ELEMENTS OF
BAYESIAN
STATISTICS

MONOGRAPHS AND TEXTBOOKS IN PURE AND APPLIED MATHEMATICS

30. *J. S. Golan*, Localization of Noncommutative Rings (1975)
31. *G. Klambauer*, Mathematical Analysis (1975)
32. *M. K. Agoston*, Algebraic Topology: A First Course (1976)
33. *K. R. Goodearl*, Ring Theory: Nonsingular Rings and Modules (1976)
34. *L. E. Mansfield*, Linear Algebra with Geometric Applications: Selected Topics (1976)
35. *N. J. Pullman*, Matrix Theory and Its Applications (1976)
36. *B. R. McDonald*, Geometric Algebra Over Local Rings (1976)
37. *C. W. Groetsch*, Generalized Inverses of Linear Operators: Representation and Approximation (1977)
38. *J. E. Kuczkowski and J. L. Gersting*, Abstract Algebra: A First Look (1977)
39. *C. O. Christenson and W. L. Voxman*, Aspects of Topology (1977)
40. *M. Nagata*, Field Theory (1977)
41. *R. L. Long*, Algebraic Number Theory (1977)
42. *W. F. Pfeffer*, Integrals and Measures (1977)
43. *R. L. Wheeden and A. Zygmund*, Measure and Integral: An Introduction to Real Analysis (1977)
44. *J. H. Curtiss*, Introduction to Functions of a Complex Variable (1978)
45. *K. Hrbacek and T. Jech*, Introduction to Set Theory (1978)
46. *W. S. Massey*, Homology and Cohomology Theory (1978)
47. *M. Marcus*, Introduction to Modern Algebra (1978)
48. *E. C. Young*, Vector and Tensor Analysis (1978)
49. *S. B. Nadler, Jr.*, Hyperspaces of Sets (1978)
50. *S. K. Segal*, Topics in Group Rings (1978)
51. *A. C. M. van Rooij*, Non-Archimedean Functional Analysis (1978)
52. *L. Corwin and R. Szczarba*, Calculus in Vector Spaces (1979)
53. *C. Sadosky*, Interpolation of Operators and Singular Integrals: An Introduction to Harmonic Analysis (1979)
54. *J. Cronin*, Differential Equations: Introduction and Quantitative Theory (1980)
55. *C. W. Groetsch*, Elements of Applicable Functional Analysis (1980)
56. *I. Vaisman*, Foundations of Three-Dimensional Euclidean Geometry (1980)
57. *H. I. Freedman*, Deterministic Mathematical Models in Population Ecology (1980)
58. *S. B. Chae*, Lebesgue Integration (1980)
59. *C. S. Rees, S. M. Shah, and C. V. Stanojević*, Theory and Applications of Fourier Analysis (1981)
60. *L. Nachbin*, Introduction to Functional Analysis: Banach Spaces and Differential Calculus (R. M. Aron, translator) (1981)
61. *G. Orzech and M. Orzech*, Plane Algebraic Curves: An Introduction Via Valuations (1981)
62. *R. Johnsonbaugh and W. E. Pfaffenberger*, Foundations of Mathematical Analysis (1981)
63. *W. L. Voxman and R. H. Goetschel*, Advanced Calculus: An Introduction to Modern Analysis (1981)
64. *L. J. Corwin and R. H. Szcarba*, Multivariable Calculus (1982)
65. *V. I. Istrătescu*, Introduction to Linear Operator Theory (1981)
66. *R. D. Järvinen*, Finite and Infinite Dimensional Linear Spaces: A Comparative Study in Algebraic and Analytic Settings (1981)

67. *J. K. Beem and P. E. Ehrlich*, Global Lorentzian Geometry (1981)
68. *D. L. Armacost*, The Structure of Locally Compact Abelian Groups (1981)
69. *J. W. Brewer and M. K. Smith, eds.*, Emmy Noether: A Tribute to Her Life and Work (1981)
70. *K. H. Kim*, Boolean Matrix Theory and Applications (1982)
71. *T. W. Wieting*, The Mathematical Theory of Chromatic Plane Ornaments (1982)
72. *D. B. Gauld*, Differential Topology: An Introduction (1982)
73. *R. L. Faber*, Foundations of Euclidean and Non-Euclidean Geometry (1983)
74. *M. Carmeli*, Statistical Theory and Random Matrices (1983)
75. *J. H. Carruth, J. A. Hildebrant, and R. J. Koch*, The Theory of Topological Semigroups (1983)
76. *R. L. Faber*, Differential Geometry and Relativity Theory: An Introduction (1983)
77. *S. Barnett*, Polynomials and Linear Control Systems (1983)
78. *G. Karpilovsky*, Commutative Group Algebras (1983)
79. *F. Van Oystaeyen and A. Verschoren*, Relative Invariants of Rings: The Commutative Theory (1983)
80. *I. Vaisman*, A First Course in Differential Geometry (1984)
81. *G. W. Swan*, Applications of Optimal Control Theory in Biomedicine (1984)
82. *T. Petrie and J. D. Randall*, Transformation Groups on Manifolds (1984)
83. *K. Goebel and S. Reich*, Uniform Convexity, Hyperbolic Geometry, and Nonexpansive Mappings (1984)
84. *T. Albu and C. Năstăsescu*, Relative Finiteness in Module Theory (1984)
85. *K. Hrbacek and T. Jech*, Introduction to Set Theory, Second Edition, Revised and Expanded (1984)
86. *F. Van Oystaeyen and A. Verschoren*, Relative Invariants of Rings: The Noncommutative Theory (1984)
87. *B. R. McDonald*, Linear Algebra Over Commutative Rings (1984)
88. *M. Namba*, Geometry of Projective Algebraic Curves (1984)
89. *G. F. Webb*, Theory of Nonlinear Age-Dependent Population Dynamics (1985)
90. *M. R. Bremner, R. V. Moody, and J. Patera*, Tables of Dominant Weight Multiplicities for Representations of Simple Lie Algebras (1985)
91. *A. E. Fekete*, Real Linear Algebra (1985)
92. *S. B. Chae*, Holomorphy and Calculus in Normed Spaces (1985)
93. *A. J. Jerri*, Introduction to Integral Equations with Applications (1985)
94. *G. Karpilovsky*, Projective Representations of Finite Groups (1985)
95. *L. Narici and E. Beckenstein*, Topological Vector Spaces (1985)
96. *J. Weeks*, The Shape of Space: How to Visualize Surfaces and Three-Dimensional Manifolds (1985)
97. *P. R. Gribik and K. O. Kortanek*, Extremal Methods of Operations Research (1985)
98. *J.-A. Chao and W. A. Woyczynski, eds.*, Probability Theory and Harmonic Analysis (1986)
99. *G. D. Crown, M. H. Fenrick, and R. J. Valenza*, Abstract Algebra (1986)
100. *J. H. Carruth, J. A. Hildebrant, and R. J. Koch*, The Theory of Topological Semigroups, Volume 2 (1986)

101. *R. S. Doran and V. A. Belfi*, Characterizations of C*-Algebras: The Gelfand-Naimark Theorems (1986)

102. *M. W. Jeter*, Mathematical Programming: An Introduction to Optimization (1986)

103. *M. Altman*, A Unified Theory of Nonlinear Operator and Evolution Equations with Applications: A New Approach to Nonlinear Partial Differential Equations (1986)

104. *A. Verschoren*, Relative Invariants of Sheaves (1987)

105. *R. A. Usmani*, Applied Linear Algebra (1987)

106. *P. Blass and J. Lang*, Zariski Surfaces and Differential Equations in Characteristic p > 0 (1987)

107. *J. A. Reneke, R. E. Fennell, and R. B. Minton*. Structured Hereditary Systems (1987)

108. *H. Busemann and B. B. Phadke*, Spaces with Distinguished Geodesics (1987)

109. *R. Harte*, Invertibility and Singularity for Bounded Linear Operators (1988).

110. *G. S. Ladde, V. Lakshmikantham, and B. G. Zhang*, Oscillation Theory of Differential Equations with Deviating Arguments (1987)

111. *L. Dudkin, I. Rabinovich, and I. Vakhutinsky*, Iterative Aggregation Theory: Mathematical Methods of Coordinating Detailed and Aggregate Problems in Large Control Systems (1987)

112. *T. Okubo*, Differential Geometry (1987)

113. *D. L. Stancl and M. L. Stancl*, Real Analysis with Point-Set Topology (1987)

114. *T. C. Gard*, Introduction to Stochastic Differential Equations (1988)

115. *S. S. Abhyankar*, Enumerative Combinatorics of Young Tableaux (1988)

116. *H. Strade and R. Farnsteiner*, Modular Lie Algebras and Their Representations (1988)

117. *J. A. Huckaba*, Commutative Rings with Zero Divisors (1988)

118. *W. D. Wallis*, Combinatorial Designs (1988)

119. *W. Więsław*, Topological Fields (1988)

120. *G. Karpilovsky*, Field Theory: Classical Foundations and Multiplicative Groups (1988)

121. *S. Caenepeel and F. Van Oystaeyen*, Brauer Groups and the Cohomology of Graded Rings (1989)

122. *W. Kozlowski*, Modular Function Spaces (1988)

123. *E. Lowen-Colebunders*, Function Classes of Cauchy Continuous Maps (1989)

124. *M. Pavel*, Fundamentals of Pattern Recognition (1989)

125. *V. Lakshmikantham, S. Leela, and A. A. Martynyuk*, Stability Analysis of Nonlinear Systems (1989)

126. *R. Sivaramakrishnan*, The Classical Theory of Arithmetic Functions (1989)

127. *N. A. Watson*, Parabolic Equations on an Infinite Strip (1989)

128. *K. J. Hastings*, Introduction to the Mathematics of Operations Research (1989)

129. *B. Fine*, Algebraic Theory of the Bianchi Groups (1989)

130. *D. N. Dikranjan, I. R. Prodanov, and L. N. Stoyanov*, Topological Groups: Characters, Dualities, and Minimal Group Topologies (1989)

131. *J. C. Morgan II,* Point Set Theory (1990)
132. *P. Biler and A. Witkowski,* Problems in Mathematical Analysis (1990)
133. *H. J. Sussmann,* Nonlinear Controllability and Optimal Control (1990)
134. *J.-P. Florens, M. Mouchart, and J. M. Rolin,* Elements of Bayesian Statistics (1990)

Other Volumes in Preparation

ELEMENTS OF BAYESIAN STATISTICS

JEAN PIERRE FLORENS

University of Social Sciences of Toulouse
Toulouse, France

MICHEL MOUCHART

JEAN-MARIE ROLIN

Catholic University of Louvain
Louvain-la-Neuve
Belgium

CRC Press
Taylor & Francis Group
Boca Raton London New York

CRC Press is an imprint of the
Taylor & Francis Group, an **informa** business

CRC Press
Taylor & Francis Group
6000 Broken Sound Parkway NW, Suite 300
Boca Raton, FL 33487-2742

First issued in paperback 2019

© 1990 by Taylor & Francis Group, LLC
CRC Press is an imprint of Taylor & Francis Group, an Informa business

ISBN-13: 978-0-8247-8123-1 (hbk)
ISBN-13: 978-0-367-40321-8 (pbk)

Library of Congress Cataloging-in-Publication Data

Florens, J. P.
     Elements of Bayesian statistics / Jean-Pierre Florens, Michel Mouchart, Jean-Marie Rolin.
     p.   cm. -- (Monographs and textbooks in pure and applied mathematics ; 134)
     Includes bibliographical references.
     ISBN 0-8247-8123-6 (alk. paper)
     1. Bayesian statistical decision theory. I. Mouchart, Michel.
II. Rolin, J. III. Title. IV. Series.
QA279.5.F56 1990   89-77562
519.5'42--dc20   CIP

Visit the Taylor & Francis Web site at
http://www.taylorandfrancis.com

and the CRC Press Web site at
http://www.crcpress.com

To Nicole, Nicole,
and Marie-Jeanne

Preface

The Mathematical Structure of a Bayesian Experiment

Statistical theory is usually based on a mathematical structure defined by a family of (sampling) probabilities on a measurable space (*viz.* the sample space). Convenience or decision-theoretic considerations typically lead to introduce a "parameter space" whose elements index the family of sampling probabilities. Endowing the parameter space with a σ-field which makes the sampling probabilities measurable is not restrictive (as long as a particular structure, such as separability, is not imposed on that σ-field). The approach to statistical methods underlying such a mathematical structure will be called *"classical"* (or *"sampling theory"*) and we shall use the following notation : $(S, \mathcal{S})$ and $(A, \mathcal{A})$ denote the sample space and the parameter space respectively and $P^a(X)$ stands for the sampling probability of $X \in \mathcal{S}$ corresponding to the parameter $a \in A$.

As far as no other restriction than making the sampling probabilities measurable is imposed on the σ-field $\mathcal{A}$, the structure just described accommodates for so-called parametric as well as for nonparametric or semi-parametric methods; more specifically the elements of A may be taken as probability measures on the sample space as well as finite dimensional characteristics of a (proper) subset of all probability measures on the sample space.

In *Bayesian methods*, the preceding structure is enriched by a probability μ — to be called "a priori" — on the parameter space. This induces a unique probability Π on the product space $(A \times S, \mathcal{A} \otimes \mathcal{S})$ which defines a *Bayesian experiment*. Thus the *prior probability* μ appears as the marginal probability of Π on $(A, \mathcal{A})$ while the *sampling probabilities* P^a constitute (regular, by construction) versions of the conditional probabilities of Π (on $(S, \mathcal{S})$) given $\mathcal{A}$. Under reasonably unrestrictive conditions, the product probability Π admits a dual decomposition into a marginal probability P on $(S, \mathcal{S})$, to be called a *predictive probability*, and regular versions of the conditional probabilities given $\mathcal{S}$ to be momentarily denoted μ^s and to be called "*posterior probabilities*". Thus the structure of a *Bayesian Experiment* may be described as follows:

$$\mathcal{E} = (A \times S, \ \mathcal{A} \otimes \mathcal{S}, \ \Pi = \mu \otimes P^a = P \otimes \mu^s).$$

The object of this monograph is a systematic study of such a structure.

The originality of the Bayesian approach does therefore not lie in the mathematical structure of the model (in particular, the structure of product space is not even essential) but rather in the interpretation of its constitutive elements. Thus the σ-fields $\mathcal{A}$ and $\mathcal{S}$ formalize the unobservable (or unknown) events and the observable (or known) events respectively. For $\mathcal{A}$-measurable random variables a, and $\mathcal{S}$-measurable random variables s, marginal and conditional expectations will be a center of interest. In particular, comparing $E(a)$ and $E(a \mid \mathcal{S})$ will be useful to describe the learning process, while comparing $E(s)$ and $E(s \mid \mathcal{A})$ will help to characterize the sampling process. A Bayesian experiment will therefore be analyzed from three points of view: one is based on the joint probability Π, another one is based on the decomposition of that joint probability into a marginal probability on $(A, \mathcal{A})$ (the prior probability) and the conditional probabilities given $\mathcal{A}$ (the sampling probabilities) and a last one is based on the decomposition of Π into a marginal probability on $(S, \mathcal{S})$ (the predictive probability) and the conditional probabilities given $\mathcal{S}$ (the posterior probabilities).

From a strictly formal point of view, the properties of a Bayesian experiment may therefore be analyzed as properties of a decomposition (or desintegration) of a probability measure into a marginal (or trace) probability and a conditional component with respect to a given sub-σ-field.

On Bayesian and Decision Theoretic Approach

The main stream of thought in this monograph is Bayesian but not decision-theoretic oriented, as far as we never consider explicitly a completely specified decision problem, i.e., a structure with a decision space and a loss function.

More specifically, although the first objective of this monograph is a systematic exposition of the Bayesian model, we also endeavour to get a deeper understanding of the role of the prior probability by examining, for instance, how far a given property would be robust to a given modification of the prior probability or by comparing concepts and results in a Bayesian and in a sampling-theory approach; in particular we give several theorems concerning the equivalence of the two approaches.

The decision-theoretic flavour is introduced by means of σ-fields on the parameter space representing "parameters of interest". Typically, such a σ-field would be the smallest one making the loss function measurable (for every decision) and would naturally be the first object of inference. Heuristically we retain, from a decision problem, the characterization of "what it depends on" without taking care of the form of such dependence.

Prerequisite Knowledge and Intended Audience

We find it tempting to mimic the introduction of Bourbaki's treatise by stating "we take statistical analysis at its starting point; thus, reading this monograph does not require, in principle, any previous knowledge ... in statistics". As a matter of fact, three domains of knowledge may be required to ensure an easy access to this monograph. Firstly, we systematically rely on σ-fields and measurability arguments. This requires a knowledge of the basis of abstract *probability theory* such as exposited in Chung

(1968) or Metivier (1972). We also refer to specific results in Neveu (1964), in Breiman (1968) or in Dellacherie and Meyer (1975, 1980) when we try to avoid repeating a more systematic exposition of the results we use. More precisely, the strictly Bayesian theory in its general formulation does not require more than a good understanding of conditional expectation but formulation in terms of densities or comparison between sampling theory and Bayesian theory are mathematicallly more demanding because such topics require the introduction of more structure. In order to make this monograph reasonably self-contained, we have briefly recalled, in the preliminary Chapter 0, some elements of probability theory as they are necessary for the reading of the first six chapters. The last three chapters use some less elementary results that are recalled when they come into use. Secondly, this monograph does not systematically treat usual *Bayesian models* such as normal or binomial models under natural conjugate prior distributions. Those models are introduced as examples only and their presentation will generally be rather sketchy. Our objective is basically to handle general properties of the Bayesian statistical structure; for a more detailed treatment of Bayesian models the reader is referred to classical textbooks as Berger (1980), Box and Tiao (1973), De Groot (1970), Good (1950, 1965), Leamer (1978), Lindley (1975), Raiffa and Schlaifer (1961) or Zellner (1971). Neither does this monograph explicitly treat foundational aspects of Bayesian methods, but our choice of topics and techniques has clearly been guided by epistemological considerations handled in De Finetti (1974), or Savage (1954); for a recent appraisal of the state of the art, see Shafer (1986). Let us mention that although those references on Bayesian models and on foundations of Bayesian methods are not a prerequisite for this monograph, they nevertheless provide an important guideline for the motivation underlying the topics treated. Thirdly some familiarity with *sampling-theory concepts* such as sufficiency, ancillarity or estimability would definitely benefit our reader; for these topics, Barra (1971), Cox and Hinkley (1974), Kendall and Stuart (1952, 1961 and 1966), or Zacks (1971) are most useful references insofar as they provide motivation and point of comparison for the concepts we develop. Let us nevertheless mention that our text has been made reasonably self-sufficient by remembering, when appropriate, the corresponding sampling-theory definitions and results.

These flexible prerequisites make about two-thirds of this monograph suitable as a reference text for a graduate course dealing with the mathematical foundations of Bayesian methods. This suggestion is also based on the fact that the present version has benefited from earlier presentations at various places, in particular at the Statistical Department of Carnegie Mellon University (U.S.A.), and to graduate students of the Universities of Aix-Marseille II (France), Granada (Spain), Louvain-la-Neuve (Belgium), Paris VI, Louis Pasteur at Strasbourg and Toulouse (France).

Content

The main theme of this monograph is the theory of reduction of a Bayesian Experiment considered as a (unique) probability measure on a product space (parameter space × sample space). The main question to be considered may be phrased as follows: given that a Bayesian Experiment may be "reduced" by marginalization or by conditioning on the parameter space and/or on the sample space, how far does such a reduction lose no "relevant" information? Clearly such questions are to be handled in terms of ancillarity and of sufficiency but the Bayesian approach allows for a symmetric treatment on the parameter and on the sample space. Also those questions will require a formalization of the ideas "learning by observing" and "without losing information". So we also consider questions such as: what, in the parameter space, receives or does not receive information from the sample? What, in the sample space, provides or does not provide, information? More than a specific field in statistical methodology the theory of reduction actually develops a general approach to face the statistical analysis of empirical data.

That providing tools helpful for the specification of a statistical model is at the center of our concern may be appreciated from the following consideration: We systematically draw our results from hypotheses about some structural properties on distribution rather than from specific distributional assumptions. Thus our results concern classes of models. Our suggestion is actually to start the specification of a model by structural hypotheses (such as prior independence, admissibility of a conditioning, an invariance property), before eventually considering, in a parametric framework, specific distributional hypotheses.

Let us now briefly sketch how the material of this book has been allocated to ten chapters. A preliminary Chapter 0 introduces some notation
and definitions and, more specifically, recalls the main probabilistic tools to
be used in the sequel. Chapter 1 introduces the basic structure which will
be systematically investigated, *viz.*, the structure of a Bayesian experiment.
The basic concepts of different reduced experiments are introduced and alternative interpretations are discussed. Chapter 2 presents the basic theory
of sufficiency and ancillarity in an unreduced Bayesian experiment, i.e., in
an experiment where a "complete" observation is available and where the
parameter of interest is identical to the complete parameter of the model.
These two basic topics are again treated, for the case of a reduced experiment, when introduced are the concepts of mutual sufficiency, mutual
exogeneity and (Bayesian) cut in a framework of joint reduction, in Chapter 3. Chapter 4 analyzes minimal sufficiency along with related topics;
there, identification and exact estimability are treated in that framework;
these concepts will also be essential in asymptotic theory. The relationships between sufficiency and ancillarity have retained a wide attention in
the sampling theory literature and constitute the topic of Chapter 5. These
first five chapters refer to what may be called "one-shot analysis". The next
two chapters handle the sequential nature of empirical observations. This
leads to enrich the basic structure by a filtration (an increasing sequence of
σ-fields) on the sample space. This new structure will allow us, in Chapter 6,
to analyze problems of sequential reduction and, in Chapter 7, asymptotic
questions. Chapter 6 centers on the sequential (observationwise) decomposition of the learning process and its reductions but also with a particular
attention on models where conditioning on the sample space calls for a
different, eventually sequential treatment of the conditioning σ-field, introducing the concepts of transitivity and of non-causality. Two main themes
are treated in Chapter 7: on one side, the relationship between (admissible)
reductions in finite sample and reductions in the asymptotic model; on the
other side, the concept of (Bayesian) exact estimability as a formalization
of the idea of consistency of Bayesian methods. Chapters 8 and 9 conclude this monograph by considering invariant experiments, first in general
(Chapter 8), next in the analysis of stochastic processes (Chapter 9). In
Chapter 8 we conclude a general approach to invariance. Thus the basic
structure of a Bayesian experiment is enriched by the introduction of a fam-

ily of operators acting on the product space (parameters and observations), eventually introducing experiments that are invariant for a monoid or for a group of transformations. For such experiments, we obtain useful reductions and analyze some asymptotic properties. Finally, Chapter 9 may be viewed as an application of both Chapters 7 and 8. More specifically, the general approch to invariance is applied to the particular case of stationary and of exchangeable processes; in that context, asymptotic properties of posterior expectations in Markov, AR, MA and ARMA processes are derived.

Method of Presentation

We have systematically resorted to the use of *conditional independence* as a unifying principle for the presentation of this monograph. This concept is widely used in probability theory, in particular for the analysis of Markov processes. In statistics, Hall, Wijsman and Ghosh (1965) relied heavily on this concept for their presentation and mention, in their introduction, earlier unpublished work by Ch. Stein; see also LeCam (1964). In the late sixties, a strong group of French statisticians at the "Institut Statistique de l'Université de Paris" (ISUP) made a systematic use of this concept (M. Littaye, M. Oheix, J.L. Petit, B. Van Cutsem, Fr. Martin, J.P. Raoult, G. Romier, etc...) in a series of papers, partly assembled in a special issue of the *Annales de l'Institut Henri Poincaré* (1969) (see also Fr. Martin, J.L. Petit, M. Littaye (1973)). In 1973, T. Speed, in a private communication, drew our attention to these contributions and commented on a long bibliography on sufficiency that he was collecting with D. Basu (Basu and Speed (1974)). Later on, A.P. Dawid (1978, 1979) also pointed out the interest of this concept for Bayesian analysis.

As a general rule, the probabilistic tools are progressively introduced as they are used. Thus almost all chapters contain a nonstatistical section introducing topics of probability theory; those sections present results either with proofs or with references to textbooks; in general our "statistical" results will be presented, often without proofs, as simple corollaries of probabilistic results. This method of presentation has the advantage of alleviating the statistical theory from too much "technical" digression. This will be emphasized by calling "Propositions" those results that do not require explicit proofs for being simple corollaries of previous "Theorems".

However, it cannot be inferred that this monograph is a simple application of probability theory. Indeed statistical motivations lead to develop concepts and definitions that are apparently of minor interest in probability theory. This implies that when we abstract statistical concepts related to statistics or parameters as arbitrary sub-σ-fields on a probability space, we obtain new probabilistic definitions and results.

The basic theory is couched in terms of σ-fields; at this level no density is used. However each chapter provides sections where the theory is repeated in terms of densities for dominated experiments. Similarly, although this monograph basically presents the Bayesian theory, each chapter also contains sections where the relationships with the analogue sampling theory concepts are commented. Although we hope to have shown, in this monograph, that the basic Bayesian theory may be presented in relatively simple terms, the reader should nevertheless be warned that the sections with densities (i.e., dominated experiments) and the sections on the classical (or sampling-theory) approach are more mathematically demanding.

Origin

Our interest in the mathematical foundations of Bayesian experiments dates back to the early seventies when, taking advantage of a one-year visit of the first author to CORE, we were puzzled by some intricacies of a general version of the linear model (Florens, Mouchart and Richard (1974, 1976, 1979)). This motivated our interest for a better understanding of the mathematics underlying the Bayesian methods (Mouchart (1976)). This book reports an effort to systematize more than 10 years of work on that field. Many results have been previously reported in published papers and in unpublished Discussion Papers (particularly from CORE). In the introductory section of each chapter we recall our previous work relative to the topic to be handled along with some perspective on the literature. It should nevertheless be mentioned that most chapters contain large portions of hitherto unpublished material. Apart from new results, materials have been included that should make this text open to a wider audience than the typical journal reader.

Acknowledgement

When concluding a work the span of which has been so many years, it is fitting to think gratefully of so many individuals who have made this work possible, even if their number makes an exhaustive list unmanageable.

Jacques Drèze has introduced one of us to the beauty of Bayesian thinking and communicated to all of us his most uncommon enthusiasm for research, rigor and sheer curiosity for the unknown and the misunderstood. Deep is our debt to Jacques as a teacher, as a colleague and as a kingpin of CORE (Center of Operations Research and Econometrics). Jacques Voranger guided the first steps of one of us in statistics (that could only be Bayesian) and in decision theory. Last but not least, this book would simply not have come to light if Jean-Pierre Raoult had not been the doctorate promotor of one of us. His careful reading and his advice were a decisive incentive to proceed toward the publication of our first results. He also launched the idea of the "Rencontre Franco-Belge de Statisticiens", so fruitful a forum for discussion. Within the doctoral program of Jean-Pierre Florens, Claude Dellacherie had a decisive influence through his guidance into a more analytical approach to probability theory.

The development of the works underlying this book has benefited, at every stage, from an exceptional scientific environment. In a sense, this book is a product of CORE; we particularly appreciated its most stimulating atmosphere materialized, in particular, by many insightful discussions with Jean-François Mertens. We also thank the Department of Mathematics, at Louvain-la-Neuve, where José Paris, head of the Unit for Probability Theory and Statistical Analysis, showed his continuing interest for our work and also provided appreciated research facilities. In France, GREQE (Groupe de Recherches en Economie Quantitative et Econométrie de Marseille, France) and, later on, GREMAQ (Groupe de Recherche en Economie Mathématique et Quantitative de Toulouse, France) completed in a most decisive way these favourable environments and we want to thank all the colleagues and administrative members of those centers.

Many stays in foreign institutions and participation in scientific meetings provided opportunities for exposing our ideas and receiving stimulating comments. From too long a list for a complete edition, we want to explicitly thank some individuals with whom we had particularly stimulating discussions and who were instrumental for inviting (at least) one of us to visit their department and eventually to share the stimulus of their environment: Ph. Nanopoulos (University of Strasbourg), J. Kadane and M. H. de Groot (Carnegie Mellon), P. Diaconis (Stanford), K. P. S. Bashkara Rao (Indian Statistical Institute, Calcutta), A. Zellner (University of Chicago), E. Moreno (University of Granada) along with the participants of the Valencia Meetings of Bayesian Statistics, the International Study Year on Bayesian Statistics (at the University of Warwick) and the Franco-Belgian Meetings of Statisticians.

Many ideas presented in this book have been elaborated jointly through research works with several colleagues. In particular, Jean-François Richard has been very closely associated to much of the research reported here; also, several results mentioned in this book have been first worked out in papers co-authored with David Hendry, Renzo Orsi or Léopold Simar. Daniela Cocchi, Anne Feyssolle, Charles Lai Tong, Hugo Roche, Solange Scotto, Velayoudoun Marimoutou and Michele Ruggiero have been demanding Ph.D. students whose questions and requirements of clarification had an important catalytic and progress-provoking role.

At the stage of final presentation of our work, Joseph Hakizamungu's work has been decisive in the preparation of the index. Vicky Barham had the desperate job of improving our linguistic talent. The late Elisabeth Pecquereau was an example of cheerful cooperation to transform unreadable cryptics into elegantly typed mathematics and Mariette Huysentruyt has been in charge of the final presentation of the manuscript into a camera ready proof. Any one knowing the authors recognizes that the merits of Elisabeth and Mariette cannot be scaled on any bounded instrument. Sheila Verkaeren was brilliant in coping with the very difficult task of synchronizing the typing work within a permanently overloaded administrative staff. Finally we have always found a totally cooperative attitude at Marcel Dekker.

Jean-Pierre Florens
Michel Mouchart
Jean-Marie Rolin

Contents

Notation

$\mathcal{M}$: σ-field, 2

$(M, \mathcal{M})$:   Measurable space, 2

$\mathcal{M}_1 \subset \mathcal{M}$:   sub-$\sigma$-field, 2

$\mathcal{M}_0 = \{\phi, M\}$:   Trivial σ-field, 2

$A \cap \mathcal{M}$:   Trace of $\mathcal{M}$ on A, 3

$\cap_{t \in T} \mathcal{M}_t$:   Intersection of σ-fields, 3

$\sigma(\mathcal{C})$: σ-field generated by $\mathcal{C}$, 3

$\cup_{t \in T} \mathcal{M}_t$:   Union of σ-fields, 3

$\bigvee_{t \in T} \mathcal{M}_t = \sigma(\cup_{t \in T} \mathcal{M}_t)$:   Wedge of σ-fields, 3

$\mathcal{B}$:   Borel σ-field of $I\!\!R$, 4

$x' \underset{\mathcal{M}}{\sim} x$, 4

$\mathbf{1}_A$:   Indicator function, 4

$A_x^{\mathcal{M}}$:   Atom of x in $\mathcal{M}$, 4

$\overset{\bullet}{M} = M/\mathcal{M}$:   Quotient set, 5

$\overset{\bullet}{\mathcal{M}}$:   Quotient σ-field, 5

$(\overset{\bullet}{M}, \overset{\bullet}{\mathcal{M}})$: Quotient measurable space, 5

$f^{-1}(\mathcal{N}) = \sigma(f)$: σ-field generated by f, 5

$f : (M, \mathcal{M}) \to (N, \mathcal{N})$: Measurable function, 5

$\bar{f}(\mathcal{M}) = \{B \subset N : f^{-1}(B) \in \mathcal{M}\}$, 6

$m : (M, \mathcal{M}) \longrightarrow (\mathbb{R}, \mathcal{B})$: Random variable, Borel function, 6

$[\mathcal{M}], [\mathcal{M}]^+, [\mathcal{M}]_\infty$: Set of random variables, of positive random variables, and of bounded random variables respectively, 6

$\pi_t : \mathsf{X}_{t \in T} M_t \to M_t$: Coordinate map, 10

$\bigotimes_{t \in T} \mathcal{M}_t$: Product σ-field of the $\mathcal{M}_t$, 10

$(M^T, \mathcal{M}^T)$, 10

$\mathcal{M}_1 \otimes \mathcal{M}_2$, 10

$\mathcal{B}^d = \mathcal{B}(\mathbb{R}^d)$: Borel σ-field of $\mathbb{R}^d$, 11

$\mu : \mathcal{M} \to \overline{\mathbb{R}}^+$: Measure (positive), 13
$(M, \mathcal{M}, \mu)$: Measure space, 13

$I(m \mathbf{1}_A) = \int_A m \, d\mu = \int_A m(x) \, \mu(dx)$: Integral, 14

$\nu \ll \mu$: Absolutely continuous, 14

$\mu \perp \nu$: Mutually singular , 14

$d\nu/d\mu$: Radon-Nikodym derivative, 15

P: Probability, 15

$(M, \mathcal{M}, P)$: Probability space, 15

$E(m) = \int m \, dP$: Expectation (mathematical), 15

$\overline{\mathcal{M}}_0$: Completed trivial σ-field, 15, 69

$\overline{\mathcal{N}}$: Completed sub-σ-field, 16, 69

$\overline{\mathcal{N}}^P$: Completion of $\mathcal{N}$ by the P-null sets, 16

$\|m\|_p$: p-semi-norms, 16

$[\mathcal{M}]_p$, 16

$m = m'$ a.s.P: Almost sure equality, 17

L_p, 17

$P^{\bullet}(\bullet): (M, \mathcal{M}) \relbar\joinrel\prec (N, \mathcal{N})$: Transition, 17

$Q \otimes P^x$, 18

$\mathcal{N}m$: Conditional expectation of m with respect to $\mathcal{N}$, 18

$E(m) = \mathcal{M}_0 m$: Expectation, 19

$P^{\mathcal{N}}(A)$: Conditional probability, 21

$P_{\mathcal{N}}(A) = P(A)$: Trace of P on $\mathcal{N}$, Marginal probability, 21

$P_{\mathcal{M}_1}^{\mathcal{M}_2}(A)$: Conditional trace probability, 21

$(S, \mathcal{S})$: Sample space, 26

$\mathcal{E} = \{(S, \mathcal{S}), P^a : a \in A\}$: Statistical experiment, 26

$(A, \mathcal{A})$: Parameter space, 26

P^a: Sampling probabilities, 26

$\mu = \Pi_{\mathcal{A}}$: Restriction of Π to $\mathcal{A}$, 26

$P = \Pi_{\mathcal{S}}$: Restriction of Π to $\mathcal{S}$, 26

$\mathcal{A} \vee \mathcal{S}$: σ-field generated by $(\mathcal{A} \times S) \cup (A \times \mathcal{S})$, 27

$\mathcal{E} = (A \times S, \mathcal{A} \vee \mathcal{S}, \Pi)$: Bayesian experiment, 27

$\mathcal{A}m$: Sampling expectation, 27

$\mathcal{S}m$: Posterior expectation, 27

$\mathcal{E} = (A \times S, \mathcal{A} \vee \mathcal{S}, \Pi = \mu \otimes P^{\mathcal{A}} = P \otimes \mu^{\mathcal{S}})$: Regular Bayesian experiment, 27

μ: Prior probability, 27

P: Predictive probability, 27

$\mu^{\mathcal{S}}$: Posterior probabilities, 27

$P^{\mathcal{A}}$: Sampling probabilities, 27

P_*: Privileged dominating probability, 28

$g(a, s) = \frac{d\Pi}{d(\mu \otimes P)}$, 30

$\mathcal{E}^{\mathcal{M}}$:   Measurable family of experiments, 36

$\mathcal{BEE}(\alpha, \sigma)$ Bayesian exponential experiment associated to α and σ, 41

$\mathcal{I} = \{\phi, A \times S\}$, 46

$\mathcal{E}_{\mathcal{M}}$: Marginal experiment, 47

$\mathcal{E}^{\mathcal{M}}$:   Conditional experiment, 47

$P^{\mathcal{A}} = \Pi^{\mathcal{A}}_{\mathcal{S}},\ \mu^{\mathcal{S}} = \Pi^{\mathcal{S}}_{\mathcal{A}},\ P^{\mathcal{B}}_{\mathcal{T}} = \Pi^{\mathcal{B}}_{\mathcal{T}},\ \mu^{\mathcal{T}}_{\mathcal{B}} = \Pi^{\mathcal{T}}_{\mathcal{B}}$, 47

$\mathcal{E}_{\mathcal{A} \vee \mathcal{T}} = (A \times S,\ \mathcal{A} \vee \mathcal{T},\ \Pi_{\mathcal{A} \vee \mathcal{T}})$:   Bayesian experiment marginal on $\mathcal{T}$, 48

$\mathcal{E}_{\mathcal{B} \vee \mathcal{S}} = (A \times S,\ \mathcal{B} \vee \mathcal{S},\ \Pi_{\mathcal{B} \vee \mathcal{S}})$:   Bayesian experiment marginal on $\mathcal{B}$, 49

$\mathcal{E}^{\mathcal{T}} = (A \times S,\ \mathcal{A} \vee \mathcal{S},\ \Pi^{\mathcal{T}})$:   Bayesian experiment conditional on $\mathcal{T}$, 51

$\mathcal{E}^{\mathcal{B}} = (A \times S,\ \mathcal{A} \vee \mathcal{S},\ \Pi^{\mathcal{B}})$:   Bayesian experiment conditional on $\mathcal{B}$, 52

$\mathcal{E}^{\mathcal{M}}_{\mathcal{B} \vee \mathcal{T}}$:   General reduced Bayesian experiment, 54

$\tilde{\mathcal{N}}m$:   conditional expectation of m given $\mathcal{N}$ with respect to $\mu \otimes P$, 59

$g_{\mathcal{N}}$, 60

$g^{\mathcal{M}}_{\mathcal{N}}$, 62

$\mathcal{M}_1 \perp\!\!\!\perp \mathcal{M}_2 \mid \mathcal{M}_3$ (equivalent to $\mathcal{M}_1 \perp\!\!\!\perp \mathcal{M}_2 \mid \mathcal{M}_3; P$): $\mathcal{M}_1$ and $\mathcal{M}_2$ are independent conditionally on $\mathcal{M}_3$, 67

$\mathcal{M}_1 \perp\!\!\!\perp \mathcal{M}_2$ (equivalent to $\mathcal{M}_1 \perp\!\!\!\perp \mathcal{M}_2 \mid \mathcal{M}_0$):   (Marginal) Independence of $\mathcal{M}_1$ and $\mathcal{M}_2$, 67

$\mathcal{T}^a s$:   Conditional expectation of s given $\mathcal{T}$ with respect to P^a, 87

$\mathcal{T}_* s$:   Conditional expectation of s given $\mathcal{T}$ with respect to a privileged dominating probability P_*, 89

$\mathcal{E}^{\mu}_{\mathcal{B} \vee \mathcal{S}};\ \Pi^{\mu}_{\mathcal{B} \vee \mathcal{S}}$, 102

$\mathcal{M}_2 \mathcal{M}_1$:   Projection of $\mathcal{M}_1$ on $\mathcal{M}_2$, 143

$g^a_* = dP^a / dP_*$, 156

$\mathcal{S}_1 = \sigma\{g^a_* : a \in A\}$:   Sampling minimal sufficient statistic, 156

$\mathcal{A}_1$:   Sampling minimal sufficient parameter, 156

$\widehat{x} = (\widehat{x}_n, n \in I\!N)$:   Coordinate process, 409

$\widehat{a}$, 409

$\widehat{\mathcal{X}}_n = \sigma(\widehat{x}_n)$, 409

$\tau(u)_n = u_{n+1}$:   Shift operator, 410

$\mathcal{U}_\Gamma$: σ-field of shift-invariant sets, 411

$\mathcal{X}_\Gamma$:   Shift-invariant events of the stochastic process, 411

$\widehat{\mathcal{X}}_\Gamma$:   Shift invariant events of the coordinate process, 411

$\sigma(u)_n = u_{\sigma(n)} \quad \forall\, n < k$:   Finite permutation operator, 413

Σ_k, Σ, 413

$\mathcal{U}_\Sigma$: σ-field of symmetric sets, 414

$\mathcal{X}_\Sigma$:   Symmetric events of the stochastic process, 414

$\widehat{\mathcal{X}}_\Sigma$:   Symmetric events of the coordinate process, 414

ELEMENTS OF
BAYESIAN
STATISTICS

0

Basic Tools and Notation
from Probability Theory

0.1 Introduction

This monograph uses three kinds of probabilistic tools:

(i) The basic results of probability theory, up to the properties of the conditional expectation with respect to a σ-field; they are essentially assumed to be known by the reader but some are briefly recalled in this preliminary chapter.

(ii) Some more original expositions of several probabilistic concepts particularly powerful in statistics (such as conditional independence, measurable separability, weak and strong identification among σ-fields, projection of σ-fields). These concepts are presented in separate sections of the chapter in which they are first used. Those probabilistic sections are identified by a slightly different notation; they refer to an abstract probability space $(M, \mathcal{M}, P)$ whereas in the statistical sections M is the Cartesian product $A \times S$.

(iii) Some more advanced but standard results of probability theory (such as martingale theory, invariance, ergodicity) are used in the last three chapters. These results are not reviewed in this chapter; they are recalled, and where useful reexpressed using our notation, as they become necessary.

This preliminary chapter thus lists the main concepts used (at least in the six first chapters), recalls the usual definitions, and introduces the notation employed throughout this book. Our terminology follows, in most cases, Dellacherie and Meyer (1975). In actual fact, relatively few concepts are necessary to follow our exposition; they essentially relate to the theory of the conditional expectation, and to the alternative characterizations of measurability. In this domain we apply some results beyond the range of elementary textbooks; these results are contained in this chapter. However, the basic properties of the concepts used here are not recalled, and the reader is referred to the textbook she(he) prefers.

The books on probability theory principally used are Metivier (1972), Neveu (1970), and Dellacherie and Meyer (1975 and 1980); Billingsley (1979), Chung (1968) and Breiman (1968) are also convenient references for most of the topics covered.

0.2   Measurable Spaces

0.2.1   σ-Fields

We first recall the notion of a σ-field of subsets of a set.

0.2.1 Definition. Let M be a set and $\mathcal{M}$ be a family of subsets of M. $\mathcal{M}$ is a *σ-field* if:

(i) $\phi \in \mathcal{M}$,
(ii) $A \in \mathcal{M} \Rightarrow A^c \in \mathcal{M}$,
(iii) $A_i \in \mathcal{M}$ $\forall i \in I$ I countable $\Rightarrow \cup_{i \in I} A_i \in \mathcal{M}$.   ∎

Using the algebraic properties of set theory, it follows from this definition that $\mathcal{M}$ is closed for every countable sequence of set operations (union, intersection, difference, symmetric difference). In particular, $M \in \mathcal{M}$, and if $A_i \in \mathcal{M}$ $\forall i \in I$ and I is countable, then $\cap_{i \in I} A_i \in \mathcal{M}$.

A *measurable space* is a pair $(M, \mathcal{M})$ where M is a set and $\mathcal{M}$ is a σ-field of subsets of M. The elements of $\mathcal{M}$ are called *measurable sets* or *events*. A *sub-σ-field* $\mathcal{M}_1$ of $\mathcal{M}$ is a σ-field of subsets of M belonging to $\mathcal{M}$. The smallest σ-field defined on M is $\mathcal{M}_0 = \{\phi, M\}$ and is called the *trivial σ-field*. It is a sub-σ-field of any σ-field of subsets of M. The largest σ-field defined on M -is the family of all subsets of M. For any $A \subset M$, the family of subsets of A, $A \cap \mathcal{M} = \{A \cap B : B \in \mathcal{M}\}$, is a σ-field on A but not on M and is called the *trace of $\mathcal{M}$ on A*.

0.2.2 Proposition. If T is an arbitrary index set and if, $\forall\, t \in T$, $\mathcal{M}_t$ is a σ-field of subsets of M, then $\cap_{t \in T} \mathcal{M}_t$ is a σ-field and is the largest σ-field contained in each $\mathcal{M}_t$. ∎

0.2.3 Definition. If $\mathcal{C}$ is any collection of subsets of M, the intersection of all the σ-fields of $\mathcal{M}$ containing $\mathcal{C}$ (which is a nonempty family) is, by Proposition 0.2.2, the smallest σ-field containing $\mathcal{C}$ and is called the *σ-field generated by $\mathcal{C}$* and denoted by $\sigma(\mathcal{C})$. ∎

As an example, the σ-field generated by a nontrivial subset of M — i.e., $A \subset M$ with $\phi \neq A \neq M$ — is given by: $\sigma(\{A\}) = \{\phi, M, A, A^c\}$. Definition 0.2.3 permits the smallest σ-field containing every element of an arbitrary collection of σ-fields to be defined. In general, if T is an arbitrary index set, $\cup_{t \in T} \mathcal{M}_t$ is not a σ-field. This fact motivates the following definition.

0.2.4 Definition. If T is an arbitrary index set and if, $\forall\, t \in T$, $\mathcal{M}_t$ is a σ-field of subsets of M, then $\sigma(\cup_{t \in T} \mathcal{M}_t)$ is the smallest σ-field containing each $\mathcal{M}_t$. It is denoted by $\bigvee_{t \in T} \mathcal{M}_t$ and is called the *wedge of the σ-fields $\mathcal{M}_t$*. ∎

0.2.5 Proposition. Let

$$\mathcal{P} = \{\cap_{t \in S} A_t \ : \ S \subset T, \ \mathrm{card}(S) < \infty, \quad A_t \in \mathcal{M}_t \quad \forall\, t \in S\},$$

i.e., the collection of all finite intersections of elements of $\cup_{t \in T} \mathcal{M}_t$. Then $\bigvee_{t \in T} \mathcal{M}_t = \sigma(\mathcal{P})$. ∎

The following two definitions introduce often suitable properties of a σ-field.

0.2.6 Definition. A σ-field $\mathcal{M}$ of subsets of M is *separable* if there exists a countable collection of subsets of M generating $\mathcal{M}$, i.e., $\exists\, \mathcal{C} \subset \mathcal{M}$ where $\mathcal{C}$ is countable such that $\mathcal{M} = \sigma(\mathcal{C})$. ∎

A useful example of separable σ-field is provided by the *Borel σ-field of a Polish space*. If M is a *topological space*, the σ-field generated by its open sets is called its *Borel σ-field*. It is also the σ-field generated by the closed sets. If M is *second countable*, i.e., if there exists a countable base for the open sets, then the Borel σ-field is separable since it is the σ-field generated by this countable base. A *Polish space* is a topological space for which there exists a distance which is compatible with its topology and which makes it a *complete separable metric space*. Recall that a metric space is separable if there exists a dense countable subset. Therefore the Borel σ-field of a Polish space is generated by the open balls with rational radius and center belonging to a dense countable subset. In particular, the Borel σ-field of the real line $I\!R$, denoted by $\mathcal{B}$, is generated by the collection of open intervals with rationals endpoints.

A σ-field $\mathcal{M}$ on M generates the following *equivalence relation*:

$$(0.2.1) \qquad x' \underset{\mathcal{M}}{\sim} x \quad \Leftrightarrow \quad \mathbf{1}_A(x') = \mathbf{1}_A(x) \quad \forall\, A \in \mathcal{M},$$

where $\mathbf{1}_A$ is the *indicator function* of the set A, i.e., $\mathbf{1}_A(x) = 1$ if $x \in A$ and $\mathbf{1}_A(x) = 0$ if $x \notin A$; we shall also use $\mathbf{1}_{\{x \in A\}}$ instead of $\mathbf{1}_A(x)$. The *atoms* of $\mathcal{M}$ are the equivalence classes defined by the equivalence relation (0.2.1). We denote by $A_x^{\mathcal{M}}$ the atom of x in $\mathcal{M}$, i.e.,

$$(0.2.2) \qquad \begin{aligned} A_x^{\mathcal{M}} &= \{x' \in M : \mathbf{1}_A(x') = \mathbf{1}_A(x) \quad \forall\, A \in \mathcal{M}\}, \\ &= \cap\{A \in \mathcal{M} \; : \; x \in A\}. \end{aligned}$$

Note that for any $A \in \mathcal{M}$, $x \in A$ implies $A_x^{\mathcal{M}} \subset A$. So every measurable set is a union of atoms of $\mathcal{M}$. In general, the atoms of a σ-field are not measurable sets. Note, however, that $\{A \subset M \; : \; \mathbf{1}_A(x') = \mathbf{1}_A(x)\}$ is a σ-field for any $x, x' \in M$; therefore if $\mathcal{M} = \sigma(\mathcal{C})$, we have:

$$(0.2.3) \qquad A_x^{\mathcal{M}} = \bigcap_{C \in \mathcal{C}} \{x' \in M \; : \; \mathbf{1}_C(x') = \mathbf{1}_C(x)\},$$

and thus if $\mathcal{M}$ is separable, the atoms are measurable.

0.2.7 Definition. A σ-field $\mathcal{M}$ of subsets of M is *separating* if all its atoms are singletons, i.e.,

$$A_x^{\mathcal{M}} = \{x\} \qquad\qquad \forall\, x \in M. \qquad\qquad \blacksquare$$

Note that if the singletons are measurable, i.e., $\{x\} \in \mathcal{M} \quad \forall x \in M$, then $\mathcal{M}$ is separating. On the other hand, $\mathcal{M}$ is separating if and only if $\forall x' \neq x$, $\exists A \in \mathcal{M}$ such that $x \in A$ and $x' \notin A$. Hence if $\mathcal{M}$ is separating, the singletons are not necessarily measurable. But if $\mathcal{M}$ is both separable and separating the singletons are measurable.

To any $(M, \mathcal{M})$, we can associate the quotient measurable space $(\overset{\bullet}{M}, \overset{\bullet}{\mathcal{M}})$ where $\overset{\bullet}{M}$ is the quotient set for equivalence relation (0.2.1) i.e., the set of atoms of M:

$$(0.2.4) \qquad \overset{\bullet}{M} = M/\mathcal{M} = \{A_x^{\mathcal{M}} \ : \ x \in M\},$$

and $\overset{\bullet}{\mathcal{M}}$, the quotient σ-field, is the image of $\mathcal{M}$ by the canonical application which associates its atom to any element of M, i.e.,

$$(0.2.5) \qquad \overset{\bullet}{\mathcal{M}} = \{\overset{\bullet}{A} \ : \ A \in \mathcal{M}\},$$

where

$$(0.2.6) \qquad \overset{\bullet}{A} = \{A_x^{\mathcal{M}} \ : \ x \in A\}.$$

Note that $\overset{\bullet}{\mathcal{M}}$ is always separating and if $\mathcal{M}$ is separable, $\overset{\bullet}{\mathcal{M}}$ is separable.

0.2.2 Measurable Functions

A. Measurability in General

Let M be a set, $(N, \mathcal{N})$ be a measurable space and f be a function defined on M with values in N. The family

$$(0.2.7) \qquad f^{-1}(\mathcal{N}) = \{f^{-1}(B) \ : \ B \in \mathcal{N}\}$$

is a σ-field on M called the σ-*field generated by* f and often denoted by $\sigma(f)$.

0.2.8 Definition. Let $(M, \mathcal{M})$ and $(N, \mathcal{N})$ be two measurable spaces and $f : M \to N$; then f is *measurable* if $f^{-1}(\mathcal{N}) \subset \mathcal{M}$, i.e., if $f^{-1}(\mathcal{N})$ is a sub-σ-field of $\mathcal{M}$. We shall often condense this property by writing $f : (M, \mathcal{M}) \to (N, \mathcal{N})$. ∎

Thus $f^{-1}(\mathcal{N})$ is the smallest σ-field on M which makes $f : M \to (N, \mathcal{N})$ measurable. On the other hand the largest σ-field on N which makes $f :$ $(M, \mathcal{M}) \to N$ measurable is given by:

$$(0.2.8) \qquad \bar{f}(\mathcal{M}) = \{B \subset N : f^{-1}(B) \in \mathcal{M}\}.$$

For example, a constant function is measurable for any $\mathcal{M}$ and $\mathcal{N}$ and generates the trivial σ-field $\mathcal{M}_0$. Reciprocally, if $\mathcal{N}$ is separating, a function generating the trivial σ-field $\mathcal{M}_0$ is constant. Note that measurability is preserved under the product of composition.

0.2.9 Proposition. Let $(M, \mathcal{M}), (N, \mathcal{N}), (S, \mathcal{S})$ be three measurable spaces. If $f : M \to N$ and g : N $\to$ S are both measurable; then $g \circ f : M \to S$ is measurable.   ∎

A useful criteria for measurability is provided by the following proposition.

0.2.10 Proposition. Let M, N be two sets, $\mathcal{C}$ be a collection of subsets of N and $f : M \to N$, then $\sigma[f^{-1}(\mathcal{C})] = f^{-1}[\sigma(\mathcal{C})]$.   ∎

A measurable function, m, defined on $(M, \mathcal{M})$ with values in $(\mathbb{R}, \mathcal{B})$ is called a *random variable* or a *Borel function*. The set of random variables defined on $(M, \mathcal{M})$ is denoted by $[\mathcal{M}]$. The set of positive (respectively bounded) random variables is denoted by $[\mathcal{M}]^+$ (respectively $[\mathcal{M}]_\infty$). Note that, by the above proposition, a real valued function m on $(M, \mathcal{M})$ will be a random variable as soon as $m^{-1}(] - \infty, a]) \in \mathcal{M}$ $\forall a \in Q$ i.e., for all rational numbers.

A random variable is called *simple* if it takes only a finite number of distinct values. For such a random variable, there exists a finite measurable partition of M, i.e., $\{A_i : i \in I\} \subset \mathcal{M}$ -and a finite number of distinct real values $\{c_i : i \in I\}$ such that:

$$(0.2.9) \qquad m = \sum_{i \in I} c_i 1_{A_i}.$$

For $m \in [\mathcal{M}]^+$, let

$$(0.2.10) \quad m_n = \sum_{1 \leq k \leq n2^n - 1} \frac{k}{2^n} \, 1_{\left\{\frac{k}{2^n} < m \leq \frac{k+1}{2^n}\right\}} + n \, 1_{\{m > n\}}.$$

Then m_n is a sequence of simple random variables increasing to m pointwise and $\sigma(m_n) \subset \sigma(m) \; \forall \, n \in I\!N_0$. This is the key fact to prove the following theorem due to Doob. It is a very powerful tool for characterizing the measurability of a random variable with respect to a sub-σ-field generated by a measurable function.

0.2.11 Theorem. Let M be a set, $(N, \mathcal{N})$ be a measurable space and $f : M \to N$. Then the following properties are equivalent:

(i) $m \in [f^{-1}(\mathcal{N})]$ (resp. $[f^{-1}(\mathcal{N})]^+$, resp. $[f^{-1}(\mathcal{N})]_\infty$),

(ii) $\exists \, n \in [\mathcal{N}]$ (resp. $[\mathcal{N}]^+$, resp. $[\mathcal{N}]_\infty$) such that $m = n \circ f$. ∎

For a proof see, e.g., Metivier (1972) Appendix 1 or Neveu (1970) Proposition II.2.5 or Dellacherie and Meyer (1975) Theorem I.18. As remarked in this last reference this theorem may be extended by considering measurable functions with values in a Polish space instead of random variables.

B. Measurability through Atoms: Blackwell Theorem

Another way of proving the measurability of a function relies on the atoms of σ-fields. Indeed, if $(M, \mathcal{M})$ and $(N, \mathcal{N})$ are two measurable spaces and f is a function defined on M with values in N, the definition 0.2.8 of the measurability of f clearly implies that:

$$(0.2.11) \qquad A_x^{\mathcal{M}} \subset A_x^{f^{-1}(\mathcal{N})} \qquad \forall \, x \in M,$$

i.e., any atom of $f^{-1}(\mathcal{N})$ is a union of atoms of $\mathcal{M}$. But

$$(0.2.12) \qquad A_x^{f^{-1}(\mathcal{N})} = f^{-1}\left[A_{f(x)}^{\mathcal{N}} \right].$$

Therefore if $\mathcal{N}$ is separating,

$$(0.2.13) \qquad A_x^{\mathcal{M}} \subset f^{-1}[\{f(x)\}],$$

or, equivalently, f is constant on the atoms of $\mathcal{M}$. The Blackwell Theorem (in the terminology of Dellacherie and Meyer (1975)) states that under some technical assumptions, this is a sufficient condition for the measurability of f. On this topic, see also Bhaskara Rao and Rao (1981).

0.2.12 Definition. A σ-field $\mathcal{M}$ of subsets of M is a *Blackwell σ-field* if $\mathcal{M}$ is separable and if $\forall \, m \in [\mathcal{M}]$ and $\forall \, A \in \mathcal{M}$, $m(A)$ is an analytic set of $\mathbb{R}$. A *Souslin σ-field* is a separating Blackwell σ-field. $\blacksquare$

An *analytic set* of $\mathbb{R}$ is the projection on $\mathbb{R}$ of a Borel set of $\mathbb{R}^2$. Remark that any Borel set is an analytic set. Remark also that a separable sub-σ-field of a Blackwell σ-field is a Blackwell σ-field. Note that if $\mathcal{M}$ is a Blackwell σ-field then $\overset{\bullet}{\mathcal{M}}$ is a Souslin σ-field. The concept of a Souslin σ-field is slightly more restrictive than the concept of a separable and separating σ-field; this is shown by the following two theorems.

0.2.13 Theorem. A σ-field $\mathcal{M}$ of subsets of M is separable and separating if and only if there exist $B \subset \mathbb{R}$ and $\varphi : (M, \mathcal{M}) \to (B, B \cap \mathcal{B})$ bijective and bimeasurable. $\blacksquare$

0.2.14 Theorem. A σ-field $\mathcal{M}$ of subsets of M is a Souslin σ-field if and only if there exist B, an analytic subset of $\mathbb{R}$, and $\varphi : (M, \mathcal{M}) \to (B, B \cap \mathcal{B})$ bijective and bimeasurable. $\blacksquare$

For the proofs of these theorems see, e.g., Dellacherie and Meyer (1975) Theorem I.11. and Theorem III.25. More regularity than a Souslin σ-field is sometimes required to obtain deeper results. This leads to the notion of a Standard Borel σ-field.

0.2.15 Definition. A σ-field $\mathcal{M}$ of subsets of M is a *Standard Borel* or *Lusin σ-field* if and only if there exist $B \in \mathcal{B}$ (a Borel set) and a mapping $\varphi : (M, \mathcal{M}) \to (B, B \cap \mathcal{B})$ bijective and bimeasurable.

Therefore any standard Borel σ-field is a Souslin σ-field and any Souslin σ-field is both separable and separating. Note also that the Borel σ-field of a Polish space is a Standard Borel σ-field (see Dellacherie and Meyer (1975) III.17, III.20 and III.21). We now state the *Blackwell Theorem*.

0.2.16 Theorem. Let $\mathcal{M}$ be a Blackwell σ-field of subsets of M, $\mathcal{M}_1$ -be a sub-σ-field of $\mathcal{M}$ and $\mathcal{M}_2$ be a separable sub-σ-field of $\mathcal{M}$. Then the following properties are equivalent:

(i) $\mathcal{M}_1 \subset \mathcal{M}_2$,

(ii) $A_x^{\mathcal{M}_2} \subset A_x^{\mathcal{M}_1} \ \forall \ x \in M$, i.e., any atom of $\mathcal{M}_1$ is a union of atoms of $\mathcal{M}_2$. ∎

For a proof see, e.g., Dellacherie and Meyer (1975) Theorem III.26. This theorem entails the following important corollary.

0.2.17 Corollary. Let $\mathcal{M}$ be a Blackwell σ-field of subsets of M, $\mathcal{M}_1$ be a separable sub-σ-field of $\mathcal{M}$, $(N,\mathcal{N})$ be a measurable space and $f : (M,\mathcal{M}) \rightarrow (N,\mathcal{N})$. Then the following properties are equivalent:

(i) $f^{-1}(\mathcal{N}) \subset \mathcal{M}_1$, i.e., f is $\mathcal{M}_1$-measurable,

(ii) $x' \underset{\mathcal{M}_1}{\sim} x \Rightarrow f(x') \underset{\mathcal{N}}{\sim} f(x)$.

In particular if $\mathcal{N}$ is separating, the following properties are equivalent:

(i) $f^{-1}(\mathcal{N}) \subset \mathcal{M}_1$, i.e., f is $\mathcal{M}_1$-measurable,

(ii) $f(A_x^{\mathcal{M}_1}) = \{f(x)\}$, i.e., f is constant on the atoms of $\mathcal{M}_1$.

Therefore if $m \in [\mathcal{M}]$, the following properties are equivalent:

(i) $m \in [\mathcal{M}_1]$,

(ii) $m(A_x^{\mathcal{M}_1}) = \{m(x)\}$, i.e., m is constant on the atoms of $\mathcal{M}_1$. ∎

For random variables, the property $m \in [\mathcal{M}_1]$ always implies that m is constant on the atoms of $\mathcal{M}_1$. However, the reciprocal condition requires that $m \in [\mathcal{M}]$, where $\mathcal{M}$ is a Blackwell σ-field and $\mathcal{M}_1$ is a separable sub-σ-field of $\mathcal{M}$.

0.2.3   Product of Measurable Spaces

Let us consider $\{(M_t,\mathcal{M}_t) : t \in T\}$ an arbitrary family of measurable spaces. Let $\tilde{M} = \mathsf{X}_{t \in T} \, M_t$, i.e.,

$$(0.2.14) \qquad x \in \tilde{M} \quad \Longleftrightarrow \quad x = (x_t \ : \ t \in T) \quad x_t \in M_t \quad \forall \, t \in T,$$

and $\forall\, t \in T$, $\pi_t : \tilde{M} \to M_t$ the *coordinate map*, i.e.,

$$(0.2.15) \qquad\qquad\qquad \pi_t(x) = x_t \qquad \forall\, t \in T.$$

Then the *product σ-field* of the $\mathcal{M}_t$, denoted by $\bigotimes_{t\in T} \mathcal{M}_t$, is the σ-field of subsets of $\tilde{M}$ defined by:

$$(0.2.16) \qquad\qquad\qquad \bigotimes_{t\in T} \mathcal{M}_t = \bigvee_{t\in T} \pi_t^{-1}(\mathcal{M}_t).$$

In particular, if $M_t = M$ and $\mathcal{M}_t = \mathcal{M}$ $\forall\, t \in T$, we denote the product of the measurable spaces $\{(M_t, \mathcal{M}_t) : t \in T\}$ by $(M^T, \mathcal{M}^T)$ and M^T may be viewed as the set of all functions defined on T with values in M.

The *product σ-field* $\bigotimes_{t\in T} \mathcal{M}_t$ is also generated as follows: a *cylinder set* A is defined as a set of the form:

$$
\begin{aligned}
A \ &= \ \{x \in \tilde{M} : \pi_{t_j}(x) \in B_{t_j}, \quad \forall\, 1 \le j \le k\}, \\
&= \ \bigcap_{1\le j\le k} \{x \in \tilde{M} : x_{t_j} \in B_{t_j}\},
\end{aligned}
$$

where $B_{t_j} \in \mathcal{M}_{t_j}$ $\forall\, 1 \le j \le k$. By Proposition 0.2.5, the σ-field $\bigotimes_{t\in T} \mathcal{M}_t$ as defined in (0.2.16) is equal to the σ-field generated by the class of all cylinder sets.

The following proposition addresses the measurability of functions with values in a product σ-field.

0.2.18 Proposition. Let $(N, \mathcal{N})$ be a measurable space, T be an arbitrary index set, and $\{(M_t, \mathcal{M}_t),\ t \in T\}$ be a family of measurable spaces. Then f, defined on $(N, \mathcal{N})$ with values in $(\bigtimes_{t\in T} M_t, \bigotimes_{t\in T} \mathcal{M}_t)$, is measurable if and only if $\pi_t \circ f \ :\ N \to M_t$ is measurable $\forall\, t \in T$. ∎

Let us consider the particular case of two measurable spaces $(M_1, \mathcal{M}_1)$ and $(M_2, \mathcal{M}_2)$. On $M_1 \times M_2$, $\mathcal{M}_1$ and $\mathcal{M}_2$ induce the following σ-fields, called *σ-fields of cylinder sets*.

$$(0.2.17) \qquad\qquad \pi_1^{-1}(\mathcal{M}_1) \ = \ \{A \times M_2 \ : A \in \mathcal{M}_1\},$$
$$(0.2.18) \qquad\qquad \pi_2^{-1}(\mathcal{M}_2) \ = \ \{M_1 \times B \ : B \in \mathcal{M}_2\}.$$

Then, by definition,

$$(0.2.19) \qquad\qquad \mathcal{M}_1 \otimes \mathcal{M}_2 \ = \ \pi_1^{-1}(\mathcal{M}_1) \vee \pi_2^{-1}(\mathcal{M}_2),$$

and by Proposition 0.2.5, it is the case that

$$(0.2.20) \qquad \mathcal{M}_1 \otimes \mathcal{M}_2 \;=\; \sigma[\{A \times B : A \in \mathcal{M}_1, \; B \in \mathcal{M}_2\}],$$

i.e., $\mathcal{M}_1 \otimes \mathcal{M}_2$ is also the σ-field on $M_1 \times M_2$ generated by the rectangles.

Warning. *Throughout this book, we will identify $\mathcal{M}_i$ and $\pi_i^{-1}(\mathcal{M}_i)$ for $i = 1, 2$, and therefore $\mathcal{M}_1 \otimes \mathcal{M}_2$ will be identified to $\mathcal{M}_1 \vee \mathcal{M}_2$.* ∎

This construction is trivially extended to a product of a finite number of measurable spaces. The most familiar example is provided by $I\!\!R^d$. With the usual Euclidian metric, $I\!\!R^d$ is a Polish space. Using Proposition 0.2.5, the Borel σ-field on $I\!\!R^d$, $\mathcal{B}(I\!\!R^d)$, may be shown to be equal to $\mathcal{B}^d$, i.e., the product σ-field of the Borel σ-fields on $I\!\!R$. A measurable function defined on $(M, \mathcal{M})$ with values in $(I\!\!R^d, \mathcal{B}^d)$, $m = (m_1, m_2, \ldots, m_d)'$, is called a *random vector*. By Proposition 0.2.18, this is equivalent to the property that $m_i \in [\mathcal{M}] \; \forall \, 1 \le i \le d$.

0.2.4 Monotone Class Theorems

In this section we present theorems useful for proving that a property satisfied by a class of subsets is also satisfied by the σ-field generated by this class. These theorems are usually called *Monotone Class Theorems*. Before reviewing these theorems, some new definitions are required.

0.2.19 Definitions. A family $\mathcal{C}$ of subsets of M is called

(i)   an *algebra* or a *field* if
   1. $\phi \in \mathcal{C}$,
   2. $A \in \mathcal{C} \;\; \Rightarrow \;\; A^c \in \mathcal{C}$,
   3. $A_1, A_2 \in \mathcal{C} \;\; \Rightarrow \;\; A_1 \cup A_2 \in \mathcal{C}$;

(ii)   a *monotone class* if, $A_n \in \mathcal{C} \;\; \forall \, n \in I\!\!N$,
   1. $A_n \subset A_{n+1} \;\; \forall \, n \in I\!\!N \;\; \Rightarrow \;\; \cup_{n \in I\!\!N} A_n \in \mathcal{C}$,
   2. $A_{n+1} \subset A_n \;\; \forall \, n \in I\!\!N \;\; \Rightarrow \;\; \cap_{n \in I\!\!N} A_n \in \mathcal{C}$;

(iii)   a *π-system* if
   $$A_1, A_2 \in \mathcal{C} \;\; \Rightarrow \;\; A_1 \cap A_2 \in \mathcal{C};$$

(iv) a *d-system* if
   1. $M \in \mathcal{C}$,
   2. $A_1, A_2 \in \mathcal{C}$ $A_1 \subset A_2$ $\Rightarrow$ $A_2 - A_1 \in \mathcal{C}$,
   3. $A_n \in \mathcal{C}$ $\forall\, n \in I\!N$, $A_n \subset A_{n+1}$ $\forall\, n \in I\!N$ $\Rightarrow$ $\cup_{n \in I\!N} A_n \in \mathcal{C}$.   ∎

Hence a field is closed for a finite sequence of set operations, a monotone class is closed for monotone sequences of subsets and a π-system is closed for finite intersections. It is important to note that the class of all cylinder sets, defined in Section 0.2.3, is a π-system. The following theorem will be useful for subsequent analysis.

0.2.20 Theorem.

(i)   If a monotone class of subsets of M contains a field of subsets of M it also contains the σ-field generated by this field.

(ii)   If a d-system of subsets of M contains a π-system of subsets of M it also contains the σ-field generated by this π-system.   ∎

For a proof of (i), the reader is referred to: Dellacherie and Meyer (1975) Theorem I.19 and Chung (1968) Theorem 2.1.2 and Neveu (1970) Propositions 1.4.1 and 1.4.2; and for a proof of (ii), see: Neveu (1970) Exercise 1.4.5, Breiman (1968) Proposition 2.23, and Blumenthal and Getoor (1968) Chapter 0 Theorem 2.2.

We now provide a functional form of the Monotone Class Theorem, which is used throughout this monograph.

0.2.21 Theorem. Let M be a set and $\mathcal{C}$ a π-system of subsets of M. Let $\mathcal{L}$ be a vector space of real-valued functions on M. If

(i) $1_A \in \mathcal{L}$ $\forall\, A \in \mathcal{C}$,

(ii) $1_M \in \mathcal{L}$,

(iii) $f_n \in \mathcal{L}$ $\forall\, n \in I\!N$, $0 \leq f_n \leq f_{n+1}$ $\forall\, n \in I\!N$, $f = \sup_n f_n$ bounded
       $\Rightarrow f \in \mathcal{L}$.
       Then $[\sigma(\mathcal{C})]_\infty \subset \mathcal{L}$.   ∎

For a proof see, e.g., Blumenthal and Getoor (1968) Chapter 0, Theorem 2.3. The main argument is based on the fact that $\{A \subset M : 1_A \in \mathcal{L}\}$ is a

d-system since $A_1 \subset A_2$ implies $\mathbf{1}_{A_2 - A_1} = \mathbf{1}_{A_2} - \mathbf{1}_{A_1}$ and $A_n \uparrow \cup_{n \in \mathbb{N}} A_n = A$ implies $\mathbf{1}_{A_n} \uparrow \mathbf{1}_A$. This theorem is used mainly in connection with Proposition 0.2.5, which states that $\bigvee_{t \in T} \mathcal{M}_t = \sigma(\mathcal{P})$ where $\mathcal{P}$ is a π-system.

0.3 Probability Spaces

0.3.1 Measures and Integrals

We first recall the definition of a measure.

0.3.1 Definition. Let $(M, \mathcal{M})$ be a measurable space and $\mu : \mathcal{M} \rightarrow \overline{\mathbb{R}}^+$. Then,

(i) μ is a *(positive) measure* on $(M, \mathcal{M})$ if μ is *countably additive*, i.e., $A_i \in \mathcal{M}$ $\forall i \in I$, I countable, and $A_i \cap A_{i'} = \phi$ $\forall i \neq i'$ imply that $\mu(\cup_{i \in I} A_i) = \sum_{i \in I} \mu(A_i)$,

(ii) the measure μ is *finite* if $\mu(M) < \infty$,

(iii) the measure μ is *σ-finite* if
$\exists A_n \in \mathcal{M}$, $n \in \mathbb{N}$, $M = \cup_{n \in \mathbb{N}} A_n$ such that $\mu(A_n) < \infty \, \forall \, n \in \mathbb{N}$. ∎

Note that the countable additivity of μ implies that $\mu(\phi) = 0$. The triple $(M, \mathcal{M}, \mu)$ is then called a *measure space*. The integral is constructed on this measure space (see, for instance, Metivier (1972) Chap. II-4, Billingsley (1979) Chap. 3, Sections 15 et 16).

0.3.2 Theorem. Let $(M, \mathcal{M}, \mu)$ be a σ-finite measure space. There exists a unique mapping I $: [\mathcal{M}]^+ \rightarrow \overline{\mathbb{R}}^+$ satisfying:

(i) $\text{I}(c_1 m_1 + c_2 m_2) = c_1 \, \text{I}(m_1) + c_2 \, \text{I}(m_2)$ $\forall \, m_i \in [\mathcal{M}]^+$, $c_i \in \mathbb{R}^+$
$i = 1, 2$,

(ii) $m_n \in [\mathcal{M}]^+$ $m_n \uparrow m$ $\Rightarrow$ $\text{I}(m_n) \uparrow \text{I}(m)$,

(iii) $\text{I}(\mathbf{1}_A) = \mu(A)$ $\forall \, A \in \mathcal{M}$.

This mapping, called the *integral*, is also written as $\text{I}(m) = \int m \, d\mu$. ∎

The proof is based on (0.2.10). Indeed under (i) (ii) and (iii),

(0.3.1) $I(m) = \sup_n I(m_n),$ and

(0.3.2) $I(m_n) = \sum_{1 \leq k \leq n2^n - 1} \frac{k}{2^n} \mu\left[\left\{\frac{k}{2^n} < m \leq \frac{k+1}{2^n}\right\}\right] + n\,\mu[\{m > n\}].$

Note also the following alternative notation:

(0.3.3) $I(m\mathbf{1}_A) = \int_A m\,d\mu = \int_A m(x)\,\mu(dx).$

A further useful theorem is the *Lebesgue Decomposition Theorem*.

0.3.3 Theorem. Let $(M, \mathcal{M})$ be a measurable space and μ and ν be two σ-finite measures on $(M, \mathcal{M})$. Then there exist $N \in \mathcal{M}$ with $\mu(N) = 0$ and $m \in [\mathcal{M}]^+$ such that:
(i) $\nu(A) = \nu(A \cap N) + \int_A m\,d\mu \quad \forall\, A \in \mathcal{M}.$

Moreover, if $N' \in \mathcal{M}$ with $\mu(N') = 0$ and $m' \in [\mathcal{M}]^+$ also satisfy (i), then,
(ii) $\nu(N \Delta N') = 0,$

(iii) $\mu[\{m' \neq m\}] = 0.$ ■

For a proof see, e.g., Chow and Teicher (1978) Section 6.5. and Neveu (1970) Proposition IV-1-3.

0.3.4 Definitions. Let μ and ν be two σ-finite measures on the measurable space $(M, \mathcal{M})$.
(i) ν is *absolutely continuous* with respect to μ, and is denoted by $\nu \ll \mu$, if $A \in \mathcal{M}$ and $\mu(A) = 0$ imply $\nu(A) = 0.$

(ii) μ and ν are *mutually singular*, and is denoted by $\mu \perp \nu$, if there exists $N \in \mathcal{M}$ such that $\mu(N) = 0$ and $\nu(N^c) = 0.$ ■

With these definitions, Theorem 0.3.3. may be expressed as:

0.3.5 Corollary. Let μ and ν be two σ-finite measures on the same measurable space $(M, \mathcal{M})$. Then there exist two unique σ-finite measures λ_1 and λ_2 such that

(i) $\nu = \lambda_1 + \lambda_2$,

(ii) $\lambda_1 \perp \mu$,

(iii) $\lambda_2 \ll \mu$. ∎

The Radon-Nikodym Theorem is a corollary of Theorem 0.3.3:

0.3.6 Corollary. Let μ and ν be two σ-finite measures on the measurable space $(M, \mathcal{M})$. Then the following properties are equivalent:

(i) $\nu \ll \mu$,

(ii) $\exists\, m \in [\mathcal{M}]^+$ essentially unique such that $\nu(A) = \int_A m \, d\mu \;\; \forall\, A \in \mathcal{M}$. ∎

Essential uniqueness refers to property (iii) of Theorem 0.3.3. A commonly used notation for m is $d\nu/d\mu$ and it is called a *Radon-Nikodym derivative* of ν with respect to μ.

0.3.2 Probabilities. Expectations. Null Sets

Let us first recall the definitions of probability and expectation.

0.3.7 Definition. Let $(M, \mathcal{M})$ be a measurable space. A *probability* is a measure P such that $P(M) = 1$. The triple $(M, \mathcal{M}, P)$ is then called a *probability space* and the integral associated to P is called the (mathematical) *expectation* and is denoted indifferently as $E(m) = \int m \, dP$ $m \in [\mathcal{M}]^+$. ∎

The properties of probability and expectation may be found in Metivier (1972), Chung (1968) and Neveu (1970).

In the probability space $(M, \mathcal{M}, P)$, we denote by $\overline{\mathcal{M}}_0$ the sub-σ-field of $\mathcal{M}$ defined by

$$(0.3.4) \qquad \overline{\mathcal{M}}_0 = \{A \in \mathcal{M} \; : \; P(A)^2 = P(A)\}.$$

Thus, $\overline{\mathcal{M}}_0$ is generated by the family of sets of probability zero, called *null sets*, and is called the *completed trivial σ-field*. We also denote, for $\mathcal{N}$ a sub-σ-field of $\mathcal{M}$,

$$(0.3.5) \qquad \overline{\mathcal{N}} = \mathcal{N} \vee \overline{\mathcal{M}}_0.$$

This is the smallest σ-field containing $\mathcal{N}$ and the (measurable) null sets and it is usually called the *completed sub-σ-field* $\overline{\mathcal{N}}$. Remark that this definition of (measurable) completion should be distinguished from the Lebesgue completion by all the subsets of the (measurable) null sets (see Neveu (1970) I-4). Thus, in our definition, the completion $\overline{\mathcal{M}}$ of $\mathcal{M}$ is equal to $\mathcal{M}$. If we want to make explicit the role of the probability P we write $\overline{\mathcal{N}}^P$ the completion of $\mathcal{N}$ by the P-null sets. In some cases it will be useful to increase a sub-σ-field $\mathcal{M}_1$ of $\mathcal{M}$ with the null sets of a second sub-σ-field $\mathcal{M}_2$, i.e., $\mathcal{M}_1 \vee (\mathcal{M}_2 \cap \overline{\mathcal{M}}_0)$. When $\mathcal{M}_1$ is a sub-σ-field of $\mathcal{M}_2$, the following relation obtains:

$$(0.3.6) \qquad \mathcal{M}_1 \vee (\mathcal{M}_2 \cap \overline{\mathcal{M}}_0) = \overline{\mathcal{M}}_1 \cap \mathcal{M}_2 \quad \forall \, \mathcal{M}_1 \subset \mathcal{M}_2 \subset \mathcal{M}.$$

A measurable function defined on a probability space permits the transfer of probability from one measurable space to another.

0.3.8 Proposition. Let $(M, \mathcal{M}, P)$ be a probability space, $(N, \mathcal{N})$ be a measurable space and $f : (M, \mathcal{M}) \rightarrow (N, \mathcal{N})$. If $Q = P \circ f^{-1}$ then Q is a probability on $(N, \mathcal{N})$ called the *image under f of the probability P*. Moreover for all $n \in [\mathcal{N}]^+$, we have equivalently:

(i) $E_Q(n) = E_P(n \circ f)$,

(ii) $\int_N n \; d(P \circ f^{-1}) = \int_M (n \circ f) \; dP.$ ∎

For a random variable $m \in [\mathcal{M}]$, the p-semi-norms for $p \in (0, \infty]$ are defined by

$$(0.3.7) \qquad \|m\|_p = \{E[| \, m \, |^p]\}^{1/p} \qquad p \in (0, \infty),$$

$$(0.3.8) \qquad \|m\|_\infty = \inf\{c \; : \; P(| \, m \, | > c) = 0\},$$

and the corresponding vector spaces usually denoted by $\mathcal{L}_p(M, \mathcal{M}, P)$ or more simply by $\mathcal{L}_p$ will hereafter be denoted by $[\mathcal{M}]_p$, i.e.,

$$(0.3.9) \qquad [\mathcal{M}]_p = \{m \in [\mathcal{M}] \; : \; \|m\|_p < \infty\}.$$

The advantage of this notation is that it stresses the measurability property: if N is a sub-σ-field of $\mathcal{M}$, $[N]_p$ represents the linear subspace of $[\mathcal{M}]_p$ of N-measurable random variables. Note that the same notation $[\mathcal{M}]_\infty$ is used for bounded and for almost surely bounded random variables. Recall that

$$(0.3.10) \qquad [\mathcal{M}]_{p_2} \subset [\mathcal{M}]_{p_1} \quad \forall\, p_1 < p_2.$$

The expectation may be extended as a linear functional on $[\mathcal{M}]_1$ by setting

$$(0.3.11) \qquad E(m) = E(m^+) - E(m^-),$$

where $m^+ = \max(m, 0)$ and $m^- = (-m)^+$.

Two random variables, $m, m' \in [\mathcal{M}]$, are said to be *almost surely equal* if $P[\{m \neq m'\}] = 0$. This property will be denoted as $m = m'$ a.s. or $m = m'$ a.s.P if we want to make explicit the role of the probability P. This property is in fact an equivalence relation. Recall that the quotient space under this equivalence relation on $\mathcal{L}_p$, denoted as

$$(0.3.12) \qquad L_p = \mathcal{L}_p/\sim,$$

is a Banach space $\forall\, p \in [1, \infty]$, and that L_2 is a Hilbert space. If $\frac{1}{p} + \frac{1}{q} = 1$ and $p \in [1, \infty)$, then the dual space of L_p, denoted as L_p^*, is isomorphic to L_q.

0.3.3 Transition and Product Probability

Transition is a primordial concept in Bayesian statistics.

0.3.9 Definition. Let $(M, \mathcal{M})$ and $(N, \mathcal{N})$ be two measurable spaces. A *transition* (or *transition probability*, or *Markov kernel*) from $(M, \mathcal{M})$ to $(N, \mathcal{N})$ is a mapping $P^\bullet(\bullet) : M \times \mathcal{N} \to [0, 1]$ such that

(i) $P^\bullet(A) \in [\mathcal{M}] \quad \forall\, A \in \mathcal{N}$,

(ii) $P^x(\bullet)$ is a probability on $(N, \mathcal{N}) \quad \forall\, x \in M$.

We denote such a transition by $P^\bullet(\bullet) : (M, \mathcal{M}) \longrightarrow\!\!\!< (N, \mathcal{N})$. ∎

A transition from $(M, \mathcal{M})$ to $(N, \mathcal{N})$ and a probability on $(M, \mathcal{M})$ jointly define a probability on the product space:

0.3.10 Theorem. Let us consider a probability space $(M, \mathcal{M}, Q)$ and a transition $P^{\bullet}(\bullet) : (M, \mathcal{M}) \longrightarrow< (N, \mathcal{N})$. Then there exists a unique probability denoted by $Q \otimes P^x$ on $(M \times N, \mathcal{M} \otimes \mathcal{N})$ such that

$$(Q \otimes P^x)(A \times B) = \int_A P^x(B) \, Q(dx) \quad \forall \, A \in \mathcal{M} \quad \forall \, B \in \mathcal{N}.$$

Moreover if $h \in [\mathcal{M} \otimes \mathcal{N}]^+$ we have successively that

$\int_N h(x, y) P^x(dy) \in [\mathcal{M}]^+$

$\int_{M \times N} h \, d(Q \otimes P^x) = \int_M Q(dx) \int_N h(x, y) \; P^x(dy).$ ∎

For a proof see Neveu (1970) III.2. In particular, if $P^x = P \quad \forall \, x \in M$, the transition reduces to a probability on $(N, \mathcal{N})$ and the last property of Theorem 0.3.10 reduces to Fubini Theorem.

0.3.4   Conditional Expectation

In this section, we review the most important properties of the conditional expectation. The conditional expectation will be the most important tool used in this monograph.

0.3.11 Definition. Let $(M, \mathcal{M}, P)$ be a probability space and $\mathcal{N}$ be a sub-σ-field of $\mathcal{M}$. For any $m \in [\mathcal{M}]^+$, we define a *conditional expectation of m with respect to $\mathcal{N}$*, denoted by $\mathcal{N}m$, as any $n \in [\mathcal{N}]^+$ such that $E[mr] = E[nr] \quad \forall \, r \in [\mathcal{N}]^+$. ∎

Remark that n is essentially unique, in the sense that if n and n' satisfy Definition 0.3.11 then $n = n'$ a.s.P; and if n satisfies Definition 0.3.11, $n' = n$ a.s.P and $n' \in [\mathcal{N}]$, then n' also satisfies Definition 0.3.11.

Warning.   *$\mathcal{N}m$ might be interpreted as an equivalence class for almost sure equality. We will not adopt this interpretation here. Since $\mathcal{N}m$ is only almost surely defined, any equality involving conditional expectations are then almost sure equalities but, in the following, we lighten the notation* *by deleting the "a.s." for those equalities where the reference probability is not ambiguous. Such equalities will thus be verified almost surely by any version of these conditional expectations.* ∎

The existence of a conditional expectation may be based on the Radon-Nikodym Theorem as in Neveu (1970) IV.3. Indeed, if we define, for $m \in [\mathcal{M}]_\infty^+$, $\lambda(B) = E[m\mathbf{1}_B]$ and $\mu(B) = P(B)$, then λ and μ are finite measures on $(M, \mathcal{N})$ and $\lambda \ll \mu$. It may also relies on the Riesz Theorem about linear functionals on Hilbert spaces as, e.g., in Dellacherie and Meyer (1975) II.3. This approach to conditional expectation motivates the notation $\mathcal{N}m$ introduced by Hunt (1957), rather than the usual notation $E(m \mid \mathcal{N})$. We have adopted Hunt's notation because of its compactness and because, in the Hilbert case, $\mathcal{N}m$ actually represents the orthogonal projection of the random variable $m \in [\mathcal{M}]_2^+$ onto the subspace $[\mathcal{N}]_2^+$. In Chapter 4 this notation is extended to the concept of projection of a σ-field onto another σ-field.

If $\mathcal{N} = \mathcal{M}_0 = \{\phi, M\}$, i.e., the trivial σ-field, $\mathcal{M}_0 m$ is a constant and is equal to $E(m)$. Therefore, for the sake of coherence, we use the notation $\mathcal{M}_0 m$ rather than $E(m)$. Thus, using our notation, the definition of the conditional expectation of m with respect to $\mathcal{N}$ may be rewritten as

$$(0.3.13) \qquad \mathcal{N}m \in [\mathcal{N}]^+,$$

$$(0.3.14) \qquad \mathcal{M}_0(m \cdot r) = \mathcal{M}_0(\mathcal{N}m \cdot r) \quad \forall\, r \in [\mathcal{N}]^+.$$

The computation of conditional expectations relies on the following proposition, the proof of which is a direct application of the Monotone Class Theorem.

0.3.12 Proposition. Let $(M, \mathcal{M}, P)$ be a probability space, $\mathcal{N}$ be a sub-σ-field of $\mathcal{M}$ and $\mathcal{C}$ be a π-system of subsets of M such that $\mathcal{N} = \sigma(\mathcal{C})$. If $m \in [\mathcal{M}]_1^+$ and $n \in [\mathcal{N}]_1^+$ satisfy:

(i) $\mathcal{M}_0 m = \mathcal{M}_0 n,$

(ii) $\mathcal{M}_0(m\mathbf{1}_A) = \mathcal{M}_0(n\mathbf{1}_A) \quad \forall\, A \in \mathcal{C},$

then,

(iii) $n = \mathcal{N}m$ a.s.P.   ∎

We now list the basic properties of conditional expectations. For a proof see, e.g., Metivier (1972) IV.2. or Neveu (1970) IV.3. The first set of properties are the same as those of the expectation except that they are only almost surely satisfied.

0.3.13 Proposition. Let $(M, \mathcal{M}, P)$ be a probability space and $\mathcal{N}$ be a sub-σ-field of $\mathcal{M}$. Then:

(i) $\mathcal{N}(cm) = c\mathcal{N}(m)$ $\forall\, m \in [\mathcal{M}]^+$ $\forall\, c \in \mathbb{R}^+$,

(ii) $\mathcal{N}(m_1 + m_2) = \mathcal{N}m_1 + \mathcal{N}m_2$ $\forall\, m_i \in [\mathcal{M}]^+$ $i = 1, 2$,

(iii) $m_i \in [\mathcal{M}]^+$ $i = 1, 2$ and $m_1 \leq m_2$ imply $0 \leq \mathcal{N}m_1 \leq \mathcal{N}m_2$,

(iv) $m_n \in [\mathcal{M}]^+$ $\forall\, n \in I\!N$ and $m_n \uparrow m$ imply $\mathcal{N}m = \sup_n \mathcal{N}m_n$.   ∎

In contrast, the following set of properties are particular to conditional expectations.

0.3.14 Proposition. Let $(M, \mathcal{M}, P)$ be a probability space.

(i)   If $\mathcal{N}$ is a sub-σ-field of $\mathcal{M}$, then: $\forall\, m \in [\mathcal{M}]^+$, $\forall\, n \in [\mathcal{N}]^+$,
$\mathcal{N}(mn) = n \cdot \mathcal{N}m$.

In particular, $\mathcal{N}(n) = n$ and so $\mathcal{N}(\mathbf{1}_M) = 1$.

(ii) If $\mathcal{N}_i \subset \mathcal{M}$ $i = 1, 2$ and $\mathcal{N}_1 \subset \mathcal{N}_2$ then,
$\mathcal{N}_2(\mathcal{N}_1 m) = \mathcal{N}_1(\mathcal{N}_2 m) = \mathcal{N}_1 m$ $\forall\, m \in [\mathcal{M}]^+$.

In particular, $\mathcal{M}_0(\mathcal{N}m) = \mathcal{M}_0 m$ $\forall\, \mathcal{N} \subset \mathcal{M}$, $\forall\, m \in [\mathcal{M}]^+$.   ∎

Concerning the completion of sub-σ-fields by null sets, it is easily verified that:

(0.3.15) $\overline{\mathcal{N}}m = \mathcal{N}m$ a.s.P

although the equivalence class corresponding to $\overline{\mathcal{N}}m$ is larger than that of $\mathcal{N}m$.

As for the expectation, the conditional expectation may be defined on $[\mathcal{M}]_1$ by setting:

(0.3.16) $$\mathcal{N}m = \mathcal{N}m^+ - \mathcal{N}m^-.$$

Under this extension, the conditional expectation with respect to $\mathcal{N}$ becomes a linear operator from $[\mathcal{M}]_1$ into $[\mathcal{N}]_1$. Most of the properties of the conditional expectations (Proposition 0.3.13 (i) (ii) (iii) and Proposition 0.3.14 (i) and (ii)) remain true provided that they make sense, i.e., $+\infty - \infty$ must be avoided. The *Jensen inequality* is useful in this context.

0.3.15 Proposition. Let $(M, \mathcal{M}, P)$ be a probability space and $\mathcal{N}$ be a sub-σ-field of $\mathcal{M}$. If $m \in [\mathcal{M}]_1$ and $\varphi : \mathbb{R} \to \mathbb{R}$ is a convex function such that $\varphi(m) \in [\mathcal{M}]_1$, then $\varphi(\mathcal{N}m) \le \mathcal{N}[\varphi(m)]$. ∎

The conditional probability is defined through the conditional expectation.

0.3.16 Definition. Let $(M, \mathcal{M}, P)$ be a probability space and $\mathcal{N}$ be a sub-σ-field of $\mathcal{M}$, the *conditional probability* of $A \in \mathcal{M}$ with respect to $\mathcal{N}$, denoted by $P^{\mathcal{N}}(A)$, is defined by $P^{\mathcal{N}}(A) = \mathcal{N}(\mathbf{1}_A)$. ∎

We shall also use the notation $P_{\mathcal{N}}$ to denote the restriction of P to a sub-σ-field $\mathcal{N}$ of $\mathcal{M}$, i.e.,

$$P_{\mathcal{N}}(A) = P(A) \qquad \forall\, A \in \mathcal{N}.$$

$P_{\mathcal{N}}$ is also called the *trace* of P on $\mathcal{N}$ or the *marginal probability* on $\mathcal{N}$. Combining these two notations we obtain for any $\mathcal{M}_i$, $i = 1, 2$, sub-σ-fields of $\mathcal{M}$:

$$P_{\mathcal{M}_1}^{\mathcal{M}_2}(A) = P^{\mathcal{M}_2}(A) \quad \forall\, A \in \mathcal{M}_1.$$

Note that $\mathcal{M}_2$ need not be a sub-σ-field of $\mathcal{M}_1$, but both $\mathcal{M}_1$ and $\mathcal{M}_2$ are sub-σ-fields of $\mathcal{M}$.

Recall that, in general, a conditional probability need not be a transition. Indeed, the countable additivity property

(0.3.17) $$P^{\mathcal{N}}(\cup_n A_n) = \sum_n P^{\mathcal{N}}(A_n),$$

for $A_n \in \mathcal{M}$, $n \in \mathbb{N}$, $A_n \cap A_{n'} = \phi$ $\forall n \neq n'$, is only almost sure. Therefore, by the uncountability of all these properties, the null sets where it fails may pile down to give a nonnull set.

0.3.17 Definition. For $\mathcal{M}_1$ and $\mathcal{M}_2$, sub-σ-fields of $\mathcal{M}$, one says that there exists a *regular version of the conditional probability* on $\mathcal{M}_1$ given $\mathcal{M}_2$, if there exists a transition $P^{\bullet}(\bullet)$: $(M, \mathcal{M}_2) \longrightarrow\!\!\!\!\!< (M, \mathcal{M}_1)$ such that

$$P^{\mathcal{M}_2}_{\mathcal{M}_1}(A) = P^x(A) \quad \forall A \in \mathcal{M}_1. \qquad \blacksquare$$

$P^{\bullet}(\bullet)$ is also called a *desintegration* of P on $\mathcal{M}_1$ given $\mathcal{M}_2$. Jirina Theorem identifies a condition which, if satisfied, implies the existence of a regular conditional probability.

0.3.18 Theorem. Let $(M, \mathcal{M}, P)$ be a probability space and $\mathcal{M}_1$ be a Standard Borel sub-σ-field of $\mathcal{M}$, then for any $\mathcal{M}_3 \subset \mathcal{M}_1$ and $\mathcal{M}_2 \subset \mathcal{M}$, there exists a regular conditional probability on $\mathcal{M}_3$ given $\mathcal{M}_2$. $\qquad \blacksquare$

For a proof and more general results see, e.g., Blackwell and Dubbins (1975), Hoffman and Jørgensen (1971), Dellacherie and Meyer (1975) III.3, Neveu (1970), Corollary of Proposition V.4.4, or Parthasarathy (1977), Section 46.

0.3.5 Densities

Let $(M, \mathcal{M}, P)$ be a probability space. If μ is a σ-finite measure on $(M, \mathcal{M})$ such that $P \ll \mu$, then by the theorem of Radon-Nikodym, we know this is equivalent to stating that there exists $f \in [\mathcal{M}]^+_1$ such that

$$(0.3.18) \qquad P(A) = \int_A f \, d\mu \qquad \forall A \in \mathcal{M}.$$

In the special case of a probability space such an f is called a *density*, instead of a Radon-Nikodym derivative.

Consider now a transition $P^{\bullet}(\bullet)$: $(M, \mathcal{M}) \longrightarrow\!\!\!\!\!< (N, \mathcal{N})$ such that

$$(0.3.19) \qquad P^x \ll \mu \qquad \forall x \in M.$$

Thus each P^x admits a density $f_x \in [\mathcal{N}]_1^+$. In a Bayesian framework, it may be suitable to have versions of these densities such that, considered as defined on $M \times N$, it is bimeasurable, i.e., $f_\bullet(\bullet) \in [\mathcal{M} \otimes \mathcal{N}]$. The following theorem gives conditions implying the existence of such a bimeasurable selection of those densities.

0.3.19 Theorem. Let $(N, \mathcal{N})$ and $(M, \mathcal{M})$ be measurable spaces with $\mathcal{N}$ separable. Let $P^\bullet(\bullet) : (M, \mathcal{M}) \longrightarrow\!\!\!< (N, \mathcal{N})$ be a transition and Q a probability on $(N, \mathcal{N})$ such that

(i) $P^x \ll Q \quad \forall\, x \in M$.

Then there exists a function $f \in [\mathcal{M} \otimes \mathcal{N}]_1^+$ such that

(ii) $P^x(A) = \int_A f(x, y)\ Q(dy) \quad \forall\, A \in \mathcal{N}$.   ■

For a proof see, e.g., Dellacherie and Meyer (1980) V.5.

1

Bayesian Experiments

1.1 Introduction

In this chapter, we introduce the formal structure to be analyzed in this monograph. This structure relies on the basic concept of a "Bayesian experiment". Whereas in sampling theory the basic concept is that of a "statistical experiment", i.e., a family of probability measures on a given sample space, the Bayesian experiment is characterized by a unique probability measure on the product of the parameter space and the sample space. This structure is first introduced in its full generality, i.e., without imposing regularity conditions. Conditions for the existence of a regular conditional probability (such as the posterior probability) or for the existence of a measure dominating every sampling probability are discussed subsequently. Examples are given to show that, in order to be dealt within a single framework, standard situations do indeed require a fully general formulation of the Bayesian experiment. The last section of this chapter handles the reduction of Bayesian experiments, i.e., the problem of reducing a probability measure on a product space and, which permits the introduction of the notions of marginal and of conditional experiments; we insist both on the symmetry between parameters and observations in the reduction process and on the care to be taken for those concepts of parameters and of observations in the treatment of a conditional experiment.

Unlike most other chapters, this one does not include a section on some

topic in probability theory as it does not make use of probabilistic techniques
beyond those reviewed in the preliminary Chapter 0. It should nonetheless
be mentioned that in the sections on dominated experiments the treatment
of densities is more detailed than in most statistical textbooks; it also somewhat idiosyncratic.

Sections 1.2 and 1.4 are based on Florens (1974) and Florens and Mouchart
(1977), (1986a), but subsections 1.4.4 and 1.4.5 derive from later work. Section 1.3 is based on Florens and Mouchart (1986b).

1.2   The Basic Concepts of Bayesian Experiments

1.2.1   General Definitions

We start from a *statistical experiment* $\mathcal{E}$ defined as:

$$(1.2.1) \qquad\qquad \mathcal{E} = \{(S, \mathcal{S}), P^a : a \in A\}$$

where $(S, \mathcal{S})$ is a measurable space (the *sample space*) and $\{P^a : a \in A\}$ is a
family of probability measures on the sample space indexed by a *parameter*
a belonging to a *parameter space* A. The probability measures P^a are called
the *sampling probabilities*. A probability measure Π on the product space
$A \times S$ may be constructed by endowing the parameter space with a prob-
ability measure μ on $(A, \mathcal{A})$ where the σ-field $\mathcal{A}$ makes $P^a(X)$ measurable
for any $X \in \mathcal{S}$ and by extending to $\mathcal{A} \otimes \mathcal{S}$ (in a unique way) the function Π
defined on $\mathcal{A} \times \mathcal{S}$ as follows:

$$(1.2.2) \qquad \Pi(E \times X) = \int_E P^a(X)\mu(da) \qquad E \in \mathcal{A}, \quad X \in \mathcal{S}.$$

The measure constructed from (1.2.2) is denoted as:

$$(1.2.3) \qquad\qquad \Pi = \mu \otimes P^{\mathcal{A}}.$$

We shall also denote by P, the marginal measure on $(S, \mathcal{S})$:

$$(1.2.4) \qquad\qquad P(X) = \Pi(A \times X) \qquad X \in \mathcal{S}.$$

Note that by the construction (1.2.2), P^a becomes a regular version of the
restriction to $\mathcal{S}$ of the conditional probability Π given $\mathcal{A}$ (thus we wrote, in
(1.2.3) $P^{\mathcal{A}}$ instead of P^a), and that μ becomes the restriction of Π to $\mathcal{A}$, or

the marginal measure on $(A, \mathcal{A})$. (On the constructions (1.2.2) and (1.2.3) see Theorem 0.3.10.)

Remark. Strictly speaking, we identify μ, a probability measure on $(A, \mathcal{A})$, with the restriction of Π to the σ-field of cylinders $\mathcal{A} \times S$. More generally, we shall systematically identify the sub-σ-fields $\mathcal{B} \subset \mathcal{A}$ (respectively, $\mathcal{T} \subset \mathcal{S}$) with the sub-$\sigma$-fields of the corresponding cylinders $\mathcal{B} \times S$ (respectively, $A \times \mathcal{T}$); similarly, we identify the product σ-fields $\mathcal{A} \otimes \mathcal{S}$ with $\mathcal{A} \vee \mathcal{S}$, the σ-field generated by $(\mathcal{A} \times S) \cup (A \times \mathcal{S})$. The central feature of the Bayesian model is that it views a statistical experiment as a (unique) probability measure on the product space $A \times S$; more formally we define:

1.2.1 Definition. *A Bayesian experiment* is defined by the following probability space:

$$(1.2.5) \qquad \mathcal{E} = (A \times S, \mathcal{A} \vee \mathcal{S}, \Pi)$$

where $(A, \mathcal{A})$ is called the *parameter space* and $(S, \mathcal{S})$ is called the *sample space*. The restriction of Π on $(A, \mathcal{A})$ is called the *prior probability* and the restriction of Π on $(S, \mathcal{S})$ is called the *predictive probability*. For any integrable random variable m defined on $A \times S$, a conditional expectation given $\mathcal{A}$, denoted $\mathcal{A}m$, is called a *sampling expectation* and a conditional expectation given $\mathcal{S}$, denoted $\mathcal{S}m$, is called a *posterior expectation*. A *statistic* is either a $\mathcal{S}$-measurable function defined on $A \times S$ with value in some measurable space or, more simply, a sub-σ-field of $\mathcal{S}$. Similarly, a *subparameter*, or more simply a *parameter*, is a sub-σ-field of $\mathcal{A}$. ∎

Besides the decomposition of Π as in (1.2.3), rendering this concept operational often requires the existence of the converse decomposition of Π into a marginal probability P on $(S, \mathcal{S})$, and a regular conditional probability given $\mathcal{S}$, represented by a transition which is denoted $\mu^{\mathcal{S}}$; this observation motivates the following definition:

1.2.2 Definition. *A regular Bayesian experiment* is defined as the following (desintegrable) Bayesian experiment:

$$(1.2.6) \qquad \mathcal{E} = (A \times S, \ \mathcal{A} \vee \mathcal{S}, \ \Pi = \mu \otimes P^{\mathcal{A}} = P \otimes \mu^{\mathcal{S}})$$

where μ and P are respectively the *prior* and the *predictive probabilities*, and

the transitions $P^{\mathcal{A}} : (A, \mathcal{A}) \longrightarrow\!\!< (S, \mathcal{S})$ and $\mu^{\mathcal{S}} : (S, \mathcal{S}) \longrightarrow\!\!< (A, \mathcal{A})$
are called the *sampling* and the *posterior probabilities*, respectively.   ∎

1.2.2 Dominated Experiments

The statistical literature often uses arguments in terms of density, and
the objective of this section is to introduce those density-based arguments
relevant to Bayesian experiments. In a sampling theory framework, a *dominated* statistical experiment is a statistical experiment such that there exists
a dominating σ-finite measure λ, i.e., λ is such that $\forall a \in A : P^a \ll \lambda$. In
such a case, the theorem of Halmos and Savage (1949) says that when a statistical experiment is dominated by some measure λ, it is also dominated by
a "privileged" dominating probability; more precisely if $P^a \ll \lambda$ $\forall a \in A$,
there exists $A_0 \subset A$ countable such that $P^a \ll P_* \ll \lambda$ $\forall a \in A$ for
any P_* of the form $P_* = \sum_{a \in A_0} \alpha(a) P^a$ where $\alpha(a) > 0$ $\forall a \in A_0$ and
$\sum_{a \in A_0} \alpha(a) = 1$

In a Bayesian framework, an interesting domination property arises
when $\Pi \ll \mu \otimes P$, i.e., Π is dominated by the independent product of its
marginals (on A and on S). This occurs, in particular, when the sampling
probabilities are dominated, (i.e., $P^a \ll \lambda$ $\forall a \in A$) and when there exists a
bimeasurable Radon-Nikodym density, i.e., there exists an $\mathcal{A} \vee \mathcal{S}$-measurable
function $f(a, s)$ such that:

$$\frac{dP^a}{d\lambda} = f(a, s) \quad \text{a.e.}\lambda \quad \forall a \in A.$$

In this situation Π is also dominated by $\mu \otimes \lambda$, since by Fubini Theorem:

$$\Pi(E \times X) = \int_{E \times X} f(a, s) \mu(da) \lambda(ds).$$

Furthermore, there also exists an $\mathcal{A} \vee \mathcal{S}$-measurable function $g_*(a, s)$ that,
for each $a \in A$ is a density of P^a with respect to P_* and such that $f(a, s) =
g_*(a, s) f_*(s)$ a.e.$\mu \otimes \lambda$ where $f_*(s) = \sum_{a \in A_0} \alpha(a) f(a, s)$. As the Bayesian
analogue of the Halmos-Savage Theorem says that $\Pi \ll \mu \otimes \lambda$ implies $\Pi \ll
\mu \otimes P$, in a Bayesian experiment, the predictive probability P plays a role
similar to that of a privileged dominating probability, even when such a
P_* does not exist. (An example of this use of P will be shown up in the
theory of sufficiency in Section 2.3.7). This assertion is established in the
next theorem.

1.2.3 Theorem. If $\Pi \ll \mu \otimes \lambda$ where λ is a σ-finite measure on $(S, \mathcal{S})$, then $\Pi \ll \mu \otimes P$.

Proof. Let $h(a, s)$ be a positive $\mathcal{A} \vee \mathcal{S}$-measurable function such that:

$$(*) \qquad \int_{A \times S} h(a, s)\mu(da)P(ds) = 0$$

We have to show that this implies:

$$(**) \qquad \int_{A \times S} h(a, s)\, f(a, s)\mu(da)\lambda(ds) = 0$$

Remembering that

$$P(X) = \Pi(A \times X) = \int_{A \times X} f(a, s)\mu(da)\lambda(ds)$$

we obtain, using the Fubini theorem:

$$P(X) = \int_X k(s)\lambda(ds)$$

where

$$k(s) = \int_A f(a, s)\mu(da).$$

Therefore $(*)$ may be written as:

$$\int_{A \times S} h(a, s)\, k(s)\, \mu(da)\lambda(ds) = 0.$$

Hence there exists $S_0 \in \mathcal{S}$ such that $\lambda(S_0) = 0$ and

$$k(s) \int_A h(a, s)\mu(da) = 0 \quad \forall s \in S_0^c.$$

Therefore $h(a, s) = 0 \quad$ a.s.$\mu \quad \forall s \in S_0^c \cap \{k > 0\}$. As well, from the definition of $k(s)$, we also have $f(a, s) = 0 \quad$ a.s.$\mu \quad \forall s \in S_0^c \cap \{k = 0\}$. Therefore $h(a, s)f(a, s) = 0 \quad$ a.s.$\mu \quad \forall s \in S_0^c$. Hence, by Fubini Theorem, $(*)$ implies $(**)$ because:

$$\int_{A \times S} h(a, s)f(a, s)\mu(da)\lambda(ds) = \int_{S_0^c} \lambda(ds) \int_A h(a, s)f(a, s)\mu(da) = 0. \quad \blacksquare$$

Thus, a Radon-Nikodym derivative of Π with respect to $\mu \otimes P$ may be taken as:

$$(1.2.7) \qquad \frac{d\Pi}{d(\mu \otimes P)} = \frac{f(a, s)}{k(s)} = g(a, s) \quad \text{when } k(s) > 0$$

$$= 1 \qquad\qquad \text{when } k(s) = 0,$$

where $k(s)$ is defined as in Theorem 1.2.3. Note that any such Radon-Nikodym derivative also has the property that:

$$\int_A g(a, s)\mu(da) = 1 \qquad a.s.P$$
$$\int_S g(a, s)P(ds) = 1 \qquad a.s.\mu$$

These observations motivate the following definition:

1.2.4 Definition. A Bayesian experiment is *dominated* if the associated probability Π is dominated by $\mu \otimes P$. We shall denote g(a,s) a version of $d\Pi/d(\mu \otimes P)$. ∎

Again, by Fubini Theorem, if a Bayesian experiment is dominated, it is also regular; indeed (1.2.2) may be written as:

$$(1.2.8) \qquad \Pi(E \times X) = \int_X dP \int_E g d\mu$$
$$= \int_E d\mu \int_X g dP \qquad E \in \mathcal{A}, \quad X \in \mathcal{S}.$$

Therefore, a regular version of the conditional probabilities of Π given $\mathcal{A}$ and given $\mathcal{S}$ may be specified by setting:

$$(1.2.9) \qquad P^{\mathcal{A}}(X) = \int_X g dP \qquad X \in \mathcal{S}$$

$$(1.2.10) \qquad \mu^{\mathcal{S}}(E) = \int_E g d\mu \qquad E \in \mathcal{A}$$

If μ is represented by a density $h(a)$ w.r.t. some measure μ_0 on $(A, \mathcal{A})$ — i.e., $d\mu/d\mu_0 = h$ a.s.μ_0 — $\mu^{\mathcal{S}}$ may also be specified in such a way that it is a.s. dominated by μ_0; indeed its density $h(a|s)$ follows from (1.2.7) and (1.2.10) and is nothing else than the *Bayes Theorem* expressed in terms of densities:

$$(1.2.11) \qquad \frac{d\mu^{\mathcal{S}}}{d\mu_0} = h(a|s) = \frac{h(a)f(a, s)}{\int_A h(a)f(a, s)\mu_0(da)}$$

The relationship between dominance in sampling theory — i.e., $P^a \ll \lambda$ $\forall\, a \in A$ — and Bayesian dominance — i.e., $\Pi \ll \mu \otimes P$ — may be interpreted as follows. Suppose we build a Bayesian experiment, as in (1.2.2), from a given specification of P^a and μ. If Π is dominated by $\mu \otimes P$ this does not imply that P^a is dominated; but there does exist a family $\tilde{P}^a$, e.g., $\tilde{P}^a = \int g dP$ as in (1.2.9), such that $\Pi = \mu \otimes \tilde{P}^a$ and $\tilde{P}^a \ll P$ $\forall a \in A$. Note that for such a family of $\tilde{P}^a$, $P^a(X) = \tilde{P}^a(X)$ a.s. μ for any $X \in \mathcal{S}$, but the exceptional set may depend on X. Conversely, if P^a is dominated and even if there exists a bimeasurable version $f(a, s)$ of the Radon-Nikodym derivative, this does not imply that P^a is dominated by P, but that P^a may be replaced by an almost equivalent family $\tilde{P}^a$ as above.

Note finally that taking advantage of the symmetry (w.r.t. A and S) of the Bayesian experiment, dominance of Π by $\mu \otimes P$ is also obtained when the posterior probabilities μ^S are dominated in such a way that there exist bimeasurable Radon-Nikodym derivatives, in which case the above analysis may be symmetrically replicated.

Remark. To calculate the Radon-Nikodym derivative of Π with respect to $\mu \otimes P$, a probability, rather than with respect to the product of reference measures (such as the Lebesgue or some counting measure) is as natural, in a Bayesian framework, as expressing, in hypothesis testing, a likelihood ratio as a Radon-Nikodym derivative of a (sampling) probability with respect to another (sampling) probability. Indeed, in sampling theory, it is known that likelihood ratios lead to the neglect of irrelevant factors in the likelihood function. Similarly, in a Bayesian framework, if one defines $f^*(a, s) = d\Pi/d(\mu_0 \otimes \lambda)$ where μ_0 (respectively, λ) is considered as a "natural" reference measure on $(A, \mathcal{A})$ (resp $(S, \mathcal{S})$) dominating μ (respectively, P), one obtains

$$(1.2.12) \qquad f^*(a, s) = h(a)k(s)g(a, s) \quad \text{a.s.} \mu_0 \otimes \lambda$$

where $h(a) = d\mu/d\mu_0$ and $k(s) = dP/d\lambda$ are two scale factors providing no information with regards to the stochastic association within $A \times S$.

1.2.3 Three Remarks on Regular and Dominated Experiments

(i) In general, the existence of a regular conditional probability is a problem concerning the regularity of a measurable space. Thus, for instance, if both $(A, \mathcal{A})$ and $(S, \mathcal{S})$ are Polish spaces, any probability on $(A \times S, \mathcal{A} \vee \mathcal{S})$ defines a regular Bayesian experiment (see Theorem 0.3.18).

(ii) When the sampling probabilities P^a are dominated by a σ-finite measure λ, Doob Theorem implies that if $\mathcal{S}$ is separable there exists an $\mathcal{A} \vee \mathcal{S}$-

measurable function $f(a,s)$ such that: $\forall a \in A$, $f(a,s) = dP^a/d\lambda$ a.e. λ (see Theorem 0.3.19). As seen above, this is sufficient to define $g(a,s)$ as in (1.2.7) for any probability measure μ on $(A,\mathcal{A})$, so the thus-defined Bayesian experiment is dominated and, consequently, regular.

(iii) The motivation for defining general Bayesian experiments that are neither dominated nor (even) regular is not purely esthetic. Firstly, it is worthwhile showing that basic concepts such as sufficiency or identification may be defined without relying on so-called "technical" (or regularity) conditions. Secondly, certain questions both in asymptotic theory and in nonparametric methods naturally involve nondominated sampling probabilities, and it is therefore useful to adopt a unique framework capable of handling these questions. This will be illustrated by some examples in Section 1.3.

1.2.4   A Remark Regarding the Interpretation of Bayesian Experiments

As stated above, a usual interpretation of a Bayesian experiment consists of considering $(A,\mathcal{A})$ as a space of unknown parameters and μ as representing a typically subjective prior information: This is motivated, for instance, by a decision-theoretic argument after Savage's *Foundations of Statistics* (Savage (1954)). The structure of a unique probability measure on a product space allows one other interpretations. Basically, the distinction between the spaces A and S may be based on an informational criterion; in particular A may involve not only unknown (and unobservable) parameters but also latent (or unobservable) variables and/or future observations of observable variables. In such a case, the probability μ on $(A,\mathcal{A})$ may be interpreted partly as purely subjective and partly as representation of some physical process; this flexibility in the interpretation does not affect the mathematical structure thus far developed.

An example of this flexibility may be sketched in the framework of prediction problems. Let $(Z,\mathcal{Z})$ be a measurable space of future observations. In the general case, the probability Π on $A \times S$ induced by a Bayesian experiment may be extended to a probability Π_* on $A \times S \times Z$ by introducing a sampling transition $Q^{A\vee S} : (A \times S, \mathcal{A} \vee \mathcal{S}) \longrightarrow\!\!\!< (Z,\mathcal{Z})$ in such a way that $\Pi_* = (\mu \otimes P^A) \otimes Q^{A\vee S}$. One may consider R, the restriction of Π_* on $\mathcal{S} \vee \mathcal{Z}$. It should be noted that in the probability space $(S \times Z, \mathcal{S} \vee \mathcal{Z}, R)$, the subjective probability μ on $(A, \mathcal{A})$ has been utilized essentially as a helping device to assign a probability on $\mathcal{S} \vee \mathcal{Z}$; this is consistent with de Finetti's approach to coherent prediction (de Finetti (1937)). But note also that the probability space $(S \times \mathcal{Z}, \mathcal{S} \vee \mathcal{Z}, R)$ has the same mathematical structure as a Bayesian experiment where $(Z, \mathcal{Z})$ serves as a parameter space. This means that the analysis of Bayesian experiments presented below may be reinterpreted in a purely predictive framework; in this case, the marginal probability on $(Z, \mathcal{Z})$ is a mixture of subjective prior probability on $(A, \mathcal{A})$ and of sampling probability transitions ($P^{\mathcal{A}}$ and $Q^{\mathcal{A} \vee \mathcal{S}}$).

Another example of flexible interpretation is found in the field of hierarchical models, particularly in superpopulation models with finite populations. In such applications, a vector of parameters $b = (b_i : i \in \mathcal{N})$ characterizes the elements of a finite population $(\mathcal{N})$, and a model, with parameters c, characterizes the way the b_i's have been generated. Interest lies in some functions of b and/or of c. The model generating b, given c, may receive either a sampling interpretation or a purely subjective interpretation, and the b_i's may or may not be observable. The distinction between "parameter" and "observation" may therefore seem artificial in such contexts; nevertheless, the basic structure of a unique probability on a product space remains the basic mathematical object of interest (see, e.g., Cassel, Särndal and Wretman (1977), Cocchi and Mouchart (1986, 1989); see also Scott and Smith (1971, 1973), Smith (1983), Sugden (1979, 1985), and Sugden and Smith (1984)). Note that these kinds of models (where the b_i's may be in very large number) provides one of many motivations to "reduce" a Bayesian experiment into a lower-dimensional one. This will be the object of Section 1.4.

1.2.5   A Remark on Sampling Theory and Bayesian Methods

As indicated above, in a sampling theory framework, a statistical experiment, such as (1.2.1), merely specifies a family of sampling probabilities. Therefore, both in estimation and in prediction, the only possible object of interest is the study of the sampling properties of some function defined on the sample space: there is, strictly speaking, no way of escaping conditioning on the parameter space. It should nevertheless be pointed out that there is some flexibility in the hierarchical model when latent variables or incidental parameters can be integrated out conditionally one some hyperparameters (or superpopulation parameters).

In contrast, in a Bayesian experiment, the unique probability measure on $A \times S$ (or on $(A \times S \times \mathcal{Z})$) allows for a great diversity of interests. It is still fruitful to ask questions about sampling properties such as: What is the sampling distribution of the posterior expectation ? and does it converge (in some suitable sense) to the "true" parameter ? These questions are, in most Bayesian works, considered of rather minor interest. What is usually considered as "genuinely Bayesian" is to take full advantage of the complete probabilistic structure to analyze either the predictive distributions (marginal distribution on S or on $\mathcal{Z}$ given S) by integrating out unknown parameters (and, possibly, unobservable variables) or the posterior distribution on A given S, where conditioning is done on already observed variables only.

In other words, the basic distinction between sampling theory and Bayesian methods is the following: Sampling theory develops properties conditionally on unobserved parameters whereas Bayesian methods condition only on already observed variables and integrate out irrelevant unobserved parameters or variables.

1.2.6 A Remark Regarding So-called "Improper" Prior Distributions

In this monograph we do not introduce prior measures on $(A, \mathcal{A})$ that are not probability measures. It should, however, be mentioned that if the prior measure is σ-finite (but not finite), a σ-finite measure on the product space $A \times S$ may be defined as in (1.1.3). However, the marginal measure P on $(S, \mathcal{S})$ may fail to be σ-finite, and consequently the measure transition μ^S may fail to exist due to the nonapplicability of the Radon-Nikodym Theorem (see Corollary 0.3.6 or, for more details, see Mouchart (1976)). If P is σ-finite, μ^S may be defined as a probability transition, but several theorems presented below are no longer true; in particular, one has to take care of so-called "marginalization paradoxes" (see, e.g., Dawid, Stone and Zidek (1973)).

We also do not address the problem of specifying a Bayesian experiment, but we consider only a given experiment and analyze its reductions.

Thus, the problem of specifying a prior distribution for a given sampling process does not receive systematic analysis. In particular, the question of defining and specifying so-called "noninformative" prior distributions will be dealt with only as far as this question is related to the analysis of invariant Bayesian experiments, which is studied in Chapter 8. Readers interested in more detailed analysis of "improper" or "noninformative" prior distribution may consult, *inter alii*, Bernardo (1979), Hartigan (1964), Jeffreys (1961), Villegas (1971, 1972, 1977 a and b) or Zellner (1971).

1.2.7 Families of Bayesian Experiments

As stated above, a Bayesian experiment is defined with reference to a unique prior probability μ on $(A, \mathcal{A})$. Thus we do not deal systematically with families of prior probabilities. In particular, we do not develop an analysis in terms of the family of probabilities Π^μ on $A \times S$, indexed by the set of *all* probabilities on $(A, \mathcal{A})$. We shall nevertheless take two steps in that direction. The first step is the following proposition, which is motivated by the fact that in the subsequent analysis, the null sets of the prior probability play an important role. This proposition establishes a dominance property between two Bayesian experiments constructed from the same sampling process.

1.2.5 Proposition. Let $\Pi_i = \mu_i \otimes P^{\mathcal{A}} = P_i \otimes \mu_i^{\mathcal{A}} \quad i = 1, 2$. Then:

$$\mu_1 \ll \mu_2 \Rightarrow \quad (i) \quad P_1 \ll P_2$$
$$(ii) \quad \Pi_1 \ll \Pi_2$$
$$(iii) \quad \frac{d\Pi_1}{d\Pi_2} \in [\mathcal{A}]^+ \qquad \blacksquare$$

The proof is obvious, after noticing that for any $X \in \mathcal{S}, P_i(X)$ is weighted average of nonnegative $P^{\mathcal{A}}(X)$ with μ_i as weighting function. Note that the converse implication (in particular, of (i)) is false. This proposition may clearly be restated by considering two probabilities $\Pi_i(i = 1, 2)$ with common posterior probabilities $\mu^{\mathcal{S}}$ and different predictive probabilities such that $P_1 \ll P_2$. This would clearly imply $d\Pi_1/d\Pi_2$ is $\mathcal{S}$-measurable.

The second step in the direction of a family of prior probabilities is taken in the next definition, although the concept of a measurable family of experiments has much wider use.

1.2.6 Definition. Consider a measurable space $(M, \mathcal{M})$. One may define a *measurable family of experiments* $\mathcal{E}^{\mathcal{M}}$ on $(A \times S, \mathcal{A} \vee \mathcal{S})$ by giving transition probabilities $\Pi^{\mathcal{M}}, \mu^{\mathcal{M}}, P^{\mathcal{A} \vee \mathcal{M}}, P^{\mathcal{M}}$ and $\mu^{\mathcal{S} \vee \mathcal{M}}$ (which are respectively defined on $M \times (\mathcal{A} \vee \mathcal{S}), M \times \mathcal{A}, (A \times M) \times \mathcal{S}, M \times \mathcal{S}$ and $(S \times M) \times \mathcal{A}$) such that

$$\Pi^{\mathcal{M}} = \mu^{\mathcal{M}} \otimes P^{\mathcal{A} \vee \mathcal{M}} = P^{\mathcal{M}} \otimes \mu^{\mathcal{S} \vee \mathcal{M}}. \qquad \blacksquare$$

Recall that transition probabilities play a fundamental role in the study of conditional probabilities. Similarly, the concept of a measurable family of experiments will supply the natural ground for the analysis of conditional experiments in Section 1.4. One situation (see Lindley and Smith (1972)) in which this concept is useful is found when $(M, \mathcal{M})$ represents a (possibly parametric) family of prior probabilities on $(A, \mathcal{A})$ for one then has: $P^{\mathcal{A} \vee \mathcal{M}} = P^{\mathcal{A}}$; this kind of structure is investigated in particular detail in the next chapter.

1.3   Some Examples of Bayesian Experiments

In this section, some well-known Bayesian experiments are discussed to illustrate the diversity of situations in which the Bayesian statistician must act. We want to show, by way of these examples, that in spite of strikingly different levels of mathematical difficulty, there is an identical structure underlying all these situations, viz., a unique probability measure on a product space $(A \times S, \mathcal{A} \vee \mathcal{S})$ such that the description of such a structure does not require technicalities (such as topological requirements) other than a description of the sets which are going to be considered as "measurable", and a description of the probability measure Π on this product space through its dual decomposition $\mu \otimes P^{\mathcal{A}}$ and $P \otimes \mu^{\mathcal{S}}$.

The first example has a double objective: we want to provide an example where it is natural to think in terms of probability measures rather than in terms of densities, and we want to show the flexibility possible in the interpretation of the basic spaces A and S. In particular, the basic structure may also be used in a purely predictive setup where A represents future (or unknown) observations, and S represents actual observations.

The second example also has a double motivation. This is a situation where reasoning in terms of densities is most natural, and where the symmetry between the parameters and the observations is very striking.

The last two examples are situations where reasoning in terms of densities are definitely unsuitable; in Example 3, this is because the sampling probabilities are not dominated, and in Example 4 this is because the parameter space is "too big".

These last two examples are slightly more technical in nature, and are merely sketched out in this section. Example 3 is treated more thoroughly in Chapters 7 and 9, whereas Example 4 will be revisited in Chapter 8.

Example 1. (*A normal experiment*)
Suppose first that the statistical experiment specifies that the observation s is a p-dimensional vector generated according to a normal distribution with unknown expectation a and known covariance matrix V. We write:

$$(s|a) \sim N_p(a, V).$$

Let the prior distribution on a also be normal:

$$a \sim N_p(m_0, V_0).$$

The Bayesian experiment is therefore represented as a $2p$-dimensional normal vector:

$$\begin{pmatrix} a \\ s \end{pmatrix} \sim N_{2p} \left[\begin{pmatrix} m_0 \\ m_0 \end{pmatrix}, \begin{pmatrix} V_0 & V_0 \\ V_0 & V_0 + V \end{pmatrix} \right].$$

The Bayesian statistician is typically interested either in predicting s before observing it; this will be done by means of the marginal distribution of s:

$$s \sim N_p(m_0, V_0 + V)$$

or in revising his/her opinions on a after observing s; this will be done by means of the conditional distribution of a given s, which is also p-variate normal with moments:

$$
\begin{aligned}
E(a \mid s) &= (V_0^{-1} + V^{-1})^{-1}(V_0^{-1}m_0 + V^{-1}s) \\
V(a \mid s) &= [V_0^{-1} + V^{-1}]^{-1}.
\end{aligned}
$$

A first extension of this Bayesian experiment is to consider a prediction problem. Let z be a "future" observation generated as s and independently of s:

$$(z \mid a, s) \sim N_p(a, V).$$

We then have a $3p$-variate model:

$$\begin{pmatrix} a \\ s \\ z \end{pmatrix} \sim N_{3p} \left[\begin{pmatrix} m_0 \\ m_0 \\ m_0 \end{pmatrix}, \begin{pmatrix} V_0 & V_0 & V_0 \\ V_0 & V_0 + V & V_0 \\ V_0 & V_0 & V_0 + V \end{pmatrix} \right].$$

In a purely predictive approach, one could consider that the main object of interest is the pair (s, z), and that the parameter a has been introduced only to facilitate the probability assignment for (s, z); here the object of interest becomes a *Bayesian predictive experiment*:

$$\begin{pmatrix} s \\ z \end{pmatrix} \sim N_{2p} \left[\begin{pmatrix} m_0 \\ m_0 \end{pmatrix}, \begin{pmatrix} V_0 + V & V_0 \\ V_0 & V_0 + V \end{pmatrix} \right].$$

The prior predictive distribution of z is evidently the same as the predictive distribution of s in the original Bayesian experiment: $z \sim N(m_0, V_0 + V)$. The prediction of z after observing s will be realized through the distribution of $(z \mid s)$ which is also normal with moments:

$$\begin{aligned} E(z \mid s) &= E(a \mid s) = [V_0^{-1} + V^{-1}]^{-1}[V_0^{-1} m_0 + V^{-1} s] \\ V(z \mid s) &= V_0 + V - V_0(V_0 + V)^{-1} V_0 = V + (V_0^{-1} + V^{-1})^{-1}. \end{aligned}$$

It is useful to notice that the transformation, in distribution, of $z \to (z \mid s)$ in the Bayesian predictive experiment is formally the same as the transformation $a \to (a \mid s)$ in the original experiment.

Another extension of this original Bayesian experiment would be to decompose the probability on a with respect to another parameter c such that $(s \mid a, c) \sim (s \mid a)$. Suppose, for instance:

$$\begin{aligned} c &\sim N_k(c_0, C_0) \\ (a \mid c) &\sim N_p(b_0 + B_0 c, A_0) \end{aligned}$$

where C_0 (respectively, A_0) is a $k \times k$ (respectively, $p \times p$) covariance matrix, c_0 (respectively, b_0) is a $k \times 1$ (respectively, $p \times 1$) vector and B_0 is a $p \times k$

matrix. In order to obtain the same marginal distribution on a as in the original experiment, we assume:

$$
\begin{aligned}
m_0 &= b_0 + B_0 c_0 \\
V_0 &= A_0 + B_0 C_0 B_0'.
\end{aligned}
$$

This kind of structure may arise in hierarchical models, in which case c is called a "hyperparameter" and its role may be to facilitate the specification of the prior probability on $(A, \mathcal{A})$.

In all of these situations there is a basic model represented by:

$$
\begin{pmatrix} c \\ a \\ s \end{pmatrix} \sim N_{2p+k} \left[\begin{pmatrix} c_0 \\ m_0 \\ m_0 \end{pmatrix}, \begin{pmatrix} C_0 & C_0 B_0' & C_0 B_0' \\ B_0 C_0 & V_0 & V_0 \\ B_0 C_0 & V_0 & V_0 + V \end{pmatrix} \right].
$$

In some contexts, attention may be concentrated on the probability on (a, s), as already treated in the original Bayesian experiment, but in other contexts attention is instead focused on the probability on (c, s):

$$
\begin{pmatrix} c \\ s \end{pmatrix} \sim N_{k+p} \left[\begin{pmatrix} c_0 \\ m_0 \end{pmatrix} \begin{pmatrix} C_0 & C_0 B_0' \\ B_0 C_0 & V_0 + V \end{pmatrix} \right].
$$

This is again the same structure as that of the original Bayesian experiment.

Let finally consider an i.i.d. sampling of a normal experiment with unknown variance. Here, $a = (m, V) \in A = \mathbb{R}^p \times C_p$ where C_p is the cone of the $p \times p$ positive definite matrices and $s = (s, \dots, s_n) \in S = \mathbb{R}^{np}$ (depending on the context we shall interpret s as either an np-vector or an $n \times p$-matrix). Here $\mathcal{A}$ and $\mathcal{S}$ are the usual Borel σ-fields of their corresponding Euclidian spaces. The sampling probabilities are described by

$$
(s_i \mid a) \sim i.N(m, V)
$$

and for the prior specification we shall consider an Inverted-Wishart distribution for V, i.e., $V \sim IW_p(\nu_0, V_0)$. For the (prior) conditional distribution of $(m \mid V)$ we again consider an normal distribution with a variance proportional to V, i.e., $(m \mid V) \sim N(m_0, n_0^{-1}V)$. In terms of density (with respect to the Lebesgue measure):

$$\begin{aligned}
\mu(a) &= \mu_1(V) \times \mu_2(m|V) \\
&= \left[\left\{2^{\nu_0 p/2}\pi^{(p-1)p/4}\prod_{i=1}^{p}\Gamma\left((\nu_0+1-i)/2\right)\right\}^{-1}\right. \\
&\quad \times |V_0|^{\nu_0/2}|V|^{-(\nu_0+p+1)/2}\exp\left\{-\tfrac{1}{2}trV^{-1}V_0\right\}\bigr] \\
&\quad \times \left[(2\pi)^{-p/2}|n_0^{-1}V|^{-1/2}\ \exp-\tfrac{1}{2}n_0(m-m_0)'V^{-1}(m-m_0)\right].
\end{aligned}$$

with $\nu_0 > p$, $V_0 \in C_p$, $m_0 \in {I\!\!R}^p$ et $n_0 > 0$.

Note that the prior marginal distribution of m is p-variate Student: $m \sim S_p(m_0, n_0^{-1}V_0, \nu_0 - p + 1)$ with density

$$\begin{aligned}
\mu_3(m) &= \pi^{-p/2}\frac{\Gamma((\nu_0+1)/2)}{\Gamma((\nu_0-p+1)/2)}|n_0^{-1}V_0|^{-1/2} \\
&\quad \times \left[1 + n_0(m-m_0)'V_0^{-1}(m-m_0)\right]^{-(\nu_0+1)/2}.
\end{aligned}$$

It may be helpful, for the interpretation of this prior specification, to recall its moments:

$$\begin{aligned}
\text{if } \nu_0 > \quad p: & \quad E(m|V) = m_0, \quad V(m\,|\,V) = n_0^{-1}V \\
\text{if } \nu_0 > \quad p: & \quad E(m) = m_0 \\
\text{if } \nu_0 > \quad p+1: & \quad E(V) = \frac{1}{\nu_0-p-1}V_0 \\
\text{if } \nu_0 > \quad p+1: & \quad V(m) = \frac{1}{\nu_0-p-1}n_0^{-1}V_0.
\end{aligned}$$

Let us now describe the dual decomposition of the probability Π on $A \times S$. It may be checked (see, e.g., De Groot (1970), Chap. 9 or Press (1972), Section 6.2) that the predictive distribution of s, considered as an $n \times p$-matrix, is matrix-Student: $s \sim S_{n\times p}(i_n m_0', V_0, I_n+n_0^{-1}i_n i_n', \nu_0)$, where i_n is an n-vector with all elements equal to 1, and with density

$$p(s) = K\left|V_0 + (s - i_n m_0')'\left[I_n - (n_0+n)^{-1}i_n i_n'\right](s - i_n m_0')\right|^{-(\nu_0+n)/2}.$$

where:

$$\begin{aligned}
K &= \left[\pi^{np/2}\prod_{j=1}^{p}\Gamma\left((\nu_0+1-j)/2\right)\Big/\Gamma\left((\nu_0+n+1-j)/2\right)\right]^{-1} \\
&\quad \times |V_0|^{\nu_0/2}\left|I_n - (n_0+n)^{-1}i_n i_n'\right|^{p/2}.
\end{aligned}$$

Furthermore, the posterior distribution of $a = (m, V)$ has the same structure as its prior: $(V\,|\,s)$ is again Inverted-Wishart $(V\,|\,s) \sim IW_p(\nu_*, V_*)$

and $(m \mid V, s)$ is again normal $(m \mid V, s) \sim N_p(m_*, n_*^{-1}V)$ where the parameters of the posterior distributions are computed as follows:

$$
\begin{aligned}
\nu_* &= \nu_0 + n \\
V_* &= V_0 + (s - i_n m_0')' \left[I_n - (n_0 + n)^{-1} i_n i_n' \right] (s - i_n m_0') \\
m_* &= (n_0 + n)^{-1}(n_0 m_0 + n\bar{s}) \qquad \text{where} \quad \bar{s}' = \frac{1}{n} i_n' s \\
n_* &= n_0 + n.
\end{aligned}
$$

The posterior marginal distribution of $(m \mid s)$ and the posterior moments may be computed in exactly the same manner as a priori, replacing formally the subindices 0 by $*$. This prior specification — called an Inverted Wishart-Normal Distribution — has different, and interesting properties. Firstly, it is "closed" under normal sampling in the sense that the transformation "prior-to-posterior" is only a matter of revising (a finite number of) parameters while keeping a same analytical form of the densities, this is a property of being "natural-conjugate" to the normal sampling (for more detail, see, e.g., Raiffa and Schlaifer (1961)). ∎

Example 2. (*A canonical Bayesian exponential experiment*)
In this example, densities are the natural way to represent probabilities, and the symmetry between the parameters and the observations is at an extreme. Let $A = S = I\!R^p$ and $\mathcal{A} = \mathcal{S} = \mathcal{B}^p$. Also let α and σ be two σ-finite measures, on $(A, \mathcal{A})$ and $(S, \mathcal{S})$ respectively, such that $\int \exp(a's)d(\alpha \otimes \sigma) = 1$ where $a's$ is the inner product. (Thus, a more general set-up would be to take A and S in duality for a given scalar product). Note that the main requirement on α and σ is to make the function $\exp(a's)$ integrable: normalization to 1 is a matter of notational convenience only. The canonical Bayesian exponential experiment associated to α and σ is defined as:

$$
\mathcal{BEE}(\alpha, \sigma) = \left\{ A \times S, \ \mathcal{A} \vee \mathcal{S}, \ d\Pi / d(\alpha \otimes \sigma) = e^{a's} \right\},
$$

i.e., Π is characterized by

$$
\Pi(E \times X) = \int_{E \times X} e^{a's} d(\alpha \otimes \sigma) \qquad E \in \mathcal{A} \quad S \in \mathcal{S}.
$$

In other words, $e^{a's}$ is the density of Π with respect to the measure $\alpha \otimes \sigma$ (which would make $\mathcal{A}$ and $\mathcal{S}$ independent in probability under a suitable normalization of $\alpha \otimes \sigma$, as soon as $\alpha \otimes \sigma$ is a finite measure). Let us now describe the dual decomposition of Π in terms of densities. We first denote the logarithm of the Laplace transform of α and σ as:

$$L(a) = \ln \int_S e^{a's} d\sigma \qquad M(s) = \ln \int_A e^{a's} d\alpha.$$

Note that L (respectively, M) is well defined because $\int e^{a's} d\sigma > 0 \quad \forall a$. We may now describe the prior and the predictive probability as:

$$\frac{d\mu}{d\alpha} = e^{L(a)} \qquad \frac{dP}{d\sigma} = e^{M(s)}.$$

This exponential structure is dominated (and regular), and the sampling and the posterior probabilities both admit a representation in terms of densities:

$$\frac{dP^a}{d\sigma} = e^{a's - L(a)} \qquad \frac{d\mu^s}{d\alpha} = e^{a's - M(s)}.$$

We may also write the density of Π with respect to $\mu \otimes P$ rather than with respect to $\alpha \otimes \sigma$:

$$\frac{d\Pi}{d(\mu \otimes P)} = e^{a's - M(s) - L(a)}.$$

In many cases, the measures α and σ are themselves defined through densities with respect to "natural" measures α_0 and σ_0 (such as the Lebesgue measure or the counting measure). More specifically let:

$$\frac{d\alpha}{d\alpha_0} = e^{f(a)} \qquad \frac{d\sigma}{d\sigma_0} = e^{g(s)}.$$

The prior and predictive densities now become:

$$\frac{d\mu}{d\alpha_0} = e^{f(a) + L(a)} \qquad \frac{dP}{d\sigma_0} = e^{g(s) + M(s)}.$$

Also the usual form of the exponential family (for the sampling distribution) is obtained as:

$$\frac{dP^a}{d\sigma_0} = e^{a's - L(a) + g(s)}.$$

Similarly,

$$\frac{d\mu^s}{d\alpha_0} = e^{a's - M(s) + f(a)}$$

$$\frac{d\Pi}{d(\alpha_0 \otimes \sigma_0)} = e^{a's + f(a) + g(s)}.$$

Suppose now that the differentiation and integration may be commuted in $\int \frac{d}{ds} \exp(a's - M(s))d\alpha$ and $\int \frac{d}{da} \exp(a's - L(a))d\sigma$ (for sufficient conditions see, e.g., Monfort (1982), Barra (1971)), then we obtain:

$$E(s \mid a) = \frac{d}{da}L(a) \qquad E(a \mid s) = \frac{d}{ds}M(s)$$

$$V(s \mid a) = \frac{d^2}{dada'}L(a) \qquad V(a \mid s) = \frac{d^2}{dsds'}M(s).$$

As an example consider the case of normal sampling with known (unit) variance along with a normal prior distribution (with expectation equal to m_0 and variance equal to b_0^{-1}). In such a case, one can easily check that $s \sim N(m_0, b_0^{-1} + 1)$ and $(a \mid s) \sim N[(b_0 + 1)^{-1}(s + b_0 m_0), (b_0 + 1)^{-1}]$. Here both α_0 and σ_0 are the Lebesgue measure (denoted as λ).

More specifically, when

$$\frac{dP^a}{d\lambda} = (2\pi)^{-1/2} \exp -\frac{1}{2}(s - a)^2$$

$$\frac{d\mu}{d\lambda} = (2\pi)^{-1/2} b_0^{1/2} \exp -\frac{b_0}{2}(a - m_0)^2$$

the joint density becomes

$$\frac{d\Pi}{d(\lambda \otimes \lambda)} = \exp\left\{ as - \frac{1}{2}s^2 - \frac{1}{2}a^2(b_0 + 1) \right.$$
$$+ \left. ab_0 m_0 - \frac{1}{2}b_0 m_0^2 - \ln(2\pi) + \frac{1}{2}\ln b_0 \right\}$$

and the reference measure may be taken as:

$$\frac{d\sigma}{d\lambda} = e^{g(s)} = \exp\left(-\frac{1}{2}s^2 - \frac{1}{2}\ln 2\pi \right)$$

$$\frac{d\alpha}{d\lambda} = e^{f(a)}$$

$$= \exp\left(-\frac{1}{2}a^2(b_0 + 1) - \frac{1}{2}b_0 m_0^2 + ab_0 m_0 - \frac{1}{2}\ln(2\pi) + \frac{1}{2}\ln b_0 \right)$$

or, with $\varphi(z)$ denoting the value of a standardized normal density at z:

$$g(s) = \ln \varphi(s)$$

$$f(a) = \ln \varphi[(b_0 + 1)^{\frac{1}{2}}(a - b_0(b_0 + 1)^{-1}m_0)]$$

$$-\frac{1}{2}[m_0^2 b_0(b_0 + 1)^{-1} - \ln[b_0(b_0 + 1)^{-1}]]$$

or with $e_0 = b_0(b_0 + 1)^{-1} \in (0, 1)$,

$$f(a) = \ln \varphi[(1 - e_0)^{-\frac{1}{2}}(a - e_0 m_0)] - \frac{1}{2}[e_0 m_0^2 - \ln e_0].$$

Thus we obtain

$$L(a) = \ln \frac{d\mu}{d\lambda} - f(a) = \frac{1}{2}a^2.$$

It can be checked that:

$$E(s \mid a) = \frac{d}{da}L(a) = a; \qquad V(s \mid a) = \frac{d^2 L(a)}{da} = 1.$$

Similarly:

$$M(s) = \ln \frac{dP}{d\lambda} - g(s)$$

$$= \frac{1}{2}\ln(b_0^{-1} + 1) + \frac{1}{2}(b_0^{-1} + 1)^{-1}[s^2 b_0^{-1} + 2m_0 s - m_0^2],$$

and it can be checked that:

$$E(a \mid s) = \frac{d}{ds}M(s) = \frac{1}{2}(b_0^{-1} + 1)^{-1}[2sb_0^{-1} + 2m_0]$$

$$= \frac{s + b_0 m_0}{b_0 + 1}$$

$$V(a \mid s) = \frac{d^2}{ds^2}M(s) = \frac{1}{b_0 + 1}.$$

Until now the presentation has treated the parameter and the sample spaces symmetrically. Let us now consider the sampling process as given, as also the measure σ (on $(S, \mathcal{S})$) and the function $L(a)$. One may look for a prior specification which would "naturally conjugate" with the given sampling process. Diaconis and Ylvisaker (1979) have proposed the following analysis. Suppose that σ, and therefore $L(a)$, are such that for some (n_0, m_0), and for some reference measure α_0, the integral $\int e^{n_0(a'm_0 - L(a))} d\alpha_0$ converges; denote that positive quantity as $[k_0(n_0, m_0)]^{-1}$. Then the prior specification

$$\frac{d\mu}{d\alpha_0} = k_0(n_0, m_0)e^{n_0(a'm_0 - L(a))}$$

has the property that the posterior density has the form

$$\frac{d\mu^s}{d\alpha_0} = k_0(n_*, m_*)e^{n_*(a'm_* - L(a))}$$

where $n_* = n_0 + 1$ and $m_* = (n_0 + 1)^{-1}(n_0 m_0 + s)$. This choice for the prior specification corresponds to

$$
\begin{aligned}
f(a) &= \ln \frac{d\alpha}{d\alpha_0} = n_0 m_0' a - (n_0 + 1)L(a) + \ln k_0(n_0, m_0) \\
M(s) &= -\ln k_0(n_*, m_*) + \ln k_0(n_0, m_0).
\end{aligned}
$$

The extension of this experiment to i.i.d. sampling is presented in Chapter 2. ∎

Example 3. (*An asymptotic experiment*)
Consider the simple univariate normal model, with known variance, from an asymptotic point of view. More specifically let $(s_j \mid a) \sim i.N(a, 1)$ $j \in I\!N$ along with $a \sim N(0, 1)$. Thus $A = I\!R$ and $S = I\!R^{I\!N}$, i.e., $s = (s_1, s_2 \ldots)$ and $\mathcal{A} \vee \mathcal{S}$ is the usual Borel σ-field (generated by the cylinders); μ is the standardized normal probability distribution on $I\!R$ while $P^{\mathcal{A}}$ is an independent Gaussian process (with easily found mean and covariance function) for each $a \in A$. Thus $\Pi = \mu \otimes P^{\mathcal{A}}$ is built without difficulty. Similarly the predictive process P (on $I\!R^{I\!N}$) is an exchangeable Gaussian (the structure of exchangeability is investigated more systematically in Chapter 9) process (obtained as the projective limit of the finite dimensional symmetric distributions of $(s_1 \ldots s_n)$ with $E(s_j) = 0$, $V(s_j) = 2$, $\text{Cov}(s_j, s_k) = 1$ (for $j \neq k$)). The strong law of large numbers implies that the posterior transition $\mu^{\mathcal{S}}$ may be characterized as follows: $\mu^{\mathcal{S}} = \delta_t$ a.s. where δ_x is the degenerate probability giving mass 1 to the point x and $t = \limsup \bar{s}_n$ where $\bar{s}_n$ is the sample mean, i.e., $\bar{s}_n = n^{-1} \sum_{1 \leq j \leq n} s_j$; in other words, the posterior distribution gives mass 1 to the value of a corresponding to the limiting value of the sample mean. We would like to draw attention to the fact that this Bayesian experiment is genuinely simple: the probabilities and transitions $\mu, P^{\mathcal{A}}, P, \mu^{\mathcal{S}}$ and Π are easily described even though the sampling probabilities are not dominated (once A is not countable). This is an example of a situation in which reasoning in terms of density is unsuitable despite the essential simplicity of the model.

Example 4. (*A non parametric experiment*)
For simplicity, let $(S, \mathcal{S})$ be the real line along with its Borel sets and $A = \{a : \mathcal{S} \to [0, 1]\} = [0, 1]^{\mathcal{S}}$. For $\mathcal{A}$, we take the product σ-field, (i.e.,

the σ-field generated by all the cylinders based on a Borel set of $[0, 1]$ for a finite number of coordinates, see Section 0.2.3). For the prior probability, Ferguson (1973, 1974) (see also Cifarelli, Muliere and Scarsini (1981), Mouchart and Simar (1984b), and Rolin (1983)) has shown that a Dirichlet process offers a workable alternative. This process may be built as the limit of the projective system defined as follows: for every $(T_1 \ldots T_m)$, a measurable finite partition of S, the random vector $(a(T_1) \ldots a(T_m))$ is distributed as an m-dimensional Dirichlet distribution with parameters $(a_0(T_1) \ldots a_0(T_m))$ where a_0 is a finite (i.e., turns out that, with a Dirichlet prior, a is (almost surely) a probability measure on $(S, \mathcal{S})$. For the sampling transitions we simply consider a unit size random sample: $(s \mid a) \sim a$. Note that the product σ-field, $\mathcal{A}$ on A, makes the functions $a(T)$ measurable for every $T \in \mathcal{S}$, and therefore the product probability $\Pi = \mu \otimes P^{\mathcal{A}}$ is a well-defined probability on $(A \times S, \mathcal{A} \vee \mathcal{S})$.

In the dual decomposition $\Pi = P \otimes \mu^{\mathcal{S}}$, we have $P = P_0$, where P_0 is defined by $a_0 = n_0 P_0$, and the posterior transition is again a Dirichlet process with parameters $a_0 + \delta_s$, where δ_s is the probability measure giving mass 1 to the point $\{s\}$.   ∎

1.4   Reduction of Bayesian Experiments

1.4.1   Introduction

We first adapt to Bayesian experiments some of the notation presented in Chapter 0.

Let $\mathcal{M} \subset \mathcal{A} \vee \mathcal{S}$ be a sub-σ-field of $\mathcal{A} \vee \mathcal{S}$. We write $m \in [\mathcal{M}]$ to denote that m is a real-valued function defined on $A \times S$ and is $\mathcal{M}$-measurable. If it is also non-negative, we write $m \in [\mathcal{M}]^+$; if $|m|^p$ is integrable, we write $m \in [\mathcal{M}]_p$, and if m is bounded we write $m \in [\mathcal{M}]_\infty$. For any $\mathcal{A} \vee \mathcal{S}$-measurable integrable or positive function n, its conditional expectation given $\mathcal{M}$ is written $\mathcal{M}n$. The trivial σ-field $\{\phi, A \times S\}$ is denoted $\mathcal{I}$; therefore, the mathematical expectation of n is denoted $\mathcal{I}n$.

One goal of statistical theory is to reduce a given problem to its essential elements, so as to both simplify and reduce the burden of specification and the volume of the computations involved in the treatment of statistical data. In this section we essentially consider the problem of reducing a given Bayesian experiment $\mathcal{E}$. There are two natural ways of reducing the probability Π characterizing $\mathcal{E}$: marginalizing it, (i.e., considering a restriction of Π), or conditioning it, (i.e., considering a desintegration of Π), in both cases with respect to a given sub-σ-field $\mathcal{M} \subset \mathcal{A} \vee \mathcal{S}$. Thus, $\Pi_\mathcal{M}$, the restriction of Π to $\mathcal{M}$ defines a marginal experiment $\mathcal{E}_\mathcal{M}$, and $\Pi^\mathcal{M}$, the conditional probability Π given $\mathcal{M}$, defines a conditional experiment $\mathcal{E}^\mathcal{M}$. Under the usual regularity conditions (ensuring the existence of a regular version of the conditional probability), $\mathcal{E}^\mathcal{M}$ represents a measurable family of experiments (in the sense of Definition 1.2.6, the specification $\mathcal{M} \subset \mathcal{A} \vee \mathcal{S}$ was not necessary). Note that when Π is not reduced, i.e., is a probability on $\mathcal{A} \vee \mathcal{S}$, we drop the lower index (and write Π rather than $\Pi_{\mathcal{A} \vee \mathcal{S}}$) when there is no danger of confusion. Remember that $\Pi_\mathcal{A}$ is identified with μ, and $\Pi_\mathcal{S}$ with P, i.e., when μ (resp P) is a probability on $\mathcal{A}$ (respectively, $\mathcal{S}$) we also drop the lower index; furthermore $\Pi_\mathcal{S}^\mathcal{A}$ (respectively, $\Pi_\mathcal{A}^\mathcal{S}, \Pi_\mathcal{T}^\mathcal{B}, \Pi_\mathcal{B}^\mathcal{T}$) is denoted as $P^\mathcal{A}$ (respectively, $\mu^\mathcal{S}, P_\mathcal{T}^\mathcal{B}, \mu_\mathcal{B}^\mathcal{T}$).

For reduction by marginalization, the most useful cases are obtained when $\mathcal{M}$ has a product structure: $\mathcal{M} = \mathcal{B} \vee \mathcal{S}$ (with $\mathcal{B} \subset \mathcal{A}$) or $\mathcal{M} = \mathcal{A} \vee \mathcal{T}$ (with $\mathcal{T} \subset \mathcal{S}$). For expository purposes, these reductions are analyzed for, and motivated on, both the sample and parameter spaces.

Finally, recall that a sub-σ-field $\mathcal{T} \subset \mathcal{S}$ is typically the σ-field generated by some statistic, say t, defined on the sample space; a sub-σ-field $\mathcal{B} \subset \mathcal{A}$ is typically the σ-field generated by some function, say b, defined on the parameter space (see Bahadur (1955b)).

1.4.2 Marginal Experiments

A. Marginalization on the Sample Space

Suppose we start by specifying a Bayesian experiment on $A \times S$ where S is the σ-field generated by an unreduced observation s. Then the statistician may want to restrict her/his attention to a statistic $t \in [\mathcal{S}]$, either because only t is actually observable, or because s would be too expensive to observe competely, or because the statistician thinks, rightly or wrongly, that "information would not be lost" by observing t only. These reflections motivate the following definition.

1.4.1 Definition. The *Bayesian experiment marginal on* $T, \mathcal{E}_{A \vee T}$, is defined as follows:

$$(1.4.1) \qquad\qquad \mathcal{E}_{A \vee T} = (A \times S, \ \mathcal{A} \vee \mathcal{T}, \ \Pi_{A \vee T})$$

where $\Pi_{A \vee T}$ is the restriction of Π on $\mathcal{A} \vee \mathcal{T}$. If $\mathcal{E}_{A \vee T}$ is *regular* we also have:

$$(1.4.2) \qquad\qquad \Pi_{A \vee T} \ = \ \mu \otimes P_T^{\mathcal{A}} = P_T \otimes \mu^T \qquad\qquad \blacksquare$$

Example. Consider again a simple univariate normal model with known variance: $s = (s_1, \ldots, s_n)$ and $(s_i \mid a) \sim i.N(a, 1)$. Suppose now that only the sign of s_i is observable. Here $S = \mathbb{R}^n$, and $\mathcal{S}$ is its Borel σ-field. The actual observation may be represented by a statistic $t = (t_1, \ldots, t_n)$ such as $t_i = \mathbf{1}_{(0,\infty)}(s_i)$, but the sub-$\sigma$-field $\mathcal{T}$ of $\mathcal{S}$, associated to t, is clearly invariant under any recoding of t; this justifies our consideration of marginal probabilities on a sub-σ-field rather than Π transformed under a measurable mapping. Note also that the $(t_i \mid a)$ are independently distributed as a Bernoulli variable with parameter $1 - \Phi(-a)$ (where $\Phi(\cdot)$ is the cumulative distribution function of a standardized normal variable) whereas the prior probability on $(A, \mathcal{A})$ is the same in the unreduced experiment $\mathcal{E}$ as in the marginal experiment $\mathcal{E}_{A \vee T}$. $\qquad\qquad \blacksquare$

B. Marginalization on the Parameter Space

In a statistical analysis attention is often focused on some functions of the parameter. For instance, in a decision-theoretic approach, the loss-function may merely depend on some functions of the parameters. In such circumstances $\mathcal{B}$, a sub-σ-field of $\mathcal{A}$, is said to represent the decision parameters (or the parameters of interest) if it is the smallest sub-σ-field of $\mathcal{A}$ which makes the loss-function measurable for every decision. Consequently, it may seem natural to restrict the Bayesian experiment to the σ-field $\mathcal{B}$ on A only: this leads to the elimination, by integration, of the so-called "nuisance" parameters (see also Basu (1977) and Dawid (1980)). Another reason for undertaking the same type of reduction may be that the statistician thinks, rightly or wrongly, that a function b, generating the σ-field $\mathcal{B}$, is "sufficient" to describe the sampling process, and that the observation will give information about b only.

1.4.2 Definition. The *Bayesian experiment marginal on* $\mathcal{B}$, $\mathcal{E}_{\mathcal{B}\vee\mathcal{S}}$, is defined as follows:

(1.4.3) $$\mathcal{E}_{\mathcal{B}\vee\mathcal{S}} = (A \times S, \ \mathcal{B} \vee \mathcal{S}, \ \Pi_{\mathcal{B}\vee\mathcal{S}})$$

where $\Pi_{\mathcal{B}\vee\mathcal{S}}$ is the restriction of Π on $\mathcal{B}\vee\mathcal{S}$. If $\mathcal{E}_{\mathcal{B}\vee\mathcal{S}}$ is *regular* we also have:

(1.4.4) $$\Pi_{\mathcal{B}\vee\mathcal{S}} = \mu_{\mathcal{B}} \otimes P^{\mathcal{B}} = P \otimes \mu_{\mathcal{B}}^{\mathcal{S}}. \qquad \blacksquare$$

Note that $P^{\mathcal{B}}$ is sometimes called *the marginalized sampling probability* and may be computed, in the regular case, as follows (see also Raiffa and Schlaifer (1961), Section 2.1.1):

(1.4.5) $$P^{\mathcal{B}}(X) = \int_A P^{\mathcal{A}}(X)d\mu^{\mathcal{B}} \qquad X \in \mathcal{S}.$$

Indeed, for any $t \in [\mathcal{S}]^+$, Proposition 0.3.14(ii) implies that $\mathcal{B}t = \mathcal{B}(\mathcal{A}t)$.

Example. Consider a simple univariate normal model with unknown variance: $s = (s_1, \ldots, s_n)$ and $(s_i \mid a) \sim i.N(m, v)$ where $a = (m, v)$. Suppose first that m is the only parameter of interest. Thus we want to reduce the original model on (m, v, s) into a marginal model on (m, s). If, for example, $(v \mid m)$ is distributed as an Inverted-Wishart variable — $(v \mid m) \sim IW(\nu_0, v_0(m))$ —, the marginal sampling probability is Student — $(s \mid m) \sim S_n(\nu_0, m i_n, v_0(m)I_n)$ — thus $(s \mid m)$ is a random vector of exchangeable, uncorrelated, but not independent variables and the prior distribution in the marginal model on (m, s) is evidently the same as the marginal prior distribution on m in the unreduced model on (m, v, s). Suppose now that v is the only parameter of interest, and that $(m \mid v)$ is a normal distribution: $(m \mid v) \sim N(m_0, n_0^{-1}v)$. Then the marginal sampling probability becomes $(s \mid v) \sim N_n(m_0 i_n, v[I_n + n_0^{-1}i_n i_n'])$, i.e., an exchangeable (equicorrelated) normal vector. Again, the prior distribution in the marginal model on (v, s) is the same as the marginal prior distribution on v in the unreduced model on (m, v, s). $\qquad \blacksquare$

C. Joint Marginalization

There are many circumstances, in particular in sequential analysis, where marginal reductions are operated simultaneously on the sample space and on the parameter space. So, considering two sub-σ-fields $\mathcal{B} \subset \mathcal{A}$ and $\mathcal{T} \subset \mathcal{S}$, we may define:

(1.4.6) $$\mathcal{E}_{\mathcal{B}\vee\mathcal{T}} = (A \times S, \ \mathcal{B} \vee \mathcal{T}, \ \Pi_{\mathcal{B}\vee\mathcal{T}})$$

and, in the regular case, we have:

$$(1.4.7) \qquad \Pi_{\mathcal{B}\vee\mathcal{T}} = \mu_{\mathcal{B}} \otimes P_{\mathcal{T}}^{\mathcal{B}} = P_{\mathcal{T}} \otimes \mu_{\mathcal{B}}^{\mathcal{T}}.$$

In the marginal experiment $\mathcal{E}_{\mathcal{B}\vee\mathcal{T}}$, a statistic is either a sub-σ-field of $\mathcal{T}$ or a $\mathcal{T}$-measurable function defined on S, a subparameter is either a sub-σ-field of $\mathcal{B}$ or a $\mathcal{B}$-measurable function defined on A. Note that any marginal experiment has exactly the same mathematical structure as a general Bayesian experiment.

Example. As an example of joint marginalization, suppose that in Example 1 of Section 1.3 the loss-function depends on a linear combination of the expectations only; i.e., consider the subparameter $b = \sum_i \alpha_i a_i = \alpha' a$. It then seems reasonable to reduce the observation s to the same linear combination, viz., $t = \sum_i \alpha_i s_i = \alpha' s$; defining the sub-$\sigma$-field: $\mathcal{B} = \sigma(b)$ and $\mathcal{T} = \sigma(t)$, the marginal experiment $\mathcal{E}_{\mathcal{B}\vee\mathcal{T}}$ is characterized by the joint probability:

$$\begin{pmatrix} b \\ t \end{pmatrix} \sim N_2 \left[\begin{pmatrix} \alpha' m_0 \\ \alpha' m_0 \end{pmatrix}, \begin{pmatrix} \alpha' V_0 \alpha & \alpha' V_0 \alpha \\ \alpha' V_0 \alpha & \alpha'(V_0 + V)\alpha \end{pmatrix} \right].$$

In this reduction, the sampling probabilities are represented by a normal distribution with parameters:

$$E(t \mid b) = b \qquad\qquad V(t \mid b) = \alpha' V \alpha$$

and the posterior distribution is also normal with parameters:

$$E(b \mid t) = (1 - p)\alpha' m_0 + pt$$

$$V(b \mid t) = \left[\frac{1}{\alpha' V_0 \alpha} + \frac{1}{\alpha' V \alpha} \right]^{-1}$$

where

$$p = \frac{\alpha' V_0 \alpha}{\alpha'(V_0 + V)\alpha}. \qquad\qquad\blacksquare$$

In the next chapter, we shall see that this joint marginalization of the original experiment is indeed "natural" in the following sense. If $b = \alpha' a$ is the only parameter of interest it will be shown, in a sense to be made precise, that the marginal experiment $\mathcal{E}_{\mathcal{B}\vee\mathcal{T}}$ "does not lose relevant information".

1.4.3 Conditional Experiment

Recall that conditional probabilities are defined a.s. only; and according to the convention introduced in Chapter 0, we omit the a.s. proviso in all equalities involving a conditional probability.

A. Conditioning on the Sample Space

The concept of conditional experiment may be motivated by the following example.

Example. Let us consider a regression model $(y \mid X, a) \sim N_n(Xb, vI)$ where y is an $(n \times 1)$-vector, X is an $(n \times k)$-matrix, b is an $(k \times 1)$-vector of regression coefficient, v is a positive real number. This model may be embedded in a Bayesian experiment with observation $s = (y, X)$ and parameter $a = (b, v, c)$, where c are parameters characterizing the marginal sampling process generating X. The nature of a regression model is to operate an analysis "conditionally" on the statistic X; this is an analysis where the sampling process generating X is not specified explicitly. Other examples will also be encountered in sequential analysis (Chap. 6.). ∎

1.4.3 Definition. The *Bayesian experiment conditional on* T, $\mathcal{E}^T$, is defined as:

$$(1.4.8) \qquad \mathcal{E}^T = (A \times S, \ \mathcal{A} \vee \mathcal{S}, \ \Pi^T)$$

where Π^T is a conditional probability of Π given T. The conditional experiment $\mathcal{E}^T$ is said to be *regular* if there exists a regular version of Π^T such that there exist regular versions of μ^S and $P^{A \vee T}$, in this case we then have:

$$(1.4.9) \qquad \Pi^T = \mu^T \otimes P^{A \vee T} = P^T \otimes \mu^S \qquad \blacksquare$$

Equalities (1.4.9) are based on the following identities:

$$(1.4.10) \quad \Pi[E \times (X \cap Y)] = \int_Y \Pi^T(E \times X) dP_T$$

$$= \int_Y dP_T \int_X \mu^S(E) dP^T$$

$$= \int_Y dP_T \int_E P^{A \vee T}(X) d\mu^T$$

$$E \in \mathcal{A}, \ X \in \mathcal{S}, \ Y \in \mathcal{T}.$$

Equalities (1.4.10) come from the fact that for any $a \in [\mathcal{A}]^+$ and $s \in [\mathcal{S}]^+$, one has, using Proposition 0.3.14(ii), $T(as) = T(s \cdot Sa) = T[a \cdot (\mathcal{A} \vee T)s]$.

Even though the last two equalities are true in general, the existence of a regular version of Π^T implies the existence of a regular version of μ^T and P^T and justifies the integrations in (1.4.10). A given version of Π^T gives a T-measurable family of probabilities on $\mathcal{A} \vee \mathcal{S}$ to which one may associate a measurable family of regular experiments once the conditional probabilities $P^{\mathcal{A} \vee T}$ and $\mu^{\mathcal{S}}$ admit a regular version. Note also that $P^{\mathcal{A} \vee T}$ is a sampling probability conditional on T.

B. Conditioning on the Parameter Space

The Bayesian treatment of exact restrictions on the parameter space leads naturally to the concept of a conditional experiment $\mathcal{E}^{\mathcal{B}}$. Classically, an exact restriction is a surface in the parameter space. Imagine, for example, that s is distributed as a bivariate normal: $(s \mid a) \sim N_2(m, V)$ where $a = (m, V)$. Suppose now that $m = (m_1, m_2)$ has a continuous probability distribution and that one wishes to corporate the information "m_1 and m_2 have the same value". It is well known, by the Borel-Kolmogorov paradox, — see Kolmogorov (1950) — that the distribution of m_1 conditionally on $m_1 - m_2 = 0$ or on $m_1 m_2^{-1} = 1$ is *not* the same. This may be viewed as two different disintegrations of μ or of Π: the σ-field generated by $m_1 - m_2$ is clearly different from that generated by $m_1 m_2^{-1}$ and represents a different structure of information, (i.e., a different set of "possible messages"). A restriction on the parameter space should therefore be handled through a sub-σ-field of the parameter space rather than through a subset of the parameter space (see Mouchart and Orsi (1976, 1986), and Mouchart and Roche (1987)). This approach indeed leads to a coherent treatment of restrictions defined by a surface of zero prior probability, and motivates the following definition:

1.4.4 Definition. The *Bayesian experiment conditional on* $\mathcal{B}$, is defined as:

$$(1.4.11) \qquad\qquad \mathcal{E}^{\mathcal{B}} \; = \; (A \times S, \; \mathcal{A} \vee \mathcal{S}, \; \Pi^{\mathcal{B}})$$

where $\Pi^{\mathcal{B}}$ is a conditional probability of Π given $\mathcal{B}$. The conditional experiment $\mathcal{E}^{\mathcal{B}}$ is said to be *regular* if there exists a regular version of $\Pi^{\mathcal{B}}$ such that there exists a regular version of $\mu^{\mathcal{B} \vee \mathcal{S}}$; in this case we have:

$$(1.4.12) \qquad \Pi^{\mathcal{B}} = \mu^{\mathcal{B}} \otimes P^{\mathcal{A}} = P^{\mathcal{B}} \otimes \mu^{\mathcal{B} \vee \mathcal{S}}. \qquad \blacksquare$$

As in (1.4.9), the above equalities are based on the following identities:

$$(1.4.13) \quad \Pi[(E \cap F) \times T] = \int_F \Pi^{\mathcal{B}}(E \times T) d\mu_{\mathcal{B}}$$
$$= \int_F d\mu_{\mathcal{B}} \int_E P^{\mathcal{A}}(T) d\mu^{\mathcal{B}}$$
$$= \int_F d\mu_{\mathcal{B}} \int_T \mu^{\mathcal{B} \vee \mathcal{S}}(E) dP^{\mathcal{B}}$$
$$E \in \mathcal{A}, \ F \in \mathcal{B}, \ T \in \mathcal{S}.$$

Again, (1.4.13) follows from the fact that, by Proposition 0.3.14(ii), for any $a \in [\mathcal{A}]^+$ and $s \in [\mathcal{S}]^+$, $\mathcal{B}(as) = \mathcal{B}(a \cdot \mathcal{A}s) = \mathcal{B}[s \cdot (\mathcal{B} \vee \mathcal{S})a]$. The integrals in (1.4.13) are justified by the fact that the existence of a regular version of $\Pi^{\mathcal{B}}$ implies the existence of a regular version of $\mu^{\mathcal{B}}$ and $P^{\mathcal{B}}$. Note also that $P^{\mathcal{A}}$ has been assumed to be a transition of sampling probability. The next proposition makes the role of $\mu_{\mathcal{B}}$ more precise for the construction of the conditional experiment $\mathcal{E}^{\mathcal{B}}$: it shows that $\mathcal{E}^{\mathcal{B}}$ is robust with respect to changes of $\mu_{\mathcal{B}}$ into an equivalent $\mu'_{\mathcal{B}}$.

1.4.5 Proposition. Let us consider $\mathcal{E}$ and $\mathcal{E}'$, two Bayesian experiments on $(A \times S, \mathcal{A} \vee \mathcal{S})$ with probabilities Π and Π' such that

(i) $\quad \Pi \sim \Pi'$

(ii) $\quad \dfrac{d\Pi'}{d\Pi} \in [\mathcal{B}]^+.$

Then $\mathcal{E}^{\mathcal{B}} = \mathcal{E}'^{\mathcal{B}}$. $\qquad \blacksquare$

C. Combined Reductions

Joint conditioning on the sample space and on the parameter space may also be defined as follows:

$$(1.4.14) \qquad \mathcal{E}^{\mathcal{B} \vee \mathcal{T}} = (A \times S, \mathcal{A} \vee \mathcal{S}, \Pi^{\mathcal{B} \vee \mathcal{T}})$$

where, in the regular case, the conditional probability $\Pi^{\mathcal{B} \vee \mathcal{T}}$ may be decomposed as:

$$(1.4.15) \qquad \Pi^{\mathcal{B} \vee \mathcal{T}} = \mu^{\mathcal{B} \vee \mathcal{T}} \otimes P^{\mathcal{A} \vee \mathcal{T}} = P^{\mathcal{B} \vee \mathcal{T}} \otimes \mu^{\mathcal{B} \vee \mathcal{S}}.$$

Subsequently, we shall also encounter various combinations of reduction by marginalization *and* reduction by conditioning. In general, for $\mathcal{M} \subset \mathcal{A} \vee \mathcal{S}$, $\mathcal{E}_{\mathcal{B} \vee T}^{\mathcal{M}}$ is defined in the regular case, using:

$$(1.4.16) \qquad \Pi_{\mathcal{B} \vee T}^{\mathcal{M}} = \mu_{\mathcal{B}}^{\mathcal{M}} \otimes P_T^{\mathcal{B} \vee \mathcal{M}} = P_T^{\mathcal{M}} \otimes \mu_{\mathcal{B}}^{\mathcal{M} \vee T}$$

D. Parameters and Observations in Conditional Experiments

The analysis of conditional experiments requires that some care be taken with regards to the concepts of a parameter and an observation. Let $\mathcal{M}$ be any sub-σ-field of $\mathcal{A} \vee \mathcal{S}$. In $\mathcal{E}^{\mathcal{M}}$, a statistic is basically a function defined on the available information, i.e., on the observation *and* on the conditioning variables. This leads to the following definitions:

1.4.6 Definition. A sub-σ-field $\mathcal{N}$ is an $\mathcal{E}^{\mathcal{M}}$-*statistic* if $\mathcal{N} \subset \mathcal{M} \vee \mathcal{S}$ and $\overline{\mathcal{M}} \cap (\mathcal{M} \vee \mathcal{S})$ is called the *trivial* $\mathcal{E}^{\mathcal{M}}$-statistic.   ∎

1.4.7 Definition. An $\mathcal{E}^{\mathcal{M}}$-statistic, $\mathcal{N}$, which is also an $\mathcal{E}$-statistic, (i.e., $\mathcal{N} \subset \mathcal{S}$) is called a *uniform $\mathcal{E}^{\mathcal{M}}$-statistic*.   ∎

1.4.8 Definition. An $\mathcal{E}^{\mathcal{M}}$-statistic, $\mathcal{N}$, is a *strong $\mathcal{E}^{\mathcal{M}}$-statistic* if $\mathcal{M} \subset \mathcal{N}$ (and therefore $\mathcal{M} \subset \mathcal{N} \subset \mathcal{M} \vee \mathcal{S}$).   ∎

Note that a statistic $\mathcal{N} \subset \mathcal{M} \vee \mathcal{S}$ generates a family of σ-fields on the sample space as follows. For any $V \in \mathcal{N}$, $a \in A$, define a section of V at a as follows: $V_a = \{s \in S \mid (a, s) \in V\}$. Clearly, $V_a \in \mathcal{S}$ (see, e.g., Neveu (1974), Chap. III-2) and $\mathcal{N}_a = \{V_a \mid V \in \mathcal{N}\}$ is a sub-σ- field of $\mathcal{S}$. In the uniform case, this family collapses into a unique σ-field $T \subset \mathcal{S}$.

Example. Let us illustrate these concepts by the following example. Let $(s_i \mid a) \sim i.N(m, v)$ with $a = (m, v)$ and consider $\mathcal{E}^{\mathcal{M}}$ where $\mathcal{M} = \sigma(m) \in \mathcal{A}$ is the σ-field generated by m ; let $\mathcal{N}_1 = \sigma(\sum_i (s_i - m)^2)$ and, consider also $\mathcal{N}_2 = \sigma(\sum_i (s_i - m)^2, m)$ and $\mathcal{N}_3 = \sigma(\sum_i s_i, \sum_i s_i^2)$. $\mathcal{N}_j (j = 1, 2, 3)$ are each $\mathcal{E}^{\mathcal{M}}$-statistics, but only $\mathcal{N}_3$ is a uniform $\mathcal{E}^{\mathcal{M}}$-statistic. In $\mathcal{E}^{\mathcal{M}}$, (i.e., once m is known), $\mathcal{N}_1$ and $\mathcal{N}_2$ give the very same information: $\mathcal{N}_2$ may be considered as a redefinition of $\mathcal{N}_1$ so as to incorporate $\mathcal{M}$; $\mathcal{N}_2$ is thus a strong $\mathcal{E}^{\mathcal{M}}$-statistic. For each value of m, $\mathcal{N}_1$ (or $\mathcal{N}_2$) generates a family of different σ-fields on $\mathcal{S}$ while $\mathcal{N}_3$ generates only one such σ-field on $\mathcal{S}$. ∎

Similarly, in $\mathcal{E}^{\mathcal{M}}$ relevant functions of the parameters may depend on both $\mathcal{A}$ and $\mathcal{M}$. For example, the regression model $y = X\beta + \epsilon$ may be viewed as an experiment conditionalized on X and the function $X\beta$ may be considered as a parameter of interest; note also that in heteroscedastic models, the sampling variance of ϵ may, for example, be a function of $X\beta$ or of X.

1.4.9 Definition. A σ-field $\mathcal{L}$ is an $\mathcal{E}^{\mathcal{M}}$-*parameter* iff $\mathcal{L} \subset \mathcal{A} \vee \mathcal{M}$ and $(\mathcal{A} \vee \mathcal{M}) \cap \overline{\mathcal{M}}$ is called the *trivial* $\mathcal{E}^{\mathcal{M}}$-parameter. ∎

1.4.10 Definition. An $\mathcal{E}^{\mathcal{M}}$-parameter, $\mathcal{L}$, which is also an $\mathcal{E}$-parameter, (i.e., $\mathcal{L} \subset \mathcal{A}$) is called a *uniform $\mathcal{E}^{\mathcal{M}}$-parameter.* ∎

1.4.11 Definition. An $\mathcal{E}^{\mathcal{M}}$-parameter, $\mathcal{L}$, is a *strong $\mathcal{E}^{\mathcal{M}}$-parameter* iff $\mathcal{M} \subset \mathcal{L}$ (and therefore $\mathcal{M} \subset \mathcal{L} \subset \mathcal{A} \vee \mathcal{M}$). ∎

Remarks.

(i) Clearly a statistic (respectively, parameter) in $\mathcal{E}$ is a uniform $\mathcal{E}^{\mathcal{M}}$- statistic (respectively, $\mathcal{E}^{\mathcal{M}}$-parameter) for any $\mathcal{M}$. Note also that if $\mathcal{M}$ is trivial, — i.e., $\mathcal{M} = \mathcal{I} = \{\phi, A \times S\}$ — then $\mathcal{E}^{\mathcal{I}} = \mathcal{E}$, and any $\mathcal{E}^{\mathcal{I}}$-statistic is both strong and uniform.

(ii) If $\mathcal{M} \subset \mathcal{A}$ (respectively, $\mathcal{M} \subset \mathcal{S}$), an $\mathcal{E}^{\mathcal{M}}$-parameter (respectively, $\mathcal{E}^{\mathcal{M}}$-statistic) is necessarily uniform.

1.4.4 Complementary Reductions

An important theme in the Bayesian literature is the analysis of the process of "learning by observing", viz., the transformation "prior-to-posterior": $\mu \to \mu^{\mathcal{S}}$. The concept of complementary reductions corresponds to a decomposition of this learning process either on the sample space or on the parameter space.

A. Decomposition on the Sample Space

Let T be a sub-σ-field of S. The transformation of μ into μ^S may be decomposed into two steps. The first step is to obtain μ^T by revising μ in the light of partial sample information T; this is done in $\mathcal{E}_{A \vee T}$. The second step is to obtain μ^S by revising μ^T in the light of the complete sample information; this is done in $\mathcal{E}_{A \vee S}^{T}$. This decomposition is illustrated in the following diagram:

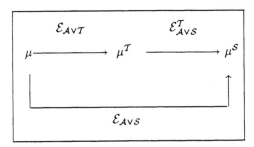

B. Decomposition on the Parameter Space

Let B be a sub-σ-field of A. Let us decompose μ into $\mu_B \otimes \mu^B$ and μ^S into $\mu_B^S \otimes \mu^{B \vee S}$. In $\mathcal{E}_{B \vee S}$, μ_B is revised by the result of a sampling process, characterized by P^B, and in $\mathcal{E}^B$, μ^B is revised by the unreduced sampling process, characterized by P^A. This may be illustrated as follows:

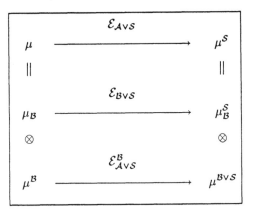

These decompositions motivate the following definition:

1.4.12 Definition. For any sub-σ-field $\mathcal{B} \subset \mathcal{A}$ the reductions $\mathcal{E}_{\mathcal{B} \vee \mathcal{S}}$ and $\mathcal{E}_{\mathcal{A} \vee \mathcal{S}}^{\mathcal{B}}$ are called *complementary*, as are the reductions $\mathcal{E}_{\mathcal{A} \vee \mathcal{T}}$ and $\mathcal{E}_{\mathcal{A} \vee \mathcal{S}}^{\mathcal{T}}$ for any sub-σ-field $\mathcal{T} \subset \mathcal{S}$. ∎

Example. Consider the Bayesian experiment $\mathcal{E}$ defined as follows: Let $A = S = I\!\!R^2$ and $\mathcal{A} = \mathcal{S} = \mathcal{B}^2$ (the Borel σ-field of $I\!\!R^2$) along with Π be defined through: $(s \mid a) \sim N_2(a, V_1)$ and $a \sim N_2(a_0, V_0)$ where V_0 and V_1 are known positive definite symmetric 2×2 matrices. Consider $\mathcal{B} = \sigma(\alpha'a)$ where α is a vector of given constants. Thus a pair of complementary reductions is $\mathcal{E}_{\mathcal{B} \vee \mathcal{S}}$ and $\mathcal{E}_{\mathcal{A} \vee \mathcal{S}}^{\mathcal{B}}$. In $\mathcal{E}_{\mathcal{B} \vee \mathcal{S}}$, the prior probability is given by $\alpha'a \sim N(\alpha'a_0, \alpha'V_0\alpha)$ and the marginalized sampling probabilities are

$$(s \mid \alpha'a) \sim N_2[a_0 + V_0\alpha(\alpha'V_0\alpha)^{-1}\alpha'(a - a_0), \quad V_1 + V_\alpha],$$

where $V_\alpha = V(a \mid \alpha'a) = V_0 - V_0\alpha(\alpha'V_0\alpha)^{-1}\alpha'V_0$. In this experiment, the prior probability is transformed into

$$(\alpha'a \mid s) \sim N[\alpha'a_0 + \alpha'V_0(V_0 + V_1)^{-1}(s - a_0), \quad \alpha'(V_0^{-1} + V_1^{-1})^{-1}\alpha].$$

In the complementary reduction, $\mathcal{E}_{\mathcal{A} \vee \mathcal{S}}^{\mathcal{B}}$, the prior distribution $(a \mid \alpha'a) \sim N_2(a_\alpha, V_\alpha)$, where $a_\alpha = E(a \mid \alpha'a) = a_0 + V_0\alpha(\alpha'V_0\alpha)^{-1}\alpha'(a - a_0)$, is revised into

$$(a \mid \alpha'a, s) \sim N_2[a_\alpha + V_\alpha(V_1 + V_\alpha)^{-1}(s - a_\alpha), V_\alpha - V_\alpha(V_1 + V_\alpha)^{-1}V_\alpha].$$

Similarly, consider $\mathcal{T} = \sigma(\beta's)$ where β is a vector of given constants. In $\mathcal{E}_{\mathcal{A} \vee \mathcal{T}}$ the prior distribution $a \sim N_2(a_0, V_0)$ is revised into $(a \mid \beta's) \sim N_2(a_\beta, V_\beta)$, where

$$
\begin{aligned}
a_\beta &= E(a \mid \beta's) = a_0 + V_0\beta[\beta'(V_0 + V_1)\beta]^{-1}\beta'(s - a_0) \\
V_\beta &= V(a \mid \beta's)V_0 - V_0\beta[\beta'(V_0 + V_1)\beta]^{-1}\beta'V_0.
\end{aligned}
$$

In the complementary reduction $\mathcal{E}_{\mathcal{A} \vee \mathcal{S}}^{\mathcal{T}}$ the prior distribution is the posterior distribution of $\mathcal{E}_{\mathcal{A} \vee \mathcal{T}}$: $(a \mid \beta's) \sim N_2(a_\beta, V_\beta)$, and is revised into

$$(a \mid \beta's, s) \sim (a \mid s) \sim N_2[(V_0^{-1} + V_1^{-1})^{-1}(V_0^{-1}a_0 + V_1^{-1}s), (V_0^{-1} + V_1^{-1})^{-1}]. \quad ∎$$

C. Complementary Reductions in Conditional Experiments

The above definitions correspond to decompositions of an unreduced experiment $\mathcal{E}_{A \vee S}$. Identical decompositions may be operated on marginal experiments $\mathcal{E}_{A \vee T}$, $\mathcal{E}_{B \vee S}$ or $\mathcal{E}_{B \vee T}$ since they are all characterized by a unique probability on a product space. Some care must be taken when defining complementary reductions of a conditional experiment, since conditional experiments are characterized, in the regular case, by a probability transition rather than a unique probability.

Let us consider arbitrary σ-fields $B \subset A$, $T \subset S$ and $M \subset A \vee S$. From the discussion on parameters and statistics in conditional experiments (Section 1.4.3.D), one may identify the conditional experiments $\mathcal{E}_{B \vee T}^{M}$ and $\mathcal{E}_{(B \vee M) \vee (M \vee T)}^{M}$. In other words, in conditional experiments, the product structure $B \vee T$ is considered unaffected if the parameters involve observable variables included in the conditioning σ-field, or if the statistics involve parameters included in the conditioning σ-field.

The decomposition of $\mathcal{E}_{B \vee T}^{M}$ on the sample space may be operated through a statistic of this conditional experiment, $N \subset M \vee T$. This decomposition is illustrated in the following diagram:

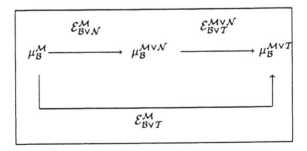

Similarly, the decomposition of $\mathcal{E}_{B \vee T}^{M}$ on the parameter space may be operated through a parameter of the conditional experiment, $\mathcal{L} \subset B \vee M$. This decomposition is illustrated in the following diagram:

$$
\begin{array}{ccc}
& \mathcal{E}^{\mathcal{M}}_{\mathcal{B}\vee\mathcal{T}} & \\
\mu^{\mathcal{M}}_{\mathcal{B}} & \longrightarrow & \mu^{\mathcal{M}\vee\mathcal{T}}_{\mathcal{B}} \\
\| & & \| \\
& \mathcal{E}^{\mathcal{M}}_{\mathcal{L}\vee\mathcal{T}} & \\
\mu^{\mathcal{M}}_{\mathcal{L}} & \longrightarrow & \mu^{\mathcal{M}\vee\mathcal{T}}_{\mathcal{L}} \\
\otimes & & \otimes \\
& \mathcal{E}^{\mathcal{L}\vee\mathcal{M}}_{\mathcal{B}\vee\mathcal{T}} & \\
\mu^{\mathcal{L}\vee\mathcal{M}}_{\mathcal{B}} & \longrightarrow & \mu^{\mathcal{L}\vee\mathcal{M}\vee\mathcal{T}}_{\mathcal{B}}
\end{array}
$$

As before, for any $\mathcal{L} \subset \mathcal{B} \vee \mathcal{M}$, $\mathcal{E}^{\mathcal{M}}_{\mathcal{L}\vee\mathcal{T}}$ and $\mathcal{E}^{\mathcal{L}\vee\mathcal{M}}_{\mathcal{B}\vee\mathcal{T}}$ are called complementary reductions on the parameter space of $\mathcal{E}^{\mathcal{M}}_{\mathcal{B}\vee\mathcal{T}}$, and for any $\mathcal{N} \subset \mathcal{M} \vee \mathcal{T}$, $\mathcal{E}^{\mathcal{M}}_{\mathcal{B}\vee\mathcal{N}}$ and $\mathcal{E}^{\mathcal{M}\vee\mathcal{N}}_{\mathcal{B}\vee\mathcal{T}}$ are called complementary reductions on the sample space of $\mathcal{E}^{\mathcal{M}}_{\mathcal{B}\vee\mathcal{T}}$.

Let us note that when $\mathcal{M}$ is trivial, i.e., $\mathcal{M} = \mathcal{I}$, the two decompositions illustrated in the above diagrams coincide with the decompositions presented in Sections 1.4.4A and 1.4.4. B.

1.4.5 Dominance in Reduced Experiments

If $\mathcal{E}$ is dominated in the sense of Definition 1.2.4, i.e., Π is dominated by $\mu \otimes P$, we shall show that for every $\mathcal{M}$ and $\mathcal{N}$ sub-σ-fields of $\mathcal{A} \vee \mathcal{S}$, $\Pi^{\mathcal{M}}_{\mathcal{N}}$ is dominated in a particular sense by $(\mu \otimes P)^{\mathcal{M}}_{\mathcal{N}}$.

A. Notation

We first recall some notation introduced in Chapter 0. As before, for any integrable random variable m defined on $A \times S$, we denote by $\mathcal{N}m$ the conditional expectation of m given $\mathcal{N}$ with respect to Π. Furthermore, we denote by $\tilde{\mathcal{N}}m$ the conditional expectation of m given $\mathcal{N}$ with respect to $\mu \otimes P$. Thus, *the symbol "~" refers to conditional expectations with respect* to $\mu \otimes P$. In particular the expectation of m with respect to $\mu \otimes P$ will be denoted by $\tilde{\mathcal{I}}m$.

Remember that all equalities for conditional expectations are almost sure equalities. More precisely, it may be noted that it is almost sure with respect to the trace of the probability on the largest σ-field involved in the equality. When the underlying probability is not ambiguous, we skip this "a.s." proviso, since conditional expectations are only defined almost surely. Such an omission should thus not endanger the correct understanding of the formulae and theorems. In this section, the proviso a.s.Π will be omitted as before, whereas the proviso a.s.$\mu \otimes P$ will not be omitted.

B. Dominance in Marginal Experiments

B.1. Marginal Densities

It is readily seen that, for any sub-σ-field $\mathcal{N}$ of $\mathcal{A} \vee \mathcal{S}$, $\Pi_{\mathcal{N}}$ is dominated by $(\mu \otimes P)_{\mathcal{N}}$. If g denotes a version of $d\Pi/d(\mu \otimes P)$ and $g_{\mathcal{N}}$ a version of $d\Pi_{\mathcal{N}}/d(\mu \otimes P)_{\mathcal{N}}$, we have the following identity:

$$(1.4.17) \qquad\qquad g_{\mathcal{N}} = \tilde{\mathcal{N}}g \qquad \text{a.s. } \mu \otimes P.$$

This comes from the fact that, for any $n \in [\mathcal{N}]^+$, by definition of $g_{\mathcal{N}}$, $\mathcal{I}n = \tilde{\mathcal{I}}(n \cdot g_{\mathcal{N}})$ and, also, $\mathcal{I}n = \tilde{\mathcal{I}}(n \cdot g) = \tilde{\mathcal{I}}(n \cdot \tilde{\mathcal{N}}g)$. These relationships lead to the identity (1.4.17).

In particular, since for any $a \in [\mathcal{A}]^+$ (respectively, $s \in [\mathcal{S}]^+$), $\mathcal{I}a = \tilde{\mathcal{I}}a$ (respectively, $\mathcal{I}s = \tilde{\mathcal{I}}s$), this implies $g_{\mathcal{A}} = 1$ a.s. $\mu \otimes P$ (respectively, $g_{\mathcal{S}} = 1$ a.s. $\mu \otimes P$). From (1.4.17), we also obtain that for any $\mathcal{B}$ sub-σ-field of $\mathcal{A}$, $g_{\mathcal{B}} = 1$ a.s. $\mu \otimes P$, and for any $\mathcal{T}$ sub-σ-field of $\mathcal{S}$, $g_{\mathcal{T}} = 1$ a.s. $\mu \otimes P$.

B.2. Dominance in Marginal Experiments

If $\mathcal{E}$ is dominated, and if $\mathcal{B}$ is a sub-σ-field of $\mathcal{A}$ and $\mathcal{T}$ a sub-σ-field of $\mathcal{S}$, the marginal experiment $\mathcal{E}_{\mathcal{B}\vee\mathcal{T}}$ is dominated in the sense of Definition 1.2.4, i.e., it is dominated by the product of its prior and predictive probabilities; indeed, one has clearly, $(\mu \otimes P)_{\mathcal{B}\vee\mathcal{T}} = \mu_{\mathcal{B}} \otimes P_{\mathcal{T}}$ and, therefore, $g_{\mathcal{B}\vee\mathcal{T}}$ is a version of $d\Pi_{\mathcal{B}\vee\mathcal{T}}/d(\mu_{\mathcal{B}} \otimes P_{\mathcal{T}})$. Thus, the same argument as in Section 1.2.2 leads to the following result:

1.4.13 Proposition. If $\mathcal{E}$ is dominated (in the sense of Definition 1.2.4), then, for any sub-σ-fields $\mathcal{B} \subset \mathcal{A}$ and $\mathcal{T} \subset \mathcal{S}$, the marginal experiment $\mathcal{E}_{\mathcal{B} \vee \mathcal{T}}$ is regular, i.e., the conditional probabilities $P_{\mathcal{T}}^{\mathcal{B}}$ and $\mu_{\mathcal{B}}^{\mathcal{T}}$ are regular and may be written as:

$$(1.4.18) \qquad P_{\mathcal{T}}^{\mathcal{B}}(X) = \int_X g_{\mathcal{B} \vee \mathcal{T}} \; dP \qquad X \in \mathcal{T}$$

$$(1.4.19) \qquad \mu_{\mathcal{B}}^{\mathcal{T}}(E) = \int_E g_{\mathcal{B} \vee \mathcal{T}} \; d\mu \qquad E \in \mathcal{B} \qquad\qquad \blacksquare$$

B.3. Marginal Densities in the Regular Case

Clearly with respect to $\mu \otimes P$, $\mathcal{A}$ and $\mathcal{S}$ are independent σ-fields. (The concept of independence among σ-fields is treated more systematically in Section 2.2). If $\mathcal{B}$ is a sub-σ-field of $\mathcal{A}$ and $\mathcal{T}$ a sub- σ-field of $\mathcal{S}$, we can see that for any $a \in [\mathcal{A}]^+$ and $s \in [\mathcal{S}]^+$

$$(1.4.20) \qquad \widetilde{\mathcal{B} \vee \mathcal{T}}(as) = \tilde{\mathcal{B}}a \cdot \tilde{\mathcal{T}}s = \mathcal{B}a \cdot \mathcal{T}s$$

since Π and $\mu \otimes P$ have the same marginals.

From this equation, we obtain that if μ, (respectively, P) admits a regular conditional probability given $\mathcal{B}$ (respectively, $\mathcal{T}$), one has:

$$(1.4.21) \qquad (\mu \otimes P)^{\mathcal{B} \vee \mathcal{T}} = \mu^{\mathcal{B}} \otimes P^{\mathcal{T}}.$$

From this, we obtain that $g_{\mathcal{B} \vee \mathcal{T}}$ may be computed as follows:

$$(1.4.22) \qquad g_{\mathcal{B} \vee \mathcal{T}} = \int_{A \times S} g(a, s) \mu^{\mathcal{B}}(da) P^{\mathcal{T}}(ds).$$

In particular,

$$g_{\mathcal{B} \vee \mathcal{S}} = \int_A g(a, s) \mu^{\mathcal{B}}(da)$$

$$g_{\mathcal{A} \vee \mathcal{T}} = \int_S g(a, s) P^{\mathcal{T}}(ds)$$

Example. We consider a probit type model for which we analyze only a sample of size one: $(s \mid a) \sim N(a, 1)$ with the prior probability $a \sim N(0, 1)$. Let λ be the Lebesgue measure on $\mathbb{R}$. We first notice that

$$\frac{d\Pi}{d(\mu \otimes \lambda)} = (2\pi)^{-\frac{1}{2}} \exp[-\frac{1}{2}(s - a)^2]$$

thus, it may also be checked that:

$$\frac{d\Pi}{d(\mu \otimes P)} = 2^{\frac{1}{2}} \exp[-\frac{1}{2}(a^2 - 2as + \frac{1}{2}s^2)]$$

because $s \sim N(0,2)$. If $t = \mathbf{1}_{[0,\infty[}(s)$, and $T = \sigma(t)$, P_T gives a probability equal to $\frac{1}{2}$ at $]-\infty, 0[$ and at $[0,\infty[$ and P^T is characterized by:

$$\frac{dP^T}{d\lambda} = [2f(s \mid 0,2)\mathbf{1}_{[0,\infty[}(s)]^t [2f(s \mid 0,2)\mathbf{1}_{]-\infty,0[}(s)]^{1-t}$$

where f is the density with respect to λ of the $N(0,2)$ distribution.

$$\frac{d\Pi_{A\vee T}}{d(\mu \otimes P_T)} = 2\Phi(-a)^{1-t}\{1 - \Phi(-a)\}^t$$

and this density may be computed equivalently by the conditional expectation of $d\Pi/d(\mu \otimes P)$ given $A \vee T$, or by integrating $d\Pi/d(\mu \otimes P)$ with respect to P^T ∎

C. Dominance in Conditional Experiments

C.1. Conditional Densities

We now consider $\mathcal{N}$ and $\mathcal{M}$, two sub-σ-fields of $\mathcal{A} \vee \mathcal{S}$. For any $n \in [\mathcal{N}]^+$ and any $m \in [\mathcal{M}]^+$, the expectation of the product $n \cdot m$ may be written as:

$$(1.4.23) \qquad \mathcal{I}(n \cdot m) = \tilde{\mathcal{I}}[n \cdot m \cdot g_{\mathcal{N}\vee\mathcal{M}}] = \tilde{\mathcal{I}}[m \cdot \tilde{\mathcal{M}}(n \cdot g_{\mathcal{N}\vee\mathcal{M}})].$$

But it is also true that

$$(1.4.24) \qquad \mathcal{I}(n \cdot m) = \mathcal{I}[m \cdot \mathcal{M}n] = \tilde{\mathcal{I}}[m \cdot \mathcal{M}n \cdot g_{\mathcal{M}}];$$

hence, for any $m \in [\mathcal{M}]^+$,

$$(1.4.25) \qquad \tilde{\mathcal{I}}[m \cdot \tilde{\mathcal{M}}(n \cdot g_{\mathcal{N}\vee\mathcal{M}})] = \tilde{\mathcal{I}}[m \cdot \mathcal{M}n \cdot g_{\mathcal{M}}],$$

and this implies that, for any $n \in [\mathcal{N}]^+$,

$$(1.4.26) \qquad \tilde{\mathcal{M}}(n \cdot g_{\mathcal{N}\vee\mathcal{M}}) = \mathcal{M}n \cdot g_{\mathcal{M}} \quad \text{a.s.}\mu \otimes P.$$

Now $\Pi[g_{\mathcal{M}} = 0] = \tilde{\mathcal{I}}[g_{\mathcal{N}\vee\mathcal{M}} \mathbf{1}_{\{g_{\mathcal{M}}=0\}}] = \tilde{\mathcal{I}}[g_{\mathcal{M}} \mathbf{1}_{\{g_{\mathcal{M}}=0\}}] = 0$. So $g_{\mathcal{M}}$ is positive a.s.Π and $\{g_{\mathcal{M}} = 0\} \subset \{g_{\mathcal{N}\vee\mathcal{M}} = 0\}$ a.s.$\mu \otimes P$. Noticing that $g_{\mathcal{M}} \in [\mathcal{M}]^+$, and if we define

$$(1.4.27) \qquad g_{\mathcal{N}}^{\mathcal{M}} = \frac{g_{\mathcal{N}\vee\mathcal{M}}}{g_{\mathcal{M}}} \qquad \text{if } g_{\mathcal{M}} > 0$$

$$\qquad\qquad\qquad = 1 \qquad\qquad \text{if } g_{\mathcal{M}} = 0$$

we obtain the following identity: for any $n \in [\mathcal{N}]^+$

$$(1.4.28) \qquad \mathcal{M}n = \tilde{\mathcal{M}}[n \cdot g_{\mathcal{N}}^{\mathcal{M}}] \qquad \text{a.s.}\Pi.$$

This tells us that $\Pi_{\mathcal{N}}^{\mathcal{M}}$ is in some sense dominated by $(\mu \otimes P)_{\mathcal{N}}^{\mathcal{M}}$ since a conditional expectation with respect to Π may be replaced by a conditional expectation with respect to $\mu \otimes P$, using the "density" $g_{\mathcal{N}}^{\mathcal{M}}$.

It is important to note that $g_{\mathcal{N}}^{\mathcal{M}} \in [\mathcal{N} \vee \mathcal{M}]^+$, and that we may rewrite (1.4.27) as:

$$(1.4.29) \qquad g_{\mathcal{N}\vee\mathcal{M}} = g_{\mathcal{M}} \cdot g_{\mathcal{N}}^{\mathcal{M}} \qquad \text{a.s.}\mu \otimes P.$$

C.2. Dominance in Conditional Experiments

If $\mathcal{E}$ is dominated and $\mathcal{M}$ is a sub-σ-field of $\mathcal{A} \vee \mathcal{S}$, it is not clear whether the conditional experiment $\mathcal{E}^{\mathcal{M}}$ will be dominated in the sense of Definition 1.2.4; indeed, if there exists a regular version of $(\mu \otimes P)^{\mathcal{M}}$, then $\Pi^{\mathcal{M}}$ will also admit a regular version, and thus also $\mu^{\mathcal{M}}$ and $P^{\mathcal{M}}$; but, in general, one would not have $(\mu \otimes P)^{\mathcal{M}} = \mu^{\mathcal{M}} \otimes P^{\mathcal{M}}$.

In the special case where $\mathcal{M} = \mathcal{B} \vee \mathcal{T}$, with $\mathcal{B} \subset \mathcal{A}$ and $\mathcal{T} \subset \mathcal{S}$ such that there exist regular versions of $\mu^{\mathcal{B}}$ and $P^{\mathcal{T}}$, then $(\mu \otimes P)^{\mathcal{B}\vee\mathcal{T}}$ and $\Pi^{\mathcal{B}\vee\mathcal{T}}$ will also admit regular versions and, as seen earlier, $(\mu \otimes P)^{\mathcal{B}\vee\mathcal{T}} = \mu^{\mathcal{B}} \otimes P^{\mathcal{T}}$. In this situation $\mu^{\mathcal{B}\vee\mathcal{T}}$ (respectively, $P^{\mathcal{B}\vee\mathcal{T}}$) is regular and dominated by $\mu^{\mathcal{B}}$ (respectively, $P^{\mathcal{T}}$); $g_{\mathcal{A}}^{\mathcal{B}\vee\mathcal{T}}$ (respectively, $g_{\mathcal{S}}^{\mathcal{B}\vee\mathcal{T}}$) is a version of $d\mu^{\mathcal{B}\vee\mathcal{T}}/d\mu^{\mathcal{B}}$ (respectively, $dP^{\mathcal{B}\vee\mathcal{T}}/dP^{\mathcal{T}}$). Indeed if $E \in \mathcal{A}$,

$$(1.4.30) \qquad \begin{aligned} \mu^{\mathcal{B}\vee\mathcal{T}}(E) &= \Pi^{\mathcal{B}\vee\mathcal{T}}(E \times S) &= \widetilde{\mathcal{B}\vee\mathcal{T}}[g_{\mathcal{A}}^{\mathcal{B}\vee\mathcal{T}}\mathbf{1}_{E\times S}] \\ &= \int_{E\times S} g_{\mathcal{A}}^{\mathcal{B}\vee\mathcal{T}} d\mu^{\mathcal{B}} dP^{\mathcal{T}} &= \int_E g_{\mathcal{A}}^{\mathcal{B}\vee\mathcal{T}} d\mu^{\mathcal{B}}, \end{aligned}$$

where the last equality comes from the fact that $g_{\mathcal{A}}^{\mathcal{B}\vee\mathcal{T}} \in [\mathcal{A} \vee \mathcal{T}]^+$. Similarly, if $X \in \mathcal{S}$,

$$(1.4.31) \qquad P^{\mathcal{B}\vee\mathcal{T}}(X) = \int_X g_{\mathcal{S}}^{\mathcal{B}\vee\mathcal{T}} dP^{\mathcal{T}}.$$

Hence, $\mu^{\mathcal{B}\vee\mathcal{T}} \otimes P^{\mathcal{B}\vee\mathcal{T}}$ is regular and dominated by $\mu^{\mathcal{B}} \otimes P^{\mathcal{T}}$, and $g_{\mathcal{A}}^{\mathcal{B}\vee\mathcal{T}} \cdot g_{\mathcal{S}}^{\mathcal{B}\vee\mathcal{T}}$ is a version of $d(\mu^{\mathcal{B}\vee\mathcal{T}} \otimes P^{\mathcal{B}\vee\mathcal{T}})/d(\mu^{\mathcal{B}} \otimes P^{\mathcal{T}})$. Now, if we define

$$(1.4.32) \qquad h^{\mathcal{B}\vee\mathcal{T}} = \frac{g_{\mathcal{A}\vee\mathcal{S}}^{\mathcal{B}\vee\mathcal{T}}}{g_{\mathcal{A}}^{\mathcal{B}\vee\mathcal{T}} \cdot g_{\mathcal{S}}^{\mathcal{B}\vee\mathcal{T}}} = \frac{g \cdot g_{\mathcal{B}\vee\mathcal{T}}}{g_{\mathcal{A}\vee\mathcal{T}} \cdot g_{\mathcal{B}\vee\mathcal{S}}}$$

on $\{g_{\mathcal{A}\vee\mathcal{T}} > 0\} \cap \{g_{\mathcal{B}\vee\mathcal{S}} > 0\}$ and 1 elsewhere, we obtain the following result.

1.4.14 Theorem. If $\mathcal{E}$ is dominated, then for any sub-σ-field $\mathcal{B} \subset \mathcal{A}$ and $\mathcal{T} \subset \mathcal{S}$ such that there exist regular versions of $\mu^{\mathcal{B}}$ and $P^{\mathcal{T}}$, the conditional experiment $\mathcal{E}^{\mathcal{B} \vee \mathcal{T}}$ is dominated and regular, i.e., the conditional probabilities $P^{\mathcal{A} \vee \mathcal{T}}$ and $\mu^{\mathcal{B} \vee \mathcal{S}}$ are regular and may be written as

$$(1.4.33) \qquad P^{\mathcal{A} \vee \mathcal{T}}(X) = \int_X h^{\mathcal{B} \vee \mathcal{T}} dP^{\mathcal{B} \vee \mathcal{T}}$$

$$= \int_X g_{\mathcal{A} \vee \mathcal{S}}^{\mathcal{A} \vee \mathcal{T}} dP^{\mathcal{T}} \qquad X \in \mathcal{S}$$

$$(1.4.34) \qquad \mu^{\mathcal{B} \vee \mathcal{S}}(E) = \int_E h^{\mathcal{B} \vee \mathcal{T}} d\mu^{\mathcal{B} \vee \mathcal{T}}$$

$$= \int_E g_{\mathcal{A} \vee \mathcal{S}}^{\mathcal{B} \vee \mathcal{S}} d\mu^{\mathcal{B}} \qquad E \in \mathcal{A}$$

Proof. If $m \in [\mathcal{A} \vee \mathcal{S}]^+$,

$$\widetilde{\mathcal{B} \vee \mathcal{T}}\left[m \cdot h^{\mathcal{B} \vee \mathcal{T}} \cdot g_{\mathcal{A}}^{\mathcal{B} \vee \mathcal{T}} \cdot g_{\mathcal{S}}^{\mathcal{B} \vee \mathcal{T}}\right] = \widetilde{\mathcal{B} \vee \mathcal{T}}\left[m \cdot g_{\mathcal{A} \vee \mathcal{S}}^{\mathcal{B} \vee \mathcal{T}} \cdot 1_{\{g_{\mathcal{A}}^{\mathcal{B} \vee \mathcal{T}} > 0\} \cap \{g_{\mathcal{S}}^{\mathcal{B} \vee \mathcal{T}} > 0\}}\right]$$

$$= \widetilde{\mathcal{B} \vee \mathcal{T}}\left[m \cdot g_{\mathcal{A} \vee \mathcal{S}}^{\mathcal{B} \vee \mathcal{T}}\right] = (\mathcal{B} \vee \mathcal{T})(m).$$

Since, as seen before, $\{g > 0\} \subset \{g_{\mathcal{A} \vee \mathcal{T}} > 0\} \cap \{g_{\mathcal{B} \vee \mathcal{S}} > 0\}$ a.s.$\mu \otimes P$. ∎

As a final remark, note that, in particular:

$$(1.4.35) \qquad h^{\mathcal{B}} = g_{\mathcal{A} \vee \mathcal{S}}^{\mathcal{B} \vee \mathcal{S}} = \frac{g}{g_{\mathcal{B} \vee \mathcal{S}}}$$

$$(1.4.36) \qquad h^{\mathcal{T}} = g_{\mathcal{A} \vee \mathcal{S}}^{\mathcal{A} \vee \mathcal{T}} = \frac{g}{g_{\mathcal{A} \vee \mathcal{T}}}.$$

2

Admissible Reductions:
Sufficiency and Ancillarity

2.1 Introduction

In this chapter we formalize the idea that a given reduction of a Bayesian experiment "does not lose useful information". This concept leads to the definition of admissible reductions. Finally, we characterize the admissibility of reductions effectuated either by marginalization or by conditioning.

Given that the object of Bayesian inference is the transformation of "prior to posterior" probabilities, a Bayesian experiment is said to be "*totally non-informative*" if the transformation prior to posterior is trivial, i.e., if the posterior probabilities are almost all equal to the prior probability. It will therefore be said that a reduced experiment $\mathcal{E}_1$ "*loses no information*" with respect to the unreduced experiment $\mathcal{E}$ if the complementary reduction $\mathcal{E}_2$ (in the sense of Section 1.4.4) is totally non-informative. Indeed in this case the revision of the prior probability is entirely determined in the reduction $\mathcal{E}_1$. Therefore, in such a situation, $\mathcal{E}_1$ is called an *admissible reduction* of $\mathcal{E}$. As we shall see later on, $\mathcal{E}$ may be either an unreduced experiment $\mathcal{E}_{\mathcal{A} \vee \mathcal{S}}$ or an already reduced experiment, such as $\mathcal{E}_{\mathcal{B} \vee \mathcal{S}}$, where $\mathcal{B}$ represents the parameter of interest. In this chapter, we only consider reduction of an unreduced experiment $\mathcal{E}_{\mathcal{A} \vee \mathcal{S}}$.

A parameter (or a statistic) is said to be *sufficient* if the corresponding marginal reduction is admissible. A parameter (or a statistic) is said to be *ancillary* if the corresponding conditional reduction is admissible. In other words, sufficiency and ancillarity appear as concepts characterizing the admissibility of reductions by marginalization and by conditioning respectively.

In a Bayesian framework, those concepts are naturally rendered operational by using the tool of stochastic independence both in marginal and in conditional terms. For this reason, we first review in the next section the main properties of conditional independence; that section is based on Mouchart and Rolin (1984b) but Sections 2.2.5 and 2.2.6 are essentially new. The same material has also been partly presented in Van Putten and Van Schuppen (1985); for a table of correspondence of results, see Mouchart and Rolin (1985); on the same topic, see also Dawid (1979a, 1979b, 1980b) and Pitman and Speed (1973). Section 2.3 is based on Florens and Mouchart (1977, 1986a), but Sections 2.3.6 and 2.3.7 are essentially new.

2.2   Conditional Independence

2.2.1   Notation

This section handles a topic in probability theory. For this reason, we recall the general notation introduced in Chapter 0, viz., $(M, \mathcal{M}, P)$ is an abstract probability space; thus M need not be equal to $A \times S$. Sub-σ-fields of $\mathcal{M}$ will be denoted by $\mathcal{M}_i$; in particular $\mathcal{M}_0$ is always the trivial σ-field $\{\phi, M\}$. Also, $m \in [\mathcal{M}_i]^+$ indicates that m is a positive real-valued $\mathcal{M}_i$-measurable function defined on M and $\mathcal{M}_i m$ denotes the conditional expectation of m with respect to P, given $\mathcal{M}_i$, i.e., $\mathcal{M}_i m = E(m \mid \mathcal{M}_i)$.

The concepts and equalities introduced below are, in general, stated in terms of positive real-valued functions and are extended to integrable (real-valued) functions using standard arguments. Finally, let us also recall that equalities involving conditional expectations should be read as almost sure equalities.

2.2.2 Definition of Conditional Independence

The following theorem defines and completely characterizes the concept of conditional independence:

2.2.1 Theorem. The following conditions are equivalent:

(i) $\mathcal{M}_3(m_1 m_2) = \mathcal{M}_3 m_1 \cdot \mathcal{M}_3 m_2 \quad \forall\, m_i \in [\mathcal{M}_i]^+, i = 1, 2;$

(ii) $(\mathcal{M}_2 \vee \mathcal{M}_3) m_1 = \mathcal{M}_3 m_1 \qquad \forall\, m_1 \in [\mathcal{M}_1]^+;$

(iii) $\mathcal{M}_2(\mathcal{M}_3 m_{13}) = \mathcal{M}_2 m_{13} \quad \forall\, m_{13} \in [\mathcal{M}_1 \vee \mathcal{M}_3]^+.$

Proof. The equivalence of (i) and (ii) are standard (see, e.g., Dellacherie and Meyer (1975), Chap. 2). The equivalence of (i) and (iii) is less well known, and follows from the following argument

$$\mathcal{M}_0(m_1 m_2 m_3) = \mathcal{M}_0[m_2 \cdot \mathcal{M}_3(m_1 m_3)] \qquad \forall\, m_i \in [\mathcal{M}_i]^+ \quad (i = 1, 2, 3)$$

if and only if

$$\mathcal{M}_0(m_1 m_2 m_3) = \mathcal{M}_0[\mathcal{M}_3 m_1 \cdot \mathcal{M}_3 m_2 \cdot m_3] \qquad \forall\, m_i \in [\mathcal{M}_i]^+ \quad (i = 1, 2, 3)$$

since

$$\mathcal{M}_0[m_2 \cdot \mathcal{M}_3(m_1 m_3)] = \mathcal{M}_0[\mathcal{M}_3 m_2 \cdot m_1 m_3] = \mathcal{M}_0[\mathcal{M}_3 m_1 \cdot \mathcal{M}_3 m_2 \cdot m_3].$$

The proof is then completed by making use of the monotone class Theorem 0.2.21, to extend these equalities from the products $m_1 \cdot m_3$ to all functions $m_{13} \in [\mathcal{M}_1 \vee \mathcal{M}_3]$. ∎

2.2.2 Definition. Under any one of the conditions of Theorem 2.2.1, we say that $\mathcal{M}_1$ and $\mathcal{M}_2$ are *independent conditionally on* $\mathcal{M}_3$ and we write $\mathcal{M}_1 \perp\!\!\!\perp \mathcal{M}_2 \mid \mathcal{M}_3$. If we want to make explicit the role of the probability P in this concept, we write $\mathcal{M}_1 \perp\!\!\!\perp \mathcal{M}_2 \mid \mathcal{M}_3; P$. ∎

Remarks.

1) Clearly when $\mathcal{M}_3 = \mathcal{M}_0$, Definition 2.2.2. corresponds to the usual (or "marginal") independence of σ-fields, and we write $\mathcal{M}_1 \perp\!\!\!\perp \mathcal{M}_2$ rather

than $\mathcal{M}_1 \perp\!\!\!\perp \mathcal{M}_2 \mid \mathcal{M}_0$. Note that conditions (i) and (ii) generalize, for the case of non-trivial $\mathcal{M}_3$, the familiar concept of stochastic independence (among σ-fields) whereas condition (iii) is genuinely a conditional concept in the sense that it boils down to condition (ii) for trivial $\mathcal{M}_3$.

2) It is clear from (i) that the concept of conditional independence is symmetric in $\mathcal{M}_1$ and $\mathcal{M}_2$. This allows us to add conditions (ii bis) and (iii bis) obtained by interchanging $\mathcal{M}_1$ and $\mathcal{M}_2$. In the sequel, we omit such obvious duplications.

3) Condition (iii) of Theorem 2.2.1 cannot be weakened into

$$\mathcal{M}_2(\mathcal{M}_3 m_1) = \mathcal{M}_2 m_1 \qquad \forall\, m_1 \in [\mathcal{M}_1]^+$$

(as done in Martin, Petit and Littaye (1973)); this is shown by the following counterexample. Take three measurable sets $A_i \in \mathcal{M}$, each with positive probability, pairwise independent but not jointly independent, and define $\mathcal{M}_i = \sigma(A_i)$. For any $m \in [\mathcal{M}_1]^+$, pairwise independence implies $\mathcal{M}_2(\mathcal{M}_3 m) = \mathcal{M}_2 m$ whereas, in general, $\mathcal{M}_3(m_1 m_2) = \mathcal{M}_3 m_1 \cdot \mathcal{M}_3 m_2$ for $m_1 \in [\mathcal{M}_1]^+$, and $m_2 \in [\mathcal{M}_2]^+$ is not implied by pairwise independence.

As $\mathcal{M}_3 \subset \mathcal{M}_1$ is equivalent to $\mathcal{M}_1 \vee \mathcal{M}_3 = \mathcal{M}_1$, Theorem 2.2.1 trivially implies the following corollary.

2.2.3 Corollary. If $\mathcal{M}_3 \subset \mathcal{M}_1$, then the following properties are equivalent:

(i) $\mathcal{M}_1 \perp\!\!\!\perp \mathcal{M}_2 \mid \mathcal{M}_3$,

(ii) $\mathcal{M}_1 m_2 = \mathcal{M}_3 m_2 \quad \forall\, m_2 \in [\mathcal{M}_2]^+$,

(iii) $\mathcal{M}_2(\mathcal{M}_3 m_1) = \mathcal{M}_2 m_1 \qquad \forall\, m_1 \in [\mathcal{M}_1]^+$.   ∎

The next corollary provides elementary properties of conditional independence. These properties follow in a straightforward way from elementary properties of conditional expectations.

2.2.4 Corollary.
(i) $\mathcal{M}_1 \subset \mathcal{M}_3 \;\Rightarrow\; \mathcal{M}_1 \perp\!\!\!\perp \mathcal{M}_2 \mid \mathcal{M}_3 \qquad \forall\, \mathcal{M}_2$,

(ii) $\mathcal{M}_1 \perp\!\!\!\perp \mathcal{M}_2 \mid \mathcal{M}_3$ and $\mathcal{M}_5 \subset \mathcal{M}_1 \;\Rightarrow\; \mathcal{M}_5 \perp\!\!\!\perp \mathcal{M}_2 \mid \mathcal{M}_3$.   ∎

2.2.3   Null Sets and Completion

In such topics as minimal sufficiency or asymptotic theory, null sets may play a crucial role, and the literature is plagued with errors due to the overlooking or misuse of null sets (see, e.g.,Basu (1955) and (1958), and Dawid (1979b)). In this section we introduce the measurable completion of σ-fields. This tool will usually allow a suitable treatment of null sets; more particularly, it provides a characterization of conditional independence in terms of a measurability property.

As in Section 0.3.2, the *completed trivial σ-field* $\overline{\mathcal{M}_0}$ is defined by:

$$(2.2.1) \qquad \overline{\mathcal{M}_0} = \{A \in \mathcal{M} : P(A)^2 = P(A)\}$$

and for $\mathcal{M}_i$, a sub-σ-field of $\mathcal{M}$, we define the *completed σ-field* $\overline{\mathcal{M}_i}$ as:

$$(2.2.2) \qquad \overline{\mathcal{M}_i} = \mathcal{M}_i \vee \overline{\mathcal{M}_0}.$$

Let us remark that we complete sub-σ-fields by measurable sets only and not by subsets of measurable null sets as is usually done (in Lebesgue completion). In this way, we avoid the danger of losing the separability of a σ-field.

Let us recall two basic properties of the conditional expectation:

$$(2.2.3) \qquad \overline{\mathcal{M}_i} m \;=\; \mathcal{M}_i m \qquad \forall\, m \in [\mathcal{M}]^+$$

$$(2.2.4) \qquad m \in [\overline{\mathcal{M}_i}]^+ \;\Leftrightarrow\; m \in [\mathcal{M}]^+ \;\text{ and }\; m = \mathcal{M}_i m.$$

The main properties used in operating with completion of σ-fields are summarized in the following lemma.

2.2.5 Lemma.

(i) $\overline{\mathcal{M}_1 \vee \mathcal{M}_2} = \overline{\mathcal{M}_1} \vee \overline{\mathcal{M}_2} = \overline{\mathcal{M}_1} \vee \mathcal{M}_2 = \mathcal{M}_1 \vee \overline{\mathcal{M}_2};$

(ii) $\overline{\mathcal{M}_1 \cap \mathcal{M}_2} \subset \overline{\mathcal{M}_1} \cap \overline{\mathcal{M}_2} = \overline{\mathcal{M}_1 \cap \mathcal{M}_2} = \mathcal{M}_1 \cap \overline{\mathcal{M}_2};$

(iii) $\mathcal{M}_1 \subset \mathcal{M}_2 \Rightarrow \overline{\mathcal{M}_1} \cap \mathcal{M}_2 = \mathcal{M}_1 \vee (\overline{\mathcal{M}_0} \cap \mathcal{M}_2).$ ∎

Properties (i) and (ii) are fairly standard. Property (iii) will be used repeatedly. More insight into this property may be obtained by first noticing that $\mathcal{M}_2 \cap \overline{\mathcal{M}_0}$ represents the σ-field of the trivial sets of $\mathcal{M}_2$. Therefore, when $\mathcal{M}_1$ is included in $\mathcal{M}_2$, the completion of $\mathcal{M}_1$ by the null sets of $\mathcal{M}_2$ is equal to the sets of $\mathcal{M}_2$ that are almost surely equal to sets of $\mathcal{M}_1$.

We are now in a position of characterizing conditional independence as a measurability property.

2.2.6 Theorem. The following properties are equivalent:

(i) $\mathcal{M}_1 \perp\!\!\!\perp \mathcal{M}_2 \mid \mathcal{M}_3$.

(ii) $m_2 \in [\mathcal{M}_2]^+ \Rightarrow (\mathcal{M}_1 \vee \mathcal{M}_3)m_2 \in [\overline{\mathcal{M}_3}]^+$.

Proof. Using the Condition (ii) in Theorem 2.2.1 and the basic property (2.2.4), the proof follows from the following identity which is an application of Proposition 0.3.14(ii): $\forall\, m_2 \in [\mathcal{M}_2]^+$,

$$(\mathcal{M}_1 \vee \mathcal{M}_3)m_2 \;=\; \mathcal{M}_3[(\mathcal{M}_1 \vee \mathcal{M}_3)m_2] \;=\; \mathcal{M}_3 m_2. \qquad \blacksquare$$

Using Lemma 2.2.5, conditional independence may be extended to completed σ-fields as follows.

2.2.7 Corollary. The following properties are equivalent:

(i) $\mathcal{M}_1 \perp\!\!\!\perp \mathcal{M}_2 \mid \mathcal{M}_3$,

(ii) $\mathcal{M}_1 \perp\!\!\!\perp \mathcal{M}_2 \mid \overline{\mathcal{M}_3}$,

(iii) $\overline{\mathcal{M}_1} \perp\!\!\!\perp \overline{\mathcal{M}_2} \mid \mathcal{M}_3$. $\qquad\qquad\qquad\qquad\qquad\qquad\blacksquare$

Remember that for marginal independence the completed trivial σ-field is the largest σ-field that is, independent of itself. The next corollary extends this property to conditional independence.

2.2.8 Corollary. The following properties are equivalent:

(i) $\mathcal{M}_1 \subset \overline{\mathcal{M}_3}$,

(ii) $\mathcal{M}_1 \perp\!\!\!\perp \mathcal{M}_1 \mid \mathcal{M}_3$,

(iii) $m_1 \in [\mathcal{M}_1]^+ \Rightarrow \mathcal{M}_1(\mathcal{M}_3 m_1) = m_1$.

Proof. The equivalence of (i) and (ii) follows directly from Theorem 2.2.6. That (i) $\Rightarrow$ (iii) is clear. That (iii) $\Rightarrow$ (i) relies on the following identity for any $m_1 \in [\mathcal{M}_1]_\infty$:

$$\mathcal{M}_0\{(m_1-\mathcal{M}_3 m_1)^2\} = \mathcal{M}_0\{m_1(m_1-\mathcal{M}_3 m_1)\} = \mathcal{M}_0\{m_1[m_1-\mathcal{M}_1(\mathcal{M}_3 m_1)]\}$$

which is equal to zero under (iii).   ∎

Another facet of Corollary 2.2.8 is revealed by considering a conditional version of the well-known property that any event common to two independent σ-fields is trivial. More specifically, we have:

2.2.9 Corollary. If $\mathcal{M}_1 \perp\!\!\!\perp \mathcal{M}_2 \mid \mathcal{M}_3$, then $\overline{\mathcal{M}_1} \cap \overline{\mathcal{M}_2} \subset \overline{\mathcal{M}_3}$.   ∎

We shall see that this corollary, elementary from a probabilistic point of view, plays a crucial role in asymptotic theory.

2.2.4   Basic Properties of Conditional Independence

The main tool supplied by conditional independence is indicated in the next theorem:

2.2.10 Theorem. The following properties are equivalent.

(i) $\mathcal{M}_1 \perp\!\!\!\perp \mathcal{M}_2 \mid \mathcal{M}_3$ and $\mathcal{M}_1 \perp\!\!\!\perp \mathcal{M}_4 \mid \mathcal{M}_2 \vee \mathcal{M}_3$,

(ii) $\mathcal{M}_1 \perp\!\!\!\perp (\mathcal{M}_2 \vee \mathcal{M}_4) \mid \mathcal{M}_3$,

(iii) $\mathcal{M}_1 \perp\!\!\!\perp \mathcal{M}_4 \mid \mathcal{M}_3$ and $\mathcal{M}_1 \perp\!\!\!\perp \mathcal{M}_2 \mid \mathcal{M}_4 \vee \mathcal{M}_3$.

Proof. By the symmetry of $\mathcal{M}_2$ and $\mathcal{M}_4$, it suffices to prove the equivalence of (i) and (ii).

(ii) $\Rightarrow$ (i): the proof follows from a simple property of conditional expectations: if $m \in [\mathcal{M}]^+$ is such that $\mathcal{M}_1 m = \mathcal{M}_2 m$ and if $\mathcal{M}_1 \subset \mathcal{M}_2$, then for any $\mathcal{M}_3$ such that $\mathcal{M}_1 \subset \mathcal{M}_3 \subset \mathcal{M}_2$, we have $\mathcal{M}_3 m = \mathcal{M}_1 m$. Hence, if (ii) is true, then $\forall m_1 \in [\mathcal{M}_1]^+, (\mathcal{M}_2 \vee \mathcal{M}_4 \vee \mathcal{M}_3) m_1 = \mathcal{M}_3 m_1 = (\mathcal{M}_2 \vee \mathcal{M}_3) m_1$. The last equality is due to the fact that $\mathcal{M}_3 \subset \mathcal{M}_2 \vee \mathcal{M}_3 \subset \mathcal{M}_2 \vee \mathcal{M}_4 \vee \mathcal{M}_3$; these two equalities are clearly equivalent to (i).

(i) $\Rightarrow$ (ii): $\forall m_1 \in [\mathcal{M}_1]^+ : (\mathcal{M}_2 \vee \mathcal{M}_4 \vee \mathcal{M}_3) m_1 = (\mathcal{M}_2 \vee \mathcal{M}_3) m_1 = \mathcal{M}_3 m_1$ is implied by (i). Therefore $\{(\mathcal{M}_2 \vee \mathcal{M}_4) \vee \mathcal{M}_3\} m_1 = \mathcal{M}_3 m_1$ which is (ii). ∎

Next corollary shows that the property $\mathcal{M}_1 \perp\!\!\!\perp \mathcal{M}_2 \mid \mathcal{M}_3$ remains true if $\mathcal{M}_1$ (or $\mathcal{M}_2$) is enlarged in the direction of $\mathcal{M}_3$, or if $\mathcal{M}_3$ is enlarged in the direction of $\mathcal{M}_1$ and/or of $\mathcal{M}_2$.

2.2.11 Corollary. If $\mathcal{M}_1 \perp\!\!\!\perp \mathcal{M}_2 \mid \mathcal{M}_3, \mathcal{M}_5 \subset \mathcal{M}_1 \vee \mathcal{M}_3, \mathcal{M}_4 \subset \mathcal{M}_2 \vee \mathcal{M}_3$, then

(i) $(\mathcal{M}_1 \vee \mathcal{M}_5) \perp\!\!\!\perp (\mathcal{M}_2 \vee \mathcal{M}_4) \mid \mathcal{M}_3$,

(ii) $\mathcal{M}_1 \perp\!\!\!\perp \mathcal{M}_2 \mid \mathcal{M}_3 \vee \mathcal{M}_4 \vee \mathcal{M}_5$.

Proof. Indeed, $\mathcal{M}_4 \subset \mathcal{M}_2 \vee \mathcal{M}_3$ implies, by Corollary 2.2.4(i), that $\mathcal{M}_1 \perp\!\!\!\perp \mathcal{M}_4 \mid \mathcal{M}_2 \vee \mathcal{M}_3$; along with $\mathcal{M}_1 \perp\!\!\!\perp \mathcal{M}_2 \mid \mathcal{M}_3$ this gives, by Theorem 2.2.10, the result for $\mathcal{M}_5 = \mathcal{M}_0$. The proof is completed by repeating the same argument for an arbitrary $\mathcal{M}_5 \subset \mathcal{M}_1 \vee \mathcal{M}_3$. ∎

Theorem 2.2.10 shows which pairs of conditions of conditional independence (viz., (i) or (iii)) do produce (and are actually equivalent to) the "larger" condition (ii). This is indeed a crucial property, and is used repeatedly subsequently. In this chapter, as well as in Chapters 3 and 4, this theorem will have a central role in the development of the theory of admissible reductions. When studying optimal reductions, we shall show that pairs of conditions other than (i) or (iii) may also imply, but not be equivalent to, the "larger" condition (ii), provided some supplementary conditions are granted: these new conditions are presented in Chapter 5.

If we wish to decrease the σ-fields in $\mathcal{M}_1 \perp\!\!\!\perp \mathcal{M}_2 \mid \mathcal{M}_3$, the elementary property (ii) of Corollary 2.2.4 allows us to replace $\mathcal{M}_1$ or $\mathcal{M}_2$ by any of their respective sub-σ-fields; the next theorem characterizes a method for decreasing the conditioning σ-field.

2.2.12 Theorem.

If $\mathcal{M}_3 \vee \mathcal{M}_4 \subset \mathcal{M}_1$,

$\mathcal{M}_1 \perp\!\!\!\perp \mathcal{M}_2 \mid \mathcal{M}_3$ and $\mathcal{M}_1 \perp\!\!\!\perp \mathcal{M}_2 \mid \mathcal{M}_4$,

then

(i) $\mathcal{M}_1 \perp\!\!\!\perp \mathcal{M}_2 \mid \overline{\mathcal{M}_3} \cap \overline{\mathcal{M}_4}$,

(ii) $\mathcal{M}_1 \perp\!\!\!\perp \mathcal{M}_2 \mid \overline{\mathcal{M}_3} \cap \mathcal{M}_4$.

Proof. (i) is an immediate consequence of Theorem 2.2.6. (ii) then follows from Lemma 2.2.5 (ii) and Corollary 2.2.7.   ∎

Note that (i) and (ii) are clearly equivalent. Also, if $\mathcal{M}_3 \vee \mathcal{M}_4 \subset \mathcal{M}_1$, then any one of (i) or (ii) implies the other two conditions of this theorem. Furthermore, taking $\mathcal{M}_1 = \mathcal{M}_3 \vee \mathcal{M}_4$ leads to the following corollary:

2.2.13 Corollary. For any sub-σ-fields $\mathcal{M}_i$ $(i = 2, 3, 4)$, the following properties are equivalent:

(i) $\mathcal{M}_2 \perp\!\!\!\perp \mathcal{M}_3 \mid \mathcal{M}_4$ and $\mathcal{M}_2 \perp\!\!\!\perp \mathcal{M}_4 \mid \mathcal{M}_3$;

(ii) $\mathcal{M}_2 \perp\!\!\!\perp (\mathcal{M}_3 \vee \mathcal{M}_4) \mid \overline{\mathcal{M}_3} \cap \overline{\mathcal{M}_4}$.   ∎

2.2.5   Conditional Independence and Densities

We first characterize conditional independence in terms of densities. The basic idea is to start with a probability for which conditional independence is easily verified and to then characterize, in terms of the corresponding densities, conditional independence with respect to another probability dominated by the first one.

2.2.14 Theorem. Let P' be a probability on $(M, \mathcal{M})$ such that $P' \ll P$, and let $g = dP'/dP$. If $\mathcal{M}_1 \perp\!\!\!\perp \mathcal{M}_2 \mid \mathcal{M}_3; P$, then the following properties are equivalent:

(i) $\mathcal{M}_1 \perp\!\!\!\perp \mathcal{M}_2 \mid \mathcal{M}_3; P'$,

(ii) $g_{\mathcal{M}_1 \vee \mathcal{M}_2}^{\mathcal{M}_3} = g_{\mathcal{M}_1}^{\mathcal{M}_3} \cdot g_{\mathcal{M}_2}^{\mathcal{M}_3}$, on $\{g_{\mathcal{M}_3} > 0\}$,

(iii) $g_{\mathcal{M}_1}^{\mathcal{M}_2 \vee \mathcal{M}_3} = g_{\mathcal{M}_1}^{\mathcal{M}_3}$, on $\{g_{\mathcal{M}_2 \vee \mathcal{M}_3} > 0\}$,

(iv) $g_{\mathcal{M}_1 \vee \mathcal{M}_2 \vee \mathcal{M}_3} = g_{\mathcal{M}_1 \vee \mathcal{M}_3} \cdot g_{\mathcal{M}_2 \vee \mathcal{M}_3} / g_{\mathcal{M}_3}$ on $\{g_{\mathcal{M}_3} > 0\}$.

Proof. As indicated in Section 1.4.5.C., one has, in general,

$$g_{\mathcal{M}_1 \vee \mathcal{M}_2 \vee \mathcal{M}_3} = g_{\mathcal{M}_3} \cdot g_{\mathcal{M}_2}^{\mathcal{M}_3} \cdot g_{\mathcal{M}_1}^{\mathcal{M}_2 \vee \mathcal{M}_3}.$$

It is then clear that (ii), (iii) and (iv) are equivalent. Now, by Formula (1.4.28), if $\mathcal{N}'m$ denotes a conditional expectation of m, given $\mathcal{N}$, with respect to P', one then has, in general, for any $m_i \in [\mathcal{M}_i]^+$ $(i = 1, 2)$:

$$\mathcal{M}_3'(m_1 m_2) = \mathcal{M}_3\big(m_1 \cdot m_2 \cdot g_{\mathcal{M}_1 \vee \mathcal{M}_2}^{\mathcal{M}_3}\big) \qquad \text{a.s.} P',$$

$$\mathcal{M}_3'(m_1) \quad = \mathcal{M}_3\big(m_1 \cdot g_{\mathcal{M}_1}^{\mathcal{M}_3}\big) \qquad \text{a.s.} P',$$

$$\mathcal{M}_3'(m_2) \quad = \mathcal{M}_3\big(m_2 \cdot g_{\mathcal{M}_2}^{\mathcal{M}_3}\big) \qquad \text{a.s.} P'.$$

By conditional independence with respect to P, and using Corollary 2.2.11, we have, a.s.P:

$$\mathcal{M}_3\big(m_1 \cdot g_{\mathcal{M}_1}^{\mathcal{M}_3}\big) \cdot \mathcal{M}_3\big(m_2 \cdot g_{\mathcal{M}_2}^{\mathcal{M}_3}\big) = \mathcal{M}_3\big(m_1 \cdot m_2 \cdot g_{\mathcal{M}_1}^{\mathcal{M}_3} \cdot g_{\mathcal{M}_2}^{\mathcal{M}_3}\big).$$

Therefore,

$$\mathcal{M}_3'(m_1 \cdot m_2) = \mathcal{M}_3'm_1 \cdot \mathcal{M}_3'm_2 \qquad \text{a.s.} P'$$

is equivalent to:

$$\mathcal{M}_3\big(m_1 \cdot m_2 \cdot g_{\mathcal{M}_1 \vee \mathcal{M}_2}^{\mathcal{M}_3}\big) = \mathcal{M}_3\big(m_1 \cdot m_2 \cdot g_{\mathcal{M}_1}^{\mathcal{M}_3} \cdot g_{\mathcal{M}_2}^{\mathcal{M}_3}\big) \qquad \text{a.s.} P'.$$

This is equivalent to:

$$g_{\mathcal{M}_3} \cdot g_{\mathcal{M}_1 \vee \mathcal{M}_2}^{\mathcal{M}_3} = g_{\mathcal{M}_3} \cdot g_{\mathcal{M}_1}^{\mathcal{M}_3} \cdot g_{\mathcal{M}_2}^{\mathcal{M}_3} \qquad \text{a.s.} P,$$

and by definition of conditional densities, this is clearly equivalent to

$$g_{\mathcal{M}_1 \vee \mathcal{M}_2}^{\mathcal{M}_3} = g_{\mathcal{M}_1}^{\mathcal{M}_3} \cdot g_{\mathcal{M}_2}^{\mathcal{M}_3} \qquad \text{a.s.} P'. \qquad \blacksquare$$

This theorem shows that conditional independence may be characterized in terms of densities as well as in terms of conditional expectations, as we have done until now. This is indeed in the spirit of Subsection 1.4.5., where we write marginal densities as conditional expectations, and conditional densities as ratios of marginal densities.

We prefer, however, to continue in terms of conditional expectations because, firstly, a dominated measure for which conditional independence is easily verified and densities easily computed is not always available and, secondly, because we feel that null sets are more easily taken into account in terms of conditional expectations. Indeed, in the case of densities, one must first choose a version of g on $\mathcal{M}_1 \vee \mathcal{M}_2 \vee \mathcal{M}_3$ for this density is defined a.s.P only; then the marginal densities, $g_{\mathcal{M}_i}$, are specified a.s.$P_{\mathcal{M}_i}$, and the conditional densities, $g_{\mathcal{M}_i}^{\mathcal{M}_j}$, are defined a.s.$P_{\mathcal{M}_i \vee \mathcal{M}_j}$. For this reason, both the statement and the proof of this seemingly simple theorem are rather involved in terms of null sets, sometimes with respect to P, and sometimes with respect to P'. For example, Property (iii) in the theorem has to be verified a.s.P on the set $\{g_{\mathcal{M}_2 \vee \mathcal{M}_3} > 0\}$ but is typically false on the set $\{g_{\mathcal{M}_2 \vee \mathcal{M}_3} = 0\} \cap \{g_{\mathcal{M}_3} > 0\}$.

2.2.15 Corollary. Let g be defined as in Theorem 2.2.14. If

(i) $\mathcal{M}_1 \perp\!\!\!\perp \mathcal{M}_2 \mid \mathcal{M}_3; P$,

(ii) $g \in [\overline{\mathcal{M}_2 \vee \mathcal{M}_3}]^+$,

then

(iii) $\mathcal{M}_1 \perp\!\!\!\perp \mathcal{M}_2 \mid \mathcal{M}_3; P'$.

Proof. By the definition of the marginal density (1.4.17) and by the measurability of g, we easily obtain:

$$g_{\mathcal{M}_1 \vee \mathcal{M}_2 \vee \mathcal{M}_3} = g_{\mathcal{M}_2 \vee \mathcal{M}_3} = g \qquad \text{a.s.}P.$$

On the other hand by conditional independence with respect to P,

$$g_{\mathcal{M}_1 \vee \mathcal{M}_3} = (\mathcal{M}_1 \vee \mathcal{M}_3)g = \mathcal{M}_3 g = g_{\mathcal{M}_3} \qquad \text{a.s.}P.$$

The first relation implies $g_{\mathcal{M}_1 \vee \mathcal{M}_2}^{\mathcal{M}_3} = g_{\mathcal{M}_2}^{\mathcal{M}_3}$ a.s.P' and the second implies $g_{\mathcal{M}_1}^{\mathcal{M}_3} = 1$ a.s.P'. This clearly implies $g_{\mathcal{M}_1 \vee \mathcal{M}_2}^{\mathcal{M}_3} = g_{\mathcal{M}_1}^{\mathcal{M}_3} \cdot g_{\mathcal{M}_2}^{\mathcal{M}_3}$ a.s.P'. ∎

This corollary shows that conditional independence may be checked by a measurability of the density g. Generally, that is easily verified. The condition $g \in \left[\overline{\mathcal{M}_2 \vee \mathcal{M}_3}\right]^+$ is equivalent to $g_{\mathcal{M}_1}^{\mathcal{M}_2 \vee \mathcal{M}_3} = 1$, which means that P and P' do not differ on their conditional probabilities given $\mathcal{M}_2 \vee \mathcal{M}_3$. In other words conditional independence with respect to P is not altered if P is modified in such a way that only its restriction on $\mathcal{M}_2 \vee \mathcal{M}_3$ is affected, and if the modified restriction $P'_{\mathcal{M}_2 \vee \mathcal{M}_3}$ is dominated by $P_{\mathcal{M}_2 \vee \mathcal{M}_3}$.

2.2.6   Conditional Independence as Point Properties

Theorem 2.2.6 characterizes conditional independence as a measurability property. We now consider the problem of verifying such a measurability condition. The easiest case arises when the conditioning σ-field $\mathcal{M}_3$ is generated by a measurable function $f : (M, \mathcal{M}) \rightarrow (N, \mathcal{N})$, where $(N, \mathcal{N})$ is an arbitrary measurable space, i.e., $\mathcal{M}_3 = f^{-1}(\mathcal{N})$. In such a situation, a simple corollary of Theorem 0.2.11 gives the following result.

2.2.16 Proposition. If $\mathcal{M}_3 = f^{-1}(\mathcal{N})$, then the following two properties are equivalent:

(i) $\mathcal{M}_1 \perp\!\!\!\perp \mathcal{M}_2 \mid \mathcal{M}_3$;

(ii) $\forall\, m_2 \in [\mathcal{M}_2]^+$, there exists $n \in [\mathcal{N}]^+$ such that $n \circ f$ is a version of $(\mathcal{M}_1 \vee \mathcal{M}_3) m_2$.   ∎

Without this assumption on $\mathcal{M}_3$, we have to introduce more technical conditions so as to be able to use Blackwell Theorem (see Theorem 0.2.16). Let us recall that the *atoms* of a σ-field $\mathcal{M}_3$ — see (0.2.2) — are the equivalence classes of the equivalence relation between points of M defined as

$$x \underset{\mathcal{M}_3}{\sim} x' \Leftrightarrow 1_X(x) = 1_X(x') \qquad \forall\, X \in \mathcal{M}_3,$$

i.e., two points of M are equivalent for $\mathcal{M}_3$ if they are not separated by an $\mathcal{M}_3$-measurable set. Blackwell Theorem provides a characterization of conditional independence through the constancy, on the atoms of the conditioning σ-field, of the conditional expectations.

2.2.17 Proposition. If $\mathcal{M}_1$ is a Blackwell σ-field and if $\mathcal{M}_3 \subset \mathcal{M}_1$ is a separable σ-field, then the two following properties are equivalent:

(i) $\mathcal{M}_1 \perp\!\!\!\perp \mathcal{M}_2 \mid \mathcal{M}_3$,

(ii) $\forall\, m \in [\mathcal{M}_2]^+$ there exists a version of $\mathcal{M}_1 m$ that is, constant on the atoms of $\mathcal{M}_3$. ∎

Note that the condition of the separability of the conditioning σ-field is crucial for the validity of Proposition 2.2.17. Indeed, as a counterexample, let M be the real line, $\mathcal{M}_1$ its Borel σ-field, $\mathcal{M}_2$ any σ-field on $(I\!R)$, and let $\mathcal{M}_3$ be the (non separable) σ-field generated by the singletons. In this case, $\mathcal{M}_1$ is a Blackwell σ-field, $\mathcal{M}_3$ a (strict) sub-σ-field of $\mathcal{M}_1$, and any function on M is constant on the atoms of $\mathcal{M}_3$, even when the property $\mathcal{M}_1 \perp\!\!\!\perp \mathcal{M}_2 \mid \mathcal{M}_3$ is not verified.

2.3 Admissible Reductions of an Unreduced Experiment

2.3.1 Introduction

We are now in a position to render operational, in terms of stochastic independence, the concepts of ancillarity, and of sufficiency, i.e., the property of a σ-field that makes admissible the corresponding reduction by conditioning or by marginalization. In this section, we consider admissible reductions of an unreduced experiment $\mathcal{E} = \mathcal{E}_{A \vee S}$. In the next chapter we shall consider admissible reductions of already reduced experiments either by marginalization (such as $\mathcal{E}_{A \vee T}$ or $\mathcal{E}_{B \vee S}$) in Section 3.2. or by conditioning (such as $\mathcal{E}_{A \vee S}^{\mathcal{M}}$) in Section 3.3. It should be noted that Sections 2.3 and 3.2 are introduced for expository purposes only: Section 3.3. presents a general theory which encompasses both Sections 2.3 and 3.2.

The first step consists of rendering precise the concept of a totally non-informative experiment and of an admissible reduction. It should now be clear that the property "the posterior probability is almost surely equal to the prior probability" is a concept of stochastic independence between observation and parameter. This remark motivates the following definition for the case of a general Bayesian experiment $\mathcal{E}_{B \vee T}^{\mathcal{M}}$ which includes, as particular cases, $\mathcal{M} = \mathcal{I} = \{\phi,\, A \times S\}$ or $\mathcal{B} = \mathcal{A}$ or $\mathcal{T} = \mathcal{S}$.

2.3.1 Definition. The Bayesian experiment $\mathcal{E}_{\mathcal{B} \vee \mathcal{T}}^{\mathcal{M}}$ is *totally non-informative* if and only if any one of the following equivalent conditions holds:

(i) $\mathcal{B} \perp\!\!\!\perp \mathcal{T} \mid \mathcal{M}$;

(ii) $\mu_{\mathcal{B}}^{\mathcal{M}}(E) = \mu_{\mathcal{B}}^{\mathcal{M} \vee \mathcal{T}}(E) \quad \forall\, E \in \mathcal{B}$;

(iii) $P_{\mathcal{T}}^{\mathcal{M}}(Y) = P_{\mathcal{T}}^{\mathcal{B} \vee \mathcal{M}}(Y) \quad \forall\, Y \in \mathcal{T}$.

A reduction of a given Bayesian experiment is an *admissible reduction* if the complementary reduction, in the sense of Section 1.4.4, is totally non-informative.   ∎

Similarly to Definition 2.3.1 we define the concepts of ancillarity and sufficiency by means of three equivalent conditions. The first one, expressed in terms of stochastic independence, leads to the explicit identification of those reductions which are totally non-informative and those which are admissible. The second equivalent condition is expressed in terms of the property of distributions on the parameter space; this provides a characteristic property of the "prior-to-posterior" transformation. The third equivalent condition is expressed in terms of a characterization of the sampling process, and is useful for comparisons with the corresponding sampling theory concepts and for rendering the role of the prior probability more precise. The fact that these three conditions are equivalent is, using Monotone Class Theorems (see Section 0.2.4), a trivial consequence of the equivalent forms of conditional independence as in Theorem 2.2.1.

2.3.2   Admissible Reductions on the Sample Space

Let us consider a statistic $\mathcal{T}$, i.e., a sub-σ-field of $\mathcal{S}$.

2.3.2 Definition. $\mathcal{T}$ is $\mathcal{E}$-*sufficient* if and only if $\mathcal{E}^{\mathcal{T}}$ is totally non-informative, i.e., if and only if one of the following equivalent conditions holds

(i) $\mathcal{A} \perp\!\!\!\perp \mathcal{S} \mid \mathcal{T}$,

(ii) $\mu^{\mathcal{S}}(E) = \mu^{\mathcal{T}}(E), \quad \forall\, E \in \mathcal{A}$,

(iii) $P^{\mathcal{A} \vee \mathcal{T}}(X) = P^{\mathcal{T}}(X), \quad \forall\, X \in \mathcal{S}$.   ∎

2.3.3 Definition. $\mathcal{T}$ is $\mathcal{E}$-*ancillary* if and only if $\mathcal{E}_{\mathcal{A} \vee \mathcal{T}}$ is totally non-informative, i.e., if and only if one of the following equivalent conditions holds

(i) $\mathcal{A} \perp\!\!\!\perp \mathcal{T}$,

(ii) $\mu^{\mathcal{T}}(E) = \mu(E), \quad \forall\, E \in \mathcal{A},$

(iii) $P_{\mathcal{T}}^{\mathcal{A}}(X) = P_{\mathcal{T}}(X), \quad \forall\, X \in \mathcal{T}.$ ∎

2.3.3   Admissible Reductions on the Parameter Space

Consider a parameter $\mathcal{B}$, i.e., a sub-σ-field of $\mathcal{A}$.

2.3.4 Definition. $\mathcal{B}$ is $\mathcal{E}$-*sufficient* if and only if $\mathcal{E}^{\mathcal{B}}$ is totally non-informative, i.e., if and only if one of the following equivalent conditions holds:

(i) $\mathcal{A} \perp\!\!\!\perp \mathcal{S} \mid \mathcal{B}$,

(ii) $\mu^{\mathcal{B} \vee \mathcal{S}}(E) = \mu^{\mathcal{B}}(E), \quad \forall\, E \in \mathcal{A},$

(iii) $P^{\mathcal{A}}(X) = P^{\mathcal{B}}(X), \quad \forall\, X \in \mathcal{S}.$ ∎

2.3.5 Definition. $\mathcal{B}$ is $\mathcal{E}$-*ancillary* if and only if $\mathcal{E}_{\mathcal{B} \vee \mathcal{S}}$ is totally non-informative, i.e., if and only if one of the following equivalent conditions holds

(i) $\mathcal{B} \perp\!\!\!\perp \mathcal{S}$,

(ii) $\mu_{\mathcal{B}}^{\mathcal{S}}(E) = \mu_{\mathcal{B}}(E), \quad \forall\, E \in \mathcal{B},$

(iii) $P^{\mathcal{B}}(X) = P(X), \quad \forall\, X \in \mathcal{S}.$ ∎

2.3.4   Some Comments on the Definitions

(i) The Bayesian framework allows us to provide a set of homogeneous definitions that reveal, in particular, the completely parallel treatment of sufficiency in the sample space and sufficiency in the parameter space. Barankin (1960), along with Picci (1977), take advantage of this formal

analogy between the two concepts of sufficiency. Similarly, we have introduced a definition of ancillarity on the parameter space that is completely
parallel to the definition of ancillarity on the sample space.

(ii) All of the Conditions (ii) and (iii) of the definitions above are equivalent to measurability conditions (Theorem 2.2.6), and may be checked using
Propositions 2.2.16 and 2.2.17. In particular, the sufficiency of $T \subset S$ may
be characterized as follows: in the first case, if T is generated by a statistic
f defined on S with values in some measurable space $(N, \mathcal{N})$, T is sufficient
if and only if $\forall\, a \in [\mathcal{A}]^+$, Sa depends only on f; in the second case, if
S is a Blackwell σ-field and if T is separable, T is sufficient if and only if
$\forall\, a \in [\mathcal{A}]^+$, Sa is constant on the atoms of T.

(iii) With respect to the concept of sufficiency on the sample space (respectively, on the parameter space), Condition (ii) says that the conditional
reduction $\mathcal{E}^{T}_{\mathcal{A} \vee S}$ (respectively, $\mathcal{E}^{\mathcal{B}}_{\mathcal{A} \vee S}$) is totally non-informative (i.e., "prior
= posterior" a.s.) and, accordingly, that the marginal reduction $\mathcal{E}_{\mathcal{A} \vee T}$ (re-
spectively, $\mathcal{E}_{\mathcal{B} \vee S}$) is an admissible reduction of $\mathcal{E}_{\mathcal{A} \vee S}$. On the sample space,
this also means that in the process of "learning by observing" (i.e., in the
transformation $\mu \to \mu^S$), it is "sufficient" to observe T instead of S, when T
is sufficient, in the sense that the inference will be the same (i.e., same prior
and same posterior) in the reduced experiment $\mathcal{E}_{\mathcal{A} \vee T}$ as in the unreduced
experiment $\mathcal{E}_{\mathcal{A} \vee S}$. Note that Condition (iii) also says that the reason for
which $\mathcal{E}^{T}_{\mathcal{A} \vee S}$ is totally non-informative is precisely that its sampling transition "does not depend on $\mathcal{A}$": this reasoning will be deepened in Section
2.3.7 when comparing this Bayesian concept and the corresponding one in
sampling theory. On the parameter space the sufficiency of $\mathcal{B}$ means that $\mathcal{B}$
is "sufficient" to describe the sampling process (Condition (iii)) and, consequently, that the observation S brings information on $\mathcal{B}$ only in the sense
that, conditionally on $\mathcal{B}$, observation brings no information (Condition (ii)).

(iv) For the concept of ancillarity on the sample space (respectively,
on the parameter space), Conditions (ii) says that the marginal reduction
$\mathcal{E}_{\mathcal{A} \vee T}$ (respectively, $\mathcal{E}_{\mathcal{B} \vee S}$) is totally non-informative, and accordingly, the
conditional reduction $\mathcal{E}^{T}_{\mathcal{A} \vee S}$ (respectively, $\mathcal{E}^{\mathcal{B}}_{\mathcal{A} \vee S}$) is an admissible reduction
of $\mathcal{E}_{\mathcal{A} \vee S}$. On the sample space, this means that in the process of "learning-
by-observing", the random character of T is "ancillary", i.e., of no actual
relevance, and may in fact be neglected in the sense of specifying the sam-

pling process "as if" T were not random; this is indeed characteristic of $\mathcal{E}^T_{A \vee S}$. Condition (iii) also shows that $\mathcal{E}_{A \vee T}$ is totally non-informative because its sampling transition "does not depend on $\mathcal{A}$". On the parameter space the ancillarity of $\mathcal{B}$ means that the sampling process brings no information on $\mathcal{B}$ (Condition (ii)). Note that the Condition (iii) is genuinely Bayesian, as it depends crucially on the conditional prior distribution μ^B, which is the only part of μ to be revised by the observation $\mathcal{S}$.

Let us now illustrate these concepts by some examples. We first note that the Bayesian concepts of sufficiency and ancillarity on the sample space are, up to μ-null sets, the same as the corresponding concepts in sampling theory. In Section 2.3.7, we shall make this assertion more precise; nevertheless, in broad terms it can be said that any example of sufficient or of ancillary statistics, in a sampling theory framework, may also be used as an example in a Bayesian framework. Here we shall discuss some examples concerning the role of the μ-null sets. (Discrete examples are given in Rolin (1985)).

Example 1. Consider a univariate normal experiment: $s = (s_1, \cdots, s_n)$ with $(s_i \mid a) \sim i.N(m, v)$, where $a = (m, v)$. In a sampling theory framework, $t = (t_1, t_2)$ with $t_1 = \sum_i s_i$ and $t_2 = \sum_i s_i^2$, is known to be sufficient whereas t_1 is sufficient when the variance v is known, i.e., when $(s_i \mid a) \sim i.N(m, v_0)$, where v_0 is a fixed number. In a Bayesian framework, the case of known variance is equivalently treated either by considering the sampling process $(s_i \mid a) \sim i.N(m, v_0)$ with some prior distribution on $A = \mathbb{R}$, or by considering the sampling process $(s_i \mid a) \sim i.N(m, v)$ with a prior distribution on $A = \mathbb{R} \times \mathbb{R}^+$ such that $\mu(\{v = v_0\}) = 1$; in both cases t_1 may be shown to be sufficient. ∎

Example 2. The role of the prior information may also be illustrated within the Example 1 of Section 1.3, viz., a p-dimensional normal experiment with known variance: $(s \mid a) \sim N_p(a, V)$ and $a \sim N_p(m_0, V_0)$ where V and V_0 are $p \times p$ positive semidefinite symmetric matrices of known elements. Let $T = \sigma(Ts)$ where T is an $(r \times p)$-matrix of known elements. Here $t = Ts$ is a sufficient statistic if and only if $\text{cov}(a, s \mid t) = 0$, i.e., if and only if $V_0 - V_0 T'[T(V_0 + V)T']^{-1}T(V_0 + V) = 0$. Such a condition clearly links the prior information (V_0) and the sampling process (V). Let us consider, in particular, the i.i.d. case; now the prior information is such that all components of a are almost surely equal, i.e., $m_0 = \bar{m} i_p$ and $V_0 = v_0 i_p i_p'$ (thus V_0 is of rank 1 and p is here the sample size). Furthermore, the sampling process is such that $V = v I_p$ (where v is a known positive number). It is now easy to check that the sum of the observations $t = i_p' s$ is sufficient, indeed:

$$
\begin{aligned}
\mathrm{cov}(a, s \mid i_p' s) &= v_0 i_p i_p' - v_0 i_p i_p' i_p [i_p'(v_0 i_p i_p' + v I_p) i_p]^{-1} i_p'(v_0 i_p i_p' + v I_p) \\
&= 0
\end{aligned}
$$

Note that the sufficiency of $i_p' s$ does not require that $m_0 = \bar{m} i_p$.

Let us now turn to the conditions for the ancillarity of $\mathcal{T} = \sigma(Ts)$. Here $t = Ts$ is ancillary if and only if $\mathrm{cov}(a, t) = 0$, i.e., if and only if $V_0 T' = 0$. In this (very) particular case, such a condition depends only on the structure of the prior information. For instance, in the i.i.d. case (i.e., $V_0 = v_0 i_p i_p'$), t is ancillary when $T = I_p - \frac{1}{p} i_p i_p'$ ($t = Ts$ is now the vector of the deviations of the observations from the empirical average). Note that any function of Ts would also be ancillary.

Let us now consider $\mathcal{B} = \sigma(Ba)$, where B is an $r \times p$-matrix of known elements. In this situation $b = Ba$ is a sufficient parameter if and only if $\mathrm{cov}(a, s \mid b) = 0$ which is equivalent to $V(a \mid b) = 0$ because, in such a situation, $E(s \mid a) = a$. In other words, b is sufficient if a is a.s. a function of b. This condition is trivially satisfied for any regular matrix B, but otherwise depends on the structure of the prior information (viz., V_0). For any B, one has $V(a \mid b) = V_0 - V_0 B'(BV_0 B')^+ BV_0$ (where A^+ denotes the Moore-Penrose inverse of A) (see Marsaglia (1964)). In particular, if V_0 has the form $V_0 = v_0 i_p i_p'$ this condition is satisfied for any $1 \times p$ matrix B such that $Bi_p \neq 0$. On the other hand, $b = Ba$ is an ancillary parameter if and only if $\mathrm{cov}(b, s) = 0$, which is equivalent to $\mathrm{cov}(a, b) = 0$, (because, again, $E(s \mid a) = a$), i.e., that b is a.s. constant or $BV_0 = 0$. In particular, if V_0 has the form $V_0 = v_0 i_p i_p'$, this condition is satisfied for any matrix B such that $Bi_p = 0$.   ∎

2.3.5 Elementary Properties of Sufficiency and Ancillarity

Using Corollaries 2.2.11(ii) and 2.2.4(ii) we obtain the following monotonicity properties when considering two statistics T and U, or two parameters B and C.

2.3.6 Proposition.
(i)   If T is $\mathcal{E}$-sufficient and $T \subset U$ then U is $\mathcal{E}$-sufficient,

(ii) If T is $\mathcal{E}$-ancillary and $U \subset T$ then U is $\mathcal{E}$-ancillary,

(iii) If B is $\mathcal{E}$-sufficient and $B \subset C$ then C is $\mathcal{E}$-sufficient,

(iv) If B is $\mathcal{E}$-ancillary and $C \subset B$ then C is $\mathcal{E}$-ancillary.   ■

These properties imply that one will typically look for minimal sufficient statistics or parameters, and maximal ancillary statistics or parameters. This problem is treated in Chapter 4.

The role of the prior probability may be understood by asking which modifications of the prior probability guard intact a given property of sufficiency or of ancillarity. This aspect of robustness with respect to the prior specification is the object of the next proposition, the proof of which is a simple application of Corollary 2.2.15.

2.3.7 Proposition. Let us consider $\mathcal{E}$ and $\mathcal{E}'$, two Bayesian experiments on $(A \times S,\ \mathcal{A} \vee \mathcal{S})$ defined by their probabilities Π and Π'. If
(i) $\Pi' \ll \Pi$,

(ii) $d\Pi'/d\Pi \in [\mathcal{A}]^{+}$.

Then
(iii) If $T \subset S$ is an $\mathcal{E}$-sufficient statistic, it is also $\mathcal{E}'$-sufficient;

(iv) If $T \subset S$ is an $\mathcal{E}$-ancillary statistic, it is also $\mathcal{E}'$-ancillary;

(v)   If $B \subset A$ is an $\mathcal{E}$-sufficient parameter, it is also $\mathcal{E}'$-sufficient;

(vi) If $\mathcal{B} \subset \mathcal{A}$ is an $\mathcal{E}$-ancillary parameter, and if $d\Pi'/d\Pi \in [\overline{\mathcal{B}}]^+$,
then it is also $\mathcal{E}'$-ancillary.   ∎

In other words, for a given sampling process, equivalent prior probabilities give the same sufficient σ-fields both on the sample space and on the
parameter space and the same ancillary σ-fields on the sample space. The
situation is quite different for ancillarity on the parameter space since this
last property is retained when the prior probability is modified only on its
restriction to the ancillary σ-field.

2.3.6   Sufficiency and Ancillarity in a Dominated Experiment

We now express ancillarity and sufficiency, for the case of a dominated
Bayesian experiment, as properties of the densities.

2.3.8 Proposition. Let $\mathcal{E} = (A \times S, \mathcal{A} \vee \mathcal{S}, \Pi)$ be a dominated Bayesian
experiment and $g = d\Pi/d(\mu \otimes P)$. Then

(i) T is an $\mathcal{E}$-sufficient statistic if and only if $g_{\mathcal{S}}^{\mathcal{A} \vee T} = 1$;

(ii) T is an $\mathcal{E}$-ancillary statistic if and only if $g_{\mathcal{A} \vee T} = 1$;

(iii) $\mathcal{B}$ is an $\mathcal{E}$-sufficient parameter if and only if $g_{\mathcal{A}}^{\mathcal{B} \vee \mathcal{S}} = 1$;

(iv) $\mathcal{B}$ is an $\mathcal{E}$-ancillary parameter if and only if $g_{\mathcal{B} \vee \mathcal{S}} = 1$.   ∎

The proof of this proposition follows in a straightforward way from
Theorem 2.2.14 after noting that $\mathcal{A}$ and $\mathcal{S}$ are evidently independent for
the probability $\mu \otimes P$. Thus any condition of the type $\mathcal{A} \perp\!\!\!\perp \mathcal{S} \mid T; \mu \otimes P$
or $\mathcal{B} \perp\!\!\!\perp \mathcal{S}; \mu \otimes P$ is trivially true . Let us note that these characterizations
in terms of densities may take various forms. For example, T will be an
$\mathcal{E}$-sufficient statistic if and only if $g_{\mathcal{A}}^{\mathcal{S}} = g_{\mathcal{A}}^{T}$ or, equivalently, $g_{\mathcal{A} \vee \mathcal{S}}^{T} = g_{\mathcal{A}}^{T}$ or
equivalently $g_{\mathcal{A} \vee \mathcal{S}} = g_{\mathcal{A} \vee T}$. However, the simplest property which must be
verified is a measurability property of the density, such as the one given in
Corollary 2.2.15. This leads to the following proposition.

2.3.9 Proposition. Let $\mathcal{E} = (A \times S, \mathcal{A} \vee \mathcal{S}, \Pi)$ be a dominated Bayesian
experiment, and let $g = d\Pi/d(\mu \otimes P)$. Then

(i) T is an $\mathcal{E}$-sufficient statistic if and only if $g \in [\overline{\mathcal{A} \vee T}^{\mu \otimes P}]$,

(ii) $\mathcal{B}$ is an $\mathcal{E}$-sufficient parameter if and only if $g \in [\overline{\mathcal{B} \vee \mathcal{S}}^{\mu \otimes P}]$.   ∎

Here $\overline{\mathcal{A} \vee T}^{\mu \otimes P}$ denotes $\mathcal{A} \vee T$ completed by the $\mu \otimes P$-null sets.

Proposition 2.3.8 may be viewed as a Bayesian analogue of the Neyman's factorization theorem in sampling theory — see, e.g., Barra (1971), Chap. 2, Theorem 1. Consider indeed, as was done in the remark at the end of Section 1. 2. 2, the Radon-Nikodym derivative $f^*(a, s)$ of Π with respect to the product of two reference measures μ_0 on $(A, \mathcal{A})$ and λ on $(S, \mathcal{S})$. Furthermore, when T is generated by a function $t : (S, \mathcal{S}) \to (U, \mathcal{U})$ — i.e., $T = t^{-1}(\mathcal{U})$ — then there exists, by Theorem 0.2.11, $\tilde{g} : A \times U \to \mathbb{R}_+$ such that $g(a, s) = \tilde{g}(a, t(s))$ a.s.$(\mu \otimes P)$, in which case (1.2.12) takes the familiar form of Neyman's factorization:

$$(2.3.1) \qquad f^*(a, s) = h(a)\, k(s)\, \tilde{g}(a, t(s)) \quad \text{a.s.}(\mu_0 \otimes \lambda)$$

This is illustrated in the following example.

Example. Consider a simple random sample of size n in the canonical Bayesian exponential experiment (as in Example 2 of Section 1.3). Thus $s = (s_1, \cdots, s_n)$ and $(s_i \mid a)$ are i.i.d. with density

$$\frac{dP_i^a}{d\sigma_0} = \exp[a's_i - L(a) + g(s_i)],$$

where σ_0 is the reference measure on the sample space of one observation. Let λ be the reference measure on the sample space for n observations (i.e., $\lambda = \sigma_0^n$), the data density of $s = (s_1, \cdots, s_n)$ is

$$\frac{dP^a}{d\lambda} = \exp[a' \sum_i s_i - nL(a) + \sum_i g(s_i)],$$

and the predictive density of s is written as:

$$
\begin{aligned}
\frac{dP}{d\lambda} &= \frac{d \int P^a \mu(da)}{d\lambda} = \int \frac{dP^a}{d\lambda} \mu(da) \\
&= \exp\{\sum_i g(s_i)\} \int \exp[a' \sum_i s_i - nL(a) + L(a)]\, \alpha(da) \\
&= \exp\{\sum_i g(s_i) + M_n(\sum_i s_i)\}.
\end{aligned}
$$

We recall that, as defined in Section 1.3 (Example 2), $L(a) = \ln d\mu/d\alpha$ and M_n is defined as:

$$M_n(s) = \ln \int \exp[a's - (n-1)L(a)]\ \alpha(da)$$

(and, therefore $M_1(s) = M(s) = \ln \int \exp(a's)d\alpha$ as defined in Section 1.3).

Let Π_n be the probability measure on the σ-field generated by $(a, s) = (a, s_1, \ldots, s_n)$. The sufficiency of $t = \sum_i s_i$ may be expressed using a Bayesian version of Neyman's factorization theorem on the density of Π_n with respect to the reference measure $\alpha_0 \otimes \lambda$:

$$\frac{d\Pi_n}{d(\alpha_0 \otimes \lambda)} = f^*(a, s) = \exp\{a't - (n-1)L(a) + f(a) + \sum_i g(s_i)\}$$

(thus, in terms of the notation in (2.3.1), $\ln h(a) = f(a) - (n-1)L(a)$, $\ln k(s) = \sum_i g(s_i)$ and $\ln \tilde{g}(a, t) = a't$). Alternatively, it may be interpreted using the property (i) of Proposition (2.3.8):

$$\frac{d\Pi_n}{d(\mu \otimes P)} = g(a, s)$$

$$= \frac{d\Pi_n/d(\alpha_0 \otimes \lambda)}{d(\mu \otimes P)/d(\alpha_0 \otimes \lambda)}$$

$$= \frac{\exp\{a't - (n-1)L(a) + f(a) + \sum_i g(s_i)\}}{\exp\{f(a) + L(a) + \sum_i g(s_i) + M_n(t)\}}$$

$$= \exp\{a't - M_n(t) - nL(a)\}.$$

Let us note that the marginal experiment $\mathcal{E}_{A \vee T}$, characterized by the image probability denoted by $\Pi_{a,t}$, is also a canonical Bayesian exponential experiment provided the measure on $(A, \mathcal{A})$ is allowed to depend on the sample size. More specifically, recall from Example 2, Section 1.3, that

$$\frac{dP_i^a}{d\sigma} = \exp[a's_i - L(a)]$$

and therefore:

$$\frac{d\Pi_n}{d(\alpha \otimes \sigma^n)} = \exp[a' \sum_i s_i - (n-1)L(a)].$$

Therefore, if we define

$$\frac{d\alpha_n}{d\alpha} = \exp -(n-1)L(a),$$

we obtain:

$$\frac{d\Pi_n}{d(\alpha_n \otimes \sigma^n)} = \exp[a' \sum_i s_i]$$

which implies

$$\frac{d\Pi_{a,t}}{d(\alpha_n \otimes \sigma_t)} = e^{a't}$$

where σ_t is the image measure of σ^n by $t = \sum_i s_i$, i.e., $\sigma_t = \sigma^{*n}$, the n-th convolution of σ. Thus $\mathcal{E}_{A \vee T} = \mathcal{BEE}(\alpha_n, \sigma^{*n})$.

2.3.7 Sampling Theory and Bayesian Methods

In this section, we compare the sampling theory concepts of sufficiency and ancillarity in the sample space with the corresponding Bayesian concepts. Similar comparisons could be made in the parameter space using the duality properties. Here, Bayesian sufficiency and Bayesian ancillarity refer to Definitions 2.3.2. to 2.3.5.

Let us consider a statistical experiment $\mathcal{E} = \{(S, \mathcal{S}), P^a, a \in A\}$ and $\mathcal{A}$, a σ-field on A, which makes $P^a(X)$ measurable for any $X \in \mathcal{S}$. For $s \in [\mathcal{S}]^+$ and $\mathcal{T}$ a sub-σ-field of $\mathcal{S}$, we denote by $\mathcal{T}^a s$ the conditional expectation, given $\mathcal{T}$, of s with respect to the sampling probability P^a. In particular, $\mathcal{I}^a s$ denotes the expectation of s with respect to P^a.

It is often desirable for $\mathcal{T}^a s$ to be bimeasurable. However, this is not in general implied by the fact that $P^a(X)$ is measurable for any $X \in \mathcal{S}$. This therefore requires a supplementary assumption which is described in the following definition.

2.3.10 Definition. We shall say that a statistic $\mathcal{T} \subset \mathcal{S}$ is *regular* if, for each $s \in [\mathcal{S}]^+$, $\mathcal{T}^a s \in [\mathcal{A} \vee \mathcal{T}]$ (more precisely, if there exists an $\mathcal{A} \vee \mathcal{T}$-measurable function which is a version of $\mathcal{T}^a s$ for each $a \in A$). ∎

Note that this condition of regularity is weaker than the condition of separability. Indeed, for any $s \in [\mathcal{S}]_\infty^+$ define a family of measures on $(S, \mathcal{S})$ as follows: $dQ^{a,s}/dP^a = s$. Clearly, $\mathcal{T}^a s = dQ_\mathcal{T}^{a,s}/dP_\mathcal{T}^a$ and, by Theorem 0.3.19, the separability of $\mathcal{T}$ ensures the existence of a bimeasurable version of this Radon-Nikodym derivative. But the condition of the regularity of $\mathcal{T}$ is more general, in the sense that, if A is finite, then any sub-σ-field of $\mathcal{S}$ is regular.

A. Sufficiency

Let μ be a probability on $(A, \mathcal{A})$, and let $\mathcal{E} = (A \times S, \mathcal{A} \vee S, \Pi)$ be the corresponding Bayesian experiment as defined in Section 1.2.1. It is then clear that if $T \subset S$ is regular, $T^a s$ can be chosen as a version of $(\mathcal{A} \vee T)s$ for any $s \in [S]^+$.

In sampling theory a sufficient statistic is a sub-σ-field of S such that, given this σ-field, there exists a common version of the sampling expectations, of any S-measurable function. More precisely, following Halmos and Savage (1949) — see also Bahadur (1955a) and Barra (1971) — we have:

2.3.11 Definition. A sub-σ-field $T \subset S$ is a *sampling sufficient statistic* if for any $s \in [S]^+$, there exists $t \in [T]^+$ such that $\forall\, a \in A, T^a s = t$ a.s.P^a. ∎

Note that, in general, this definition does not possess the monotonicity property of the corresponding Bayesian definition, in other words, T sufficient and $\mathcal{U} \supset T$ do not imply $\mathcal{U}$ sufficient (see Burkholder (1961)). This clearly shows that the two definitions are not equivalent without further assumptions. However, we have the following result:

2.3.12 Theorem. Let $T \subset S$ be a sampling sufficient statistic. Then, for any prior probability μ, T is also Bayesian sufficient.

Proof. If $s \in [S]^+$ and $t \in [T]^+$ are such that $T^a s = t$ a.s.P^a for any $a \in A$, it follows that for any prior probability μ, $(\mathcal{A} \vee T)s = t$ a.s.Π, and so $(\mathcal{A} \vee T)s \in [\overline{T}]^+$. ∎

To obtain the converse implication, we need to place some assumptions on the prior probablity μ.

2.3.13 Definition. We will say that the prior probability μ is *regular for the family* $\{P^a, a \in A\}$ if $s \in [S]_\infty$ and $\mathcal{I}^a s = 0$ a.s.μ imply $\mathcal{I}^a s = 0$ $\forall\, a \in A$. ∎

Two special cases of such a property are the following:

(i) if A is countable, a prior probablity μ will be regular if it gives positive mass to each point of A;

(ii) if A is a topological space, and if $\{P^a, a \in A\}$, the sampling probabilities are such that for any $s \in [\mathcal{S}]_\infty$, $\mathcal{I}^a s$ is continuous on A, then a prior probability μ will be regular if it gives positive probability to each open (measurable) subset of A.

This leads to the following result:

2.3.14 Theorem. Let μ be a regular prior probablity on $(A, \mathcal{A})$. If $\mathcal{T} \subset \mathcal{S}$ is a Bayesian sufficient statistic, then $\mathcal{T}$ is a sampling sufficient statistic.

Proof. If $\mathcal{T}$ is Bayesian sufficient, then $\forall s \in [\mathcal{S}]_\infty^+$, $(\mathcal{A} \vee \mathcal{T})s = \mathcal{T}s$ a.s.Π. Hence, $\forall b \in [\mathcal{A}]_\infty^+$, $\forall t \in [\mathcal{T}]_\infty^+$, we obtain successively:

$$\mathcal{I}(b \cdot t \cdot s) = \mathcal{I}(b \cdot t \cdot \mathcal{T}s),$$

$$\int_A b \cdot \mathcal{I}^a(t \cdot s)d\mu = \int_A b \cdot \mathcal{I}^a(t \cdot \mathcal{T}s)d\mu.$$

It follows that, $\forall s \in [\mathcal{S}]_\infty^+$, $\forall t \in [\mathcal{T}]_\infty^+$

$$\mathcal{I}^a(t \cdot s) = \mathcal{I}^a(t \cdot \mathcal{T}s) \quad \text{a.s.}\mu.$$

Hence $\mathcal{I}^a(t \cdot s) = \mathcal{I}^a(t \cdot \mathcal{T}s) \; \forall \, a \in A$ since μ is regular. This implies that, $\forall \, a \in A$ and $\forall \, s \in [\mathcal{S}]_\infty^+$, $\mathcal{T}^a s = \mathcal{T}s$ a.s.P^a. ∎

Suppose now that the statistical experiment is classically dominated, as in Section 1.2.2. By the factorization theorem, sufficiency becomes a measurability property and thus recovers the property of monotonicity exactly as in a Bayesian experiment where, even in the undominated case, sufficiency is a measurability property. Moreover in this situation, the definition of sufficiency is restated in terms of a *"privileged"* dominating probability P_* (see Section 1.2.2). A statistic $\mathcal{T} \subset \mathcal{S}$ is then sufficient if and only if $\forall \, s \in [\mathcal{S}]^+$, $\mathcal{T}^a s = \mathcal{T}_* s$ a.s.P^a $\forall \, a \in A$, where $\mathcal{T}_* s$ is the conditional expectation, given $\mathcal{T}$, of s with respect to P_* (see Halmos and Savage (1949 — proof of Theorem 1)). Let us compare this equivalent definition with the Bayesian definition: $\forall \, s \in [\mathcal{S}]^+$, $(\mathcal{A} \vee \mathcal{T})s = \mathcal{T}s$ a.s.Π; recall that if $\mathcal{T}$ is regular, $(\mathcal{A} \vee \mathcal{T})s = \mathcal{T}^a s$ a.s. Π. We see that in a Bayesian experiment, the predictive probability plays a role similar to that of a privileged dominating measure even if the Bayesian experiment is not dominated.

Remark. It should be clear that the existence of a prior probablity μ regular for the family $\{P^a : a \in A\}$ implies that for each $a \in A$, P^a is dominated by the predictive probability P (indeed, in Definition 2.3.12, let s be the indicator function of any P-null set); the statistical experiment is therefore classically dominated. Furthermore, the predictive probability is equivalent to any privileged dominating probability P_* since they are both weighted averages of elements of $\{P^a : a \in A\}$.

A further step would be to particularize these concepts to a family of experiments associated with a family of prior probabilities and identical sampling process. In this approach, Martin, Petit and Littaye (1973) have shown that, in the dominated case, classical sufficiency (on the sample space) is equivalent to Bayesian sufficiency (on the sample space) for the family of all prior probabilities, and in the undominated case, classical pairwise sufficiency (on the sample space), (see e.g., Halmos and Savage (1949)) is equivalent to Bayesian sufficiency (on the sample space) for the family of purely atomic prior probabilities; further results may be found in Ramamoorthy (1980a, 1980b) and Roy and Ramamoorthy (1979). Note that Bayesian sufficiency may also be seen as a classical pairwise sufficiency; this fact is shown in the next theorem.

2.3.15 Theorem. If $T \subset S$ is a Bayesian sufficient statistic, then $A \vee T$ is sufficient in the classical experiment $(A \times S, A \vee S, \{\Pi, \mu \otimes P\})$; the converse is also true if Π is dominated by $\mu \otimes P$.

Proof. $\forall s \in [S]^+$, $Ts = \tilde{T}s$ a.s.P and hence a.s.$(\mu \otimes P)$. But $(\widetilde{A \vee T})s = \tilde{T}s$ a.s.$(\mu \otimes P)$. If T is sufficient, $(A \vee T)s = Ts$ a.s.Π and, consequently, Ts is a common version of $(\widetilde{A \vee T})s$ and $(A\vee T)s$. However, if t is a common version of $(\widetilde{A \vee T})s$ and of $(A \vee T)s$, then $t = \tilde{T}s = Ts$ a.s.$(\mu \otimes P)$, hence a.s.Π by domination, and so $(A \vee T)s = Ts$ a.s.Π.   ∎

B. Ancillarity

We now compare ancillarity in sampling theory and in a Bayesian framework. Recall the sampling theory definition of an ancillary statistic.

2.3.16 Definition. A sub-σ-field $T \subset S$ is a *sampling ancillary statistic* if for any $t \in [T]^+$, there exists a constant k in $\mathbb{R}^+$ such that $\mathcal{I}^a t = k$ $\forall a \in A$.   ∎

Let μ be a probability measure on $(A, \mathcal{A})$ and $\mathcal{E} = (A \times S, \mathcal{A} \vee \mathcal{S}, \Pi)$ the corresponding Bayesian experiment. Using Monotone Class Theorems, it is clear that $\forall\, s \in [\mathcal{S}]^+$, $\mathcal{I}^a s \in [\mathcal{A}]^+$ and so $\mathcal{I}^a s$ may be chosen as a version of $\mathcal{A}s$ for any $s \in [\mathcal{S}]^+$ i.e., $\mathcal{A}s = \mathcal{I}^a s$ a.s.Π. This gives the following results.

2.3.17 Theorem. If $T \subset S$ is a sampling ancillary statistic, then for any prior probablity μ, T is also Bayesian ancillary. ∎

2.3.18 Theorem. Let μ be a regular prior probablity on $(A, \mathcal{A})$. If $T \subset S$ is a Bayesian ancillary statistic, then T is a sampling ancillary statistic.

Proof. If T is Bayesian ancillary, for any $t \in [T]_\infty^+$, $\mathcal{A}t = \mathcal{I}t$ a.s.Π and so $\mathcal{I}^a t = \mathcal{I}t$ a.s.Π and this is equivalent to $\mathcal{I}^a(t - \mathcal{I}t) = 0$ a.s.μ. Since μ is regular this implies $\mathcal{I}^a t = \mathcal{I}t$ $\forall\, a \in A$. ∎

Similarly to Proposition 2.3.14, the Bayesian conception of ancillarity may be seen as classical pairwise ancillarity in the sense that $T \subset S$ is a ancillary statistic in the Bayesian sense if and only if $\mathcal{A} \vee T$ is ancillary in the classical experiment $(A \times S, \mathcal{A} \vee \mathcal{S}, \{\Pi, \mu \otimes P\})$.

2.3.8 A First Result on the Relations between Sufficiency and Ancillarity

The definitions of sufficiency and ancillarity expressed in terms of conditional independence allows the basic properties of conditional independence (Section 2.2.4) to be used, giving rise to the following result:

2.3.19 Theorem. In a Bayesian experiment $\mathcal{E} = (A \times S,\ \mathcal{A} \vee \mathcal{S},\ \Pi)$

(i) Any statistic $\mathcal{U}$ predictively independent of a sufficient statistic T is ancillary. Moreover $\mathcal{U}$ and T are sampling independent.

(ii) Any parameter $\mathcal{C}$ a priori independent of a sufficient parameter $\mathcal{B}$ is ancillary. Moreover $\mathcal{C}$ and $\mathcal{B}$ are a posteriori independent.

Proof. By duality it suffices to prove (i). Now, by assumption, $\mathcal{U} \perp\!\!\!\perp \mathcal{T}$ and $\mathcal{A} \perp\!\!\!\perp \mathcal{S} \mid \mathcal{T}$ (by sufficiency of $\mathcal{T}$). By Corollary 2.2.4(ii), $\mathcal{A} \perp\!\!\!\perp \mathcal{U} \mid \mathcal{T}$. By Theorem 2.2.10, $\mathcal{U} \perp\!\!\!\perp \mathcal{T}$ and $\mathcal{A} \perp\!\!\!\perp \mathcal{U} \mid \mathcal{T}$ is equivalent to $\mathcal{U} \perp\!\!\!\perp (\mathcal{A} \vee \mathcal{T})$ and is also equivalent to $\mathcal{U} \perp\!\!\!\perp \mathcal{A}$ and $\mathcal{U} \perp\!\!\!\perp \mathcal{T} \mid \mathcal{A}$.   ∎

In sampling theory, this kind of results has received a considerable attention (see, e.g., Barra (1971), Basu (1955, 1958), Koehn and Thomas (1975), Soler (1970)). A sampling theory analogue of Theorem 2.3.18 may be stated as follows:

2.3.20 Theorem. In a dominated experiment, any statistic $\mathcal{U}$ independent of a sufficient statistic, with respect to a privileged dominating probablity, is ancillary and independent of this sufficient statistic with respect to each sampling probability.

Proof. Recall that $T_* s$ denotes the conditional expectation of s, given $\mathcal{T}$, with respect to privileged dominating probability P_*. By sufficiency, $\forall s \in [\mathcal{S}]^+$, $T^a s = T_* s$ a.s.P^a $\forall a \in A$ and by independence, $\forall u \in [\mathcal{U}]^+$, $T_* u = \mathcal{I}_* u$ a.s.P_*. Hence for any $u \in [\mathcal{U}]^+$, $T^a u = \mathcal{I}_* u$ a.s.P^a. $\forall a \in A$. Thus $\forall u \in [\mathcal{U}]^+$, $\forall t \in [\mathcal{T}]^+$, $\mathcal{I}^a(ut) = \mathcal{I}^a(\mathcal{I}_* u \cdot t) = \mathcal{I}_* u \cdot \mathcal{I}^a t$. This implies $\mathcal{I}_* u = \mathcal{I}^a u$, thus $\mathcal{U}$ is ancillary and $\mathcal{I}^a(ut) = \mathcal{I}^a u \cdot \mathcal{I}^a t$, i.e., $\mathcal{U} \perp\!\!\!\perp \mathcal{T}$ with respect to P^a $\forall a \in A$.   ∎

Note that Basu Theorems consider the following different problems. Let $\mathcal{T}$ be a sufficient statistic ($\mathcal{A} \perp\!\!\!\perp \mathcal{S} \mid \mathcal{T}$). Is a statistic $\mathcal{U}$, independent of $\mathcal{T}$ in the sampling process ($\mathcal{U} \perp\!\!\!\perp \mathcal{T} \mid \mathcal{A}$), ancillary ($\mathcal{U} \perp\!\!\!\perp \mathcal{A}$) ? Is an ancillary statistic $\mathcal{U}$ ($\mathcal{U} \perp\!\!\!\perp \mathcal{A}$) independent of $\mathcal{T}$ in the sampling process ($\mathcal{U} \perp\!\!\!\perp \mathcal{T} \mid \mathcal{A}$) ?

The answer is known to be negative without further requirements as illustrated in the next example. The Bayesian versions of these problems are treated in Chapter 5.

Example. Let $S = \{s_1, s_2\}$ and $A = \{a_1, a_2, a_3\}$. Suppose that Π is defined by Table 1 and $\mathcal{B}$ is the σ-field generated by the function $b(a_1) = b(a_2) = b_1$ and $b(a_3) = b_2$ where $b_1 \neq b_2$. That $\mathcal{B}$ is ancillary is easily seen from Table 2. but clearly $\mathcal{B}$ is not independent of $\mathcal{A}$ which is sufficient and is not independent of any other sufficient parameter. Therefore the converse of Theorem 2.3.19 is not true.

Table 1. Table 2.

s	$p(s_i \mid a_j)$		
	a_1	a_2	a_3
s_1	$\frac{2}{5}$	$\frac{1}{4}$	$\frac{1}{3}$
s_2	$\frac{3}{5}$	$\frac{3}{4}$	$\frac{2}{3}$
$\mu(a_j)$	$\frac{5}{12}$	$\frac{4}{12}$	$\frac{3}{12}$

s	$p(s_i \mid b_j)$	
	b_1	b_2
s_1	$\frac{1}{3}$	$\frac{1}{3}$
s_2	$\frac{2}{3}$	$\frac{2}{3}$
$\mu(b_j)$	$\frac{3}{4}$	$\frac{1}{4}$

■

The next proposition is genuinely Bayesian insofar as it has no sampling theory counterpart.

2.3.21 Theorem. In a Bayesian experiment $\mathcal{E} = (A \times S, \mathcal{A} \vee \mathcal{S}, \Pi)$

(i) Any statistic $\mathcal{U}$ independent of a sufficient parameter $\mathcal{B}$ is ancillary,

(ii) Any parameter $\mathcal{C}$ independent of a sufficient statistic $\mathcal{T}$ is ancillary.

Proof. By duality, it suffices to prove (i). By assumptions we have: $\mathcal{U} \perp\!\!\!\perp \mathcal{B}$ and $\mathcal{A} \perp\!\!\!\perp \mathcal{S} \mid \mathcal{B}$. This implies, by Corollary 2.2.4(ii), $\mathcal{A} \perp\!\!\!\perp \mathcal{U} \mid \mathcal{B}$. Now, by Theorem 2.2.10, $\mathcal{U} \perp\!\!\!\perp \mathcal{B}$ and $\mathcal{A} \perp\!\!\!\perp \mathcal{U} \mid \mathcal{B}$ is equivalent to $\mathcal{U} \perp\!\!\!\perp \mathcal{A}$. ■

3

Admissible Reductions in
Reduced Experiments

3.1 Introduction

An implicit conclusion of Chapter 1 is that any statistical model actually used in practice should be viewed as a reduction of a larger model, this reduction being obtained by (i) marginalization on actually available and seemingly relevant observations, (ii) marginalization on parameters deemed to be of some interest, and (iii) conditioning on functions of both parameters and observations. In particular, one undertakes conditioning given "reasonable" assumptions about parameter values and about variables which should be regarded as "exogenous", i.e., as if the marginal processes of data generation were not worthy of complete specification. Thus any model in actual use may be viewed as a model of the form $\mathcal{E}_{\mathcal{B}\vee\mathcal{T}}^{\mathcal{M}}$, a reduction of an unreduced model $\mathcal{E}_{\mathcal{A}\vee\mathcal{S}}$ (with $\mathcal{B} \subset \mathcal{A}$, $\mathcal{T} \subset \mathcal{S}$ and $\mathcal{M} \subset \mathcal{A} \vee \mathcal{S}$); in other words, instead of specifying a probability $\Pi_{\mathcal{A}\vee\mathcal{S}}$ on a "huge" space $\mathcal{A} \vee \mathcal{S}$, the statistician typically specifies a "smaller" probability measure $\Pi_{\mathcal{B}\vee\mathcal{T}}^{\mathcal{M}}$, thereby economizing on the specification of the two associated submodels, namely, $\Pi_{\mathcal{M}}$ and $\Pi_{\mathcal{A}\vee\mathcal{S}}^{\mathcal{B}\vee\mathcal{M}\vee\mathcal{T}}$. The motivation for such a reduction is that the unreduced model $\mathcal{E}_{\mathcal{A}\vee\mathcal{S}}$ would typically be unmanageable and, further, that the more detailed specification of $\mathcal{E}_{\mathcal{M}}$ and of $\mathcal{E}^{\mathcal{B}\vee\mathcal{M}\vee\mathcal{T}}$ would not only not be worth the marginal cost, but would also probably aggravate the problem of specification errors. Thus the reduction $\mathcal{E}_{\mathcal{B}\vee\mathcal{T}}^{\mathcal{M}}$ is viewed not as an admissible reduction of $\mathcal{E}_{A\vee S}$ but, rather, as an operational starting point; any loss of information related to this reduction being accepted *ab initio*.

In view of this state of affairs, two types of questions naturally emerge. How can $\mathcal{E}_{B\vee T}^{\mathcal{M}}$ be simplified further without additional loss of information? And when is a pair of experiments a *jointly admissible* reduction of $\mathcal{E}_{A\vee S}$? The response to these questions is the object of this chapter.

The first question concerns the admissible reduction of $\mathcal{E}_{B\vee T}^{\mathcal{M}}$; having accepted a loss of information in the reduction of $\mathcal{E}_{A\vee S}$ to $\mathcal{E}_{B\vee T}^{\mathcal{M}}$, one seeks to further simplify $\mathcal{E}_{B\vee T}^{\mathcal{M}}$ without additional loss of information. Technically, this is the question of ancillarity and sufficiency in reduced experiments, and is studied in Sections 3.2 and 3.3. For expository purposes, we have chosen to examine firstly the admissibility of the reduction $\mathcal{E}_{B\vee T}$ where the original experiment is reduced by marginalization only, and thereafter the reduction $\mathcal{E}_{B\vee T}^{\mathcal{M}}$ where the original experiment is reduced both by marginalization and by conditioning. In addition to providing a smoother development of the theory than a single exposition with regards to $\mathcal{E}_{B\vee T}^{\mathcal{M}}$ only, the presentation regarding $\mathcal{E}_{B\vee T}$ will also provide results that are easily put to use (both in applied work and in the theory to be exposed in later chapters).

The second kind of question is rather different in nature. In Section 1.4.4. we decomposed the learning process (i.e., the transformation $\mu \rightarrow \mu^S$) either on the sample space or on the parameter space. Let us now consider a joint decomposition of the learning process, as illustrated below:

$$
\begin{array}{ccccc}
\mu_A & \xrightarrow{\hspace{1.5cm}\mathcal{E}_{A\vee S}\hspace{1.5cm}} & & & \mu_A^S \\[4pt]
\| & & & & \| \\[4pt]
\mu_B & \xrightarrow{\mathcal{E}_{B\vee T}} & \mu_B^T & \xrightarrow{\mathcal{E}_{B\vee S}^T} & \mu_B^S \\[4pt]
\otimes & & & & \otimes \\[4pt]
\mu_A^B & \xrightarrow{\mathcal{E}_{A\vee T}^B} & \mu_A^{B\vee T} & \xrightarrow{\mathcal{E}_{A\vee S}^{B\vee T}} & \mu_A^{B\vee S}
\end{array}
$$

We may now ask whether the pair of experiments $(\mathcal{E}_{\mathcal{B} \vee \mathcal{T}}, \mathcal{E}_{\mathcal{A} \vee \mathcal{S}}^{\mathcal{B} \vee \mathcal{T}})$ constitutes a *jointly admissible* reduction of $\mathcal{E}_{\mathcal{A} \vee \mathcal{S}}$, (i.e., whether both $\mathcal{E}_{\mathcal{B} \vee \mathcal{S}}^{\mathcal{T}}$ and
$\mathcal{E}_{\mathcal{A} \vee \mathcal{T}}^{\mathcal{B}}$ are totally noninformative), or whether the pair $\mathcal{E}_{\mathcal{A} \vee \mathcal{T}}^{\mathcal{B}}, \mathcal{E}_{\mathcal{B} \vee \mathcal{S}}^{\mathcal{T}}$ constitute a jointly admissible reduction of $\mathcal{E}_{\mathcal{A} \vee \mathcal{S}}$ (i.e., whether both $\mathcal{E}_{\mathcal{B} \vee \mathcal{T}}$ and
$\mathcal{E}_{\mathcal{A} \vee \mathcal{S}}^{\mathcal{B} \vee \mathcal{T}}$ are totally noninformative). These questions are the object of Section
3.4. Note, in particular, that in the first case one would carry out reduc-
tions by marginalization (or conditioning) on both the parameter and the
sample space whereas in the second case on would carry out reductions by
marginalization on the one space and conditioning on the other space.

It should be stressed why the questions treated in Section 3.4 are different in nature from the questions treated in Chapter 2, and also from those
treated in Sections 3.2 and 3.3. Indeed, the properties of sufficiency and ancillarity in a given experiment (unreduced or already reduced) are relevant
mostly at the stage of inference or prediction once the data is collected, and
the experiment is specified; at that level, a property of sufficiency and/or
of ancillarity is a property of a given probability measure (on $\mathcal{A} \vee \mathcal{S}$), and
is useful mainly for simplifying the calculations required by the inference
and/or the prediction. By contrast, a property of joint admissibility, as
discussed in Section 3.4, is useful mainly at the stage of model specification.
Indeed, in practice, the specification of a Bayesian experiment, typically of
the form $\mathcal{E}_{\mathcal{B} \vee \mathcal{T}}^{\mathcal{M}}$, generally involves the following sequence of steps. A large
experiment $\mathcal{E}_{\mathcal{A} \vee \mathcal{S}}$ is first considered as a general framework of reference, in
the sense that the probability space $(\mathcal{A} \times \mathcal{S}, \mathcal{A} \vee \mathcal{S}, \Pi)$ is not specified in
detail but is only briefly sketched. Subsequently, an answer is sought to
the question: Which aspects of $\mathcal{E}_{\mathcal{A} \vee \mathcal{S}}$ should be now specified in detail before considering the stage of inference and/or of prediction? The conditions
for jointly admissible reductions characterize the structural properties of Π
(namely, conditions of stochastic independence) before its detailed analytical form is specified making it possible to decide which parts of Π should
be further specified as they are associated with totally noninformative experiments. Furthermore this section will also lead to conditions allowing for
specifying separately part of (or, all) the probability measures involved in
the two experiments constituting the pair of jointly admissible reductions of
$\mathcal{E}_{\mathcal{A} \vee \mathcal{S}}$. Finally, the structure of the loss function will often imply that only
one of the two jointly admissible reductions actually involves the parameter
of interest. As a result of adopting the appropriate specification strategy,

the experiment actually used at the stage of inference and/or of prediction will not only be much simpler than the reference experiment $\mathcal{E}_{A \vee S}$ (if this were completely specified) but will also be robust against a large class of specification errors.

Section 3.5 is, in a sense, complementary to Section 3.4, for we progressively shift emphasis from building an experiment towards comparing two given experiments with a view to deciding whether or not one is redundant with respect to the other. We shall see that the inherent symmetry (between parameters and observations) of a Bayesian experiment will provide a natural environment for a Bayesian presentation not only of sufficiency in the sense of Blackwell (1951, 1953) and LeCam (1966, 1986), but also of model choice, model specification or hypothesis testing. Note, however, that a detailed treatment of these topics would also require a treatment of approximate admissibility, a topic not treated in this monograph (interested readers are referred, in particular, to Csiszar (1967a), Torgersen (1972, 1976, 1981), and Florens (1983). In other words, Section 3.5 merely sketches a general framework which handles, from a Bayesian point of view, a large class of problems in statistical methodology, which a rapidly growing literature suggest are of fundamental import.

Florens and Mouchart (1977) has been the starting point for Sections 3.2, 3.3, and 3.4 but the sections on dominated experiments (3.3.6 and 3.4.4) were worked out later. The Section 3.2.4 on partial sufficiency is based on Mouchart and Rolin (1984a). This presentation, along with its examples, also relies on Florens and Mouchart (1985a) and (1986a). Most of Section 3.5 sketches work under progress, the first steps of which have been reported in Florens, Mouchart and Scotto (1983), Florens and Scotto (1984), and Florens and Mouchart (1985c). Steps toward a Bayesian approach to hypothesis testing and encompassing are not reported in this chapter, but may be found in Florens and Mouchart (1988) and Florens (1988) respectively. The motivation underlying those works is found in the appeal for a better understanding of exogeneity in econometric modelling expressed in Florens, Mouchart and Richard (1974), and which led to a systematic investigation of the linear model with error in the variables (Florens, Mouchart and Richard (1979)), which is the source of the examples of this chapter.

Let us conclude this introduction by remarking that this chapter does not contain a strictly probabilistic section. Thus, at the technical level, this chapter makes use of exactly the same tools as the first two chapters. Furthermore, a discussion of the relationship between Bayesian and sampling theory is confined to some remarks, the general discussion in Section 2.3.7 continuing to serve as the general framework of reference.

3.2 Admissible Reduction in Marginal Experiments

3.2.1 Introduction

In this section, we shall see that, because a marginal experiment has exactly the same structure as an unreduced experiment — viz., a unique probability Π on a product space — the concepts of ancillarity and of sufficiency are essentially the same. We shall therefore present the basic definitions with only very few comments. Subsequently, we analyze some relationships between admissible reductions in unreduced and in marginal experiments, and pay particular attention to the problem of sufficiency with respect to a sub-parameter for the class of all prior probabilities.

3.2.2 Basic Concepts

We consider two sub-σ-fields $\mathcal{B} \subset \mathcal{A}$ and $\mathcal{T} \subset \mathcal{S}$. The marginal experiment $\mathcal{E}_{\mathcal{B} \vee \mathcal{T}}$ was defined in Formula (1.4.6). Now, the Definitions 2.3.2 to 2.3.5 may be used in order to define sufficiency or ancillarity, in the experiment $\mathcal{E}_{\mathcal{B} \vee \mathcal{T}}$, of any sub-$\sigma$-fields $\mathcal{C} \subset \mathcal{B}$ or $\mathcal{U} \subset \mathcal{T}$. For instance, the condition:

$$(3.2.1) \qquad \mathcal{B} \perp\!\!\!\perp \mathcal{T} \mid \mathcal{U}$$

is, as in Definition 2.3.2, equivalent to each one of the following two:

$$(3.2.2) \qquad \mu^{\mathcal{T}}(E) = \mu^{\mathcal{U}}(E) \quad \forall\, E \in \mathcal{B}$$

$$(3.2.3) \qquad P^{\mathcal{B} \vee \mathcal{U}}(X) = P^{\mathcal{U}}(X) \quad \forall\, X \in \mathcal{T}$$

and means that $\mathcal{U}$ is $\mathcal{E}_{\mathcal{B} \vee \mathcal{T}}$-sufficient or, equivalently, that $\mathcal{E}^{\mathcal{U}}_{\mathcal{B} \vee \mathcal{T}}$ is totally noninformative. Similar comments can be made with respect to conditions such as $\mathcal{B} \perp\!\!\!\perp \mathcal{T} \mid \mathcal{C}$, i.e., $\mathcal{E}_{\mathcal{B} \vee \mathcal{T}}$-sufficiency of $\mathcal{C}$, $\mathcal{B} \perp\!\!\!\perp \mathcal{U}$, i.e., $\mathcal{E}_{\mathcal{B} \vee \mathcal{T}}$-ancillarity of $\mathcal{U}$, or $C \perp\!\!\!\perp T$, i.e., $\mathcal{E}_{\mathcal{B}\vee T}$-ancillarity of C. It should be stressed that this analysis remains consistent with $\mathcal{B} = \mathcal{A}$ and/or $T = \mathcal{S}$.

Apart from this formal identity of concepts in $\mathcal{E}_{\mathcal{A}\vee\mathcal{S}}$ and $\mathcal{E}_{\mathcal{B}\vee T}$, it should be noted that as the condition $C \perp\!\!\!\perp T$ is symmetric in C and T, this not only means that $\mathcal{E}_{C\vee T}$ is totally noninformative, but also means (equivalently) that C is $\mathcal{E}_{\mathcal{A}\vee T}$-ancillary, or that T is $\mathcal{E}_{C\vee\mathcal{S}}$-ancillary. This is a first example of a condition leading to an admissibility property in two (reduced) experiments. This topic is pursued more systematically in Section 3.4. The concern of the next definition is precisely to stress the reciprocal character of the ancillarity of C with respect to T, and of T with respect to C.

3.2.1 Definition. The parameter $C(\subset \mathcal{A})$ and the statistic $T(\subset \mathcal{S})$ are *mutually ancillary* if and only if they are independent, i.e., $C \perp\!\!\!\perp T$. ∎

In general, the conditions for admissible reductions of experiments marginalized on a sub-σ-field C of the parameter space depend crucially on the (conditional) prior probability μ^C. The following example illustrates this point.

Example. Let $s = (x_1, ..., x_n)$ with $x_i = (y_i, z_i)'$ and

$$a = \begin{pmatrix} a_{11} & a_{12} \\ a_{12} & a_{22} \end{pmatrix}.$$

Suppose $(x_i \mid a) \sim i.N_2(0, a)$ and a is a priori distributed as an Inverse-Wishart distribution. If we define $c = a_{12}a_{22}^{-1}$ and $t = (z_1, \ldots, z_n)$, then $C = \sigma(c)$ and $T = \sigma(t)$ are mutually ancillary; indeed, a basic property of this prior distribution is $c \perp\!\!\!\perp a_{22}$; this implies, by Theorem 2.2.10, that $c \perp\!\!\!\perp t$ because $c \perp\!\!\!\perp t \mid a_{22}$ (indeed $(t \mid a) \sim N_n(0, a_{22}I_n)$). Note, however, that such a mutual ancillarity would no longer be true if the prior distribution on a did not render c and a_{22} independent. ∎

3.2.3 Sufficiency and Ancillarity in Unreduced and in Marginal Experiments

As already mentioned in Section 2.3.8, the relationship between ancillarity and sufficiency has drawn a good deal of attention in the statistical literature. The results on the unreduced experiment $\mathcal{E}_{AVS}$, discussed in Section 2.3.8, also hold within any of the marginal experiments $\mathcal{E}_{AVT}$, $\mathcal{E}_{BVS}$ or $\mathcal{E}_{BVT}$; they are not repeated here, and are left as an exercise for the reader. Rather, the object of this section is to examine the relationships between sufficiency (or ancillarity) in a marginal and in an unreduced experiment. These relationships we now report are fairly trivial consequences of the monotonocity properties of conditional independence (as given in Section 2.2) and are presented not merely for their own interest, but also as a first step toward the analysis of jointly admissible reductions in Section 3.4.

By Corollary 2.2.4, one has the following implication:

$$(3.2.4) \qquad A \perp\!\!\!\perp S \mid T \;\Rightarrow\; B \perp\!\!\!\perp U \mid T \qquad \forall\, U \subset S \;\; \forall\, B \subset A.$$

Therefore, if T is $\mathcal{E}$-sufficient then T is $\mathcal{E}_{BVU}$-sufficient for any $B \subset A$ and any U such that $T \subset U \subset S$. Similarly, as

$$(3.2.5) \qquad A \perp\!\!\!\perp T \;\Rightarrow\; B \perp\!\!\!\perp T \qquad \forall\, B \subset A$$

one has: if T is $\mathcal{E}$-ancillary, then B and T are mutually ancillary for any $B \subset A$, and T is $\mathcal{E}_{BVU}$-ancillary for any $B \subset A$ and U such that $T \subset U \subset S$. Also, as:

$$(3.2.6) \qquad A \perp\!\!\!\perp S \mid B \;\Rightarrow\; C \perp\!\!\!\perp T \mid B \quad \forall\, C \subset A \;\; \forall\, T \subset S$$

one has: if B is $\mathcal{E}$-sufficient, then B is $\mathcal{E}_{CVT}$-sufficient for any $B \subset C \subset A$ and $T \subset S$. Finally, as

$$(3.2.7) \qquad B \perp\!\!\!\perp S \;\Rightarrow\; B \perp\!\!\!\perp T \qquad \forall\, T \subset S$$

one has: if B is $\mathcal{E}$-ancillary, then B and T are mutually ancillary for any $T \subset S$, and B is $\mathcal{E}_{CVT}$-ancillary for any C such that $B \subset C \subset A$ and any $T \subset S$.

Clearly, the converse of these implications is false. In particular, to be sufficient in a marginal experiment, a sub-σ-field has to make a smaller class of functions measurable than in a complete experiment. A sub-σ-field may therefore be sufficient in a marginal experiment but not in the associated unreduced experiment, as illustrated by the following example.

Example. (See Florens, Mouchart and Richard (1974)). Let $s = (s_1, \ldots, s_n)$ with $(s_i \mid a) \sim i.N(m_i, v)$ where $a = (m_1, \ldots, m_i, \ldots, m_n, v)$. For the prior distribution, define $m = (m_1, \ldots, m_i, \ldots, m_n)$ and suppose $m \perp\!\!\!\perp v$, $m_i \sim i.N(0,1)$ and v has an arbitrary distribution (continuous on $\mathbb{R}_+$). If we consider $\mathcal{B} = \sigma(v)$ and $T = (\sum_{1 \le i \le n} s_i^2)$, then T is $\mathcal{E}_{\mathcal{B}\vee\mathcal{S}}$-sufficient (because $(s_i \mid v) \sim i.N(0, v+1)$ but is clearly not $\mathcal{E}$-sufficient.   ∎

3.2.4   A Remark on "Partial" Sufficiency

Sufficient statistics for the marginal experiment $\mathcal{E}_{\mathcal{B}\vee\mathcal{S}}$ are relevant when $\mathcal{B}$ is the only parameter of interest (i.e., $\mathcal{B}$ makes the loss-function measurable for any decisions) and the "other" parameters are nuisance parameters (on nuisance parameter from a Bayesian point of view, see, e.g., Dawid (1980a), or for a sampling approach, see Godambe (1976, 1980)). This suggests that it is appropriate to speak of "partial" sufficiency in such circumstances. Furthermore, a recurrent theme in the literature has been to examine the relationship between classical sufficiency and "Bayes sufficiency for all prior distribution" (see Section 2.3.7). These two considerations lead Kolmogorov (1942) to suggest the following definition of partial sufficiency.

3.2.2 Definition. A statistic $T \subset \mathcal{S}$ is *K-sufficient* for $\mathcal{B} \subset \mathcal{A}$ if, for every prior probability μ on $(A, \mathcal{A})$, T is $\mathcal{E}^\mu_{\mathcal{B}\vee\mathcal{S}}$-sufficient where $\mathcal{E}^\mu_{\mathcal{B}\vee\mathcal{S}}$ is defined as follows: $(A \times S, \mathcal{B}\vee\mathcal{S}, \Pi^\mu_{\mathcal{B}\vee\mathcal{S}})$ and $\Pi^\mu_{\mathcal{B}\vee\mathcal{S}}$ is the trace on $\mathcal{B}\vee\mathcal{S}$ of $\Pi^\mu_{\mathcal{A}\vee\mathcal{S}} = \mu \otimes P^{\mathcal{A}}$. When $\mathcal{B} = \mathcal{A}$, we simply say that T is K-sufficient.   ∎

In Section 2.3.7, we investigated the relationship between Bayesian sufficiency for a particular prior probability, K-sufficiency (i.e., Bayesian sufficiency for any prior probability) and sampling sufficiency. Now, we consider the relationship between K-sufficiency (for the unreduced parameter $\mathcal{A}$) and K-sufficiency for a subparameter $\mathcal{B} \subset \mathcal{A}$. In this respect, Hajek (1965) — see also Basu (1977) — asserted this definition to be void in view of the following assertion:

Assertion.   (Hajek (1965)). If $\mathcal{B}$ is nontrivial (i.e., $\mathcal{B} \ne \mathcal{I} = \{\phi, A \times S\}$) and T is K-sufficient for $\mathcal{B}$ then T is K-sufficient.

Martin, Petit and Littaye (1973) gave a counterexample to Hajek's assertion and proved that if the sampling probabilities are all equivalent (i.e., have the same null sets) then Hajek's Theorem is correct; the main argument is that in this case, K-sufficiency is equivalent to pairwise sufficiency

(Martin, Petit and Littaye (1973 - Propr. I-7); see also Speed (1976)). Consider, however, the following finite example:

Example.

	Sampling Probabilities $p(s_i \mid a_j)$			
	a_1	a_2	a_3	a_4
s_1	.15	.10	.20	.25
s_2	α	.30	.10	0
s_3	.45	.30	.60	.75
s_4	$.40 - \alpha$	.30	.10	0

where $0 \leq \alpha \leq .40$

Consider the σ-fields $\mathcal{B} = \sigma(b)$ and $\mathcal{T} = \sigma(t)$ defined by the following functions:

$$(3.2.8) \qquad \begin{aligned} b(a_j) &= b_1 & j &= 1, 2, 3 \\ &= b_2 & j &= 4 \end{aligned}$$

$$(3.2.9) \qquad \begin{aligned} t(s_i) &= t_1 & i &= 1, 3 \\ &= t_2 & i &= 2, 4 \end{aligned}$$

It may be checked that $\mathcal{T}$ is K-sufficient for $\mathcal{B}$ for any value of α (even for $\alpha = 0$) while $\mathcal{T}$ is K-sufficient if and only if $\alpha = .2$. ∎

This example not only illustrates the non-necessity of the condition of equivalent sampling probabilities, but also shows that the failure in the proof of Hajek's Assertion is due to the non-respect of the following condition:

(H) $p(s_i \mid a_j) = p(s_{i'} \mid a_j) = 0$ implies:

$$p(s_i \mid a_{j'}) = \lambda(s_i, s_{i'})p(s_{i'} \mid a_{j'}) \text{ for any } a_{j'}.$$

In other words, condition (H) says that in the matrix of the sampling probabilities $p(s_i|a_j)$, any two zeros in the same column should always be associated to proportional rows. In view of the following theorem this condition turns out to also be sufficient.

3.2.3 Theorem. In the discrete case, Condition (H) is necessary and sufficient to ensure that any statistic K-sufficient for $\mathcal{B}$ (nontrivial) is also K-sufficient.

Proof. *(i) Condition (H) is necessary.*

If $p(s_i \mid a_{j_0}) = p(s_{i'} \mid a_{j_0}) = 0$, consider the following statistic t:

$$(3.2.10) \qquad\qquad t(s) \;=\; t_1 \text{ if } s = s_i \text{ or } s_{i'}$$
$$= \; t_2 \text{ otherwise.}$$

Then $\mathcal{T} = \sigma(t)$ is K-sufficient for a nontrivial subparameter defined by:

$$(3.2.11) \qquad\qquad b(a) \;=\; b_1 \text{ if } a = a_{j_0}$$
$$= \; b_2 \text{ otherwise.}$$

Indeed, $p^\mu(b_1 \mid s_i) = p^\mu(b_1 \mid s_{i'}) = 0$ for any prior probability μ; but if $\mathcal{T}$ is also K-sufficient the two rows associated with s_i and $s_{i'}$, should be proportional.

(ii) Condition (H) is sufficient.

We show that under (H), for any non K-sufficient statistic $\mathcal{T}$ and any nontrivial parameter $\mathcal{B}$, there exists a prior probability such that $\mathcal{T}$ is not Bayesian sufficient for that $\mathcal{B}$ and that prior probability. Indeed, if $\mathcal{T} = \sigma(t)$ is not K-sufficient, there exist $(s_i, s_{i'}, a_j, a_{j'})$ such that $t(s_i) = t(s_{i'})$ and $p(s_i \mid a_j)p(s_{i'} \mid a_{j'}) \neq p(s_i \mid a_{j'})p(s_{i'} \mid a_j)$. If $b(a_j) \neq b(a_{j'})$ then $\mathcal{T}$ is not sufficient for $\mathcal{B} = \sigma(b)$ once $\mu(a_j) = \mu(a_{j'}) = \frac{1}{2}$. If $b(a_j) = b(a_{j'})$, there exists a_m such that $b(a_m) \neq b(a_j)$, because $\mathcal{B}$ is not trivial. Note that $(s_i, s_{i'}, a_j, a_{j'})$ are such that

$$[p(s_i \mid a_j) + p(s_{i'} \mid a_j)][p(s_i \mid a_{j'}) + p(s_{i'} \mid a_{j'})] > 0$$

and $p(s_i \mid a_j)/p(s_{i'} \mid a_j) \neq p(s_i \mid a_{j'})/p(s_{i'} \mid a_{j'})$

on $\overline{I\!R}^+$ (where $a/0$ is defined as ∞ for $a > 0$). Furthermore, under (H), either $p(s_i \mid a_m) + p(s_{i'} \mid a_m) = 0$ and

$$p(s_i \mid a_j)p(s_{i'} \mid a_{j'}) = p(s_i \mid a_{j'})p(s_{i'} \mid a_j)$$

or $p(s_i \mid a_m) + p(s_{i'} \mid a_m) > 0$ but then the ratio $p(s_i \mid a_m)/p(s_{i'} \mid a_m)$ is well defined and cannot be equal to both $p(s_i \mid a_j)/p(s_{i'} \mid a_j)$ and $p(s_i \mid a_{j'})/p(s_{i'} \mid a_{j'})$ and, therefore, $\mathcal{T}$ is not sufficient for $\mathcal{B}$ (either for $\mu(a_j) = \mu(a_m) = \frac{1}{2}$ or for $\mu(a_{j'}) = \mu(a_m) = \frac{1}{2}$). ∎

In terms of conditional independence, the problem of equivalence between K-sufficiency for a nonconstant subparameter and K-sufficiency may be viewed as follows. If $\mathcal{B}$ and $\mathcal{S}$ are independent conditionally on $\mathcal{T}$ (i.e., $\mathcal{B} \perp\!\!\!\perp \mathcal{S} \mid \mathcal{T}$ for any prior probability), under what conditions does this imply that $\mathcal{A} \supset \mathcal{B}$ and $\mathcal{S}$ are independent conditionally on $\mathcal{T}$ (i.e., $\mathcal{A} \perp\!\!\!\perp \mathcal{S} \mid \mathcal{T}$ for any prior probability)? Clearly, in the general case, the crucial point is "for any prior probability" because, by Theorem 2.2.10, one such condition, for a particular prior probability, would be $\mathcal{A} \perp\!\!\!\perp \mathcal{S} \mid \mathcal{B} \vee \mathcal{T}$.

Condition (H), a condition on the sampling probabilities, provides the result for any nontrivial subparameter $\mathcal{B}$. An alternative approach would be to look either for a property linking a statistic $\mathcal{T}$ and a subparameter $\mathcal{B}$ or for a property, stronger than non-triviality, which should be possessed by a subparameter $\mathcal{B}$, in order for Hajek's Assertion to be true. Recently, Cano, Hernandez and Moreno (1988) tackled the case of an uncountable space $(S, \mathcal{S})$ when the sampling probabilities are dominated and the σ-field $\mathcal{A}$ on the parameter space is separating. In such a context, they modified our condition (H) into a similar necessary and sufficient condition for a given statistic $\mathcal{T}$ and a given nontrivial parameter $\mathcal{B}$.

3.3 Admissible Reductions in Conditional Experiments

3.3.1 Introduction

We now extend the concepts of sufficiency and of ancillarity to an experiment $\mathcal{E}_{\mathcal{B}\vee\mathcal{T}}^{\mathcal{M}}$, already reduced, by both conditioning and marginalization, where $\mathcal{B} \subset \mathcal{A}$, $\mathcal{T} \subset \mathcal{S}$ and $\mathcal{M} \subset \mathcal{A} \vee \mathcal{S}$. This is the more general setting. Indeed, the definitions provided in this section generalize those provided previously by setting $\mathcal{M} = \mathcal{I}$, and also provide definitions for an experiment reduced by conditioning only, by setting $\mathcal{B} = \mathcal{A}$ and $\mathcal{T} = \mathcal{S}$.

As mentioned in the introduction to this chapter, the general experiment $\mathcal{E}_{\mathcal{B}\vee\mathcal{T}}^{\mathcal{M}}$ is the natural set-up for almost any statistical analysis, and from a technical point of view, once the extension of the concepts of parameters and of statistics to conditional models (given in Section 1.4.3.D) is recalled, the formulation of the concepts of ancillarity and of sufficiency at this level of generality involves an essentially trivial extension of th conceptual apparatus elaborated for the unreduced experiment $\mathcal{E}_{\mathcal{A}\vee\mathcal{S}}$. Accordingly, we do not give extensive comments of these concepts so as to avoid largely trivial duplication of the discussions in Chapter 2. Instead, we shall limit ourselves to presenting some simple examples at the end of Section 3.3.2, after the general definitions; it should nevertheless be stressed that all forthcoming chapters make extensive use of the concepts developed in this section, and it may be viewed as a rich source of illustrations of the usefulness of these concepts (see also, e.g., Kalbfleisch (1975), and Sprott (1975)).

3.3.2   Reductions in the Sample Space

Recall that in an experiment such as $\mathcal{E}_{\mathcal{B}\vee\mathcal{T}}^{\mathcal{M}}$ (see Section 1.4.3.D), a statistic $\mathcal{N}$ is a sub-σ-field of $\mathcal{M}\vee\mathcal{T}$. Following the general principle that a reduction is admissible if the complementary reduction is totally noninformative, i.e., the prior probability is a.s. equal to the posterior probability, we obtain, using the diagrams of Subsection 1.4.4.C, the following definitions:

3.3.1 Definition. $\mathcal{N}$ is an $\mathcal{E}_{\mathcal{B}\vee\mathcal{T}}^{\mathcal{M}}$-*sufficient statistic* if and only if:

(i) $\mathcal{N} \subset \mathcal{M} \vee \mathcal{T}$;

(ii) $\mathcal{B} \perp\!\!\!\perp \mathcal{T} | \mathcal{M} \vee \mathcal{N}.$ ∎

3.3.2 Definition. $\mathcal{N}$ is an $\mathcal{E}_{\mathcal{B}\vee\mathcal{T}}^{\mathcal{M}}$-*ancillarity statistic* if and only if:

(i) $\mathcal{N} \subset \mathcal{M} \vee \mathcal{T}$;

(ii) $\mathcal{B} \perp\!\!\!\perp \mathcal{N} | \mathcal{M}.$ ∎

Recall also that a conditional independence property may be written in various ways. For instance, $\mathcal{N}$ is a $\mathcal{E}_{\mathcal{B}\vee\mathcal{T}}^{\mathcal{M}}$-sufficient statistic if and only if

$$(3.3.1) \qquad \mu_{\mathcal{B}}^{\mathcal{M}\vee\mathcal{N}\vee\mathcal{T}}(E) = \mu_{\mathcal{B}}^{\mathcal{M}\vee\mathcal{N}}(E) \quad \forall\, E \in \mathcal{B}$$

or, equivalently, if and only if

$$(3.3.2) \qquad P_{\mathcal{T}}^{\mathcal{B}\vee\mathcal{M}\vee\mathcal{N}}(X) = P_{\mathcal{T}}^{\mathcal{M}\vee\mathcal{N}}(X) \quad \forall\, X \in \mathcal{T}.$$

3.3.3   Reductions in the Parameter Space

Recall that in an experiment $\mathcal{E}_{\mathcal{B}\vee\mathcal{T}}^{\mathcal{M}}$ a parameter $\mathcal{L}$ is a sub-σ-field of $\mathcal{B} \vee \mathcal{M}$. As before we obtain the following definitions:

3.3.3 Definition. $\mathcal{L}$ is an $\mathcal{E}_{\mathcal{B}\vee\mathcal{T}}^{\mathcal{M}}$-*sufficient parameter* if and only if:

(i) $\mathcal{L} \subset \mathcal{B} \vee \mathcal{M}$

(ii) $\mathcal{B} \perp\!\!\!\perp \mathcal{T} \mid \mathcal{L} \vee \mathcal{M}.$ ∎

3.3.4 Definition. $\mathcal{L}$ is an $\mathcal{E}_{\mathcal{B}\vee\mathcal{T}}^{\mathcal{M}}$-*ancillary parameter* if and only if:

(i) $\mathcal{L} \subset \mathcal{B} \vee \mathcal{M};$

(ii) $\mathcal{L} \perp\!\!\!\perp \mathcal{T} \mid \mathcal{M}.$ ∎

Let us illustrate these concepts by two examples.

Example 1. (Sufficiency in Regression Models)

Let $s_i' = (y_i, z_i') \in \mathbb{R}^{1+p}$ $i = 1, \ldots, n$, $s' = (s_1', \ldots, s_n')$, $y' = (y_1, \ldots, y_n)$, $z' = (z_1', \ldots, z_n')$, and $Z = (z_1, \ldots, z_n)'$, an $n \times p$ matrix. Now suppose that $(z_i \mid a) \sim i.N(m, \Sigma)$, and $(y_i \mid Z, a) \sim i.N(\beta' z_i, \sigma^2)$, where $a = (m, \Sigma, \beta, \sigma^2)$ and, in particular, $\beta \in \mathbb{R}^p$. The regression model of $(y \mid z)$ may be defined as $\mathcal{E}_{\mathcal{B}\vee\mathcal{T}}^{\mathcal{M}}$ where $\mathcal{M} = \sigma(Z)$, $\mathcal{T} = \sigma(y)$, and, in particular, $\mathcal{B} = \sigma(\beta, \sigma^2)$. A natural sufficient statistic is $\mathcal{N} = \sigma(Z'y, y'y)$ where, clearly, $\mathcal{N} \subset \mathcal{M} \vee \mathcal{T}$. Although $\mathcal{B} = \sigma(\beta, \sigma^2)$ is often a natural parameter, it is sometimes more suitable to use $\mathcal{L} = \sigma(Z\beta, \sigma^2)$, which is obviously included in $\mathcal{B} \vee \mathcal{M}$, as a sufficient parameter.   ∎

Example 2. (Sufficiency in Censored Data)

Let $s_i' = (y_i, z_i) \in \mathbb{R}^2$ $i = 1, \ldots, n$, $s' = (s_1', \ldots, s_n')$. Let

$$\left(s_i \mid a \right) \sim i.N_2 \left[\binom{m}{0}, I_2 \right]$$

where $\mathcal{A} = \sigma(m)$, and y_i is observed only if it is smaller than z_i. Thus we define $x = (x_1, \ldots, x_n)'$ with $x_i = \min(y_i, z_i)$. The experiment to be analysed will often have the form $\mathcal{E}^{\mathcal{M}}_{\mathcal{B} \vee \mathcal{T}}$, where $\mathcal{M} = \sigma(z_1, \ldots, z_n)$, $\mathcal{B} = \mathcal{A}$, and $\mathcal{T} = \sigma(x_1, \ldots, x_n)$. For this reduced experiment, a sufficient statistic may be built as follows: define $w = (w_1, \ldots, w_n)'$ with $w_i = 1$ if $x_i < z_i$, and $w_i = 0$ if $x_i = z_i$, i.e., $w_i = 1_{\{x_i < z_i\}}$; one may check that $\mathcal{N} = \sigma(w, w'x)$ is an $\mathcal{E}^{\mathcal{M}}_{\mathcal{B} \vee \mathcal{T}}$-sufficient statistic. Note again that $\mathcal{N} \subset \mathcal{M} \vee \mathcal{T}$. ∎

3.3.4   Elementary Properties

In the reduced experiment $\mathcal{E}^{\mathcal{M}}_{\mathcal{B} \vee \mathcal{T}}$, the monotonocity properties of sufficient and ancillary σ-fields of an unreduced experiment (Proposition 2.3.6) are preserved. This fact is made explicit in next proposition, the proof of which is a simple application of Corollaries 2.2.11(ii) and 2.2.4(ii).

3.3.5 Proposition. In a reduced experiment $\mathcal{E}^{\mathcal{M}}_{\mathcal{B} \vee \mathcal{T}}$,

(i)   An $\mathcal{E}^{\mathcal{M}}_{\mathcal{B} \vee \mathcal{T}}$-statistic (respectively parameter) including an $\mathcal{E}^{\mathcal{M}}_{\mathcal{B} \vee \mathcal{T}}$-sufficient statistic (respectively parameter) is also $\mathcal{E}^{\mathcal{M}}_{\mathcal{B} \vee \mathcal{T}}$-sufficient.

(ii) An $\mathcal{E}^{\mathcal{M}}_{\mathcal{B} \vee \mathcal{T}}$-statistic (respectively parameter) included in an ancillary $\mathcal{E}^{\mathcal{M}}_{\mathcal{B} \vee \mathcal{T}}$-statistic (respectively parameter) and is also $\mathcal{E}^{\mathcal{M}}_{\mathcal{B} \vee \mathcal{T}}$-ancillary. ∎

The property of robustness of a sufficient or ancillary σ-field for a suitable modification of the prior probability is also preserved. This fact is made explicit in next proposition, the proof of which is based on Corollary 2.2.15.

3.3.6 Proposition. Let us consider $\mathcal{E}$ and $\mathcal{E}'$, two Bayesian experiments on $(A \times S, \mathcal{A} \vee \mathcal{S})$ defined by their probabilities Π and Π'. Let $\mathcal{B} \subset \mathcal{A}, \mathcal{T} \subset \mathcal{S}$ and $\mathcal{M} \subset \mathcal{A} \vee \mathcal{S}$. If

(i) $\Pi' \ll \Pi$

(ii) $\frac{d\Pi'}{d\Pi} \in \left[\overline{\mathcal{B} \vee \mathcal{M}}\right]^+$.

Then:

(iii) An $\mathcal{E}^{\mathcal{M}}_{\mathcal{B}\vee\mathcal{T}}$-sufficient statistic (respectively parameter) is also $\mathcal{E}'^{\mathcal{M}}_{\mathcal{B}\vee\mathcal{T}}$-sufficient;

(iv) An $\mathcal{E}^{\mathcal{M}}_{\mathcal{B}\vee\mathcal{T}}$-ancillary statistic is also $\mathcal{E}'^{\mathcal{M}}_{\mathcal{B}\vee\mathcal{T}}$-ancillary;

(v)   If $\mathcal{L}$ is an $\mathcal{E}^{\mathcal{M}}_{\mathcal{B}\vee\mathcal{T}}$-ancillary parameter and if furthermore $\frac{d\Pi'}{d\Pi} \in \left[\overline{\mathcal{L} \vee \mathcal{M}}\right]^+$, then $\mathcal{L}$ is also an $\mathcal{E}'^{\mathcal{M}}_{\mathcal{B}\vee\mathcal{T}}$-ancillary parameter.   ∎

Note that if assumption (ii) is replaced by $\left(d\Pi'/d\Pi\right) \in \left[\overline{\mathcal{M} \vee \mathcal{T}}\right]^+$, conclusion (iii) still holds, but conclusion (iv) becomes: An $\mathcal{E}^{\mathcal{M}}_{\mathcal{B}\vee\mathcal{T}}$-ancillary parameter is also $\mathcal{E}'^{\mathcal{M}}_{\mathcal{B}\vee\mathcal{T}}$-ancillary and conclusion (v) becomes: If $\mathcal{N}$ is an $\mathcal{E}^{\mathcal{M}}_{\mathcal{B}\vee\mathcal{T}}$-ancillary statistic, and if furthermore $\left(d\Pi'/d\Pi\right) \in \left[\overline{\mathcal{M} \vee \mathcal{N}}\right]^+$, then $\mathcal{N}$ is also an $\mathcal{E}'^{\mathcal{M}}_{\mathcal{B}\vee\mathcal{T}}$-ancillary statistic. It may be useful to recall that, as in Proposition 2.3.7, the condition $\left(d\Pi'/d\Pi\right) \in \left[\overline{\mathcal{B} \vee \mathcal{M}}\right]^+$ (respectively, $\left(d\Pi'/d\Pi\right) \in \left[\overline{\mathcal{M} \vee \mathcal{T}}\right]^+$) means that Π' differs from Π only in its marginal part of $\mathcal{B} \vee \mathcal{M}$ (respectively, $\mathcal{M} \vee \mathcal{T}$).

3.3.5   Relationships between Sufficiency and Ancillarity

Note that a conditional independence property such as $\mathcal{B} \perp\!\!\!\perp \mathcal{T} \mid \mathcal{M}\vee\mathcal{N}$, where $\mathcal{B} \subset \mathcal{A}$, $\mathcal{T} \subset \mathcal{S}$, $\mathcal{M} \subset \mathcal{A} \vee \mathcal{S}$, $\mathcal{N} \subset \mathcal{M} \vee \mathcal{T}$ may receive several interpretations. Indeed, it means that $\mathcal{N}$ is an $\mathcal{E}^{\mathcal{M}}_{\mathcal{B}\vee\mathcal{T}}$-sufficient statistic, and also that $\mathcal{T}$ is a uniform $\mathcal{E}^{\mathcal{M}\vee\mathcal{N}}_{\mathcal{B}\vee\mathcal{S}}$-ancillary statistic and that $\mathcal{B}$ is a uniform $\mathcal{E}^{\mathcal{M}\vee\mathcal{N}}_{\mathcal{A}\vee\mathcal{T}}$-ancillary parameter. Another interesting conditional independence property is $\mathcal{A} \perp\!\!\!\perp \mathcal{S} \mid \mathcal{B}\vee\mathcal{T}$, where $\mathcal{B} \subset \mathcal{A}$ and $\mathcal{T} \subset \mathcal{S}$. It means both that $\mathcal{B}$ is a uniform $\mathcal{E}^{\mathcal{T}}$-sufficient parameter and that $\mathcal{T}$ is a uniform $\mathcal{E}^{\mathcal{B}}$-sufficient statistic, and consequently that $\mathcal{E}^{\mathcal{B}\vee\mathcal{T}}$ is totally noninformative.

An interesting property of the stability of sufficiency in conditional experiments may be derived from Corollary 2.2.11, viz., any $\mathcal{E}$-sufficient statistic (respectively, parameter) is also a uniform $\mathcal{E}^{\mathcal{B}\vee\mathcal{T}}$-sufficient statistic (respectively, parameter) for any $\mathcal{B} \subset \mathcal{A}$ and $\mathcal{T} \subset \mathcal{S}$. Furthermore, Theorem 2.2.10 entails the following result:

3.3.7 Proposition.

(i) $T \subset S$ is $\mathcal{E}$-sufficient (respectively, $\mathcal{E}$-ancillary) if and only if there exists $\mathcal{B} \subset \mathcal{A}$ such that T is $\mathcal{E}_{\mathcal{B} \vee S}$-sufficient (respectively, $\mathcal{E}_{\mathcal{B} \vee S}$-ancillary), and $\mathcal{E}^{\mathcal{B}}$-sufficient (respectively, $\mathcal{E}^{\mathcal{B}}$-ancillary).

(ii) $\mathcal{B} \subset \mathcal{A}$ is $\mathcal{E}$-sufficient (respectively, $\mathcal{E}$-ancillary) if and only if there exists $T \subset S$ such that $\mathcal{B}$ is $\mathcal{E}_{\mathcal{A} \vee T}$-sufficient (respectively, $\mathcal{E}_{\mathcal{A} \vee T}$-ancillary), and $\mathcal{E}^{T}$ sufficient (respectively, $\mathcal{E}^{T}$-ancillary).   ∎

Using the same theorem, Theorems 2.3.19 and 2.3.20 may be extended to the general reduced experiment $\mathcal{E}_{\mathcal{B} \vee T}^{\mathcal{M}}$.

3.3.8 Theorem. In $\mathcal{E}_{\mathcal{B} \vee T}^{\mathcal{M}}$,

(i)   any statistic (respectively, parameter) independent, given $\mathcal{M}$, of an $\mathcal{E}_{\mathcal{B} \vee T}^{\mathcal{M}}$-sufficient statistic (respectively, parameter) is $\mathcal{E}_{\mathcal{B} \vee T}^{\mathcal{M}}$-ancillary and is moreover independent of this sufficient statistic (respectively, parameter) given $\mathcal{B} \vee \mathcal{M}$ (respectively, given $\mathcal{M} \vee T$);

(ii) any statistic (respectively, parameter) independent, given $\mathcal{M}$, of an $\mathcal{E}_{\mathcal{B} \vee T}^{\mathcal{M}}$-sufficient parameter (respectively, statistic) is $\mathcal{E}_{\mathcal{B} \vee T}^{\mathcal{M}}$-ancillary.   ∎

Proof. We prove (i) for a statistic only. In that case, the theorem may be restated as follows. Let $\mathcal{N}_i \subset \mathcal{M} \vee T$, $i = 1, 2$. If $\mathcal{B} \perp\!\!\!\perp T \mid \mathcal{M} \vee \mathcal{N}_2$, and $\mathcal{N}_1 \perp\!\!\!\perp \mathcal{N}_2 \mid \mathcal{M}$, then $\mathcal{B} \perp\!\!\!\perp \mathcal{N}_1 \mid \mathcal{M}$, and $\mathcal{N}_1 \perp\!\!\!\perp \mathcal{N}_2 \mid \mathcal{B} \vee \mathcal{M}$. It suffices to note that, since $\mathcal{N}_1 \subset \mathcal{M} \vee T$, $\mathcal{B} \perp\!\!\!\perp T \mid \mathcal{M} \vee \mathcal{N}_2$ implies $\mathcal{B} \perp\!\!\!\perp \mathcal{N}_1 \mid \mathcal{M} \vee \mathcal{N}_2$ (Corollary 2.2.11 (i) and Corollary 2.2.4 (ii)). By Theorem 2.2.10, when $\mathcal{N}_1 \perp\!\!\!\perp \mathcal{N}_2 \mid \mathcal{M}$, this last relation is equivalent to $\mathcal{N}_1 \perp\!\!\!\perp (\mathcal{B} \vee \mathcal{N}_2) \mid \mathcal{M}$, and this is equivalent to the conclusion. The proof of (ii) follows exactly the same approach.   ∎

Note that, when $\mathcal{N}_1$ and $\mathcal{N}_2$ are uniform statistics (namely, $\mathcal{N}_i \subset T$, $i = 1, 2$) it is enough to require $\mathcal{N}_2$ to be $\mathcal{E}_{\mathcal{B} \vee \mathcal{N}_1 \vee \mathcal{N}_2}^{\mathcal{M}}$-sufficient.

3.3.6 Sufficiency and Ancillarity in a Dominated Reduced Experiment

It is sometimes useful to deduce a conditional independence property from arguments based on the factorization of the density. In this section, considerable use is made of Theorem 2.2.14. Since by construction, $A \perp\!\!\!\perp S; \mu \otimes P$, we can obtain sufficiency and ancillarity with respect to $\mu \otimes P$ only if the conditioning σ-field $\mathcal{M}$ is of the form $\mathcal{C} \vee \mathcal{U}$ where $\mathcal{C} \subset \mathcal{A}$ and $\mathcal{U} \subset \mathcal{S}$ (Corollary 2.2.11). The next theorem describes the factorization properties of the density $g_{\mathcal{BVT}}^{\mathcal{M}}$ of the experiment $\mathcal{E}_{\mathcal{BVT}}^{\mathcal{M}}$, that are implications of the existence of sufficient and ancillary statistics and parameters.

3.3.9 Proposition. Consider the experiment $\mathcal{E}_{\mathcal{BVT}}^{\mathcal{M}}$ where $\mathcal{B} \subset \mathcal{A}$, $\mathcal{T} \subset \mathcal{S}$, $\mathcal{M} = \mathcal{C} \vee \mathcal{U}$ with $\mathcal{C} \subset \mathcal{A}$ and $\mathcal{U} \subset \mathcal{S}$. Then

(i) $\mathcal{N}$ is an $\mathcal{E}_{\mathcal{BVT}}^{\mathcal{M}}$-sufficient statistic if and only if
$$g_{\mathcal{BVT}}^{\mathcal{M}} = g_{\mathcal{BVN}}^{\mathcal{M}} \cdot g_{\mathcal{T}}^{\mathcal{MVN}};$$

(ii) $\mathcal{N}$ is an $\mathcal{E}_{\mathcal{BVT}}^{\mathcal{M}}$-ancillary statistic if and only if
$$g_{\mathcal{BVT}}^{\mathcal{M}} = g_{\mathcal{B}}^{\mathcal{M}} \cdot g_{\mathcal{N}}^{\mathcal{M}} \cdot g_{\mathcal{T}}^{\mathcal{BVMVN}};$$

(iii) $\mathcal{L}$ is an $\mathcal{E}_{\mathcal{BVT}}^{\mathcal{M}}$-sufficient parameter if and only if
$$g_{\mathcal{BVT}}^{\mathcal{M}} = g_{\mathcal{LVT}}^{\mathcal{M}} \cdot g_{\mathcal{B}}^{\mathcal{LVM}};$$

(iv) $\mathcal{L}$ is an $\mathcal{E}_{\mathcal{BVT}}^{\mathcal{M}}$-ancillary parameter if and only if
$$g_{\mathcal{BVT}}^{\mathcal{M}} = g_{\mathcal{L}}^{\mathcal{M}} \cdot g_{\mathcal{T}}^{\mathcal{M}} \cdot g_{\mathcal{B}}^{\mathcal{LVMVT}}. \qquad \blacksquare$$

Although sometimes useful, these decompositions of the densities are nenetheless complicated and, in particular, verification requires integrations on some parameters and/or statistics. Additional assumptions may simplify this process. That is the subject of the next section.

3.4   Jointly Admissible Reductions

In Sections 3.2 and 3.3 we analyzed successive reductions of experiments. In this section, we analyze joint reductions of experiments, i.e., reductions on both sample and parameter spaces. In contrast with the analysis above, both reductions are introduced simultaneously; this approach is motivated by the search both for computational simplicity and for robustness of the specifications.

When the jointly admissible reductions are obtained by marginalization on both the sample space and the parameter space, this is referred to as *mutual sufficiency*; when they are obtained by marginalization on one space and conditioning on the other space, this is referred to as *mutual exogeneity*. The concept of a *Bayesian cut* combines both possibilities.

As the concept of an unreduced experiment is actually relative to a given experiment, the definitions provided in Section 3.4.5 are for the most general form of a reduced experiment $\mathcal{E}_{\mathcal{B}\vee\mathcal{T}}^{\mathcal{M}}$. For expository reasons, however, we introduce the concepts and provide both proofs and comments for the simpler case where $\mathcal{M} = \mathcal{I}$, $\mathcal{B} = \mathcal{A}$, and $\mathcal{T} = \mathcal{S}$. The proofs for the general case are straightforward extensions of this simple case. The more general set-up is necessary, however, when dealing, for example, with sequential problems as in Chapter 6. In Section 3.4.6, some examples will reveal the relationships between the three basic concepts of mutual sufficiency, mutual exogeneity, and the Bayesian cut; while in Sections 3.4.1 to 3.4.3, each example illustrates only one of these concepts.

3.4.1   Mutual Sufficiency

If $\mathcal{B} \subset \mathcal{A}$ represents the only parameter of interest, we have seen that, without losing relevant information, the experiment $\mathcal{E}$ may be reduced to $\mathcal{E}_{\mathcal{B}\vee\mathcal{T}}$ if $\mathcal{T}$ is an $\mathcal{E}_{\mathcal{B}\vee\mathcal{S}}$-sufficient statistic. The transformation $\mu_{\mathcal{B}} \rightarrow \mu_{\mathcal{B}}^{\mathcal{S}}$ will, in general, require the integration of $P^{\mathcal{A}}$ with respect to $\mu^{\mathcal{B}}$ so as to obtain $P^{\mathcal{B}}$. The next theorem states when such an integration may be avoided. However, when $\mathcal{B} \subset \mathcal{A}$ represents the only parameter of interest, it often happens that one may find a statistic $\mathcal{T}$ such that the sampling distribution of $\mathcal{T}$ depends on $\mathcal{B}$ only. Unless a further condition is introduced, this does not imply, in general, that statistical inference about $\mathcal{B}$ (thus, its posterior distribution) depends on $\mathcal{T}$ only. The next definition formalizes this double requirement.

3.4.1 Theorem. Let $\mathcal{E} = (A \times S, \mathcal{A} \vee \mathcal{S}, \Pi)$ be a Bayesian experiment. For $\mathcal{B} \subset \mathcal{A}$ and $\mathcal{T} \subset \mathcal{S}$, the following properties are equivalent:

(i) $\mathcal{B} \perp \mathcal{S} \mid \mathcal{T}$ and $\mathcal{A} \perp \mathcal{T} \mid \mathcal{B}$

(ii) $\mathcal{T}a = \mathcal{S}(\mathcal{B}a)$ $\qquad \forall\, a \in [\mathcal{A}]^+$

(iii) $\mathcal{B}s = \mathcal{A}(\mathcal{T}s)$ $\qquad \forall\, s \in [\mathcal{S}]^+.$

Proof. By symmetry it suffices to prove the equivalence of (i) and (ii). By Theorem 2.2.1. (iii), $\mathcal{A} \perp \mathcal{T} \mid \mathcal{B}$ implies $\mathcal{T}a = \mathcal{T}(\mathcal{B}a)$, and by Theorem 2.2.1(ii) $\mathcal{B} \perp \mathcal{S} \mid \mathcal{T}$ implies $\mathcal{T}(\mathcal{B}a) = \mathcal{S}(\mathcal{B}a)$. Hence (i) implies (ii). Furthermore, applying (ii) for $a \in [\mathcal{B}]^+ \subset [\mathcal{A}]^+$ yields $\mathcal{B} \perp \mathcal{S} \mid \mathcal{T}$; hence $\forall a \in [\mathcal{A}]^+$, $\mathcal{S}(\mathcal{B}a) = \mathcal{T}(\mathcal{B}a)$, and applying again (ii) yields $\mathcal{T}a = \mathcal{T}(\mathcal{B}a)$, i.e., by Theorem 2.2.1(iii), $\mathcal{A} \perp \mathcal{T} \mid \mathcal{B}$. ∎

3.4.2 Definition. Under any one of the conditions of Theorem 3.4.1., we define $\mathcal{B}$ and $\mathcal{T}$ to be *mutually sufficient*. ∎

Example. Consider $x = (y, z)'$ and $(x \mid a) \sim N_2[(\sin \varphi, \cos \varphi)', \eta^{-1} I_2]$ with $a = (\varphi, \eta) \in A = (-\pi \ +\pi] \times \mathbb{R}_0^+$. Let the prior distribution be an independent product of a gamma distribution for η and a uniform distribution for φ. It is rather easily checked that the posterior distribution of η depends on $r^2 = y^2 + z^2$ only (i.e., $\eta \perp x \mid r^2$), while the sampling distribution of r^2 (being $(1/\eta)\chi_2^2(\eta)$) depends on η only (i.e., $a \perp r^2 \mid \eta$); therefore η and r^2 are mutually sufficient. It is shown, in Chapter 8, that this example crucially depends on the fact that the prior distribution (on φ) displays a suitable invariance property on a compact parameter space. ∎

Remark. The concept of mutual sufficiency in the form of condition (i) in Theorem 3.4.1. is used by Martin, Petit and Littaye (1973), Definition III-1, for a family of prior distributions. ∎

The property of mutual sufficiency means that the pair of reductions $\mathcal{E}^{\mathcal{B} \vee \mathcal{T}}$ and $\mathcal{E}_{\mathcal{B} \vee \mathcal{T}}$ is jointly admissible in the sense that the pair $\mathcal{E}_{\mathcal{B} \vee \mathcal{S}}^{\mathcal{T}}$ and $\mathcal{E}_{\mathcal{A} \vee \mathcal{T}}^{\mathcal{B}}$ are both totally noninformative. It is therefore natural to analyze some implications of mutual sufficiency for both $\mathcal{E}_{\mathcal{B} \vee \mathcal{T}}$ and $\mathcal{E}^{\mathcal{B} \vee \mathcal{T}}$.

We first consider the marginal reduction $\mathcal{E}_{\mathcal{B}\vee\mathcal{T}}$. The reduction by joint marginalization on both the parameters and the observations has been abundantly studied in the statistical literature, in particular after Barnard's (1963) work on sufficiency. A key issue is the elimination, in a sampling theory framework, of nuisance parameters; "in the absence of any information on these parameters", the condition $\mathcal{B} \perp\!\!\!\perp \mathcal{S} \mid \mathcal{T}$ may be viewed as a Bayesian condition, (on the relationship between the prior distribution and the sampling process). This leads to the easy elimination of the nuisance parameter when the sampling distribution of $\mathcal{T}$ depends on $\mathcal{B}$ only. Whereas the construction of a statistic $\mathcal{T}$, mutually sufficient with a given parameter $\mathcal{B}$, may be difficult in general, it is shown, in Chapter 8, that invariance arguments arise quite naturally in such an undertaking; such arguments have been used, namely, in fiducial theory (see, e.g., Fraser (1968)), to justify more or less heuristic inferences on $\mathcal{B}$ based on $\mathcal{T}$ only. From an analytical point of view, recall that in the regular case — see (1.4.7) — one has, in general:

$$(3.4.1) \qquad \Pi_{\mathcal{B}\vee\mathcal{T}} = \mu_{\mathcal{B}} \otimes P_{\mathcal{T}}^{\mathcal{B}} = P_{\mathcal{T}} \otimes \mu_{\mathcal{B}}^{\mathcal{T}}$$

where

$$(3.4.2) \qquad P_{\mathcal{T}}^{\mathcal{B}}(X) = \int_A P_{\mathcal{T}}^{\mathcal{A}}(X) d\mu^{\mathcal{B}} \qquad \forall \, X \in \mathcal{T}.$$

Now, if only the first conditional independence relation of condition (i) of Theorem 3.4.1. — viz., $\mathcal{T}$ is $\mathcal{E}_{\mathcal{B}\vee\mathcal{S}}$-sufficient — were true, $P_{\mathcal{T}}^{\mathcal{B}}$ would require an integration as in (3.4.2). But the second conditional independence relation of condition (i) of Theorem 3.4.1. says that

$$(3.4.3) \qquad P_{\mathcal{T}}^{\mathcal{B}}(X) = P_{\mathcal{T}}^{\mathcal{A}}(X) \qquad \forall \, X \in \mathcal{T}.$$

The integration in (3.4.2) may thus be avoided in the course of the transformation $\mu_{\mathcal{B}} \to \mu_{\mathcal{B}}^{\mathcal{T}}$. Clearly, for a given sampling transition $P_{\mathcal{S}}^{\mathcal{A}}$, the relation $\mathcal{B} \perp\!\!\!\perp \mathcal{S} \mid \mathcal{T}$ depends on $\mu^{\mathcal{B}}$ only, whereas the relation $\mathcal{A} \perp\!\!\!\perp \mathcal{T} \mid \mathcal{B}$ does not depend on the prior probability (see 3.4.3). However, these two relations do not imply any restriction on $\Pi_{\mathcal{B}\vee\mathcal{T}}$, and are therefore properties of $\Pi^{\mathcal{B}\vee\mathcal{T}}$ only. In many instances there will be a $\mathcal{B}$ and $\mathcal{T}$ mutually sufficient for a family M of prior distributions. If this is the case then only $\mu_{\mathcal{B}}$ should be specified for the analysis of the transformation $\mu_{\mathcal{B}} \to \mu_{\mathcal{B}}^{\mathcal{S}}$. Furthermore, by Corollary 2.2.15, if $\mu_{\mathcal{B}}$ is modified into $\mu_{\mathcal{B}}'$, mutual sufficiency will not be affected provided $\mu_{\mathcal{B}}' \ll \mu_{\mathcal{B}}$.

Mutual sufficiency also has useful implications on $\mathcal{E}^{BvT}$. Suppose, for instance, that B is introduced to operate with exact restrictions, on the parameter space. Then, the condition $A \perp\!\!\!\perp T \mid B$ means that T is ancillary in $\mathcal{E}^B$, and $\mathcal{E}^B$ may therefore be admissibly reduced to $\mathcal{E}^{BvT}$. From an analytical point of view, recall that in the regular case, one has, in general,

$$(3.4.4) \qquad \Pi^{BvT} = \mu^{BvT} \otimes P^{AvT} = P^{BvT} \otimes \mu^{BvS}.$$

Now, $A \perp\!\!\!\perp T \mid B$ implies that:

$$(3.4.5) \qquad \mu^{BvT}(E) = \mu^B(E) \qquad \forall\, E \in A,$$

i.e., the transformation $\mu^B \to \mu^{BvS}$ may be obtained from P^{AvT}. If, furthermore, $B \perp\!\!\!\perp S \mid T$, i.e., B is $\mathcal{E}^T$-ancillary, then one has

$$(3.4.6) \qquad P^{BvT}(X) = P^T(X) \qquad \forall\, X \in S,$$

and in such a case (3.4.4) is now written as

$$(3.4.7) \qquad \Pi^{BvT} = \mu^B \otimes P^{AvT} = P^T \otimes \mu^{BvS}.$$

The concept of mutual sufficiency is weaker than the requirement that both B and T be $\mathcal{E}$-sufficient. Indeed, as a corollary of Theorem 2.2.10, one has the following link between these concepts:

3.4.3 Proposition. For $B \subset A$ and $T \subset S$, the following conditions are equivalent:
(i) $A \perp\!\!\!\perp S \mid B$ and $A \perp\!\!\!\perp S \mid T$;

(ii) $B \perp\!\!\!\perp S \mid T$ and $A \perp\!\!\!\perp T \mid B$ and $A \perp\!\!\!\perp S \mid B \vee T$. ∎

Note that if B and T were both $\mathcal{E}$-sufficient, one would have

$$(3.4.8) \qquad \Pi^{BvT} = \mu^B \otimes P^T = P^T \otimes \mu^B.$$

This is precisely the meaning of Proposition 3.4.3: if B and T are both $\mathcal{E}$-sufficient, this is equivalent to B and T be mutually sufficient, and to the conditional experiment $\mathcal{E}^{BvT}$ being totally noninformative, namely, $A \perp\!\!\!\perp S \mid B \vee T$. Thus the $\mathcal{E}_{AvS}$-sufficiency of both B and T would make $\mathcal{E}_{BvT}$ an admissible reduction of $\mathcal{E}_{AvS}$, whereas the mutual sufficiency of B and T make $\mathcal{E}_{BvT}$, and $\mathcal{E}^{BvT}$ a jointly admissible pair of reductions of $\mathcal{E}_{AvS}$.

3.4.2   Mutual Exogeneity

In the previous section, mutual sufficiency leads to the consideration of joint reduction by marginalization (or conditioning) on both spaces A and S. The concept of mutual exogeneity, to be defined, leads to the consideration of pairs of joint reductions involving marginalization on one space and conditioning on the other.

3.4.4 Definition. In a Bayesian experiment $\mathcal{E} = (A \times S, \mathcal{A} \vee \mathcal{S}, \Pi)$, $\mathcal{C} \subset \mathcal{A}$, and $\mathcal{T} \subset \mathcal{S}$, are *mutually exogenous* if:

(i) $\mathcal{C} \perp\!\!\!\perp \mathcal{T}$;

(ii) $\mathcal{A} \perp\!\!\!\perp \mathcal{S} \mid \mathcal{C} \vee \mathcal{T}$.   ∎

In other words, $\mathcal{C}$ and $\mathcal{T}$ are mutually exogenous if $\mathcal{C}$ and $\mathcal{T}$ are mutually ancillary (condition (i)), and $\mathcal{C}$ is an $\mathcal{E}^{\mathcal{T}}$-sufficient parameter, or, equivalently $\mathcal{T}$ is an $\mathcal{E}^{\mathcal{C}}$-sufficient statistic (condition (ii)).

Example. Consider $s = (y, z)'$ and $a = (b, c)'$. Suppose that

$$(s \mid a) \sim N_2\left[\begin{pmatrix} b+c \\ b-c \end{pmatrix} ; \begin{pmatrix} 2 & 1 \\ 1 & 1 \end{pmatrix}\right],$$
$$c \sim N(0,1), \quad (b \mid c) \sim N(c,1).$$

Then z and c are mutually ancillary since $(z \mid c) \sim N(0,2)$ and, since $(y \mid b, c, z) \sim N(z + 2c, 1)$, c is a sufficient parameter in $\mathcal{E}^{\mathcal{T}}$, where $\mathcal{T}$ is the σ-field generated by z.   ∎

Thus, the concept of mutual exogeneity means that the pair of reductions $\mathcal{E}^{\mathcal{C}}_{\mathcal{A} \vee \mathcal{T}}$ and $\mathcal{E}^{\mathcal{T}}_{\mathcal{C} \vee \mathcal{S}}$ is jointly admissible in the sense that the pairs $\mathcal{E}_{\mathcal{C} \vee \mathcal{T}}$ and $\mathcal{E}^{\mathcal{C} \vee \mathcal{T}}$ are both totally noninformative. It is therefore natural to analyze (some) implications of mutual exogeneity on $\mathcal{E}^{\mathcal{C}}_{\mathcal{A} \vee \mathcal{T}}$ and on $\mathcal{E}^{\mathcal{T}}_{\mathcal{C} \vee \mathcal{S}}$.

First consider $\mathcal{E}^{\mathcal{T}}_{\mathcal{C} \vee \mathcal{S}}$. Recall that, in the regular case (1.4.9), one has, in general:

(3.4.9) $\Pi^{\mathcal{T}}_{\mathcal{C} \vee \mathcal{S}} = \mu^{\mathcal{T}}_{\mathcal{C}} \otimes P^{\mathcal{C} \vee \mathcal{T}} = P^{\mathcal{T}} \otimes \mu^{\mathcal{S}}_{\mathcal{C}}.$

If $\mathcal{C}$ and $\mathcal{T}$ are mutually exogenous, condition (i) is equivalent to:

(3.4.10) $\mu^{\mathcal{T}}_{\mathcal{C}}(E) = \mu_{\mathcal{C}}(E) \qquad \forall E \in \mathcal{C},$

and condition (ii) is equivalent to:

$$(3.4.11) \qquad P^{\mathcal{A}\vee\mathcal{T}}(X) = P^{\mathcal{C}\vee\mathcal{T}}(X) \qquad \forall\, X \in \mathcal{S}.$$

Thus, we obtain:

$$(3.4.12) \qquad \Pi^{\mathcal{T}}_{\mathcal{C}\vee\mathcal{S}} = \mu_{\mathcal{C}} \otimes P^{\mathcal{A}\vee\mathcal{T}} = P^{\mathcal{T}} \otimes \mu^{\mathcal{S}}_{\mathcal{C}},$$

i.e., the transformation $\mu_{\mathcal{C}} \to \mu^{\mathcal{S}}_{\mathcal{C}}$ is obtained from $P^{\mathcal{A}\vee\mathcal{T}}$ without integrating this latter with respect to $\mu^{\mathcal{C}\vee\mathcal{T}}$. Note also that, in view of (3.4.11), condition (ii) does not depend on the prior probability. Therefore, $\mu^{\mathcal{C}}$ is used merely to check condition (i). In other words, if only $\mathcal{C}$ is of interest, and if $\mathcal{C}$ and $\mathcal{T}$ are mutually exogenous, the process generating $\mathcal{T}$ becomes irrelevant, and the data-generating process may be directly conditioned on $\mathcal{T}$ (without integration with respect to $\mu^{\mathcal{C}}$). This is precisely the meaning of the so-called "exogenous variables" used in econometrics models (see, e.g., Koopmans (1950), or Malinvaud (1978)). This Bayesian analogue insists that the exogeneity concept is relative to a parameter of interest.

Now consider $\mathcal{E}^{\mathcal{C}}_{\mathcal{A}\vee\mathcal{T}}$. In the regular case — see (1.4.12) — one has, in general,

$$(3.4.13) \qquad \Pi^{\mathcal{C}}_{\mathcal{A}\vee\mathcal{T}} = \mu^{\mathcal{C}} \otimes P^{\mathcal{A}}_{\mathcal{T}} = P^{\mathcal{C}}_{\mathcal{T}} \otimes \mu^{\mathcal{C}\vee\mathcal{T}}.$$

If $\mathcal{C}$ and $\mathcal{T}$ are mutually exogenous, condition (i) is equivalent to:

$$(3.4.14) \qquad P^{\mathcal{C}}_{\mathcal{T}}(X) = P_{\mathcal{T}}(X) \qquad \forall\, X \in \mathcal{T}$$

and condition (ii) is equivalent to:

$$(3.4.15) \qquad \mu^{\mathcal{C}\vee\mathcal{S}}(E) = \mu^{\mathcal{C}\vee\mathcal{T}}(E) \qquad \forall\, E \in \mathcal{A}.$$

Finally, under mutual exogeneity we obtain:

$$(3.4.16) \qquad \Pi^{\mathcal{C}}_{\mathcal{A}\vee\mathcal{T}} = \mu^{\mathcal{C}} \otimes P^{\mathcal{A}}_{\mathcal{T}} = P_{\mathcal{T}} \otimes \mu^{\mathcal{C}\vee\mathcal{S}},$$

i.e., the transformation $\mu^{\mathcal{C}} \to \mu^{\mathcal{C}\vee\mathcal{S}}$ is obtained from $P^{\mathcal{A}}_{\mathcal{T}}$ only (and not from $P^{\mathcal{A}}_{\mathcal{S}}$). In other words, if $\mathcal{C}$ is used to represent exact restrictions on the parameter space, and if $\mathcal{C}$ and $\mathcal{T}$ are mutually exogenous, the observations on the sample space may be admissibly marginalized on $\mathcal{T}$, and the prediction on $\mathcal{T}$ is not affected by the use of this exact restriction.

Note, finally, that for a given $\mathcal{C}$, by Corollary 2.2.15, mutual exogeneity (as mutual sufficiency) is robust with respect to a modification of $\mu_{\mathcal{C}}$ into $\mu'_{\mathcal{C}}$ if $\mu'_{\mathcal{C}} \ll \mu_{\mathcal{C}}$.

3.4.3   Bayesian Cut

The notion of a Bayesian cut combines the concepts of mutual sufficiency and of mutual exogeneity in such a way as to provide a complete decomposition of a Bayesian experiment. It is also the Bayesian analogue of the concept of a cut as introduced by Barndorff-Nielsen (1973, 1978).

3.4.5 Definition. Let $\mathcal{E} = (A \times S, A \vee S, \Pi)$ be a Bayesian experiment; let also $B \subset A$, $C \subset A$, and $T \subset S$ be sub-σ-fields. Then (B, C, T) operates a *Bayesian cut* on $\mathcal{E}$ if:

(i) $B \perp\!\!\!\perp C$;

(ii) $A \perp\!\!\!\perp T \mid B$;

(iii) $A \perp\!\!\!\perp S \mid C \vee T$.   ∎

The basic idea of a Bayesian cut is that the sub-σ-fields of A which make P_T^A and $P_S^{A \vee T}$ measurable are *a priori* independent; in other words, conditions (ii) and (iii) may actually be viewed as definitions of the subparameters B and C, thus leaving condition (i) as the proper characterization of a Bayesian cut. Unless otherwise indicated, any references to a cut in the sequel, should be interpreted as a Bayesian cut. The main properties of a cut are summarized in the following theorem:

3.4.6 Theorem. If (B, C, T) operates a cut on $\mathcal{E}$, then:

(i) $B \vee C$ is $\mathcal{E}$-sufficient, i.e., $A \perp\!\!\!\perp S \mid B \vee C$;

(ii) B and T are mutually sufficient, i.e., $A \perp\!\!\!\perp T \mid B$ and $B \perp\!\!\!\perp S \mid T$;

(iii) C and T are mutually exogenous, i.e., $C \perp\!\!\!\perp T$ and $A \perp\!\!\!\perp S \mid C \vee T$;

(iv) B and C are a posteriori independent, i.e., $B \perp\!\!\!\perp C \mid T$ and $B \perp\!\!\!\perp C \mid S$.

Proof. In Definition 3.4.5., (ii) implies $A \perp\!\!\!\perp T \mid B \vee C$ and (iii) implies $A \perp\!\!\!\perp S \mid B \vee C \vee T$ (Corollary 2.2.11 (ii)). These two conditional independence relations are equivalent to $A \perp\!\!\!\perp S \mid B \vee C$. This proves (i). Now $A \perp\!\!\!\perp T \mid B$ implies $C \perp\!\!\!\perp T \mid B$, and, along with $B \perp\!\!\!\perp C$, this is equivalent to $C \perp\!\!\!\perp (T \vee B)$ which implies $C \perp\!\!\!\perp T$, and $B \perp\!\!\!\perp C \mid T$. This proves (iii) and

the first part of (iv). This also implies, by Theorem 2.2.10, that $B \perp\!\!\!\perp C \mid T$.
Since $A \perp\!\!\!\perp S \mid C \vee T$ implies $B \perp\!\!\!\perp S \mid C \vee T$, applying Theorem 2.2.10 again
gives $B \perp\!\!\!\perp (S \vee C) \mid T$. This is equivalent to $B \perp\!\!\!\perp S \mid T$ and $B \perp\!\!\!\perp C \mid S$, and
this proves (ii) and the second part of (iv).   ■

In consequence, if B represents the parameter of interest, T is $\mathcal{E}_{B \vee S}$-
sufficient and the transformation $\mu_B \to \mu_B^T$ does not require integration of
the sampling probabilities P_T^A with respect to μ^B (or, equivalently, with
respect to μ_C because $B \perp\!\!\!\perp C$). Likewise, if C represents the parameter
of interest, T is $\mathcal{E}_{C \vee S}$-ancillary and the transformation $\mu_C \to \mu_C^S$ does not
require integration of the sampling probabilities $P^{A \vee T}$ with respect to $\mu^{C \vee T}$.
In any case, inferences on B and C are completely separated in the sense that
B and C are independent both *a priori* and *a posteriori*, and the revision
of the prior probabilities is based on two different aspects of the sampling
process — those characterized by P_T^B and by $P_S^{C \vee T}$ respectively.

The structure of a cut is not affected by a modification of the prior
probability μ into μ' if $\mu' \ll \mu$, and if this modification preserves the independence of B and C. This offers an easy characterization of the class of
prior distributions compatible with a cut.

Given the concepts and results developed thus far, we can now respond
to the following question: How far can we reduce a sampling process by conditioning on a statistic T without losing information about the parameters
of interest C? If all the parameters are of interest (i.e., $C = A$), conditioning
on T will be admissible if T is ancillary, in the sense of Chapter 2; this condition is identical in a sampling theory framework and in a Bayesian approach,
up to the μ-null sets. If, however, the parameters of interest define a strict
sub-σ-field of A, i.e., $C \subset A$ and $C \neq A$, the condition of mutual ancillarity
between C and T is the minimal condition which ensures the admissibility of
conditioning on T for inference on C. This condition, however, depends crucially on the specific form of μ^C, the prior distribution conditional on C, and
has no true equivalent for sampling theory. It is, in particular, difficult to
describe a modification of μ^C that preserves a property of mutual ancillarity.
Robustness with respect to the specification of μ^C and ease of computation
are improved if one furthermore requires C to be an $\mathcal{E}^T$-sufficient parameter.
This is precisely the condition of mutual exogeneity but once again there is
no fully equivalent concept in sampling theory. With the stronger condition

of a Bayesian cut, it is straightforward to characterize the modifications of μ that preserve the condition of a Bayesian cut: one needs merely to preserve the prior independence between C and B, where B is a parameter sufficient to describe the sampling process generating T; this requirement is very close to the condition of being variation-free (i.e., factorization of the parameter space) used in the sampling theory concept of a cut.

Finally, there exists some kind of converse to Theorem 3.4.6 which links the three types of joint reductions when A takes the form $B \vee C$. This gives the following proposition, the proof of which is a direct application of Theorem 2.2.10:

3.4.7 Proposition. Let $A = B \vee C$ and $T \subset S$.

(i)   If C and T are mutually exogenous and if $B \perp\!\!\!\perp C \mid T$, then (B, C, T) operates a cut on $\mathcal{E}$;

(ii)   If B and T are mutually sufficient and B and C are independent both *a priori* and *a posteriori* — i.e., $B \perp\!\!\!\perp C$ and $B \perp\!\!\!\perp C \mid S$ — then (B, C, T) operates a cut on $\mathcal{E}$.   ∎

3.4.4   Joint Reductions in a Dominated Experiment

Consider a dominated Bayesian experiment $\mathcal{E} = (A \times S, A \vee S, \Pi)$ and

$$(3.4.17) \qquad\qquad g_{A\vee S} = \frac{d\Pi}{d(\mu \otimes P)}.$$

In this situation mutual sufficiency, mutual exogeneity, and the Bayesian cut may be characterized by a factorization of density.

3.4.8 Theorem. In a dominated Bayesian experiment $\mathcal{E}$, $B \subset A$, and $T \subset S$ are mutually sufficient if and only if:

$$(3.4.18) \qquad\qquad g_B^S = g_T^A \qquad \text{a.s.}\,\mu \otimes P.$$

Proof. In view of Theorem 2.2.14, $B \perp\!\!\!\perp S \mid T$ and $A \perp\!\!\!\perp T \mid B$ are equivalent to $g_B^S = g_B^T$ and $g_T^A = g_T^B$ or, equivalently, $g_{B\vee S} = g_{B\vee T}$ and $g_{A\vee T} = g_{B\vee T}$.

This implies $g_{BVS} = g_{AVT}$, which is equivalent to (3.4.18). Now, if (3.4.18) holds, it suffices to show that $g_{BVT} = g_{AVT}$ to complete the proof. But since $A \perp\!\!\!\perp S; \mu \otimes P$, we have $(A \vee T) \perp\!\!\!\perp (B \vee S) \mid B \vee T; \mu \otimes P$ and so, using 2.2.1 (ii), we obtain

$$g_{BVT} = (\widetilde{B \vee T})(g_{BVS}) = (\widetilde{A \vee T})(g_{BVS}) = (\widetilde{A \vee T})(g_{AVT}) = g_{AVT}. \quad \blacksquare$$

3.4.9 Theorem. In a dominated Bayesian experiment $\mathcal{E}$, $\mathcal{C} \subset \mathcal{A}$, and $\mathcal{T} \subset \mathcal{S}$ are mutually exogenous if and only if:

$$(3.4.19) \qquad\qquad g_{\mathcal{C}}^{\mathcal{S}} = g_{\mathcal{S}}^{A \vee T} \qquad \text{a.s.} \mu \otimes \text{P}.$$

Proof. By Theorem 2.2.14, $\mathcal{C} \perp\!\!\!\perp \mathcal{T}$ and $\mathcal{A} \perp\!\!\!\perp \mathcal{S} \mid \mathcal{C} \vee \mathcal{T}$ are equivalent to $g_{CVT} = 1$, and $g_{AVS} = g_{AVT} \cdot g_{CVS}/g_{CVT}$. This implies $g_{AVS} = g_{AVT} \cdot g_{CVS}$ which is clearly equivalent to (3.4.19). To finish the proof, it suffices to show that (3.4.19) implies $g_{CVT} = 1$. Now, since $(A \vee T) \perp\!\!\!\perp (C \vee S) \mid C \vee T; \mu \otimes P$, using 2.2.1 (i), we obtain:

$$g_{CVT} = (\widetilde{C \vee T})(g_{AVS}) = (\widetilde{C \vee T})[g_{AVT} \cdot g_{CVS}]$$

$$= (\widetilde{C \vee T})(g_{AVT}) \cdot (\widetilde{C \vee T})(g_{CVS})$$

$$= g_{CVT} \cdot g_{CVT} = g_{CVT}^2.$$

Since $\widetilde{\mathcal{I}}(g_{CVT}) = 1$, it follows that g_{CVT} is the indicator function of a set of probability one, i.e., $g_{CVT} = 1$ a.s. $\mu \otimes P$. $\quad\blacksquare$

3.4.10 Theorem. Let $\mathcal{E}$ be a dominated Bayesian experiment and let $\mathcal{B} \subset \mathcal{A}, \mathcal{C} \subset \mathcal{A}$, and $\mathcal{T} \subset \mathcal{S}$. If $\mathcal{B} \perp\!\!\!\perp \mathcal{C}$, $(\mathcal{B}, \mathcal{C}, \mathcal{T})$ operates a cut if and only if:

$$(3.4.20) \qquad\qquad g_{\mathcal{A}}^{\mathcal{S}} = g_{\mathcal{T}}^{\mathcal{B}} \cdot g_{\mathcal{S}}^{C \vee T} \qquad \text{a.s.} \mu \otimes \text{P}.$$

Proof. By Theorem 2.2.14, $\mathcal{A} \perp\!\!\!\perp \mathcal{T} \mid \mathcal{B}$ and $\mathcal{A} \perp\!\!\!\perp \mathcal{S} \mid \mathcal{C} \vee \mathcal{T}$ are equivalent to $g_{AVT} = g_{BVT}$, and $g_{AVS} = g_{AVT} \cdot g_{CVS}/g_{CVT}$. This implies $g_{AVS} = g_{BVT} \cdot g_{CVS}/g_{CVT}$ which is equivalent to (3.4.20). To finish the proof, it suffices to show that $(3.4.20)$ implies $g_{A \vee T} = g_{B \vee T}$. Now, since $\mathcal{A} \perp\!\!\!\perp (\mathcal{C} \vee \mathcal{S}) \mid \mathcal{C} \vee \mathcal{T}; \mu \otimes P$, we obtain, by Theorem 2.2.1 (ii):

$$
\begin{aligned}
g_{A \vee T} &= (\widetilde{A \vee T})(g_{A \vee S}) = (\widetilde{A \vee T}) \left[\frac{g_{B \vee T} \cdot g_{C \vee S}}{g_{C \vee T}} \right] \\[2mm]
&= \frac{g_{B \vee T}}{g_{C \vee T}} (\widetilde{A \vee T})(g_{C \vee S}) \\[2mm]
&= \frac{g_{B \vee T}}{g_{C \vee T}} (\widetilde{C \vee T})(g_{C \vee S}) = \frac{g_{B \vee T} \cdot g_{C \vee T}}{g_{C \vee T}} = g_{B \vee T}. \qquad \blacksquare
\end{aligned}
$$

3.4.5   Joint Reductions in a Conditional Experiment

If the Bayesian experiment $\mathcal{E} = (A \times S, \mathcal{A} \vee \mathcal{S}, \Pi)$ has been reduced both by conditioning and by marginalization into $\mathcal{E}^{\mathcal{M}}_{\mathcal{B} \vee \mathcal{T}}$ where $\mathcal{B} \subset \mathcal{A}, \mathcal{T} \subset \mathcal{S}$ and $\mathcal{M} \subset \mathcal{A} \vee \mathcal{S}$, we may generalize the concepts of mutual sufficiency, mutual exogeneity and the Bayesian cut as done below. Note, however, that this section is rather sketchy: the extension of the main concepts and results from $\mathcal{E}_{\mathcal{A} \vee \mathcal{S}}$ to $\mathcal{E}^{\mathcal{M}}_{\mathcal{B} \vee \mathcal{T}}$ is formally straightforward, and so the extension of the comments and motivations is consequently rather trivial, and indeed superfluous. It should nevertheless be emphasized that, because most statistical models used in empirical work are conditional, the framework developed in this section is actually the most relevant in practice and, in particular, underlies the whole of Chapter 6.

3.4.11 Definition. In the experiment $\mathcal{E}^{\mathcal{M}}_{\mathcal{B} \vee \mathcal{T}}$, a parameter $\mathcal{K} \subset \mathcal{B} \vee \mathcal{M}$ and a statistic $\mathcal{N} \subset \mathcal{M} \vee \mathcal{T}$ are *mutually sufficient* if:

(i) $\mathcal{K} \perp\!\!\!\perp \mathcal{T} \mid \mathcal{M} \vee \mathcal{N}$

(ii) $\mathcal{B} \perp\!\!\!\perp \mathcal{N} \mid \mathcal{K} \vee \mathcal{M}$.

3.4.12 Definition. In the experiment $\mathcal{E}^{\mathcal{M}}_{\mathcal{B} \vee \mathcal{T}}$, a parameter $\mathcal{L} \subset \mathcal{B} \vee \mathcal{M}$ and a statistic $\mathcal{N} \subset \mathcal{M} \vee \mathcal{T}$ are *mutually exogenous* if:

(i) $\mathcal{L} \perp\!\!\!\perp \mathcal{N} \mid \mathcal{M}$

(ii) $\mathcal{B} \perp\!\!\!\perp \mathcal{T} \mid \mathcal{L} \vee \mathcal{M} \vee \mathcal{N}$.

3.4.13 Definition. In the experiment $\mathcal{E}^{\mathcal{M}}_{\mathcal{B} \vee \mathcal{T}}$, the parameters $\mathcal{K} \subset \mathcal{B} \vee \mathcal{M}$ and $\mathcal{L} \subset \mathcal{B} \vee \mathcal{M}$, and the statistic $\mathcal{N} \subset \mathcal{M} \vee \mathcal{T}$ operate a *Bayesian cut* if:

(i) $\mathcal{K} \perp\!\!\!\perp \mathcal{L} \mid \mathcal{M}$

(ii) $\mathcal{B} \perp\!\!\!\perp \mathcal{N} \mid \mathcal{K} \vee \mathcal{M}$

(iii) $\mathcal{B} \perp\!\!\!\perp \mathcal{T} \mid \mathcal{L} \vee \mathcal{M} \vee \mathcal{N}.$ ∎

The theorems linking these concepts are easily extended to the conditional case.

3.4.14 Proposition. In the experiment $\mathcal{E}^{\mathcal{M}}_{\mathcal{B}\vee\mathcal{T}}$, a parameter $\mathcal{K} \subset \mathcal{B} \vee \mathcal{M}$ and a statistic $\mathcal{N} \subset \mathcal{M} \vee \mathcal{T}$ are $\mathcal{E}^{\mathcal{M}}_{\mathcal{B}\vee\mathcal{T}}$-sufficient if and only if they are mutually sufficient and $\mathcal{B} \perp\!\!\!\perp \mathcal{T} \mid \mathcal{K} \vee \mathcal{M} \vee \mathcal{N}.$ ∎

3.4.15 Proposition. In the experiment $\mathcal{E}^{\mathcal{M}}_{\mathcal{B}\vee\mathcal{T}}$, if $\mathcal{K} \subset \mathcal{B} \vee \mathcal{M}$, $\mathcal{L} \subset \mathcal{B} \vee \mathcal{M}$, and $\mathcal{N} \subset \mathcal{M} \vee \mathcal{T}$ operate a cut, then:

(i) $\mathcal{K} \vee \mathcal{L}$ is $\mathcal{E}^{\mathcal{M}}_{\mathcal{B}\vee\mathcal{T}}$-sufficient, i.e., $\mathcal{B} \perp\!\!\!\perp \mathcal{T} \mid \mathcal{K} \vee \mathcal{L} \vee \mathcal{M};$

(ii) $\mathcal{K}$ and $\mathcal{T}$ are mutually sufficient;

(iii) $\mathcal{L}$ and $\mathcal{T}$ are mutually exogenous;

(iv) $\mathcal{K}$ and $\mathcal{L}$ are a posteriori independent, i.e., $\mathcal{K} \perp\!\!\!\perp \mathcal{L} \mid \mathcal{M} \vee \mathcal{T}.$ ∎

3.4.16 Proposition. Let $\mathcal{B}\vee\mathcal{M} = \mathcal{K}\vee\mathcal{L}\vee\mathcal{M}$ where $\mathcal{K} \subset \mathcal{B}\vee\mathcal{M}$, $\mathcal{L} \subset \mathcal{B}\vee\mathcal{M}$, and $\mathcal{N} \subset \mathcal{M} \vee \mathcal{T}.$

(i)   If $\mathcal{L}$ and $\mathcal{N}$ are mutually exogenous in $\mathcal{E}^{\mathcal{M}}_{\mathcal{B}\vee\mathcal{T}}$ and if $\mathcal{K} \perp\!\!\!\perp \mathcal{L} \mid \mathcal{M} \vee \mathcal{N}$, then $(\mathcal{K}, \mathcal{L}, \mathcal{N})$ operates a cut on $\mathcal{E}^{\mathcal{M}}_{\mathcal{B}\vee\mathcal{T}}.$

(ii) If $\mathcal{K}$ and $\mathcal{N}$ are mutually sufficient in $\mathcal{E}^{\mathcal{M}}_{\mathcal{B}\vee\mathcal{T}}$, and if $\mathcal{K}$ and $\mathcal{L}$ are independent, given $\mathcal{M}$, both *a priori* and *a posteriori*, i.e., $\mathcal{K} \perp\!\!\!\perp \mathcal{L} \mid \mathcal{M}$ and $\mathcal{K} \perp\!\!\!\perp \mathcal{L} \mid \mathcal{M} \vee \mathcal{T}$, then $(\mathcal{K}, \mathcal{L}, \mathcal{N})$ operates a cut on $\mathcal{E}^{\mathcal{M}}_{\mathcal{B}\vee\mathcal{T}}.$ ∎

The definitions receive the same interpretations as in Sections 3.4.1 to 3.4.3. Suppose $\mathcal{K}$ and $\mathcal{N}$ are mutually sufficient. If $\mathcal{K}$ is the parameter of interest, the inference on $\mathcal{K}$, i.e., the transformation $\mu_{\mathcal{K}}^{\mathcal{M}} \to \mu_{\mathcal{K}}^{\mathcal{M}\vee\mathcal{T}}$ only will depend upon $P_{\mathcal{N}}^{\mathcal{B}\vee\mathcal{M}}$. If $\mathcal{K}$ represents an exact restriction, the inference on $\mathcal{B}$, i.e., the transformation $\mu_{\mathcal{B}}^{\mathcal{K}\vee\mathcal{M}} \to \mu_{\mathcal{B}}^{\mathcal{K}\vee\mathcal{M}\vee\mathcal{T}}$ will depend upon $P_{\mathcal{T}}^{\mathcal{B}\vee\mathcal{M}\vee\mathcal{N}}$ only.

Suppose $\mathcal{L}$ and $\mathcal{N}$ are mutually exogenous. If $\mathcal{L}$ represents the parameter of interest, the inference on $\mathcal{L}$, i.e., the transformation $\mu_{\mathcal{L}}^{\mathcal{M}} \to \mu_{\mathcal{L}}^{\mathcal{M}\vee\mathcal{T}}$ may depend upon $P_{\mathcal{T}}^{\mathcal{B}\vee\mathcal{M}\vee\mathcal{N}}$ only. If $\mathcal{L}$ represents an exact restriction, the inference on $\mathcal{B}$, i.e., the transformation $\mu_{\mathcal{B}}^{\mathcal{L}\vee\mathcal{M}} \to \mu_{\mathcal{B}}^{\mathcal{L}\vee\mathcal{M}\vee\mathcal{T}}$ will depend upon $P_{\mathcal{N}}^{\mathcal{B}\vee\mathcal{M}}$ only.

Further, if $(\mathcal{K}, \mathcal{L}, \mathcal{N})$ operates a cut on $\mathcal{E}_{\mathcal{B}\vee\mathcal{T}}^{\mathcal{M}}$, the inference on $\mathcal{K}$ and $\mathcal{L}$ will be completely separated in the sense that $\mathcal{K}$ and $\mathcal{L}$ are independent both *a priori* and *a posteriori*. The inferences on $\mathcal{K}$ and $\mathcal{L}$, i.e., the transformations $\mu_{\mathcal{K}}^{\mathcal{M}} \to \mu_{\mathcal{K}}^{\mathcal{M}\vee\mathcal{T}}$ and $\mu_{\mathcal{L}}^{\mathcal{M}} \to \mu_{\mathcal{L}}^{\mathcal{M}\vee\mathcal{T}}$, will be carried out using two disjoint aspects of the sampling process — those characterized by $P_{\mathcal{N}}^{\mathcal{B}\vee\mathcal{M}}$ and by $P_{\mathcal{T}}^{\mathcal{B}\vee\mathcal{M}\vee\mathcal{N}}$, respectively.

Finally, if $\mathcal{E} = (A \times S, \mathcal{A} \vee \mathcal{S}, \Pi)$ is a dominated Bayesian experiment, Theorems 3.4.8, 3.4.9. and 3.4.10. may be extended to the case of a general reduced experiment $\mathcal{E}_{\mathcal{B}\vee\mathcal{T}}^{\mathcal{M}}$ where $\mathcal{M} = \mathcal{C}\vee\mathcal{U}$ with $\mathcal{C} \subset \mathcal{A}$ and $\mathcal{U} \subset \mathcal{S}$. Indeed, factorizations of the densities impose conditional independence relations on $\mu \otimes P$. In fact, $\mathcal{B} \perp\!\!\!\perp \mathcal{T} \mid \mathcal{M}$; $\mu \otimes P$ can be deduced from $\mathcal{A} \perp\!\!\!\perp \mathcal{S}$; $\mu \otimes P$ only if $\mathcal{M}$ has the form $\mathcal{C}\vee\mathcal{U}$. With this caveat, the extensions of Theorems 3.4.8, 3.4.9 and 3.4.10 are rather straightforward, and are summarized as follows:

3.4.17 Proposition. Let $\mathcal{E} = (A \times S, \mathcal{A} \vee \mathcal{S}, \Pi)$ be a dominated Bayesian experiment, and let $g = d\Pi/d(\mu \otimes P)$. Let $\mathcal{B} \vee \mathcal{C} \subset \mathcal{A}$, $\mathcal{T} \vee \mathcal{U} \subset \mathcal{S}$ and $\mathcal{M} = \mathcal{C} \vee \mathcal{U}$. Let $\mathcal{K}$ and $\mathcal{L}$ be $\mathcal{E}_{\mathcal{B}\vee\mathcal{T}}^{\mathcal{M}}$-parameters and let $\mathcal{N}$ be an $\mathcal{E}_{\mathcal{B}\vee\mathcal{T}}^{\mathcal{M}}$-statistic. Then

(i) $\mathcal{K}$ and $\mathcal{N}$ are mutually sufficient if and only if

$$(3.4.21) \qquad\qquad g_{\mathcal{K}}^{\mathcal{M}\vee\mathcal{T}} = \frac{g_{\mathcal{K}}^{\mathcal{M}}}{g_{\mathcal{N}}^{\mathcal{M}}} \cdot g_{\mathcal{N}}^{\mathcal{B}\vee\mathcal{M}};$$

(ii) $\mathcal{L}$ and $\mathcal{N}$ are mutually exogenous if and only if

(3.4.22)
$$g_{\mathcal{L}}^{\mathcal{MVT}} = \frac{g_{\mathcal{L}}^{\mathcal{M}}}{g_{\mathcal{T}}^{\mathcal{MVN}}} \cdot g_{\mathcal{T}}^{\mathcal{BVMVN}};$$

(iii) If $\mathcal{K} \perp\!\!\!\perp \mathcal{L} \mid \mathcal{M}$, $(\mathcal{L}, \mathcal{K}, \mathcal{N})$ operates a cut if and only if

(3.4.23)
$$g_{\mathcal{T}}^{\mathcal{BVM}} = g_{\mathcal{N}}^{\mathcal{KVM}} \cdot g_{\mathcal{T}}^{\mathcal{LVMVN}}. \qquad\blacksquare$$

3.4.6 Some Examples

In this section we illustrate the different types of joint reductions and some of the links between them.

Example 1. (The finite case). Let us consider bivariate parameters and observations: $a = (b, c)$ and $s = (t, u)$. Suppose now that each coordinate may assume two values only: $b \in \{b_1, b_2\}, c \in \{c_1, c_2\}, t \in \{t_1, t_2\}$ and $u \in \{u_1, u_2\}$. The Bayesian model characterized by the joint probability $\pi(a, s)$ is suitably defined by the following 15 numbers (assumed to be different from zero):

$$
\begin{aligned}
\mu(c_1) &= \mu_o \\
\mu(b_1 \mid c_i) &= \mu_i & i &= 1, 2 \\
p(t_1 \mid c_i, b_j) &= p_{ij} & i, j &= 1, 2 \\
p(u_1 \mid t_i, c_j, b_k) &= q_{ijk} & i, j, k &= 1, 2.
\end{aligned}
$$

Mutual ancillarity between c and t is equivalent to $p(t_1 \mid c_1) = p(t_1 \mid c_2)$, i.e.,

(R1)
$$\mu_1 p_{11} + (1 - \mu_1) p_{12} = \mu_2 p_{21} + (1 - \mu_2) p_{22}.$$

For mutual exogeneity, the supplementary condition $(a \perp\!\!\!\perp s \mid c, t)$ is equivalent to the four equalities:

(R2)
$$q_{ij1} = q_{ij2} \qquad\qquad i, j = 1, 2.$$

For a cut, it is necessary to verify condition (R2) plus the following three:

(R3)
$$\mu_1 = \mu_2 \qquad\qquad (\text{i.e., } b \perp\!\!\!\perp c)$$

(R4)
$$p_{1j} = p_{2j} \qquad j = 1, 2 \qquad (\text{i.e., } a \perp\!\!\!\perp t \mid b).$$

Even under prior independence ($R3$), it should be clear that mutual exogeneity ($R1 + R2$) does not imply a cut ($R2 + R3 + R4$) but it may be checked that the two conditions of posterior independence ($b \perp\!\!\!\perp c \mid t$), along with mutual ancillarity, are equivalent to ($R3+R4$): posterior independence ($b \perp\!\!\!\perp c \mid t$) plus mutual exogeneity do indeed imply a cut.   ∎

Example 2. Contrary to Example 1, we now exhibit a case where mutual exogeneity is equivalent to a cut. Let the prior distribution be: $a = (b, c) \sim N(0, \Sigma)$, and the sampling process generate $s = (t, u)$ is as follows: $(s \mid a) \sim N(a, V)$, where Σ and V are known 2×2 symmetric positive definite matrices. Clearly, $\pi(a, s)$ is given by:

$$\begin{pmatrix} a \\ s \end{pmatrix} \sim N\left(0, \begin{pmatrix} \Sigma & \Sigma \\ \Sigma & \Sigma + V \end{pmatrix} \right).$$

Here cov $(b, c) = 0$ (i.e., prior independence) is equivalent to $\mathrm{cov}(c, t) = 0$ (mutual ancillarity of c and t); note that it is also equivalent to the mutual ancillarity of b and u. Since b is clearly sufficient in the marginal experiment $\mathcal{E}_{a,t}$ (indeed $(t \mid a) \sim N(b, v_{11})$ implies $a \perp\!\!\!\perp t \mid b$) we conclude that, in this case, the mutual exogeneity of c and t (in this case , cov $(c, t) = 0$ and $\mathrm{cov}(b, u \mid c, t) = 0$) does imply that there is a cut. Note also that, after some manipulations, one may verify that $\mathrm{cov}(b, u \mid c, t) = 0$ is equivalent, in this example, to the independence of u and t in the sampling process (i.e., cov $(u, t \mid a) = 0$).   ∎

In the next two examples we construct regression models from a larger model involving also the data generating processs of the regressors. In Example 3, the cut does not impose any restriction on the parameter space but merely involves a simultaneous decomposition of the parameters and of the sampling process. In Example 4, the cut does place restrictions on the parameter space. This example also leads to the analysis of so-called incomplete simultaneous equation models.

Example 3. Let $s = (x_1, x_2, \ldots, x_n)$ where $x_i' = (y_i', z_i') \in \mathbb{R}^m$, $y_i \in \mathbb{R}^\ell$, $z_i \in \mathbb{R}^k$ $(k + \ell = m)$. Let Σ be a $m \times m$ symmetric positive definite matrix and $a = (\sigma_{ij} : 1 \le i \le j \le m)$. We suppose that $(x_i \mid a) \sim i.N(0, \Sigma)$. If we set $b_1 = \Sigma_{yz} \Sigma_{zz}^{-1}, b_2 = \Sigma_{yy} - \Sigma_{yz} \Sigma_{zz}^{-1} \Sigma_{zy}$ and $c = \Sigma_{zz}$, and if we suppose that a priori $\Sigma \sim I.W.(\nu_0, S_0)$, then $b = (b_1, b_2)$ and c are independent. Moreover if we set $y = (y_1, y_2, \ldots, y_n)'$ and $z = (z_1, z_2, \ldots, z_n)'$, then (b, c, z) operates a cut since $(z_i \mid a) \sim i.N(0, c)$ and $(y_i \mid a, z) \sim i.N(b_1 z_i, b_2)$. Note that the structure of the cut is preserved if the prior distribution is modified while maintaining the prior independence between b and c. Suppose now that b alone is of interest; the conditional model $(y \mid a, z)$ would be the only model of interest even if the marginal model generating z was modified in such a way that its new parameters, say c_*, were independent of b. In this example, the cut places no restriction on the parameter Σ. ∎

Example 4. In the previous example, the x_i's are i.i.d. conditionally on a. Let us now consider the linear model $(x_i \mid a) \sim i.N(\xi_i, \Sigma)$ where Σ is a $m \times m$ symmetric positive definitive matrix, $\xi_i \in \mathbb{R}^m$, $1 \le i \le n$, and $a = (\Sigma, \theta, \xi_i : 1 \le i \le n)$. Suppose that $A_\theta \xi_i = 0 \quad \forall\, 1 \le i \le n$ where A_θ is a $p \times m$ matrix with rank p which is identified by θ (i.e., known if θ is known). The ξ_i's are generally called incidental parameters. As in the previous example, x_i is partitioned into y_i and z_i, as is Σ. We also partition $\xi = (\xi_1, \xi_2, \ldots, \xi_n)'$ into (ξ_y, ξ_z), and A_θ into (B_θ, C_θ). We now define $\eta_i = E(y_i \mid a, z_i) \in \mathbb{R}^\ell$ and $\eta = (\eta_1, \eta_2, \ldots, \eta_n)'$. Even if we decompose the parameter $a = (\Sigma, \theta, \xi)$ into $b = (b_2, \theta, \eta)$ where $b_2 = \Sigma_{yy} - \Sigma_{yz} \Sigma_{zz}^{-1} \Sigma_{zy}$, $c = (\Sigma_{zz}, \xi_z)$, (b, c, z) operates a cut only under an (exogeneity) assumption placed on the parameter a, viz., $B_\theta b_1 + C_\theta = 0$ where $b_1 = \Sigma_{yz} \Sigma_{zz}^{-1}$ or, equivalently, if (conditionally on a), z_i and $A_\theta x_i$ are independent. Under this additional assumption, the conditional model is written as

$$(y_i \mid a, z_i) \sim i.N(\eta_i, b_2)$$

$$B_\theta \eta_i + C_\theta z_i = 0 \text{ a.s.}$$

If B_θ is square $(p = \ell)$, this model is equivalent to a simultaneous equations model $B_\theta y_i + C_\theta z_i = u_i$ and $(u_i \mid a) \sim i.N(0, b_2)$. In this case, the incidental parameters η_i are eliminated by the identity $\eta_i = -B_\theta^{-1} C_\theta z_i$; otherwise (i.e., $p < m$), this model corresponds to a so-called "incomplete model". Note that the incompleteness of the model involves the presence of $(n - p)$ incidental parameters (for more details, see Florens, Mouchart and Richard (1979), Griliches (1974), Lindley and El-Sayad (1968) Neyman and Scott (1948, 1951)). ∎

3.5   Comparison of Experiments

Consider two Bayesian experiments:

$$\mathcal{E}_i = (A \times S, \mathcal{B}_i \vee \mathcal{T}_i, \Pi^i) \quad i = 1, 2.$$

Note that assuming a common product space $A \times S$ is not restrictive once we allow for different σ-fields $\mathcal{B}_i$ and $\mathcal{T}_i$. For instance, if the sample spaces of those two experiments has "nothing in common", one could formally embed their respective sample spaces into a common space S such as $S = \mathcal{T}_1 \times \mathcal{T}_2$, and accordingly embed their natural σ-fields of observations (i.e., of observable events) into that common S.

If however, we want to "compare" $\mathcal{E}_1$ and $\mathcal{E}_2$, they should share in common either the parameters (i.e., $\mathcal{B}_1 = \mathcal{B}_2$), or the observations (i.e., $\mathcal{T}_1 = \mathcal{T}_2$). These two cases correspond to different problems. Thus, when $\mathcal{B}_1 = \mathcal{B}_2 = \mathcal{A}$, we handle a comparison on the sample space and have in mind comparisons of experiments in the sense of Blackwell (1951), LeCam (1966) (see also Sacksteder (1967) and Heyer (1982)), i.e., the choice between two experimental designs in order to infer on a given parameter and, when $\mathcal{T}_1 = \mathcal{T}_2 = \mathcal{S}$, we handle comparisons on the parameter space and have in mind choices of models to "explain" a given observation (and also predict future observations to be generated by the same "unknown" model). In both cases we are interested in the question of whether experiment $\mathcal{E}_2$ is, in some sense, redundant with respect to experiment $\mathcal{E}_1$. In order to stress the difference of contexts, we shall say that $\mathcal{E}_1$ is "sufficient" for $\mathcal{E}_2$ when $\mathcal{B}_1 = \mathcal{B}_2$, and that $\mathcal{E}_1$ "encompasses" $\mathcal{E}_2$ when $\mathcal{T}_1 = \mathcal{T}_2$.

The basic idea for comparing $\mathcal{E}_1$ and $\mathcal{E}_2$ is to look for an embedding of $\mathcal{E}_1$ into an extended experiment $\mathcal{E}$ in such a way that its characteristic σ-field ($\mathcal{T}_1$ when comparing on the sample space or $\mathcal{B}_1$ when comparing on the parameter space) is sufficient in the extended experiment $\mathcal{E}$, and that $\mathcal{E}_2$ is partially embedded into $\mathcal{E}$. This will lead to three levels of comparison according to the degree of embedding of $\mathcal{E}_2$ into $\mathcal{E}$ and to be qualified as weak, coherent, and strong.

3.5.1   Comparison on the Sample Space: Sufficiency

In Section 3.2.3, we interpreted a condition of the type $\mathcal{A} \perp\!\!\!\perp \mathcal{T}_2 \mid \mathcal{T}_1$ when $\mathcal{T}_1 \subset \mathcal{T}_2 \subset \mathcal{S}$ as "$\mathcal{T}_1$ is $\mathcal{E}_{\mathcal{A} \vee \mathcal{T}_2}$-sufficient". Noting that, in general, this condition is equivalent to $\mathcal{A} \perp\!\!\!\perp (T_1 \vee T_2) \mid T_1$, we may interpret $\mathcal{A} \perp\!\!\!\perp T_2 \mid T_1$ as "T_1 is $\mathcal{E}_{\mathcal{A} \vee T_1 \vee T_2}$-sufficient" even if T_1 is not included in T_2.

Let us now reverse that analysis and consider two experiments:

$$(3.5.1) \qquad \mathcal{E}_i = (A \times S,\ \mathcal{A} \vee T_i,\ \Pi^i) \quad i = 1, 2.$$

Here, the important feature is that both experiments share a common parameter space $(A, \mathcal{A})$; thus, in the regular case, Π^i has the form:

$$(3.5.2) \qquad \Pi^i = \mu_{\mathcal{A}}^i \otimes P_{T_i}^{i,\mathcal{A}}.$$

We want to make precise the idea that $\mathcal{E}_1$ is sufficient for $\mathcal{E}_2$ in the sense that the information on $\mathcal{A}$ contained in $\mathcal{E}_2$ is already contained in $\mathcal{E}_1$.

3.5.1 Definition. $\mathcal{E}_1$ is *weakly* (respectively, *coherently*, *strongly*) *sufficient* for $\mathcal{E}_2$ if there exists a probability Π on $\mathcal{A} \vee T_1 \vee T_2$ such that:

(i) $\Pi_{\mathcal{A} \vee T_1} = \Pi^1$.

(ii) $\mathcal{A} \perp\!\!\!\perp T_2 \mid T_1; \Pi$.

(iii-w) *(weakly)*: $\forall\, t_2 \in [T_2]^+$ $\exists\, b \in [\mathcal{A}]^+$ such that $\mathcal{A}^2 t_2 = b$ and $\mathcal{A} t_2 = b$.

(iii-c) *(coherently)*: weakly along with $P_{T_2} = P_{T_2}^2$.

(iii-s) *(strongly)*: $\Pi_{\mathcal{A} \vee T_2} = \Pi^2$. $\qquad\qquad\blacksquare$

Here $\mathcal{A}^2 t_2$ denotes the conditional expectation of t_2 given $\mathcal{A}$ computed from the probability Π^2; hence $\mathcal{A}^2 t_2$ is a class of random variables defined up to μ^2-equivalence (where μ^2 is the restriction of Π^2 to $\mathcal{A}$), while $\mathcal{A} t_2$ being constructed from Π is defined up to a μ^1-equivalence (because of condition (i)). Finally, condition (iii-w) means that there exists a common version, in $\mathcal{E}$ and in $\mathcal{E}^2$, of the conditional expectation of any T_2-measurable (and positive) function given $\mathcal{A}$.

Thus the Bayesian experiment $\mathcal{E}_1$ is sufficient for $\mathcal{E}_2$ if it can be embedded (condition (i)) in an extended experiment $\mathcal{E}$ on $\mathcal{A} \vee T_1 \vee T_2$ in such a way that its observations T_1 are $\mathcal{E}$-sufficient (condition (ii)), and that $\mathcal{E}_2$ also is (more or less) embedded in $\mathcal{E}$ (conditions (iii)). In a sense, conditions (i) and

(ii) are somewhat technical in nature (i.e., apart from pathological cases, they can always be met) and the very idea of sufficiency among experiments lies in the juxtaposition of condition (ii) and one of the conditions (iii).

More specifically, we know, from Theorem 2.2.1-(iii), that condition (ii) above implies that $\mathcal{A}t_2 = \mathcal{A}[T_1 t_2] = \mathcal{A}^1[T_1 t_2] \ \forall \ t_2 \in [T_2]^+$. Thus in the regular case, and when S factorizes into $S = T_1 \times T_2$ in such a way that both T_1 and T_2 are cylinders σ-fields based respectively on T_1 and T_2, the embedding experiment $\mathcal{E}$ satisfying conditions (i) and (ii) will typically be build from a transition defined on $T_1 \times T_2$ and denoted $Q_{T_2}^{T_1}$ as follows:

$$(3.5.3) \qquad\qquad \Pi = \mu_{\mathcal{A}}^1 \otimes P_{T_1}^{1,\mathcal{A}} \otimes Q_{T_2}^{T_1}$$

and the problem of sufficiency becomes the question of whether there exists a transition $Q_{T_2}^{T_1}$ satisfying one of the conditions (iii). Note also that when Π has the structure (3.5.3), the sampling probabilities of $\mathcal{E}$ on T_2 may be represented in the form:

$$(3.5.4) \qquad\qquad P_{T_2}^{\mathcal{A}}(X_2) = \int Q_{T_2}^{T_1}(X_2) \ dP_{T_1}^{1,\mathcal{A}} \quad X_2 \in T_2.$$

Let us now consider the three levels (conditions (iii-w,c, and s)) for sufficiency. Weak sufficiency — condition (iii-w) — means that the embedding probability Π should be such that there exists a common version $\mathcal{A}t_2$ and $\mathcal{A}^2 t_2$ (for any $t_2 \in [T_2]^+$) and therefore, in the regular case (3.5.3), a common version of $P_{T_2}^{\mathcal{A}}$ — as defined in (3.5.4) and of $P_{T_2}^{2,\mathcal{A}}$. In particular, when the two prior probabilities $\mu_{\mathcal{A}}^1$ and $\mu_{\mathcal{A}}^2$ are equivalent, all the versions of the conditional expectations $\mathcal{A}t_2$ and $\mathcal{A}^2 t_2$ are both a.s.$\mu_{\mathcal{A}}^1$ and a.s.$\mu_{\mathcal{A}}^2$ equivalent under weak sufficiency, and therefore, in the regular case, the sampling probabilities of $\mathcal{E}$ and $\mathcal{E}^2$ coincide on T_2:

$$(3.5.5) \ P_{T_2}^{2,\mathcal{A}}(X_2) = \int Q_{T_2}^{T_1}(X_2) \ dP_{T_1}^{1,\mathcal{A}} \quad \text{a.s.}\mu_{\mathcal{A}}^i \quad (i = 1, 2) \quad \forall \ X_2 \in T_2.$$

This interpretation of weak sufficiency shows that it can also be viewed in general as a sampling sufficiency in the same spirit as Theorem 2.3.15. More specifically we have:

3.5.2 Theorem. Under conditions (i) and (ii) of Definition 3.5.1, condition (iii-w) is equivalent to the sufficiency of $\mathcal{A}$ in the classical experiment $(\mathcal{A} \times S, \mathcal{A} \vee T_2, \{\Pi_{\mathcal{A} \vee T_2}, \Pi^2\})$.

Proof. The classical sufficiency of $\mathcal{A}$ means that for any $m \in [\mathcal{A} \vee T_2]^+$, there exists a common version of $\mathcal{A}m$ and of $\mathcal{A}^2 m$. By a monotone class argument, this is equivalent to require the existence of a common version for any m of the form $m = a \cdot t_2$ with $a \in [\mathcal{A}]^+$ and $t_2 \in [T_2]^+$ but this is equivalent to condition (iii-w) because $\mathcal{A}(a \cdot t_2) = a \, \mathcal{A}t_2$, and $\mathcal{A}^2(a \cdot t_2) = a \cdot \mathcal{A}^2 t_2$. ∎

Finally weak sufficiency may also be viewed, under a very weak regularity condition — condition (o) in next theorem — as a Bayesian sufficiency in an extended Bayesian experiment.

3.5.3 Theorem. If there exists a probability Π^{*2} on $\mathcal{A} \vee T_1 \vee T_2$ such that:

(o) $\Pi^{*2}_{\mathcal{A} \vee T_2} = \Pi^2$

and if $\mathcal{E}_1$ is weakly sufficient for $\mathcal{E}_2$, then there exists a probability Π^* on $\mathcal{J} \vee \mathcal{A} \vee T_1 \vee T_2$ where $\mathcal{J}$ is the σ-field of all subsets of $J = \{1, 2\}$ identified, as usual, with $\mathcal{J} \times A \times S$, such that:

(∗i) $\Pi^*(j) > 0 \quad j = 1, 2.$

(∗ii) $\Pi^{*j}_{\mathcal{A} \vee T_j} = \Pi^j \quad j = 1, 2.$

(∗iii) $\mathcal{A} \perp\!\!\!\perp T_2 \mid T_1; \Pi^{*1}.$

(∗iv) $\mathcal{J} \perp\!\!\!\perp T_2 \mid \mathcal{A}; \Pi^*.$

Conversely, the existence of such a probability Π^* implies that $\mathcal{E}_1$ is weakly sufficient for $\mathcal{E}_2$.

Proof. We first prove that, under (o), the weak sufficiency of $\mathcal{E}_1$ for $\mathcal{E}_2$ implies the existence of a measure Π^* satisfying the four conditions (∗): For (∗i), take any number α such that $0 < \alpha < 1$ and define: $\Pi^*(1) = \alpha$ and $\Pi^*(2) = 1 - \alpha$. Next, for $j = 1$, take $\Pi^{*1} = \Pi$ where Π is the probability involved in Definition 3.5.1: this guarantees (∗ii) for $j = 1$ and (∗iii); and for $j = 2$ take the Π^{*2} implied in condition (o): this guarantees (∗ii) for $j = 2$. Finally (∗iv) is a direct implication of Theorem 3.5.2 because the finite character of J entails both Theorems 2.3.12 and 2.3.14. It remains to remark that Theorem 3.5.2 also renders trivial the fact that the existence of a probability Π^* on $\mathcal{J} \vee \mathcal{A} \vee \mathcal{T}_1 \vee \mathcal{T}_2$ satisfying the four conditions $(*)$ implies the weak sufficiency of $\mathcal{E}_1$ for $\mathcal{E}_2$.   ∎

In this theorem, condition (o) is a (very weak) regularity condition to be discussed below in a "technical digression", the random element $j \in J = \{1,2\}$ should be viewed as a parameter labelling the two experiments $\mathcal{E}_i$ and $\Pi^*(j)$ as an (implicit) prior probability on these experiments. Now Theorem 3.5.3 says that weak sufficiency among experiments may also be viewed as the $\mathcal{E}^*$-sufficiency of the parameters common to the two experiments (i.e., $\mathcal{A}$) when we only observe the data available from the second experiment (i.e., $\mathcal{T}_2$), where $\mathcal{E}^*$ is a Bayesian experiment extending to the labels of the given experiments $\mathcal{E}_1$ and $\mathcal{E}_2$, the extended experiment underlying Definition 3.5.1.

Let us now turn to the coherent sufficiency. Among the extensions of Π^1 on $\mathcal{A} \vee \mathcal{T}_1$ into a probability Π on $\mathcal{A} \vee \mathcal{T}_1 \vee \mathcal{T}_2$, it may seem desirable to constraint Π to have the same predictive probability on $\mathcal{T}_2$ as Π^2: this is condition (iii-c). It is particularly desirable, in a subjective approach to the concept of probability, when the Bayesian experiments $\mathcal{E}^1$ and $\mathcal{E}^2$ come from a same person (rather than from two "experts") willing to be coherent in his probability assignment. Note, however, that the condition $P_{\mathcal{T}_2} = P_{\mathcal{T}_2}^2$, along with conditions (i) and (ii) does not imply condition (iii-w).

Clearly strong sufficiency (condition iii-s) implies that both experiments $\mathcal{E}^1$ and $\mathcal{E}^2$ share a common prior probability $\mu = \mu^1 = \mu^2$ on $\mathcal{A}$ and that $\mathcal{E}^1$ can be extended into an experiment $\mathcal{E}$ in $\mathcal{A} \vee \mathcal{T}_1 \vee \mathcal{T}_2$ that has the same sampling probabilities and the same predictive probability on $\mathcal{T}_2$ as $\mathcal{E}^2$; in other words $\mathcal{E}^1$ and $\mathcal{E}^2$ may be viewed as two marginal reductions of an experiment $\mathcal{E}$ for which $\mathcal{T}_1$ is sufficient. Clearly, this implies coherent and weak sufficiency of $\mathcal{E}^1$ for $\mathcal{E}^2$.

Technical Digression.   The above discussion involves the difficulty of extending a probability on a given σ-field into a larger σ-field. More specifically, let $(M, \mathcal{M})$ be a measurable space, and $\mathcal{M}_i$ $(i = 1, 2)$ be two sub-σ-fields of $\mathcal{M}$. A first question is: Under which condition a probability on $(M, \mathcal{M}_1)$ can be extended to a probability on $(M, \mathcal{M}_1 \vee \mathcal{M}_2)$? This is clearly not always possible: for instance, a uniform distribution on the Borel sets of the unit interval cannot be extended to a probability on all the subsets of the unit interval; but we have mentioned in (3.5.3) that if M does factorize in such a way that the σ-fields $\mathcal{M}_i$ $(i = 1, 2)$ are cylinders based on the factors, this will always be possible by Theorem 0.3.10.

Now, strong sufficiency raises a more complex question: given two probabilities P_i on $(M, \mathcal{M}_i)$ ($i = 1, 2$), does it exist a probability P on $(M, \mathcal{M}_1 \vee \mathcal{M}_2)$ admitting P_i as marginals (i.e., such that $P_{\mathcal{M}_i} = P_i$, $i = 1, 2$)? This problem has been addressed in Strassen (1965), and Diaconis and Zabell(1982) have shown that a necessary condition is:

$$(3.5.6) \quad \forall\, X_i \in \mathcal{M}_i\ (i = 1, 2)\ :\ X_1 \cap X_2 = \phi \Rightarrow P_1(X_1) + P_2(X_2) \leq 1.$$

Note that this condition is trivially met when $\mathcal{M}_i$ are cylinders σ-fields based on factors of M because in such a case $X_1 \cap X_2 = \phi$ implies that $X_1 = \phi$ or $X_2 = \phi$. ∎

Let us compare Definition 3.5.1 with previous works of Blackwell and LeCam. Blackwell (1951)'s original definition may be translated as follows:

Definition A. Consider two statistical experiments:

$$\mathcal{E}_i = \{(T_i, \mathcal{T}_i), P^{i,a} : a \in A\} \quad i = 1, 2.$$

$\mathcal{E}_1$ is *sufficient* for $\mathcal{E}_2$ if and only if there exists a transition probability $\Lambda : (T_1, \mathcal{T}_1) \longrightarrow\!\!\!< (T_2, \mathcal{T}_2)$ such that

$$(3.5.7) \qquad P^{2,a}(X) = \int_{T_1} \Lambda^t(X)\ \ P^{1,a}(dt) \quad \forall\, X \in \mathcal{T}_2.$$

∎

This definition may be generalized (see LeCam (1966)) as follows:

Definition B. Consider two statistical experiments:

$$\mathcal{E}_i = \{(S, \mathcal{T}_i), P^{i,a} : a \in A\} \quad i = 1, 2.$$

$\mathcal{E}_1$ is *sufficient* for $\mathcal{E}_2$ if and only if there exists a family, $\{P^a : a \in A\}$ of probabilities on $\mathcal{T}_1 \vee \mathcal{T}_2$ such that $P^{i,a}$ are the restrictions of P^a on $\mathcal{T}_i$, $i = 1, 2$ and $\mathcal{T}_1$ is sufficient for the experiment $\mathcal{E} = \{S, (\mathcal{T}_1 \vee \mathcal{T}_2), P^a : a \in A\}$. ∎

Repeating the arguments in Section 2.3.7, and taking a σ-field $\mathcal{A}$ on A that makes $P^a(X)$ $\mathcal{A}$-measurable for any $X \in T_1 \vee T_2$, we have the following relationships:

(i) If $\mathcal{E}_1$ is sufficient for $\mathcal{E}_2$ in the sense of Definition B, then for any prior probabilities μ_1 and μ_2 on $(A, \mathcal{A})$, $\mathcal{E}_1$ is weakly sufficient for $\mathcal{E}_2$ in the sense of Definition 3.5.1. Moreover, if $\mu_1 = \mu_2$, $\mathcal{E}_1$ is strongly sufficient for $\mathcal{E}_2$.

(ii) If $\mathcal{E}_1$ is strongly sufficient for $\mathcal{E}_2$ in the sense of Definition 3.5.1, if the involved probability Π admits a regular conditional probability on $T_1 \vee T_2$ given $\mathcal{A}$, (see Theorem 0.3.18) and if μ is regular, then $\mathcal{E}_1$ is sufficient for $\mathcal{E}_2$ in the sense of Definition B.

Let us illustrate these concepts by an example.

Example. Let us consider $A = \mathbb{R}_0^+$, $S = \mathbb{R}^2$, $(t_1 \mid a) \sim N(0, a)$ and $(t_2 \mid a) \sim N(0, a + 1)$. To show that for any prior probability on a, $\mathcal{E}_1$ is sufficient for $\mathcal{E}_2$ in the sense of Definition B, one may remark that the family of probabilities

$$\left(\begin{pmatrix} t_1 \\ t_2 \end{pmatrix} \mid a \right) \sim N_2 \left(0, \begin{bmatrix} a & a \\ a & a+1 \end{bmatrix} \right),$$

clearly extends the two given families of $(t_1 \mid a)$ and $(t_2 \mid a)$, that its conditional probability of $(t_2 \mid t_1, a)$ does not depend on a (namely, $(t_2 \mid t_1, a) \sim N(t_1, 1)$) and, therefore, the conditions of Definition B are verified.   ∎

Note, finally, that Definition 3.5.1 is easily adapted to deal with the presence of nuisance parameters: it suffices to consider a sub-σ-field $\mathcal{B} \subset \mathcal{A}$ instead of $\mathcal{A}$. Similar problems as in Section 3.2.4 may be handled, but, in general, a comparison of experiments with nuisance parameter in a sampling theory framework raises conceptual difficulties (see, e.g., Goel and De Groot (1979)).

3.5.2 Comparison on the Parameter Space: Encompassing

A dual analysis of Section 3.5.1 can be made for the comparison of two Bayesian experiments characterized by a same sample space but different parameter spaces. Consider specifically:

$$(3.5.8) \qquad \mathcal{E}_i = (A \times S, \, \mathcal{B}_i \vee S, \, \Pi^i)$$

where, in the regular case, Π^i has the form

$$(3.5.9) \qquad \Pi^i = \mu^i_{\mathcal{B}_i} \otimes P^{i,\mathcal{B}_i}_S.$$

3.5.4 Definition. $\mathcal{E}_1$ *encompasses weakly* (respectively, *coherently, strongly*) $\mathcal{E}_2$ if there exists a probability Π on $\mathcal{B}_1 \vee \mathcal{B}_2 \vee S$ such that

(i) $\Pi_{\mathcal{B}_1 \vee S} = \Pi^1$;

(ii) $\mathcal{B}_2 \perp\!\!\!\perp S \mid \mathcal{B}_1, \Pi$;

(iii-w) (*weakly*) $\forall\, b_2 \in [\mathcal{B}_2]^+ \; \exists\, t \in [S]^+$ such that $Sb_2 = t$ and $S^2 b_2 = t$;

(iii-c) (*coherently*) weakly along with: $\mu_{\mathcal{B}_2} = \mu^2_{\mathcal{B}_2}$;

(iii-s) (*strongly*) $\Pi_{\mathcal{B}_2 \vee S} = \Pi^2$. ∎

Under the same notation as in Definition 3.5.1, condition (iii-w) means that for any (positive) $\mathcal{B}_2$-measurable function (of the parameter), the classes of their posterior expectations admit a common version under Π and under Π^2, remembering that Sb_2 is defined up to a P^1_S-a.s.-equivalence while $S^2 b_2$ is defined up to a P^2_S-a.s.-equivalence.

We shall not repeat dual versions of Theorems 3.5.2 and 3.5.3, but, rather concentrate the attention of reinterpreting the analysis of Section 3.5.1 in the regular case (i.e., all conditional probabilities admitting a regular version) with a parameter space factorizable into $A = B_1 \times B_2$ in such a way that the $\mathcal{B}_i$'s are cylinders σ-fields based on the factors B_i. In this context, the embedding experiment $\mathcal{E}$ satisfying conditions (i) and (ii) will typically be built from a transition defined on $B_1 \times B_2$, and denoted $\nu^{\mathcal{B}_1}_{\mathcal{B}_2}$ as follows:

$$(3.5.10) \qquad \Pi = P^1_S \otimes \mu^{1,S}_{\mathcal{B}_1} \otimes \nu^{\mathcal{B}_1}_{\mathcal{B}_2}$$

while the experiments $\mathcal{E}_i$ are also representables as:

$$(3.5.11) \qquad \Pi^i = P^i_S \otimes \mu^{i,S}_{\mathcal{B}_i} \qquad i = 1, 2.$$

Note that the posterior probabilities in the extended experiment have the form:

$$(3.5.12) \qquad \mu_{\mathcal{B}_2}^{\mathcal{S}}(E_2) = \int \nu_{\mathcal{B}_2}^{\mathcal{B}_1}(E_2) \; d\mu_{\mathcal{B}_1}^{1,\mathcal{S}} \qquad \forall \; E_2 \in \mathcal{B}_2,$$

since by (ii) in Definition 3.5.4, we have $\mathcal{S}b_2 = \mathcal{S}(\mathcal{B}_1 b_2) = \mathcal{S}^1(\mathcal{B}_1 b_2)$ $\forall \; b_2 \in [\mathcal{B}_2]^+$.

The problem of encompassing is the question of whether there exists a transition $\nu_{\mathcal{B}_2}^{\mathcal{B}_1}$ satisfying one of the conditions (iii). Essentially, (weak) encompassing is the existence of a common version of $\mu_{\mathcal{B}_2}^{2,\mathcal{S}}$ and $\mu_{\mathcal{B}_2}^{\mathcal{S}}$: this is the Bayesian formulation of the idea that any inference possible in $\mathcal{E}_2$ (i.e., $\mu_{\mathcal{B}_2}^{2,\mathcal{S}}$) may be reconstructed from $\mathcal{E}_1$ without retrieving the sample. This idea may be viewed as a statistical approach to the principle that, in the scientific world, a new theory ($\mathcal{E}_1$) should, at least, be able to explain the phenomena explained by an older theory ($\mathcal{E}_2$). This concern has been repeatedly raised in D. Hendry's econometric work (Hendry and Anderson (1977) and Davidson, Hendry, Sbra and Yeo (1978) and formulated in Hendry and Richard (1982, 1983, 1987). The Bayesian formulation adopted here follows that of Florens, Mouchart and Scotto (1983), and Florens and Mouchart (1985).

The transition $\nu_{\mathcal{B}_2}^{\mathcal{B}_1}$ is the key to reinterpret the inferences of the experiment $\mathcal{E}_2$ from the experiment $\mathcal{E}_1$ and should therefore be viewed as a Bayesian analogue to the pseudo-true value which is the limit (a.s. or in probability) or the expectation of the estimator of a parameter of one model under another model. This idea has a long history in statistics (Cox (1961,1962), Berk (1966, 1970), Huber (1967), Akaike (1974)); and in econometrics Pesaran (1974), Hausman (1978), Sawa (1978), Hausman and Taylor (1981), White (1982), Gouriéroux, Monfort and Trognon (1983,1984), Mizon and Richard (1986)).indexMizon and Richard Note that the classical pseudo-true value is a function of the parameter (of the "true" model) while its Bayesian analogue is a transition (from the parameter of the encompassing experiment to that of the encompassed experiment).

Coherent encompassing means that in the regular case as in (3.5.9), one requires:

$$(3.5.13) \qquad \mu_{\mathcal{B}_2}^{2}(E_2) = \int \nu_{\mathcal{B}_2}^{\mathcal{B}_1}(E_2) \; d\mu_{\mathcal{B}_1}^{1} \qquad \forall \; E_2 \in \mathcal{B}_2,$$

i.e., a coherence of the prior distribution on $\mathcal{B}_2$ in the experiment $\mathcal{E}_2$ and in the extension of $\mathcal{E}_1$ into $\mathcal{E}$; this is also Condition (3.5.12) transported *a priori*. In the general case, as in Definition 3.5.4, it follows from (i) that $\mu_{B_1 \cap B_2} = \mu^1_{B_1 \cap B_2}$ while conditions (iii-c) imply $\mu_{B_1 \cap B_2} = \mu^2_{B_1 \cap B_2}$; therefore coherent encompassing implies that μ^1 and μ^2 must coincide on $B_1 \cap B_2$. The restrictiveness of this supplementary condition depends, among others, on the interpretation of the parameters; when $\mathcal{E}_1$ and $\mathcal{E}_2$ refer to two different sampling schemes, the crucial question is whether one considers the meaning, and eventually the prior probability, of the parameters, independent or not of the model.

Strong encompassing implies not only weak and coherent encompassing but also a same predictive probability on both experiments $\mathcal{E}_1$ and $\mathcal{E}_2$. This restriction would typically be unacceptable in most asymptotic i.i.d. models because a common predictive probability on $I\!\!R^\infty$ would imply a unique decomposition into a prior probability and an i.i.d. sampling (see, e.g., Hewitt and Savage (1955), Chow and Teicher (1978), or Dellacherie and Meyer (1980, Chap. 5)). Even in small samples, this requirement is in opposition with the idea of comparing two models on the basis of their predictive abilities, a clearly natural idea. In particular, in the light of Theorem 3.5.3, transposed to the comparison on the parameter space, it may be shown (Florens and Mouchart (1985)) that a common predictive probability implies that the model labels are not identifiable and, that the model labels are exactly estimable only if the predictive probabilities are mutually singular; this last aspect will be deepened in Chapter 6. Thus it should be clear that for the encompassing among experiments, the strong concept (i.e., condition (iii-s)) is definitely less palatable than for the sufficiency among experiments.

Let us conclude this brief presentation by mentioning that the concept of encompassing offers one possible avenue for a Bayesian theory of hypothesis testing, once is accepted the idea that the null and the alternative hypothesis may be associated to two experiments, being the two hypotheses nested or not. In this approach, testing is viewed as a problem of approximate encompassing in the sense of looking for a transition $\nu^{B_1}_{B_2}$ satisfying "as well as possible" the conditions of Definition 3.5.4. The idea is that if experiment $\mathcal{E}_1$ (exactly) encompasses experiment $\mathcal{E}_2$, the first one will be preferred to the second one. When this is not the case, one may look for a transition $\nu^{B_1}_{B_2}$ minimizing, in $P^1_{T_1}$, either a distance between posterior expectations of a parameter of interest, or a divergence between posterior distributions when these are evaluated in $\mathcal{E}_2$ and in $\mathcal{E}$. This is a Bayesian analogue of proposals made by Mizon and Richard (1986) in a sampling theory framework; preliminary results of this Bayesian approach may be found in Florens and Mouchart (1988). Details and applications of encompassing analysis are given in Florens and Richard (1989), and Florens, Hendry and Richard (1989).

4

Optimal Reductions: Maximal Ancillarity and Minimal Sufficiency

4.1 Introduction

In this chapter we seek to identify the "optimal" reduction of a given experiment. Clearly, such an optimal reduction should be derived from monotonicity properties of sufficiency and ancillarity.

As discussed in Section 2.3.5, ancillarity is preserved for any sub-σ-field of an ancillary one. One is therefore interested in identifying a maximal ancillary σ-field. Unfortunately, stochastic independence is not preserved by the wedge operation ($\vee$), and, consequently, neither is ancillarity. In Section 4.2. we make precise the concept of maximal ancillarity in both marginal and conditional experiments and we prove its existence using Zorn Lemma. Maximal ancillary σ-fields are not unique; but sufficient conditions to ensure that an ancillary σ-field is maximal are provided in Chapter 5.

In Section 2.3.5 it was also mentioned that, in a Bayesian experiment, any σ-field containing a sufficient σ-field is itself sufficient. Interest is therefore focused on the identification of a minimal sufficient σ-field. Theorem 2.2.12 showed that the intersection of two sufficient completed σ-field is itself sufficient. The minimal sufficient σ-field may therefore be defined through the intersection of all completed sufficient σ-fields. Consequently, minimal sufficient σ-fields are essentially unique.

In the next sections we take a more constructive approach: we construct the minimal sufficient σ-field using the concept of the projection among σ-fields. We first define and collect the main properties of the projection among σ-fields in Section 4.3; we then use this operation in Section 4.4 to construct the minimal sufficient σ-field. Next we introduce, in Section 4.5, a new probabilistic tool, namely, the concept of identification among σ-fields. We then specialize on the parameter space the analysis of Section 4.4 and consider, in Section 4.6, the general problem of identification as a problem of minimal sufficiency on the parameter space.

We conclude this chapter by describing in Section 4.7 an ideal (or limit) experiment, called "totally informative" where observations give a perfect information on parameters. This will provide a Bayesian concept of exact estimability, that is introduced, in a punctual form, in Section 4.8.

Section 4.2 on maximal ancillarity essentially gives a set of definitions in the framework of this monograph but most substantial results on this topic are relegated to Chapter 5 because they require several results of this chapter along with new probabilistic tools developed in Chapter 5. Section 4.3 on the projections of σ-fields presents and extends the results contained in Mouchart and Rolin (1984b). The first steps toward the analysis of minimal sufficiency in Section 4.4 were set out in Florens (1974) and elaborated in Florens and Mouchart (1977). The work on the identification among σ-fields, started in Mouchart and Rolin (1984b), has been applied in Section 4.5 to the identification of parameters in Bayesian statistics after a series of papers — Florens (1974), Florens and Mouchart (1977, 1980, 1986a); its relationship with sampling theory was presented in Florens, Mouchart and Rolin (1985). Finally, the section on totally informative experiments elaborates on ideas first sketched in Florens and Rolin (1984).

4.2 Maximal Ancillarity

As mentioned in the introduction, ancillarity is not preserved by wedge operation. Here is a standard example of such a situation

Example. Let $s = (x_1, x_2, \ldots, x_n)$ where $x_i = (y_i, z_i)$ and let $a \in]-1, 1[$. Suppose that $(x_i \mid a) \sim i.N_2(0, \Sigma_a)$ where $\Sigma_a = \begin{pmatrix} 1 & a \\ a & 1 \end{pmatrix}$. Then each of $y = (y_1, \ldots, y_n)$ or $z = (z_1, \ldots, z_n)$ are separately ancillary whatever is the prior distribution on a. But clearly $\sigma(y) \vee \sigma(z) = \sigma(s)$ is not ancillary. (For a discussion of this example in the sampling theory framework see, e.g., Cox (1971) and Cox and Hinkley (1974), Example 2.30). ∎

Therefore, one cannot speak of a unique maximal ancillary σ-field. We may however prove the existence of extremal, for the inclusion, ancillary σ-fields.

4.2.1 Theorem. The family of ancillary parameters (respectively, statistics) in an unreduced Bayesian experiment $\mathcal{E} = (A \times S, \ \mathcal{A} \vee \mathcal{S}, \ \Pi)$, i.e., $\{\mathcal{B} \subset \mathcal{A} : \mathcal{B} \perp\!\!\!\perp \mathcal{S}\}$, (respectively, $\{\mathcal{T} \subset \mathcal{S} : \mathcal{A} \perp\!\!\!\perp \mathcal{T}\}$) admits maximal elements.

Proof. This family is nonempty since it contains the trivial sub-σ-field of $\mathcal{A}$, i.e., $\overline{\mathcal{I}} \cap \mathcal{A}$, since trivially $\overline{\mathcal{I}} \perp\!\!\!\perp \mathcal{S}$. If T is an arbitrary ordered set and $\{\mathcal{B}_t : t \in T\}$ is a chain of ancillary parameters, i.e., $t \leq t' \ \Rightarrow \ \mathcal{B}_t \subset \mathcal{B}_{t'}$, it admits a maximal element $\mathcal{B}_T = \sigma\{\bigcup_{t \in T} \mathcal{B}_t\}$. Indeed, $\forall \ E \in \bigcup_{t \in T} \mathcal{B}_t$, $\mathcal{S}1_E = \mathcal{I}1_E$. By an application of the monotone class theorem, $\mathcal{S}1_E = \mathcal{I}1_E$ $\forall \ E \in \mathcal{B}_T$, i.e., $\mathcal{B}_T$ is ancillary. By Zorn Lemma, this proves the existence of maximal elements in the ancillary parameters. ∎

It should be noticed that any maximal ancillary parameter (respectively, statistics) contains the trivial parameter $\overline{\mathcal{I}} \cap \mathcal{A}$ (respectively, the trivial statistics $\overline{\mathcal{I}} \cap \mathcal{S}$). Indeed, by Lemma 2.2.5(iii), if $\mathcal{B}$ is an ancillary parameter, then $\mathcal{B} \vee (\overline{\mathcal{I}} \cap \mathcal{A}) = \overline{\mathcal{B}} \cap \mathcal{A}$ is still an ancillary parameter. We will, however, restrict our attention to ancillary σ-fields which the following definitions identify as essentially maximal.

4.2.2 Definition. A parameter $\mathcal{B}$ is $\mathcal{E}$-*maximal ancillary* if

(i) $\mathcal{B}$ is $\mathcal{E}$-ancillary (i.e., $\mathcal{B} \perp\!\!\!\perp \mathcal{S}$);

(ii) if $\mathcal{C}$ is an $\mathcal{E}$-ancillary parameter and $\mathcal{C} \supset \mathcal{B}$ then $\mathcal{C} \subset \overline{\mathcal{B}} \cap \mathcal{A}$. ∎

4.2.3 Definition. A statistic T is $\mathcal{E}$-*maximal ancillary* if

(i) T is $\mathcal{E}$-ancillary (i.e., $A \perp\!\!\!\perp T$);

(ii) if U is an $\mathcal{E}$-ancillary statistic and $U \supset T$ then $U \subset \overline{T} \cap S$.   ∎

According to these definitions, B (respectively, T is an $\mathcal{E}$-maximal ancillary parameter (respectively, statistic) if and only if $\overline{B} \cap A$ (respectively, $\overline{T} \cap S$) is a maximal element in the sense of Theorem 4.2.1.

These definitions extend naturally to conditional experiments as follows.

4.2.4 Definition. A parameter $\mathcal{L}$ is $\mathcal{E}_{\mathcal{B} \vee T}^{\mathcal{M}}$-*maximal ancillary* if and only if

(i) $\mathcal{L}$ is an $\mathcal{E}_{\mathcal{B} \vee T}^{\mathcal{M}}$-ancillary parameter (i.e., $\mathcal{L} \subset \mathcal{B} \vee \mathcal{M}$ and $\mathcal{L} \perp\!\!\!\perp T \mid \mathcal{M}$);

(ii) If $\mathcal{K}$ is an $\mathcal{E}_{\mathcal{B} \vee T}^{\mathcal{M}}$-ancillary parameter and $\mathcal{K} \supset \mathcal{L}$
      then $\mathcal{K} \subset \overline{\mathcal{L} \vee \mathcal{M}} \cap (\mathcal{B} \vee \mathcal{M})$.   ∎

4.2.5 Definition. A statistic $\mathcal{N}$ is $\mathcal{E}_{\mathcal{B} \vee T}^{\mathcal{M}}$-*maximal ancillary* if and only if

(i) $\mathcal{N}$ is an $\mathcal{E}_{\mathcal{B} \vee T}^{\mathcal{M}}$-ancillary statistic (i.e., $\mathcal{N} \subset \mathcal{M} \vee T$ and $\mathcal{B} \perp\!\!\!\perp \mathcal{N} \mid \mathcal{M}$);

(ii) If $\mathcal{R}$ is an $\mathcal{E}_{\mathcal{B} \vee T}^{\mathcal{M}}$-ancillary statistic and $\mathcal{R} \supset \mathcal{N}$
      then $\mathcal{R} \subset \overline{\mathcal{M} \vee \mathcal{N}} \cap (\mathcal{M} \vee T)$.   ∎

Note that $\mathcal{L}$ is an $\mathcal{E}_{\mathcal{B} \vee T}^{\mathcal{M}}$-maximal ancillary parameter if and only if $\overline{\mathcal{L} \vee \mathcal{M}} \cap (\mathcal{B} \vee \mathcal{M})$ is a maximal element in the family of $\mathcal{E}_{\mathcal{B} \vee T}^{\mathcal{M}}$-ancillary parameters. The existence of these maximal elements is assured by a conditional version of Theorem 4.2.1.

As maximal ancillary parameters (or statistics) are not unique, the practical problem is to determine whether a given ancillary parameter (or statistic) is actually maximal. Such a characterization of maximality requires tools developed in Chapter 5.

4.3 Projections of σ-Fields

4.3.1 Introduction

As in Section 2.2, we now handle a topic of general probability theory and we adopt the same notation: $(M, \mathcal{M}, P)$ is an abstract probability space, $\mathcal{M}_i$'s are sub-σ-fields of $\mathcal{M}$, $\mathcal{M}_0 = \{\phi, M\}$ is the trivial σ-field and $\overline{\mathcal{M}_0}$ is the completed trivial σ-field.

The motivation to introduce the projection of σ-fields is to construct the "smallest" sub-σ-field of a σ-field $\mathcal{M}_2$ conditionally on which $\mathcal{M}_2$ becomes independent of another given σ-field $\mathcal{M}_1$.

4.3.2 Definition and Elementary Properties

The projection of a σ-field $\mathcal{M}_1$ on a σ-field $\mathcal{M}_2$ is the smallest sub-σ-field of $\mathcal{M}_2$ with respect to which conditional expectations of $\mathcal{M}_1$-measurable functions given $\mathcal{M}_2$ are measurable. More precisely:

4.3.1 Definition. For any two sub-σ-fields $\mathcal{M}_1$, $\mathcal{M}_2$ of $\mathcal{M}$ we define the *projection of $\mathcal{M}_1$ on $\mathcal{M}_2$*, which we denote by $\mathcal{M}_2\mathcal{M}_1$, as follows:

$$(4.3.1) \qquad \mathcal{M}_2\mathcal{M}_1 = \sigma\{\mathcal{M}_2 m_1 \; : \; m_1 \in [\mathcal{M}_1]^+\}. \qquad \blacksquare$$

Remarks

(i) This definition has to be interpreted as the σ-field generated by *every version* of the conditional expectation of every positive $\mathcal{M}_1$-measurable function.

(ii) This operation is crucially dependent on the probability P and, in particular, on the P-null sets.

4.3.2 Proposition. *Elementary Properties of Projections of σ-Fields*

(i) $\overline{\mathcal{M}_0} \cap \mathcal{M}_2 \subset \overline{\mathcal{M}_1} \cap \mathcal{M}_2 \subset \mathcal{M}_2\mathcal{M}_1 \subset \mathcal{M}_2$;

(ii) $\mathcal{M}_2 \subset \overline{\mathcal{M}_1} \; \Rightarrow \; \mathcal{M}_2\mathcal{M}_1 = \mathcal{M}_2$;

(iii) $\mathcal{M}_1 \subset \overline{\mathcal{M}}_2 \Rightarrow \mathcal{M}_2\mathcal{M}_1 = \overline{\mathcal{M}}_1 \cap \mathcal{M}_2$;

(iv) $\mathcal{M}_3 \subset \overline{\mathcal{M}}_1 \Rightarrow \mathcal{M}_2\mathcal{M}_3 \subset \mathcal{M}_2\mathcal{M}_1$;

(v) $\mathcal{M}_4 \subset \overline{\mathcal{M}}_2 \Rightarrow \mathcal{M}_4\mathcal{M}_1 \subset \mathcal{M}_4(\mathcal{M}_2\mathcal{M}_1)$;

(vi) $\mathcal{M}_2\overline{\mathcal{M}}_1 = \mathcal{M}_2\mathcal{M}_1$;

(vii) $\overline{\mathcal{M}}_2\mathcal{M}_1 = \overline{\mathcal{M}_2\mathcal{M}_1}$;

(viii) $\overline{\mathcal{M}_2\mathcal{M}_1} \cap \mathcal{M}_2 = \mathcal{M}_2\mathcal{M}_1$.   ∎

Remarks

(i) Note that, due to Property (i), $\mathcal{M}_2\mathcal{M}_1$ contains *all* the null sets of $\mathcal{M}_2$, i.e., $\overline{\mathcal{M}}_0 \cap \mathcal{M}_2$. This is reflected in Property (viii), if we recall (Lemma 2.2.5(iii)) that $\overline{\mathcal{M}_2\mathcal{M}_1} \cap \mathcal{M}_2 = (\mathcal{M}_2\mathcal{M}_1) \vee (\overline{\mathcal{M}}_0 \cap \mathcal{M}_2)$. Property (ii) says that if we project a σ-field on one of its sub-σ-fields we obtain exactly this sub-σ-field. On the other hand, by Property (iii), the projection on a σ-field of one of its sub-σ-fields amounts to the completion of this sub-σ-field by the null sets of the greater σ-field.

(ii) It is important to remark that *projections of σ-fields is a nonassociative operation* and henceforth requires the use of parentheses.

(iii) Properties (i) through (iv) are analogous to properties of projections on linear spaces (up to null sets). This justifies the term of "projection" for this operation among σ-fields. Note, however, that in case of linear spaces Property (v) would be an equality; nonetheless, in the following example of projections among σ-fields, only the strict inclusion is verified.

Example. Let $s = (x, y)', a = (b, c, V)$ and suppose that

$$(s \mid a) \sim N_2 \left[\begin{pmatrix} b \\ cb \end{pmatrix}, V \right]$$
$$(b \mid c, V) \sim N(0, 1).$$

If $\mathcal{S} = \sigma(s)$ and $\mathcal{A} = \sigma(a)$, then $\mathcal{A}\mathcal{S} = \mathcal{A}$. Nevertheless,

$$(s \mid c, V) \sim N_2 \left[\begin{pmatrix} 0 \\ 0 \end{pmatrix}, \begin{pmatrix} v_{11} + 1 & v_{12} + c \\ v_{12} + c & v_{22} + c^2 \end{pmatrix} \right].$$

Therefore if $\mathcal{B} = \sigma(c, V)$ and $\mathcal{C} = \sigma(v_{11}, v_{12} + c, v_{22} + c^2)$, then $\mathcal{BS} = \mathcal{C} \subset \mathcal{B}$ and $\mathcal{C} \neq \mathcal{B}$. This shows that $\mathcal{BS} = \mathcal{C}$ is strictly included in $\mathcal{B}(\mathcal{AS}) = \mathcal{BA} = \mathcal{B}$. The last equality is justified by the elementary Property 4.3.2(ii). ∎

4.3.3 Projections and Conditional Independence

We now respond to the question posed in Section 4.2.1 and show that the projection of $\mathcal{M}_1$ on $\mathcal{M}_2$ is the smallest sub-σ-field of $\mathcal{M}_2$ conditionally on which $\mathcal{M}_1$ and $\mathcal{M}_2$ are independent.

4.3.3 Theorem. For any sub-σ-fields $\mathcal{M}_1$ and $\mathcal{M}_2$,

(i) $\mathcal{M}_1 \perp\!\!\!\perp \mathcal{M}_2 \mid \mathcal{M}_2\mathcal{M}_1$;

(ii) $\mathcal{M}_4 \subset \mathcal{M}_2$ and $\mathcal{M}_1 \perp\!\!\!\perp \mathcal{M}_2 \mid \mathcal{M}_4 \Rightarrow \mathcal{M}_2\mathcal{M}_1 \subset \overline{\mathcal{M}}_4$.

Proof. For any σ-field $\mathcal{M}_4 \subset \mathcal{M}_2$, Theorem 2.2.6 implies that:

$$\mathcal{M}_1 \perp\!\!\!\perp \mathcal{M}_2 \mid \mathcal{M}_4 \text{ if and only if } \mathcal{M}_2 m \in [\overline{\mathcal{M}}_4]^+, \ \forall \ m \in [\mathcal{M}_1]^+.$$

This implies $\mathcal{M}_2\mathcal{M}_1 \subset \overline{\mathcal{M}}_4$ and the proof is completed by noticing that, trivially,

$$\mathcal{M}_2 m \in [\mathcal{M}_2\mathcal{M}_1]^+ \quad \forall \ m \in [\mathcal{M}_1]^+. \qquad ∎$$

This theorem shows that $\mathcal{M}_2\mathcal{M}_1$ is the intersection of all sub-σ-fields of $\mathcal{M}_2$ containing the null sets of $\mathcal{M}_2$ and conditionally on which $\mathcal{M}_1$ and $\mathcal{M}_2$ are independent. $\mathcal{M}_2\mathcal{M}_1$ may then be written as follows

$$(4.3.2) \qquad \mathcal{M}_2\mathcal{M}_1 = \bigcap_{\mathcal{N} \in \mathcal{F}_{2.1}} (\overline{\mathcal{N}} \cap \mathcal{M}_2)$$

where $\mathcal{F}_{2.1} = \{\mathcal{N} \subset \mathcal{M}_2 : \mathcal{M}_1 \perp\!\!\!\perp \mathcal{M}_2 \mid \mathcal{N}\}$. It may be noticed that Mac-Kean (1963) defined the projection of σ-fields through the right-hand side of (4.3.2), using Lebesgue completion instead of measurable completion as in this monograph.

Note also that, in the language of system theory, the projection $\mathcal{M}_1\mathcal{M}_2$ (respectively, $\mathcal{M}_2\mathcal{M}_1$) gives *the minimal internal splitting σ-field of $\mathcal{M}_1$*

and $\mathcal{M}_2$ which is included in $\mathcal{M}_1$ (respectively, $\mathcal{M}_2$) (see, e.g., Lindquist and Picci (1979) and Lindquist, Picci and Ruckebusch (1979)).

We now investigate further the relation between conditional independence and projection. We first provide an alternative characterization of conditional independence in terms of projections; this characterization is a straightforward application of Theorem 2.2.6.

4.3.4 Proposition. The following properties are equivalent:

(i) $\mathcal{M}_1 \perp\!\!\!\perp \mathcal{M}_2 \mid \mathcal{M}_3$;

(ii) $(\mathcal{M}_1 \vee \mathcal{M}_3)\mathcal{M}_2 \subset \overline{\mathcal{M}}_3$;

(iii) $(\mathcal{M}_2 \vee \mathcal{M}_3)\mathcal{M}_1 \subset \overline{\mathcal{M}}_3$.

In particular $\mathcal{M}_1 \perp\!\!\!\perp \mathcal{M}_2$ if and only if $\mathcal{M}_2\mathcal{M}_1 \subset \overline{\mathcal{M}}_0$ or if and only if $\mathcal{M}_1\mathcal{M}_2 \subset \overline{\mathcal{M}}_0$.   ∎

Thus, Proposition 4.3.4. characterizes conditional independence by inclusions of projections. The following Corollary 4.3.5 relies on Theorem 2.2.10 to obtain several equivalent decompositions of a given conditional independence property between σ-fields into conditional independences involving projections either on or of the conditioning σ-field. This corollary is repeatedly used in deducing new conditional independences from a given one. In particular, in Corollary 4.3.6, a given conditional independence is shown to provide a refinement of the elementary Properties 4.3.2. in the form of specific inclusions or equalities among projections of σ-fields.

4.3.5 Corollary. The following properties are equivalent:

(i) $\mathcal{M}_1 \perp\!\!\!\perp \mathcal{M}_2 \mid \mathcal{M}_3$;

(ii) $(\mathcal{M}_1 \vee \mathcal{M}_3) \perp\!\!\!\perp \mathcal{M}_2 \mid \mathcal{M}_3\mathcal{M}_2$;

(iii)   1. $\mathcal{M}_1 \perp\!\!\!\perp \mathcal{M}_2\mathcal{M}_3 \mid \mathcal{M}_3$;

        2. $(\mathcal{M}_1 \vee \mathcal{M}_3) \perp\!\!\!\perp \mathcal{M}_2 \mid \mathcal{M}_2\mathcal{M}_3$;

(iv) 1. $\mathcal{M}_1 \perp\!\!\!\perp \mathcal{M}_2 \mid \mathcal{M}_3\mathcal{M}_1 \vee \mathcal{M}_3\mathcal{M}_2$;

2. $(\mathcal{M}_1 \vee \mathcal{M}_2) \perp\!\!\!\perp \mathcal{M}_3 \mid \mathcal{M}_3\mathcal{M}_1 \vee \mathcal{M}_3\mathcal{M}_2$;

(v) 1. $\mathcal{M}_1\mathcal{M}_3 \perp\!\!\!\perp \mathcal{M}_2\mathcal{M}_3 \mid \mathcal{M}_3$;

2. $\mathcal{M}_1\mathcal{M}_3 \perp\!\!\!\perp \mathcal{M}_2 \mid \mathcal{M}_2\mathcal{M}_3$;

3. $\mathcal{M}_1 \perp\!\!\!\perp \mathcal{M}_2\mathcal{M}_3 \mid \mathcal{M}_1\mathcal{M}_3$;

4. $\mathcal{M}_1 \perp\!\!\!\perp \mathcal{M}_2 \mid \mathcal{M}_1\mathcal{M}_3 \vee \mathcal{M}_2\mathcal{M}_3$;

5. $(\mathcal{M}_1 \vee \mathcal{M}_2) \perp\!\!\!\perp \mathcal{M}_3 \mid \mathcal{M}_1\mathcal{M}_3 \vee \mathcal{M}_2\mathcal{M}_3$.

Proof. All equivalences stated in the proof come from Theorem 2.2.10 and, unless otherwise stated, asserted but unproved conditional independences come from Theorem 4.3.3.

(a) (i) is equivalent to (ii) since $\mathcal{M}_3 \perp\!\!\!\perp \mathcal{M}_2 \mid \mathcal{M}_3\mathcal{M}_2$ is always true.

(b) (i) is equivalent to (iii)1 along with $\mathcal{M}_1 \perp\!\!\!\perp \mathcal{M}_2 \mid \mathcal{M}_3 \vee \mathcal{M}_2\mathcal{M}_3$. But this is equivalent to (iii)2 since $\mathcal{M}_3 \perp\!\!\!\perp \mathcal{M}_2 \mid \mathcal{M}_2\mathcal{M}_3$ is always true. Hence, (i) is equivalent to (iii).

(c) (ii) is equivalent to the following two conditional independences:

(1) $(\mathcal{M}_1 \vee \mathcal{M}_3\mathcal{M}_1) \perp\!\!\!\perp \mathcal{M}_2 \mid \mathcal{M}_3\mathcal{M}_2$

(2) $\mathcal{M}_3 \perp\!\!\!\perp \mathcal{M}_2 \mid \mathcal{M}_1 \vee \mathcal{M}_3\mathcal{M}_1 \vee \mathcal{M}_3\mathcal{M}_2$.

But (1) is equivalent to (iv)1 since, by Corollary 2.2.11,

$$\mathcal{M}_3 \perp\!\!\!\perp \mathcal{M}_2 \mid \mathcal{M}_3\mathcal{M}_2 \quad \Rightarrow \quad \mathcal{M}_3\mathcal{M}_1 \perp\!\!\!\perp \mathcal{M}_2 \mid \mathcal{M}_3\mathcal{M}_2.$$

On the other hand, (2) is equivalent to (iv)2, since by Corollary 2.2.11,

$$\mathcal{M}_3 \perp\!\!\!\perp \mathcal{M}_1 \mid \mathcal{M}_3\mathcal{M}_1 \quad \Rightarrow \quad \mathcal{M}_3 \perp\!\!\!\perp \mathcal{M}_1 \mid \mathcal{M}_3\mathcal{M}_1 \vee \mathcal{M}_3\mathcal{M}_2.$$

Hence (ii) is equivalent to (iv).

(d) (iii)1 is equivalent to (v)1 along with $\mathcal{M}_1 \perp\!\!\!\perp \mathcal{M}_2\mathcal{M}_3 \mid \mathcal{M}_3 \vee \mathcal{M}_1\mathcal{M}_3$. But this is equivalent to $\mathcal{M}_1 \perp\!\!\!\perp (\mathcal{M}_3 \vee \mathcal{M}_2\mathcal{M}_3) \mid \mathcal{M}_1\mathcal{M}_3$ since it always true that $\mathcal{M}_1 \perp\!\!\!\perp \mathcal{M}_3 \mid \mathcal{M}_1\mathcal{M}_3$. This last conditional independence is equivalent to (v)3 along with

(3) $\mathcal{M}_1 \perp\!\!\!\perp \mathcal{M}_3 \mid \mathcal{M}_1\mathcal{M}_3 \vee \mathcal{M}_2\mathcal{M}_3$.

On the other hand (iii)2 is equivalent to (v)2 along with

(4) $(\mathcal{M}_1 \vee \mathcal{M}_3) \perp\!\!\!\perp \mathcal{M}_2 \mid \mathcal{M}_1\mathcal{M}_3 \vee \mathcal{M}_2\mathcal{M}_3$

and this is equivalent to (v)4 along with

(5) $\mathcal{M}_3 \perp\!\!\!\perp \mathcal{M}_2 \mid \mathcal{M}_1\mathcal{M}_3 \vee \mathcal{M}_2\mathcal{M}_3 \vee \mathcal{M}_1$.

It can be checked that (3) and (5) are equivalent to (v)5. Therefore (iii) is equivalent to (v).   ∎

Note that, from Proposition 4.3.4, Corollary 4.3.5. might also be presented as equivalences between sets of inclusions among various projections. In this spirit, the next corollary refines some of the most relevant inclusions of projections of σ-fields implied by Corollary 4.3.5, which are viewed as implications of conditional independence.

4.3.6 Corollary. If $\mathcal{M}_1 \perp\!\!\!\perp \mathcal{M}_2 \mid \mathcal{M}_3$, then

(i) $\mathcal{M}_3(\mathcal{M}_1 \vee \mathcal{M}_2) = \mathcal{M}_3\mathcal{M}_1 \vee \mathcal{M}_3\mathcal{M}_2$;

(ii) $(\mathcal{M}_1 \vee \mathcal{M}_2)\mathcal{M}_3 \subset \overline{\mathcal{M}_1\mathcal{M}_3 \vee \mathcal{M}_2\mathcal{M}_3}$;

(iii) $\mathcal{M}_3\mathcal{M}_2 \subset (\mathcal{M}_1 \vee \mathcal{M}_3)\mathcal{M}_2 \subset \overline{\mathcal{M}_3\mathcal{M}_2}$;

(iv) $\mathcal{M}_2(\mathcal{M}_1 \vee \mathcal{M}_3) = \mathcal{M}_2\mathcal{M}_3$;

(v) $\mathcal{M}_2\mathcal{M}_1 = \mathcal{M}_2(\mathcal{M}_1\mathcal{M}_3) \subset \mathcal{M}_2(\mathcal{M}_3\mathcal{M}_1) \subset \mathcal{M}_2\mathcal{M}_3$.

Proof.

(i) Clearly $\mathcal{M}_3\mathcal{M}_1 \vee \mathcal{M}_3\mathcal{M}_2 \subset \mathcal{M}_3(\mathcal{M}_1 \vee \mathcal{M}_2)$. Combining elementary Property 4.3.2 (viii), and Proposition 4.3.4, we successively obtain:

$$\mathcal{M}_3(\mathcal{M}_1 \vee \mathcal{M}_2) \subset \mathcal{M}_3\mathcal{M}_1 \vee \mathcal{M}_3\mathcal{M}_2$$
$$\Leftrightarrow \quad \mathcal{M}_3(\mathcal{M}_1 \vee \mathcal{M}_2) \subset \overline{\mathcal{M}_3\mathcal{M}_1 \vee \mathcal{M}_3\mathcal{M}_2}$$
$$\Leftrightarrow \quad (\mathcal{M}_1 \vee \mathcal{M}_2) \perp\!\!\!\perp \mathcal{M}_3 \mid \mathcal{M}_3\mathcal{M}_1 \vee \mathcal{M}_3\mathcal{M}_2.$$

Therefore (i) is equivalent to (iv)2 in Corollary 4.3.5.

(ii) By Proposition 4.3.4, (ii) is equivalent to

$$(\mathcal{M}_1 \vee \mathcal{M}_2) \perp\!\!\!\perp \mathcal{M}_3 \mid \mathcal{M}_1\mathcal{M}_3 \vee \mathcal{M}_2\mathcal{M}_3$$

which is (v)5 in Corollary 4.3.5.

(iii) Theorem 2.2.1 (ii) defining conditional independence implies (iii). Note that by Proposition 4.3.4, (iii) is actually equivalent to $\mathcal{M}_1 \perp\!\!\!\perp \mathcal{M}_2 \mid \mathcal{M}_3$, i.e., (i) in Corollary 4.3.5.

(iv) Clearly, $\mathcal{M}_2\mathcal{M}_3 \subset \mathcal{M}_2(\mathcal{M}_1 \vee \mathcal{M}_3)$. Combining elementary Property 4.3.2 (viii), and Proposition 4.3.4, we successively obtain:

$$\mathcal{M}_2(\mathcal{M}_1 \vee \mathcal{M}_3) \subset \mathcal{M}_2\mathcal{M}_3$$
$$\Leftrightarrow \quad \mathcal{M}_2(\mathcal{M}_1 \vee \mathcal{M}_3) \subset \overline{\mathcal{M}_2\mathcal{M}_3}$$
$$\Leftrightarrow \quad (\mathcal{M}_1 \vee \mathcal{M}_3) \perp\!\!\!\perp \mathcal{M}_2 \mid \mathcal{M}_2\mathcal{M}_3.$$

Therefore (iv) is equivalent to (iii)2 in Corollary 4.3.5.

(v) Clearly $\mathcal{M}_2(\mathcal{M}_1\mathcal{M}_3) \subset \mathcal{M}_2\mathcal{M}_1$ and $\mathcal{M}_2(\mathcal{M}_3\mathcal{M}_1) \subset \mathcal{M}_2\mathcal{M}_3$; by Corollary 4.3.5(ii) and (iii)2 (with permutation of $\mathcal{M}_1$ and $\mathcal{M}_2$), we obtain: $\mathcal{M}_1 \perp\!\!\!\perp \mathcal{M}_2 \mid \mathcal{M}_3\mathcal{M}_1$ and $\mathcal{M}_1 \perp\!\!\!\perp \mathcal{M}_2 \mid \mathcal{M}_1\mathcal{M}_3$. By (iv), we already know that:

$$\mathcal{M}_1 \perp\!\!\!\perp \mathcal{M}_2 \mid \mathcal{M}_3\mathcal{M}_1 \Rightarrow \mathcal{M}_2(\mathcal{M}_1 \vee \mathcal{M}_3\mathcal{M}_1) = \mathcal{M}_2(\mathcal{M}_3\mathcal{M}_1).$$

Therefore we conclude that $\mathcal{M}_2\mathcal{M}_1 \subset \mathcal{M}_2(\mathcal{M}_3\mathcal{M}_1)$. Moreover,

$$\mathcal{M}_1 \perp\!\!\!\perp \mathcal{M}_2 \mid \mathcal{M}_1\mathcal{M}_3 \Rightarrow \mathcal{M}_2(\mathcal{M}_1 \vee \mathcal{M}_1\mathcal{M}_3) = \mathcal{M}_2(\mathcal{M}_1\mathcal{M}_3),$$

i.e., $\mathcal{M}_2\mathcal{M}_1 = \mathcal{M}_2(\mathcal{M}_1\mathcal{M}_3)$.

It is interesting to remark that (v) is therefore equivalent to

(vbis) 1. $\quad \mathcal{M}_1 \perp\!\!\!\perp \mathcal{M}_2 \mid \mathcal{M}_2(\mathcal{M}_1\mathcal{M}_3)$;

2. $\quad \mathcal{M}_1 \perp\!\!\!\perp \mathcal{M}_2 \mid \mathcal{M}_2(\mathcal{M}_3\mathcal{M}_1)$. ∎

It should be noted that in Corollary 4.3.6, Property (iii) actually implies that $(\mathcal{M}_1 \vee \mathcal{M}_3)\mathcal{M}_2 \subset \overline{\mathcal{M}_3}$, i.e., Condition (ii) of Proposition 4.3.4 and is therefore equivalent to (and not only implied by) $\mathcal{M}_1 \perp\!\!\!\perp \mathcal{M}_2 \mid \mathcal{M}_3$.

Like Corollary 4.3.6, the next theorem exhibits equalities among projections implied by a double conditional independence: this gives procedures to simplify the computation of the minimal conditioning σ-field; furthermore, due to its symmetric form, this theorem imposes, as a corollary, some structural properties on projections.

4.3.7 Theorem. If

(i) $\mathcal{M}_1 \perp\!\!\!\perp \mathcal{M}_2 \mid \mathcal{M}_3$ and $\mathcal{M}_3 \subset \mathcal{M}_1,$

(ii) $\mathcal{M}_1 \perp\!\!\!\perp \mathcal{M}_2 \mid \mathcal{M}_4$ and $\mathcal{M}_4 \subset \mathcal{M}_2,$

then

(iii) $\mathcal{M}_3\mathcal{M}_4 \subset \mathcal{M}_1\mathcal{M}_2 \subset \overline{\mathcal{M}_3\mathcal{M}_4}.$

More precisely,

(iv) $\mathcal{M}_3\mathcal{M}_4 = (\mathcal{M}_1\mathcal{M}_2) \cap \mathcal{M}_3 = \mathcal{M}_3(\mathcal{M}_1\mathcal{M}_2);$

(v) $\mathcal{M}_1\mathcal{M}_2 = \overline{\mathcal{M}_3\mathcal{M}_4} \cap \mathcal{M}_1 = \mathcal{M}_1(\mathcal{M}_3\mathcal{M}_4).$

Proof. Since (i) and (ii) imply $\mathcal{M}_1 \perp\!\!\!\perp \mathcal{M}_4 \mid \mathcal{M}_3$, we obtain, by Corollary 4.3.6(iii), that: $\mathcal{M}_3\mathcal{M}_4 \subset \mathcal{M}_1\mathcal{M}_4 \subset \overline{\mathcal{M}_3\mathcal{M}_4}$. But from (ii) and Corollary 4.3.6(iv), one has $\mathcal{M}_1\mathcal{M}_4 = \mathcal{M}_1\mathcal{M}_2$. This proves (iii). To prove (iv) and (v), it suffices to remark that (iii) implies $\overline{\mathcal{M}_3\mathcal{M}_4} = \overline{\mathcal{M}_1\mathcal{M}_2}$ and to use 4.3.2(iii) and (viii). ∎

This theorem may be phrased as follows: the only difference between $\mathcal{M}_1\mathcal{M}_2$ and $\mathcal{M}_3\mathcal{M}_4$ is a matter of null sets of $\mathcal{M}_1$. Note also that, by symmetry, (iii) may also be expressed in terms of $\mathcal{M}_2\mathcal{M}_1$ and $\mathcal{M}_4\mathcal{M}_3$.

The next corollary is obtained by choosing a $\mathcal{M}_3$ ($= \mathcal{M}_1\mathcal{M}_2$) and a $\mathcal{M}_4$ ($= \mathcal{M}_2\mathcal{M}_1$) so as to make conditions (i) and (ii) in Theorem 4.3.7 trivial. Consequently, this corollary provides structural properties of the projections.

4.3.8 Corollary. For any sub-σ-fields $\mathcal{M}_1$ and $\mathcal{M}_2$ of $\mathcal{M}$

$$\mathcal{M}_1\mathcal{M}_2 = (\mathcal{M}_1\mathcal{M}_2)\mathcal{M}_2 = \mathcal{M}_1(\mathcal{M}_2\mathcal{M}_1) = (\mathcal{M}_1\mathcal{M}_2)(\mathcal{M}_2\mathcal{M}_1).$$ ∎

This corollary, along with the elementary Properties 4.3.2, imply that, regardless of how big they are, expressions involving only projection operations of two sub-σ-fields $\mathcal{M}_1$ and $\mathcal{M}_2$ may always be simplified into either $\mathcal{M}_2\mathcal{M}_1$ or $\mathcal{M}_1\mathcal{M}_2$. We now turn to a further structural property of the projection of σ-fields, namely, that if $\mathcal{M}_3 \subset \mathcal{M}_1$ and $m \in [\mathcal{M}_1]^+$, $(\mathcal{M}_2 \vee \mathcal{M}_3)m$ will depend only on countably many conditional expectations $\mathcal{M}_2 m_i$ with $m_i \in [\mathcal{M}_1]^+$, and on $\mathcal{M}_3$-measurable functions.

4.3.9 Theorem. If $\mathcal{M}_3 \subset \mathcal{M}_1$ then:

(i) $(\mathcal{M}_2 \vee \mathcal{M}_3)\mathcal{M}_1 \subset (\mathcal{M}_2\mathcal{M}_1) \vee \overline{\mathcal{M}_3}$;

(ii) $\mathcal{M}_1(\mathcal{M}_2 \vee \mathcal{M}_3) = (\mathcal{M}_1\mathcal{M}_2) \vee \mathcal{M}_3$.

Proof. Clearly, $\mathcal{M}_3 \subset \mathcal{M}_1$ implies that $\mathcal{M}_2 \perp\!\!\!\perp \mathcal{M}_3 \mid \mathcal{M}_1$. Therefore by Corollary 4.3.6(i) and (ii), we successively obtain

$$\begin{aligned}
\mathcal{M}_1(\mathcal{M}_2 \vee \mathcal{M}_3) &= \mathcal{M}_1\mathcal{M}_2 \vee \mathcal{M}_1\mathcal{M}_3 \\
(\mathcal{M}_2 \vee \mathcal{M}_3)\mathcal{M}_1 &\subset \overline{\mathcal{M}_2\mathcal{M}_1 \vee \mathcal{M}_3\mathcal{M}_1}.
\end{aligned}$$

Since $\mathcal{M}_3 \subset \mathcal{M}_1$, $\mathcal{M}_3\mathcal{M}_1 = \mathcal{M}_3$, by elementary Property 4.3.2(ii), and $\mathcal{M}_1\mathcal{M}_3 = \overline{\mathcal{M}}_3 \cap \mathcal{M}_1 = \mathcal{M}_3 \vee (\mathcal{M}_1 \cap \overline{\mathcal{M}}_0)$ by elementary Property 4.3.2(iii) and Lemma 2.2.5(iii). The proof is completed by noticing that, by elementary Property 4.3.2(i), $\mathcal{M}_1 \cap \overline{\mathcal{M}}_0 \subset \mathcal{M}_1\mathcal{M}_2$. ∎

It is sometimes interesting to identify a minimal conditioning σ-field conditionally on which two σ-fields $\mathcal{M}_1$ and $\mathcal{M}_2$ are independent in a restrained class of such σ-fields. We may require, for instance, that it contains a third σ-field $\mathcal{M}_3$. A solution to this problem is given in the next proposition.

4.3.10 Proposition. For any sub-σ-fields $\mathcal{M}_1$, $\mathcal{M}_2$, and $\mathcal{M}_3$:

(i) $\mathcal{M}_1 \perp\!\!\!\perp \mathcal{M}_2 \mid (\mathcal{M}_2 \vee \mathcal{M}_3)(\mathcal{M}_1 \vee \mathcal{M}_3)$;

(ii)　$\mathcal{M}_3 \subset \mathcal{M}_4 \subset \mathcal{M}_2 \vee \mathcal{M}_3$ and $\mathcal{M}_1 \perp\!\!\!\perp \mathcal{M}_2 \mid \mathcal{M}_4$
　　　$\Rightarrow (\mathcal{M}_2 \vee \mathcal{M}_3)(\mathcal{M}_1 \vee \mathcal{M}_3) \subset \overline{\mathcal{M}}_4.$

Moreover

$$(\mathcal{M}_2 \vee \mathcal{M}_3)(\mathcal{M}_1 \vee \mathcal{M}_3) = \{(\mathcal{M}_2 \vee \mathcal{M}_3)\mathcal{M}_1\} \vee \mathcal{M}_3. \qquad \blacksquare$$

Another interesting feature of projections of σ-fields is that they simplify the search for conditional independence, since it suffices to check the desired conditional independence on smaller σ-fields. More precisely:

4.3.11 Theorem. Let $\mathcal{M}_1$ and $\mathcal{M}_2$ be sub-σ-fields of $\mathcal{M}$. Then for any $\mathcal{M}_3 \subset \mathcal{M}_1$, $\mathcal{M}_4 \subset \mathcal{M}_2$, $\mathcal{M}_5 \subset \mathcal{M}_1$:

(i)　$\mathcal{M}_3 \perp\!\!\!\perp \mathcal{M}_2\mathcal{M}_1 \mid \mathcal{M}_4 \vee \mathcal{M}_5 \Leftrightarrow \mathcal{M}_3 \perp\!\!\!\perp \mathcal{M}_2 \mid \mathcal{M}_4 \vee \mathcal{M}_5;$

(ii)　$\mathcal{M}_3 \perp\!\!\!\perp \mathcal{M}_1\mathcal{M}_2 \mid \mathcal{M}_4 \vee \mathcal{M}_5 \Rightarrow \mathcal{M}_3 \perp\!\!\!\perp \mathcal{M}_2 \mid \mathcal{M}_4 \vee \mathcal{M}_5.$

Moreover

(iii)　$\mathcal{M}_1 \perp\!\!\!\perp \mathcal{M}_2 \mid \mathcal{M}_4 \vee \mathcal{M}_5 \Leftrightarrow \mathcal{M}_1\mathcal{M}_2 \perp\!\!\!\perp \mathcal{M}_2\mathcal{M}_1 \mid \mathcal{M}_4 \vee \mathcal{M}_5.$

Proof. Since $\mathcal{M}_1 \perp\!\!\!\perp \mathcal{M}_2 \mid \mathcal{M}_1\mathcal{M}_2$ and $\mathcal{M}_1 \perp\!\!\!\perp \mathcal{M}_2 \mid \mathcal{M}_2\mathcal{M}_1$, always hold by Theorem 4.3.3, we obtain from Corollary 2.2.11(ii), that

$$\mathcal{M}_3 \perp\!\!\!\perp \mathcal{M}_2 \mid \mathcal{M}_1\mathcal{M}_2 \vee \mathcal{M}_4 \vee \mathcal{M}_5$$
$$\mathcal{M}_3 \perp\!\!\!\perp \mathcal{M}_2 \mid \mathcal{M}_2\mathcal{M}_1 \vee \mathcal{M}_4 \vee \mathcal{M}_5.$$

So by Theorem 2.2.10,

$$\mathcal{M}_3 \perp\!\!\!\perp \mathcal{M}_2\mathcal{M}_1 \mid \mathcal{M}_4 \vee \mathcal{M}_5 \quad \Leftrightarrow \quad \mathcal{M}_3 \perp\!\!\!\perp (\mathcal{M}_2 \vee \mathcal{M}_2\mathcal{M}_1) \mid \mathcal{M}_4 \vee \mathcal{M}_5,$$

and this is equivalent to $\mathcal{M}_3 \perp\!\!\!\perp \mathcal{M}_2 \mid \mathcal{M}_4 \vee \mathcal{M}_5$. Similarly

$$\mathcal{M}_3 \perp\!\!\!\perp \mathcal{M}_1\mathcal{M}_2 \mid \mathcal{M}_4 \vee \mathcal{M}_5 \quad \Leftrightarrow \quad \mathcal{M}_3 \perp\!\!\!\perp (\mathcal{M}_2 \vee \mathcal{M}_1\mathcal{M}_2) \mid \mathcal{M}_4 \vee \mathcal{M}_5,$$

and this implies $\mathcal{M}_3 \perp\!\!\!\perp \mathcal{M}_2 \mid \mathcal{M}_4 \vee \mathcal{M}_5$. To complete the proof it suffices to remark that (iii) is obtained by symmetrizing (i). 　　■

4.4 Minimal Sufficiency

In this section, we use projections among σ-fields to construct a minimal sufficient statistic and a minimal sufficient parameter.

4.4.1 Minimal Sufficiency in Unreduced and in Marginal Experiments

We now apply the results of Section 4.3 to the unreduced Bayesian experiment $\mathcal{E} = \mathcal{E}_{A \vee S}$. Minimality with respect to set-inclusion requires some care; this fact inspires the following definition:

4.4.1 Definition. A parameter $\mathcal{B} \subset \mathcal{A}$ is $\mathcal{E}$-*minimal sufficient* if and only if:

(i) $\mathcal{B}$ is $\mathcal{E}$-sufficient (i.e., $\mathcal{A} \perp\!\!\!\perp \mathcal{S} \mid \mathcal{B}$);

(ii) $\mathcal{C} \subsetneq \mathcal{A}$ and $\mathcal{E}$-sufficient $\Rightarrow$ $\mathcal{B} \subset \overline{\mathcal{C}} \cap \mathcal{A}$. ∎

Remember that $\overline{\mathcal{B}} \cap \mathcal{A}$ is the smallest sub-σ-field of $\mathcal{A}$ containing $\mathcal{B}$ and all the null sets of $\mathcal{A}$ while $\overline{\mathcal{B}}$ is, in general, not a sub-σ-field of $\mathcal{A}$. Next, from Definition 4.4.1, two sub-σ-fields of $\mathcal{A}$ can be $\mathcal{E}$-minimal sufficient parameters even when one is strictly contained in the other one. This motivates the following definition.

4.4.2 Definition. A parameter $\mathcal{B} \subset \mathcal{A}$, is $\mathcal{A}$-*complete* if $\overline{\mathcal{B}} \cap \mathcal{A} = \mathcal{B}$, i.e., if $\mathcal{B}$ contains all the null sets of $\mathcal{A}$. ∎

It is clear that two $\mathcal{A}$-complete $\mathcal{E}$-minimal sufficient parameters coincide. Hence the elementary Property 4.3.2(i) and Theorem 4.3.3 guarantee the existence of a unique $\mathcal{A}$-complete $\mathcal{E}$-minimal sufficient parameter.

4.4.3 Proposition. In the Bayesian experiment $\mathcal{E} = \{A \times S, \mathcal{A} \vee \mathcal{S}, \Pi\}$ the parameter $\mathcal{AS}$ is the unique $\mathcal{A}$-complete $\mathcal{E}$-minimal sufficient parameter. Therefore a parameter $\mathcal{B} \subset \mathcal{A}$ is $\mathcal{E}$-minimal sufficient if and only if $\overline{\mathcal{B}} \cap \mathcal{A} = \mathcal{AS}$. ∎

By duality we obtain the same concepts on the sample space.

4.4.4 Definition. A statistic $T \subset S$ is $\mathcal{E}$-*minimal sufficient* if:

(i) T is $\mathcal{E}$-sufficient (i.e., $A \perp\!\!\!\perp S \mid T$);

(ii) $\mathcal{U} \subset S$ and $\mathcal{E}$-sufficient $\Rightarrow$ $T \subset \overline{\mathcal{U}} \cap S.$ ∎

4.4.5 Definition. A statistic $T \subset S$ is S-*complete* if $\overline{T} \cap S = T$, i.e., if T contains all the null sets of S.   ∎

4.4.6 Proposition. In the Bayesian experiment $\mathcal{E} = \{A \times S,\ A \vee S,\ \Pi\}$ the statistic SA is the unique S-complete $\mathcal{E}$-minimal sufficient statistic. Therefore a statistic $T \subset S$ is $\mathcal{E}$-minimal sufficient if and only if $\overline{T} \cap S = SA.$ ∎

An identical analysis of minimal sufficiency in the marginal experiment $\mathcal{E}_{B \vee T}$ can be done by substituting B to A and T to S. So, in $\mathcal{E}_{B \vee T}$, BT is the B-complete $\mathcal{E}_{B \vee T}$-minimal sufficient parameter and TB is the T-complete $\mathcal{E}_{B \vee T}$-minimal sufficient statistic.

4.4.2   Elementary Properties of Minimal Sufficiency

The search of the $\mathcal{E}$-minimal sufficient parameter and statistic may be simplified if we already know a sufficient parameter B and a sufficient statistic T using Theorem 4.3.7.

4.4.7 Proposition. If B is an $\mathcal{E}$-sufficient parameter and T is an $\mathcal{E}$-sufficient statistic, then:

(i) $BT \subset AS \subset \overline{BT};$

(ii) $TB \subset SA \subset \overline{TB}.$ ∎

If we modify the prior distribution μ in μ' in such a way that $\mu' \ll \mu$, we have an inclusion among minimal sufficient sub-σ-fields.

4.4.8 Theorem. Let $\mathcal{E}$ and $\mathcal{E}'$ be two Bayesian experiments on $(A \times S, A \vee S)$ defined by their probabilities Π and Π'. If

(i) $\Pi' \ll \Pi$,

(ii) $\dfrac{d\Pi'}{d\Pi} \in [A]^+$,

then the $\mathcal{E}'$-minimal sufficient parameter (resp., statistic) is included in the $\mathcal{E}$-minimal sufficient parameter (resp., statistic) completed by the null sets with respect to Π' of A (resp., of S).

Proof. Let us denote by $\overline{\mathcal{I}}^{\pi'}$ the null sets of $A \vee S$ with respect to Π' and by $(AS)_{\pi'}$ the projection of S on A with respect to Π'. Under (ii), by Corollary 2.2.15, $A \perp\!\!\!\perp S \mid (AS)_\pi; \Pi$ implies $A \perp\!\!\!\perp S \mid (AS)_\pi; \Pi'$. Therefore, $(AS)_{\pi'} \subset (AS)_\pi \vee (A \cap \overline{\mathcal{I}}^{\pi'})$ by Theorem 4.3.3(ii). By symmetry, $(SA)_{\pi'} \subset (SA)_\pi \vee (S \cap \overline{\mathcal{I}}^{\pi'})$. ∎

4.4.3   Minimal Sufficiency in a Dominated Experiment

In a dominated Bayesian experiment $\mathcal{E} = (A \times S,\ A \vee S,\ \Pi)$ with $\Pi \ll \mu \otimes P$ the search for the minimal sufficient parameter and the minimal sufficient statistic may be considerably simplified by looking at measurability properties of the density $g_{A\vee S} = d\Pi/(d\mu \otimes P)$.

4.4.9 Theorem. In a dominated Bayesian experiment $\mathcal{E} = \{A \times S, A \vee S, \Pi\}$, a parameter $B \subset A$ and a statistic $T \subset S$ are $\mathcal{E}$-sufficient if and only if $g_{A\vee S} \in \overline{B \vee T}^{\mu \otimes P}$.

Proof. If $g_{A\vee S} \in \overline{B \vee T}$, then $g_{A\vee S} \in \overline{B \vee S}$ and $g_{A\vee S} \in \overline{A \vee T}$ and, by Proposition 2.3.9, this is equivalent to saying that B is $\mathcal{E}$-sufficient and T is $\mathcal{E}$-sufficient. However, by Proposition 2.3.8(i) and (iii), B and T are $\mathcal{E}$-sufficient if and only if $g_{A\vee S} = g_{A\vee T}$ and $g_{A\vee S} = g_{B\vee S}$. Now, since $A \perp\!\!\!\perp S; \mu \otimes P$ implies $(A \vee T) \perp\!\!\!\perp (B \vee S) \mid B \vee T; \mu \otimes P$ we have, successively,

$$
\begin{aligned}
g_{B\vee T} &= (\widetilde{B \vee T})(g_{A\vee S}) = (\widetilde{B \vee T})(g_{A\vee T}) = (\widetilde{B \vee S})(g_{A\vee T}) \\
&= (\widetilde{B \vee S})(g_{B\vee S}) = g_{B\vee S} = g_{A\vee S}.
\end{aligned}
$$

Hence, $g_{A\vee S} = g_{B\vee T}$ a.s. $\mu \otimes P$. ∎

This result provides the following characterization of the $\mathcal{E}$-minimal sufficient parameter and of the $\mathcal{E}$-minimal sufficient statistic.

4.4.10 Proposition. Let $\mathcal{E} = \{A \times S, \mathcal{A} \vee \mathcal{S}, \Pi\}$ be a dominated Bayesian experiment. Then:

(i) $g_{\mathcal{A} \vee \mathcal{S}} \in \overline{\mathcal{A}\mathcal{S} \vee \mathcal{S}\mathcal{A}}^{\mu \otimes P}$;

(ii)   If $g_{\mathcal{A} \vee \mathcal{S}} \in \overline{\mathcal{B} \vee \mathcal{T}}^{\mu \otimes P}$,   where $\mathcal{B} \subset \mathcal{A}$ and $\mathcal{T} \subset \mathcal{S}$;
       then $\mathcal{A}\mathcal{S} \subset \overline{\mathcal{B}} \cap \mathcal{A}$ and $\mathcal{S}\mathcal{A} \subset \overline{\mathcal{T}} \cap \mathcal{S}$.   ∎

4.4.4   Sampling Theory and Bayesian Methods

Let us consider a statistical experiment $\mathcal{E} = \{(S, \mathcal{S}); P^a \ a \in \mathcal{A}\}$ and $\mathcal{A}$ a σ-field on A which makes $P^a(X)$ measurable for any $X \in \mathcal{S}$. We use the notation introduced in Section 2.3.7. In sampling theory, a minimal sufficient statistic is known to exist only in the dominated case, whereas it always exists in a Bayesian experiment (see, e.g., Barra (1971), Chap. II, Section 5; see also Pitcher (1957, 1965) . Indeed, if g_*^a is a version of dP^a/dP_* where P_* is a privileged dominating probability, it is known that $\mathcal{T}$ is a sufficient statistic if and only if $g_*^a \in \overline{\mathcal{T}}^{P_*} \ \forall \ a \in A$. Therefore if $\mathcal{S}_1 = \sigma\{g_*^a : a \in A\}$, $\mathcal{S}_1$ is minimal sufficient in the sense that $\mathcal{S}_1 \subset \overline{\mathcal{T}}^{P_*}$ $\forall \ \mathcal{T}$ sufficient.

4.4.11 Theorem. If μ is a regular probability on $(A, \mathcal{A})$ in the sense of Section 2.3.7, then the $\mathcal{S}$-complete (with respect to Π) $\mathcal{E}$-minimal sufficient statistic is equal to $\mathcal{S}_1$ completed by the null sets of $\mathcal{S}$ with respect to P_*, i.e., $\mathcal{S}\mathcal{A} = \overline{\mathcal{S}}_1^{P_*}$.

Proof. By Theorem 2.3.12, $\mathcal{S}\mathcal{A} \subset \overline{\mathcal{S}}_1^{\Pi} \cap \mathcal{S}$. Now, if μ is regular this implies that the predictive probability P is equivalent to any privileged dominating probability $P_*(P \sim P_*)$ and so $\overline{\mathcal{S}}_1^{\Pi} \cap \mathcal{S} = \overline{\mathcal{S}}_1^P = \overline{\mathcal{S}}_1^{P_*}$. However, by Theorem 2.3.14, $\mathcal{S}_1 \subset \overline{\mathcal{S}\mathcal{A}}^{P_*}$. But, as before, $\overline{\mathcal{S}\mathcal{A}}^{P_*} = \overline{\mathcal{S}\mathcal{A}}^{\Pi} \cap \mathcal{S} = \mathcal{S}\mathcal{A}$, and so $\overline{\mathcal{S}}_1^{P_*} = \mathcal{S}\mathcal{A}$.   ∎

In a sampling theory framework, minimal sufficiency on the parameter space is introduced as follows (see, e.g. Barankin (1961)). For any $X \in \mathcal{S}$, let $\mathcal{A}_X$ be the σ-field on A generated by the mapping $a \longrightarrow P^a(X)$ and let $\mathcal{A}_1 = \bigvee_{X \in \mathcal{S}} \mathcal{A}_X$. Then $\mathcal{A}_1$ is the minimal sufficient parameter, i.e., the smallest sub-σ-field of $\mathcal{A}$ which makes the sampling probabilities $P^a(X)$ measurable $\forall\, X \in \mathcal{S}$. We then have the following result.

4.4.12 Theorem. For any prior probability μ on $(A, \mathcal{A})$ the $\mathcal{A}$-complete (with respect to Π) $\mathcal{E}^\mu$-minimal sufficient parameter is equal to $\mathcal{A}_1$ completed by the null sets of $\mathcal{A}$ with respect to μ, i.e., $\mathcal{A}\mathcal{S} = \overline{\mathcal{A}}_1^\mu$.

Proof. With the notation of Section 2.3.7, $\mathcal{A}s = \mathcal{I}^a s$ a.s.$\Pi\ \forall\, s \in [\mathcal{S}]^+$. Hence $\mathcal{A}_1 \subset \overline{\mathcal{A}\mathcal{S}}^\Pi \cap \mathcal{A} = \mathcal{A}\mathcal{S}$. However, $\mathcal{A}s \in \overline{\mathcal{A}}_1^\Pi \cap \mathcal{A}\ \ \forall s \in \mathcal{S}^+$, i.e., $\mathcal{A}\mathcal{S} \subset \overline{\mathcal{A}}_1^\Pi \cap \mathcal{A} = \overline{\mathcal{A}}_1^\mu$. ∎

4.4.5 Minimal Sufficiency in Conditional Experiment

Minimal sufficiency in a conditional experiment $\mathcal{E}^{\mathcal{M}}_{\mathcal{B}\vee\mathcal{T}}$ requires some care because, in general, the intersection of two $\mathcal{E}^{\mathcal{M}}_{\mathcal{B}\vee\mathcal{T}}$-sufficient statistics need not be $\mathcal{E}^{\mathcal{M}}_{\mathcal{B}\vee\mathcal{T}}$-sufficient; indeed, although $\mathcal{N}_i \subset \mathcal{M} \vee \mathcal{T}$ $(i = 1, 2)$ and $\mathcal{B} \perp\!\!\!\perp \mathcal{T} \mid \mathcal{M} \vee \mathcal{N}_i$ do imply, by Theorem 2.2.12, that

$$\mathcal{B} \perp\!\!\!\perp \mathcal{T} \mid \overline{(\mathcal{M} \vee \mathcal{N}_1)} \cap \overline{(\mathcal{M} \vee \mathcal{N}_2)},$$

it should be stressed that $\overline{(\mathcal{M} \vee \mathcal{N}_1)} \cap \overline{(\mathcal{M} \vee \mathcal{N}_2)}$ does not have the form $\overline{\mathcal{M} \vee (\mathcal{N}_1 \cap \mathcal{N}_2)}$. This is precisely why the attention is limited to strong $\mathcal{E}^{\mathcal{M}}_{\mathcal{B}\vee\mathcal{T}}$-statistics $\mathcal{N}_i$, i.e., $\mathcal{M} \subset \mathcal{N}_i \subset \mathcal{M} \vee \mathcal{T}$.

We first note that Theorem 4.3.9(ii) implies that for any $\mathcal{M}$:

$$(4.4.1) \qquad (\mathcal{M} \vee \mathcal{T})(\mathcal{B} \vee \mathcal{M}) = [(\mathcal{M} \vee \mathcal{T})\mathcal{B}] \vee \mathcal{M};$$

in other words, the projection of $\mathcal{B} \vee \mathcal{M}$ on $\mathcal{M} \vee \mathcal{T}$ is a strong $\mathcal{E}^{\mathcal{M}}_{\mathcal{B}\vee\mathcal{T}}$-statistic. Now Proposition 4.3.10 provides, through the next proposition, the natural construction of minimal sufficient σ-fields in conditional experiments.

4.4.13 Proposition. In the conditional experiment $\mathcal{E}^{\mathcal{M}}_{\mathcal{B}\vee\mathcal{T}}$:

(i) $(\mathcal{M} \vee \mathcal{T})(\mathcal{B} \vee \mathcal{M})$ is the sufficient statistic which is minimal among the $(\mathcal{M} \vee \mathcal{T})$-complete strong $\mathcal{E}^{\mathcal{M}}_{\mathcal{B}\vee\mathcal{T}}$-statistics;

(ii) $(\mathcal{B} \vee \mathcal{M})(\mathcal{M} \vee \mathcal{T})$ is the sufficient parameter which is minimal among the $(\mathcal{B} \vee \mathcal{M})$-complete strong $\mathcal{E}^{\mathcal{M}}_{\mathcal{B}\vee\mathcal{T}}$-parameters. ∎

4.4.6   Optimal Mutual Sufficiency

If we wish to identify the "optimal" joint reduction of a given experiment, only mutual sufficiency will be considered, due to the non-uniqueness and the nonoperational character of maximal ancillarity; indeed, the structure of mutual exogeneity and of Bayesian cut both involve ancillarity conditions.

We first note that if $(\mathcal{B}, \mathcal{T})$ is mutually sufficient we may, fixing $\mathcal{B}$, optimize $\mathcal{T}$ or, fixing $\mathcal{T}$, optimize $\mathcal{B}$.

4.4.14 Theorem. If $(\mathcal{B}, \mathcal{T})$ are mutually sufficient, then

(i) $(\mathcal{B}, \mathcal{SB})$ are mutually sufficient and $\mathcal{SB} \subset \overline{\mathcal{T}} \cap \mathcal{S}$;

(ii) $(\mathcal{AT}, \mathcal{T})$ are mutually sufficient and $\mathcal{AT} \subset \overline{\mathcal{B}} \cap \mathcal{A}$.

Proof. The mutual sufficiency of $(\mathcal{B}, \mathcal{T})$ is equivalent to $\mathcal{A} \perp\!\!\!\perp \mathcal{T} \mid \mathcal{B}$ and $\mathcal{B} \perp\!\!\!\perp \mathcal{S} \mid \mathcal{T}$. Clearly, $\mathcal{B} \perp\!\!\!\perp \mathcal{S} \mid \mathcal{SB}$. Now, $\mathcal{B} \perp\!\!\!\perp \mathcal{S} \mid \mathcal{T}$ is equivalent to $\mathcal{SB} \subset \overline{\mathcal{T}}$, and then $\mathcal{A} \perp\!\!\!\perp \mathcal{T} \mid \mathcal{B}$ implies $\mathcal{A} \perp\!\!\!\perp \mathcal{SB} \mid \mathcal{B}$, i.e., $(\mathcal{B}, \mathcal{SB})$ are mutually sufficient. Similarly, $(\mathcal{AT}, \mathcal{T})$ are mutually sufficient.   ∎

Simultaneous optimization on $\mathcal{B}$ and $\mathcal{T}$ is not relevant without constraint since the solution is trivial, i.e., $\overline{\mathcal{I}} \cap \mathcal{A}$ and $\overline{\mathcal{I}} \cap \mathcal{S}$.

Now if $\mathcal{B}$ represents a parameter of interest, it is not always possible to exhibit a statistic $\mathcal{T}$ such that $\mathcal{B}$ and $\mathcal{T}$ are mutually sufficient; in particular, $\mathcal{B}$ and $\mathcal{S}$ are not mutually sufficient unless $\mathcal{B}$ is sufficient. It is therefore meaningful to look for a pair of σ-fields $\mathcal{B}_1$ and $\mathcal{T}$ such that $\mathcal{B} \subset \mathcal{B}_1$ and $\mathcal{B}_1$ and $\mathcal{T}$ are mutually sufficient. Clearly, $\mathcal{A}$ and $\mathcal{S}$ would satisfy these conditions but one would like to exhibit a pair of smallest ones. This is the goal of the next theorem:

4.4.15 Theorem. Let $\mathcal{B}$ be a sub-σ-field of $\mathcal{A}$ in a Bayesian experiment $\mathcal{E}$. Then there exists a couple of σ-fields $\mathcal{B}_1 \subset \mathcal{A}$ and $\mathcal{T}_1 \subset \mathcal{S}$ such that:

(i) $\mathcal{B} \subset \mathcal{B}_1$;

(ii) $\mathcal{B}_1$ and $\mathcal{T}_1$ are mutually sufficient;

(iii) $\mathcal{C}$ and $\mathcal{T}$ mutually sufficient and $\mathcal{B} \subset \mathcal{C}$ imply $\mathcal{B}_1 \subset \overline{\mathcal{C}} \cap \mathcal{A}$ and $\mathcal{T}_1 \subset \overline{\mathcal{T}} \cap \mathcal{S}$.

The pair $(\mathcal{B}_1, \mathcal{T}_1)$ will be called $\mathcal{B}$-*optimal mutually sufficient.*

Proof. Let us consider:

$$\Phi = \{(\mathcal{C}, \mathcal{U} \mid \mathcal{C} = \overline{\mathcal{C}} \cap \mathcal{A} \supset \mathcal{A}\mathcal{U} \vee \mathcal{B}, \mathcal{U} = \overline{\mathcal{U}} \cap \mathcal{S} \supset \mathcal{S}\mathcal{C}\}.$$

Note that Φ is not empty because $(\mathcal{A}, \mathcal{S}) \in \Phi$ and that each pair $(\mathcal{C}, \mathcal{U})$ in Φ are mutually sufficient by Theorem 4.3.3. Thus we define:

$$\Phi_{\mathcal{A}} = \{\mathcal{C} \mid \exists \,\mathcal{U} : (\mathcal{C}, \mathcal{U}) \in \Phi\} \qquad \mathcal{B}_1 = \cap\{\mathcal{C} \mid \mathcal{C} \in \Phi_{\mathcal{A}}\}$$
$$\Phi_{\mathcal{S}} = \{\mathcal{U} \mid \exists \,\mathcal{C} : (\mathcal{C}, \mathcal{U}) \in \Phi\} \qquad \mathcal{T}_1 = \cap\{\mathcal{U} \mid \mathcal{U} \in \Phi_{\mathcal{S}}\}.$$

Clearly, $\mathcal{B}_1 = \overline{\mathcal{B}}_1 \cap \mathcal{A}$, $\mathcal{B} \subset \mathcal{B}_1$ and $\mathcal{T}_1 = \overline{\mathcal{T}}_1 \cap \mathcal{S}$. We only have to show that $(\mathcal{B}_1, \mathcal{T}_1) \in \Phi$. Indeed,

$$\forall \, \mathcal{C} \in \Phi_{\mathcal{A}} \quad \exists \, \mathcal{U} \quad \mathcal{A}\mathcal{T}_1 \subset \mathcal{A}\mathcal{U} \subset \mathcal{C}.$$

Hence $\mathcal{A}\mathcal{T}_1 \subset \mathcal{B}_1$. Similarly $\mathcal{S}\mathcal{B}_1 \subset \mathcal{T}_1$. ∎

Note that the proof does not give a construction of $\mathcal{B}_1$ and $\mathcal{T}_1$. However, according to Theorem 4.4.14, $\mathcal{B}_1$ and $\mathcal{T}_1$ must obey some relations which may be helpful in searching for them.

4.4.16 Corollary. If $(\mathcal{B}_1, \mathcal{T}_1)$ are $\mathcal{B}$-optimal mutually sufficient, then:
(i) $\mathcal{T}_1 = \mathcal{S}\mathcal{B}_1$;

(ii) $\mathcal{B}_1 = \mathcal{B} \vee \mathcal{A}\mathcal{T}_1$. ∎

In applications, it often occurs that the determination of $\mathcal{B}_1$ and $\mathcal{T}_1$ is facilitated due to the following argument: If $\mathcal{B}$ is the parameter of interest, then $\mathcal{T}_1 = \mathcal{S}\mathcal{B}$ is the $\mathcal{E}_{\mathcal{B}}$-minimal sufficient statistic. Now $\mathcal{B}_1 = \mathcal{A}\mathcal{T}_1 = \mathcal{A}(\mathcal{S}\mathcal{B})$ is the $\mathcal{E}_{\mathcal{T}_1}$-minimal sufficient parameter. If it can be proved that $\mathcal{B} \subset \mathcal{B}_1$ and $\mathcal{B}_1 \perp\!\!\!\perp \mathcal{S} \mid \mathcal{T}_1$, then $(\mathcal{B}_1, \mathcal{T}_1)$ are the $\mathcal{B}$-optimal mutually sufficient σ-fields. Indeed, if $(\mathcal{C}, \mathcal{U})$ is mutually sufficient with $\mathcal{B} \subset \mathcal{C}$, then

$\mathcal{C} \perp\!\!\!\perp \mathcal{S} \mid \mathcal{U}$ implies $\mathcal{B} \perp\!\!\!\perp \mathcal{S} \mid \mathcal{U}$, and therefore, $\mathcal{T}_1 = \mathcal{SB} \subset \overline{\mathcal{U}} \cap \mathcal{S}$. By the same token, $\mathcal{A} \perp\!\!\!\perp \mathcal{U} \mid \mathcal{C}$ implies $\mathcal{A} \perp\!\!\!\perp \mathcal{T}_1 \mid \mathcal{C}$, and therefore, $\mathcal{B}_1 = \mathcal{AT}_1 \subset \overline{\mathcal{C}} \cap \mathcal{A}$.

Note that $\mathcal{B}_1$ is equal to $\mathcal{B}$ if and only if there exists $\mathcal{T} \subset \mathcal{S}$ such that $(\mathcal{B}, \mathcal{T})$ is mutually sufficient. But this is not generally true, as illustrated by the following example.

Example. Consider again the example provided in Section 3.4.1, i.e., $x = (y, z)'$ and

$$(x \mid a) \sim N_2[(\sin \varphi, \cos \varphi)', \eta^{-1} I_2]$$

with

$$a = (\varphi, \eta) \in A = (-\pi, \pi] \times \mathbb{R}_0^+.$$

Let $\varphi \perp\!\!\!\perp \eta$, where η follows a gamma distribution and φ follows a uniform distribution. Recall that $\eta \perp\!\!\!\perp x \mid r^2$ and $a \perp\!\!\!\perp r^2 \mid \eta$ since $(r^2 \mid a) \sim \frac{1}{\eta} \chi_2^2(\eta)$. Now, if $\mathcal{B} = \sigma\{\eta\}$, and $\mathcal{S} = \sigma\{x\}$, $\mathcal{T}_1 = \mathcal{SB} = \sigma\{r^2\}$, and $\mathcal{B}_1 = \mathcal{AT}_1 = \mathcal{B}$. Therefore, since $\mathcal{B}_1 \perp\!\!\!\perp \mathcal{S} \mid \mathcal{T}_1$, by the remark following Corollary 4.4.16 it follows that (η, r^2) is η-optimal mutually sufficient.   ∎

4.5   Identification among σ-Fields

In a Bayesian experiment, the sample information bears on a sufficient parameter only: conditionally on a sufficient parameter, the Bayesian experiment is totally non-informative. This raises the following question: why should one introduce a σ-field larger than the minimal sufficient parameter, i.e., the smallest σ-field that makes the sampling probabilities measurable? In other words, why should $\mathcal{A}$ be different from $\mathcal{AS}$? In nonexperimental fields (such as econometrics — see, e.g., Drèze (1974), Fisher (1966), or Hannan (1971)) such models are often introduced. There, it is standard procedure to introduce redundant parametrization as an early stage of model building or as a support for relevant prior information or because the parameter of interest (making e.g., the loss function measurable) is larger than the minimal sufficient parameter. In experimental fields, it may be the case that the experimental design will not provide information on all the parameters of a theoretically relevant model.

A model for which the parametrization is redundant (i.e., $\mathcal{AS}$ is strictly contained in $\mathcal{A}$) will be called non-identified. Otherwise, i.e., if $\mathcal{AS} = \mathcal{A}$, the model will be said identified. The object of the next section is to study the relations existing between $\mathcal{A}$ and $\mathcal{AS}$. We first present the abstract probability theory underlying the statistical analysis of identification that is presented in Section 4.6.

As in Sections 2.2 and 4.3, we handle a topic of general probability theory and so adopt the same notation; in particular, $(M, \mathcal{M}, P)$ is an abstract probability space.

4.5.1 Definition. $\mathcal{M}_1$ is *identified* by $\mathcal{M}_2$ *conditionally on* $\mathcal{M}_3$ or, equivalently, $\mathcal{M}_2$ *identifies* $\mathcal{M}_1$ *conditionally on* $\mathcal{M}_3$ if

$$(\mathcal{M}_1 \vee \mathcal{M}_3)(\mathcal{M}_2 \vee \mathcal{M}_3) = \mathcal{M}_1 \vee \mathcal{M}_3.$$

In this case we write $\mathcal{M}_1 < \mathcal{M}_2 \mid \mathcal{M}_3$. If $\mathcal{M}_3 = \mathcal{M}_0$ we say $\mathcal{M}_1$ is *identified* by $\mathcal{M}_2$, and write $\mathcal{M}_1 < \mathcal{M}_2$. If we want to make explicit the role of the probability P, we write $\mathcal{M}_1 < \mathcal{M}_2 \mid \mathcal{M}_3; P$ and $\mathcal{M}_1 < \mathcal{M}_2; P$. ∎

Note that, by Definition 4.5.1, $\mathcal{M}_1 < \mathcal{M}_2 \mid \mathcal{M}_3$ is equivalent to $(\mathcal{M}_1 \vee \mathcal{M}_3) < (\mathcal{M}_2 \vee \mathcal{M}_3)$. Consequently, theorems stated conditionally on $\mathcal{M}_3$ need be proved for $\mathcal{M}_3 = \mathcal{M}_0$ only.

4.5.2 Proposition. *Elementary Properties of Identification*

(i) $\mathcal{M}_1 < \mathcal{M}_2 \mid \mathcal{M}_3 \Leftrightarrow \overline{\mathcal{M}}_1 < \mathcal{M}_2 \mid \mathcal{M}_3 \Leftrightarrow \mathcal{M}_1 < \overline{\mathcal{M}}_2 \mid \mathcal{M}_3$
$\Leftrightarrow \mathcal{M}_1 < \mathcal{M}_2 \mid \overline{\mathcal{M}}_3$;

(ii) $\mathcal{M}_1 \subset \overline{\mathcal{M}_2 \vee \mathcal{M}_3} \Rightarrow \mathcal{M}_1 < \mathcal{M}_2 \mid \mathcal{M}_3$;

(iii) $\mathcal{M}_2 \subset \overline{\mathcal{M}_1 \vee \mathcal{M}_3}$ and $\mathcal{M}_1 < \mathcal{M}_2 \mid \mathcal{M}_3 \Rightarrow \overline{\mathcal{M}_1 \vee \mathcal{M}_3} = \overline{\mathcal{M}_2 \vee \mathcal{M}_3}$;

(iv) $\mathcal{M}_2 \subset \overline{\mathcal{M}_4 \vee \mathcal{M}_3}$ and $\mathcal{M}_1 < \mathcal{M}_2 \mid \mathcal{M}_3 \Rightarrow \mathcal{M}_1 < \mathcal{M}_4 \mid \mathcal{M}_3$;

(v) $(\mathcal{M}_1 \vee \mathcal{M}_4) < \mathcal{M}_2 \mid \mathcal{M}_3 \Rightarrow \mathcal{M}_1 < \mathcal{M}_2 \mid \mathcal{M}_3 \vee \mathcal{M}_4$
$\Rightarrow \mathcal{M}_1 < (\mathcal{M}_2 \vee \mathcal{M}_4) \mid \mathcal{M}_3$.

Proof. Only elementary Property (v) requires a proof. Take $\mathcal{M}_3 = \mathcal{M}_0$. Clearly, by elementary Property (iv), $\mathcal{M}_1 \vee \mathcal{M}_4 < \mathcal{M}_2$ implies that:

$$\mathcal{M}_1 \vee \mathcal{M}_4 < \mathcal{M}_2 \vee \mathcal{M}_4, \quad \text{i.e.,} \quad \mathcal{M}_1 < \mathcal{M}_2 \mid \mathcal{M}_4.$$

Now, if $\mathcal{M}_1 \vee \mathcal{M}_4 < \mathcal{M}_2 \vee \mathcal{M}_4$, then

$$\mathcal{M}_1 \subset \mathcal{M}_1 \vee \mathcal{M}_4 = (\mathcal{M}_1 \vee \mathcal{M}_4)(\mathcal{M}_2 \vee \mathcal{M}_4) \subset \overline{\mathcal{M}_4} \vee \mathcal{M}_1(\mathcal{M}_2 \vee \mathcal{M}_4)$$

by Theorem 4.3.9(i). Next, $\mathcal{M}_1 \perp\!\!\!\perp (\mathcal{M}_2 \vee \mathcal{M}_4) \mid \mathcal{M}_1(\mathcal{M}_2 \vee \mathcal{M}_4)$ implies, by Corollary 2.2.11 and Corollary 2.2.7, that

$$\mathcal{M}_1 \perp\!\!\!\perp \{\overline{\mathcal{M}_4} \vee \mathcal{M}_1(\mathcal{M}_2 \vee \mathcal{M}_4)\} \mid \mathcal{M}_1(\mathcal{M}_2 \vee \mathcal{M}_4).$$

Therefore, $\mathcal{M}_1 \perp\!\!\!\perp \mathcal{M}_1 \mid \mathcal{M}_1(\mathcal{M}_2 \vee \mathcal{M}_4)$ and by Corollary 2.2.8, this is equivalent to $\mathcal{M}_1 \subset \overline{\mathcal{M}_1(\mathcal{M}_2 \vee \mathcal{M}_4)}$ and, therefore, $\mathcal{M}_1 = \mathcal{M}_1(\mathcal{M}_2 \vee \mathcal{M}_4)$ by elementary Property 4.3.2(viii), i.e., $\mathcal{M}_1 < \mathcal{M}_2 \vee \mathcal{M}_4$. ∎

Note that elementary Property (iv) amounts to saying that the identifying σ-field can be increased. However, the identified σ-field cannot be decreased: i.e., $\mathcal{M}_5 \subset \overline{\mathcal{M}_1 \vee \mathcal{M}_3}$ and $\mathcal{M}_1 < \mathcal{M}_2 \mid \mathcal{M}_3$ does not imply, in general, that $\mathcal{M}_5 < \mathcal{M}_2 \mid \mathcal{M}_3$. Indeed, consider the following example. Let $\{A_i : i = 1, 2, 3, 4\}$ be a measurable partition of M such that $P(A_i) = \frac{1}{4}$; take $\mathcal{M}_1 = \sigma\{A_1 \cup A_3, A_2, A_4\}$, $\mathcal{M}_2 = \sigma\{A_1 \cup A_2, A_3 \cup A_4\}$, $\mathcal{M}_3 = \mathcal{M}_0$ and $\mathcal{M}_5 = \sigma\{A_1 \cup A_3, A_2 \cup A_4\}$. Clearly, $\mathcal{M}_1\mathcal{M}_2 = \mathcal{M}_1$ but $\mathcal{M}_5\mathcal{M}_2 = \mathcal{M}_0$.

The next two theorems and their corollaries examine the relationships which link identification and conditional identification under some conditional independence conditions. In essence, the first theorem states the conditions under which identification implies conditional identification. The second theorem states conditions under which the identifying σ-field may be reduced and, as a corollary, it states when conditional identification implies identification.

4.5.3 Theorem. Let $\mathcal{M}_i$ $(i = 1, 2, 3, 4)$ be sub-σ-fields of $\mathcal{M}$. If

(i) $\mathcal{M}_4 \perp\!\!\!\perp \mathcal{M}_2 \mid \mathcal{M}_1 \vee \mathcal{M}_3$,

(ii) $\mathcal{M}_1 < \mathcal{M}_2 \mid \mathcal{M}_3$,

then

(iii) $\mathcal{M}_1 < \mathcal{M}_2 \mid \mathcal{M}_3 \vee \mathcal{M}_4$.

Note that (i) is trivially satisfied if $\mathcal{M}_4 \subset \overline{\mathcal{M}_1 \vee \mathcal{M}_3}$.

Proof. Take $\mathcal{M}_3 = \mathcal{M}_0$. Now, by Theorem 4.3.9 (ii),

$$\overline{(\mathcal{M}_1 \vee \mathcal{M}_4)(\mathcal{M}_2 \vee \mathcal{M}_4)} = \overline{\mathcal{M}_4 \vee \overline{(\mathcal{M}_1 \vee \mathcal{M}_4)\mathcal{M}_2}}.$$

Under (i), by Corollary 4.3.6(iii), $\overline{(\mathcal{M}_1 \vee \mathcal{M}_4)\mathcal{M}_2} = \overline{\mathcal{M}_1\mathcal{M}_2}$ and, by (ii), $\overline{\mathcal{M}_1\mathcal{M}_2} = \overline{\mathcal{M}_1}$. Therefore

$$\overline{(\mathcal{M}_1 \vee \mathcal{M}_4)(\mathcal{M}_2 \vee \mathcal{M}_4)} = \overline{\mathcal{M}_1 \vee \mathcal{M}_4},$$

and the result follows from elementary Property 4.3.2(viii). ∎

4.5.4 Theorem. Let $\mathcal{M}_i$ $(i = 1, 2, 3, 4)$ be sub-σ-fields of $\mathcal{M}$. If

(i) $\mathcal{M}_4 \perp\!\!\!\perp \mathcal{M}_1 \mid \mathcal{M}_2 \vee \mathcal{M}_3$,

(ii) $\mathcal{M}_1 < (\mathcal{M}_2 \vee \mathcal{M}_4) \mid \mathcal{M}_3$,

then

(iii) $\mathcal{M}_1 < \mathcal{M}_2 \mid \mathcal{M}_3$.

Note that (i) is trivially satisfied if $\mathcal{M}_4 \subset \overline{\mathcal{M}_2 \vee \mathcal{M}_3}$.

Proof. Take $\mathcal{M}_3 = \mathcal{M}_0$. Under (i), $\mathcal{M}_1(\mathcal{M}_2 \vee \mathcal{M}_4) = \mathcal{M}_1\mathcal{M}_2$ by Corollary 4.3.6(iv). But, under (ii), $\mathcal{M}_1(\mathcal{M}_2 \vee \mathcal{M}_4) = \mathcal{M}_1$ and, therefore, $\mathcal{M}_1\mathcal{M}_2 = \mathcal{M}_1$, i.e., $\mathcal{M}_1 < \mathcal{M}_2$. ∎

4.5.5 Corollary. Let $\mathcal{M}_i$ $(i = 1, 2, 3, 4)$ be sub-σ-fields of $\mathcal{M}$. If

(i) $\mathcal{M}_4 \perp\!\!\!\perp \mathcal{M}_1 \mid \mathcal{M}_2 \vee \mathcal{M}_3$.

(ii) $\mathcal{M}_1 < \mathcal{M}_2 \mid \mathcal{M}_3 \vee \mathcal{M}_4$, then

(iii) $\mathcal{M}_1 < \mathcal{M}_2 \mid \mathcal{M}_3.$

Note that (i) is trivially satisfied if $\mathcal{M}_4 \subset \overline{\mathcal{M}_2 \vee \mathcal{M}_3}$.

Proof. Indeed, by elementary Property 4.5.2(v), $\mathcal{M}_1 < \mathcal{M}_2 \mid \mathcal{M}_3 \vee \mathcal{M}_4$ implies $\mathcal{M}_1 < (\mathcal{M}_2 \vee \mathcal{M}_4) \mid \mathcal{M}_3$, and the result follows from Theorem 4.5.4.   ∎

Combining Theorems 4.5.3 and 4.5.4 together with Corollary 4.5.5 entails the following corollary.

4.5.6 Corollary. Let $\mathcal{M}_i (i = 1, 2, 3, 4)$ be sub-σ-fields of $\mathcal{M}$. If

(o) $\mathcal{M}_4 \perp\!\!\!\perp (\mathcal{M}_1 \vee \mathcal{M}_2) \mid \mathcal{M}_3$

the following identification properties are equivalent:

(i) $\mathcal{M}_1 < \mathcal{M}_2 \mid \mathcal{M}_3;$

(ii) $\mathcal{M}_1 < (\mathcal{M}_2 \vee \mathcal{M}_4) \mid \mathcal{M}_3;$

(iii) $\mathcal{M}_1 < \mathcal{M}_2 \mid \mathcal{M}_3 \vee \mathcal{M}_4.$ ∎

Restricting the conditioning σ-field in Theorem 4.5.4 to be the minimal one we obtain, as a corollary:

4.5.7 Corollary. Let $\mathcal{M}_5 \subset \mathcal{M}_1, \mathcal{M}_6 \subset \mathcal{M}_2$. Then, for any $\mathcal{M}_4 \subset \mathcal{M}_2$,

(i) $\mathcal{M}_4 < \mathcal{M}_1 \mid \mathcal{M}_5 \vee \mathcal{M}_6 \Leftrightarrow \mathcal{M}_4 < \mathcal{M}_1 \mathcal{M}_2 \mid \mathcal{M}_5 \vee \mathcal{M}_6;$

(ii) $\mathcal{M}_4 < \mathcal{M}_1 \mid \mathcal{M}_5 \vee \mathcal{M}_6 \Rightarrow \mathcal{M}_4 < \mathcal{M}_2 \mathcal{M}_1 \mid \mathcal{M}_5 \vee \mathcal{M}_6.$

Proof. By Corollary 2.2.11 and Theorem 4.3.3, we successively have:

$$\mathcal{M}_1 \perp\!\!\!\perp \mathcal{M}_2 \mid \mathcal{M}_1\mathcal{M}_2 \vee \mathcal{M}_5 \vee \mathcal{M}_6$$
$$\mathcal{M}_1 \perp\!\!\!\perp \mathcal{M}_2 \mid \mathcal{M}_2\mathcal{M}_1 \vee \mathcal{M}_5 \vee \mathcal{M}_6.$$

Therefore $\mathcal{M}_4 < \mathcal{M}_1 \mid \mathcal{M}_5 \vee \mathcal{M}_6$ implies $\mathcal{M}_4 < \mathcal{M}_1\mathcal{M}_2 \mid \mathcal{M}_5 \vee \mathcal{M}_6$ by Theorem 4.5.4, and the reverse implication is trivial. On the other hand, by elementary Property 4.5.2(iv), $\mathcal{M}_4 < \mathcal{M}_1 \mid \mathcal{M}_5 \vee \mathcal{M}_6$ implies $\mathcal{M}_4 < \mathcal{M}_1 \vee \mathcal{M}_2\mathcal{M}_1 \mid \mathcal{M}_5 \vee \mathcal{M}_6$ and, by Theorem 4.5.4, this implies $\mathcal{M}_4 < \mathcal{M}_2\mathcal{M}_1 \mid \mathcal{M}_5 \vee \mathcal{M}_6$. ∎

4.6 Identification in Bayesian Experiments

4.6.1 Identification in a Reduced Experiment

Let us consider a general experiment $\mathcal{E}_{\mathcal{B}\vee\mathcal{T}}^{\mathcal{M}}$ where $\mathcal{B} \subset \mathcal{A}, \mathcal{T} \subset \mathcal{S}$ and $\mathcal{M} \subset \mathcal{A} \vee \mathcal{S}$. We have seen, in Proposition 4.4.13, that the minimal sufficient $(\mathcal{B} \vee \mathcal{M})$-complete strong $\mathcal{E}_{\mathcal{B}\vee\mathcal{T}}^{\mathcal{M}}$-parameter is given by $(\mathcal{B} \vee \mathcal{M})(\mathcal{M} \vee \mathcal{T})$. In line with the introduction to Section 4.5, we are lead to provide the following definition of identification.

4.6.1 Definition. The experiment $\mathcal{E}_{\mathcal{B}\vee\mathcal{T}}^{\mathcal{M}}$ is *identified* if $\mathcal{B} < \mathcal{T} \mid \mathcal{M}$, i.e., if $(\mathcal{B} \vee \mathcal{M})(\mathcal{M} \vee \mathcal{T}) = \mathcal{B} \vee \mathcal{M}$. In particular, the unreduced experiment $\mathcal{E}$ is *identified* if $\mathcal{A} < \mathcal{S}$, i.e., if $\mathcal{A}\mathcal{S} = \mathcal{A}$. ∎

More generally, we may introduce the concept of identification for an $\mathcal{E}_{\mathcal{B}\vee\mathcal{T}}^{\mathcal{M}}$-parameter, i.e., a sub-$\sigma$- field $\mathcal{L} \subset \mathcal{B} \vee \mathcal{M}$.

4.6.2 Definition. $\mathcal{L}$ is an $\mathcal{E}_{\mathcal{B}\vee\mathcal{T}}^{\mathcal{M}}$-*identified parameter* if the experiment $\mathcal{E}_{\mathcal{L}\vee\mathcal{T}}^{\mathcal{M}}$ is identified, i.e., $\mathcal{L} < \mathcal{T} \mid \mathcal{M}$ or $(\mathcal{L} \vee \mathcal{M})(\mathcal{M} \vee \mathcal{T}) = \mathcal{L} \vee \mathcal{M}$. In particular, $\mathcal{B}$ is an $\mathcal{E}$-identified parameter if $\mathcal{E}_{\mathcal{B}\vee\mathcal{S}}$ is identified, i.e., $\mathcal{B} < \mathcal{S}$ or $\mathcal{B}\mathcal{S} = \mathcal{B}$. ∎

Note that identification appears as a problem of minimal sufficiency on the parameter space rather than on the sample space. This formal analogy has already been remarked among others by Barankin (1961), Kadane (1974) and Picci (1977).

Note also that the concept of identification makes this property dependent on the prior probability through its null sets (see the remarks following 4.3.1 and 4.3.2); the following example illustrates this fact.

Example. Let $A = \mathbb{R}$, $\mathcal{A}$ be the Borel sets of $\mathbb{R}$. Let us assume $(x_i \mid a) \sim i.N(|a|, 1)$ and $s = (x_1, x_2, \ldots, x_n)'$. If μ is a prior probability then, clearly, $\mathcal{AS} = \overline{\mathcal{A}}_1^\mu$ where $\mathcal{A}_1 = \sigma\{|a|\} = \{E \in \mathcal{A} : -E = E\}$. If μ is equivalent to the Lebesgue measure, the experiment is not identified but it will be identified if μ is equivalent to the Lebesgue measure on $\mathbb{R}^+$ and if $\mu(\mathbb{R}^-) = 0$. ∎

From the elementary Properties 4.5.2(iv), we obtain the following result.

4.6.3 Proposition. The experiment $\mathcal{E}_{\mathcal{BVT}}^{\mathcal{M}}$ is identified if and only if there exists an $\mathcal{E}_{\mathcal{BVT}}^{\mathcal{M}}$-statistic $\mathcal{N} \subset \mathcal{M} \vee \mathcal{T}$ such that $\mathcal{E}_{\mathcal{BVN}}^{\mathcal{M}}$ is identified. ∎

From the remark following 4.5.2, note that even if $\mathcal{E}_{\mathcal{BVT}}^{\mathcal{M}}$ is identified, a subsequent marginalization on a statistic $\mathcal{N}$, $\mathcal{E}_{\mathcal{BVN}}^{\mathcal{M}}$, or on a parameter $\mathcal{L}$, $\mathcal{E}_{\mathcal{LVT}}^{\mathcal{M}}$, need not be identified. This is obvious for $\mathcal{E}_{\mathcal{BVN}}^{\mathcal{M}}$. This is even more striking for $\mathcal{E}_{\mathcal{LVT}}^{\mathcal{M}}$, but recall that the sampling probabilities of an experiment marginalized on a parameter are obtained by averaging the sampling probabilities; this integration may evidently lead to the loss of identification. This problem has received a good deal of attention in particular applications in which some additional assumptions guarantee the identification of marginal models (see, e.g., Elbers and Ridders (1982) for such results in proportional hazard models). In particular, a sub-parameter of an identified parameter is not necessarily identified. We have already provided one counterexample (following 4.5.2); here is another:

Example. Let $s = (x_1, \ldots, x_n)$, $x_i = (x_{1i}, x_{2i})'$, $a = (a_0, a_1, \ldots, a_n, \Sigma)$ where Σ is a positive definite 2×2 symmetric matrix. Suppose

$$(a_i \mid a_0, \Sigma) \sim i.N(0; 1), \qquad (x_i \mid a) \sim i.N_2\left[\begin{pmatrix} a_i \\ a_0 a_i \end{pmatrix}, \Sigma\right],$$

and let (a_0, Σ) have a prior distribution equivalent to the Lebesgue measure. This model is clearly identified. If we marginalize this model on (a_0, Σ), we obtain

$$(x_i \mid a_0, \Sigma) \sim i.N_2 \left[\begin{pmatrix} 0 \\ 0 \end{pmatrix}; \begin{pmatrix} \sigma_{11} + 1 & \sigma_{12} + a_0 \\ \sigma_{12} + a_0 & \sigma_{22} + a_0^2 \end{pmatrix} \right].$$

This marginal model is not identified.   ∎

It is interesting to remark that the identification of $\mathcal{L}$ in $\mathcal{E}_{\mathcal{BVT}}^{\mathcal{M}}$ depends crucially on $\mu_{\mathcal{B}}^{\mathcal{LVM}}$. If $(a_i \mid a_0, \Sigma)$ were not normal in the above example, identification would not be lost in the marginal experiment (see, e.g., Reiersøl (1950)).

We now turn to the relationship between sufficiency and identification.

On the parameter space, a sufficient parameter is not necessarily identified; but it is identified if and only if this parameter is almost surely equal to the minimal sufficient parameter.

4.6.4 Theorem. An $\mathcal{E}_{\mathcal{BVT}}^{\mathcal{M}}$-sufficient parameter, $\mathcal{L}$, is identified if and only if $\mathcal{L} \vee \mathcal{M}$ is almost surely equal to the minimal complete strong $\mathcal{E}_{\mathcal{BVT}}^{\mathcal{M}}$-sufficient parameter, i.e.,

$$\overline{\mathcal{L} \vee \mathcal{M}} \cap (\mathcal{B} \vee \mathcal{M}) = (\mathcal{B} \vee \mathcal{M})(\mathcal{M} \vee \mathcal{T}).$$

Proof. $\mathcal{L}$ is $\mathcal{E}_{\mathcal{BVT}}^{\mathcal{M}}$-sufficient is equivalent to $(\mathcal{B} \vee \mathcal{M}) \perp\!\!\!\perp (\mathcal{M} \vee \mathcal{T}) \mid \mathcal{M} \vee \mathcal{L}$. By Corollary 4.3.6(ii), $\overline{(\mathcal{L} \vee \mathcal{M})(\mathcal{M} \vee \mathcal{T})} = \overline{(\mathcal{B} \vee \mathcal{M})(\mathcal{M} \vee \mathcal{T})}$. If $\mathcal{L}$ is identified, $(\mathcal{L} \vee \mathcal{M})(\mathcal{M} \vee \mathcal{T}) = \mathcal{L} \vee \mathcal{M}$. Therefore, by elementary Property 4.3.2(viii), we obtain the result. However, if $\overline{\mathcal{L} \vee \mathcal{M}} = \overline{(\mathcal{B} \vee \mathcal{M})(\mathcal{M} \vee \mathcal{T})}$, $\mathcal{L}$ is clearly sufficient and, by the elementary Property 4.3.2(vii) and Corollary 4.3.8, $\overline{\mathcal{L} \vee \mathcal{M}}$ is identified as is $\mathcal{L}$ by elementary Property 4.5.2(i).   ∎

As we have seen, an $\mathcal{E}_{\mathcal{BVT}}^{\mathcal{M}}$-parameter $\mathcal{L}$ contained in $(\mathcal{B} \vee \mathcal{M})(\mathcal{M} \vee \mathcal{T})$ is not necessarily identified. Conversely, an $\mathcal{E}_{\mathcal{BVT}}^{\mathcal{M}}$-identified parameter $\mathcal{L}$ is not necessarily contained in $(\mathcal{B} \vee \mathcal{M})(\mathcal{M} \vee \mathcal{T})$.

Example. Let $S = \{s_1, s_2\}$ and $A = \{a_1, a_2, a_3\}$. Take

$$p(s_1 \mid a_1) = \frac{1}{2},$$
$$p(s_1 \mid a_2) = p(s_1 \mid a_3) = \frac{1}{3}.$$

Then if $\mu(a_i) > 0, i = 1, 2, 3$, we obtain

$$\mathcal{AS} = \mathcal{B}_1 = \sigma(\{a_1\}, \quad \{a_2, a_3\}).$$

If we take $\mathcal{B}_2 = \sigma(\{a_1, a_2\}, \{a_3\})$, then $\mathcal{B}_2\mathcal{S} = \mathcal{B}_2$ and, clearly,
$\mathcal{B}_2 \not\subset \mathcal{B}_1 = \mathcal{AS}$.   ∎

On the sample space a sufficient statistic permits the verification of identification.

4.6.5 Theorem. In $\mathcal{E}_{\mathcal{BVT}}^{\mathcal{M}}$, a parameter $\mathcal{L}$ is identified if and only if $\mathcal{L}$ is identified by a sufficient statistic $\mathcal{N}$.

Proof. $\mathcal{B} \perp\!\!\!\perp \mathcal{T} \mid \mathcal{M} \vee \mathcal{N}$, since $\mathcal{N}$ is sufficient, and this is equivalent to $(\mathcal{B}\vee\mathcal{M}) \perp\!\!\!\perp (\mathcal{M}\vee\mathcal{T}) \mid \mathcal{M}\vee\mathcal{N}$. This implies that $(\mathcal{L}\vee\mathcal{M}) \perp\!\!\!\perp (\mathcal{M}\vee\mathcal{T}) \mid \mathcal{M}\vee\mathcal{N}$, and the result follows from Theorem 4.5.4.   ∎

As a simple corollary to Theorem 4.5.3 and Corollary 4.5.5, we obtain the next proposition which renders explicit the relationship between identification of an experiment and identification in a further conditioning of this experiment.

4.6.6 Proposition.

(i)   If $\mathcal{E}_{\mathcal{BVT}}^{\mathcal{M}}$ is identified,

then $\mathcal{E}_{\mathcal{BVT}}^{\mathcal{LVM}}$ is identified for any $\mathcal{E}_{\mathcal{BVT}}^{\mathcal{M}}$-parameter $\mathcal{L}$;

(ii)   If $\mathcal{E}_{\mathcal{BVT}}^{\mathcal{MVN}}$ is identified for some $\mathcal{E}_{\mathcal{BVT}}^{\mathcal{M}}$-statistic $\mathcal{N}$,

then $\mathcal{E}_{\mathcal{BVT}}^{\mathcal{M}}$ is identified.   ∎

When an experiment is not identified, a basic objective of the theory of identification is to look for a parameter such that the experiment conditioned on this parameter is identified. When this parameter represents exact restrictions, they are called "identifying restrictions" (see, e.g., Kadane (1974)). Hence the following definition:

4.6.7 Definition. An $\mathcal{E}^{\mathcal{M}}_{\mathcal{B}\vee\mathcal{T}}$-parameter $\mathcal{L}$ is *identifying* for $\mathcal{E}^{\mathcal{M}}_{\mathcal{B}\vee\mathcal{T}}$ if $\mathcal{E}^{\mathcal{L}\vee\mathcal{M}}_{\mathcal{B}\vee\mathcal{T}}$ is identified. ∎

Example. Let $s = (x_1, \ldots, x_n)$ and $a = (a_1, a_2)$. Suppose that the prior probability μ is equivalent to the Lebesgue measure on $\mathbb{R}^2$, and that $(x_i \mid a) \sim i.N(a_1 + a_2, 1)$. This experiment is clearly not identified. The σ-field $\mathcal{B}$ generated by the linear function $b = \lambda_1 a_1 + \lambda_2 a_2$ (λ_j known, $j = 1, 2$) is identifying unless $\lambda_1 = \lambda_2$. Indeed let $\mathcal{C}$ be the σ-field generated by $a_1 + a_2$. Clearly, $\mathcal{A}\mathcal{S} = \overline{\mathcal{C}} \cap \mathcal{A}$ and $\mathcal{B} \vee \mathcal{C} = \mathcal{A}$. Thus, by Theorem 4.3.9(ii), $\mathcal{A}(\mathcal{S} \vee \mathcal{B}) = \mathcal{B} \vee \mathcal{A}\mathcal{S} = \mathcal{A}$. ∎

Let us now look at the robustness of identification for a proper modification of the prior probability. Consider $\mathcal{E}$ and $\mathcal{E}'$, two Bayesian experiments on $(A \times S, \mathcal{A} \vee \mathcal{S})$ defined by their probabilities Π and Π'. If $\Pi' \ll \Pi$ and $d\Pi'/d\Pi \in [\mathcal{A}]^+$, and if $\mathcal{E}'$ is identified, we cannot conclude that $\mathcal{E}$ is identified because Theorem 4.4.8 only implies that $\mathcal{A}$ is equal to $\mathcal{A}\mathcal{S}$ completed by the null sets of $\mathcal{A}$ with respect to Π'; this equality still involves Π' since the class of null sets with respect to Π' is larger than the class of null sets with respect to Π in general. This is not so if the two prior probabilities are equivalent.

4.6.8 Proposition. Let $\mathcal{E}$ and $\mathcal{E}'$ be two Bayesian experiments defined on $(A \times S, \mathcal{A} \vee \mathcal{S})$ by their probabilities Π and Π'. Let $\mathcal{B} \subset \mathcal{A}$, $\mathcal{T} \subset \mathcal{S}$, $\mathcal{M} \subset \mathcal{A} \vee \mathcal{S}$. If

(i) $\Pi' \sim \Pi$,

(ii) $\dfrac{d\Pi'}{d\Pi} \in [\overline{\mathcal{B} \vee \mathcal{M}}]^+$,

then $\mathcal{E}^{\mathcal{M}}_{\mathcal{B}\vee\mathcal{T}}$ is identified if and only if $\mathcal{E}'^{\mathcal{M}}_{\mathcal{B}\vee\mathcal{T}}$ is identified. ∎

Even though the problem of identification is usually studied with reference to the parameter space, we can take advantage of the complete duality of the parameter space and the sample space of a Bayesian experiment to define the concept of an identified statistic:

4.6.9 Definition. $\mathcal{N}$ is an $\mathcal{E}^{\mathcal{M}}_{\mathcal{B}\vee\mathcal{T}}$-*identified statistic* if $\mathcal{N} < \mathcal{B} \mid \mathcal{M}$, i.e., $(\mathcal{M} \vee \mathcal{N})(\mathcal{B} \vee \mathcal{M}) = \mathcal{M} \vee \mathcal{N}$.   ∎

This definition may be useful when searching for the minimal sufficient statistic, as illustrated by the following proposition which is the dual of Theorem 4.6.4.

4.6.10 Proposition. An $\mathcal{E}^{\mathcal{M}}_{\mathcal{B}\vee\mathcal{T}}$-sufficient statistic $\mathcal{N}$ is identified if and only if $\mathcal{M} \vee \mathcal{N}$ is almost surely equal to the minimal complete strong $\mathcal{E}^{\mathcal{M}}_{\mathcal{B}\vee\mathcal{T}}$-sufficient statistic, i.e.,

$$\overline{\mathcal{M} \vee \mathcal{N}} \cap (\mathcal{M} \vee \mathcal{T}) = (\mathcal{M} \vee \mathcal{T})(\mathcal{B} \vee \mathcal{M}).$$ ∎

We conclude this section by studying the relationship between joint reductions and identification. There is not much to say in the case of mutual sufficiency and mutual exogeneity. Indeed, even if the unreduced experiment is identified, the definitions of mutual sufficiency and of mutual exogeneity do not ensure any minimal character to the parameter of interest.

The situation is quite different for the cut. Let $\mathcal{E} = \{A \times S, \mathcal{A} \vee \mathcal{S}, \Pi\}$ be an unreduced Bayesian experiment, and suppose that $\mathcal{B} \subset \mathcal{A}, \mathcal{C} \subset \mathcal{A}$ and $\mathcal{T} \subset \mathcal{S}$ operate a cut. If the unreduced model is identified, i.e., $\mathcal{A} = \mathcal{B}\vee\mathcal{C}$ and $(\mathcal{B}\vee\mathcal{C})\mathcal{S} = \mathcal{B}\vee\mathcal{C}$, the next theorem will show that $\mathcal{E}_{\mathcal{B}\vee\mathcal{T}}$ is also identified, i.e., $\mathcal{B}\mathcal{T} = \mathcal{B}$, but, in general, $\mathcal{E}^{\mathcal{T}}_{\mathcal{C}\vee\mathcal{S}}$ need not be identified, i.e., $(\mathcal{C} \vee \mathcal{T})\mathcal{S} \neq \mathcal{C} \vee \mathcal{T}$ because the minimal $\mathcal{E}^{\mathcal{T}}_{\mathcal{C}\vee\mathcal{S}}$-sufficient parameter is not necessarily uniform. Consider, for instance, the following example:

Example. Let $s = (x_1, x_2, \ldots, x_n)$ with $x_i = (x_{1i}, x_{2i})'$. Let also $a = (a_1, a_2, \ldots, a_n, \sigma_{11}, \sigma_{12}, \sigma_{22})$ with $a_i = (b_i, c_i)'$. Suppose that $(x_i \mid a) \sim i.N_2(a_i, \Sigma)$ and that the prior probability on a is equivalent to the Lebesgue measure. Clearly, $\mathcal{A}\mathcal{S} = \mathcal{A}$ (the $2n + 3$ components of a are identified). Now, $(x_{1i} \mid a) \sim i.N(b_i, \sigma_{11})$ and $(x_{2i} \mid a, x_1) \sim i.N[m_i, d]$ where $m_i = c_i + (\sigma_{12}/\sigma_{11})(x_{1i} - b_i)$ and $d = \sigma_{22} - (\sigma_{12}^2/\sigma_{11})$. Define $\mathcal{T} = \sigma\{x_{11}, \ldots, x_{1n}\}$, $\mathcal{B} = \sigma\{b_1, \ldots, b_n, \sigma_{11}\}$, $\mathcal{C} = \sigma\{c_1 - (\sigma_{12}/\sigma_{11})b_1, \ldots, c_n - (\sigma_{12}/\sigma_{22})b_n, d\}$, and $\mathcal{N} = \sigma\{m_1, \ldots, m_n, d\}$. Clearly, $\mathcal{A} = \mathcal{B}\vee\mathcal{C}$ and $(\mathcal{B}, \mathcal{C}, \mathcal{T})$ operates a cut if $\mathcal{B} \perp\!\!\!\perp \mathcal{C}$, and we also have $\mathcal{B}\mathcal{T} = \mathcal{B}$, but $(\mathcal{C} \vee \mathcal{T})\mathcal{S} = \mathcal{N} \vee \mathcal{T}$ which is strictly included in $\mathcal{C} \vee \mathcal{T}$, and $\mathcal{N}$ is not a uniform parameter.   ∎

4.6.11 Theorem. Let $\mathcal{E} = (A \times S, \mathcal{A} \vee \mathcal{S}, \Pi)$ be an identified Bayesian experiment. Suppose $\mathcal{B} \subset \mathcal{A}$, $\mathcal{C} \subset \mathcal{A}$, and $\mathcal{T} \subset \mathcal{S}$ operate a cut, and $\mathcal{A} = \mathcal{B} \vee \mathcal{C}$. Then $\mathcal{E}_{\mathcal{B} \vee \mathcal{T}}$ is identified.

Proof. Since $\mathcal{E}$ is identified, and $\mathcal{A} = \mathcal{B} \vee \mathcal{C}$, $\mathcal{B} \vee \mathcal{C} < \mathcal{S}$. By the elementary Property 4.5.2 (v), this implies that $\mathcal{B} < \mathcal{C} \vee \mathcal{S}$. Now, since $\mathcal{B} \perp\!\!\!\perp \mathcal{S} \mid \mathcal{C} \vee \mathcal{T}$, by Theorem 4.5.4, this implies $\mathcal{B} < \mathcal{C} \vee \mathcal{T}$. But $\mathcal{B} \perp\!\!\!\perp \mathcal{C}$ and $\mathcal{C} \perp\!\!\!\perp \mathcal{T} \mid \mathcal{B}$ is equivalent to $\mathcal{C} \perp\!\!\!\perp (\mathcal{B} \vee \mathcal{T})$; this consequently implies that $\mathcal{B} \perp\!\!\!\perp \mathcal{C} \mid \mathcal{T}$. Therefore, by another application of Theorem 4.5.4, $\mathcal{B} < \mathcal{C} \vee \mathcal{T}$ implies $\mathcal{B} < \mathcal{T}$. ∎

4.6.2 Sampling Theory and Bayesian Methods

A. Identification in Unreduced Experiments

In a sampling theory framework, one starts from a statistical experiment:

$$\mathcal{E} = \{(S, \mathcal{S}), P^a : a \in A\}.$$

A widely accepted definition of identification is the following (see, e.g., Fisher (1966), Rothenberg (1971), LeCam and Schwartz (1960), Bunke and Bunke (1974) or Deistler and Seifert (1978)):

4.6.12 Definition. The statistical experiment is *s-identified* (s for "sampling") if the mapping $a \to P^a$ is injective (i.e., if: $a \neq a' \Rightarrow P^a \neq P^{a'}$). ∎

Remark. An equivalent definition may be stated as follows. One first introduces an equivalence relation on $S : a \underset{s}{\sim} a'$, read as a and a' are *observationally equivalent*, and defined as: $a \underset{s}{\sim} a' \Leftrightarrow P^a = P^{a'}$. Denote the quotient space for this equivalence as A/s. The statistical experiment is then identified if $A/s = A/o$ where A/o is the trivial partition: $A/o = \{\{a\} \mid a \in A\}$.

In a decision theory context, the parameter of interest is naturally introduced as a σ-field that makes both the loss function and the family of mappings $a \longrightarrow P^a(X)$, $X \in \mathcal{S}$ measurable. This leads to this other concept of identification (see, e.g., Neveu (1970), example in Section III.2).

4.6.13 Definition. Let $\mathcal{A}$ be a σ-field on A representing the parameter of interest. Let $\mathcal{A}_1$ be the smallest sub-σ-field of $\mathcal{A}$ that makes the mappings

$a \longrightarrow P^a(X)$, $X \in \mathcal{S}$, measurable (see Theorem 4.4.12). The statistical experiment $\mathcal{E}$ is *c-identified* (c for "classical") if $\mathcal{A} = \mathcal{A}_1$.   ∎

The third definition has often been suggested as a Bayesian counterpart to Definition 4.6.12 (see, e.g., Schönfeld (1975), Deistler and Seifert (1978)). This concept requires that the statistical experiment $\mathcal{E}$ is endowed with a prior probability μ on $(A, \mathcal{A})$ where $\mathcal{A}$ makes the mappings $a \longrightarrow P^a(X), X \in \mathcal{S}$, measurable.

4.6.14 Definition. The statistical experiment $\mathcal{E}$ is *a.s.s-identified* (a.s.s. for "almost sure sampling") if $\exists A_0 \in \mathcal{A}$ such that $\mu(A_0) = 0$ and the statistical experiment $\mathcal{E}_0 = \{(S, \mathcal{S}), P^a : a \in A - A_o\}$ is s-identified (equivalently, if the mapping $a \longrightarrow P^a$ is injective on $A - A_0$).   ∎

For the sake of discussion, we recall the definition of the previous section:

4.6.15 Definition. The Bayesian experiment $\mathcal{E} = \{A \times S, \mathcal{A} \vee \mathcal{S}, \Pi\}$ is *b-identified* (b for Bayesian) if $\mathcal{A} = \mathcal{AS}$.   ∎

We now examine the relationships between these four concepts of identification. These concepts rely on different levels of specification: the first three concepts are based on a family of sampling probabilities without (Definition 4.6.12) or with (Definitions 4.6.13 and 4.6.14) a measurable structure on the parameter space, and without (Definitions 4.6.12 and 4.6.13) or with (Definition 4.6.14) a prior probability on that measurable structure. Note also that the fourth concept does not formally impose any regularity condition on the Bayesian experiment.

Let us consider the relationships between the two "non-Bayesian" concepts. This involves the concept of a *separating* σ-field (Definition 0.2.7), i.e., a σ-field whose atoms are singletons or, in other words, $A/\mathcal{A} = A/o$.

4.6.16 Theorem. *c-identification implies s-identification if and only if $\mathcal{A}$ is separating.*

Proof. Recall that for any σ-field $\mathcal{B} \subset \mathcal{A}$, the $\mathcal{B}$-measurable functions are constant on the $\mathcal{B}$-atoms. In particular, if $\mathcal{B} \supset \mathcal{A}_1$, $P^a(X)$ is constant on the

$\mathcal{B}$-atoms; in other words, the elements of the $\mathcal{B}$-atoms are observationally equivalent. We now prove that if $\mathcal{B} = \mathcal{A}_1$, the $\mathcal{A}_1$-atoms are exactly the equivalence classes of A/s. Indeed, if we denote by $\mathcal{A}_X$, as in Section 4.4.4, the σ-field generated by the mapping $a \to P^a(X)$, we have, by Proposition 0.2.5, $P^a = P^{a'}$ (i.e., $a \underset{s}{\sim} a'$) $\Leftrightarrow a \underset{\mathcal{A}_X}{\sim} a' \; \forall \, X \in \mathcal{S} \Leftrightarrow a \underset{\mathcal{A}_1}{\sim} a'$, because $\mathcal{A}_1 = \bigvee_{X \in \mathcal{S}} \mathcal{A}_X$. It is therefore clear that : $\mathcal{A} = \mathcal{A}_1 \Rightarrow A/s = A/o$ if and only if $A/\mathcal{A} = A/o$. $\blacksquare$

4.6.17 Theorem. If $\mathcal{A}$ is a Blackwell σ-field (see Definition 0.2.12) and $\mathcal{S}$ is separable then s-identification implies c-identification:

$$A/s = A/o \Rightarrow \mathcal{A} = \mathcal{A}_1.$$

Proof.

(i) We first prove that if $\mathcal{S}$ is separable, then $\mathcal{A}_1$ is also separable. Indeed, consider a countable π-system of sets $\{X_n\}$ generating $\mathcal{S}$. Then $\mathcal{A}_{X_n}$ is separable as it is the inverse image of Borel sets in $[0, 1]$ by the mapping $a \to P^a(X_n)$. Since $\{X \subset S : a \to P^a(X)$ is Borel measurable$\}$ is a d-system, by Theorem 0.2.20(ii), it contains $\mathcal{S}$, and therefore $\mathcal{A}_1 = \bigvee_n \mathcal{A}_{X_n}$, therefore $\mathcal{A}_1$ is separable.

(ii) s-identification implies that $\mathcal{A}_1$ is separating because $a \underset{s}{\sim} a' \Leftrightarrow a \underset{\mathcal{A}_1}{\sim} a'$ (see proof of Theorem 4.6.16). Therefore $\mathcal{A}$ is also separating, as $\mathcal{A}_1 \subset \mathcal{A}$.

(iii) As $\mathcal{A}_1$ is a separable sub-σ-field of a separating σ-field $\mathcal{A}$, the result follows from Blackwell's Theorem (see Theorem 0.2.16).

It should be noticed that the separability of $\mathcal{S}$ is merely used to establish the separability of $\mathcal{A}_1$ without describing $\mathcal{A}_1$ explicitly. Recall that a *Souslin σ-field* is a separating Blackwell σ-field (Definition 0.2.12). As the proof of Theorem 4.6.17 shows, in part (ii), that $\mathcal{A}$ is separating, we may synthesize Theorems 4.6.16 and 4.6.17 as follows:

4.6.18 Theorem. If $\mathcal{A}$ is a Souslin σ-field and $\mathcal{S}$ is separable, s-idenfication and c-identification are equivalent. $\blacksquare$

We next consider the relationship between the two "Bayesian" concepts.

4.6.19 Theorem. If $\mathcal{A}$ is both separable and separating, b-identification implies $a.s.s.$-identification.

Proof.

(i) $\mathcal{A} = \overline{\mathcal{A}}_1^\mu$ because, by Theorem 4.4.12, $\mathcal{AS} = \overline{\mathcal{A}}_1^\mu$. Let us consider $\{E_n\}$, a countable algebra generating $\mathcal{A}$. Then $\forall\, n\; \exists\; F_n \in \mathcal{A}_1$ such that $m(E_n \Delta F_n) = 0$. We define $A_0 = \cup_n (E_n \Delta F_n)$. Clearly $\mu(A_0) = 0$.

(ii) $\mathcal{A}$ separable and separating $\Rightarrow \forall\, a,\, a' \in A,\, a \neq a'\; \exists n$ such that $a \in E_n$ and $a' \in E_n^c$. Otherwise $1_{E_n}(a) = 1_{E_n}(a')\; \forall\, n$. This implies, by (0.2.3), $a \underset{\mathcal{A}}{\sim} a'$, i.e., $a = a'$ since $\mathcal{A}$ is separating. Recall that the separability of $\mathcal{A}$ implies that the $\mathcal{A}$-atoms are measurable sets (see Section 0.2).

(iii) Now if $a \in A - A_0$ and $a' \underset{s}{\sim} a$ (i.e., $a' \underset{\mathcal{A}_1}{\sim} a$), and $a' \neq a$, then $a \in E_n \cap F_n$ and $a' \in E_n^c \cap F_n$, which implies that $a' \in A_0$. Therefore $A - A_0$ contains only one point of each $\mathcal{A}_1$-atoms, i.e., the mapping $a \longrightarrow P^a$ is injective on $A - A_0$. ∎

4.6.20 Theorem. If $\mathcal{A}$ is a Blackwell σ-field and $\mathcal{S}$ is separable, then $a.s.s.$-identification implies b-identification.

Proof. Let us consider the trace of $\mathcal{A}$ on $A - A_0$:

$$\mathcal{A}^- = \{E \cap (A - A_0) \mid E \in \mathcal{A}\} = \mathcal{A} \cap (A - A_0)$$

which is a Blackwell σ-field on $A - A_0$, and which makes the mappings $a \longrightarrow P^a(X),\, X \in \mathcal{S}$, measurable on $A - A_0$. Let $\mathcal{A}_1^-$ be the smallest σ-field on $A - A_0$ such that the mapping $a \longrightarrow P^a(X),\, X \in \mathcal{S}$, are measurable on $A - A_0$; then, as before, $\mathcal{A}_1^-$ is separable and by Theorem 4.6.17, $\mathcal{A}^- = \mathcal{A}_1^-$. Moreover, remark that

$$\mathcal{A}_1^- = \{E \cap (A - A_0) \mid E \in \mathcal{A}_1\}.$$

Therefore $\mathcal{A}$ and $\mathcal{A}_1$ have the same trace on a set of measure 1; therefore $\mathcal{A}$ and $\mathcal{A}_1$ are almost surely equal, i.e., $\mathcal{A} = \overline{\mathcal{A}}_1^\mu$. ∎

Similarly to the "non-Bayesian" concepts, one may summarize Theorems 4.6.19 and 4.6.20 as follows:

4.6.21 Theorem. If $\mathcal{A}$ is a Souslin σ-field and $\mathcal{S}$ is separable, *a.s.s.*-identification and *b*-identification are equivalent. ∎

Let us now consider the links between "Bayesian" and "non-Bayesian" concepts. Clearly:

(i) *s*-identification implies *a.s.s.*-identification for any μ;

(ii) *c*-identification implies *b*-identification for any μ.

Converse results may be sought for either a specific μ or for a family of μ. Consider the following example:

Example. Let $(s \mid a) \sim N(a, 1)$ if $a^2 \neq 1$ and $(s \mid a) \sim N(1, 1)$ if $a^2 = 1$. This model is not identified in any non-Bayesian terms because $(-1) \underset{s}{\sim} (+1)$. If the prior is such that $\mu(\{-1\}) \cdot \mu(\{+1\}) = 0$, then this model is *a.s.s.*-identified with $A_0 = \{-1\}$ or $A_0 = \{+1\}$ and, therefore, by Theorem 4.6.21, is *b*-identified for any prior such that $\mu(\{-1\}) \cdot \mu(\{+1\}) = 0$. ∎

This example shows that no "gentle" condition on a specific μ will ensure that *a.s.s.*-identification or *b*-identification implies *s*-identification or *c*-identification.

B. Identification in Marginal Experiments

Identification theory is a domain in which Bayesian and classical methods treat nuisance parameters differently. Consider a loss function $\ell(a, d)$ where d is a decision. Parameters of interest may be defined either using the equivalence relationship:

$$a \underset{\ell}{\sim} a' \quad \Leftrightarrow \quad \forall \, d : \ell(a, d) = \ell(a', d),$$

or using the smallest sub-σ-field $\mathcal{B}$ of $\mathcal{A}$ that makes $\ell(a, d)$ measurable for all d:

$$\mathcal{B} = \bigvee_d \sigma\{\ell(\cdot, d)\}.$$

The same argument as in the proof of Theorem 4.6.16 shows that these two approaches are equivalent in the sense that

$$A/\ell = A/\mathcal{B}.$$

We may now generalize the "non-Bayesian" concepts of identification by considering only the parameters of interest. Essentially, we are formalizing the intuitive idea that an "identified subparameter should be a function of any sufficient parameter" (see, e.g., Fisher (1966), Malinvaud (1978)). This entails the following definitions:

4.6.22 Definition. A statistical experiment is:

(i) ℓ-s-*identified* if $A/s \leq A/\ell$ (equivalently if $a \underset{s}{\sim} a' \Rightarrow a \underset{\ell}{\sim} a'$)

(ii) ℓ-c-*identified* if $\mathcal{B} \subset \mathcal{A}_1$.   ■

Then one gets the following results:

(a)   By Theorem 4.6.16, ℓ-c-identification implies ℓ-s-identification for any loss function ℓ;

(b)   by Theorem 4.6.17, if $\mathcal{A}$ is a Blackwell σ-field and $\mathcal{S}$ is separable, then ℓ-s-identification and ℓ-c-identification are equivalent.

The classical condition $\mathcal{B} \subset \mathcal{A}_1$ implies, in a Bayesian experiment defined through any μ on $(A, \mathcal{A})$ that

$$\mathcal{B} \subset \overline{\mathcal{A}}_1^\mu = \mathcal{AS}.$$

This property should not however be taken as an alternative Bayesian concept of partial identification. Indeed, as mentioned in Section 4.4, if $\mathcal{B}$ is the only parameter of interest, the relevant experiment is $\mathcal{E}_{\mathcal{B} \vee \mathcal{S}}$ and the identification condition in this experiment is $\mathcal{B} = \mathcal{BS}$. This is so because when $\mathcal{B}$ is the only parameter of interest, inference bears only on the transformation $\mu_{\mathcal{B}} \to \mu_{\mathcal{B}}^{\mathcal{S}}$. The examples of Section 4.6 have shown that, in general, there is non-implication between the two conditions $\mathcal{B} \subset \mathcal{AS}$ and $\mathcal{B} = \mathcal{BS}$.

4.7   Exact and Totally Informative Experiments

This section is mainly motivated by Bayesian asymptotic theory but as a preliminary, we consider a somewhat broader question: to what extent does an observation give "perfect" information about a parameter? The natural answer to this question is: it does once the parameter is known perfectly after (almost) any observation of the sample. Consider the general reduced experiment $\mathcal{E}^{\mathcal{M}}_{\mathcal{B}\vee\mathcal{T}}$. A parameter $\mathcal{L} \subset \mathcal{B} \vee \mathcal{M}$ is known perfectly if $\mathcal{L} \subset \overline{\mathcal{M} \vee \mathcal{T}}$. Note that $\mathcal{L} \subset \overline{\mathcal{M} \vee \mathcal{T}}$ is equivalent to $\mathcal{L} \perp\!\!\!\perp \mathcal{L} \mid \mathcal{M} \vee \mathcal{T}$ by Corollary 2.2.8. However, $\mathcal{B} \perp\!\!\!\perp \mathcal{L} \mid \mathcal{M} \vee \mathcal{T}$ is implied by $\mathcal{L} \subset \overline{\mathcal{M} \vee \mathcal{T}}$ and, by Corollary 2.2.11, implies $\mathcal{L} \perp\!\!\!\perp \mathcal{L} \mid \mathcal{M} \vee \mathcal{T}$.

By symmetry this justifies the following definitions:

4.7.1 Definition. In the general reduced Bayesian experiment $\mathcal{E}^{\mathcal{M}}_{\mathcal{B}\vee\mathcal{T}}$,

(i)   A parameter $\mathcal{L} \subset \mathcal{B} \vee \mathcal{M}$ is *exactly estimable* if $\mathcal{L} \subset \overline{\mathcal{M} \vee \mathcal{T}}$ or, equivalently, $\mathcal{L} \perp\!\!\!\perp \mathcal{B} \mid \mathcal{M} \vee \mathcal{T}$.

(ii) A statistic $\mathcal{N} \subset \mathcal{M} \vee \mathcal{T}$ is *exactly estimating* if $\mathcal{N} \subset \overline{\mathcal{B} \vee \mathcal{M}}$ or, equivalently, $\mathcal{N} \perp\!\!\!\perp \mathcal{T} \mid \mathcal{B} \vee \mathcal{M}$. ∎

For expository reasons, we provide comments with respect to the unreduced experiment $\mathcal{E} = \mathcal{E}_{\mathcal{A}\vee\mathcal{S}}$ only.

In $\mathcal{E}_{\mathcal{A}\vee\mathcal{S}}$, a parameter $\mathcal{C} \subset \mathcal{A}$ is exactly estimable if $\mathcal{C} \subset \overline{\mathcal{S}}$. Thus, any sub-$\sigma$-field of $\mathcal{A} \cap \overline{\mathcal{S}}$ is exactly estimable. It should be stressed that in $\mathcal{C} \subset \overline{\mathcal{S}}$, the completion of $\mathcal{S}$ is made w.r.t. Π, i.e., $\overline{\mathcal{S}}$ is generated by $\mathcal{S}$ and all the Π-null sets of the product space $\mathcal{A} \vee \mathcal{S}$. Exact estimability means that, for any $c \in [\mathcal{C}]^+$, the posterior expectation $\mathcal{S}c$ is equal to c a.s. Π; equivalently if c is the indicator function of an event $C \in \mathcal{C}$, one has: $\mathcal{S}1_C = 1_C$ a.s.Π. This is also: $\mathcal{S}1_C$ is equal to 0 or 1 a.s. P. In other words, $\mathcal{C} \subset \overline{\mathcal{S}}$ formalizes the idea that $\mathcal{C}$ is "known perfectly" after the observation of the sample.

Similarly, a statistic $\mathcal{U} \subset \mathcal{S}$ is exactly estimating if $\mathcal{U} \subset \overline{\mathcal{A}}$. Thus any sub-$\sigma$-field of $\overline{\mathcal{A}} \cap \mathcal{S}$ is exactly estimating. By the same argument as above, for any event $U \in \mathcal{U}$, its sampling probability $\mathcal{A}1_U$ is equal to 0 or 1 a.s.μ. Heuristically, any $u \in [\mathcal{U}]$ is a.s. constant in the sampling.

As shown in the following theorem, the exactly estimable parameters are in complete duality with the exactly estimating statistics.

4.7.2 Theorem. In a general reduced Bayesian experiment $\mathcal{E}^{\mathcal{M}}_{\mathcal{B}\vee\mathcal{T}}$, a parameter $\mathcal{L} \subset \mathcal{B} \vee \mathcal{M}$ is exactly estimable if and only if there exists an exactly estimating statistic $\mathcal{N} \subset \mathcal{M} \vee \mathcal{T}$ such that $\overline{\mathcal{L}} = \overline{\mathcal{N}}$.

Proof.

(i) It suffices to pose $\mathcal{N} = \overline{\mathcal{L}} \cap (\mathcal{M} \vee \mathcal{T})$. Clearly, $\mathcal{N}$ is is an $\mathcal{E}^{\mathcal{M}}_{\mathcal{B} \vee \mathcal{T}}$-statistic, i.e., $\mathcal{N} \subset \mathcal{M} \vee \mathcal{T}$; furthermore, $\mathcal{N}$ is exactly estimating because $\mathcal{N} \subset \overline{\mathcal{L}} \subset \overline{\mathcal{B} \vee \mathcal{M}}$ and $\overline{\mathcal{N}} = \overline{\mathcal{L}}$ because $\overline{\mathcal{N}} = \overline{\mathcal{L}} \cap \overline{\mathcal{M} \vee \mathcal{T}}$ (Lemma 2.2.5(ii)). Then $\overline{\mathcal{N}} = \overline{\mathcal{L}}$ since $\overline{\mathcal{L}} \subset \overline{\mathcal{M} \vee \mathcal{T}}$.

(ii) Reciprocally, if $\overline{\mathcal{L}} = \overline{\mathcal{N}} \subset \overline{\mathcal{M} \vee \mathcal{T}}$ then $\mathcal{L} \subset \mathcal{B} \vee \mathcal{M}$ is exactly estimating. ∎

In the asymptotic case a parameter $b \in [\mathcal{A}]$ will be exactly estimable if there exists a strongly consistent sequence of estimators of b, i.e., if $\exists \, t_n \in [\mathcal{S}]$, $n \in \mathbb{N}$, such that $t_n \to b$ a.s.II. In sampling theory, it is known that a necessary condition for the existence of such a sequence is the identifiability of the parameter.

The Bayesian analogue of this result, given in the next proposition, is a simple consequence of the elementary Property 4.5.2(ii)

4.7.3 Proposition. In $\mathcal{E}^{\mathcal{M}}_{\mathcal{B} \vee \mathcal{T}}$,

(i)   Any exactly estimable parameter, $\mathcal{L} \subset (\mathcal{B} \vee \mathcal{M}) \cap \overline{(\mathcal{M} \vee \mathcal{T})}$, is identified, i.e., $\mathcal{L} < \mathcal{T} \mid \mathcal{M}$.

(ii) Any exactly estimating statistic, $\mathcal{N} \subset \overline{(\mathcal{B} \vee \mathcal{M})} \cap (\mathcal{M} \vee \mathcal{T})$, is identified, i.e., $\mathcal{N} < \mathcal{B} \mid \mathcal{M}$. ∎

Clearly, in the general reduced experiment $\mathcal{E}^{\mathcal{M}}_{\mathcal{B} \vee \mathcal{T}}$,

$$\mathcal{L}_{\max} = (\mathcal{B} \vee \mathcal{M}) \cap \overline{(\mathcal{M} \vee \mathcal{T})}$$

is the unique maximal exactly estimable parameter and

$$\mathcal{N}_{\max} = \overline{(\mathcal{B} \vee \mathcal{M})} \cap (\mathcal{M} \vee \mathcal{T})$$

is the unique maximal exactly estimating statistic. Furthermore, by Lemma 2.2.5(ii), we have:

$$\overline{\mathcal{L}}_{\max} = \overline{\mathcal{N}}_{\max} = \overline{(\mathcal{B} \vee \mathcal{M})} \cap \overline{(\mathcal{M} \vee \mathcal{T})}.$$

The elementary Property 4.3.2(i) provides two sets of inclusions;these are primordial to understanding the possible informational content of any Bayesian experiment.

4.7.4 Proposition. In the general reduced experiment $\mathcal{E}^{\mathcal{M}}_{\mathcal{B}\vee\mathcal{T}}$,

(i) The trivial parameter, i.e., $(\mathcal{B}\vee\mathcal{M})\cap\overline{\mathcal{M}}$, is exactly estimable and any exactly estimable parameter is included in the minimal sufficient strong parameter, i.e.,
$$(\mathcal{B}\vee\mathcal{M})\cap\overline{\mathcal{M}}\subset(\mathcal{B}\vee\mathcal{M})\cap\overline{(\mathcal{M}\vee\mathcal{T})}\subset(\mathcal{B}\vee\mathcal{M})(\mathcal{M}\vee\mathcal{T})\subset\mathcal{B}\vee\mathcal{M};$$

(ii) the trivial statistic, i.e., $\overline{\mathcal{M}}\cap(\mathcal{M}\vee\mathcal{T})$ is exactly estimating and any exactly estimating statistic is included in the minimal sufficient strong statistic, i.e.,
$$\overline{\mathcal{M}}\cap(\mathcal{M}\vee\mathcal{T})\subset\overline{(\mathcal{B}\vee\mathcal{M})}\cap(\mathcal{M}\vee\mathcal{T})\subset(\mathcal{M}\vee\mathcal{T})(\mathcal{B}\vee\mathcal{M})\subset\mathcal{M}\vee\mathcal{T}.\ \blacksquare$$

Recall that equalities in the last inclusions in the above two relations amount to $\mathcal{E}^{\mathcal{M}}_{\mathcal{B}\vee\mathcal{T}}$-identifications, i.e., $\mathcal{B}<\mathcal{T}\mid\mathcal{M}$ and $\mathcal{T}<\mathcal{B}\mid\mathcal{M}$. Note that there is no relationship between those two properties. Equality for the first inclusions means that only trivial parameters (respectively, trivial statistics) are exactly estimable (respectively, exactly estimating). Note that these two properties are in fact equivalent since, by Lemma 2.2.5(ii), each of these is equivalent to $\overline{(\mathcal{B}\vee\mathcal{M})}\cap\overline{(\mathcal{M}\vee\mathcal{T})}=\overline{\mathcal{M}}$, i.e., $\overline{\mathcal{L}}_{\max}=\overline{\mathcal{N}}_{\max}=\overline{\mathcal{M}}$. This situation is analyzed more deeply in Chapter 5.

The next theorem shows that equalities for the middle inclusions are also equivalent, i.e., equality between the maximal exactly estimable parameter and the minimal sufficient strong parameter is equivalent to equality between the maximal exactly estimating statistic and the minimal sufficient strong statistic.

4.7.5 Theorem. In the general reduced Bayesian experiment $\mathcal{E}^{\mathcal{M}}_{\mathcal{B}\vee\mathcal{T}}$, the following conditions are equivalent:

(i) $(\mathcal{B}\vee\mathcal{M})\cap\overline{(\mathcal{M}\vee\mathcal{T})}=(\mathcal{B}\vee\mathcal{M})(\mathcal{M}\vee\mathcal{T});$

(ii) $\overline{(\mathcal{B}\vee\mathcal{M})}\cap(\mathcal{M}\vee\mathcal{T})=(\mathcal{M}\vee\mathcal{T})(\mathcal{B}\vee\mathcal{M}).$

Proof. By Theorem 4.3.3(i) and Corollary 2.2.11,

$$(\mathcal{B} \vee \mathcal{M})(\mathcal{M} \vee \mathcal{T}) \subset (\mathcal{B} \vee \mathcal{M}) \cap \overline{(\mathcal{M} \vee \mathcal{T})}$$

implies

$$(\mathcal{B} \vee \mathcal{M}) \perp\!\!\!\perp (\mathcal{M} \vee \mathcal{T}) \mid (\mathcal{B} \vee \mathcal{M}) \cap \overline{(\mathcal{M} \vee \mathcal{T})}.$$

By Corollary 2.2.7 and Lemma 2.2.5(ii), this is equivalent to

$$(\mathcal{B} \vee \mathcal{M}) \perp\!\!\!\perp (\mathcal{M} \vee \mathcal{T}) \mid \overline{(\mathcal{B} \vee \mathcal{M})} \cap \overline{(\mathcal{M} \vee \mathcal{T})}.$$

But, by Theorem 4.3.3(ii) and by Lemma 2.2.5(ii),

$$(\mathcal{B} \vee \mathcal{M}) \perp\!\!\!\perp (\mathcal{M} \vee \mathcal{T}) \mid (\mathcal{B} \vee \mathcal{M}) \cap \overline{(\mathcal{M} \vee \mathcal{T})}$$

implies

$$(\mathcal{B} \vee \mathcal{M})(\mathcal{M} \vee \mathcal{T}) \subset \overline{(\mathcal{B} \vee \mathcal{M})} \cap \overline{(\mathcal{M} \vee \mathcal{T})}$$

and, therefore,

$$(\mathcal{B} \vee \mathcal{M})(\mathcal{M} \vee \mathcal{T}) \subset (\mathcal{B} \vee \mathcal{M}) \cap \overline{(\mathcal{M} \vee \mathcal{T})}.$$

Therefore (i) is equivalent to

$$(\mathcal{B} \vee \mathcal{M}) \perp\!\!\!\perp (\mathcal{M} \vee \mathcal{T}) \mid \overline{(\mathcal{B} \vee \mathcal{M})} \cap \overline{(\mathcal{M} \vee \mathcal{T})}$$

and is equivalent to (ii) by symmetry. ∎

Let us comment on Propositions 4.7.3, 4.7.4 and Theorem 4.7.5 in the unreduced experiment $\mathcal{E}_{A \vee S}$. It was mentioned, in Section 4.6.1, that for a parameter $\mathcal{C} \subset \mathcal{A}$ of $\mathcal{E}_{A \vee S}$ there is, in general, no relationship between $\mathcal{C} \subset \mathcal{AS}$ and $\mathcal{C} = \mathcal{CS}$. We notice that, if $\mathcal{C}$ is exactly estimable, i.e., $\mathcal{C} \subset \mathcal{A} \cap \overline{\mathcal{S}}$, then, from Proposition 4.7.3, $\mathcal{C} = \mathcal{CS}$ and, from Proposition 4.7.4, $\mathcal{C} \subset \mathcal{AS}$. Therefore, if $\mathcal{C}$ is exactly estimable these two properties are simultaneously satisfied. Note also that, in view of Theorem 4.4.12, $\mathcal{C} \subset \mathcal{AS}$ is a.s. equivalent to the sampling definition of an identified subparameter, while $\mathcal{C} = \mathcal{CS}$ is a genuinely Bayesian property since it requires integration on $\mathcal{A}$ conditionally on $\mathcal{C}$ in order to be verified. Note also that if the minimal sufficient parameter $\mathcal{AS}$ is exactly estimable, i.e., $\mathcal{AS} = \mathcal{A} \cap \overline{\mathcal{S}}$, the Bayesian equivalent definition of an identified subparameter, i.e., $\mathcal{C} \subset \mathcal{AS}$, implies that this parameter is exactly estimable and therefore is identified, i.e., $C \subset \mathcal{A} \cap \overline{\mathcal{S}}$ and $C = CS$. But it has not been shown that if the minimal sufficient parameter is exactly estimable, i.e., $\mathcal{AS} = \mathcal{A} \cap \overline{\mathcal{S}}$, then any identified parameter, i.e., $C \subset \mathcal{A}$ and $C = CS$, would be exactly estimable, i.e., $C \subset \mathcal{A} \cap \overline{\mathcal{S}}$, unless, of course, the Bayesian experiment $\mathcal{E}_{\mathcal{A} \vee \mathcal{S}}$ is identified, i.e., $\mathcal{A} = \mathcal{AS}$.

Proposition 4.7.4. states that the best one can hope for is to estimate exactly the functions of the minimal sufficient parameter, i.e., $\mathcal{AS}$. This, along with Theorem 4.7.5, leads to the following definition:

4.7.6 Definition. The general reduced Bayesian experiment $\mathcal{E}_{\mathcal{B} \vee \mathcal{T}}^{\mathcal{M}}$ is *exact* if any one of the following equivalent conditions holds:

(i) $(\mathcal{B} \vee \mathcal{M})(\mathcal{M} \vee \mathcal{T}) \subset \overline{\mathcal{M} \vee \mathcal{T}}$;

(ii) $(\mathcal{M} \vee \mathcal{T})(\mathcal{B} \vee \mathcal{M}) \subset \overline{\mathcal{B} \vee \mathcal{M}}$,

i.e., the minimal sufficient strong parameter (respectively, statistic) is exactly estimable (respectively, estimating). ∎

The following proposition, whose proof is straightforward, completely characterizes the structure of an exact Bayesian experiment.

4.7.7 Proposition. If the general reduced Bayesian experiment $\mathcal{E}_{\mathcal{B} \vee \mathcal{T}}^{\mathcal{M}}$ is exact, then

$$\overline{(\mathcal{B} \vee \mathcal{M})(\mathcal{M} \vee \mathcal{T})} = \overline{(\mathcal{M} \vee \mathcal{T})(\mathcal{B} \vee \mathcal{M})} = \overline{(\mathcal{B} \vee \mathcal{M})} \cap \overline{(\mathcal{M} \vee \mathcal{T})}. \qquad \blacksquare$$

In general, the minimal sufficient parameter is not exactly estimable in a finite sample experiment. But this may be true asymptotically. But in such a situation, relations such as (i) and (ii) in Definition 4.7.6 may be difficult to verify, for they require computation of the minimal sufficient strong parameter (or statistic). We thus need a more operationally useful criterion to establish whether an experiment is exact. To such an end, the most useful result, inspired by Theorem 4.7.2, is provided by the following theorem:

4.7.8 Theorem. The general reduced Bayesian experiment $\mathcal{E}^{\mathcal{M}}_{\mathcal{B}\vee\mathcal{T}}$ is exact if and only if there exists either an exactly estimating sufficient statistic, i.e.,

(i) $\exists\,\mathcal{N}\subset\mathcal{M}\vee\mathcal{T}$ such that

      1. $\mathcal{B}\perp\!\!\!\perp\mathcal{T}\mid\mathcal{M}\vee\mathcal{N}$,

      2. $\mathcal{N}\subset\overline{\mathcal{B}\vee\mathcal{M}}$;

or, there exists an exactly estimable sufficient paramater, i.e.,

(ii) $\exists\,\mathcal{L}\subset\mathcal{B}\vee\mathcal{M}$ such that

      1. $\mathcal{B}\perp\!\!\!\perp\mathcal{T}\mid\mathcal{M}\vee\mathcal{L}$,

      2. $\mathcal{L}\subset\overline{\mathcal{M}\vee\mathcal{T}}$.

Moreover, for any $\mathcal{N}$ satisfying (i), we have

(iii) $\overline{\mathcal{M}\vee\mathcal{N}}=\overline{(\mathcal{B}\vee\mathcal{M})}\cap\overline{(\mathcal{M}\vee\mathcal{T})}$,

and, for any $\mathcal{L}$ satifying (ii), we have

(iv) $\overline{\mathcal{L}\vee\mathcal{M}}=\overline{(\mathcal{B}\vee\mathcal{M})}\cap\overline{(\mathcal{M}\vee\mathcal{T})}$.

Proof. By symmetry it suffices to prove (i). By Proposition 4.7.7 and Theorem 4.7.5, the necessity is clear by taking $\mathcal{N}=\overline{(\mathcal{B}\vee\mathcal{M})}\cap(\mathcal{M}\vee\mathcal{T})$. On the other hand, by Proposition 4.3.10 or by Proposition 4.4.13, (i) 1. implies $(\mathcal{M}\vee\mathcal{T})(\mathcal{B}\vee\mathcal{M})\subset\overline{\mathcal{M}\vee\mathcal{N}}$ and (i) 2. implies $\overline{\mathcal{M}\vee\mathcal{N}}\subset\overline{\mathcal{B}\vee\mathcal{M}}$. Therefore $(\mathcal{M}\vee\mathcal{T})(\mathcal{B}\vee\mathcal{M})\subset\overline{\mathcal{B}\vee\mathcal{M}}$, i.e., $\mathcal{E}^{\mathcal{M}}_{\mathcal{B}\vee\mathcal{T}}$ is exact. More precisely, $\overline{(\mathcal{M}\vee\mathcal{T})(\mathcal{B}\vee\mathcal{M})}\subset\overline{\mathcal{M}\vee\mathcal{N}}\subset\overline{(\mathcal{B}\vee\mathcal{M})}\cap\overline{(\mathcal{M}\vee\mathcal{T})}$ and this shows (iii), by Proposition 4.7.7. ∎

Let us provide comment on Theorem 4.7.8. in the unreduced experiment $\mathcal{E}_{\mathcal{A}\vee\mathcal{S}}$. In order to show that $\mathcal{E}_{\mathcal{A}\vee\mathcal{S}}$ is exact, i.e., $\mathcal{A}\mathcal{S}\subset\overline{\mathcal{S}}$, it suffices to exhibit a sufficient statistic, i.e., $\mathcal{T}\subset\mathcal{S}$ such that $\mathcal{A}\perp\!\!\!\perp\mathcal{S}\mid\mathcal{T}$, that is exactly estimating, i.e., $\mathcal{T} \subset \overline{\mathcal{A}}$. Let us note that any such statistic is almost surely unique since it follows from (iii), that $\overline{\mathcal{T}} = \overline{\mathcal{A}} \cap \overline{\mathcal{S}}$ or, equivalently $\overline{\mathcal{T}} \cap \mathcal{S} = \overline{\mathcal{A}} \cap \mathcal{S}$, i.e., $\mathcal{T}$ completed by the null sets of $\mathcal{S}$ is equal to the maximal exactly estimating statistic, which is also the minimal sufficient statistic $\mathcal{SA}$ since when $\mathcal{E}_{\mathcal{A} \vee \mathcal{S}}$ is exact, $\mathcal{SA} = \overline{\mathcal{A}} \cap \mathcal{S}$ and $\mathcal{AS} = \mathcal{A} \cap \overline{\mathcal{S}}$, (since $\overline{\mathcal{AS}} = \overline{\mathcal{SA}} = \overline{\mathcal{A}} \cap \overline{\mathcal{S}}$ by Proposition 4.7.7). This also implies that $\mathcal{A} \cap \overline{\mathcal{T}} = \mathcal{A} \cap \overline{\mathcal{S}} = \mathcal{AS}$.

This section has presented the fundamental theorems of Bayesian asymptotic theory (insofar as a.s. convergence is concerned). In Chapters 7, 8 and 9 we see how particular structures such as independent sampling and invariance can be used to make the criteria of exact estimability operational. Combining identification and exactness of a Bayesian experiment leads to the concept of a totally informative Bayesian experiment.

4.7.9 Definition. The reduced Bayesian experiment $\mathcal{E}_{\mathcal{B} \vee \mathcal{T}}^{\mathcal{M}}$ is *totally informative* if it is identified and exact, or equivalently, if $\mathcal{B} \subset \overline{\mathcal{M} \vee \mathcal{T}}$. ∎

The equivalence stated in the definition comes from Proposition 4.7.4(i). Indeed, identification and exactness amount to saying that the last two inclusions are actually equalities, and is therefore equivalent to

$$\mathcal{B} \vee \mathcal{M} = (\mathcal{B} \vee \mathcal{M}) \cap \overline{(\mathcal{M} \vee \mathcal{T})},$$

which is equivalent to $\mathcal{B} \subset \overline{\mathcal{M} \vee \mathcal{T}}$.

Furthermore, if $\mathcal{E}_{\mathcal{B} \vee \mathcal{T}}^{\mathcal{M}}$ is totally non-informative, i.e., $\mathcal{B} \perp \!\!\! \perp \mathcal{T} \mid \mathcal{M}$, this is equivalent, by Theorem 4.3.4 and Corollary 2.2.11, to saying that $(\mathcal{B} \vee \mathcal{M})(\mathcal{M} \vee \mathcal{T}) = (\mathcal{B} \vee \mathcal{M}) \cap \overline{\mathcal{M}}$, i.e., the minimal sufficient complete strong parameter is equal to the trivial parameter. Therefore, $\mathcal{E}_{\mathcal{B} \vee \mathcal{T}}^{\mathcal{M}}$ totally non-informative means that the first two inclusions in Proposition 4.7.4(i) are actually equalities. In other words, $\mathcal{E}_{\mathcal{B} \vee \mathcal{T}}^{\mathcal{M}}$ is totally non-informative if and only if $\mathcal{E}_{\mathcal{B} \vee \mathcal{T}}^{\mathcal{M}}$ is exact and if the maximal exactly estimable parameter is the trivial parameter. Finally, note that $\mathcal{E}_{\mathcal{B} \vee \mathcal{T}}^{\mathcal{M}}$ will be both totally informative and totally non-informative if all the inclusions in Proposition 4.7.4(i) are actually equalities; this is equivalent to saying that $\mathcal{B} \subset \overline{\mathcal{M}}$, i.e., the full parameter is trivial.

4.8   Punctual Exact Estimability

In asymptotic theory, Bayesian consistency has generally been presented as the convergence of the posterior distributions to a point mass or a Dirac distribution (see, e.g., LeCam (1986), Berk (1966, 1970), Hartigan (1983)). In this context, punctual exact estimability of a Bayesian experiment $\mathcal{E} = (A \times S, \ \mathcal{A} \vee \mathcal{S}, \ \Pi)$ may be presented as follows: There exists a measurable function $\alpha : (S, \mathcal{S}) \rightarrow (A, \mathcal{A})$ such that $\delta_{\alpha(s)}$ is a version of the posterior probability μ^s, i.e.,

$$(4.8.1) \qquad \mu^s(E) = \delta_{\alpha(s)}(E) = \mathbf{1}_E[\alpha(s)] \quad \forall \, E \in \mathcal{A}, \quad \forall \, s \in S.$$

In our conventions, this may be expressed as follows:

4.8.1 Definition. In the unreduced Bayesian experiment $\mathcal{E} = (A \times S, \ \mathcal{A} \vee \mathcal{S}, \ \Pi)$, the parameter $\mathcal{A}$ is *punctually exactly estimable* if there exists a measurable function $f : (A \times S, \mathcal{S}) \rightarrow (A \times S, \mathcal{A})$ such that

$$(4.8.2) \qquad \mathcal{S} \, \mathbf{1}_E = \mathbf{1}_E \circ f = \mathbf{1}_{f^{-1}(E)} \quad \text{a.s.} \Pi \quad \forall \, E \in \mathcal{A}.$$

This definition is in fact essentially equivalent to (4.8.1). Indeed, it suffices to define $f(a, s) = (\alpha(s), s)$.

We now compare Definition 4.8.1 with our Definition 4.7.1, i.e., $\mathcal{A} \subset \overline{\mathcal{S}}$.

4.8.2 Theorem. In the unreduced Bayesian experiment $\mathcal{E}_{\mathcal{A} \vee \mathcal{S}}$, if $\mathcal{A}$ is punctually exactly estimable then $\mathcal{A}$ is exactly estimable, i.e., $\mathcal{A} \subset \overline{\mathcal{S}}$.

Proof. From (4.8.2), it follows that $(\mathcal{S} \, \mathbf{1}_E)^2 = (\mathcal{S} \, \mathbf{1}_E)$. Therefore,

$$\mathcal{S} \, \mathbf{1}_E = \mathbf{1}_E \quad \text{a.s.} \Pi \quad \forall \, E \in \mathcal{A},$$

i.e., $\mathcal{A} \subset \overline{\mathcal{S}}$.   ∎

From the proof of Theorem 4.8.2, we see that (4.8.2) implies

$$(4.8.3) \qquad \mathbf{1}_E = \mathbf{1}_{f^{-1}(E)} \quad \text{a.s.} \Pi \quad \forall \, E \in \mathcal{A};$$

this implies

$$(4.8.4) \qquad \mu = P \circ f^{-1},$$

i.e. the prior probability is the image under f of the predictive probability (see Proposition 0.3.8). The converse of Theorem 4.8.2 requires that a regularity condition be satisfied.

4.8.3 Theorem. In the unreduced Bayesian experiment $\mathcal{E}_{\mathcal{A}\vee\mathcal{S}}$, if $\mathcal{A}$ is a Standard Borel σ-field and if $\mathcal{A}$ is exactly estimable, i.e., $\mathcal{A} \subset \overline{\mathcal{S}}$, then $\mathcal{A}$ is punctually exactly estimable.

Proof. By theorem 0.3.18, let μ^s be a regular version of the posterior probabilities, i.e., $\mathcal{S}\ \mathbf{1}_E = \mu^s(E)$ a.s.Π $\forall\ E \in \mathcal{A}$. Since $\mathcal{A} \subset \overline{\mathcal{S}}$, $\mu^s(E) = \mathbf{1}_E$ a.s.Π $\forall\ E \in \mathcal{A}$. Now, let $A_t = \varphi^{-1}((-\infty, t])$ where φ is the bijection provided by Definition 0.2.15. Since μ^s is a probability measure on $\mathcal{A}$, $\forall\ s \in S$ $\mu^s(A_t)$ is increasing and right continuous in t. Therefore, if we define:

$$t(s) = \inf\{t : \mu^s(A_t) = 1\},$$

then, clearly, $t \in [\mathcal{S}]$, since

$$\{s \in S : t(s) \le t\} = \{s \in S : \mu^s(A_t) = 1\}.$$

Define $\alpha(s) = \varphi^{-1}[t(s)]$ and $f(a, s) = (\alpha(s), s)$; by the monotone class Theorem 0.2.20(ii), we conclude that f satisfies (4.8.2), since $\forall\ t \in \mathbb{R}$, $\forall\ s \in S$, $\mu^s(A_t) = \mathbf{1}_{A_t}[\alpha(s)]$. ∎

5

Optimal Reductions: Further Results

5.1 Introduction

The goal of this chapter is to study the relationship between sufficiency and ancillarity in greater detail, paying particular attention to minimal sufficiency and maximal ancillarity. Along these lines, a first result was presented in Chapter 2 (Theorem 2.3.19) for an unreduced experiment, and in Chapter 3 (Theorem 3.3.8) for the general reduced experiment.

In sampling theory, other results have been established, most of which are due to Basu. Theorem Two in Basu (1955) states that any statistic which, in the sampling process, is independent of a sufficient statistic is ancillary provided a further condition links the parameter and the sufficient statistic. In the Bayesian framework the problem addressed by Basu Theorem may be expressed as: under what condition on $\mathcal{A}$ and $\mathcal{T}$ does $\mathcal{A} \perp\!\!\!\perp \mathcal{S} \mid \mathcal{T}$ and $\mathcal{U} \perp\!\!\!\perp \mathcal{T} \mid \mathcal{A}$ imply $\mathcal{U} \perp\!\!\!\perp \mathcal{A}$? This condition on $\mathcal{A}$ and $\mathcal{T}$ has been called the measurable separability of $\mathcal{A}$ and $\mathcal{T}$; it essentially implies that $\mathcal{A}$ and $\mathcal{T}$ have no non-trivial event in common.

Theorem One in Basu (1955) states that any ancillary statistic is independent of a sufficient statistic, in the sampling process, provided a further condition, which links the parameter and the sufficient statistic is satisfied, namely, the sufficient statistic must be boundedly complete. In the Bayesian framework this becomes: under what conditions on $\mathcal{A}$ and $\mathcal{T}$ does $\mathcal{A} \perp\!\!\!\perp \mathcal{S} \mid \mathcal{T}$ and $\mathcal{U} \perp\!\!\!\perp \mathcal{A}$, imply that $\mathcal{U} \perp\!\!\!\perp \mathcal{T} \mid \mathcal{A}$? This condition has been called the strong identifiability of T by $\mathcal{A}$. This name has been retained because it is a stronger condition than the identifiability of T by $\mathcal{A}$ as defined in Chapter 4 (Definition 4.5.1).

In this chapter, we generalize these two notions (measurable separability — Section 5.2 — and strong identifiability — Section 5.4 —) so as to provide similar results in reduced experiments. These two properties are of some independent interest. We therefore present, in these sections, a general theory of these properties and of their relationship with both conditional independence and the projection of σ-fields. We subsequently discuss their application in statistics (in Sections 5.3 and 5.5), and the relationship between these concepts and the analogous sampling theory concepts.

In view of the principle of conditioning, the choice of a suitable ancillary statistic is generally considered as important in parametric inference, and in this context, it matters to know whether or not a given ancillary statistic is maximal, i.e., whether or not it is a non-injective function of some other ancillary statistic. Recent references on this topic include Barnard and Sprott (1971), Basu (1959, 1964,1975), Becker and Gordon (1983), Cox (1971), Cox and Hinkley (1974, Chap. 2), Fraser (1973). As mentioned in Chapter 4 (Section 4.2), although one may give a definition and prove the existence (using Zorn's Lemma) of maximal ancillary σ-fields (i.e., parameters or observations)), we know of no way to construct such maximal ancillary σ-field. However, the tools developed in this chapter provide two alternative criteria of maximality. These criteria may be illustrated by the following simple example (see Basu (1959), Example f). Consider a simple observation on a normal variate with known mean; more specifically $(x \mid a) \sim N(0, a)$; consider also the ancillary statistic $y = \mathbf{1}_{\{x \geq 0\}}$. We prove that, in order to assert that y is essentially maximal ancillary, any one of the following two arguments is valid: (i) x^2 is sufficient and complete and (x^2, y) is equivalent to x; (ii) conditionally on y, x is complete. Argument (i) actually leads to a Bayesian version and an extension to the conditional experiment (such as regression-type models) of Theorem 7 in Basu (1959) providing sufficient conditions for the maximality of an ancillary statistic. Argument (ii) replaces the requirement that there exists a complete sufficient statistic by a requirement that a property of conditional completeness be verified; we shall see that this argument is actually both more general and more useful in practice than argument (i). Indeed, argument (ii) may be used when the minimal sufficient statistic is not complete whereas when the minimal sufficient statistic is complete, in view of the independence between a complete sufficient statistic and any ancillary statistic, conditioning on an ancillary statistic becomes irrelevant. Furthermore, argument (ii) may be used to resolve the apparent contradiction between the principle of sufficiency and the principle of conditioning. Indeed, once a sample has been reduced to a sufficient statistic, further conditioning on any larger (i.e., not included) ancillary statistic will be irrelevant: the best reduction is obtained through the distribution of a minimal sufficient statistic conditionally on an included ancillary statistic which makes the sufficient statistic conditionally complete; argument (ii) ensures that such an ancillary statistic is maximal within that sufficient statistic.

Sections 5.2 and 5.4 treat measurable separability and strong identification of σ-fields, the two probabilistic tools of this chapter; they make use of and extend Mouchart and Rolin (1984b) and (1986). The statistical Sections 5.3 and 5.4 are based on Mouchart and Rolin (1984b) and (1984c); however, Sections 5.3.3 and 5.5.4 on sampling theory and Bayesian methods are the fruit of more recent work.

5.2 Measurable Separability

We now handle a topic in general probability theory and adopt, as before, our usual notation; in particular, $(M, \mathcal{M}, P)$ is an abstract probability space.

The question raised by Basu Second Theorem in the Introduction 5.1 is actually equivalent to the following problem: if $\mathcal{M}_1 \perp\!\!\!\perp \mathcal{M}_2 \mid \mathcal{M}_3$ and $\mathcal{M}_1 \perp\!\!\!\perp \mathcal{M}_3 \mid \mathcal{M}_2$, under what supplementary condition is $\mathcal{M}_1 \perp\!\!\!\perp (\mathcal{M}_2 \vee \mathcal{M}_3)$? In this section we treat a somewhat more general problem by considering a conditional version of this question. This leads to the concept of *measurable separability* among σ-fields. This concept is weaker than (i.e., is implied by) independence; like independence, it has both a conditional and a marginal version. In its marginal version, it says that $\mathcal{M}_1$ and $\mathcal{M}_2$ are measurably separated if the only events in common are trivial and, in its conditional version, it corresponds to the property that events common to two σ-fields are in the conditioning σ-field.

5.2.1 Theorem. The following conditions are equivalent:

(i) $\overline{(\mathcal{M}_1 \vee \mathcal{M}_3)} \cap \overline{(\mathcal{M}_2 \vee \mathcal{M}_3)} = \overline{\mathcal{M}_3}$;

(ii) $(\mathcal{M}_1 \vee \mathcal{M}_3) \cap \left(\overline{\mathcal{M}_2 \vee \mathcal{M}_3}\right) \subset \overline{\mathcal{M}_3}$;

(iii) $m \in [\mathcal{M}_1 \vee \mathcal{M}_3]_\infty^+$ and $(\mathcal{M}_2 \vee \mathcal{M}_3)m = m$ imply $\mathcal{M}_3 m = m$.

Proof. Clearly (i) implies (ii) since for instance
$(\mathcal{M}_1 \vee \mathcal{M}_3) \cap \overline{(\mathcal{M}_2 \vee \mathcal{M}_3)} \subset \overline{(\mathcal{M}_1 \vee \mathcal{M}_3)} \cap \overline{(\mathcal{M}_2 \vee \mathcal{M}_3)}$. However, since, by
Lemma 2.2.5 (ii), $\overline{(\mathcal{M}_1 \vee \mathcal{M}_3)} \cap \overline{(\mathcal{M}_2 \vee \mathcal{M}_3)} = (\mathcal{M}_1 \vee \mathcal{M}_3) \cap \overline{(\mathcal{M}_2 \vee \mathcal{M}_3)}$,
(ii) implies that $\overline{(\mathcal{M}_1 \vee \mathcal{M}_3)} \cap \overline{(\mathcal{M}_2 \vee \mathcal{M}_3)} \subset \overline{\mathcal{M}_3}$. Therefore (ii) implies
(i) since, clearly, $\overline{\mathcal{M}_3} \subset \overline{(\mathcal{M}_1 \vee \mathcal{M}_3)} \cap \overline{(\mathcal{M}_2 \vee \mathcal{M}_3)}$. And (iii) is equivalent
to (ii) by (2.2.4).   ∎

5.2.2 Definition. Under any one of the conditions of Theorem 5.2.1,
$\mathcal{M}_1$ and $\mathcal{M}_2$ are *measurably separated conditionally on* $\mathcal{M}_3$ and we write
$\mathcal{M}_1 \parallel \mathcal{M}_2 \mid \mathcal{M}_3$. If we want to make explicit the role of the probability P in
this concept, we also write $\mathcal{M}_1 \parallel \mathcal{M}_2 \mid \mathcal{M}_3; P$. When $\mathcal{M}_3 = \mathcal{M}_0$ we simply
say that $\mathcal{M}_1$ and $\mathcal{M}_2$ are *measurably separated* and write $\mathcal{M}_1 \parallel \mathcal{M}_2$.   ∎

It is clear, from condition (i) in Theorem 5.2.1, that the concept of measurable separability is symmetric in $\mathcal{M}_1$ and $\mathcal{M}_2$; thus, one could formally
add conditions (ii bis) and (iii bis), which would be obtained by interchanging $\mathcal{M}_1$ and $\mathcal{M}_2$.

Let us also remark that events in $(\mathcal{M}_1 \vee \mathcal{M}_3) \cap \overline{(\mathcal{M}_2 \vee \mathcal{M}_3)}$ may be
characterized in several ways. Obviously,

(5.2.1) $A \in (\mathcal{M}_1 \vee \mathcal{M}_3) \cap \overline{(\mathcal{M}_2 \vee \mathcal{M}_3)}$

is equivalent to

(5.2.2) $A \in \mathcal{M}_1 \vee \mathcal{M}_2$ and $\exists B \in \mathcal{M}_2 \vee \mathcal{M}_3$
                                   such that $1_A = 1_B$ a.s.

By (2.2.4), this is also equivalent to

(5.2.3) $A \in \mathcal{M}_1 \vee \mathcal{M}_3$ and $(\mathcal{M}_2 \vee \mathcal{M}_3)1_A = 1_A$.

This clearly implies that

(5.2.4) $A \in \mathcal{M}_1 \vee \mathcal{M}_3$ and $\{(\mathcal{M}_2 \vee \mathcal{M}_3)\, 1_A\}^2 = (\mathcal{M}_2 \vee \mathcal{M}_3)\, 1_A,$

i.e., the expectation of the indicator function of A conditionally on $\mathcal{M}_2 \vee \mathcal{M}_3$ is a $\{0,1\}$-valued function, i.e., is itself an indicator function. But this is equivalent to

(5.2.5) $A \in \mathcal{M}_1 \vee \mathcal{M}_3$ and $(\mathcal{M}_2 \vee \mathcal{M}_3)\left[\{1_A - (\mathcal{M}_2 \vee \mathcal{M}_3)\, 1_A\}^2\right] = 0,$

i.e., the variance of the indicator function of A conditionally on $\mathcal{M}_2 \vee \mathcal{M}_3$ is equal to zero. This evidently implies (5.2.3). Therefore Properties (5.2.2) through (5.2.5) are equivalent characterizations of events in $(\mathcal{M}_1 \vee \mathcal{M}_3) \cap \overline{(\mathcal{M}_2 \vee \mathcal{M}_3)}$.

Therefore, if $\mathcal{M}_1 \parallel \mathcal{M}_2 \mid \mathcal{M}_3$, any set A satisfying one of the properties (5.2.2) through (5.2.5) also satisfies:

(5.2.6) $\exists\, C \in \mathcal{M}_3$ such that $1_A = 1_C$ a.s.

or, equivalently,

(5.2.7) $\mathcal{M}_3 1_A = 1_A$

or, equivalently,

(5.2.8) $(\mathcal{M}_3 1_A)^2 = \mathcal{M}_3 1_A$

or, equivalently,

(5.2.9) $\mathcal{M}_3 \left[\{1_A - \mathcal{M}_3 1_A\}^2\right] = 0.$

To gain more insight into the basic idea of measurable separability, consider x_i, $i = 1, 2, 3$, real random variables and $\mathcal{M}_i = \sigma\{x_i\}$. Using Theorem 0.2.11, in this case $\mathcal{M}_1 \parallel \mathcal{M}_2 \mid \mathcal{M}_3$ may also be expressed as follows: If their exist two (bounded) Borel measurable functions on $\mathbb{R}^2$, f and g such that $f(x_1, x_3) = g(x_2, x_3)$ a.s., then there exists a (bounded) Borel measurable function on $\mathbb{R}$, h, such that $f(x_1, x_3) = h(x_3)$ a.s. In particular, $\mathcal{M}_1 \parallel \mathcal{M}_2$ is equivalent to the following statement: If there exist two Borel measurable functions on $\mathbb{R}$, f and g, such that $f(x_1) = g(x_2)$ a.s., then there exists a constant $c \in \mathbb{R}$ such that $f(x_1) = c$ a.s. Such a condition is not satisfied if the support of x_1 and x_2 has the shape of Figure 1, where the events $\{x_1 \in A\}$ and $\{x_2 \in B\}$ are the same a.s. for the joint probability. Thus, the condition of measurable separability seeks to preclude such pathologies and, as will be shown in examples later on in this chapter, validates simple reasoning in terms of densities which would, otherwise, give invalid results.

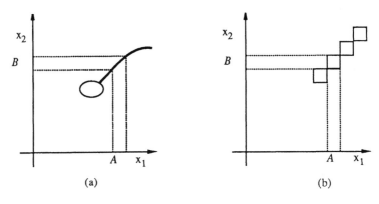

(a)   (b)

Figure 1

It should be clear from Theorem 5.2.1 and from these remarks that measurable separability depends on the probability P through its null set only. More specifically we have the following property of robustness with respect to P:

5.2.3 Proposition. Let P and P' be two probability measures on $(M, \mathcal{M})$. If $P \sim P'$, then $\mathcal{M}_1 \parallel \mathcal{M}_2 \mid \mathcal{M}_3; P \iff \mathcal{M}_1 \parallel \mathcal{M}_2 \mid \mathcal{M}_3; P'$.   ∎

We now collect a series of general properties of measurable separability. They do not require proof since they are simple consequences of the definition and of Lemma 2.2.5. They are organized according to their similarity to the corresponding properties in the case of conditional independence.

5.2.4 Proposition. *Elementary properties of measurable separability:*

(i) $\mathcal{M}_1 \parallel \mathcal{M}_2 \mid \mathcal{M}_3 \Longleftrightarrow \overline{\mathcal{M}_1} \parallel \mathcal{M}_2 \mid \mathcal{M}_3 \Longleftrightarrow \mathcal{M}_1 \parallel \mathcal{M}_2 \mid \overline{\mathcal{M}_3}$;

(ii) $\mathcal{M}_1 \subset \overline{\mathcal{M}_3} \Longrightarrow \mathcal{M}_1 \parallel \mathcal{M}_2 \mid \mathcal{M}_3 \quad \forall \mathcal{M}_2$;

(iii) $\mathcal{M}_1 \parallel \mathcal{M}_2 \mid \mathcal{M}_3 \Longleftrightarrow (\mathcal{M}_1 \vee \mathcal{M}_3) \parallel (\mathcal{M}_2 \vee \mathcal{M}_3) \mid \mathcal{M}_3$;

(iv) $\mathcal{M}_1 \parallel \mathcal{M}_2 \mid \mathcal{M}_3$ and $\mathcal{M}_4 \subset \overline{\mathcal{M}_2 \vee \mathcal{M}_3} \Longrightarrow \mathcal{M}_1 \parallel \mathcal{M}_4 \mid \mathcal{M}_3$. ■

The next theorem is similar to Theorem 2.2.10 insofar as it makes a characteristic property of measurable separability explicit.

5.2.5 Theorem. If

(i) $\mathcal{M}_1 \parallel \mathcal{M}_2 \mid \mathcal{M}_3$,

(ii) $\mathcal{M}_1 \parallel \mathcal{M}_4 \mid \mathcal{M}_2 \vee \mathcal{M}_3$,

then

(iii) $\mathcal{M}_1 \parallel (\mathcal{M}_2 \vee \mathcal{M}_4) \mid \mathcal{M}_3$.

Proof. Indeed,

$$
\begin{aligned}
\overline{\mathcal{M}_1 \vee \mathcal{M}_3} \quad \cap \quad & \overline{\mathcal{M}_2 \vee \mathcal{M}_4 \vee \mathcal{M}_3} \\
= \quad & \overline{\mathcal{M}_1 \vee \mathcal{M}_3} \cap \left\{ \overline{\mathcal{M}_1 \vee \mathcal{M}_2 \vee \mathcal{M}_3} \cap \overline{\mathcal{M}_4 \vee \mathcal{M}_2 \vee \mathcal{M}_3} \right\} \\
= \quad & \overline{\mathcal{M}_1 \vee \mathcal{M}_3} \cap \overline{\mathcal{M}_2 \vee \mathcal{M}_3} \qquad \text{by (ii)} \\
= \quad & \overline{\mathcal{M}_3} \qquad \text{by (i).}
\end{aligned}
$$
■

The next proposition is similar to the first part of Corollary 2.2.11.

5.2.6 Proposition. If $\mathcal{M}_1 \parallel \mathcal{M}_2 \mid \mathcal{M}_3$ then for any $\mathcal{M}_4 \subset \overline{\mathcal{M}_2 \vee \mathcal{M}_3}$ and $\mathcal{M}_5 \subset \overline{\mathcal{M}_1 \vee \mathcal{M}_3}$,

$$(\mathcal{M}_1 \vee \mathcal{M}_5) \parallel (\mathcal{M}_2 \vee \mathcal{M}_4) \mid \mathcal{M}_3.$$
■

In view of the following theorem, measurable separability may be viewed as a necessary condition of conditional independence (this explains the notation $\|$ and $\perp$).

5.2.7 Theorem. If $\mathcal{M}_1 \perp \mathcal{M}_2 \mid \mathcal{M}_3$ then $\mathcal{M}_1 \| \mathcal{M}_2 \mid \mathcal{M}_3$.

Proof. $\mathcal{M}_1 \perp \mathcal{M}_2 \mid \mathcal{M}_3$ implies, by Corollary 2.2.11,
$(\mathcal{M}_1 \vee \mathcal{M}_3) \perp (\mathcal{M}_2 \vee \mathcal{M}_3) \mid \mathcal{M}_3$.
An application of Corollary 2.2.9 finishes the proof.   ∎

It should be noticed that measurable separability preserves most, but not all, of the properties of independence. For instance, when there is conditional independence, the condition analogous to 5.2.5(iii) is equivalent to the condition analogous to 5.2.5(i) and (ii); this was Theorem 2.2.10. When there is measurable separability, Condition 5.2.5(iii) clearly implies 5.2.5(i) but not 5.2.5(ii). Similarly, Proposition 5.2.6 is the analogue for measurable separability of Corollary 2.2.11(i) for conditional independence but the analogous statement for measurable separability of Corollary 2.2.11(ii) is not true in general. More specifically, $\mathcal{M}_1 \| \mathcal{M}_2$ and $\mathcal{M}_4 \subset \mathcal{M}_2$ do not imply $\mathcal{M}_1 \| \mathcal{M}_2 \mid \mathcal{M}_4$.

Here is an example of these two lacks of implications.

Example. Let M_i, $(i = 1, 2, 3, 4)$ be a finite partition of M with $P(M_i) > 0 \ \forall i$. Take

$$\mathcal{M}_1 = \sigma \{M_1 \cup M_3\}, \qquad \mathcal{M}_2 = \sigma \{M_1 \cup M_2, M_3\},$$
$$\mathcal{M}_3 = \mathcal{M}_0, \qquad\qquad \mathcal{M}_4 = \sigma \{M_1 \cup M_2\}.$$

Clearly, $\mathcal{M}_4 \subset \mathcal{M}_2, \mathcal{M}_1 \| \mathcal{M}_2$, but $\mathcal{M}_2 \subset \mathcal{M}_1 \vee \mathcal{M}_4$. Thus, $\mathcal{M}_1 \| \mathcal{M}_2 \mid \mathcal{M}_4$ is not verified.   ∎

The next result shows how a conditional independence property may be used to increase σ-fields without sacrificing a measurable separability property.

5.2.8 Theorem. Let

(i) $\mathcal{M}_1 \perp\!\!\!\perp \mathcal{M}_2 \mid \mathcal{M}_3 \vee \mathcal{M}_4$

(ii) $\mathcal{M}_5 \subset \mathcal{M}_1 \vee \mathcal{M}_3 \vee \mathcal{M}_4$

(iii) $\mathcal{M}_5 \parallel \mathcal{M}_4 \mid \mathcal{M}_3$

then

$$\mathcal{M}_5 \parallel (\mathcal{M}_2 \vee \mathcal{M}_4) \mid \mathcal{M}_3.$$

Proof. Clearly, (i) and (ii) imply $\mathcal{M}_5 \perp\!\!\!\perp \mathcal{M}_2 \mid \mathcal{M}_3 \vee \mathcal{M}_4$. By Theorem 5.2.7, $\mathcal{M}_5 \parallel \mathcal{M}_2 \mid \mathcal{M}_3 \vee \mathcal{M}_4$ and, along with (iii), gives the result by Theorem 5.2.5. ∎

The next corollary applies Theorem 5.2.8 to minimal conditioning σ-fields and is analogous to Theorem 4.3.11 for conditional independence.

5.2.9 Corollary. Let $\mathcal{M}_3 \vee \mathcal{M}_5 \subset \mathcal{M}_1$ and $\mathcal{M}_4 \subset \mathcal{M}_2$. Then

(i) $\mathcal{M}_3 \parallel \mathcal{M}_2\mathcal{M}_1 \mid \mathcal{M}_4 \vee \mathcal{M}_5 \iff \mathcal{M}_3 \parallel \mathcal{M}_2 \mid \mathcal{M}_4 \vee \mathcal{M}_5;$

(ii) $\mathcal{M}_3 \parallel \mathcal{M}_1\mathcal{M}_2 \mid \mathcal{M}_4 \vee \mathcal{M}_5 \implies \mathcal{M}_3 \parallel \mathcal{M}_2 \mid \mathcal{M}_4 \vee \mathcal{M}_5.$

Moreover

(iii) $\mathcal{M}_1 \parallel \mathcal{M}_2 \mid \mathcal{M}_4 \vee \mathcal{M}_5 \iff \mathcal{M}_1\mathcal{M}_2 \parallel \mathcal{M}_2\mathcal{M}_1 \mid \mathcal{M}_4 \vee \mathcal{M}_5.$

Proof. (i) and (ii) are consequences of Theorem 5.2.8, if we notice that $\mathcal{M}_1 \perp\!\!\!\perp \mathcal{M}_2 \mid \mathcal{M}_2\mathcal{M}_1 \vee \mathcal{M}_4 \vee \mathcal{M}_5$ and $\mathcal{M}_1 \perp\!\!\!\perp \mathcal{M}_2 \mid \mathcal{M}_1\mathcal{M}_2 \vee \mathcal{M}_4 \vee \mathcal{M}_5$. (iii) is obtained by symmetrizing (i). ∎

The last theorem in this section responds to the question raised at the beginning of this section and may be viewed as a variant of Theorem 2.2.10.

5.2.10 Theorem. If

(i) $\mathcal{M}_1 \perp\!\!\!\perp \mathcal{M}_2 \mid \mathcal{M}_4 \vee \mathcal{M}_3$;

(ii) $\mathcal{M}_1 \perp\!\!\!\perp \mathcal{M}_4 \mid \mathcal{M}_2 \vee \mathcal{M}_3$;

(iii) $\mathcal{M}_2 \parallel \mathcal{M}_4 \mid \mathcal{M}_3$,

then

(iv) $\mathcal{M}_1 \perp\!\!\!\perp (\mathcal{M}_2 \vee \mathcal{M}_4) \mid \mathcal{M}_3$.

Proof. (i) $\Longrightarrow \mathcal{M}_1 \perp\!\!\!\perp (\mathcal{M}_2 \vee \mathcal{M}_3) \mid \mathcal{M}_4 \vee \mathcal{M}_3$ by Corollary 2.2.11(i); similarly (ii) $\Longrightarrow \mathcal{M}_1 \perp\!\!\!\perp (\mathcal{M}_4 \vee \mathcal{M}_3) \mid \mathcal{M}_2 \vee \mathcal{M}_3$. By Corollary 2.2.13, these properties imply $\mathcal{M}_1 \perp\!\!\!\perp (\mathcal{M}_2 \vee \mathcal{M}_3 \vee \mathcal{M}_4) \mid \overline{\mathcal{M}_2 \vee \mathcal{M}_3} \cap \overline{\mathcal{M}_3 \vee \mathcal{M}_4}$. By (iii), $\mathcal{M}_1 \perp\!\!\!\perp (\mathcal{M}_2 \vee \mathcal{M}_3 \vee \mathcal{M}_4) \mid \overline{\mathcal{M}_3}$, and this is equivalent to (iv). ∎

Proposition 5.2.3 and Theorem 5.2.7 entail the following corollary which is very useful for verifying a measurable separability property as it shows that there is (conditional) measurable separability once the probability is equivalent (i.e, same null sets) to a probability that makes $\mathcal{M}_1$ and $\mathcal{M}_2$ (conditionally) independent.

5.2.11 Corollary. If $\exists P'$ probability on $(M, \mathcal{M})$ such that

(i) $P \sim P'$;

(ii) $\mathcal{M}_1 \perp\!\!\!\perp \mathcal{M}_2 \mid \mathcal{M}_3; P'$,

then for any $\mathcal{M}_5 \subset \overline{\mathcal{M}_1 \vee \mathcal{M}_3}$, and for any $\mathcal{M}_4 \subset \overline{\mathcal{M}_2 \vee \mathcal{M}_3}$,

(iii) $\mathcal{M}_1 \parallel \mathcal{M}_2 \mid \mathcal{M}_3 \vee \mathcal{M}_4 \vee \mathcal{M}_5; P$.

Proof. $\mathcal{M}_1 \perp\!\!\!\perp \mathcal{M}_2 \mid \mathcal{M}_3; P'$ implies, by Corollary 2.2.11, that

$$\mathcal{M}_1 \perp\!\!\!\perp \mathcal{M}_2 \mid \mathcal{M}_3 \vee \mathcal{M}_4 \vee \mathcal{M}_5; P'.$$

Therefore, by Theorem 5.2.7,

$$\mathcal{M}_1 \parallel \mathcal{M}_2 \mid \mathcal{M}_3 \vee \mathcal{M}_4 \vee \mathcal{M}_5; P',$$

and by Proposition 5.2.3, under (i), this implies

$$\mathcal{M}_1 \parallel \mathcal{M}_2 \mid \mathcal{M}_3 \vee \mathcal{M}_4 \vee \mathcal{M}_5; P. \qquad \blacksquare$$

5.3 Measurable Separability in Bayesian Experiments

As mentioned in the introduction to this chapter, measurable separability is both necessary and sufficient to ensure that any statistic independent of a sufficient statistic is ancillary. In fact, this condition is a very weak regularity condition which ensure, heuristically, that the parameter and the statistic are properly distinguished. Such a property is particulary desirable in finite sample as it precludes undesirable pathologies and justifies usual manipulations of sampling densities. In contrast, such a property is undesirable in asymptotic theory where instead one hopes for total informativity; this latter property appears to be antithetic to measurable separability.

We first introduce the concept of a measurably separated Bayesian experiment (Section 5.3.1) and then analyse the role of measurable separability in Basu Second Theorem both in a purely Bayesian form (Section 5.3.2) and in sampling theory (Section 5.3.3).

5.3.1 Measurably Separated Bayesian Experiment

We first define the notion of a measurably separated experiment.

5.3.1 Definition. The Bayesian experiment $\mathcal{E}^{\mathcal{M}}_{\mathcal{B}\vee\mathcal{T}}$ is *measurably separated* if $\mathcal{B} \parallel \mathcal{T} \mid \mathcal{M}$, i.e., if any one of the following three equivalent conditions holds:

(i) $\overline{(\mathcal{B} \vee \mathcal{M})} \cap \overline{(\mathcal{M} \vee \mathcal{T})} = \overline{\mathcal{M}};$

(ii) $(\mathcal{B} \vee \mathcal{M}) \cap \overline{(\mathcal{M} \vee \mathcal{T})} = (\mathcal{B} \vee \mathcal{M}) \cap \overline{\mathcal{M}};$

(iii) $\overline{(\mathcal{B} \vee \mathcal{M})} \cap (\mathcal{M} \vee \mathcal{T}) = \overline{\mathcal{M}} \cap (\mathcal{M} \vee \mathcal{T}).$

In particular, the unreduced Bayesian experiment $\mathcal{E}$ is measurably separated if $\mathcal{A} \parallel \mathcal{S}$, i.e., if any one of the following three equivalent conditions holds:

(i) $\overline{\mathcal{A}} \cap \overline{\mathcal{S}} = \overline{\mathcal{I}}$;

(ii) $\mathcal{A} \cap \overline{\mathcal{S}} = \mathcal{A} \cap \overline{\mathcal{I}}$;

(iii) $\overline{\mathcal{A}} \cap \mathcal{S} = \overline{\mathcal{I}} \cap \mathcal{S}$.   ∎

This definition may be interpreted as follows: in a measurably separated Bayesian experiment $\mathcal{E}_{\mathcal{B} \vee \mathcal{T}}^{\mathcal{M}}$, the maximal exactly estimable parameter $(\mathcal{B} \vee \mathcal{M}) \cap \overline{(\mathcal{M} \vee \mathcal{T})}$ is equal to the trivial parameter $(\mathcal{B} \vee \mathcal{M}) \cap \overline{\mathcal{M}}$, i.e., the conditioning σ-field, $\mathcal{M}$, completed by the null sets of the full parameter $\mathcal{B} \vee \mathcal{M}$, or, equivalently, the maximal exactly estimating statistic $\overline{(\mathcal{B} \vee \mathcal{M})} \cap (\mathcal{M} \vee \mathcal{T})$ is equal to the trivial statistic $\overline{\mathcal{M}} \cap (\mathcal{M} \vee \mathcal{T})$, i.e., the conditioning σ-field, $\mathcal{M}$, completed by the null sets of the full observation $\mathcal{M} \vee \mathcal{T}$.

Recall the primordial sets of inclusions for the Bayesian experiment $\mathcal{E}_{\mathcal{B} \vee \mathcal{T}}^{\mathcal{M}}$, given in Proposition 4.7.4, i.e.,

(5.3.1)
$$(\mathcal{B} \vee \mathcal{M}) \cap \overline{\mathcal{M}} \subset (\mathcal{B} \vee \mathcal{M}) \cap \overline{(\mathcal{M} \vee \mathcal{T})}$$
$$\subset (\mathcal{B} \vee \mathcal{M})(\mathcal{M} \vee \mathcal{T}) \subset (\mathcal{B} \vee \mathcal{M}),$$

(5.3.2)
$$\overline{\mathcal{M}} \cap (\mathcal{M} \vee \mathcal{T}) \subset \overline{(\mathcal{B} \vee \mathcal{M})} \cap (\mathcal{M} \vee \mathcal{T})$$
$$\subset (\mathcal{M} \vee \mathcal{T})(\mathcal{B} \vee \mathcal{M}) \subset \mathcal{M} \vee \mathcal{T}.$$

The measurable separability property of the Bayesian experiment $\mathcal{E}_{\mathcal{B} \vee \mathcal{T}}^{\mathcal{M}}$ may be seen as a property completely opposite to that of the total informativity of this experiment. Indeed, $\mathcal{E}_{\mathcal{B} \vee \mathcal{T}}^{\mathcal{M}}$ is measurably separated if any of the first inclusions in (5.3.1) and (5.3.2) is an equality, and if the maximal exactly estimable parameter is trivial. In contrast, $\mathcal{E}_{\mathcal{B} \vee \mathcal{T}}^{\mathcal{M}}$ is totally informative if the maximal exactly estimable parameter is the full parameter (see Definition 4.7.9). Recall that $\mathcal{E}_{\mathcal{B} \vee \mathcal{T}}^{\mathcal{M}}$ is exact if any of the second inclusions in (5.3.1) and (5.3.2) is an equality. Thus the general Bayesian experiment $\mathcal{E}_{\mathcal{B} \vee \mathcal{T}}^{\mathcal{M}}$ is both measurably separated and exact if and only if

(5.3.3) $$(\mathcal{B} \vee \mathcal{M}) \cap \overline{\mathcal{M}} = (\mathcal{B} \vee \mathcal{M})(\mathcal{M} \vee \mathcal{T})$$

or, equivalently, if and only if:

$$(5.3.4) \qquad \overline{\mathcal{M}} \cap (\mathcal{M} \vee \mathcal{T}) = (\mathcal{M} \vee \mathcal{T})(\mathcal{B} \vee \mathcal{M}).$$

But, by Proposition 4.3.10, Corollary 2.2.7 and elementary Properties 4.3.2(i), 5.3.3 or 5.3.4 is also equivalent to

$$(5.3.5) \qquad \mathcal{B} \perp\!\!\!\perp \mathcal{T} \mid \mathcal{M},$$

i.e., the Bayesian experiment $\mathcal{E}_{\mathcal{B}\vee\mathcal{T}}^{\mathcal{M}}$ is totally non informative. Thus, in such a situation, the trivial parameter, i.e., $(\mathcal{B} \vee \mathcal{M}) \cap \overline{\mathcal{M}}$, is both the minimal sufficient parameter and exactly estimable parameter (this is (5.3.3)); and nothing is learned about any other parameter (this is (5.3.5)). Typically, a finite sample experiment would be measurably separated whereas an asymptotic experiment would be exact.

Verification of the measurable separability of the Bayesian experiment $\mathcal{E}_{\mathcal{B}\vee\mathcal{T}}^{\mathcal{M}}$ may be simplified by using Corollary 5.2.9. This leads to the following proposition.

5.3.2 Proposition. A Bayesian experiment $\mathcal{E}_{\mathcal{B}\vee\mathcal{T}}^{\mathcal{M}}$ is measurably separated if and only if one of the following three equivalent conditions holds.

(i)   There exists an $\mathcal{E}_{\mathcal{B}\vee\mathcal{T}}^{\mathcal{M}}$-sufficient statistic $\mathcal{N}$ measurably separated from the parameter i.e., $\mathcal{B} \parallel \mathcal{N} \mid \mathcal{M}$;

(ii) The full observation $\mathcal{T}$ is measurably separated from an $\mathcal{E}_{\mathcal{B}\vee\mathcal{T}}^{\mathcal{M}}$-sufficient parameter $\mathcal{L}$, i.e., $\mathcal{L} \parallel \mathcal{T} \mid \mathcal{M}$;

(iii) The minimal $\mathcal{E}_{\mathcal{B}\vee\mathcal{T}}^{\mathcal{M}}$-sufficient strong statistic and the minimal $\mathcal{E}_{\mathcal{B}\vee\mathcal{T}}^{\mathcal{M}}$-sufficient strong parameter are measurabily separated conditionally on $\mathcal{M}$ i.e., $(\mathcal{B} \vee \mathcal{M})(\mathcal{M} \vee \mathcal{T}) \parallel (\mathcal{M} \vee \mathcal{T})(\mathcal{B} \vee \mathcal{M}) \mid \mathcal{M}$.   ∎

In the dominated case, measurable separability is generally readily verified as a consequence of the fact that $\mathcal{A} \perp\!\!\!\perp \mathcal{S}; \mu \otimes P$ is always true; then Corollary 2.2.11 and a simple application of Proposition 5.2.11 give the following result.

5.3.3 Proposition. In a dominated Bayesian experiment $\mathcal{E} = (A \times S, \mathcal{A} \vee \mathcal{S}, \Pi)$, if $\Pi \sim \mu \otimes P$ then $\forall \mathcal{B} \subset \mathcal{A}$, $\mathcal{C} \subset \mathcal{A}$, $\mathcal{T} \subset \mathcal{S}$, $\mathcal{U} \subset \mathcal{S}$, $\mathcal{E}_{\mathcal{B} \vee \mathcal{T}}^{\mathcal{C} \vee \mathcal{U}}$ is measurably separated. ∎

We conclude this section by discussing the definition of a measurably separated experiment and some of the connections between this definition and certain other previously defined characterizations of a Bayesian experiment. For expository reasons, we restrain our attention to the unreduced experiment $\mathcal{E} = \mathcal{E}_{\mathcal{A} \vee \mathcal{S}}$.

By definition, $\mathcal{E}$ is measurably separated if the maximal exactly estimable parameter $\mathcal{A} \cap \overline{\mathcal{S}}$ is the trivial sub-σ-field of $\mathcal{A}$ or equivalently if the maximal exactly estimating statistic $\overline{\mathcal{A}} \cap \mathcal{S}$ is the trivial sub-σ-field of $\mathcal{S}$. As seen in Section 4.7, or by Formula (5.2.4), a set E in $\mathcal{A} \cap \overline{\mathcal{S}}$ is entirely characterized by the fact that $E \in \mathcal{A}$ and $\left[\mu^{\mathcal{S}}(E) \right]^2 = \mu^{\mathcal{S}}(E)$, since this is equivalent to $\mu^{\mathcal{S}}(E) = 1_E$ a.s.Π. Similarly, a set X in $\overline{\mathcal{A}} \cap \mathcal{S}$ is characterized by the fact that $X \in \mathcal{S}$ and $\left[P^{\mathcal{A}}(X) \right]^2 = P^{\mathcal{A}}(X)$, since this is equivalent to $P^{\mathcal{A}}(X) = 1_X$ a.s.Π. It follows that the experiment $\mathcal{E}$ will be measurably separated if the following implication holds:

$$E \in \mathcal{A} \qquad \left[\mu^{\mathcal{S}}(E) \right]^2 = \mu^{\mathcal{S}}(E) \implies \left[\mu(E) \right]^2 = \mu(E)$$

or, equivalently,

$$X \in \mathcal{S} \qquad \left[P^{\mathcal{A}}(X) \right]^2 = P^{\mathcal{A}}(X) \implies \left[P(X) \right]^2 = P(X),$$

i.e., $\mathcal{E}$ is measurably separated if it does not exist a set $X \in \mathcal{S}$ (respectively, $E \in \mathcal{A}$) with $0 < P(X) < 1$ (respectively, $0 < \mu(E) < 1$) such that the sampling probabilities (respectively, the posterior probabilities) are indicator functions. Such a set has been called a *splitting set* by Koehn and Thomas (1975). Note also that these properties are non-symmetric but they are equivalent to a symmetric one, i.e., $\overline{\mathcal{A}} \cap \overline{\mathcal{S}} = \overline{\mathcal{I}}$. Breiman, LeCam and Schwartz (1964) obtained the same result through a somewhat more involved argument (see also Skibinski (1960), who also treated families of probabilities).

5.3.2　Basu Second Theorem

In this section, we provide a generalized Bayesian version of Basu second Theorem; this states that any statistic independent of a sufficient statistic in the sampling process is ancillary.

5.3.4 Theorem. In a measurably separated Bayesian experiment $\mathcal{E}^{\mathcal{M}}_{\mathcal{B}\vee\mathcal{T}}$,

(i) Every $\mathcal{E}^{\mathcal{M}}_{\mathcal{B}\vee\mathcal{T}}$-statistic $\mathcal{R}$ sampling independent of an $\mathcal{E}^{\mathcal{M}}_{\mathcal{B}\vee\mathcal{T}}$-sufficient statistic $\mathcal{N}$ is ancillary. Moreover $\mathcal{R}$ is predictively independent of $\mathcal{N}$.

(ii) Every $\mathcal{E}^{\mathcal{M}}_{\mathcal{B}\vee\mathcal{T}}$-parameter $\mathcal{K}$ *a posteriori* independent of an $\mathcal{E}^{\mathcal{M}}_{\mathcal{B}\vee\mathcal{T}}$-sufficient parameter $\mathcal{L}$ is ancillary. Moreover $\mathcal{K}$ is *a priori* independent of $\mathcal{L}$.

Proof. By duality it suffices to prove the theorem for a statistic. By hypothesis $\mathcal{R} \perp\!\!\!\perp \mathcal{N} \mid \mathcal{B}\vee\mathcal{M}$. The sufficiency of $\mathcal{N}$, i.e., $\mathcal{B} \perp\!\!\!\perp \mathcal{T} \mid \mathcal{M}\vee\mathcal{N}$ implies $\mathcal{B} \perp\!\!\!\perp \mathcal{R} \mid \mathcal{M} \vee \mathcal{N}$; the measurable separability of $\mathcal{E}^{\mathcal{M}}_{\mathcal{B}\vee\mathcal{T}}$, i.e., $\mathcal{B} \parallel \mathcal{T} \mid \mathcal{M}$, implies $\mathcal{B} \parallel \mathcal{N} \mid \mathcal{M}$. By Theorem 5.2.10, we obtain $\mathcal{R} \perp\!\!\!\perp (\mathcal{B} \vee \mathcal{N}) \mid \mathcal{M}$, and this implies $\mathcal{R} \perp\!\!\!\perp \mathcal{B} \mid \mathcal{M}$, i.e., $\mathcal{R}$ is ancillary and $\mathcal{R} \perp\!\!\!\perp \mathcal{N} \mid \mathcal{M}$, i.e., $\mathcal{R}$ and $\mathcal{N}$ are independent for the predictive probability. ∎

The condition of measurable separability of the experiment $\mathcal{E}^{\mathcal{M}}_{\mathcal{B}\vee\mathcal{T}}$ is not merely sufficient but also necessary.

5.3.5 Theorem. If, in a Bayesian experiment $\mathcal{E}^{\mathcal{M}}_{\mathcal{B}\vee\mathcal{T}}$, every $\mathcal{E}^{\mathcal{M}}_{\mathcal{B}\vee\mathcal{T}}$-statistic sampling independent of an $\mathcal{E}^{\mathcal{M}}_{\mathcal{B}\vee\mathcal{T}}$-sufficient statistic is ancillary, then the experiment is measurably separated.

Proof. $\mathcal{N}_{\max} = \overline{(\mathcal{B} \vee \mathcal{M})}\cap(\mathcal{M} \vee \mathcal{T})$, the maximal exactly estimating $\mathcal{E}^{\mathcal{M}}_{\mathcal{B}\vee\mathcal{T}}$-statistic, is clearly independent in the sampling process of any $\mathcal{E}^{\mathcal{M}}_{\mathcal{B}\vee\mathcal{T}}$-sufficient statistic (in particular of $\mathcal{T}$), since $\mathcal{N}_{\max} \subset \overline{\mathcal{B} \vee \mathcal{M}}$. It is therefore ancillary, i.e., $\mathcal{N}_{\max} \perp\!\!\!\perp \mathcal{B} \mid \mathcal{M}$. But this is equivalent to $\mathcal{N}_{\max} \perp\!\!\!\perp \overline{\mathcal{B} \vee \mathcal{M}} \mid \mathcal{M}$, and so $\mathcal{N}_{\max} \perp\!\!\!\perp \mathcal{N}_{\max} \mid \mathcal{M}$. By Corollary 2.2.8 this is equivalent to $\mathcal{N}_{\max} \subset \overline{\mathcal{M}}$, or equivalently, $\mathcal{N}_{\max} = \overline{\mathcal{M}}\cap(\mathcal{M}\vee\mathcal{T})$, i.e., the experiment $\mathcal{E}^{\mathcal{M}}_{\mathcal{B}\vee\mathcal{T}}$ is measurably separated. ∎

Note that this proof uses the hypothesis of Theorem 5.3.5 for $\mathcal{N}_{\max}$ only. This theorem cannot, however, be improved since once measurable separability has been verified from the ancillarity of $\mathcal{N}_{\max}$, Theorem 5.3.4 implies the ancillarity of any other $\mathcal{E}^{\mathcal{M}}_{\mathcal{B}\vee\mathcal{T}}$-statistic sampling independent of a sufficient statistic.

We may point out that in the proof of Theorem 5.3.4, we use only the measurable separability of the parameter and the sufficient statistic, i.e., $\mathcal{B} \parallel \mathcal{N} \mid \mathcal{M}$. But when $\mathcal{N}$ is sufficient Proposition 5.3.2 (i) shows that this assumption is actually equivalent to $\mathcal{B} \parallel \mathcal{T} \mid \mathcal{M}$, i.e., the measurable separability of $\mathcal{E}_{\mathcal{B} \vee \mathcal{T}}^{\mathcal{M}}$.

5.3.3 Sampling Theory and Bayesian Methods

Let us consider a statistical experiment $\mathcal{E} = \{(S, \mathcal{S}), P^a, a \in A\}$ and $\mathcal{A}$ a σ-field on A which makes $P^a(X)$ measurable for any $X \in \mathcal{S}$. We use the same notation as in Section 2.3.7.

In 1955 Basu stated that any statistic $\mathcal{U} \subset \mathcal{S}$ independent of a sufficient statistic $\mathcal{T} \subset \mathcal{S}$, i.e., $\mathcal{U} \perp\!\!\!\perp \mathcal{T}; P^a$, $\forall a \in A$, is ancillary. Basu (1958) subsequently qualified his statement by giving a sufficient condition for the theorem to be true, i.e., $\mathcal{T}$ does not contain a *splitting set*.

5.3.6 Definition. A set $X \in \mathcal{S}$ is a splitting set if there exists a set $E \in \mathcal{A}$ non-trivial (i.e., $E \neq \phi$ and $E \neq A$) such that $P^a(X) = 1_E(a)$.   ∎

We then deduce the following characterization.

5.3.7 Proposition. A statistic $\mathcal{T} \subset \mathcal{S}$ does not contain a splitting set if and only if the following implication holds:

$X \in \mathcal{T}, [P^a(X)]^2 = P^a(X),\ \forall a \in A,$ imply

either $P^a(X) = 1\ \ \forall a \in A$ or $P^a(X) = 0\ \ \forall a \in A.$ ∎

More recently Khoen and Thomas (1975) proved that this condition is also necessary, i.e., any statistic $\mathcal{U} \subset \mathcal{S}$ independent of a sufficient statistic $\mathcal{T} \subset \mathcal{S}$ is ancillary if and only if $\mathcal{T}$ does not contain a splitting set. For the sake of completeness, we reproduce the proof: Let $u \in [\mathcal{U}]^+$. Since $\mathcal{T}$ is sufficient, there exists $t \in [\mathcal{T}]^+$ such that $T^a u = t$ a.s.P^a, $\forall a \in A$. But $\mathcal{U} \perp\!\!\!\perp \mathcal{T}; P^a$, $\forall a \in A$, implies $T^a u = I^a u$ a.s.P^a, $\forall a \in A$, and so $t = I^a u$ a.s.P^a, $\forall a \in A$. Fix $a_0 \in A$ and define $X = \{t = I^{a_0} u\}$. Clearly, $[P^a(X)]^2 = P^a(X)\ \ \forall a \in A$. Now, since $P^{a_0}(X) = 1$, if $\mathcal{T}$ does not contain a splitting set $P^a(X) = 1$, $\forall a \in A$. Thus $T^a u = I^{a_0} u$ a.s.P^a,

$\forall a \in A$, and this implies $\mathcal{I}^a u = \mathcal{I}^{a_0} u$, $\forall a \in A$, i.e., $\mathcal{U}$ is ancillary. For necessity it suffices to note that, if X is a splitting set and $\mathcal{U}$ is the σ-field generated by X, then $\mathcal{U}$ is not ancillary whereas $\mathcal{U} \perp\!\!\!\perp \mathcal{S}; P^a$, $\forall a \in A$. ∎

Let us now compare the sampling theory condition, i.e., the non-existence of a splitting set in $\mathcal{T}$, with the Bayesian condition, i.e., the measurable separability of $\mathcal{A}$ and $\mathcal{T}$. Recall that for a prior probability μ, $\forall X \in \mathcal{S}$, $P^{\mathcal{A}}(X) = P^a(X)$ a.s.μ. In the previous section we saw that $\mathcal{A}$ and $\mathcal{T}$ are measurably separated if and only if the following implication holds:

$$X \in \mathcal{T}, \quad \left[P^{\mathcal{A}}(X)\right]^2 = P^{\mathcal{A}}(X) \implies [P(X)]^2 = P(X).$$

To make precise the role of the prior probability μ, this condition may be rephrased as follows:

5.3.8 Proposition. $\mathcal{A}$ and $\mathcal{T}$ are measurably separated ($\mathcal{A} \parallel \mathcal{T}$) if and only if the following implication holds:

$X \in \mathcal{T}, [P^a(X)]^2 = P^a(X)$ a.s.μ imply

either $P^a(X) = 1$ a.s.μ , or $P^a(X) = 0$ a.s.μ. ∎

In view of Propositions 5.3.7 and 5.3.8, it is clear that if μ is a prior probability such that the predictive probability dominates all the sampling probabilities, then $\mathcal{A} \parallel \mathcal{T}$ implies the non-existence of a splitting set in $\mathcal{T}$. This will be true, in particular, if μ is regular (Definition 2.3.13). If, moreover, μ possesses the property that $[P^a(X)]^2 = P^a(X)$ a.s.μ implies $[P^a(X)]^2 = P^a(X)$, $\forall a \in A$, then the non-existence of a splitting set in $\mathcal{T}$ is equivalent to the measurable separability of $\mathcal{A}$ and $\mathcal{T}$.

5.4 Strong Identification of σ-Fields

In this section, we treat a topic in general probability theory and adopt the usual notation. In particular, $(M, \mathcal{M}, P)$ is an abstract probability space.

The question raised by Basu First Theorem in the introduction is actually equivalent to the following problem: If $\mathcal{M}_1 \perp\!\!\!\perp \mathcal{M}_2 \mid \mathcal{M}_3$ and $\mathcal{M}_1 \perp\!\!\!\perp \mathcal{M}_2$, under what condition on $\mathcal{M}_3$ and $\mathcal{M}_2$ is it true that $\mathcal{M}_1 \perp\!\!\!\perp (\mathcal{M}_2 \vee \mathcal{M}_3)$? This condition is called the *strong identifiability* of $\mathcal{M}_3$ by $\mathcal{M}_2$. We will generalize this condition so as to handle a conditional version of this problem. The identifiability of $\mathcal{M}_3$ by $\mathcal{M}_2$, as defined in Chapter 4 (Definition 4.5.1), is weaker than (since implied by) the strong identification of $\mathcal{M}_3$ by $\mathcal{M}_2$. We use Section 4.5 as a guide in collecting general properties of the strong identification.

5.4.1 Definition and General Properties

For reasons explained below, we introduce the concept of strong identification in L_p-spaces.

5.4.1 Definition. $\mathcal{M}_1$ is *strongly p-identified by* $\mathcal{M}_2$ *conditionally on* $\mathcal{M}_3$ or, equivalently, $\mathcal{M}_2$ *strongly p-identifies* $\mathcal{M}_1$ *conditionally on* $\mathcal{M}_3$ if the following implication holds:

$$m \in [\mathcal{M}_1 \vee \mathcal{M}_3]_p \quad \text{and} \quad (\mathcal{M}_2 \vee \mathcal{M}_3)\, m = 0 \implies m = 0 \quad \text{a.s.}$$

In this case we write $\mathcal{M}_1 \ll_p \mathcal{M}_2 \mid \mathcal{M}_3$. If $\mathcal{M}_3 = \mathcal{M}_0$ we say that $\mathcal{M}_1$ is *strongly p-identified by* $\mathcal{M}_2$ and we write $\mathcal{M}_1 \ll_p \mathcal{M}_2$. If $p = \infty$, we simply say that $\mathcal{M}_1$ is *strongly identified by* $\mathcal{M}_2$ *conditionally on* $\mathcal{M}_3$ and write $\mathcal{M}_1 \ll \mathcal{M}_2 \mid \mathcal{M}_3$. If we want to make explicit the role of the probability P, we write $\mathcal{M}_1 \ll_p \mathcal{M}_2 \mid \mathcal{M}_3; P$.   ∎

We first make some general remarks about the above definition.

(i) In order to make the role of $\mathcal{M}_3$ more explicit, note that the implication in the definition is actually equivalent to the following:

$$m \in [\mathcal{M}_1 \vee \mathcal{M}_3]_p \text{ and } (\mathcal{M}_2 \vee \mathcal{M}_3)\, m \in \overline{\mathcal{M}_3} \implies m \in \overline{\mathcal{M}_3}.$$

(ii) Note also that $\mathcal{M}_1 \ll_p \mathcal{M}_2 \mid \mathcal{M}_3$ is equivalent to

$$(\mathcal{M}_1 \vee \mathcal{M}_3) \ll_p (\mathcal{M}_2 \vee \mathcal{M}_3).$$

Consequently properties stated conditionally on $\mathcal{M}_3$ need be proved for $\mathcal{M}_3 = \mathcal{M}_0$ only.

(iii) Remark that $\mathcal{M}_1 \ll_p \mathcal{M}_2 \mid \mathcal{M}_3$ implies $\mathcal{M}_1 \ll_{p'} \mathcal{M}_2 \mid \mathcal{M}_3$ $\forall p \le p' \le \infty$ since in that case $[\mathcal{M}_1 \vee \mathcal{M}_3]_{p'} \subset [\mathcal{M}_1 \vee \mathcal{M}_3]_p$.

With few exceptions, the most useful theorems require only the weakest version of strong identification, i.e., $p = \infty$. Therefore even if the general properties of strong identification are true for any $p \in [1, \infty]$, in the sequel they are stated and proved only for $p = \infty$.

We now collect some elementary properties of strong identification. They do not require proof as they are simple consequences of the definition.

5.4.2 Proposition. *Elementary properties of strong identification*

(i) $\mathcal{M}_1 \ll \mathcal{M}_2 \mid \mathcal{M}_3 \iff \overline{\mathcal{M}_1} \ll \mathcal{M}_2 \mid \mathcal{M}_3 \iff \mathcal{M}_1 \ll \overline{\mathcal{M}_2} \mid \mathcal{M}_3$
$\iff \mathcal{M}_1 \ll \mathcal{M}_2 \mid \overline{\mathcal{M}_3};$

(ii) $\mathcal{M}_1 \subset \overline{\mathcal{M}_2 \vee \mathcal{M}_3} \implies \mathcal{M}_1 \ll \mathcal{M}_2 \mid \mathcal{M}_3;$

(iii) $\mathcal{M}_2 \subset \overline{\mathcal{M}_1 \vee \mathcal{M}_3}$ and $\mathcal{M}_1 \ll \mathcal{M}_2 \mid \mathcal{M}_3$
$\implies \overline{\mathcal{M}_1 \vee \mathcal{M}_3} = \overline{\mathcal{M}_2 \vee \mathcal{M}_3};$

(iv) $\mathcal{M}_2 \subset \overline{\mathcal{M}_4 \vee \mathcal{M}_3}$ and $\mathcal{M}_1 \ll \mathcal{M}_2 \mid \mathcal{M}_3 \implies \mathcal{M}_1 \ll \mathcal{M}_4 \mid \mathcal{M}_3;$

(v) $\mathcal{M}_5 \subset \overline{\mathcal{M}_1 \vee \mathcal{M}_3}$ and $\mathcal{M}_1 \ll \mathcal{M}_2 \mid \mathcal{M}_3 \implies \mathcal{M}_5 \ll \mathcal{M}_2 \mid \mathcal{M}_3.$ ∎

Note that Property (v) is, in general, not true for identification (see the comment following 4.5.2).

The concept of strong identification may be interpreted in terms of the geometry of Banach spaces. Indeed, for $m \in [\mathcal{M}]$, let $\overset{\bullet}{m}$ be the equivalence class of m (in $[\mathcal{M}]$) with respect to P-almost sure equality. For $\mathcal{M}_1$, a sub-σ-field of $\mathcal{M}$, denote by $L_p\left(\overline{\mathcal{M}_1}\right)$ the linear space $\left\{ \overset{\bullet}{m} : m \in [\mathcal{M}_1]_p \right\}$ which is a closed linear subspace of the Banach space $L_p(\mathcal{M})$ (see, e.g., Neveu (1964) Section II.6 and IV.3). For any $p \in [1, \infty]$, the conditional expectation given $\overline{\mathcal{M}_1}$ may be viewed as a continuous linear operator defined on $L_p(\mathcal{M})$ onto $L_p\left(\overline{\mathcal{M}_1}\right)$; it is in fact an orthogonal projector (i.e., symmetric and idempotent) also denoted as $\overline{\mathcal{M}_1}$. Let us denote by $N_p(T)$ the null space in $L_p(\mathcal{M})$ of a continuous linear operator T, i.e., the closed linear subspace defined by $\left\{ \overset{\bullet}{m} \in L_p(\mathcal{M}) : T \overset{\bullet}{m} = 0 \right\}$. With this notation Definition 5.4.1 may be rewritten (for $1 \leq p \leq \infty$) as

$$(5.4.1) \qquad N_p\left[\overline{\mathcal{M}_2 \vee \mathcal{M}_3} \circ \overline{\mathcal{M}_1 \vee \mathcal{M}_3} \right] = N_p\left[\overline{\mathcal{M}_1 \vee \mathcal{M}_3} \right].$$

Indeed, (5.4.1) means that $\overline{\mathcal{M}_2 \vee \mathcal{M}_3}\left[\overline{\mathcal{M}_1 \vee \mathcal{M}_3}\; \overset{\bullet}{m}\right] = 0$ implies $\overline{(\mathcal{M}_1 \vee \mathcal{M}_3)}\; \overset{\bullet}{m} = 0$, since the other inclusion is trivial. This is clearly equivalent to Definition 5.4.1. Thus (5.4.1) means that strong identification may be viewed as the injectivity of the $\mathcal{M}_2 \vee \mathcal{M}_3$-conditional expectation operator, restricted to the p-integrable functions of $[\mathcal{M}_1 \vee \mathcal{M}_3]$. In the reflexive case, i.e., $p \in (1, \infty)$, this is equivalent to the surjectivity of the adjoint operator. The next theorem makes use of that feature in order to characterize strong identification by an approximation property of $\mathcal{M}_1 \vee \mathcal{M}_3$-measurable functions by a sequence of $\mathcal{M}_1 \vee \mathcal{M}_3$-conditional expectation of $\mathcal{M}_2 \vee \mathcal{M}_3$-measurable functions.

5.4.3 Theorem. For any $p \in (1, \infty)$, $\frac{1}{p} + \frac{1}{q} = 1$, $\mathcal{M}_1 \ll_p \mathcal{M}_2 \mid \mathcal{M}_3$ if and only if for each $m \in [\mathcal{M}_1 \vee \mathcal{M}_3]_q$ there exists a sequence

$$\{m_k : 1 \leq k < \infty\} \subset [\mathcal{M}_2 \vee \mathcal{M}_3]_q$$

such that $\mathcal{M}_0 \left[\mid m - (\mathcal{M}_1 \vee \mathcal{M}_3)\, m_k \mid^q\right] \longrightarrow 0$ as $k \longrightarrow \infty$.
In particular, $\mathcal{M}_1 \ll_2 \mathcal{M}_2 \mid \mathcal{M}_3$ if and only if, for each $m \in [\mathcal{M}_1 \vee \mathcal{M}_3]_2$, there exists a sequence $\{m_k : 1 \leq k < \infty\} \subset [\mathcal{M}_2 \vee \mathcal{M}_3]_2$ such that $\mathcal{M}_0 \left[\{m - (\mathcal{M}_1 \vee \mathcal{M}_3) m_k\}^2\right] \longrightarrow 0$ as $k \longrightarrow 0$.

Proof. Recall that $\mathcal{M}_1 \ll_p \mathcal{M}_2 \mid \mathcal{M}_3$ is equivalent to

$$N_p \left[\overline{\mathcal{M}_2 \vee \mathcal{M}_3} \circ \overline{\mathcal{M}_1 \vee \mathcal{M}_3}\right] = N_p \left[\overline{\mathcal{M}_1 \vee \mathcal{M}_3}\right]$$

where the two sides of the expression are closed linear subspaces of the Banach space $L_p(\mathcal{M})$. Recall that in a normed linear space the null space of a linear operator is the perpendicular space to the range of the adjoint operator defined on the dual space $L_q(\mathcal{M})$. Then, if we denote by $R_q(T)$ the range of the linear operator $T : L_q(\mathcal{M}) \longrightarrow L_q(\mathcal{M})$, i.e., $R_q(T) = \left\{T\, \overset{\bullet}{m} : \overset{\bullet}{m} \in L_q(\mathcal{M})\right\}$, then $\mathcal{M}_1 \ll_p \mathcal{M}_2 \mid \mathcal{M}_3$ is equivalent to

$$(5.4.2) \qquad R_q \left[\overline{\mathcal{M}_1 \vee \mathcal{M}_3} \circ \overline{\mathcal{M}_2 \vee \mathcal{M}_3}\right]^{\perp} = R_q \left[\overline{\mathcal{M}_1 \vee \mathcal{M}_3}\right]^{\perp}.$$

This implies that

$$(5.4.3) \qquad R_q \left[\overline{\mathcal{M}_1 \vee \mathcal{M}_3} \circ \overline{\mathcal{M}_2 \vee \mathcal{M}_3}\right]^{\perp\perp} = R_q \left[\overline{\mathcal{M}_1 \vee \mathcal{M}_3}\right]^{\perp\perp}.$$

If $1 < p < \infty$, then $L_p(\mathcal{M})$ is a reflexive Banach space: for such a space, if L is a closed linear subspace of its dual $L_q(\mathcal{M})$, then $L^{\perp\perp} = L$. Recall also that in a normed linear space, the range of a linear operator is not necessarily closed, but the range of an idempotent operator $(T^2 = T)$ is a closed linear space (since it is the null space of $I-T$). Hence, for $1 < p < \infty$, (5.4.2) is equivalent to

$$(5.4.4) \qquad R_q\left[\overline{\mathcal{M}_1 \vee \mathcal{M}_3}\right] = \overline{R_q\left[\overline{\mathcal{M}_1 \vee \mathcal{M}_3} \circ \overline{\mathcal{M}_2 \vee \mathcal{M}_3}\right]}$$

where the upper bar on the right hand side represents the closure of the linear space in the Banach space, and this implies $\mathcal{M}_1 \ll_p \mathcal{M}_2 \mid \mathcal{M}_3$ when considering the perpendicular spaces.

Thus, (5.4.4) means that for any $\overset{\bullet}{n} \in L_q(\mathcal{M})$ there exists a sequence $\left\{\overset{\bullet}{n}_k : 1 \le k < \infty\right\} \subset L_q(\mathcal{M})$ such that $\overline{\mathcal{M}_1 \vee \mathcal{M}_3}\left[\overline{\mathcal{M}_2 \vee \mathcal{M}_3}\,\overset{\bullet}{n}_k\right]$ converges to $\overline{\mathcal{M}_1 \vee \mathcal{M}_3}\,\overset{\bullet}{n}$ in $L_q(\mathcal{M})$ as $k \longrightarrow \infty$, or, for any $\overset{\bullet}{m} \in L_q\left(\overline{\mathcal{M}_1 \vee \mathcal{M}_3}\right)$ there exists a sequence $\left\{\overset{\bullet}{m}_k : 1 \le k < \infty\right\} \subset L_q\left(\overline{\mathcal{M}_2 \vee \mathcal{M}_3}\right)$ such that $\overline{\mathcal{M}_1 \vee \mathcal{M}_3}\,\overset{\bullet}{m}_k$ converges to $\overset{\bullet}{m}$ in $L_q(\mathcal{M})$ as $k \longrightarrow \infty$, this is clearly equivalent to saying that for any $m \in [\mathcal{M}_1 \vee \mathcal{M}_3]_q$ there exists a sequence $\{m_k : 1 \le k < \infty\} \subset [\mathcal{M}_2 \vee \mathcal{M}_3]_q$ such that $(\mathcal{M}_1 \vee \mathcal{M}_3)\, m_k$ converge to m in $[\mathcal{M}]_q$. ∎

The next theorem provides a sufficient condition for strong identification in the form of a representability condition of $\mathcal{M}_1$-measurable functions by means of $\mathcal{M}_1$-conditional expectation of $\mathcal{M}_2$-measurable functions.

5.4.4 Theorem. Let $p \in [1, \infty]$. If for any $m_1 \in [\mathcal{M}_1]_p$ there exists $m_2 \in [\mathcal{M}_2]_p$ such that $m_1 = \mathcal{M}_1 m_2$, then $\mathcal{M}_1 \ll_q \mathcal{M}_2$ where $1/p + 1/q = 1$.

Proof. Let $m \in [\mathcal{M}_1]_q$ such that $\mathcal{M}_2 m = 0$. Then, for any $m_1 \in [\mathcal{M}_1]_p$ representable as $m_1 = \mathcal{M}_1 m_2$ for $m_2 \in [\mathcal{M}_2]_p$,

$$\mathcal{M}_0\,(m m_1) = \mathcal{M}_0\,(m \cdot \mathcal{M}_1 m_2) = \mathcal{M}_0\,(m m_2) = \mathcal{M}_0\,(\mathcal{M}_2 m \cdot m_2) = 0,$$

and this implies $m = 0$ a.s. ∎

When $p \in (1, \infty)$, this theorem is actually a particular application of Theorem 5.4.3, letting $\mathcal{M}_3 = \mathcal{M}_0$ and $m_k = m_2 \quad \forall\, k \in I\!N$. Remark that $\mathcal{M}_1 \ll_q \mathcal{M}_2$ as soon as the above representation ($m_1 = \mathcal{M}_1 m_2$) holds merely for a linear lattice in $[\mathcal{M}_1]_p$ generating $[\mathcal{M}_1]_p$.

5.4.2 Strong Identification and Conditional Independence

We first examine the relations existing between strong identification and conditional strong identification. In essence, the first theorem states supplementary hypotheses under which the conditioning σ-field in a conditional strong identification property may be increased; the second theorem states supplementary hypotheses under which the strongly identifying σ-field may be reduced, and as a corollary, it reveals how a conditional strong identification implies a strong identification.

5.4.5 Theorem. Let $\mathcal{M}_i$ ($i = 1, 2, 3, 4$) be sub-σ-fields of $\mathcal{M}$. If

(i) $\mathcal{M}_4 \perp\!\!\!\perp \mathcal{M}_2 \mid \mathcal{M}_3 \vee \mathcal{M}_1$

(ii) $\mathcal{M}_1 \ll \mathcal{M}_2 \mid \mathcal{M}_3$,

then

(iii) $\mathcal{M}_1 \ll \mathcal{M}_2 \mid \mathcal{M}_3 \vee \mathcal{M}_4$.

Note that (i) is trivially satisfied if $\mathcal{M}_4 \subset \overline{\mathcal{M}_1 \vee \mathcal{M}_3}$.

Proof. Take $\mathcal{M}_3 = \mathcal{M}_0$. Let $m \in [\mathcal{M}_1 \vee \mathcal{M}_4]_\infty$ be such that $(\mathcal{M}_2 \vee \mathcal{M}_4) m = 0$. Then, by Theorem 2.2.1 (iii), for any $n \in [\mathcal{M}_4]_\infty$, (i) implies that $\mathcal{M}_2 [\mathcal{M}_1 (mn)] = \mathcal{M}_2 (mn)$. But $\mathcal{M}_2 (mn) = \mathcal{M}_2 [n (\mathcal{M}_2 \vee \mathcal{M}_4) m] = 0$ and so, by (ii), $\mathcal{M}_1 (mn) = 0$. Therefore for any $n \in [\mathcal{M}_4]_\infty$ and $r \in [\mathcal{M}_1]_\infty$, $\mathcal{M}_0 (mnr) = 0$ and, by Theorem 0.2.21, this shows that $m = 0$ a.s.   ∎

5.4.6 Theorem. Let $\mathcal{M}_i$ ($i = 1, 2, 3, 4$) be sub-σ-fields of $\mathcal{M}$. If

(i) $\mathcal{M}_4 \perp\!\!\!\perp \mathcal{M}_1 \mid \mathcal{M}_3 \vee \mathcal{M}_2$

(ii) $\mathcal{M}_1 \ll (\mathcal{M}_2 \vee \mathcal{M}_4) \mid \mathcal{M}_3$,

then

(iii) $\mathcal{M}_1 \ll \mathcal{M}_2 \mid \mathcal{M}_3$.

Note that (i) is trivially satisfied if $\mathcal{M}_4 \subset \overline{\mathcal{M}_2 \vee \mathcal{M}_3}$.

Proof. Take $\mathcal{M}_3 = \mathcal{M}_0$. Let $m \in [\mathcal{M}_1]_\infty$ be such that $\mathcal{M}_2 m = 0$. By Theorem 2.2.1 (ii), (i) implies that $(\mathcal{M}_2 \vee \mathcal{M}_4) m = \mathcal{M}_2 m$. Therefore $(\mathcal{M}_2 \vee \mathcal{M}_4) m = 0$ and, by (ii), this implies $m = 0$ a.s. ∎

Elementary Property 5.4.2(v) shows that $\mathcal{M}_1 \ll \mathcal{M}_2 \mid \mathcal{M}_3 \vee \mathcal{M}_4$ implies $\mathcal{M}_1 \ll (\mathcal{M}_2 \vee \mathcal{M}_4) \mid \mathcal{M}_3$. This leads to the following straightforward corollary:

5.4.7 Corollary. Let $\mathcal{M}_i$ ($i = 1, 2, 3, 4$) be sub-σ-fields of $\mathcal{M}$. If

(i) $\mathcal{M}_4 \perp\!\!\!\perp \mathcal{M}_1 \mid \mathcal{M}_3 \vee \mathcal{M}_2$

(ii) $\mathcal{M}_1 \ll \mathcal{M}_2 \mid \mathcal{M}_3 \vee \mathcal{M}_4$,

then

(iii) $\mathcal{M}_1 \ll \mathcal{M}_2 \mid \mathcal{M}_3$.

Note that (i) is trivially satisfied if $\mathcal{M}_4 \subset \overline{\mathcal{M}_2 \vee \mathcal{M}_3}$. ∎

Remark that Corollary 5.4.7 is rather trivial whereas the analogous result for identification, Corollary 4.5.5, was a consequence of elementary Property 4.5.2(v), the proof of which was rather involved. Combining Theorems 5.4.5, 5.4.6 and Corollary 5.4.7, entails the following corollary:

5.4.8 Corollary. Let $\mathcal{M}_i$ $(i = 1, 2, 3, 4)$ be sub-σ-fields of $\mathcal{M}$. If

(o) $\mathcal{M}_4 \perp\!\!\!\perp (\mathcal{M}_1 \vee \mathcal{M}_2) \mid \mathcal{M}_3$

the following strong identification properties are equivalent:

(i) $\mathcal{M}_1 \ll \mathcal{M}_2 \mid \mathcal{M}_3$;

(ii) $\mathcal{M}_1 \ll (\mathcal{M}_2 \vee \mathcal{M}_4) \mid \mathcal{M}_3$;

(iii) $\mathcal{M}_1 \ll \mathcal{M}_2 \mid \mathcal{M}_3 \vee \mathcal{M}_4$.   ∎

Restricting the conditional σ-field in Theorem 5.4.6 to be the minimal one we obtain, as a corollary:

5.4.9 Corollary. Let $\mathcal{M}_5 \subset \mathcal{M}_1$, $\mathcal{M}_6 \subset \mathcal{M}_2$. Then, for any $\mathcal{M}_4 \subset \mathcal{M}_2$,

(i) $\mathcal{M}_4 \ll \mathcal{M}_1 \mid \mathcal{M}_5 \vee \mathcal{M}_6 \iff \mathcal{M}_4 \ll \mathcal{M}_1\mathcal{M}_2 \mid \mathcal{M}_5 \vee \mathcal{M}_6$;

(ii) $\mathcal{M}_4 \ll \mathcal{M}_1 \mid \mathcal{M}_5 \vee \mathcal{M}_6 \implies \mathcal{M}_4 \ll \mathcal{M}_2\mathcal{M}_1 \mid \mathcal{M}_5 \vee \mathcal{M}_6$.

Proof. By Corollary 2.2.11 and Theorem 4.3.3, we successively have:

$$\mathcal{M}_1 \perp\!\!\!\perp \mathcal{M}_4 \mid \mathcal{M}_1\mathcal{M}_2 \vee \mathcal{M}_5 \vee \mathcal{M}_6$$

$$\mathcal{M}_1 \perp\!\!\!\perp \mathcal{M}_4 \mid \mathcal{M}_2\mathcal{M}_1 \vee \mathcal{M}_5 \vee \mathcal{M}_6.$$

Therefore $\mathcal{M}_4 \ll \mathcal{M}_1 \mid \mathcal{M}_5 \vee \mathcal{M}_6$ implies $\mathcal{M}_4 \ll \mathcal{M}_1\mathcal{M}_2 \mid \mathcal{M}_5 \vee \mathcal{M}_6$, by Theorem 5.4.6, and the reverse implication is trivial. However, by elementary Property 5.4.2(iv), $\mathcal{M}_4 \ll \mathcal{M}_1 \mid \mathcal{M}_5 \vee \mathcal{M}_6$ implies

$$\mathcal{M}_4 \ll (\mathcal{M}_1 \vee \mathcal{M}_2\mathcal{M}_1) \mid \mathcal{M}_5 \vee \mathcal{M}_6$$

and, by Theorem 5.4.6, this implies $\mathcal{M}_4 \ll \mathcal{M}_2\mathcal{M}_1 \mid \mathcal{M}_5 \vee \mathcal{M}_6$.   ∎

The next theorem shows that a conditional independence implies the transitivity of strong identification.

5.4.10 Theorem. Let $\mathcal{M}_i$ $(i = 1, 2, 3, 4)$ be sub-σ-fields of $\mathcal{M}$. If

(i) $\mathcal{M}_1 \perp\!\!\!\perp \mathcal{M}_2 \mid \mathcal{M}_3 \vee \mathcal{M}_4$,

(ii) $\mathcal{M}_1 \ll \mathcal{M}_4 \mid \mathcal{M}_3$,

(iii) $\mathcal{M}_4 \ll \mathcal{M}_2 \mid \mathcal{M}_3$,

then

(iv) $\mathcal{M}_1 \ll \mathcal{M}_2 \mid \mathcal{M}_3$.

Proof. Take $\mathcal{M}_3 = \mathcal{M}_0$. By Theorem 2.2.1 (iii), for any $m \in [\mathcal{M}_1]_\infty$, $\mathcal{M}_2 m = \mathcal{M}_2 (\mathcal{M}_4 m)$. If $\mathcal{M}_2 m = 0$, by condition (iii), $\mathcal{M}_4 m = 0$ and, by condition (ii), $m = 0$ a.s. ∎

We now investigate the connection between identification and strong identification. The next theorem exhibits an incompatibility between independence and strong identification.

5.4.11 Theorem. Let $\mathcal{M}_i$ ($i = 1, 2, 3, 4$) be sub-σ-fields of $\mathcal{M}$. If

(i) $\mathcal{M}_1 \perp\!\!\!\perp \mathcal{M}_2 \mid \mathcal{M}_3$

(ii) $\mathcal{M}_1 \ll \mathcal{M}_2 \mid \mathcal{M}_3$,

then

(iii) $\mathcal{M}_1 \subset \overline{\mathcal{M}_3}$.

Proof. Let $m \in [\mathcal{M}_1]_\infty$. Under (i), we have $(\mathcal{M}_2 \vee \mathcal{M}_3) m = \mathcal{M}_3 m$ by 2.2.1(ii). Therefore $(\mathcal{M}_2 \vee \mathcal{M}_3) \{m - \mathcal{M}_3 m\} = 0$. Clearly, we also have $m - \mathcal{M}_3 m \in [\mathcal{M}_1 \vee \mathcal{M}_3]_\infty$. So by (ii), $m - \mathcal{M}_3 m = 0$ a.s., i.e., $m \in \left[\overline{\mathcal{M}_3}\right]_\infty$. ∎

This theorem entails the fact that strong identification implies identification.

5.4.12 Theorem. Let $\mathcal{M}_i$ ($i = 1, 2, 3, 4$) be sub-σ-fields of $\mathcal{M}$. If $\mathcal{M}_1 \ll \mathcal{M}_2 \mid \mathcal{M}_3$ then $\mathcal{M}_1 < \mathcal{M}_2 \mid \mathcal{M}_3$.

Proof. Take $\mathcal{M}_3 = \mathcal{M}_0$. Clearly, $\mathcal{M}_1 \perp\!\!\!\perp \mathcal{M}_2 \mid \mathcal{M}_1\mathcal{M}_2$. By Theorem 5.4.5, with $\mathcal{M}_4 = \mathcal{M}_1\mathcal{M}_2$, $\mathcal{M}_1 \ll \mathcal{M}_2$ implies that $\mathcal{M}_1 \ll \mathcal{M}_2 \mid \mathcal{M}_1\mathcal{M}_2$. Therefore by Theorem 5.4.11, $\mathcal{M}_1 \subset \overline{\mathcal{M}_1\mathcal{M}_2}$, and this is equivalent to $\mathcal{M}_1 = \mathcal{M}_1\mathcal{M}_2$, i.e., $\mathcal{M}_1 < \mathcal{M}_2$. ∎

We now provide the condition under which $\mathcal{M}_1 \perp\!\!\!\perp (\mathcal{M}_2 \vee \mathcal{M}_3)$ when $\mathcal{M}_1 \perp\!\!\!\perp \mathcal{M}_2 \mid \mathcal{M}_3$ and $\mathcal{M}_1 \perp\!\!\!\perp \mathcal{M}_2$. This theorem may be viewed as a complement of Theorem 2.2.10.

5.4.13 Theorem. Let $\mathcal{M}_i$ ($i = 1, 2, 3, 4$) be sub-σ-fields of $\mathcal{M}$. If

(i) $\mathcal{M}_1 \perp\!\!\!\perp \mathcal{M}_2 \mid \mathcal{M}_3$,

(ii) $\mathcal{M}_1 \perp\!\!\!\perp \mathcal{M}_2 \mid \mathcal{M}_3 \vee \mathcal{M}_4$,

(iii) $\mathcal{M}_4 \ll \mathcal{M}_2 \mid \mathcal{M}_3$,

then

(iv) $\mathcal{M}_1 \perp\!\!\!\perp (\mathcal{M}_2 \vee \mathcal{M}_4) \mid \mathcal{M}_3$.

Proof. Since (ii) is equivalent to $\mathcal{M}_1 \perp\!\!\!\perp (\mathcal{M}_2 \vee \mathcal{M}_3) \mid \mathcal{M}_3 \vee \mathcal{M}_4$, by Theorem 2.2.1 (iii), $(\mathcal{M}_2 \vee \mathcal{M}_3) [(\mathcal{M}_3 \vee \mathcal{M}_4) m] = (\mathcal{M}_2 \vee \mathcal{M}_3) m$ for any $m \in [\mathcal{M}_1]_\infty$. But, by (i) and 2.2.1 (ii), $(\mathcal{M}_2 \vee \mathcal{M}_3) m = \mathcal{M}_3 m$. Therefore

$$(\mathcal{M}_2 \vee \mathcal{M}_3) [(\mathcal{M}_3 \vee \mathcal{M}_4) m - \mathcal{M}_3 m] = 0.$$

By Condition (iii) this implies $(\mathcal{M}_3 \vee \mathcal{M}_4) m = \mathcal{M}_3 m$, namely, $\mathcal{M}_1 \perp\!\!\!\perp \mathcal{M}_4 \mid \mathcal{M}_3$ and this, along with Condition (ii), is equivalent to (iv), by Theorem 2.2.10. ∎

We end this section by showing that strong identification has useful implications for the theory of projection of σ-fields under a conditional independence relation. Indeed, if $\mathcal{M}_1 \perp\!\!\!\perp \mathcal{M}_2 \mid \mathcal{M}_3$, then by Corollary 4.3.6(v) we have

$$\mathcal{M}_2\mathcal{M}_1 = \mathcal{M}_2(\mathcal{M}_1\mathcal{M}_3) \subset \mathcal{M}_2(\mathcal{M}_3\mathcal{M}_1) \subset \mathcal{M}_2\mathcal{M}_3.$$

Assumptions of strong p-identification with $p \in (1, \infty)$ will give equalities instead of inclusions in the above relation.

5.4.14 Theorem. Let $p \in (1, \infty)$ and $\mathcal{M}_i$ $(i = 1, 2, 3)$ be sub-σ-fields of $\mathcal{M}$ such that:

(o) $\mathcal{M}_1 \perp\!\!\!\perp \mathcal{M}_2 \mid \mathcal{M}_3$,

(i) if $\mathcal{M}_3\mathcal{M}_1 \ll_p \mathcal{M}_1$ then $\mathcal{M}_2\mathcal{M}_1 = \mathcal{M}_2(\mathcal{M}_3\mathcal{M}_1) \subset \mathcal{M}_2\mathcal{M}_3$;

(ii) if $\mathcal{M}_3 \ll_p \mathcal{M}_1$ then $\mathcal{M}_2\mathcal{M}_1 = \mathcal{M}_2(\mathcal{M}_3\mathcal{M}_1) = \mathcal{M}_2\mathcal{M}_3$.

Proof. Let $m \in [\mathcal{M}_3\mathcal{M}_1]_\infty$. If $\mathcal{M}_3\mathcal{M}_1 \ll_p \mathcal{M}_1$ for some $p \in (1, \infty)$ then, as a consequence of Theorem 5.4.3, there exists a sequence

$$\{m_k : 1 \le k < \infty\} \subset [\mathcal{M}_1]_q \quad \text{with} \quad 1/p + 1/q = 1$$

such that $(\mathcal{M}_3\mathcal{M}_1)\, m_k \to m$ a.s., and in $[\mathcal{M}]_q$ as $k \to \infty$. Since

$$\mathcal{M}_1 \perp\!\!\!\perp \mathcal{M}_3 \mid \mathcal{M}_3\mathcal{M}_1,$$

we have, by Theorem 2.2.1(ii), that $(\mathcal{M}_3\mathcal{M}_1)\, m_k = \mathcal{M}_3 m_k\ \forall\ k$. And so $\mathcal{M}_3 m_k \to m$ in $[\mathcal{M}]_q$ as $k \to \infty$. Now, for any $\mathcal{M}_2$, $\mathcal{M}_2(\mathcal{M}_3 m_k) \to \mathcal{M}_2 m$ in $[\mathcal{M}]_q$ as $k \to \infty$ since the conditional expectation is a continuous linear operator from $[\mathcal{M}]_q$ to $[\mathcal{M}]_q$. Then there exists a subsequence

$$\{m_{k_\ell} : 1 \le \ell \le \infty\} \subset [\mathcal{M}_1]_q$$

such that $\mathcal{M}_2(\mathcal{M}_3 m_{k_\ell}) \to \mathcal{M}_2 m$ a.s. Now, by Theorem 2.2.1(iii), if $\mathcal{M}_1 \perp\!\!\!\perp \mathcal{M}_2 \mid \mathcal{M}_3$, $\mathcal{M}_2(\mathcal{M}_3 m_{k_\ell}) = \mathcal{M}_2 m_{k_\ell}$. Hence $\mathcal{M}_2 m_{k_\ell} \to \mathcal{M}_2 m$ a.s. as $\ell \to \infty$. This implies that $\mathcal{M}_2 m \in \overline{\mathcal{M}_2\mathcal{M}_1}$. This is true for any $m \in [\mathcal{M}_3\mathcal{M}_1]_\infty$ and so $\mathcal{M}_2(\mathcal{M}_3\mathcal{M}_1) \subset \mathcal{M}_2\mathcal{M}_1$, and this is an equality, by 4.3.6(v). To prove (ii) it suffices to notice that, by the same argument, $\mathcal{M}_2 m \in \overline{\mathcal{M}_2\mathcal{M}_1}$ for any $m \in [\mathcal{M}_3]_\infty$, and so $\mathcal{M}_2\mathcal{M}_3 \subset \mathcal{M}_2\mathcal{M}_1$ which, along with $\mathcal{M}_2\mathcal{M}_1 \subset \mathcal{M}_2(\mathcal{M}_3\mathcal{M}_1) \subset \mathcal{M}_2\mathcal{M}_3$, gives the result. ∎

If we trivialize the conditional independence relation by letting $\mathcal{M}_2 \subset \mathcal{M}_3$, then we see that strong p-identification of $\mathcal{M}_3\mathcal{M}_1$ by $\mathcal{M}_1$ is a supplementary condition that provides a result similar to the three-perpendicular theorem for projections of σ-fields.

5.4.15 Corollary. Let $p \in (1, \infty)$ and $\mathcal{M}_i$ ($i = 1, 2, 3$) be sub-σ-fields of $\mathcal{M}$. If

(i) $\mathcal{M}_2 \subset \mathcal{M}_3$

(ii) $\mathcal{M}_3\mathcal{M}_1 \ll_p \mathcal{M}_1,$

then

(iii) $\mathcal{M}_2 (\mathcal{M}_3\mathcal{M}_1) = \mathcal{M}_2\mathcal{M}_1.$ ∎

We conclude this section by treating the problem of minimal splitting, i.e., the problem of characterizing the minimal σ-fields conditionally on which two given σ-fields are independent. We show that this concept is in fact intermediate between identification and strong identification.

5.4.3 Minimal Splitting

We have seen, in Theorem 4.3.3, that the projection σ-field $\mathcal{M}_1\mathcal{M}_2$ is the smallest sub-σ-field of $\mathcal{M}_1$, which makes $\mathcal{M}_1$ and $\mathcal{M}_2$ conditionally independent. It would be interesting to characterize and to construct all sub-σ-fields of $\mathcal{M}$ which are minimal in the class of the σ-fields that make $\mathcal{M}_1$ and $\mathcal{M}_2$ conditionally independent. This section treats a topic in probability theory which will not be used directly in the statistical sections. This subsection may therefore be skipped without inconvenience for what follows; it has nevertheless been included in this monograph not only for its intrinsic interest, but also because this problem has received a good deal of attention in system theory under the heading of "the σ-algebraic realization problem" (see, e.g., Lindquist, Picci and Ruckebusch (1979), Lindquist and Picci (1979), or Picci (1976)). The next definition uses the same terminology as in system theory and explains the title of this section:

5.4.16 Definition. Let $\mathcal{M}_i$ ($i = 1, 2, 3$) be three sub-σ-fields of $\mathcal{M}$.

(i) $\mathcal{M}_3$ *splits* $(\mathcal{M}_1, \mathcal{M}_2)$ if $\mathcal{M}_1 \perp\!\!\!\perp \mathcal{M}_2 \mid \mathcal{M}_3$;

(ii) $\mathcal{M}_3$ *minimally splits* $(\mathcal{M}_1, \mathcal{M}_2)$ if
   (a) $\mathcal{M}_3$ splits $(\mathcal{M}_1, \mathcal{M}_2)$
   (b) $\mathcal{M}_4 \subset \mathcal{M}_3$ splits $(\mathcal{M}_1, \mathcal{M}_2) \Longrightarrow \overline{\mathcal{M}_3} = \overline{\mathcal{M}_4}$.

(iii) $\mathcal{M}_3$ *minimally and internally splits* $(\mathcal{M}_1, \mathcal{M}_3)$ if
   (a) $\mathcal{M}_3$ minimally splits $(\mathcal{M}_1, \mathcal{M}_2)$ and
   (b) $\mathcal{M}_3 \subset \mathcal{M}_1 \vee \mathcal{M}_2$.

Thus, in this parlance, Theorem 4.3.3 says that $\mathcal{M}_1 \mathcal{M}_2$ splits minimally and internally $(\mathcal{M}_1, \mathcal{M}_2)$, and is the unique sub-σ-field to enjoy this property in the class of the sub-σ-fields of $\mathcal{M}_1$. In this subsection we wish to investigate the class of all σ-fields that minimally split $(\mathcal{M}_1, \mathcal{M}_2)$ without requiring them to be sub-σ-fields of $\mathcal{M}_1$ or of $\mathcal{M}_2$. We first characterize that class by means of necessary and of sufficient conditions. The next theorem shows that identification is a necessary condition for minimal splitting.

5.4.17 Theorem. If $\mathcal{M}_3$ minimally splits $(\mathcal{M}_1, \mathcal{M}_2)$ then $\mathcal{M}_3$ is identified both by $\mathcal{M}_1$ and by $\mathcal{M}_2$.

Proof. Corollary 4.3.5 (ii) shows that if $\mathcal{M}_3$ minimally splits $(\mathcal{M}_1, \mathcal{M}_2)$, then $\overline{\mathcal{M}_3} = \overline{\mathcal{M}_3 \mathcal{M}_2}$ therefore, by elementary Property 4.3.2(viii), $\mathcal{M}_3 = \mathcal{M}_3 \mathcal{M}_2$ and the equality $\mathcal{M}_3 = \mathcal{M}_3 \mathcal{M}_1$ is obtained by symmetry. ∎

Next we show that identification by $\mathcal{M}_1$ and strong identification by $\mathcal{M}_2$ (or vice versa) is a sufficient condition for minimal splitting.

5.4.18 Theorem. If

(i) $\mathcal{M}_3$ splits $(\mathcal{M}_1, \mathcal{M}_2)$ (i.e, $\mathcal{M}_1 \perp\!\!\!\perp \mathcal{M}_2 \mid \mathcal{M}_3$),

(ii) $\mathcal{M}_3$ is identified by $\mathcal{M}_1$ (i.e. $\mathcal{M}_3 < \mathcal{M}_1$), and

(iii) $\mathcal{M}_3$ is strongly identified by $\mathcal{M}_2$ (i.e. $\mathcal{M}_3 \ll \mathcal{M}_2$)
then $\mathcal{M}_3$ minimally splits $(\mathcal{M}_1, \mathcal{M}_2)$.

Proof. Suppose that $\mathcal{M}_4 \subset \mathcal{M}_3$ also splits $(\mathcal{M}_1, \mathcal{M}_2)$; we show that assumptions (i) to (iii) imply $\mathcal{M}_3 \subset \overline{\mathcal{M}_4}$. Indeed, by Theorem 5.4.5, $\mathcal{M}_4 \subset \mathcal{M}_3$ and (iii) $\mathcal{M}_3 \ll \mathcal{M}_2$ imply $\mathcal{M}_3 \ll \mathcal{M}_2 \mid \mathcal{M}_4$; therefore, by Theorem 5.4.13, $\mathcal{M}_1 \perp\!\!\!\perp (\mathcal{M}_2 \vee \mathcal{M}_3) \mid \mathcal{M}_4$ which implies $\mathcal{M}_1 \perp\!\!\!\perp \mathcal{M}_3 \mid \mathcal{M}_4$; hence, by Theorem 4.3.3 and because $\mathcal{M}_4 \subset \mathcal{M}_3$, we obtain $\mathcal{M}_3 \mathcal{M}_1 \subset \overline{\mathcal{M}_4}$. Since $\mathcal{M}_3$ is identified by $\mathcal{M}_1$ (i.e., $\mathcal{M}_3 < \mathcal{M}_1$), we conclude that $\mathcal{M}_3 \subset \overline{\mathcal{M}_4}$.   ∎

We know, from Theorem 5.4.12, that identification of σ-fields is weaker than strong identification. It may be conjectured that identification by one factor and strong identification by the other one are likely to be the weakest sufficient conditions for minimal splitting; such a view is justified by the following arguments: (i) in the linear theory (i.e., σ-fields replaced by linear subspaces of the Hilbert space of square-integrable zero expectation random variables and independence replaced by uncorrelatedness), the corresponding concept of identification and strong identification do coincide (see e.g., Lindquist, Picci and Ruckebusch (1979), or Mouchart and Rolin (1986), Appendix B) and recall that in the Gaussian case uncorrelatedness is equivalent to independence; (ii) if we restrict $\mathcal{M}_3$ to be a sub-σ-field of $\mathcal{M}_2$, then by Theorem 4.3.3, $\mathcal{M}_3$ minimally splits $(\mathcal{M}_1, \mathcal{M}_2)$ if and only if $\overline{\mathcal{M}_3} = \overline{\mathcal{M}_2 \mathcal{M}_1}$. However, when $\mathcal{M}_3 \subset \mathcal{M}_2$, $\mathcal{M}_3$ is trivially identified and strongly identified by $\mathcal{M}_2$ (elementary Properties 4.3.2(ii) and 5.4.2(ii)); therefore $\mathcal{M}_3 \subset \mathcal{M}_2$ minimally splits $(\mathcal{M}_1, \mathcal{M}_2)$ if and only if $\mathcal{M}_3$ splits $(\mathcal{M}_1, \mathcal{M}_2)$ and is identified by $\mathcal{M}_1$. Furthermore, we shall see that minimal splitting may in fact be seen as a property intermediate between identification and strong identification.

As a step toward the construction of such σ-fields, let us now turn to the representations of minimal splitting σ-fields. We have seen that any minimal splitting σ-field $\mathcal{M}_3$ is by necessity splitting $(\mathcal{M}_1, \mathcal{M}_2)$ and identified by $\mathcal{M}_1$ (also by $\mathcal{M}_2$). These two properties are equivalent to the representability of $\mathcal{M}_3$ as a projection of $\mathcal{M}_1$ on a σ-field containing $\mathcal{M}_2$, and is the focus of the next theorem.

5.4.19 Theorem. The following two conditions:

(i) $\mathcal{M}_3$ splits $(\mathcal{M}_1, \mathcal{M}_2)$ (i.e. $\mathcal{M}_1 \perp\!\!\!\perp \mathcal{M}_2 \mid \mathcal{M}_3$) and

(ii) $\mathcal{M}_3$ is identified by $\mathcal{M}_1$ (i.e. $\mathcal{M}_3\mathcal{M}_1 = \mathcal{M}_3$)

are equivalent to:

(iii) $\exists\, \mathcal{M}_4 \supset \mathcal{M}_2$ such that $\overline{\mathcal{M}_3} = \overline{\mathcal{M}_4\mathcal{M}_1}$.

If furthermore $\mathcal{M}_3 \subset \mathcal{M}_1 \vee \mathcal{M}_2$ (i.e., internal splitting), and $\mathcal{M}_4$ satisfies (iii) and

(iv) $\mathcal{M}_4 \subset \mathcal{M}_1 \vee \mathcal{M}_2$,

then $\mathcal{M}_4$ is almost surely unique, i.e., $\overline{\mathcal{M}_4} = \overline{\mathcal{M}_2 \vee \mathcal{M}_3}$.

Proof. To prove that (i) and (ii) imply (iii), it suffices to choose

(v) $\mathcal{M}_4 = \mathcal{M}_2 \vee \mathcal{M}_3$;

(iii) is obtained by remarking that $\overline{(\mathcal{M}_2 \vee \mathcal{M}_3)\,\mathcal{M}_1} = \overline{\mathcal{M}_3\mathcal{M}_1}$, by (i) and by Corollary 4.3.6(iii), and that $\overline{\mathcal{M}_3\mathcal{M}_1} = \overline{\mathcal{M}_3}$ by (ii). To prove that (iii) implies (i) and (ii), we first recall that $\mathcal{M}_1 \perp\!\!\!\perp \mathcal{M}_4 \mid \mathcal{M}_4\mathcal{M}_1$ for any $\mathcal{M}_4$, by Theorem 4.3.3; thus (iii) implies $\mathcal{M}_1 \perp\!\!\!\perp \mathcal{M}_4 \mid \mathcal{M}_3$ and this implies (i) because $\mathcal{M}_4 \supset \mathcal{M}_2$. Next note that by elementary Property 4.3.2(vii) and Corollary 4.3.8, (iii) successively implies

$$\overline{\mathcal{M}_3\mathcal{M}_1} = \overline{(\mathcal{M}_4\mathcal{M}_1)\,\mathcal{M}_1} = \overline{\mathcal{M}_4\mathcal{M}_1} = \overline{\mathcal{M}_3};$$

this gives (ii) by taking the intersection with $\mathcal{M}_3$ on both sides of the last equality. It remains to show that in the case of internal splitting $\mathcal{M}_4$ is essentially unique if we also require (iv). Theorem 4.3.9 (ii) and elementary Property 4.3.2(ii) successively imply

$$\mathcal{M}_2 \vee (\mathcal{M}_4\mathcal{M}_1) = \mathcal{M}_4\,(\mathcal{M}_1 \vee \mathcal{M}_2) = \mathcal{M}_4.$$

The first equality holds because $\mathcal{M}_2 \subset \mathcal{M}_4$ and the second equality because $\mathcal{M}_4 \subset \mathcal{M}_1 \vee \mathcal{M}_2$, and therefore, in view of (iii), $\overline{\mathcal{M}_4} = \overline{\mathcal{M}_2 \vee \mathcal{M}_3}$. ∎

An important corollary of this theorem is to show that minimal splitting is a condition intermediate between identification and strong identification, once it is recognized that any minimal splitting σ-field is representable as the projection of one factor. More specifically, we have:

5.4.20 Corollary. For any $\mathcal{M}_4 \supset \mathcal{M}_2 : \mathcal{M}_4\mathcal{M}_1 \ll \mathcal{M}_2 \Longrightarrow \mathcal{M}_4\mathcal{M}_1$ minimally splits $(\mathcal{M}_1, \mathcal{M}_2) \Longrightarrow \mathcal{M}_4\mathcal{M}_1 < \mathcal{M}_2$. ∎

Thus we have seen, in Theorem 5.4.17, that any minimal splitting σ-field is identified by each of its splitted σ-fields and, in Theorem 5.4.19, that this implies its double representability as a projection of one of these factors on a σ-field containing the other factor. The next theorem establishes the duality between these two representations with respect to $\mathcal{M}_1$ and $\mathcal{M}_2$, and shows that those representations may be taken as mutually symmetric and are in bijection, furthermore in the case of internal splitting, they are essentially unique and are in bijection.

5.4.21 Theorem. If $\mathcal{M}_3$ minimally splits $(\mathcal{M}_1, \mathcal{M}_2)$ then

(i) $\exists \mathcal{M}_4 \supset \mathcal{M}_2$ and $\mathcal{M}_5 \supset \mathcal{M}_1$ such that $\overline{\mathcal{M}_3} = \overline{\mathcal{M}_4\mathcal{M}_1} = \overline{\mathcal{M}_5\mathcal{M}_2}$;

(ii) (a) for given $\mathcal{M}_4$, $\mathcal{M}_5$ may be taken as $\mathcal{M}_5 = \mathcal{M}_1 \vee \mathcal{M}_4\mathcal{M}_1$,
    (b) for given $\mathcal{M}_5$, $\mathcal{M}_4$ may be taken as $\mathcal{M}_4 = \mathcal{M}_2 \vee \mathcal{M}_5\mathcal{M}_2$,

(iii) if, furthermore $\mathcal{M}_3 \subset \mathcal{M}_1 \vee \mathcal{M}_2$ (internal splitting), then the representation in (i) with $\mathcal{M}_2 \subset \mathcal{M}_4 \subset \mathcal{M}_1 \vee \mathcal{M}_2$ and $\mathcal{M}_1 \subset \mathcal{M}_5 \subset \mathcal{M}_1 \vee \mathcal{M}_2$ is essentialy unique, namely: $\overline{\mathcal{M}_4} = \overline{\mathcal{M}_2 \vee \mathcal{M}_3}$ and $\overline{\mathcal{M}_5} = \overline{\mathcal{M}_1 \vee \mathcal{M}_3}$.

Proof. (i) is clear from Theorems 5.4.17 and 5.4.19. For (ii), suppose that $\mathcal{M}_4$ satisfies (i); then, from Theorem 5.4.17, we have that

$$(\mathcal{M}_4\mathcal{M}_1)\,\mathcal{M}_2 = \mathcal{M}_4\mathcal{M}_1$$

and, by Corollary 4.3.5(ii)

$$\overline{(\mathcal{M}_4\mathcal{M}_1)\,\mathcal{M}_2} = \overline{(\mathcal{M}_1 \vee \mathcal{M}_4\mathcal{M}_1)\,\mathcal{M}_2}$$

because, from Theorem 4.3.3, we have $\mathcal{M}_1 \perp\!\!\!\perp \mathcal{M}_4 \mid \mathcal{M}_4\mathcal{M}_1$ which, along with $\mathcal{M}_2 \subset \mathcal{M}_4$, implies $\mathcal{M}_1 \perp\!\!\!\perp \mathcal{M}_2 \mid \mathcal{M}_4\mathcal{M}_1$. Therefore taking $\mathcal{M}_5 = \mathcal{M}_1 \vee \mathcal{M}_4\mathcal{M}_1$ ensures $\overline{\mathcal{M}_5\mathcal{M}_2} = \overline{\mathcal{M}_4\mathcal{M}_1}$. The very same argument applies for $\mathcal{M}_4$ taken as $\mathcal{M}_4 = \mathcal{M}_2 \vee \mathcal{M}_5 \mathcal{M}_1$ with $\mathcal{M}_5$ satisfying (i). Finally, (iii) is a direct consequence of Theorem 5.4.19 (iii). ∎

Let us now consider the construction of minimal splitting σ-fields. From Theorem 5.4.18 and Corollary 5.4.20, any σ-field $\mathcal{M}_4$ such that $\mathcal{M}_4 \supset \mathcal{M}_2$ and $\mathcal{M}_4 \mathcal{M}_1 \ll \mathcal{M}_2$ may be used because those properties are sufficient to imply that $\mathcal{M}_4 \mathcal{M}_1$ minimally splits $(\mathcal{M}_1, \mathcal{M}_2)$. As the strong identification of $\mathcal{M}_4 \mathcal{M}_1$ (by $\mathcal{M}_2$) may be difficult to check, it may be useful to first check a weaker (but more easily verified) condition as a screening test before a more complete examination; this is the goal of the following theorem:

5.4.22 Theorem. If

(i) $\mathcal{M}_4 \supset \mathcal{M}_2$

(ii) $\mathcal{M}_4 \mathcal{M}_1 \ll \mathcal{M}_2$,

then

(iii) $\mathcal{M}_1 \perp\!\!\!\perp \mathcal{M}_4 \mid \mathcal{M}_1 \mathcal{M}_2$.

If, furthermore,

(iv) $\mathcal{M}_4 \subset \mathcal{M}_1 \vee \mathcal{M}_2$ (internal splitting)

then

(v) $\mathcal{M}_4 \subset \overline{\mathcal{M}_2 \vee \mathcal{M}_1 \mathcal{M}_2}$.

Proof. We first prove an auxiliary result:

5.4.23 Lemma.   If

(vi) $\mathcal{M}_1 \perp\!\!\!\perp \mathcal{M}_2 \mid \mathcal{M}_3$,

(vii) $\mathcal{M}_3 \ll \mathcal{M}_2$,

(viii) $\mathcal{M}_5 \subset \mathcal{M}_1 \vee \mathcal{M}_3$

(ix) $\mathcal{M}_1 \perp\!\!\!\perp \mathcal{M}_2 \mid \mathcal{M}_5$,

then

(x) $\mathcal{M}_1 \perp\!\!\!\perp (\mathcal{M}_2 \vee \mathcal{M}_3) \mid \mathcal{M}_5$.

Proof of the lemma: Remark that under (viii), (vi) is equivalent to:

(xi) $(\mathcal{M}_1 \vee \mathcal{M}_5) \perp\!\!\!\perp \mathcal{M}_2 \mid \mathcal{M}_3$

and, by Theorem 2.2.10, (xi) is equivalent to the following two conditions:

(xii) $\mathcal{M}_5 \perp\!\!\!\perp \mathcal{M}_2 \mid \mathcal{M}_3$ and

(xiii) $\mathcal{M}_1 \perp\!\!\!\perp \mathcal{M}_2 \mid \mathcal{M}_3 \vee \mathcal{M}_5$.

By Theorem 5.4.5, (vii) along with (xii) implies that

(xiv) $\mathcal{M}_3 \ll \mathcal{M}_2 \mid \mathcal{M}_5$.

Finally, by Theorem 5.4.13, (ix) along with (xiii) and (xiv) implies (x).

Proof of the theorem: Take, in the lemma, $\mathcal{M}_3 = \mathcal{M}_4 \mathcal{M}_1$ and $\mathcal{M}_5 = \mathcal{M}_1 \mathcal{M}_2$. Clearly, (vi), (vii), (viii) and (ix) are satisfied; thus (x) is also satisfied and becomes:

(xv) $\mathcal{M}_1 \perp\!\!\!\perp (\mathcal{M}_2 \vee \mathcal{M}_4 \mathcal{M}_1) \mid \mathcal{M}_1 \mathcal{M}_2$.

In order to show that (iii) and (xv) are equivalent (under (i) and (ii)), we note that, by Theorem 2.2.10, (iii) is equivalent to (xv) along with

(xvi) $\mathcal{M}_1 \perp\!\!\!\perp \mathcal{M}_4 \mid \mathcal{M}_4 \mathcal{M}_1 \vee \mathcal{M}_2 \vee \mathcal{M}_1 \mathcal{M}_2$.

Clearly (xvi) always holds because of Corollary 2.2.11 and Theorem 4.3.3. Finally, in the internal case, (v) is shown by first noticing that, by Theorem 2.2.10, (iii) is also equivalent to $\mathcal{M}_1 \perp\!\!\!\perp \mathcal{M}_2 \mid \mathcal{M}_1 \mathcal{M}_2$ (which is always true, by Theorem 4.3.3) along with

(xvii) $\mathcal{M}_1 \perp\!\!\!\perp \mathcal{M}_4 \mid \mathcal{M}_1 \mathcal{M}_2 \vee \mathcal{M}_2$;

but by Corollary 2.2.11, (xvii) is also equivalent to:

(xviii) $(\mathcal{M}_1 \vee \mathcal{M}_2) \perp\!\!\!\perp \mathcal{M}_4 \mid \mathcal{M}_1\mathcal{M}_2 \vee \mathcal{M}_2$.

Now, under (iv), (xviii) implies (v) by Corollary 2.2.8. ∎

Looking at the linear theory as elaborated in Lindquist and Picci (1979, 1982) or Lindquist, Picci and Ruckebush (1979), it is natural to ask whether *any* minimal splitting σ-field is representable as $\mathcal{M}_3 = \mathcal{M}_4\mathcal{M}_1$ with $\mathcal{M}_2 \subset \mathcal{M}_4 \subset \mathcal{M}_2 \vee \mathcal{M}_1\mathcal{M}_2$. From Theorem 5.4.18 and Corollary 5.4.20 we know that conditions (i) and (ii) are sufficient to imply the minimal splitting of $(\mathcal{M}_1, \mathcal{M}_2)$ by $\mathcal{M}_4\mathcal{M}_1$ and, from Theorem 5.4.22, we know that, in the internal case, $\mathcal{M}_4 \subset \overline{\mathcal{M}_2 \vee \mathcal{M}_1\mathcal{M}_2}$. It is therefore natural to ask whether (v) might be either necessary or sufficient for minimal splitting. In view of the following two arguments, there is little hope that (v) is a sufficient condition. Firstly, we know of no case where conditional independence, along with some other non trivial condition, implies strong identification. Secondly, the fact that (i) and (ii) imply (v) is a corollary of the auxiliary lemma for the special case $\mathcal{M}_5 = \mathcal{M}_1\mathcal{M}_2 \subset \mathcal{M}_1$ whereas the fact that (i) and (ii) imply minimal splitting may also be viewed as a corollary of the same auxiliary lemma but when $\mathcal{M}_5 \subset \mathcal{M}_3 = \mathcal{M}_4\mathcal{M}_1$ (in this respect, it is indeed interesting to compare the proofs of Theorem 5.4.18 and of the auxiliary lemma).

To conclude this section, let us suggest the following algorithmic construction of a minimal splitting σ-field. Let $\mathcal{M}_1$ and $\mathcal{M}_2$ be two given σ-algebras, then:

(i) construct the projection $\mathcal{M}_1\mathcal{M}_2$;

(ii) pick $\mathcal{M}_4$ such that $\mathcal{M}_2 \subset \mathcal{M}_4$ and
    (a) (general) $\mathcal{M}_1 \perp\!\!\!\perp \mathcal{M}_4 \mid \mathcal{M}_1\mathcal{M}_2$,
    (b) (internal) $\mathcal{M}_4 \subset \mathcal{M}_2 \vee \mathcal{M}_1\mathcal{M}_2$;

(iii) construct the projection $\mathcal{M}_4\mathcal{M}_1$;

(iv) check whether $\mathcal{M}_4\mathcal{M}_1 < \mathcal{M}_2$:
    if no, $\mathcal{M}_4\mathcal{M}_1$ is not minimal splitting; go back to (ii);
    if yes, proceed to (v);

(v)   check whether $\mathcal{M}_4\mathcal{M}_1 \ll \mathcal{M}_2$:
      if yes, $\mathcal{M}_4\mathcal{M}_1$ is minimal splitting; stop.
      if no : proceed to (vi);

(vi)   check whether $\mathcal{M}_4\mathcal{M}_1 \ll \mathcal{M}_1$:
      if yes, $\mathcal{M}_4\mathcal{M}_1$ is minimal splitting; stop.
      if no, inconclusive; go back to (ii).

Note that this constructive procedure should end up by displaying *a* minimal splitting σ-algebra but does not pretend to build *any* minimal splitting σ-field.

5.5 Completeness in Bayesian Experiments

In this section we investigate the most important implications of strong identification in the general reduced experiment $\mathcal{E}_{\mathcal{B}\vee\mathcal{T}}^{\mathcal{M}}$. In the statistical literature the concept of strong identification among σ-fields is generally called a condition of completeness (see, e.g., Lehmann and Scheffé (1950), 1955)); this motivates the following definition.

5.5.1 Definition. In the Bayesian experiment $\mathcal{E}_{\mathcal{B}\vee\mathcal{T}}^{\mathcal{M}}$, a *statistic* $\mathcal{N}$ is *p-complete* if $\mathcal{N} \ll_p \mathcal{B} \mid \mathcal{M}$ and a *parameter* $\mathcal{L}$ is *p-complete* if $\mathcal{L} \ll_p \mathcal{T} \mid \mathcal{M}$. In particular, in the unreduced experiment $\mathcal{E}$, a *statistic* $\mathcal{T}$ is *p-complete* if $\mathcal{T} \ll_p \mathcal{A}$ and a *parameter* $\mathcal{B}$ is *p-complete* if $\mathcal{B} \ll_p \mathcal{S}$. When $p = \infty$, we say *complete* instead of ∞-*complete*.   ∎

Note that the "bounded completeness" condition of Lehmann and Scheffé (1955) corresponds to what we call complete (or ∞-complete).

We next examine some links between completeness and sufficiency or minimal sufficiency. Then we show how completeness provides sufficient conditions for maximal ancillarity and how an apparent paradox in the theory of conditional inference can be resolved. In the last two sections whe analyze some relationships between sampling theory and Bayesian methods.

5.5.1 Completeness and Sufficiency

As an application of Theorem 5.4.6, the following theorem shows that the completeness of a statistic does not require the identification of the experiment.

5.5.2 Theorem. In the Bayesian experiment $\mathcal{E}_{\mathcal{B}\vee\mathcal{T}}^{\mathcal{M}}$ a statistic (respectively, parameter) is complete if and only if it is strongly identified conditionally on $\mathcal{M}$ by a sufficient parameter (respectively, statistic).

Proof. By duality, it suffices to prove the result for a statistic. Let $\mathcal{N} \subset \mathcal{M} \vee \mathcal{T}$ be a complete statistic, i.e., $\mathcal{N} \ll \mathcal{B} \mid \mathcal{M}$, and let $\mathcal{L} \subset \mathcal{B} \vee \mathcal{M}$ be a sufficient parameter, i.e., $\mathcal{B} \perp\!\!\!\perp \mathcal{T} \mid \mathcal{L} \vee \mathcal{M}$. We have to show that $\mathcal{N} \ll \mathcal{B} \mid \mathcal{M}$ if and only if $\mathcal{N} \ll \mathcal{L} \mid \mathcal{M}$. Note that the sufficiency of $\mathcal{L}$ implies that $\mathcal{B} \perp\!\!\!\perp \mathcal{N} \mid \mathcal{L} \vee \mathcal{M}$. Since $\mathcal{B} \vee \mathcal{L} \vee \mathcal{M} = \mathcal{B} \vee \mathcal{M}$ we have, by Theorem 5.4.6, that $\mathcal{N} \ll \mathcal{B} \mid \mathcal{M}$, i.e., $\mathcal{N} \ll (\mathcal{B} \vee \mathcal{L}) \mid \mathcal{M}$, implies $\mathcal{N} \ll \mathcal{L} \mid \mathcal{M}$. Now, by elementary Property 5.4.2(iv), $\mathcal{N} \ll \mathcal{L} \mid \mathcal{M}$ implies $\mathcal{N} \ll \mathcal{B} \mid \mathcal{M}$. ∎

Theorem 5.4.12 provides relationships between the concepts of completeness and of identification.

5.5.3 Proposition. In the Bayesian experiment $\mathcal{E}_{\mathcal{B}\vee\mathcal{T}}^{\mathcal{M}}$

(i) any complete statistic is identified, i.e.,
$\mathcal{N} \subset \mathcal{M} \vee \mathcal{T}, \mathcal{N} \ll \mathcal{B} \mid \mathcal{M}$
$\implies \mathcal{N} < \mathcal{B} \mid \mathcal{M}$ or, equivalently, $(\mathcal{M} \vee \mathcal{N})(\mathcal{B} \vee \mathcal{M}) = \mathcal{M} \vee \mathcal{N}$;

(ii) a complete parameter is identified, i.e.,
$\mathcal{L} \subset \mathcal{B} \vee \mathcal{M}, \mathcal{L} \ll \mathcal{T} \mid \mathcal{M}$
$\implies \mathcal{L} < \mathcal{T} \mid \mathcal{M}$ or, equivalently, $(\mathcal{L} \vee \mathcal{M})(\mathcal{M} \vee \mathcal{T}) = \mathcal{L} \vee \mathcal{M}$. ∎

For expository reasons, we provide comment on Proposition 5.5.3 in the unreduced experiment $\mathcal{E}$ for the parameter case only. Recall that if the experiment is identified, i.e., $\mathcal{A} < \mathcal{S}$ or equivalently $\mathcal{A}\mathcal{S} = \mathcal{A}$, then a parameter $\mathcal{B} \subset \mathcal{A}$ is not necessarily identified, i.e., $\mathcal{B}\mathcal{S} = \mathcal{B}$ need not be true. However, in view of Proposition 5.5.3, under the stronger property that $\mathcal{A}$ is complete, i.e., $\mathcal{A} \ll \mathcal{S}$, then, by elementary Property 5.4.2(v), $\mathcal{B}$ is complete, i.e., $\mathcal{B} \ll \mathcal{S}$, and is therefore identified, i.e., $\mathcal{B}\mathcal{S} = \mathcal{B}$.

Let us now examine the links between the concepts of a complete statistic or parameter and the concepts of a sufficient statistic or parameter. Recall that in Theorem 4.6.4 and Proposition 4.6.10 we proved that a sufficient parameter (respectively, statistic) is identified if and only if it is almost surely equal to the minimal sufficient parameter (respectively, statistic). In view of Proposition 5.5.3, this entails the following result: a complete sufficient parameter (respectively, statistic) is almost surely equal to the minimal sufficient parameter (respectively, statistic). It is interesting to remark that usually in the statistical literature, the concept of identification is introduced usually in the parameter space, whereas the stronger concept of completeness is introduced only in the sample space. This is due to the fact that to show that $\mathcal{B} \subset \mathcal{A}$ is identified, i.e., $\mathcal{B}\mathcal{S} = \mathcal{B}$, requires the evaluation of sampling expectations, whereas to show that $\mathcal{B} \subset \mathcal{A}$ is complete, i.e., $b \in [\mathcal{B}]_\infty$ and $\mathcal{S}b = 0$ imply $b = 0$ a.s., requires the evaluation of posterior expectations. Similarly, to show that $\mathcal{T} \subset \mathcal{S}$ is identified, i.e. $\mathcal{T}\mathcal{B} = \mathcal{T}$, requires the evaluation of posterior expectations, whereas to show that $\mathcal{T} \subset \mathcal{S}$ is complete, i.e., $t \in [\mathcal{T}]_\infty$ and $\mathcal{A}t = 0$ imply $t = 0$ a.s., requires the evaluation of sampling expectations. Therefore, the classical result that a complete sufficient statistic is minimal sufficient requires a proof because, in the sampling theory approach, the concept of the identification of a statistic is missing. Consequently, and for the sake of comparison, we prove this result directly.

5.5.4 Theorem. In the Bayesian experiment $\mathcal{E}^{\mathcal{M}}_{\mathcal{B}\vee\mathcal{T}}$,

(i)   any complete sufficient statistic $\mathcal{N}$ (i.e., $\mathcal{B} \perp\!\!\!\perp \mathcal{T} \mid \mathcal{M} \vee \mathcal{N}$ and $\mathcal{N} \ll \mathcal{B} \mid \mathcal{M}$) is minimal sufficient in the sense that

$$\overline{\mathcal{M} \vee \mathcal{N}} \cap (\mathcal{M} \vee \mathcal{T}) = (\mathcal{M} \vee \mathcal{T})(\mathcal{B} \vee \mathcal{M});$$

(ii) any complete sufficient parameter $\mathcal{L}$ (i.e., $\mathcal{B} \perp\!\!\!\perp \mathcal{T} \mid \mathcal{L} \vee \mathcal{M}$ and $\mathcal{L} \ll \mathcal{T} \mid \mathcal{M}$) is minimal sufficient in the sense that

$$\overline{\mathcal{L} \vee \mathcal{M}} \cap (\mathcal{B} \vee \mathcal{M}) = (\mathcal{B} \vee \mathcal{M})(\mathcal{M} \vee \mathcal{T}).$$

Proof. By duality it suffices to prove the result for a statistic. Now $\mathcal{B} \perp\!\!\!\perp \mathcal{T} \mid \mathcal{M} \vee \mathcal{N}$ clearly implies $\mathcal{B} \perp\!\!\!\perp \mathcal{R} \mid \mathcal{M} \vee \mathcal{N}$ where $\mathcal{R} = (\mathcal{M} \vee \mathcal{T})(\mathcal{B} \vee \mathcal{M}) \subset \mathcal{M} \vee \mathcal{T}$. Hence, by Theorem 5.4.5, $\mathcal{N} \ll \mathcal{B} \mid \mathcal{M}$ implies $\mathcal{N} \ll \mathcal{B} \mid \mathcal{R}$ since $\mathcal{M} \subset \mathcal{R}$. However, $(\mathcal{B} \vee \mathcal{M}) \perp\!\!\!\perp (\mathcal{M} \vee \mathcal{T}) \mid \mathcal{R}$ implies $\mathcal{B} \perp\!\!\!\perp \mathcal{N} \mid \mathcal{R}$. Therefore, by Theorem 5.4.11, $\mathcal{N} \subset \overline{\mathcal{R}}$. Hence, $\overline{\mathcal{M} \vee \mathcal{N}} \cap (\mathcal{M} \vee \mathcal{T}) \subset \overline{\mathcal{R}} \cap (\mathcal{M} \vee \mathcal{T}) = \mathcal{R}$. By Proposition 4.3.10, $\mathcal{R} \subset \overline{\mathcal{M} \vee \mathcal{N}}$, and this gives the result.   ∎

The next theorem exhibits another relationship between sufficiency and conditional strong identification. It may be paraphrased as follows: if the full sample is strongly identified by the full parameter conditionally on a statistic, then the sample is equivalent to this statistic along with any sufficient statistic.

5.5.5 Theorem.

(i)   If $\mathcal{N}$ is an $\mathcal{E}^{\mathcal{M}}_{\mathcal{B}\vee\mathcal{T}}$-sufficient statistic and $\mathcal{R}$ is an $\mathcal{E}^{\mathcal{M}}_{\mathcal{B}\vee\mathcal{T}}$-statistic such that $\mathcal{T}$ is $\mathcal{E}^{\mathcal{M}\vee\mathcal{R}}_{\mathcal{B}\vee\mathcal{T}}$-complete, then $\overline{\mathcal{M}\vee\mathcal{T}} = \overline{\mathcal{M}\vee\mathcal{N}\vee\mathcal{R}}$;

(ii)   If $\mathcal{L}$ is an $\mathcal{E}^{\mathcal{M}}_{\mathcal{B}\vee\mathcal{T}}$-sufficient parameter and $\mathcal{K}$ is an $\mathcal{E}^{\mathcal{M}}_{\mathcal{B}\vee\mathcal{T}}$-parameter such that $\mathcal{B}$ is $\mathcal{E}^{\mathcal{K}\vee\mathcal{M}}_{\mathcal{B}\vee\mathcal{T}}$-complete, then $\overline{\mathcal{B}\vee\mathcal{M}} = \overline{\mathcal{K}\vee\mathcal{L}\vee\mathcal{M}}$.

Proof. By duality it suffices to prove (i) only. By Corollary 2.2.11, if $\mathcal{N}$ is sufficient, i.e., $\mathcal{B} \perp\!\!\!\perp \mathcal{T} \mid \mathcal{M}\vee\mathcal{N}$, then $\mathcal{B} \perp\!\!\!\perp \mathcal{T} \mid \mathcal{M}\vee\mathcal{N}\vee\mathcal{R}$. Since $\mathcal{N} \subset \mathcal{M}\vee\mathcal{T}$, by Theorem 5.4.5, $\mathcal{T} \ll \mathcal{B} \mid \mathcal{M}\vee\mathcal{R}$, implies $\mathcal{T} \ll \mathcal{B} \mid \mathcal{M}\vee\mathcal{N}\vee\mathcal{R}$. Therefore, by Theorem 5.4.11, $\mathcal{T} \subset \overline{\mathcal{M}\vee\mathcal{N}\vee\mathcal{R}}$ and this is equivalent to $\overline{\mathcal{M}\vee\mathcal{T}} = \overline{\mathcal{M}\vee\mathcal{N}\vee\mathcal{R}}$ ∎

Let us now consider other useful implications of strong p-identification in statistical experiments. For sake of brevity we discuss some of these implications in the unreduced experiment $\mathcal{E}$ only.

It is very often the case that in an identified experiment $\mathcal{E}$, the parameter $\mathcal{A}$ is strongly p-identified by the observation for some $p \in (1,\infty)$. By definition, this means $b \in [\mathcal{A}]_p$ and $\mathcal{S}b = 0$ imply $b = 0$ a.s. or, equivalently, there does not exist a non-trivial p-integrable parameter with 0 posterior expectation for almost all realizations of the sample. By Theorem 5.4.3, this is equivalent to saying that for any $c \in [\mathcal{A}]_q$ (i.e., q-integrable parameter) there exists a sequence of statistics $\{t_k : 1 \leq k < \infty\} \subset [\mathcal{S}]_q$ such that their sampling expectations $\mathcal{A}t_k$ converge a.s. μ and in $[\mathcal{A}]_q$ to c. This gives an interesting representation of the parameters.

In the sample space, if $\mathcal{T}$ is a statistic, $\mathcal{T} \ll_p \mathcal{A}$ means that $t \in [\mathcal{T}]_p$ and $\mathcal{A}t = 0$ implies $t = 0$ a.s. or, equivalently, that there does not exist a non-trivial p-integrable statistic with 0 sampling expectation for almost all values of the parameter. Again, this is equivalent to saying that any q-integrable statistic $u \in \mathcal{T}$ may be represented as an almost sure limit of posterior expectations of a sequence $\{b_k : 1 \leq k < \infty\}$ of q-integrable parameters.

We now consider a useful implication of Corollary 5.4.15.

5.5.6 Theorem. In the unreduced Bayesian experiment $\mathcal{E}$,

(i)   If $\mathcal{T}$ is a p-complete sufficient statistic for some $p \in (1, \infty)$, then for any $\mathcal{U} \subset \mathcal{S}$, $\mathcal{U}\mathcal{T}$ is the minimal sufficient statistic in $\mathcal{E}_{\mathcal{A} \vee \mathcal{U}}$;

(ii)   If $\mathcal{B}$ is a p-complete sufficient parameter for some $p \in (1, \infty)$, then, for any $\mathcal{C} \subset \mathcal{A}$, $\mathcal{C}\mathcal{B}$ is the minimal sufficient parameter in $\mathcal{E}_{\mathcal{C} \vee \mathcal{S}}$.

Proof. Since $\mathcal{T}$ is a complete sufficient statistic, it is almost surely equal to the minimal sufficient statistic, i.e., $\overline{\mathcal{T}} \cap \mathcal{S} = \mathcal{S}\mathcal{A}$ so $\mathcal{S}\mathcal{A} \ll_p \mathcal{A}$ by elementary Property 5.4.2(i). By Corollary 5.4.15, for any $\mathcal{U} \subset \mathcal{S}$, $\mathcal{U}(\mathcal{S}\mathcal{A}) = \mathcal{U}\mathcal{A}$. It now suffices to recall that $\mathcal{U}(\mathcal{S}\mathcal{A}) = \mathcal{U}\left(\overline{\mathcal{S}\mathcal{A}}\right) = \mathcal{U}\left(\overline{\mathcal{T}}\right) = \mathcal{U}\mathcal{T}$, by elementary Property 4.3.2(vi). The same argument may be used to prove the parameter case.   ∎

Note that the computation of $\mathcal{U}\mathcal{T}$ only involves the predictive probability and the computation of $\mathcal{C}\mathcal{B}$, the prior probability.

In the next section we examine relations with ancillarity and maximal ancillarity.

5.5.2   Completeness and Ancillarity

A first result linking completeness and ancillarity is the fact that a complete statistic does not contain a non-trivial ancillary statistic.

5.5.7 Theorem. In the Bayesian experiment $\mathcal{E}_{\mathcal{B} \vee \mathcal{T}}^{\mathcal{M}}$, any ancillary statistic (respectively, parameter) included in a complete statistic (respectively, parameter) is trivial, i.e., is included in $\overline{\mathcal{M}} \cap (\mathcal{M} \vee \mathcal{T})$ (respectively, in $(\mathcal{B} \vee \mathcal{M}) \cap \overline{\mathcal{M}}$).

Proof. By duality it suffices to prove the result for a statistic. Now if $\mathcal{N} \subset \mathcal{M} \vee \mathcal{T}$ is included in a complete statistic, $\mathcal{N}$ is also complete by elementary Property 5.4.2(v), i.e., $\mathcal{N} \ll \mathcal{B} \mid \mathcal{M}$. If, furthermore, $\mathcal{N}$ is ancillary, i.e., $\mathcal{B} \perp\!\!\!\perp \mathcal{N} \mid \mathcal{M}$, then by Theorem 5.4.11, $\mathcal{N} \subset \overline{\mathcal{M}}$. ∎

By Theorems 5.5.4 and 5.5.7, in order to show that the minimal sufficient statistic is not complete it suffices to show that it contains a non-trivial ancillary statistic. Here is a example of such a situation.

Example 1. Let $a \in A = \mathbb{R}_0^+$ and $s = (x, y)' \in S = \mathbb{R}_0^+ \times \mathbb{R}_0^+$ such that $x \perp\!\!\!\perp y \mid a$, $(x \mid a) \sim \exp(a)$ and $(y \mid a) \sim \exp(a^{-1})$. If μ is equivalent to the Lebesgue measure on $\mathbb{R}_0^+$, by Theorem 4.4.11, $\mathcal{S} = \sigma\{s\}$ is minimal sufficient. Now, if $\mathcal{U} = \sigma\{xy\}$, then clearly $\mathcal{U} \subset \mathcal{S}$ and $\mathcal{U}$ is ancillary, i.e., $\mathcal{U} \perp\!\!\!\perp A$ and, therefore, $\mathcal{S}$ is not complete. ∎

A straightforward application of Theorem 5.4.5 entails the following result concerning completeness and ancillarity.

5.5.8 Proposition.

(i) If $\mathcal{N}$ is an $\mathcal{E}_{\mathcal{B} \vee \mathcal{T}}^{\mathcal{M}}$-complete statistic and $\mathcal{R}$ is an $\mathcal{E}_{\mathcal{B} \vee \mathcal{T}}^{\mathcal{M} \vee \mathcal{N}}$-ancillary statistic, then $\mathcal{N}$ is an $\mathcal{E}_{\mathcal{B} \vee \mathcal{M}}^{\mathcal{M} \vee \mathcal{R}}$-complete statistic;

(ii) if $\mathcal{L}$ is an $\mathcal{E}_{\mathcal{B} \vee \mathcal{T}}^{\mathcal{M}}$-complete parameter and $\mathcal{K}$ is an $\mathcal{E}_{\mathcal{B} \vee \mathcal{T}}^{\mathcal{L} \vee \mathcal{M}}$-ancillary parameter, then $\mathcal{N}$ is an $\mathcal{E}_{\mathcal{B} \vee \mathcal{T}}^{\mathcal{K} \vee \mathcal{M}}$-complete parameter. ∎

Let us recall that the **Basu First Theorem** states that a complete sufficient statistic (which is therefore minimal sufficient and does not contain any non trivial ancillary statistic) is independent of any ancillary statistic in the sampling process. We now extend this result in a general reduced Bayesian experiment.

5.5.9 Theorem. In the Bayesian experiment $\mathcal{E}_{\mathcal{B} \vee \mathcal{T}}^{\mathcal{M}}$,

(i) any ancillary statistic is sampling and predictively independent of a complete sufficient statistic.

(ii) any ancillary parameter is a priori and a posteriori independent of a complete sufficient parameter.

Proof. By duality, it suffices to prove the result for a statistic only. The statement (i) may be written as follows. Let $\mathcal{N} \subset \mathcal{M} \vee \mathcal{T}$, such that $\mathcal{N} \ll \mathcal{B} \mid \mathcal{M}$ and $\mathcal{B} \perp\!\!\!\perp \mathcal{T} \mid \mathcal{M} \vee \mathcal{N}$; for any $\mathcal{R} \subset \mathcal{M} \vee \mathcal{T}$ such that $\mathcal{B} \perp\!\!\!\perp \mathcal{R} \mid \mathcal{M}$, we have $\mathcal{R} \perp\!\!\!\perp \mathcal{N} \mid \mathcal{B} \vee \mathcal{M}$ and $\mathcal{R} \perp\!\!\!\perp \mathcal{N} \mid \mathcal{M}$. Now, clearly, $\mathcal{B} \perp\!\!\!\perp \mathcal{R} \mid \mathcal{M} \vee \mathcal{N}$. This, along with $\mathcal{B} \perp\!\!\!\perp \mathcal{R} \mid \mathcal{M}$ and $\mathcal{N} \ll \mathcal{B} \mid \mathcal{M}$ imply, by Theorem 5.4.13, that $\mathcal{R} \perp\!\!\!\perp (\mathcal{B} \vee \mathcal{N}) \mid \mathcal{M}$, and the result follows from Theorem 2.2.10.   ∎

Here is an example of a non trivial application of Theorem 5.5.9:

Example 2. Let

$$a = (b,c)' \in A = I\!\!R_0^+ \times I\!\!R_0^+$$
$$\text{and} \quad s = (x_1, x_2, \ldots, x_n)' \in S = \left(I\!\!R_0^+\right)^n$$

such that $(x_i \mid a) \sim i.\Gamma(b,c)$, i.e., the sampling densities of x_i with respect to the Lebesgue measure on $I\!\!R_0^+$ are given by $p(x_i \mid a) = (b^c/\Gamma(c))x_i^{c-1}e^{-bx_i}$. If μ is equivalent to the Lebesgue measure on $I\!\!R_0^+ \times I\!\!R_0^+$, by Theorem 4.4.11 and by known results in sampling theory, we have that $\mathcal{T} = \sigma\{u,v\}$ where

$$u = (1/n) \sum_{1 \leq i \leq n} x_i \quad \text{and} \quad v = \left(\prod_{1 \leq i \leq n} x_i\right)^{1/n}$$

is a 1-complete sufficient statistic in $\mathcal{E}$, i.e., $\mathcal{A} \perp\!\!\!\perp \mathcal{S} \mid \mathcal{T}$ and $\mathcal{T} \ll_1 \mathcal{A}$. However, if $\mathcal{U} = \sigma(u)$, $\mathcal{B} = \sigma(b)$ and $\mathcal{C} = \sigma(c)$, since $\mu^{\mathcal{C}}$ is equivalent to the Lebesgue measure on $I\!\!R_0^+$, $\mathcal{U}$ is a 1-complete sufficient statistic in $\mathcal{E}_{\mathcal{A} \vee \mathcal{S}}^{\mathcal{C}}$, i.e., $\mathcal{B} \perp\!\!\!\perp \mathcal{S} \mid \mathcal{C} \vee \mathcal{U}$, and $\mathcal{U} \ll_1 \mathcal{B} \mid \mathcal{C}$. Now, since b is a scale parameter, if we define $\mathcal{W} = \sigma(w)$ where $w = vu^{-1}$, then $\mathcal{W} \perp\!\!\!\perp \mathcal{B} \mid \mathcal{C}$, i.e., $\mathcal{W}$ is $\mathcal{E}_{\mathcal{A} \vee \mathcal{S}}^{\mathcal{C}}$-ancillary. Therefore, by Theorem 5.5.9, $\mathcal{W} \perp\!\!\!\perp (\mathcal{B} \vee \mathcal{U}) \mid \mathcal{C}$ which implies $\mathcal{W} \perp\!\!\!\perp \mathcal{U} \mid \mathcal{A}$. Note that this result is non-trivial since it is not an easy matter to characterize the joint sampling distribution of the statistic $t = (u,w)'$.   ∎

Since ancillarity permits an admissible reduction by conditioning, knowing whether or not a given ancillary statistic is maximal is important. Such a criteria is given by the third theorem in Basu (1959), which may be paraphrased as follows: if the full sample is equivalent to a given ancillary statistic along with a complete sufficient statistic, then that ancillary statistic is maximal ancillary. But whereas this result is a useful characterization of maximal ancillarity, it is useless from the point of view of reduction by conditioning. Indeed, in the unreduced experiment $\mathcal{E}$, if $\mathcal{T} \subset \mathcal{S}$ is complete sufficient, and if $\mathcal{U} \subset \mathcal{S}$ is ancillary, then by Theorem 5.5.9, $\mathcal{U} \perp\!\!\!\perp (\mathcal{A} \vee \mathcal{T})$ and, therefore, $\mathcal{E}_{\mathcal{A} \vee \mathcal{T}}$ and $\mathcal{E}^{\mathcal{U}}_{\mathcal{A} \vee \mathcal{T}}$ are essentially the same, and so conditioning is irrelevant. From this standpoint, a result which is both of more practical interest and stronger than Basu Third Theorem is provided by conditional strong identification. This may be paraphrased as follows: an ancillary statistic, conditionally on which the full sample is complete, is maximal ancillary.

5.5.10 Theorem.

(i) If $\mathcal{R}$ is an $\mathcal{E}^{\mathcal{M}}_{\mathcal{B} \vee \mathcal{T}}$-ancillary statistic such that $\mathcal{T}$ is $\mathcal{E}^{\mathcal{M} \vee \mathcal{R}}_{\mathcal{B} \vee \mathcal{T}}$-complete, then $\mathcal{R}$ is an $\mathcal{E}^{\mathcal{M}}_{\mathcal{B} \vee \mathcal{T}}$-maximal ancillary statistic;

ii) if $\mathcal{K}$ is an $\mathcal{E}^{\mathcal{M}}_{\mathcal{B} \vee \mathcal{T}}$-ancillary parameter such that $\mathcal{B}$ is $\mathcal{E}^{\mathcal{K} \vee \mathcal{M}}_{\mathcal{B} \vee \mathcal{T}}$-complete, then $\mathcal{K}$ is an $\mathcal{E}^{\mathcal{M}}_{\mathcal{B} \vee \mathcal{T}}$-maximal ancillary parameter.

Proof. By duality, it suffices to prove (i). Following definition 4.2.5, let $\mathcal{P} \supset \mathcal{R}$ be an $\mathcal{E}^{\mathcal{M}}_{\mathcal{B} \vee \mathcal{T}}$-ancillary statistic, i.e., $\mathcal{B} \perp\!\!\!\perp \mathcal{P} \mid \mathcal{M}$. By Corollary 2.2.11, this implies $\mathcal{B} \perp\!\!\!\perp \mathcal{P} \mid \mathcal{M} \vee \mathcal{R}$. Now, $\mathcal{T}$ $\mathcal{E}^{\mathcal{M} \vee \mathcal{R}}_{\mathcal{B} \vee \mathcal{T}}$-complete, i.e., $\mathcal{T} \ll \mathcal{B} \mid \mathcal{M} \vee \mathcal{R}$, implies $\mathcal{P} \ll \mathcal{B} \mid \mathcal{M} \vee \mathcal{R}$ by elementary Property 5.4.2(v). Therefore, by Theorem 5.4.11, $\mathcal{P} \subset \overline{\mathcal{M} \vee \mathcal{R}}$, which means that $\mathcal{R}$ is maximal ancillary. ∎

Note that Theorem 5.5.10 does not rely on the existence of a complete sufficient statistic. We now show that Basu Third Theorem is a corollary of Theorem 5.5.10.

5.5.11 Corollary. In the Bayesian experiment $\mathcal{E}^{\mathcal{M}}_{\mathcal{B} \vee \mathcal{T}}$

(i) If $\mathcal{N}$ is a complete sufficient statistic and if $\mathcal{R}$ is an ancillary statistic such that $\overline{\mathcal{M} \vee \mathcal{T}} = \overline{\mathcal{M} \vee \mathcal{N} \vee \mathcal{R}}$, then $\mathcal{R}$ is maximal ancillary;

(ii) if $\mathcal{L}$ is a complete sufficient parameter and if $\mathcal{K}$ is an ancillary parameter such that $\overline{\mathcal{B} \vee \mathcal{M}} = \overline{\mathcal{K} \vee \mathcal{L} \vee \mathcal{M}}$, then $\mathcal{K}$ is maximal ancillary.

Proof. By Theorem 5.5.9, $\mathcal{N}$ complete sufficient and $\mathcal{R}$ ancillary imply $(\mathcal{B} \vee \mathcal{N}) \perp\!\!\!\perp \mathcal{R} \mid \mathcal{M}$. Therefore by Theorem 5.4.8 or by Proposition 5.5.8, $\mathcal{N}$ complete, i.e., $\mathcal{N} \ll \mathcal{B} \mid \mathcal{M}$, is equivalent to $\mathcal{N} \ll \mathcal{B} \mid \mathcal{M} \vee \mathcal{R}$. But this is equivalent to $\mathcal{T} \ll \mathcal{B} \mid \mathcal{M} \vee \mathcal{R}$, since $\overline{\mathcal{M} \vee \mathcal{T}} = \overline{\mathcal{M} \vee \mathcal{N} \vee \mathcal{R}}$, i.e., $\mathcal{T}$ is $\mathcal{E}^{\mathcal{M} \vee \mathcal{R}}_{\mathcal{B} \vee \mathcal{T}}$-complete and it suffices to apply Theorem 5.5.10 to obtain the result.   ∎

Let us remark upon the following: under the condition of Theorem 5.5.10, i.e., if $\mathcal{T}$ is $\mathcal{E}^{\mathcal{M} \vee \mathcal{R}}_{\mathcal{B} \vee \mathcal{T}}$-complete, then by Theorem 5.5.5, any $\mathcal{E}^{\mathcal{M}}_{\mathcal{B} \vee \mathcal{T}}$-sufficient statistic $\mathcal{N}$ is such that $\overline{\mathcal{M} \vee \mathcal{T}} = \overline{\mathcal{M} \vee \mathcal{N} \vee \mathcal{R}}$, and so $\mathcal{N}$ is $\mathcal{E}^{\mathcal{M} \vee \mathcal{R}}_{\mathcal{B} \vee \mathcal{T}}$-complete, i.e., $\mathcal{N} \ll \mathcal{B} \mid \mathcal{M} \vee \mathcal{R}$. Therefore, when there exists a complete sufficient statistic, Theorem 5.5.10 and Corollary 5.5.11 are actually equivalent; thus Theorem 5.5.10 only generalizes Corollary 5.5.11 when the minimal sufficient statistic is not complete; this often arises with an exponential family with (exact) restrictions on the parameters. But as noted before, this is the only case where further reduction by conditioning is relevant and, in such a situation, only Theorem 5.5.10 may be used to prove that an ancillary statistic is maximal.

As will be shown in later examples, if the sufficient condition for maximal ancillarity given in Theorem 5.5.10 does not seem to be minimal, it nevertheless clarifies an apparent contradiction between the principle of sufficiency and the principle of ancillarity in successive reductions of experiments; this is the object of next section.

5.5.3 Successive Reductions of a Bayesian Experiment

Let us consider the Bayesian experiment $\mathcal{E}^{\mathcal{M}}_{\mathcal{B} \vee \mathcal{T}}$. We may first marginalize $\mathcal{E}^{\mathcal{M}}_{\mathcal{B} \vee \mathcal{T}}$ on the minimal sufficient statistic, i.e.,

$$(5.5.1) \qquad\qquad \mathcal{N} = (\mathcal{M} \vee \mathcal{T})(\mathcal{B} \vee \mathcal{M})$$

and next condition $\mathcal{E}^{\mathcal{M}}_{\mathcal{B} \vee \mathcal{T}}$ on a maximal ancillary statistic $\mathcal{R}$, i.e., on a $\mathcal{R}$ such that

$$(5.5.2) \qquad\qquad \mathcal{B} \perp\!\!\!\perp \mathcal{R} \mid \mathcal{M} \qquad \mathcal{R} \subset \mathcal{M} \vee \mathcal{N}$$

to end up with $\mathcal{E}^{\mathcal{M} \vee \mathcal{R}}_{\mathcal{B} \vee \mathcal{N}}$ as a basis for inference.

However, we may first condition on a maximal ancillary statistic $\mathcal{P}$, i.e., on a $\mathcal{P}$ such that

$$(5.5.3) \qquad \mathcal{B} \perp\!\!\!\perp \mathcal{P} \mid \mathcal{M} \qquad \mathcal{P} \subset \mathcal{M} \vee \mathcal{T}$$

and afterwards marginalize $\mathcal{E}_{\mathcal{B} \vee \mathcal{T}}^{\mathcal{M} \vee \mathcal{P}}$ on the minimal sufficient statistic, i.e.,

$$(5.5.4) \qquad \mathcal{O} = (\mathcal{M} \vee \mathcal{T})(\mathcal{B} \vee \mathcal{M} \vee \mathcal{P})$$

to end up with $\mathcal{E}_{\mathcal{B} \vee \mathcal{O}}^{\mathcal{M} \vee \mathcal{P}}$ as a basis for inference.

Now clearly, by Theorem 4.3.9 (ii), one has

$$(5.5.5) \qquad \mathcal{O} = \mathcal{P} \vee \mathcal{N},$$

and so $\mathcal{E}_{\mathcal{B} \vee \mathcal{O}}^{\mathcal{M} \vee \mathcal{P}}$ is equivalent to $\mathcal{E}_{\mathcal{B} \vee \mathcal{N}}^{\mathcal{M} \vee \mathcal{P}}$ in the sense that their probabilities are the same. Therefore, these two successive admissible reductions end up with different final reductions, viz., $\mathcal{E}_{\mathcal{B} \vee \mathcal{N}}^{\mathcal{M} \vee \mathcal{R}}$ with $\mathcal{R} \subset \mathcal{M} \vee \mathcal{N}$ and $\mathcal{E}_{\mathcal{B} \vee \mathcal{N}}^{\mathcal{M} \vee \mathcal{P}}$ with $\mathcal{P} \subset \mathcal{M} \vee \mathcal{T}$. Restricting attention to a statistic $\mathcal{P}$ larger than $\mathcal{R}$, i.e.,

$$(5.5.6) \qquad \mathcal{R} \subset \mathcal{M} \vee \mathcal{P},$$

the above paradox may be paraphrased as follows: as soon as an experiment $\mathcal{E}_{\mathcal{B} \vee \mathcal{T}}^{\mathcal{M}}$ has been reduced by marginalization on the minimal sufficient statistic $\mathcal{N}$ and afterwards reduced by conditioning on a maximal ancillary statistic $\mathcal{R}$ included in this minimal sufficient statistic, is there any interest in reducing it further by conditioning on a larger maximal ancillary statistic $\mathcal{P}$ in the initial experiment? The answer is negative if the probabilities are the same, i.e.,

$$(5.5.7) \qquad (\mathcal{B} \vee \mathcal{N}) \perp\!\!\!\perp \mathcal{P} \mid \mathcal{M} \vee \mathcal{R}$$

for all statistics $\mathcal{P}$ satisfying

$$(5.5.8) \qquad \mathcal{B} \perp\!\!\!\perp \mathcal{P} \mid \mathcal{M} \vee \mathcal{R} \qquad \mathcal{P} \subset \mathcal{M} \vee \mathcal{T}.$$

Indeed, when $\mathcal{R}$ is $\mathcal{E}_{\mathcal{B} \vee \mathcal{N}}^{\mathcal{M}}$-ancillary, i.e., $\mathcal{B} \perp\!\!\!\perp \mathcal{R} \mid \mathcal{M}$, any $\mathcal{P}$ such that $\mathcal{R} \subset \mathcal{M} \vee \mathcal{P}$ will be $\mathcal{E}_{\mathcal{B} \vee \mathcal{N}}^{\mathcal{M}}$-ancillary, i.e., $\mathcal{B} \perp\!\!\!\perp \mathcal{P} \mid \mathcal{M}$, if and only if $\mathcal{P}$ is $\mathcal{E}_{\mathcal{B} \vee \mathcal{N}}^{\mathcal{M} \vee \mathcal{R}}$-ancillary, i.e., $\mathcal{B} \perp\!\!\!\perp \mathcal{P} \mid \mathcal{M} \vee \mathcal{R}$ by Theorem 2.2.10. Amazingly enough, the sufficient condition provided by Theorem 5.5.10 fo $\mathcal{R}$ to be maximal ancillary in $\mathcal{E}_{\mathcal{B} \vee \mathcal{N}}^{\mathcal{M}}$ ensures that for any $\mathcal{P}$ such that $\mathcal{B} \perp\!\!\!\perp \mathcal{P} \mid \mathcal{M} \vee \mathcal{R}$ and $\mathcal{R} \subset \mathcal{M} \vee \mathcal{P}$ we also have $(\mathcal{B} \vee \mathcal{N}) \perp\!\!\!\perp \mathcal{P} \mid \mathcal{M} \vee \mathcal{R}$. Thus in such a situation, conditioning on a larger ancillary statistic will be irrelevant. This result is stated formally in the following theorem.

5.5.12 Theorem.

(i)   Let $\mathcal{N}$ be an $\mathcal{E}_{\mathcal{BVT}}^{\mathcal{M}}$-sufficient statistic and $\mathcal{R}$ be an $\mathcal{E}_{\mathcal{BVT}}^{\mathcal{M}}$-ancillary statistic such that $\mathcal{N}$ is $\mathcal{E}_{\mathcal{BVT}}^{\mathcal{MVR}}$-complete. Then for any $\mathcal{E}_{\mathcal{BVT}}^{\mathcal{M}}$-ancillary statistic $\mathcal{P}$ such that $\mathcal{R} \subset \mathcal{M} \vee \mathcal{P}$ we also have $(\mathcal{B} \vee \mathcal{N}) \perp\!\!\!\perp \mathcal{P} \mid \mathcal{M} \vee \mathcal{R}$.

(ii)   Let $\mathcal{L}$ be an $\mathcal{E}_{\mathcal{BVT}}^{\mathcal{M}}$-sufficient parameter and $\mathcal{K}$ be an $\mathcal{E}_{\mathcal{BVT}}^{\mathcal{M}}$-ancillary parameter such that $\mathcal{L}$ is $\mathcal{E}_{\mathcal{BVT}}^{\mathcal{KVM}}$-complete. Then for any $\mathcal{E}_{\mathcal{BVT}}^{\mathcal{M}}$-ancillary parameter $\mathcal{G}$ such that $\mathcal{K} \subset \mathcal{M} \vee \mathcal{G}$ we also have
$\mathcal{G} \perp\!\!\!\perp (\mathcal{L} \vee \mathcal{T}) \mid \mathcal{K} \vee \mathcal{M}$.

Proof. It suffices to prove (i) by duality. By Corollary 2.2.11, $\mathcal{N}$ $\mathcal{E}_{\mathcal{BVT}}^{\mathcal{M}}$-sufficient, i.e., $\mathcal{B} \perp\!\!\!\perp \mathcal{T} \mid \mathcal{M} \vee \mathcal{N}$ implies that $\mathcal{B} \perp\!\!\!\perp \mathcal{T} \mid \mathcal{M} \vee \mathcal{N} \vee \mathcal{R}$. Therefore $\mathcal{N}$ is an $\mathcal{E}_{\mathcal{BVT}}^{\mathcal{MVR}}$-complete sufficient statistic. By Theorem 2.2.10, if $\mathcal{R}$ is $\mathcal{E}_{\mathcal{BVT}}^{\mathcal{M}}$-ancillary, i.e., $\mathcal{B} \perp\!\!\!\perp \mathcal{R} \mid \mathcal{M}$, any $\mathcal{P} \subset \mathcal{M} \vee \mathcal{T}$ such that $\mathcal{R} \subset \mathcal{M} \vee \mathcal{P}$, will be $\mathcal{E}_{\mathcal{BVT}}^{\mathcal{M}}$-ancillary, i.e., $\mathcal{B} \perp\!\!\!\perp \mathcal{P} \mid \mathcal{M}$, if and only if $\mathcal{B} \perp\!\!\!\perp \mathcal{P} \mid \mathcal{M} \vee \mathcal{R}$, i.e., if and only if $\mathcal{P}$ is $\mathcal{E}_{\mathcal{BVT}}^{\mathcal{MVR}}$-ancillary. Therefore, $(\mathcal{B} \vee \mathcal{N}) \perp\!\!\!\perp \mathcal{P} \mid \mathcal{M} \vee \mathcal{R}$ by Theorem 5.5.9.   ∎

We conclude this section by some examples showing the uses and the limitations of the results hereby obtained.

The first example is characterized by the fact that both Theorem 5.5.10 and Corollary 5.5.11 may be applied to prove the maximal character of the ancillary statistic.

Example 3. Let s be a real valued random variable such that $(s \mid a) \sim N(0, a)$ and $a \in \mathbb{R}_0^+ = (0, \infty)$. If the prior probability measure on $(\mathbb{R}_0^+, \mathcal{B}_0^+)$ is such that each open subset has a positive probability (in this case Bayesian sufficiency is equivalent to sampling sufficiency, by Theorems 2.3.12 and 2.3.14), then clearly s^2 is minimal sufficient. It is also complete. Further, $t = 1_{\{s \geq 0\}}$ is clearly ancillary. Remarking that (t, s^2) is equivalent to s, we may apply Corollary 5.5.11 to deduce that t is maximal ancillary. However, without proving that s^2 is both complete and sufficient, it may be seen that for any bounded Borel function f on $\mathbb{R}$, if $\mathcal{T} = \sigma(t)$ and $\mathcal{A} = \sigma(a)$ then

$$(\mathcal{A} \vee \mathcal{T})f(s) = 2t \int_0^\infty (2\pi a)^{1/2} \exp\left(-\frac{x^2}{2a}\right) f(x)dx$$

$$+2(1-t)\int_{-\infty}^{0}(2\pi a)^{1/2}\exp\left(-\frac{x^2}{2a}\right)f(x)dx.$$

So, by continuity in a and unicity of the Laplace transform, it may be seen that s is a complete statistic conditionally on t. Therefore Theorem 5.5.10 applies to deduce the maximality of t. ∎

In the next three examples, Corollary 5.5.11 is not applicable because the minimal sufficient statistic is not complete. However Theorem 5.5.10 insures the maximal character of the ancillary statistic. This shows that Theorem 5.5.10 is of wider application than Corollary 5.5.11.

Example 4. Let $a \in A = (0,1)$ and μ be a probability measure on A such that each open subset of A has positive probability. Let $N = \{N_1, N_2, N_3\}$ have a multinomial distribution; more precisely:

$$(N \mid a) \sim MN_3\left(n, \left\{\frac{a}{2}, \frac{1-a}{2}, \frac{1}{2}\right\}\right).$$

$\{N_1, N_2\}$ is clearly minimal sufficient but not complete since $N_1 + N_2$ is ancillary. Indeed, $(N_1 + N_2 \mid a) \sim Bi\left(n, 1/2\right)$, i.e., the binomial distribution. $N_1 + N_2$ is nevertheless maximal ancillary since $(N_1 \mid N_1 + N_2, a) \sim Bi(N_1 + N_2, a)$ implies that conditionally on $N_1 + N_2$, N_1 is complete. ∎

Example 5. Let $(b, c) = a \in A = (0,1) \times (0,1)$ and $\mu \otimes \nu$ the prior probability on A where μ and ν are as in Example 4 (note that $b \perp\!\!\!\perp c$). Let $N = (N_1, N_2, N_3, N_4)$ be such that

$$(N \mid a) \sim MN_4\left(n, \{b/2, (1-b)/2, c/2, (1-c)/2\}\right).$$

Clearly, (N_1, N_2, N_3) is minimal sufficient but not complete since

$$(N_1 + N_2 \mid a) \sim Bi\left(n, 1/2\right)$$

and $N_1 + N_2$ is therefore ancillary. Nevertheless, $N_1 + N_2$ is maximal ancillary. Indeed $N_1 \perp\!\!\!\perp N_3 \mid N_1 + N_2, a$ and

$$(N_1 \mid N_1 + N_2, a) \sim Bi(N_1 + N_2, b)$$

whereas

$$(N_3 \mid N_1 + N_2, a) \sim Bi(n - N_1 - N_2, c).$$

Therefore,

$$N_1 \perp\!\!\!\perp c \mid N_1 + N_2, b \quad \text{and} \quad N_3 \perp\!\!\!\perp b \mid N_1 + N_2, c.$$

However,

$$N_1 \ll b \mid N_1 + N_2 \quad \text{and} \quad N_3 \ll c \mid N_1 + N_2.$$

By repeated application of Theorem 2.2.10 we have that

$$(N_1, b) \perp\!\!\!\perp (N_3, c) \mid N_1 + N_2;$$

this implies $b \perp\!\!\!\perp N_3 \mid N_1 + N_2, N_1$ and, therefore, $N_1 \ll b \mid N_1 + N_2, N_3$ by Theorem 5.4.5. It also implies that $b \perp\!\!\!\perp c \mid N_1 + N_2, N_3$; so $N_3 \ll c \mid N_1 + N_2, b$, by Theorem 5.4.5, since $(N_1, b) \perp\!\!\!\perp (N_3, c) \mid N_1 + N_2$ implies that $N_1 \perp\!\!\!\perp c \mid N_1 + N_2, N_3, b$. By Theorem 5.4.10, we conclude that $(N_1, N_3) \ll (b, c) \mid N_1 + N_2$. This means that, conditionally on $N_1 + N_2$, (N_1, N_2, N_3) is complete.   ∎

Example 6. Let $A = \mathbb{R}$, μ a probability on A such that each open subset of A has positive probability. Let $s = \{s_i : 1 \le i \le n\}$, $s_i = (y_i, z_i)'$ and

$$(s_i \mid a) \sim i.N_2\left[\begin{pmatrix} 0 \\ 0 \end{pmatrix}; \begin{pmatrix} 1 + a^2 & a \\ a & 1 \end{pmatrix}\right].$$

Let $y = \{y_i : 1 \le i \le n\}$ and $z = \{z_i : 1 \le i \le n\}$. If $u = \sum_{1 \le i \le n} z_i^2$ and $v = \sum_{1 \le i \le n} z_i y_i$, then $\{u, v\}$ is minimal sufficient since the logarithm of the likelihood is proportional to $av - (1/2)(1 + a^2)u$. Now $\{u, v\}$ is not complete since z is clearly ancillary and so is u. But u is maximal ancillary in $\{u, v\}$ because conditionally on u, $\{u, v\}$ is complete. Indeed, it is obvious that, $\overset{\perp\!\!\!\perp}{\scriptstyle 1 \le i \le n} y_i \mid z, a$ and $(y_i \mid z, a) \sim N(az_i, 1)$. Hence $(v \mid z, a) \sim N(au, u)$, and so $(v \mid u, a) \sim N(au, u)$. Therefore, conditionally on u, v is complete.   ∎

We end with an example showing that the condition of Theorem 5.5.10 may not be necessary in the sense that the property of completeness is not satisfied by an apparently maximal ancillary statistic.

Example 7. Let $A = \mathbb{R}$, μ a probability on A such that each open subset of A has positive probability. Let $s = \{s_i : 1 \le i \le n\}$ where $(s_i \mid a) \sim i.U(a, a+1)$, i.e., is the uniform distribution on the interval $(a, a+1)$.

If $t = \min_{1 \leq i \leq n} s_i$ and $u = \max_{1 \leq i \leq n} s_i$ then $\{t, u\}$ is, clearly, minimal sufficient but not complete, since $u - t$ is ancillary. Even though $u - t$ seems to be maximal ancillary in $\{t, u\}$, Theorem 5.5.10 cannot be used to prove it for, conditionally on $u - t$, $\{t, u\}$ is not complete. Indeed, as $(t \mid u - t, a) \sim U(a, a + 1 - (u - t))$, it may be seen that if $\mathcal{T} = \sigma\{u - t\}$ and $\mathcal{A} = \sigma\{a\}$, then

$$(\mathcal{A} \vee \mathcal{T}) \left[\sin\left(\frac{2\pi t}{1 - (u - t)} \right) \mid u - t, a \right] = 0 \qquad \forall \, a \in \mathbb{R}. \qquad \blacksquare$$

5.5.4 Sampling Theory and Bayesian Methods

Let us consider a statistical experiment $\mathcal{E} = \{(S, \mathcal{S}); P^a \quad a \in A\}$ and $\mathcal{A}$, a σ-field on A which makes the mapping $a \longrightarrow P^a(X)$ measurable for any $X \in \mathcal{S}$. We again adopt the notation introduced in Section 2.3.7.

In sampling theory the definition of completeness is as follows (see, e.g., Lehmann and Scheffé (1955)):

5.5.13 Definition. A statistic $\mathcal{T} \subset \mathcal{S}$ is *p-complete* $(1 \leq p \leq \infty)$ if the following implication holds:

$$t \in [\mathcal{T}]_p, \quad \mathcal{I}^a t = 0 \quad \forall \, a \in A \Longrightarrow t = 0 \quad \text{a.s.} \quad P_a \quad \forall \, a \in A.$$

In case of doubt we say *sampling-p-complete*. $\qquad \blacksquare$

It is easy to see that this definition shares the same elementary properties as the Bayesian definition. Namely, any statistic $\mathcal{U}$ included in a p-complete statistic $\mathcal{T}$ is also p-complete. A p-complete statistic is also p'-complete for $\forall \, p \leq p' \leq \infty$. Any ancillary statistic $\mathcal{U}$ contained in a p-complete statistic $\mathcal{T}$ is trivial in the following sense: $\mathcal{U} \subset \mathcal{N}$ where $\mathcal{N} = \{X \in \mathcal{S} : P^a(X) = 1 \quad \forall \, a \in A \quad \text{or} \quad P^a(X) = 0 \quad \forall \, a \in A\}$. Indeed, for any $u \in [\mathcal{U}]_p$ there exists $c \in \mathbb{R}$ such that $\mathcal{I}^a u = c, \quad \forall \, a \in A$, since $\mathcal{U}$ is ancillary; this implies $\mathcal{I}^a(u - c) = 0, \quad \forall \, a \in A$, i.e., $u = c$ a.s. P^a, $\quad \forall \, a \in A$, since $\mathcal{U}$ is p-complete. Thus $u \in [\mathcal{N}]$.

The concept of completeness has three important applications in sampling theory. Firstly, in the theory of unbiased estimation the 2-completeness of a statistic $T \subset S$ implies that if there exist unbiased estimators of a parameter $b \in [A]$ based on this statistic, then they are essentially unique, i.e., $t_i \in [T]_2$ $(i = 1, 2)$ $\mathcal{I}^a t_i = b(a)$ $\forall a \in A$ $(i = 1, 2)$ imply $t_1 - t_2 \in [\mathcal{N}]$. This leads to the theorem of Lehmann-Scheffé about the existence of a unique uniformly minimum variance unbiased estimator.

The second application of completeness is Basu First Theorem which states that any ancillary statistic is independent of a sufficient complete statistic. For the sake of comparison, we reproduce the proof of this theorem. Let T be sufficient and complete and let $\mathcal{U}$ be ancillary. Then, for any $u \in [\mathcal{U}]_\infty$ there exists $c \in \mathbb{R}$ and $t \in [T]_\infty$ such that $\mathcal{I}^a u = c \ \forall \ a \in A$, and $T^a u = t$ a.s.P^a $\forall \ a \in A$. Now $\mathcal{I}^a(t - c) = \mathcal{I}^a T^a u - c = \mathcal{I}^a u - c = 0 \ \forall \ a \in A$. Thus $t = c$ a.s.P^a $\forall \ a \in A$ by completeness of T. Hence, $T^a u = \mathcal{I}^a u$ a.s.P^a $\forall \ u \in [\mathcal{U}]_\infty$ and $\forall \ a \in A$; this is equivalent to $\mathcal{U} \perp\!\!\!\perp T; P^a$ $\forall \ a \in A$. This is the sampling theory version of Theorem 5.5.9.

The third important application of completeness occurs in the dominated case only and is the sampling theory version of Theorem 5.5.4. It says that any sufficient complete statistic is almost surely equal to the minimal sufficient statistic S_1 introduced in Section 4.4.4. For the sake of comparison we give a proof of this theorem.

First note that in the dominated case,
$$\mathcal{N} = \mathcal{I}^* = \left\{ X \in S : P^*(X)^2 = P^*(X) \right\} \text{ where } P^* \text{ is any privileged domi-}$$
nating probability and so $\overline{T}^{P^*} = T \vee \mathcal{N}$. Note also that $T \subset S$ is p-complete if and only if $T \vee \mathcal{N}$ is p-complete. Now let T be sufficient and complete and let S_1 be the minimal sufficient statistic. If $t \in [T]_\infty$, then $S_1^a t = S_1^* t$ a.s.P^a $\forall \ a \in A$ since S_1 is sufficient. Trivially, $\mathcal{I}^a [t - S_1^* t] = 0 \ \forall \ a \in A$ and so $\mathcal{I}^a [t - S_1^* t] = 0 \ \forall \ a \in A$. Since $S_1 \subset T \vee \mathcal{N}$ (by minimal sufficiency) and $T \vee \mathcal{N}$ is complete, we have that $t = S_1^* t$ a.s. P^a $\forall \ a \in A$, i.e., a.s. P^*. Therefore, $T \subset S_1 \vee \mathcal{N}$ and so $T \vee \mathcal{N} = S_1 \vee \mathcal{N}$. ∎

Theorem 5.5.10 suggests another application of completeness in sampling theory for the characterization of maximal ancillarity. But it requires the unusual concept of conditional completeness.

5.5.14 Definition. A statistic $T \subset S$ is p-complete $(1 \le p \le \infty)$ conditionally on a statistic $\mathcal{U} \subset S$ if the following implication holds: $t \in [T \vee \mathcal{U}]_p$,

$$\mathcal{U}^a t = 0 \ \text{ a.s.} P^a \ \ \forall \, a \in A \quad \Longrightarrow t = 0 \ \text{ a.s.} P^a \ \ \forall a \in A. \qquad \blacksquare$$

Since the sampling theory version of Theorem 5.5.10 is not very well known, we state it as a theorem.

5.5.15 Theorem. In the statistical experiment $\mathcal{E} = \{(S, \mathcal{S}) ; P^a \ a \in A\}$, if $\mathcal{U}$ is an ancillary statistic such that $\mathcal{S}$ is complete conditionally on $\mathcal{U}$, then $\mathcal{U}$ is maximal ancillary.

Proof. Let $\mathcal{V} \supset \mathcal{U}$ and suppose that $\mathcal{V}$ is ancillary. We first show that $\mathcal{U}$ is sufficient for $\mathcal{V}$. Indeed, if $a_0 \in A$, $u \in [\mathcal{U}]_\infty$, $v \in [\mathcal{V}]_\infty$, then

$$\mathcal{I}^{a_0}(uv) = \mathcal{I}^{a_0}(u\,\mathcal{U}^{a_0}v).$$

But, by the ancillarity of $\mathcal{V}$,

$$\mathcal{I}^a(uv) = \mathcal{I}^{a_0}(uv) \quad \forall \, a \in A \quad \text{and}$$

$$\mathcal{I}^a(u\,\mathcal{U}^{a_0}v) = \mathcal{I}^{a_0}(u\,\mathcal{U}^{a_0}v) \quad \forall \, a \in A.$$

Therefore, $\forall \, a \in A$,

$$\mathcal{I}^a(uv) = \mathcal{I}^a(u\,\mathcal{U}^{a_0}v)$$

i.e., $\quad \mathcal{U}^a v = \mathcal{U}^{a_0}v \quad \text{a.s.} P^a \quad \forall \, a \in A.$

Therefore,

$$\mathcal{U}^a(v - \mathcal{U}^{a_0}v) = 0 \quad \text{a.s.} P^a \quad \forall \, a \in A,$$

and since $\mathcal{S}$ is complete conditionally on $\mathcal{U}$, $v = \mathcal{U}^{a_0}v \ \text{a.s.} P^a \ \ \forall \, a \in A$, i.e. $v \in [\mathcal{U} \vee \mathcal{N}]$. This means that $\mathcal{U} \vee \mathcal{N} = \mathcal{V} \vee \mathcal{N}$ or, equivalently, that $\mathcal{U}$ is maximal ancillary. $\qquad \blacksquare$

Note that the minimal sufficiency of a complete sufficient statistic requires the domination assumption in sampling theory whereas the Bayesian analogue, Theorem 5.5.4, does not require this assumption. Therefore sampling theory concept of completeness and the Bayesian concept of completeness may only be compared with respect to the dominated case. Recall that if μ is a prior probability on $(A, \mathcal{A})$ then, for any $t \in [\mathcal{S}]^+$, $\mathcal{A}t = \mathcal{I}^a t$ a.s. μ. Thus the Bayesian definition of p-completeness of a statistic in the regular case may be written as follows (in the sense of Definition 1.2.2):

5.5.16 Definition. In the regular case, a statistic $T \subset S$ is *Bayesian p-complete with respect to the prior probability* μ if the following implication holds:

$$t \in [T]_p \quad \mathcal{I}^a t = 0 \quad \text{a.s.} \mu \implies t = 0 \quad \text{a.s.} P.$$ ∎

The concept of a regular prior probability as defined in Section 2.3.7 (Definition 2.3.13) gives the following result:

5.5.17 Theorem.

i)   If μ is a regular prior probability, then any $T \subset S$ Bayesian complete with respect to μ is sampling complete;

ii)   if $T \subset S$ is sampling complete, then it is Bayesian complete with respect to μ for any regular prior probability μ.

Proof.

(i) Remember that if μ is a regular prior probability for the family $\{P^a : a \in A\}$, then this family is dominated and the predictive probability is equivalent to any privileged dominating probability. So if T is Bayesian complete then $\mathcal{I}^a t = 0$, $\forall\, a \in A$, with $t \in [T]_\infty$, implies that $t = 0$ a.s. P which then is equivalent to $t = 0$ a.s. P^a $\forall\, a \in A$.

(ii) If T is sampling complete and μ regular, $\mathcal{I}^a t = 0$ a.s. μ and $t \in [T]_\infty$ imply $\mathcal{I}^a t = 0$ $\forall\, a \in A$ and so $t = 0$ a.s. P^a $\forall\, a \in A$, which in turn implies $t = 0$ a.s. P. ∎

The rest of this chapter will be devoted to the comparison of completeness on the parameter space with another concept found in the literature, namely, identifiability of mixtures.

5.5.5   Identifiability of Mixtures

Let us consider a statistical experiment $\mathcal{E} = \{(S, \mathcal{S}) : P^a \ a \in A\}$ and $\mathcal{A}$ a σ-field on A which makes the mapping $a \longrightarrow P^a(X)$ measurable for any $X \in \mathcal{S}$. The identifiability of mixtures as defined by Teicher (1960,

1961, 1967) may be written as follows (see also Barndorff-Nielsen (1978), Chandra (1977) or Dawid (1979 a)):

5.5.18 Definition. In the statistical experiment $\mathcal{E}$, *the mixtures are identified* if the following implication holds: for any μ_1, μ_2 prior probabilities on $(A, \mathcal{A})$,

$$\int_A \mathcal{I}^a(s)\,\mu_1(da) = \int_A \mathcal{I}^a(s)\,\mu_2(da) \quad \forall s \in [\mathcal{S}]_\infty \implies \mu_1 = \mu_2. \qquad \blacksquare$$

In our Bayesian framework, the concept of completeness may be extended to hold for any prior probability μ:

5.5.19 Definition. In the regular Bayesian experiment $\mathcal{E}_{A\vee S}$, $\mathcal{A}$ is *uniformly complete* if $\mathcal{A} \ll \mathcal{S}$; $\mu \otimes P^A \ \forall \ \mu$, a prior probability on $(A, \mathcal{A})$.

5.5.20 Theorem. In the statistical experiment $\mathcal{E}$, the mixtures are identified if and only if $\mathcal{A}$ is uniformly complete.

Proof.

(i) Suppose that $\mathcal{A}$ is uniformly complete. Let μ_1 and μ_2 be two probabilities on $(A, \mathcal{A})$. If $\mu = \frac{1}{2}\mu_1 + \frac{1}{2}\mu_2$, then $\mu_i \ll \mu$ $i = 1, 2$ and if $b_i = d\mu_i/d\mu$ then $0 \le b_i \le 2$ a.s.; this implies that we can consider that $b_i \in [\mathcal{A}]_\infty$ for $i = 1, 2$. Consider the Bayesian experiment $\mathcal{E} = (A \times S, \mathcal{A} \vee \mathcal{S}, \Pi)$ where $\Pi = \mu \otimes P^A$. Now $\int_A \mathcal{I}^a(s)\mu_1(da) = \int_A \mathcal{I}^a(s)\mu_2(da)$ is equivalent to $\mathcal{I}[b_1 \mathcal{A} s] = \mathcal{I}[b_2 \mathcal{A} s]$ $\forall s \in [\mathcal{S}]_\infty$, i.e., $\mathcal{I}[b_1 s] = \mathcal{I}[b_2 s]$ $\forall a \in [\mathcal{S}]_\infty$ and this is equivalent to $\mathcal{S}b_1 = \mathcal{S}b_2$ or $\mathcal{S}(b_1 - b_2) = 0$. Since $\mathcal{A} \ll \mathcal{S}$, in the Bayesian experiment $\mathcal{E}$, then this implies $b_1 = b_2$ a.s., which is equivalent to $\mu_1 = \mu_2$.

(ii) Now suppose that the mixtures are identified in the experiment $\mathcal{E}$. Let μ be a prior probability on $(A, \mathcal{A})$ and consider the Bayesian experiment $\mathcal{E} = (A \times S, \mathcal{A} \vee \mathcal{S}, \Pi)$ with $\Pi = \mu \otimes P^A$. Let $b \in [\mathcal{A}]_\infty$ such that $\mathcal{S}b = 0$. Define $\mu_1 = \mu$ and $\mu_2(E) = \int_E (1 + \alpha b)\,d\mu$ where $\alpha^{-1} > \sup_A |b|$. Then μ_2 is a probability measure since $1 + \alpha b \ge 0$ and $\mathcal{I}b = 0$. Moreover, $\int_A \mathcal{I}^a(s)\,d\mu_1 = \int_A \mathcal{I}^a(s)\,d\mu_2$ $\forall s \in [\mathcal{S}]_\infty$ since this condition is equivalent to $\mathcal{I}s = \mathcal{I}[s(1 + \alpha b)]$ $\forall s \in [\mathcal{S}]_\infty$, and this is verified since $\mathcal{S}b = 0$ implies $\mathcal{I}(sb) = 0$ $\forall s \in [\mathcal{S}]_\infty$. Therefore by identifiability of the mixtures $\mu_1 = \mu_2$ and this is equivalent to $b = 0$ a.s.μ. Hence $\mathcal{A} \ll \mathcal{S}; \pi$. $\qquad \blacksquare$

6

Sequential Experiments

6.1 Introduction

In the next two chapters we study a richer structure than that investigated until now. More specifically, the basic structure — viz., a unique probability on a product space $\mathcal{A} \vee \mathcal{S}$ — is now endowed with an increasing sequence $(\mathcal{S}_n)$ of sub-σ-fields of $\mathcal{S}$. Typically, $\mathcal{S}_n$ is generated by the sample information available up to time n. In other words, the increasing sequence $(\mathcal{S}_n)$ — to be called a filtration — introduces time into the model. This has two consequences. Firstly, we shall compare the learning process up to time n and the sequence of learning processes associated with the transition from time $n - 1$ to time n; secondly, in the next chapter, we shall compare the learning process over a finite horizon with the learning process over an infinite horizon. Note that in Chapters 6 and 7, the index n will be considered as a fixed time. This material may be further developed by introducing random time: two natural orientations would be to consider either a stopping time (i.e., a random time associated to the filtration $\mathcal{S}_n$), or a random time independent of the filtration. These questions are not treated in this monograph although they are important to the correct handling of some problems with finite populations or with point processes.

As in previous chapters, the basic tool is the properties of conditional independence. In the next section, we initially provide some basic properties of sequences of conditional independence in the context of a general probability space. Section 6.3 describes the formal framework of a sequential Bayesian experiment and the different levels of analysis that appear to be relevant. These different levels of analysis lead, in general, to non-equivalent conditions for admissible reduction. Equivalence can nevertheless be achieved under some so-called transitivity condition, a concept introduced and analyzed in Section 6.4. Section 6.5 then presents the main theorems concerning the equivalence of admissible reductions at different levels. Finally, Section 6.6 provides a deeper discussion of the role of transitivity; in particular, it envisages possible modifications of this condition and gives a proof of its necessity. Sections 6.4 and 6.6 depend heavily on the tools developed in Chapter 5.

The research which underlies this chapter has been motivated by an inquiry into the nature of exogeneity and non-causality in dynamic models beginning with Florens (1979), and successively systematized in Florens (1980), Florens and Mouchart (1980) and Florens, Mouchart and Rolin (1980). The relationships between different forms of non-causality have been presented in Florens and Mouchart (1984, 1985b) while the role of non-causality in Markovian models has been analyzed in Florens, Mouchart and Rolin (1987). The examples have, for the most part, been drawn from two expository papers by Florens and Mouchart (1985a, 1986b). This work has also furthered the study of specification and inference in linear dynamic models, as in Florens, Mouchart and Richard (1986, 1987); these last two papers are the source of several of the examples contained in this chapter. (See, for application, Fiori, Florens and Lai Tong (1982)).   ∎

6.2 Sequences of Conditional Independences

As in Section 2.2, we handle a topic in abstract probability theory. For this reason, we adopt the same notation; in particular $(M, \mathcal{M}, P)$ is an abstract probability space.

6.2.1 Definition.

(i)   A *filtration* $(\mathcal{F}_n)_{n \in I\!N}$ in $(M, \mathcal{M})$ is an increasing sequence of sub-σ-fields of $\mathcal{M}$, i.e., $\forall\, n \geq 0 : \mathcal{F}_n \subset \mathcal{F}_{n+1} \subset \mathcal{M}$. We denote

$$\mathcal{F}_\infty = \bigvee_{n \geq 0} \mathcal{F}_n \quad \text{and write } \mathcal{F}_n \uparrow \mathcal{F}_\infty.$$

(ii) A sequence of sub-σ-fields $(\mathcal{J}_n)_{n \in I\!N}$ is *adapted* to a sequence of σ-fields $(\mathcal{K}_n)_{n \in I\!N}$ if $\forall\, n \geq 0 : \mathcal{J}_n \subset \mathcal{K}_n$.

(iii) $(\mathcal{J}_n)$ is $(\mathcal{K}_n)$-*recursive* if $(\mathcal{J}_{n+1})$ is adapted to $(\mathcal{J}_n \vee \mathcal{K}_{n+1})$, i.e.,
$$\forall\, n \geq 0 \quad \mathcal{J}_{n+1} \subset \mathcal{J}_n \vee \mathcal{K}_{n+1}. \qquad\qquad \blacksquare$$

Moreover we introduce the following *notation*, for an arbitrary sequence of sub-σ-fields of $\mathcal{M}$:

$$(6.2.1) \qquad\qquad \mathcal{J}_p^q = \bigvee_{p \leq n \leq q} \mathcal{J}_n \qquad 0 \leq p \leq q.$$

In particular $(\mathcal{J}_0^n)_{n \in I\!N}$ is a filtration in $(M, \mathcal{M})$ canonically associated to a sequence $(\mathcal{J}_n)$ and $\mathcal{J}_n^\infty = \bigvee_{p \geq n} \mathcal{J}_p$ is a decreasing sequence of σ-fields.

Example. Consider a sequence of random variables $m_n \in [\mathcal{M}]$, $n \in I\!N$, and $\mathcal{M}_n = \sigma(m_n)$ the σ-field generated by m_n. Then $(\mathcal{M}_0^n)_{n \in I\!N}$ is a filtration in $(M, \mathcal{M})$ canonically associated to the sequence $(m_n)_{n \in I\!N}$. Consider the sequence of the partial sums $s_n = \sum_{1 \leq i \leq n} m_i$ and $\mathcal{S}_n = \sigma(s_n)$. Then $(\mathcal{S}_n)_{n \in I\!N}$ is not a filtration but is $(\mathcal{M}_n)$-recursive. $\qquad\qquad \blacksquare$

The essential objective of this section is to present two theorems; the second theorem, which deals with filtrations, plays a crucial role in sequential analysis and in this respect is similar to Theorem 2.2.10 in the one-shot sampling context.

6.2.2 Theorem. Let $(\mathcal{F}_n)$ be a filtration in $(M, \mathcal{M})$, let $(\mathcal{J}_n)$ and $(\mathcal{K}_n)$ be two sequences adapted to $(\mathcal{F}_n)$, and let $\mathcal{M}_i$ $(i = 1, 2)$ be two arbitrary sub-σ-fields of $\mathcal{M}$. If

(o) $\quad (\mathcal{J}_n)$ is $(\mathcal{K}_n)$- recursive ,

then the following properties are equivalent:

(i) $\quad (\mathcal{K}_{n+1} \vee \mathcal{M}_1) \perp\!\!\!\perp \mathcal{F}_n \mid \mathcal{J}_n \vee \mathcal{M}_2 \quad \forall\, n \geq 0$

(ii) $(\mathcal{K}_{n+1}^{\infty} \vee \mathcal{M}_1) \perp \mathcal{F}_n \mid \mathcal{J}_n \vee \mathcal{M}_2 \quad \forall\, n \geq 0.$

Proof.

(ii) $\Rightarrow$ (i) is trivial because $\mathcal{K}_{n+1} \subset \mathcal{K}_{n+1}^{\infty}$.

(i) $\Rightarrow$ (ii) Note that, from Definition 6.2.1. (iii), the hypothesis (o) is equivalent to

(iii) $\mathcal{J}_{n+p} \subset \mathcal{J}_n \vee \mathcal{K}_{n+1}^{n+p} \quad \forall\, p \geq 1.$

Note also that, by a monotone class argument, (ii) is equivalent to

(iv) $(\mathcal{K}_{n+1}^{n+p} \vee \mathcal{M}_1) \perp \mathcal{F}_n \mid \mathcal{J}_n \vee \mathcal{M}_2 \quad \forall\, p \geq 1, \quad \forall\, n \geq 0.$

We now prove (iv) by induction on p. Clearly, (iv) is true for $p = 1$ by hypothesis (i). Furthermore, hypothesis (i) implies, by Corollary 2.2.11, that

(v) $(\mathcal{K}_{n+p+1} \vee \mathcal{M}_1) \perp \mathcal{F}_{n+p} \mid \mathcal{J}_{n+p} \vee \mathcal{K}_{n+1}^{n+p} \vee \mathcal{J}_n \vee \mathcal{M}_2,$

because the fact that both $(\mathcal{K}_n)$ and $(\mathcal{J}_n)$ are adapted to the filtration $(\mathcal{F}_n)$ implies that $\mathcal{K}_{n+1}^{n+p} \subset \mathcal{F}_{n+p}$ and $\mathcal{J}_n \subset \mathcal{F}_n$, and so (v) implies, by Corollary 2.2.4. and by (iii), that

(vi) $(\mathcal{K}_{n+p+1} \vee \mathcal{M}_1) \perp \mathcal{F}_n \mid \mathcal{J}_n \vee \mathcal{K}_{n+1}^{n+p} \vee \mathcal{M}_2.$

The induction hypothesis (iv) and property (vi) implies, by Theorem 2.2.10, that (iv) is satisfied for $p + 1$.

Theorem 6.2.2. shows that recursivity is a sufficient condition to ensure that a sequential property of conditional independence (i) may be extended to an infinite horizon (ii). This theorem plays a crucial role in the proof of the next theorem, which is the cornerstone of this chapter.

6.2.3 Theorem. Let $(\mathcal{F}_n)$ be a filtration in $(M, \mathcal{M})$, $(\mathcal{G}_n)$ be a filtration in $(M, \mathcal{M})$ adapted to $(\mathcal{F}_n)$, and $\mathcal{M}_i$ $(i = 1, 2)$ be two arbitrary sub-σ-fields of $\mathcal{M}$.

Then the following seven conditions are equivalent:

1. $(\mathcal{G}_\infty \vee \mathcal{M}_1) \perp\!\!\!\perp (\mathcal{F}_n \vee \mathcal{M}_2) \mid \mathcal{F}_{n-1} \vee \mathcal{G}_n$

2. $(\mathcal{G}_\infty \vee \mathcal{M}_1) \perp\!\!\!\perp (\mathcal{F}_n \vee \mathcal{M}_2) \mid \mathcal{F}_0 \vee \mathcal{G}_n$

3. $(\mathcal{G}_{n+1} \vee \mathcal{M}_1) \perp\!\!\!\perp (\mathcal{F}_n \vee \mathcal{M}_2) \mid \mathcal{F}_0 \vee \mathcal{G}_n$

4. (i) $(\mathcal{G}_{n+1} \vee \mathcal{M}_1) \perp\!\!\!\perp \mathcal{M}_2 \mid \mathcal{F}_n$

    (ii) $(\mathcal{G}_{n+1} \vee \mathcal{M}_1) \perp\!\!\!\perp \mathcal{F}_n \mid \mathcal{F}_0 \vee \mathcal{G}_n$

5. (i) $(\mathcal{G}_n \vee \mathcal{M}_1) \perp\!\!\!\perp \mathcal{M}_2 \mid \mathcal{F}_0$

    (ii) $(\mathcal{G}_{n+1} \vee \mathcal{M}_1) \perp\!\!\!\perp \mathcal{F}_n \mid \mathcal{F}_0 \vee \mathcal{G}_n \vee \mathcal{M}_2$

6. (i) $\mathcal{M}_1 \perp\!\!\!\perp (\mathcal{F}_n \vee \mathcal{M}_2) \mid \mathcal{F}_0 \vee \mathcal{G}_n$

    (ii) $\mathcal{G}_{n+1} \perp\!\!\!\perp (\mathcal{F}_n \vee \mathcal{M}_2) \mid \mathcal{F}_0 \vee \mathcal{G}_n \vee \mathcal{M}_1$

7. (i) $\mathcal{M}_1 \perp\!\!\!\perp (\mathcal{F}_n \vee \mathcal{M}_2) \mid \mathcal{F}_{n-1} \vee \mathcal{G}_n$

    (ii) $\mathcal{G}_{n+1} \perp\!\!\!\perp (\mathcal{F}_n \vee \mathcal{M}_2) \mid \mathcal{F}_0 \vee \mathcal{G}_n \vee \mathcal{M}_1$

where all properties hold for any $n \geq 0$, and where $\mathcal{G}_\infty$ is defined as in Definition 6.2.1: $\mathcal{G}_\infty = \bigvee_{n \geq 0} \mathcal{G}_n$, and with the convention, in 1 and 7(i), that $\mathcal{F}_{-1} = \mathcal{F}_0$.

Proof. $1 \Rightarrow 2$: We first prove that 1 implies:

8. $(\mathcal{G}_\infty \vee \mathcal{M}_1) \perp\!\!\!\perp (\mathcal{F}_n \vee \mathcal{M}_2) \mid \mathcal{F}_{n-k} \vee \mathcal{G}_n \quad 1 \leq k \leq n.$

This is indeed true for $k = 1$. Suppose now that (8) is true for some k; condition 1 may also be written as:

9. $(\mathcal{G}_\infty \vee \mathcal{M}_1) \perp\!\!\!\perp (\mathcal{F}_{n-k} \vee \mathcal{M}_2) \mid \mathcal{F}_{n-(k+1)} \vee \mathcal{G}_{n-k}$

and 9 implies, by Corollary 2.2.11 and because $(\mathcal{G}_n)$ is a filtration:

10. $(\mathcal{G}_\infty \vee \mathcal{M}_1) \perp\!\!\!\perp (\mathcal{F}_{n-k} \vee \mathcal{M}_2) \mid \mathcal{F}_{n-(k+1)} \vee \mathcal{G}_n.$

Condition 8 also implies, because $(\mathcal{F}_n)$ is a filtration:

11. $(\mathcal{G}_\infty \vee \mathcal{M}_1) \perp\!\!\!\perp (\mathcal{F}_n \vee \mathcal{M}_2) \mid \mathcal{F}_{n-(k+1)} \vee \mathcal{G}_n \vee \mathcal{F}_{n-k} \vee \mathcal{M}_2.$

Clearly, 10 and 11 are equivalent, by Theorem 2.2.10, to:

12. $(\mathcal{G}_\infty \vee \mathcal{M}_1) \perp\!\!\!\perp (\mathcal{F}_n \vee \mathcal{M}_2) \mid \mathcal{F}_{n-(k+1)} \vee \mathcal{G}_n.$

Therefore, 8 becomes also true for $k = n$, which is precisely 2.

$2 \Rightarrow 1$ is trivial, by Corollary 2.2.11, because $\mathcal{F}_0 \subset \mathcal{F}_{n-1} \subset \mathcal{F}_n$.

$2 \Leftrightarrow 3$ is a simple corollary of Theorem 6.2.2, where $(\mathcal{F}_n \vee \mathcal{M}_2, \mathcal{G}_n, \mathcal{G}_n, \mathcal{F}_0)$ replace $(\mathcal{F}_n, \mathcal{J}_n, \mathcal{K}_n, \mathcal{M}_2)$; indeed, $(\mathcal{G}_n)$ is trivially $(\mathcal{G}_n)$-recursive.

$3 \Leftrightarrow 4$ is a trivial application of Theorem 2.2.10; indeed that $(\mathcal{G}_n)$ is adapted to the filtration $(\mathcal{F}_n)$ implies that $\mathcal{F}_0 \vee \mathcal{F}_n \vee \mathcal{G}_n = \mathcal{F}_n$.

$3 \Leftrightarrow 5$: clearly, 3 is equivalent to 5(ii) along with

13. $(\mathcal{G}_{n+1} \vee \mathcal{M}_1) \perp\!\!\!\perp \mathcal{M}_2 \mid \mathcal{F}_0 \vee \mathcal{G}_n.$

The equivalence between 5 (i) and 13 may be seen as an application of the equivalence between 1 and 2 under the substitution: $\mathcal{G}_n$ is trivial, $\mathcal{F}_n \to \mathcal{G}_n \vee \mathcal{F}_0$, $\mathcal{M}_1$ and $\mathcal{M}_2$ are permuted.

$3 \Leftrightarrow 6$ is a trivial application of Theorem 2.2.10

$1 \Leftrightarrow 7$: indeed, 1 is equivalent to 7 (i) along with

14. $\mathcal{G}_\infty \perp\!\!\!\perp (\mathcal{F}_n \vee \mathcal{M}_2) \mid \mathcal{F}_{n-1} \vee \mathcal{G}_n \vee \mathcal{M}_1.$

The equivalence between 7 (ii) and 14 may be seen as an application of the equivalence between 1 and 3 under the substitution: $\mathcal{M}_1$ is trivial and $\mathcal{F}_n \to \mathcal{F}_n \vee \mathcal{M}_1$.   ∎

In the sequel, this theorem is used, in particular, to allow a sequential reduction of a conditioning σ-field $\mathcal{F}_n$ on its initial condition $\mathcal{F}_0$, or a sequential extension of a σ-field on a prediction space from an immediate prediction $\mathcal{G}_{n+1}$ to an infinite horizon $\mathcal{G}_{\infty}$.

It should be noted that the equivalence between 6 and 7 may be rephrased as "under 6 (ii), 6 (i) and 7 (i) are equivalent".

6.3 Sequential Experiments

6.3.1 Definition of Sequential Experiments

A sequential Bayesian experiment is a Bayesian experiment endowed with a filtration on the sample space. More specifically:

6.3.1 Definition. A *sequential Bayesian experiment* is defined by:

$$\mathcal{E} = (A \times S, \mathcal{A} \vee \mathcal{S}, \Pi, \mathcal{S}_n \uparrow \mathcal{S})$$

where $(\mathcal{S}_n)_{n \in I\!N}$ is a filtration in $(S, \mathcal{S})$ such that $\mathcal{S} = \bigvee_{n \geq 0} \mathcal{S}_n$. ∎

Typically, $\mathcal{S}_n$ represents the σ-field generated by the observations up to time n: $\mathcal{S}_n = \sigma(s_0, s_1, \ldots, s_n)$. $\mathcal{S}$ is then generated by the infinite sequence $(s_0, s_1, \ldots)$. Such a representation of $\mathcal{S}_n$ is unnecessary for the basic theory; in particular, in this chapter no assumption of stationarity is needed.

Π is often constructed from a stochastic process $(s_n)_{n \in I\!N}$. It is perhaps useful to recall two classical approaches to assigning a probability measure to an infinite family of measurable functions s_n. One approach is to define Π, through Kolmogorov's Theorem, as the projective limit of a projective system of finite-dimensional distributions. When the index set is $I\!N$, it is sufficient to specify these finite dimensional distributions for the sets $\{0, 1, \ldots, n\}$ for any $n \geq 0$ only. This approach naturally leads to a "global" analysis of the sequence of experiments $\mathcal{E}_{\mathcal{A} \vee \mathcal{S}_n}$. An alternative approach is to specify a sequence of transitions $\Pi_{\mathcal{A} \vee \mathcal{S}_n}^{\mathcal{S}_{n-1}}$ and to define Π using Ionescu Tulcea Theorem. This second approach naturally leads to a "sequential" analysis of the sequence of experiments $\mathcal{E}_{\mathcal{A} \vee \mathcal{S}_n}^{\mathcal{S}_{n-1}}$. These two modes of construction have different technical requirements that will not be made explicit here. Useful references to the literature on general stochastic processes are Blumenthal and Getoor (1968, Chap. 1), Dellacherie and Meyer (1975, Chap. 4), Neveu

(1964, Chap. 3) and Métivier (1968). In general, we take the existence of the probability measure Π on $(A \times S, \mathcal{A} \vee \mathcal{S})$ for granted, without taking care of its construction.

We systematically prefer using $I\!N = \{0, 1, 2, \ldots\}$ as the time index rather than $Z\!\!\!Z = \{\ldots - 2, -1, 0, 1, 2, \ldots\}$. The typical statistical problem consists of observing a stochastic process from a given instant, to be labelled "$n = 1$", and $\mathcal{S}_0$ is to be interpreted as the σ-field generated by the pre-sample information; in particular, if s_n is indexed in $Z\!\!\!Z$, one would define $\mathcal{S}_0 = \bigvee_{n \leq 0} \sigma(s_n)$. One motivation of doing so is to make the role of the initial condition explicit. In $Z\!\!\!Z$ one must specify a left-tail σ-field $\mathcal{S}_{-\infty} = \bigcap_{n \in Z\!\!\!Z} \mathcal{S}_n$ which plays essentially the same role as $\mathcal{S}_0$. It should also be mentioned that deducing the properties of $\mathcal{S}_{-\infty}$ from those of the sequence $(\mathcal{S}_n)$ often raises technical problems that do not arise with $\mathcal{S}_0$ and that do not appear particularly relevant in the statistical context; from time to time we shall sketch the nature of such problems.

A sequential Bayesian experiment can be analyzed from different view points:

(i) a *"global" analysis* considers the sequence of experiments

$$\mathcal{E}_{\mathcal{A} \vee \mathcal{S}_n} = (A \times S, \mathcal{A} \vee \mathcal{S}_n, \Pi_{\mathcal{A} \vee \mathcal{S}_n}).$$

Under the usual regularity conditions

$$(6.3.1) \qquad \Pi_{\mathcal{A} \vee \mathcal{S}_n} = \mu \otimes P_{\mathcal{S}_n}^{\mathcal{A}} = P_{\mathcal{S}_n} \otimes \mu^{\mathcal{S}_n}.$$

The prior measure μ is therefore common to every $\mathcal{E}_{\mathcal{A} \vee \mathcal{S}_n}$ and the compatibility conditions of a projective system require that $P_{\mathcal{S}_n}^{\mathcal{A}}$ (respectively, $P_{\mathcal{S}_n}$) are the restrictions to $\mathcal{S}_n$ of $P_{\mathcal{S}_{n'}}^{\mathcal{A}}$ (respectively, $P_{\mathcal{S}_{n'}}$) as soon as $n \leq n'$. If $\mathcal{E}_{\mathcal{A} \vee \mathcal{S}_n}$ is dominated with density $g_{\mathcal{A} \vee \mathcal{S}_n}$ then, for any $n' \leq n$, $\mathcal{E}_{\mathcal{A} \vee \mathcal{S}_{n'}}$ is also dominated with density $g_{\mathcal{A} \vee \mathcal{S}_{n'}} = (\widetilde{\mathcal{A} \vee \mathcal{S}_{n'}}) g_{\mathcal{A} \vee \mathcal{S}_n}$, i.e., the expectation of $g_{\mathcal{A} \vee \mathcal{S}_n}$, conditional on $\mathcal{A} \vee \mathcal{S}_n'$, with respect to $\mu \otimes P$. Thus $g_{\mathcal{A} \vee \mathcal{S}_n}$ is actually a martingale, a property to be used extensively in next chapter. Note that each experiment of the sequence $\mathcal{E}_{\mathcal{A} \vee \mathcal{S}_n}$ has the same structure as the basic experiment $\mathcal{E}_{\mathcal{A} \vee \mathcal{S}}$ introduced in the first chapter.

(ii) An *initial analysis* considers the sequence of experiments

$$\mathcal{E}_{\mathcal{A} \vee \mathcal{S}_n}^{\mathcal{S}_0} = (A \times S, \mathcal{A} \vee \mathcal{S}_n, \Pi_{\mathcal{A} \vee \mathcal{S}_n}^{\mathcal{S}_0}).$$

Under the usual regularity conditions,

$$(6.3.2) \qquad \Pi^{S_0}_{AVS_n} = \mu^{S_0} \otimes P^{AVS_0}_{S_n} = P^{S_0}_{S_n} \otimes \mu^{S_n}.$$

This kind of analysis is motivated by the fact that the initial conditions, formalized in S_0, typically represent presample information that deserves different treatment. In this context, one of the relevant problems is to know to what extent is the conditioning on S_0 admissible. Note that if S_0 is trivial, then initial and global analyses coincide. The search for admissible reductions of $\mathcal{E}^{S_0}_{AVS_n}$ is undertaken by setting, in the general reduced experiment $\mathcal{E}^{M}_{BVT}$, $M = S_0$, $B = A$ and $T = S_n$.

(iii) A *sequential analysis* considers the sequence of experiments

$$\mathcal{E}^{S_{n-1}}_{AVS_n} = (A \times S, \ A \vee S_n, \ \Pi^{S_{n-1}}_{AVS_n}).$$

Under the usual regularity conditions,

$$(6.3.3) \qquad \Pi^{S_{n-1}}_{AVS_n} = \mu^{S_{n-1}} \otimes P^{AVS_{n-1}}_{S_n} = P^{S_{n-1}}_{S_n} \otimes \mu^{S_n}.$$

A major concern of sequential analysis is to study the properties of the transformation $\mu^{S_{n-1}} \to \mu^{S_n}$, i.e., the learning process associated with individual observations when S_n is generated by a sequence of observations. The search for admissible reductions of $\mathcal{E}^{S_{n-1}}_{AVS_n}$ is undertaken by setting, in the general reduced experiment $\mathcal{E}^{M}_{BVT}$, $M = S_{n-1}$, $B = A$ and $T = S_n$.

6.3.2 Admissible Reductions in Sequential Experiments

The theory developed in the first five chapters may be used to characterize admissible reductions at each of these three levels of analysis. For the reader's convenience we summarize, in the next table, the principal sorts of reductions undertaken at each level; note that (T_n) is a sequence of σ-fields adapted to (S_n), and B and C are sub-σ-fields of A.

As mentioned earlier, when introducing the three levels of analysis in a sequential Bayesian experiment, the properties defining admissible reductions may be analyzed as straightforward applications of reductions in unreduced experiments (Chapter 2) at global level and reductions in conditional experiments (Section 3.3) at the initial and sequential levels. These properties are summarized in Table 6.1.

Table 6.1. Admissible reductions in sequential experiments.

Level of analysis Corresponding experiment	global $\mathcal{E}_{A\vee S_n}$	initial $\mathcal{E}_{A\vee S_n}^{S_0}$	sequential $\mathcal{E}_{A\vee S_n}^{S_{n-1}}$
Admissible reductions			
i) *On the sample space:* (T_n)			
sufficient	$A \perp\!\!\!\perp S_n|T_n$	$A \perp\!\!\!\perp S_n|S_0\vee T_n$	$A \perp\!\!\!\perp S_n|S_{n-1}\vee T_n$
ancillary	$A \perp\!\!\!\perp T_n$	$A \perp\!\!\!\perp T_n|S_0$	$A \perp\!\!\!\perp T_n|S_{n-1}$
ii) *On the parameter space:* B			
sufficient	$A \perp\!\!\!\perp S_n|B$	$A \perp\!\!\!\perp S_n|B\vee S_0$	$A \perp\!\!\!\perp S_n|B\vee S_{n-1}$
ancillary	$B \perp\!\!\!\perp S_n$	$B \perp\!\!\!\perp S_n|S_0$	$B \perp\!\!\!\perp S_n|S_{n-1}$
iii) *Joint reduction*			
mutual ancillarity			
c and (T_n)	$c \perp\!\!\!\perp T_n$	$c \perp\!\!\!\perp T_n|S_0$	$c \perp\!\!\!\perp T_n|S_{n-1}$
mutual sufficiency	$A \perp\!\!\!\perp T_n|B$	$A \perp\!\!\!\perp T_n|B\vee S_0$	$A \perp\!\!\!\perp T_n|B\vee S_{n-1}$
B and (T_n)	$B \perp\!\!\!\perp S_n|T_n$	$B \perp\!\!\!\perp S_n|S_0\vee T_n$	$B \perp\!\!\!\perp S_n|S_{n-1}\vee T_n$
mutual exogeneity	$c \perp\!\!\!\perp T_n$	$c \perp\!\!\!\perp T_n|S_0$	$c \perp\!\!\!\perp T_n|S_{n-1}$
c and (T_n)	$A \perp\!\!\!\perp S_n|c\vee T_n$	$A \perp\!\!\!\perp S_n|c\vee S_0\vee T_n$	$A \perp\!\!\!\perp S_n|c\vee S_{n-1}\vee T_n$
cut	$B \perp\!\!\!\perp c$	$B \perp\!\!\!\perp c|S_0$	$B \perp\!\!\!\perp c|S_0$
B, c and (T_n)	$A \perp\!\!\!\perp T_n|B$	$A \perp\!\!\!\perp T_n|B\vee S_0$	$A \perp\!\!\!\perp T_n|B\vee S_{n-1}$
	$A \perp\!\!\!\perp S_n|c\vee T_n$	$A \perp\!\!\!\perp S_n|c\vee S_0\vee T_n$	$A \perp\!\!\!\perp S_n|c\vee S_{n-1}\vee T_n$

Remark. These properties are supposed to hold for any $n \geq 0$ with the convention that
$$S_n = S_0 \quad \forall\, n \leq 0.$$

The new problems concern the relationships between admissible reductions at different levels of analysis. In this chapter, we investigate the relationships between initial and sequential admissibility of reductions. The limit properties — i.e., passing from $\mathcal{S}_n$ to $\mathcal{S} = \bigvee_{n \geq 0} \mathcal{S}_n$ — are analyzed in the next chapter which mainly involves global and sequential analysis. Links between global and initial levels of analysis are not analyzed systematically; this would essentially involve discussion of the interpretation of initial conditions along with some rather straightforward results of the following type: the independence between the parameters and the initial condition — i.e., $(\mathcal{B} \vee \mathcal{C}) \perp\!\!\!\perp \mathcal{S}_0$ — ensures the equivalence of the first condition of a Bayesian cut for the three levels of analysis. We do not investigate the optimality of admissible reductions in the initial and global analysis (such as minimal sufficiency).

In this chapter, we essentially consider a sequence $(\mathcal{T}_n)$ adapted to a filtration $(\mathcal{S}_n)$, and we investigate the properties of ancillarity and of sufficiency of this sequence from an initial and a sequential point of view. For this kind of problem it is natural to assume that $(\mathcal{T}_n)$ is a filtration. In particular, if one starts with an arbitrary sequence $(\mathcal{Z}_n)$ adapted to $(\mathcal{S}_n)$, the filtration $(\mathcal{T}_n)$ is canonically constructed as $\mathcal{T}_n = \bigvee_{i \leq n} \mathcal{Z}_i$ and the results of this chapter draw on the filtration $(\mathcal{T}_n)$ only. This explains why Theorems 6.2.2 and 6.2.3 are the basic tools of this chapter. Heuristically, the motivation of our results may be viewed as follows: in most applications $\mathcal{S}_n$ is generated by a vector-valued stochastic process (s_n) and $\mathcal{T}_n$ is generated by a subvector of s_n. This is at variance with Chapter 7 where $\mathcal{T}_n$ is typically generated by a statistic on $(s_1, \ldots, s_n)$ so that the sequence $(\mathcal{T}_n)$ is adapted to $(\mathcal{S}_n)$, but is no longer increasing. Consequently, this chapter uses different tools.

Before studying the links between initial and sequential admissibility, two remarks are in order:

(i) The first condition for the sequential cut (i.e., the cut in $\mathcal{E}_{\mathcal{A} \vee \mathcal{S}_n}^{\mathcal{S}_{n-1}}$) has been given as

(6.3.4) $$\mathcal{B} \perp\!\!\!\perp \mathcal{C} \mid \mathcal{S}_0$$

rather than

(6.3.5) $$\mathcal{B} \perp\!\!\!\perp \mathcal{C} \mid \mathcal{S}_{n-1}.$$

As these conditions are supposed to hold for any $n \geq 0$, (6.3.5) clearly implies (6.3.4); the next theorem shows that, in a cut, these conditions are actually equivalent.

6.3.2 Theorem. If $(\mathcal{B}, \mathcal{C}, (\mathcal{T}_n))$ operates a sequential cut, $\mathcal{B}$ and $\mathcal{C}$ are *a posteriori* independent, i.e., $\mathcal{B} \perp \mathcal{C} \mid \mathcal{S}_n$ for all n.

Proof. We prove the theorem by induction on n. For $n = 1$, the theorem is a trivial application of Theorem 3.4.15. Now, if the theorem is true for $n - 1$, we may replace (6.3.4) by (6.3.5) and the conclusion is true again by the same theorem.

Had we introduced (6.3.5) instead of (6.3.4) in the definition of a sequential cut, Theorem 6.3.2 would have been a direct consequence of Theorem 3.4.15, but (6.3.5), for any $n \geq 0$, is precisely equivalent to the conclusion of Theorem 6.3.2. This explains our preference for condition (6.3.4) in the definition of a sequential cut.

(ii) The second remark concerns the initial condition used in the second column of the Table 6.1 (level of initial analysis). It may seem more natural to introduce $\mathcal{T}_0$ rather than $\mathcal{S}_0$ in conditions involving only $\Pi_{\mathcal{A} \vee \mathcal{T}_n}$, for example, $\mathcal{A} \perp \mathcal{T}_n \mid \mathcal{S}_0$, $\mathcal{C} \perp \mathcal{T}_n \mid \mathcal{S}_0$ or $\mathcal{A} \perp \mathcal{T}_n \mid \mathcal{B} \vee \mathcal{S}_0$. There would be, in general, no implication between such a modified condition and the given one. It is nevertheless possible to use Theorem 2.2.10 in order to produce supplementary conditions ensuring the equivalence between the modified and the given condition. So, for instance:

$$(6.3.6) \quad (\mathcal{B} \vee \mathcal{S}_0) \perp \mathcal{T}_n \mid \mathcal{C} \vee \mathcal{T}_0 \quad \Leftrightarrow \quad \left\{ \begin{array}{l} \mathcal{B} \perp \mathcal{T}_n \mid \mathcal{C} \vee \mathcal{T}_0 \\ \mathcal{S}_0 \perp \mathcal{T}_n \mid \mathcal{B} \vee \mathcal{C} \vee \mathcal{T}_0 \end{array} \right\}$$

$$\Leftrightarrow \quad \left\{ \begin{array}{l} \mathcal{B} \perp \mathcal{T}_n \mid \mathcal{C} \vee \mathcal{S}_0 \\ \mathcal{S}_0 \perp \mathcal{T}_n \mid \mathcal{C} \vee \mathcal{T}_0 \end{array} \right\}$$

$$(6.3.7) \quad \mathcal{B} \perp (\mathcal{S}_0 \vee \mathcal{T}_n) \mid \mathcal{C} \vee \mathcal{T}_0 \quad \Leftrightarrow \quad \left\{ \begin{array}{l} \mathcal{B} \perp \mathcal{T}_n \mid \mathcal{C} \vee \mathcal{T}_0 \\ \mathcal{B} \perp \mathcal{S}_0 \mid \mathcal{C} \vee \mathcal{T}_n \end{array} \right\}$$

$$\Leftrightarrow \quad \left\{ \begin{array}{l} \mathcal{B} \perp \mathcal{T}_n \mid \mathcal{C} \vee \mathcal{S}_0 \\ \mathcal{B} \perp \mathcal{S}_0 \mid \mathcal{C} \vee \mathcal{T}_0 \end{array} \right\}.$$

We shall see that, in most cases, initial and sequential admissibility are equivalent only under a supplementary condition. This condition, to be called "transitivity", is analyzed in the next section, which also includes, *inter alia*, conditions of the same type as the supplementary conditions just mentioned (viz., conditions of the type $\mathcal{S}_0 \perp\!\!\!\perp \mathcal{T}_n \mid \mathcal{B} \vee \mathcal{T}_0$).

6.4 Transitivity

This section treats a topic in probability theory. Thus, $(M, \mathcal{M}, P)$ is an abstract probability space. In general, $(\mathcal{M}_n)$, $(\mathcal{N}_n)$, $(\mathcal{O}_n)$ etc. are sequences of sub-σ-fields of $\mathcal{M}$, and $\mathcal{M}_0$ is the trivial sub-σ-field of $\mathcal{M}$.

6.4.1 Basic Theory

A. In Sequences of σ-Fields

The concept of transitivity is at the core of the linkages between initial and sequential admissibility. This concept has been introduced in statistics by Bahadur (1954) and is used extensively, e.g., in Hall, Wijsman and Ghosh (1965). In addition to its usefulness in sequential analysis, this concept also has a long history in time series, particularly in the field of econometrics, where it is referred to as non-causality analysis (see, e.g., Granger (1969), Sims (1972) or Pierce and Haugh (1977)). In this section we generalize this concept slightly, in the following way:

6.4.1 Definition. $(\mathcal{N}_n)$ is $(\mathcal{M}_n)$-*transitive* for $(\mathcal{O}_n)$ if the sequence $(\mathcal{N}_n)$ is adapted to $(\mathcal{M}_n \vee \mathcal{O}_n)$ and

$$(6.4.1) \qquad \mathcal{N}_{n+1} \perp\!\!\!\perp \mathcal{O}_n \mid \mathcal{M}_n \vee \mathcal{N}_n, \quad \forall\, n \geq 0. \qquad \blacksquare$$

It should be mentioned that in the Definition 6.4.1, the condition $\mathcal{N}_n \subset \mathcal{M}_n \vee \mathcal{O}_n$ is not restrictive because, by Corollary 2.2.11, (6.4.1) is equivalent to $\mathcal{N}_{n+1} \perp\!\!\!\perp (\mathcal{N}_n \vee \mathcal{O}_n) \mid \mathcal{M}_n \vee \mathcal{N}_n$, this implies that $(\mathcal{N}_n)$ is $(\mathcal{M}_n)$-transitive for $(\mathcal{O}_n \vee \mathcal{N}_n)$.

The concept of transitivity will often be used in the following particular specification. $(\mathcal{O}_n)$ is a filtration representing the complete history of a stochastic process (for instance, $\mathcal{O}_n = \mathcal{S}_n$), $(\mathcal{N}_n)$ is also a filtration representing the complete history of a subprocess (for instance, $\mathcal{N}_n = \mathcal{T}_n$, an increasing sequence of uniform statistics) and $\mathcal{M}_n$ is either constant, in which case we write $\mathcal{M}_n = \mathcal{M}$ (for instance, $\mathcal{M}_n = \mathcal{A}$ or $\mathcal{S}_0$), or $\mathcal{M}_n = \mathcal{S}_{n-1}$ as in sequential analysis.

The next two results give equivalent forms of transitivity in particular cases and will be interpreted in the framework of non-causality. The first one is a transposition of Theorem 6.2.2 with $\mathcal{M}_1$ and $\mathcal{M}_2$ trivial and under the identification $(\mathcal{F}_n, \mathcal{J}_n, \mathcal{K}_n) \to (\mathcal{M}_n \vee \mathcal{O}_n, \mathcal{M}_n \vee \mathcal{N}_n, \mathcal{N}_n)$.

6.4.2 Proposition. If $(\mathcal{M}_n \vee \mathcal{N}_n)$ is $(\mathcal{N}_n)$-recursive and $(\mathcal{M}_n \vee \mathcal{O}_n)$ is a filtration, then $(\mathcal{N}_n)$ is $(\mathcal{M}_n)$-transitive for $(\mathcal{O}_n)$ if and only if

$$(6.4.2) \qquad \mathcal{N}_{n+1}^{\infty} \perp\!\!\!\perp \mathcal{O}_n \mid \mathcal{M}_n \vee \mathcal{N}_n, \quad \forall\, n \geq 0. \qquad\blacksquare$$

In Proposition 6.4.2, the condition of recursivity means indeed that $\mathcal{M}_{n+1} \subset \mathcal{M}_n \vee \mathcal{N}_n \vee \mathcal{N}_{n+1}$. Thus, if $\mathcal{M}_n$ is constant, this condition becomes trivial. However, if $(\mathcal{N}_n)$ is a filtration, namely, $\mathcal{N}_n \uparrow \mathcal{N}$, the condition of recursivity becomes $\mathcal{M}_{n+1} \subset \mathcal{M}_n \vee \mathcal{N}_{n+1}$, i.e., $(\mathcal{M}_n)$ is $(\mathcal{N}_n)$-recursive which implies, by induction, that $\mathcal{M}_n \subset \mathcal{M}_0 \vee \mathcal{N}_n$. Therefore, by Corollary 2.2.11, the $(\mathcal{M}_n)$-transitivity of $(\mathcal{N}_n)$ for $(\mathcal{O}_n)$, (i.e., $\mathcal{N}_{n+1} \perp\!\!\!\perp \mathcal{O}_n \mid \mathcal{M}_n \vee \mathcal{N}_n)$ implies that $\mathcal{N}_{n+1} \perp\!\!\!\perp (\mathcal{M}_n \vee \mathcal{O}_n) \mid \mathcal{M}_0 \vee \mathcal{M}_n \vee \mathcal{N}_n$ since $\mathcal{M}_0 \subset \mathcal{M}_n \vee \mathcal{O}_n$ (because, by hypothesis, $(\mathcal{M}_n \vee \mathcal{O}_n)$ is a filtration). Thus $\mathcal{N}_{n+1} \perp\!\!\!\perp (\mathcal{M}_n \vee \mathcal{O}_n) \mid \mathcal{M}_0 \vee \mathcal{N}_n$, i.e., $(\mathcal{N}_n)$ is $(\mathcal{M}_0)$-transitive for $(\mathcal{M}_n \vee \mathcal{O}_n)$.

The second result is a transposition of condition 1, 2 and 3 of Theorem 6.2.3 with the substitution $(\mathcal{F}_n, \mathcal{G}_n) \to (\mathcal{O}_n \vee \mathcal{M}, \mathcal{N}_n)$ and $\mathcal{M}_1$ and $\mathcal{M}_2$ trivial.

6.4.3 Proposition. If $(\mathcal{N}_n)$ is a filtration adapted to a filtration $(\mathcal{O}_n)$ with $\mathcal{N}_{\infty} = \bigvee_{n \geq 0} \mathcal{N}_n$, then the following conditions are equivalent:

(i) $(\mathcal{N}_n)$ is $(\mathcal{O}_0 \vee \mathcal{M})$-transitive for $(\mathcal{O}_n)$

(ii) $\mathcal{N}_{\infty} \perp\!\!\!\perp \mathcal{O}_n \mid \mathcal{M} \vee \mathcal{O}_{n-1} \vee \mathcal{N}_n \quad \forall\, n \geq 0$

(iii) $\mathcal{N}_{\infty} \perp\!\!\!\perp \mathcal{O}_n \mid \mathcal{M} \vee \mathcal{O}_0 \vee \mathcal{N}_n \quad \forall\, n \geq 0 \qquad\qquad\blacksquare$

Let us mention that condition (ii) or (iii) in Proposition 6.4.3 implies a weaker form of condition (i), namely, (i bis): $(\mathcal{N}_n)$ is $(\mathcal{O}_{n-1} \vee \mathcal{M})$-transitive for $(\mathcal{O}_n)$. This is an example of a property of transitivity given a variable conditioning σ-field, in fact being weaker than a property of transitivity given a constant conditioning σ-field. Similarly, one may also look for a condition that would make a $(\mathcal{O}_0 \vee \mathcal{M})$-transitivity equivalent to a $(\mathcal{M})$-transitivity (i.e., a condition that allows one to drop $\mathcal{O}_0$ in condition (iii) of Proposition 6.4.3). That condition is, as an application of Theorem 6.2.3, one of the following three equivalent conditions:

$$(6.4.3) \qquad \mathcal{N}_n \perp\!\!\!\perp \mathcal{O}_0 \mid \mathcal{M} \vee \mathcal{N}_0 \quad \forall\, n \geq 0$$

$$(6.4.4) \qquad \mathcal{N}_{n+1} \perp\!\!\!\perp \mathcal{O}_0 \mid \mathcal{M} \vee \mathcal{N}_n \quad \forall\, n \geq 0$$

$$(6.4.5) \qquad \mathcal{N}_\infty \perp\!\!\!\perp \mathcal{O}_0 \mid \mathcal{M} \vee \mathcal{N}_0.$$

The condition (6.4.3) was introduced in Section 6.3.2 — see (6.3.6) and (6.3.7) — and may be interpreted as a self-predictivity requirement at the initial conditions.

B. In Stochastic Processes

$(\mathcal{M}_n)$ is typically used in the sequel, as a sequence of conditioning σ-fields for a sequence of experiments $\mathcal{E}_{\mathcal{A} \vee \mathcal{S}_n}^{\mathcal{M}_n}$; consequently, $(\mathcal{M}_n)$ is adapted to $(\mathcal{A} \vee \mathcal{S}_n)$, and $(\mathcal{N}_n)$ and $(\mathcal{O}_n)$ are $\mathcal{E}_{\mathcal{A} \vee \mathcal{S}_n}^{\mathcal{M}_n}$-statistics. Thus, both sequences are adapted to $(\mathcal{M}_n \vee \mathcal{S}_n)$. When $\mathcal{M}_n = \mathcal{A}$ or $\mathcal{M}_n = \mathcal{A} \vee \mathcal{S}_0$ we handle properties of the sampling process up to null sets for the prior probability. More specifically, if Π is transformed into Π' such that $\Pi' \ll \Pi$ and $d\Pi'/d\Pi \in [A]^+$, then Corollary 2.2.15 ensures that a property of transitivity in a given experiment is preserved in the modified experiment. In most applications in the next section, $\mathcal{M}_n$ has the form $\mathcal{M}_n = \mathcal{A}$, $\mathcal{M}_n = \mathcal{A} \vee \mathcal{S}_0$, $\mathcal{M}_n = \mathcal{S}_0$, $\mathcal{M}_n = \mathcal{A} \vee \mathcal{S}_{n-1}$ or $\mathcal{M}_n = \mathcal{S}_{n-1}$.

The basic idea underlying the concept of transitivity may be viewed as sufficiency for prediction: The knowledge of $\mathcal{M}_n \vee \mathcal{N}_n$ is sufficient to predict $\mathcal{N}_{n+1}$ given the knowledge of $\mathcal{M}_n \vee \mathcal{O}_n$. This interpretation may explain why econometricians have instead used the expression "$\mathcal{O}_n$ does not cause $\mathcal{N}_n$ (given $\mathcal{M}_n$)" while in system theory the usage is "$\mathcal{N}_n$ causes $\mathcal{O}_n$ (given $\mathcal{M}_n$). A better expression would probably be "$\mathcal{N}_n$ is selfpredictive with respect to $\mathcal{O}_n$ (given $\mathcal{M}_n$)".

It may also be noted that Theorem 6.2.2 may be interpreted as a theorem about "generalized transitivity" while Proposition 6.4.2 shows that in a rather general case, the immediate self-predictivity characterizing the transitivity is actually equivalent to a property of self-predictivity over an infinite horizon.

A natural use for the above concepts may be found in the analysis of a stochastic vector-valued process $x_n = (y_n, z_n)$. The corresponding σ-fields will be denoted by script letters, e.g., $\mathcal{X}_n = \sigma(x_n)$ and $\mathcal{X}_n = \mathcal{Y}_n \vee \mathcal{Z}_n$. The role of $\mathcal{O}_n$ will be played by the σ-field generated by the observations of the x-process from time 0 to n; it will be denoted as $\mathcal{X}_0^n = \mathcal{Y}_0^n \vee \mathcal{Z}_0^n$. The role of $\mathcal{N}_n$ is played by $\mathcal{Z}_0^n$ and the sequence $\mathcal{M}_n$ is constant. Granger's (1969) concept of "y does not cause z (given $\mathcal{M}$)" may be written as:

$$(6.4.6) \qquad \mathcal{Z}_{n+1} \perp\!\!\!\perp \mathcal{Y}_0^n \mid \mathcal{Z}_0^n \vee \mathcal{M}, \quad \forall\, n \geq 0,$$

i.e., "$(\mathcal{Z}_0^n)$ is $\mathcal{M}$-transitive for $(\mathcal{X}_0^n)$" because (6.4.6) is equivalent to:

$$(6.4.7) \qquad \mathcal{Z}_0^{n+1} \perp\!\!\!\perp \mathcal{X}_0^n \mid \mathcal{Z}_0^n \vee \mathcal{M}, \quad \forall\, n \geq 0.$$

Since $\mathcal{M}_n$ is constant, the recursivity condition of Proposition 6.4.2 is trivial. Therefore, by this proposition, (6.4.7) is equivalent to:

$$(6.4.8) \qquad \mathcal{Z}_{n+1}^\infty \perp\!\!\!\perp \mathcal{Y}_0^n \mid \mathcal{Z}_0^n \vee \mathcal{M}, \quad \forall\, n \geq 0.$$

Sims's (1972) concept of "y does not cause z (given $\mathcal{M}$)" may be written as:

$$(6.4.9) \qquad \mathcal{Z}_{n+1}^\infty \perp\!\!\!\perp \mathcal{Y}_n \mid \mathcal{Z}_0^n \vee \mathcal{M}, \quad \forall\, n \geq 0$$

which is clearly implied by but not equivalent to Granger's concept. One can modify Sims's concept into:

$$(6.4.10) \qquad \mathcal{Z}_{n+1}^\infty \perp\!\!\!\perp \mathcal{Y}_n \mid \mathcal{Y}_0^{n-1} \vee \mathcal{Z}_0^n \vee \mathcal{M}, \quad \forall\, n \geq 0,$$

i.e., the prediction of y_n on the basis of $\mathcal{Y}_0^{n-1}$, the history of y, of $\mathcal{Z}_0^\infty$, the complete trajectory of z, and of $\mathcal{M}$, does not depend on the future values of the z-process, i.e., $\mathcal{Z}_{n+1}^\infty$. Proposition 6.4.3 shows that, because in (6.4.10) $\mathcal{Y}_n$ may be replaced by $\mathcal{Y}_0^n$, the modified Sims's concept (6.4.10) is equivalent to:

$$(6.4.11) \qquad \mathcal{Z}_{n+1} \perp\!\!\!\perp \mathcal{Y}_0^n \mid \mathcal{Y}_0 \vee \mathcal{Z}_0^n \vee \mathcal{M}, \quad \forall\, n \geq 0,$$

i.e., (after replacing in (6.4.11) $\mathcal{Z}_{n+1}$ by $\mathcal{Z}_0^{n+1}$) $(\mathcal{Z}_0^n)$ is $(\mathcal{M} \vee \mathcal{Y}_0)$-transitive for $(\mathcal{X}_0^n)$. Note that $\mathcal{M}$-transitivity is stronger than $(\mathcal{M} \vee \mathcal{Y}_0)$-transitivity; therefore, Granger's concept implies the modified Sims's concept but the modified Sims's concept becomes equivalent to Granger's concept under a supplementary condition given in (6.4.3), viz.:

$$(6.4.12) \qquad \mathcal{Z}_0^n \perp\!\!\!\perp \mathcal{Y}_0 \mid \mathcal{Z}_0 \vee \mathcal{M}, \quad \forall\, n \geq 0,$$

which is trivially satisfied when $\mathcal{Y}_0 \subset \mathcal{M}$.

Remark. It is interesting, in this context, to show how to modify the index set of the stochastic process from $I\!N$ to $\mathbb{Z}$. Formulae (6.4.6) to (6.4.12) are still well defined for $n \in \mathbb{Z}$ when "0" is replaced by "$-\infty$" because, in general, the sequence of σ-fields $\mathcal{Y}_m^n$ is monotone decreasing in m. So we have $\mathcal{Y}_{-\infty}^m = \bigvee_{-\infty < j \le n} \mathcal{Y}_j^n$ and $\mathcal{Y}_{-\infty} = \bigcap_n \mathcal{Y}_{-\infty}^n$. Proposition 6.4.2 is still valid and the conditions (6.4.6), (6.4.7) and (6.4.8) are still equivalent. However, Proposition 6.4.3 may no longer be used as such. The modified Sims's concept is now rewritten as:

(6.4.13) $\qquad \mathcal{Z}_{n+1}^\infty \perp\!\!\!\perp \mathcal{Y}_n \mid \mathcal{Y}_{-\infty}^{n-1} \vee \mathcal{Z}_{-\infty}^n \vee \mathcal{M}, \quad \forall\, n \in \mathbb{Z}.$

The argument $1 \Rightarrow 2$ in the proof of Theorem 6.2.3 may be repeated to prove that (6.4.13) is equivalent to:

(6.4.14) $\qquad \mathcal{Z}_{n+1}^\infty \perp\!\!\!\perp \mathcal{Y}_n \mid \bigcap_{1 \le p < \infty} (\mathcal{Y}_{-\infty}^{n-p} \vee \mathcal{Z}_{-\infty}^n \vee \mathcal{M}), \quad \forall\, n \in \mathbb{Z}.$

Note, however, that the proof of (6.4.14) requires a generalization to sequences of Theorem 2.2.12 (Theorem 7.2.7 in Chapter 7).

Therefore, if we want (6.4.13) to be equivalent to a condition analogous to (6.4.11), namely:

(6.4.15) $\qquad \mathcal{Z}_{n+1} \perp\!\!\!\perp \mathcal{Y}_{-\infty}^n \mid \mathcal{Y}_{-\infty} \vee \mathcal{Z}_{-\infty}^n \vee \mathcal{M}, \quad \forall\, n \in \mathbb{Z},$

we need the supplementary condition:

(6.4.16) $\displaystyle\bigcap_{p \ge 1} (\mathcal{Y}_{-\infty}^{n-p} \vee \mathcal{Z}_{-\infty}^n \vee \mathcal{M}) = \mathcal{Y}_{-\infty} \vee \mathcal{Z}_{-\infty}^n \vee \mathcal{M}, \quad \forall\, n \in \mathbb{Z}.$

This analysis leaves open the question of the actual importance of indexing the x-process in $\mathbb{Z}$ rather than in $I\!N$, and of whether the ad hoc assumption (6.4.16) on the left-tail is operational. More discussion of this issue may be found in Hosoya (1977), Chamberlain (1982), and Florens and Mouchart (1982). We now illustrate the role of the conditioning σ-field $\mathcal{M}$ in a property of transitivity. Indeed, the next example exhibits a situation where $\mathcal{M}$-transitivity is verified for some choices of $\mathcal{M}$ only.

Example. Let $\mathcal{S}_n$ be generated by $(x_0, x_1, \ldots, x_n)$ where $x_n = (y_n, z_n)'$. Suppose that

$$(x_{n+1} \mid a, x_0^n) \sim N_2[(az_n, az_n)', I] \quad \text{and} \quad (a \mid x_0) \sim N(0, 1).$$

Consider $\mathcal{M} = \sigma(a, x_0) = \mathcal{A} \vee \mathcal{X}_0$; $(\mathcal{Z}_0^n)$ is clearly $(\mathcal{A} \vee \mathcal{X}_0)$-transitive for $(\mathcal{Y}_0^n)$. Consider now $\mathcal{M} = \sigma(x_0) = \mathcal{X}_0$; $(\mathcal{Z}_0^n)$ is not $\mathcal{X}_0$-transitive for $(\mathcal{Y}_0^n)$. Indeed, $\mathcal{X}_0^n(z_{n+1}) = z_n \cdot \mathcal{X}_0^n a$ and $\mathcal{X}_0^n a$ does involve $\mathcal{Y}_0^n$; for instance, $\mathcal{X}_0^1 a = (1 + 2z_0^2)^{-1} z_0 (y_1 + z_1)$.   ∎

Let us now consider $x_n = (y_n, z_n, w_n)$ and extend (6.4.6) and (6.4.8) as follows:

$$\mathcal{Z}_{n+1} \perp\!\!\!\perp \mathcal{Y}_0^n \mid \mathcal{Z}_0^n \vee \mathcal{W}_0^n \vee \mathcal{M}$$

$$\mathcal{Z}_{n+1}^\infty \perp\!\!\!\perp \mathcal{Y}_0^n \mid \mathcal{Z}_0^n \vee \mathcal{W}_0^n \vee \mathcal{M}.$$

From Proposition 6.4.2, these two transitivity properties would be equivalent under the recursivity condition

$$\mathcal{W}_{n+1} \subset \mathcal{W}_0^n \vee \mathcal{Z}_0^{n+1} \vee \mathcal{M},$$

and under this condition the comment following Proposition 6.4.2 shows that the two transition properties are each equivalent to

$$\mathcal{Z}_{n+1} \perp\!\!\!\perp (\mathcal{Y}_0^n \vee \mathcal{W}_0^n) \mid \mathcal{Z}_0^n \vee \mathcal{W}_0 \vee \mathcal{M}$$

$$\mathcal{Z}_0^\infty \perp\!\!\!\perp (\mathcal{Y}_0^n \vee \mathcal{W}_0^n) \mid \mathcal{Z}_0^n \vee \mathcal{W}_0 \vee \mathcal{M}.$$

6.4.2 Markovian Property and Transitivity

A. In Sequences of σ-Fields

Let us now return to the analysis of transitivity in the general framework. If we consider a sequence $(\mathcal{N}_n)$ of σ-fields such that $(\mathcal{N}_0^n)$ is $(\mathcal{M}_n)$-transitive for $(\mathcal{O}_n)$, i.e.,

(6.4.17) $$\mathcal{N}_0^{n+1} \perp\!\!\!\perp \mathcal{O}_n \mid \mathcal{M}_n \vee \mathcal{N}_0^n,$$

or, equivalently,

(6.4.18) $$\mathcal{N}_{n+1} \perp\!\!\!\perp \mathcal{O}_n \mid \mathcal{M}_n \vee \mathcal{N}_0^n,$$

the interpretation of this last formula as a sufficiency for prediction of $\mathcal{N}_{n+1}$ shows that as n increases, we must retain more and more information, i.e., $\mathcal{N}_0^n$. In view of this fact, it is obviously interesting to establish when it would be possible to retain merely the last k preceding elements of the sequence, with fixed k for each n. This property is clearly formalized as

(6.4.19) $$\mathcal{N}_{n+1} \perp\!\!\!\perp \mathcal{O}_n \mid \mathcal{M}_n \vee \mathcal{N}_{n-k+1}^n.$$

This property actually means that $(\mathcal{N}^n_{n-k+1})$ is $(\mathcal{M}_n)$-transitive for $(\mathcal{O}_n)$. The rest of this section will be devoted to the comparison of the transitivity of $(\mathcal{N}^n_0)$ and of $(\mathcal{N}^n_{n-k+1})$. These two concepts are naturally linked through Markovian conditions on the sequence $(\mathcal{N}_n)$, which we now define:

6.4.4 Definition. $(\mathcal{N}_n)$ is $(\mathcal{M}_n)$-k-*Markovian* if

$$(6.4.20) \qquad \mathcal{N}_{n+1} \perp\!\!\!\perp \mathcal{N}^n_0 \mid \mathcal{M}_n \vee \mathcal{N}^n_{n-k+1}. \qquad \blacksquare$$

Note that if $\mathcal{M}_n = \mathcal{M}$ for all n, an application of Theorem 6.2.2 shows that (6.4.20) is actually equivalent to

$$(6.4.21) \qquad \mathcal{N}^\infty_{n+1} \perp\!\!\!\perp \mathcal{N}^n_0 \mid \mathcal{M} \vee \mathcal{N}^n_{n-k+1}.$$

A straightforward application of Theorem 2.2.10 gives the following proposition:

6.4.5 Proposition. $(\mathcal{N}^n_0)$ is $(\mathcal{M}_n)$-*transitive* for $(\mathcal{O}_n)$ and $(\mathcal{N}_n)$ is $(\mathcal{M}_n)$-k-*Markovian* if and only if $(\mathcal{N}^n_{n-k+1})$ is $(\mathcal{M}_n)$-transitive for $(\mathcal{O}_n)$ and $(\mathcal{N}_n)$ is $(\mathcal{M}_n \vee \mathcal{O}_n)$-$k$-Markovian. $\qquad \blacksquare$

This proposition asserts that the two conditions (6.4.18) and (6.4.20) are equivalent to (6.4.19), along with

$$(6.4.22) \qquad \mathcal{N}_{n+1} \perp\!\!\!\perp \mathcal{N}^n_0 \mid \mathcal{M}_n \vee \mathcal{O}_n \vee \mathcal{N}^n_{n-k+1}.$$

An immediate corollary would be that under a transitivity condition, one Markovian property implies the other, i.e., under (6.4.19), (6.4.22) implies (6.4.20) or under (6.4.18), (6.4.20) implies (6.4.22). Similarly, under a Markovian property, one transitivity condition implies the other, i.e., under (6.4.20), (6.4.18) implies (6.4.19) or under (6.4.22), (6.4.19) implies (6.4.18).

Let us recall that, in general, a subprocess of a Markovian process is not necessarily Markovian. This problem may be analyzed by assuming that in Proposition (6.4.5), $(\mathcal{O}_n)$ is also a filtration. We obtain this by associating the filtration $(\mathcal{O}^n_0)$ to the sequence $(\mathcal{O}_n)$ as follows. Clearly, if $(\mathcal{N}_n \vee \mathcal{O}_n)$ is $(\mathcal{M}_n)$-k-Markovian, i.e.,

$$(6.4.23) \qquad (\mathcal{N}_{n+1} \vee \mathcal{O}_{n+1}) \perp\!\!\!\perp (\mathcal{N}^n_0 \vee \mathcal{O}^n_0) \mid \mathcal{M}_n \vee (\mathcal{N}^n_{n-k+1} \vee \mathcal{O}^n_{n-k+1}),$$

then $(\mathcal{N}_n)$ is $(\mathcal{M}_n \vee \mathcal{O}_0^n)$-$k$-Markovian, i.e.,

$$(6.4.24) \qquad \mathcal{N}_{n+1} \perp\!\!\!\perp \mathcal{N}_0^n \mid \mathcal{M}_n \vee \mathcal{O}_0^n \vee \mathcal{N}_{n-k+1}^n.$$

Proposition 6.4.5 says that (6.4.23) also implies that $(\mathcal{N}_n)$ is $(\mathcal{M}_n)$-k-Markovian (i.e., $(\mathcal{O}_0^n)$ may be dropped from (6.4.24)) if $(\mathcal{N}_{n-k+1}^n)$ is $(\mathcal{M}_n)$-transitive for $(\mathcal{O}_0^n)$. Such a transitivity condition may be somewhat unsuitable in this context, because $(\mathcal{N}_{n-k+1}^n)$ is not a filtration but, as it is adapted to the filtration $(\mathcal{N}_0^n)$, a more satisfactory condition would be that $(\mathcal{N}_0^n)$ is $(\mathcal{M}_n)$-transitive for $(\mathcal{O}_0^n)$. Using the concepts introduced in Chapter 5, this assumption will be sufficient if we add a measurable separability condition. More precisely, as a straightforward corollary of Theorem 5.2.10, we have the following proposition:

6.4.6 Proposition. Under $\mathcal{O}_n \parallel \mathcal{N}_0^n \mid \mathcal{M}_n \vee \mathcal{N}_{n-k+1}^n$, if $(\mathcal{N}_0^n)$ is $(\mathcal{M}_n)$-transitive for $(\mathcal{O}_n)$ and $(\mathcal{N}_n)$ is $(\mathcal{M}_n \vee \mathcal{O}_n)$-$k$-Markovian, then $(\mathcal{N}_n)$ is $(\mathcal{M}_n)$-k-Markovian and $(\mathcal{N}_{n-k+1}^n)$ is $(\mathcal{M}_n)$-transitive for $(\mathcal{O}_n)$. $\blacksquare$

We now face the following questions: Are transitivity conditions really necessary to ensure that a subprocess of a Markovian process is Markovian? And further, is it true that the two transitivity conditions (6.4.18) and (6.4.19) occur only in marginally and conditionally Markovian processes, i.e., processes satisfying both (6.4.20) and (6.4.22)? Using strong identifiability conditions, the answers to these questions are affirmative and, as corollaries of Theorem 5.4.13, follow from the next two propositions:

6.4.7 Proposition. If $\mathcal{O}_n \ll \mathcal{N}_0^n \mid \mathcal{M}_n \vee \mathcal{N}_{n-k+1}^n$, then if $(\mathcal{N}_n)$ is both $(\mathcal{M}_n)$-k-Markovian and $(\mathcal{M}_n \vee \mathcal{O}_n)$-$k$-Markovian, then this implies that both $(\mathcal{N}_0^n)$ and $(\mathcal{N}_{n-k+1}^n)$ are $(\mathcal{M}_n)$-transitive for $(\mathcal{O}_n)$. $\blacksquare$

6.4.8 Proposition. If $\mathcal{N}_0^n \ll \mathcal{O}_n \mid \mathcal{M}_n \vee \mathcal{N}_{n-k+1}^n$, then if both $(\mathcal{N}_0^n)$ and $(\mathcal{N}_{n-k+1}^n)$ are $(\mathcal{M}_n)$-transitive for $(\mathcal{O}_n)$, then this implies that $(\mathcal{N}_n)$ is both $(\mathcal{M}_n)$- and $(\mathcal{M}_n \vee \mathcal{O}_n)$-$k$-Markovian. $\blacksquare$

B. In Stochastic Processes

Let us return to the analysis of a stochastic vector-valued process $x_n = (y_n, z_n)$ as in Section 6.4.1.B and assume that $(\mathcal{X}_n)$ is an $(\mathcal{M}_n)$-k-Markovian process, i.e.,

$$(6.4.25) \qquad \mathcal{X}_{n+1} \perp\!\!\!\perp \mathcal{X}_0^n \mid \mathcal{M}_n \vee \mathcal{X}_{n-k+1}^n, \quad \forall \, n \geq 0.$$

In this subsection, the sequence of σ-fields $(\mathcal{M}_n)$ may be used in rather different contexts. Firstly, when $\mathcal{M}_n$ is constant (i.e., $\mathcal{M}_n = \mathcal{M}$) it will typically either be trivial, as in a purely predictive set-up, or represent parameters and/or initial conditions. When $(\mathcal{M}_n)$ is not constant, it may be a filtration, typically in the form $\mathcal{M}_n = \mathcal{M} \vee \mathcal{W}_0^n$ where w_n is a stochastic process defined on the same basic space as x_n, but $(\mathcal{M}_n)$ may fail to be a filtration, such as, for example, when $(\mathcal{M}_n)$ is generated by a sequence of statistics (in this case the sequence $(\mathcal{M}_n)$ is typically adapted to a filtration without being itself a filtration). This flexibility in the interpretation motivates leaving an "abstract" sequence of σ-fields in the conditioning of the properties in the stochastic process context.

Now, (6.4.25) clearly implies that:

$$(6.4.26) \qquad \mathcal{Z}_{n+1} \perp\!\!\!\perp \mathcal{Z}_0^n \mid \mathcal{M}_n \vee \mathcal{Y}_0^n \vee \mathcal{Z}_{n-k+1}^n, \quad \forall \, n \geq 0,$$

i.e., $(\mathcal{Z}_n)$ is $(\mathcal{M}_n \vee \mathcal{Y}_0^n)$-$k$-Markovian. By Proposition 6.4.5, if $(\mathcal{Z}_{n-k+1}^n)$ is $(\mathcal{M}_n)$-transitive for $(\mathcal{Y}_0^n)$, i.e.,

$$(6.4.27) \qquad \mathcal{Z}_{n+1} \perp\!\!\!\perp \mathcal{Y}_0^n \mid \mathcal{M}_n \vee \mathcal{Z}_{n-k+1}^n, \quad \forall \, n \geq 0,$$

then $(\mathcal{Z}_n)$ is $(\mathcal{M}_n)$-k-Markovian, i.e.,

$$(6.4.28) \qquad \mathcal{Z}_{n+1} \perp\!\!\!\perp \mathcal{Z}_0^n \mid \mathcal{M}_n \vee \mathcal{Z}_{n-k+1}^n, \quad \forall \, n \geq 0,$$

and, moreover, $(\mathcal{Z}_0^n)$ is $(\mathcal{M}_n)$-transitive for $(\mathcal{Y}_0^n)$, i.e.,

$$(6.4.29) \qquad \mathcal{Z}_{n+1} \perp\!\!\!\perp \mathcal{Y}_0^n \mid \mathcal{M}_n \vee \mathcal{Z}_0^n, \quad \forall \, n \geq 0.$$

The transitivity condition (6.4.27) is rather unsatisfactory in the sense that it involves conditioning on $\mathcal{Z}_{n-k+1}^n$ and therefore, in a sequential specification of the process, requires integration on $\mathcal{Z}_0^{n-k}$. The transitivity condition

(6.4.29) is a more satisfactory assumption. Proposition 6.4.6 shows that under the measurable separability condition

$$(6.4.30) \qquad \mathcal{Y}_0^n \parallel \mathcal{Z}_0^n \mid \mathcal{M}_n \vee \mathcal{Z}_{n-k+1}^n$$

if $(\mathcal{Z}_0^n)$ is $(\mathcal{M}_n)$-transitive for $(\mathcal{Y}_0^n)$ — i.e., (6.4.29) — and if $(\mathcal{Z}_n)$ is $(\mathcal{M}_n \vee \mathcal{Y}_0^n)$-$k$-Markovian — i.e., (6.4.26) — then $(\mathcal{Z}_n)$ is $(\mathcal{M}_n)$-k-Markovian — i.e., (6.4.28) — and, morever $(\mathcal{Z}_{n-k+1}^n)$ is also $(\mathcal{M}_n)$-transitive for $(\mathcal{Y}_0^n)$ — i.e., (6.4.27). In order to verify condition (6.4.30) a very useful tool is provided by Corollary 5.2.11. Indeed, if there exists a probability Π' equivalent to Π such that $\mathcal{Y}_0^n \perp\!\!\!\perp \mathcal{Z}_0^n \mid \mathcal{M}_n; \Pi' \ \forall \, n \geq 0$, then (6.4.30) is readily verified.

Proposition 6.4.7 shows that under the strong identification condition

$$(6.4.31) \qquad \mathcal{Y}_0^n \ll \mathcal{Z}_0^n \mid \mathcal{M}_n \vee \mathcal{Z}_{n-k+1}^n$$

if $(\mathcal{Z}_n)$ is both $(\mathcal{M}_n)$- and $(\mathcal{M}_n \vee \mathcal{Y}_0^n)$-$k$-Markovian — i.e. conditions (6.4.28) and (6.4.26) — then $(\mathcal{Z}_0^n)$ and $(\mathcal{Z}_{n-k+1}^n)$ are both $(\mathcal{M}_n)$-transitive for $(\mathcal{Y}_0^n)$. Proposition 6.4.8 shows that under the strong identification condition

$$(6.4.32) \qquad \mathcal{Z}_0^n \ll \mathcal{Y}_0^n \mid \mathcal{M}_n \vee \mathcal{Z}_{n-k+1}^n$$

the two transitivity conditions — (6.4.27) and (6.4.29) — imply the two Markovian properties — (6.4.26 and 6.4.28).

Assumption (6.4.25) also implies that:

$$(6.4.26') \qquad \mathcal{Z}_{n+1} \perp\!\!\!\perp \mathcal{Z}_0^n \mid \mathcal{M}_n \vee \mathcal{Y}_{n-k+1}^n \vee \mathcal{Z}_{n-k+1}^n \quad \forall \, n \geq 0,$$

i.e., $(\mathcal{Z}_n)$ is $(\mathcal{M}_n \vee \mathcal{Y}_{n-k+1}^n)$-$k$-Markovian. Moreover, by Theorem 2.2.10 and Corollary 2.2.11, under Assumption (6.4.25), the two transitivity conditions — i.e., (6.4.27) and (6.4.29) — are respectively equivalent to

$$(6.4.27') \qquad \mathcal{Z}_{n+1} \perp\!\!\!\perp \mathcal{Y}_{n-k+1}^n \mid \mathcal{M}_n \vee \mathcal{Z}_{n-k+1}^n \qquad \forall \, n \geq 0$$

$$(6.4.29') \qquad \mathcal{Z}_{n+1} \perp\!\!\!\perp \mathcal{Y}_{n-k+1}^n \mid \mathcal{M}_n \vee \mathcal{Z}_0^n \qquad \forall \, n \geq 0.$$

Therefore, substituting $\mathcal{Y}_{n-k+1}^n$ to $\mathcal{Y}_0^n$, Conditions (6.4.30) and (6.4.31) may be relaxed into respectively:

$$(6.4.30') \qquad \mathcal{Y}_{n-k+1}^n \parallel \mathcal{Z}_0^n \mid \mathcal{M}_n \vee \mathcal{Z}_{n-k+1}^n$$

$$(6.4.31') \qquad \mathcal{Y}_{n-k+1}^n \ll \mathcal{Z}_0^n \mid \mathcal{M}_n \vee \mathcal{Z}_{n-k+1}^n.$$

Therefore, if $(\mathcal{X}_n)$ is $(\mathcal{M}_n)$-k-Markovian, Proposition 6.4.8 is irrelevant. Proposition 6.4.5 shows that if $(\mathcal{Z}_{n-k+1}^n)$ is $(\mathcal{M}_n)$-transitive for $(\mathcal{Y}_0^n)$, then $(\mathcal{Z}_n)$ is $(\mathcal{M}_n)$-k-Markovian and $(\mathcal{Z}_0^n)$ is $(\mathcal{M}_n)$-transitive for $(\mathcal{Y}_0^n)$. Proposition 6.4.6 shows that, under Condition (6.4.30'), $(\mathcal{Z}_0^n)$ is $(\mathcal{M}_n)$-transitive for $(\mathcal{Y}_0^n)$ implies that $(\mathcal{Z}_n)$ is $(\mathcal{M}_n)$-k-Markovian and $(\mathcal{Z}_{n-k+1}^n)$ is $(\mathcal{M}_n)$-transitive for $(\mathcal{Y}_0^n)$. Proposition 6.4.7 shows that, under Condition (6.4.31'), $(\mathcal{Z}_n)$ is $(\mathcal{M}_n)$-k-Markovian implies that both $(\mathcal{Z}_0^n)$ and $(\mathcal{Z}_{n-k+1}^n)$ are $(\mathcal{M}_n)$-transitive for $(\mathcal{Y}_0^n)$.

Note, however, that the strong identification Property (6.4.31') is not a necessary condition. There are in fact situations in which $(\mathcal{Z}_n)$ is both $(\mathcal{M}_n)$- and $(\mathcal{M}_n \vee \mathcal{Y}_0^n)$-$k$-Markovian but where it does not verify transitivity properties. Therefore, (6.4.31') is not valid in such situations. Indeed, (6.4.25) implies that

$$(6.4.33) \qquad \mathcal{Z}_{n+1} \perp\!\!\!\perp (\mathcal{Y}_0^n \vee \mathcal{Z}_0^n) \mid \mathcal{M}_n \vee \mathcal{Y}_{n-k+1}^n \vee \mathcal{Z}_{n-k+1}^n.$$

Therefore, under

$$(6.4.34) \qquad \mathcal{Y}_{n-k+1}^n \perp\!\!\!\perp (\mathcal{Y}_0^{n-k} \vee \mathcal{Z}_0^{n-k}) \mid \mathcal{M}_n \vee \mathcal{Z}_{n-k+1}^n$$

we have, using Theorem 2.2.10,

$$(6.4.35) \qquad (\mathcal{Z}_{n+1} \vee \mathcal{Y}_{n-k+1}^n) \perp\!\!\!\perp (\mathcal{Y}_0^{n-k} \vee \mathcal{Z}_0^{n-k}) \mid \mathcal{M}_n \vee \mathcal{Z}_{n-k+1}^n.$$

Therefore, by Corollary 2.2.11, we also have

$$(6.4.36) \qquad (\mathcal{Z}_{n+1} \vee \mathcal{Y}_{n-k+1}^n) \perp\!\!\!\perp (\mathcal{Y}_0^{n-k} \vee \mathcal{Z}_0^n) \mid \mathcal{M}_n \vee \mathcal{Z}_{n-k+1}^n.$$

This implies that $(\mathcal{Z}_n)$ is both $(\mathcal{M}_n)$- and $(\mathcal{M}_n \vee \mathcal{Y}_0^n)$-$k$-Markovian. However, (6.4.31') is violated since it follows, from (6.4.34), that for $m \in [\mathcal{Y}_n]_\infty$, $m - (\mathcal{M}_n \vee \mathcal{Z}_{n-k+1}^n) m \in [\mathcal{Y}_{n-k+1}^n \vee \mathcal{M}_n \vee \mathcal{Z}_{n-k+1}^n]_\infty$ and

$$(\mathcal{M}_n \vee \mathcal{Z}_0^n)\{m - (\mathcal{M}_n \vee \mathcal{Z}_{n-k+1}^n)m\} = (\mathcal{M}_n \vee \mathcal{Z}_0^n)m - (\mathcal{M}_n \vee \mathcal{Z}_{n-k+1}^n)m = 0,$$

since (6.4.34) implies that $\mathcal{Y}_n \perp\!\!\!\perp \mathcal{Z}_0^n \mid \mathcal{M}_n \vee \mathcal{Z}_{n-k+1}^n$.

The same reasoning shows that, even if $(\mathcal{X}_n)$ is $(\mathcal{M}_n)$-k-Markovian, which implies that $(\mathcal{Z}_n)$ is $(\mathcal{M}_n \vee \mathcal{Y}_0^n)$-$k$-Markovian, it may well be true that $(\mathcal{Z}_n)$ is $(\mathcal{M}_n)$-ℓ-Markovian with $\ell > k$. Indeed, by Corollary 2.2.11, (6.4.25) implies that

$$(6.4.37) \quad \mathcal{Z}_{n+1} \perp\!\!\!\perp (\mathcal{Y}_0^{n-k} \vee \mathcal{Z}_0^{n-\ell}) \mid \mathcal{M}_n \vee \mathcal{Y}_{n-k+1}^n \vee \mathcal{Z}_{n-\ell+1}^n, \quad \forall\, \ell \geq k.$$

Therefore if, for $\ell > k$,

$$(6.4.38) \qquad \mathcal{Y}_{n-k+1}^n \perp\!\!\!\perp (\mathcal{Y}_0^{n-k} \vee \mathcal{Z}_0^{n-\ell}) \mid \mathcal{M}_n \vee \mathcal{Z}_{n-\ell+1}^n,$$

we have, by Theorem 2.2.10,

$$(6.4.39) \quad (\mathcal{Z}_{n+1} \vee \mathcal{Y}_{n-k+1}^n) \perp\!\!\!\perp (\mathcal{Y}_0^{n-k} \vee \mathcal{Z}_0^{n-\ell}) \mid \mathcal{M}_n \vee \mathcal{Z}_{n-\ell+1}^n.$$

This shows that with $\ell > k$, $(\mathcal{Z}_n)$ is merely $(\mathcal{M}_n)$-ℓ-Markovian. The next four examples illustrate Propositions 6.4.5 to 6.4.8 as well as these last remarks, and provide greater insight into the role of the technical assumptions.

Example 1. Let us consider the two-dimensional process $x_n = (y_n, z_n)'$ defined by

$$(x_{n+1} \mid x_0^n) \sim N_2 \left[\begin{pmatrix} a_{11}y_n + a_{12}z_n \\ a_{21}y_n + a_{22}z_n \end{pmatrix}; \begin{pmatrix} \sigma_{11} & \sigma_{12} \\ \sigma_{12} & \sigma_{22} \end{pmatrix} \right].$$

Clearly, $(\mathcal{X}_n)$ is 1-Markovian. Now

$$(z_{n+1} \mid y_0^n, z_0^n) \sim N[a_{21}y_n + a_{22}z_n; \sigma_{22}]$$

and

$$(y_{n+1} \mid y_0^n, z_0^{n+1}) \sim$$
$$N\left[(a_{11} - \frac{\sigma_{12}}{\sigma_{22}}a_{21})y_n + (a_{12} - \frac{\sigma_{12}}{\sigma_{22}}a_{22})z_n + \frac{\sigma_{12}}{\sigma_{22}}z_{n+1}; \sigma_{11} - \frac{\sigma_{12}^2}{\sigma_{22}}\right].$$

If $a_{21} = 0$, then $\mathcal{Z}_{n+1} \perp\!\!\!\perp (\mathcal{Y}_0^n \vee \mathcal{Z}_0^n) \mid \mathcal{Z}_n$ and therefore $\mathcal{Z}_{n+1} \perp\!\!\!\perp \mathcal{Y}_0^n \mid \mathcal{Z}_0^n$, i.e., $(\mathcal{Z}_n)$ is transitive for $(\mathcal{Y}_0^n)$ and $\mathcal{Z}_{n+1} \perp\!\!\!\perp \mathcal{Z}_0^n \mid \mathcal{Z}_n$, i.e., $(\mathcal{Z}_n)$ is 1-Markovian. If $a_{21} \neq 0$ and $\sigma_{12} = (a_{11}/a_{21})\sigma_{22}$, then

$$\mathcal{Y}_{n+1} \perp\!\!\!\perp (\mathcal{Y}_0^n \vee \mathcal{Z}_0^n) \mid \mathcal{Z}_{n+1}, \mathcal{Z}_n$$

and, therefore,

$$\mathcal{Y}_{n+1} \perp\!\!\!\perp \mathcal{Y}_0^n \mid \mathcal{Z}_0^{n+1}.$$

Hence

$$(y_n \mid z_0^n) \sim N\left[(a_{12} - \frac{a_{11}}{a_{21}}a_{22})z_{n-1} + \frac{a_{11}}{a_{21}}z_n; \sigma_{11} - \frac{a_{11}^2}{a_{21}^2}\sigma_{22}\right]$$

and, therefore

$$(z_{n+1} \mid z_0^n) \sim N\left[a_{22}z_n + a_{21}E(y_n \mid z_0^n); \sigma_{22} + a_{21}^2 V(y_n \mid z_0^n)\right],$$

i.e.,

$$(z_{n+1} \mid z_0^n) \sim$$
$$N\left[(a_{11} + a_{22})z_n + (a_{21}a_{12} - a_{11}a_{22})z_{n-1}; (1 - a_{11}^2)\sigma_{22} + a_{21}^2\sigma_{11}\right].$$

This shows that in this case (Z_n) is 2-Markovian if $a_{11}a_{22} - a_{21}a_{12} \neq 0$. Conversely, if $a_{11}a_{22} - a_{21}a_{12} = 0$, we obtain $\mathcal{Y}_{n+1} \perp\!\!\!\perp (\mathcal{Y}_0^n \vee \mathcal{Z}_0^n) \mid \mathcal{Z}_{n+1}$ — this is Condition (6.4.34) — and

$$(z_{n+1} \mid z_0^n) \sim N\left[(a_{11} + a_{22})z_n; (1 - a_{11}^2)\sigma_{22} + a_{21}^2\sigma_{11}\right],$$

which shows that (Z_n) is 1-Markovian. Note that (6.4.31) is not satisfied since $\mathcal{Z}_0^n \left(y_n - (a_{11}/a_{21})z_n\right) = 0$ and, clearly, $(\mathcal{Z}_0^n)$ is not transitive for $(\mathcal{X}_0^n)$, i.e., $\mathcal{Z}_{n+1} \perp\!\!\!\perp \mathcal{Y}_0^n \mid \mathcal{Z}_0^n$ is not satisfied since

$$(z_{n+1} \mid x_0^n) \sim N(a_{21}y_n + a_{22}z_n; \sigma_{22}) \text{ and } a_{21} \neq 0. \qquad \blacksquare$$

Example 2. Consider the two-dimensional process $x_n = (y_n, z_n)'$ defined by

$$(x_{n+1} \mid x_0^n) \sim N_2\left[\begin{pmatrix} y_n \\ y_n + az_n \end{pmatrix}; \begin{pmatrix} 0 & 0 \\ 0 & \sigma^2 \end{pmatrix}\right],$$

with the initial condition

$$x_0 \sim N_2\left[\begin{pmatrix} 0 \\ 0 \end{pmatrix}; \begin{pmatrix} \sigma^2 & \sigma^2 \\ \sigma^2 & \sigma^2 \end{pmatrix}\right].$$

Clearly, $(\mathcal{X}_n)$ is 1-Markovian and $\mathcal{Z}_{n+1} \perp\!\!\!\perp \mathcal{X}_0^n \mid \mathcal{Y}_n \vee \mathcal{Z}_n$. this shows that $(\mathcal{Z}_n)$ is $(\mathcal{Y}_n)$-1-Markovian and that

$$(z_{n+1} \mid y_0^n, z_n) \sim N(y_n + az_n, \sigma^2).$$

Thus $(\mathcal{Z}_n)$ is not transitive for $(\mathcal{Y}_0^n)$. However, due to the initial conditions, it is also true that $\mathcal{Z}_{n+1} \perp\!\!\!\perp \mathcal{X}_0^n \mid \mathcal{Z}_0 \vee \mathcal{Z}_n$ and that

$$(z_{n+1} \mid z_0^n) \sim N(z_0 + az_n, \sigma^2).$$

Hence $(\mathcal{Z}_n)$ is not 1-Markovian but $(\mathcal{Z}_0^n)$ is transitive for $(\mathcal{Y}_0^n)$. Therefore the measurable separability condition (6.4.30), i.e, $\mathcal{Y}_0^n \parallel \mathcal{Z}_0^n \mid \mathcal{Z}_n$, is not satisfied; this is evident since $\mathcal{Z}_0 \vee \mathcal{Z}_n \subset \overline{(\mathcal{Y}_0^n \vee \mathcal{Z}_n)} \cap \mathcal{Z}_0^n$ and $\mathcal{Z}_0$ is not trivial. ∎

In this example, the violation of the measurable separability assumption is due to the degeneracy of distributions. Nonetheless, in most standard situations, as shown by Corollary 5.2.11, this assumption is very mild and readily verified.

Example 3. Consider the two-dimensional process $x_n = (y_n, z_n)'$ defined by

$$(x_{n+1} \mid x_0^n) \sim N_2 \left[\begin{pmatrix} by_n \\ y_n - by_{n-1} + az_n \end{pmatrix} ; \begin{pmatrix} \tau^2 & 0 \\ 0 & \sigma^2 \end{pmatrix} \right].$$

Clearly, $\mathcal{Z}_{n+1} \perp\!\!\!\perp \mathcal{X}_0^n \mid \mathcal{Y}_n, \mathcal{Y}_{n-1}, \mathcal{Z}_n$; furthermore,

$$\sigma(y_n - by_{n-1}) \perp\!\!\!\perp \mathcal{X}_0^{n-1} \quad \text{and} \quad \sigma(y_n - by_{n-1}) \perp\!\!\!\perp \mathcal{Z}_n \mid \mathcal{X}_0^{n-1}.$$

Therefore, $\sigma(y_n - by_{n-1}) \perp\!\!\!\perp (\mathcal{Z}_n \vee \mathcal{X}_0^{n-1})$ and $(z_{n+1} \mid z_0^n) \sim N_1(az_n, \sigma^2 + \tau^2)$ by Theorem 2.2.10. This shows that $(\mathcal{Z}_n)$ is both 1-Markovian and $(\mathcal{Y}_0^n)$-1-Markovian. However, the two transitivity conditions are not satisfied. Therefore the strong identification property (6.4.31), i.e., $\mathcal{Y}_0^n \ll \mathcal{Z}_0^n \mid \mathcal{Z}_n$, is not satisfied. Indeed, whereas $y_n - by_{n-1} \in [\mathcal{Y}_0^n]$ and is not trivial, $\mathcal{Z}_0^n(y_n - by_{n-1}) = 0$. ∎

Example 4. Consider a two-dimensional process $x_n = (y_n, z_n)'$ such that, $\forall n$, $\mathcal{Z}_0^n \perp\!\!\!\perp \mathcal{Y}_0^n$ and $(\mathcal{Z}_n)$ is not Markovian; for instance,

$$(z_{n+1} \mid z_0^n) \sim N_1(az_n + z_0, \sigma^2).$$

Clearly, in this case $(\mathcal{Z}_n)$ and $(\mathcal{Z}_0^n)$ are both transitive for $(\mathcal{Y}_0^n)$, whereas $(\mathcal{Z}_n)$ is not $(\mathcal{Y}_0^n)$-1-Markovian. Consequently the strong identification property assumption (6.4.32), i.e., $\mathcal{Z}_0^n \ll \mathcal{Y}_0^n \mid \mathcal{Z}_{n-k+1}^n$, is not satisfied. This is clear since, in this context, this assumption is equivalent to $\mathcal{Z}_0^n \ll \mathcal{Z}_{n-k+1}^n$ and, by elementary Property 5.4.2(iii), this implies $\overline{\mathcal{Z}_0^n} = \overline{\mathcal{Z}_{n-k+1}^n}$. ∎

6.5 Relations Among Admissible Reductions

We now analyze the conditions under which initial and sequential admissibility are equivalent. For the most part, we follow the table of Section 6.3.2. Throughout this section, $(\mathcal{T}_n)$ is a filtration adapted to the filtration $(\mathcal{S}_n)$.

6.5.1 Admissible Reductions on the Parameter Space

On the parameter space Theorem 6.2.3 ensures that initial and sequential admissibility are equivalent without imposing further conditions.

If, in Theorem 6.2.3, we take $\mathcal{M}_2$ and $\mathcal{G}_n$ to be trivial, then 7(ii) becomes trivial and the equivalence between 7(i) and 3 provides the following result under the substitutions $(\mathcal{M}_1, \mathcal{F}_n) \rightarrow (\mathcal{A}, \mathcal{B} \vee \mathcal{S}_n)$, for sufficiency, and $(\mathcal{M}_1, \mathcal{F}_n) \rightarrow (\mathcal{B}, \mathcal{S}_n)$, for ancillarity.

6.5.1 Proposition. A parameter $\mathcal{B} \subset \mathcal{A}$ is $\mathcal{E}_{\mathcal{A} \vee \mathcal{S}_n}^{\mathcal{S}_0}$-sufficient (respectively, $\mathcal{E}_{\mathcal{A} \vee \mathcal{S}_n}^{\mathcal{S}_0}$-ancillary) if and only if $\mathcal{B}$ is $\mathcal{E}_{\mathcal{A} \vee \mathcal{S}_n}^{\mathcal{S}_{n-1}}$-sufficient (respectively, $\mathcal{E}_{\mathcal{A} \vee \mathcal{S}_n}^{\mathcal{S}_{n-1}}$-ancillary). $\blacksquare$

6.5.2 Admissible Reductions on the Sample Space

Initial and sequential admissibility of reductions on the sample space are shown to be equivalent only under a supplementary condition of transitivity.

For sufficiency on the sample space we consider experiments which are both conditioned on a parameter $\mathcal{C} \subset \mathcal{A}$ and marginalized on $\mathcal{B}$. Unconditional sufficiency is straightforwardly obtained by choosing $\mathcal{C}$ so that it is trivial and $\mathcal{B} = \mathcal{A}$. Now choose $\mathcal{M}_2$ in Theorem 6.2.3 so that it is trivial; the next proposition results from the equivalence between 6 and 7 under the substitution $(\mathcal{M}_1, \mathcal{F}_n, \mathcal{G}_n) \rightarrow (\mathcal{B}, \mathcal{S}_n \vee \mathcal{C}, \mathcal{T}_n)$. This implies, in particular, that Condition 3 becomes

$$(6.5.1) \qquad (\mathcal{B} \vee \mathcal{T}_{n+1}) \perp\!\!\!\perp \mathcal{S}_n \mid \mathcal{C} \vee \mathcal{S}_0 \vee \mathcal{T}_n.$$

6.5.2 Proposition. If $(\mathcal{T}_n)$ is $(\mathcal{B} \vee \mathcal{C} \vee \mathcal{S}_0)$-transitive for $(\mathcal{S}_n)$, then $(\mathcal{T}_n)$ is $\mathcal{E}_{\mathcal{B} \vee \mathcal{S}_n}^{\mathcal{C} \vee \mathcal{S}_0}$-sufficient if and only if $(\mathcal{T}_n)$ is $\mathcal{E}_{\mathcal{B} \vee \mathcal{S}_n}^{\mathcal{C} \vee \mathcal{S}_{n-1}}$-sufficient. $\blacksquare$

For ancillarity on the sample space, let C and (T_n) be mutually ancillary conditionally on B; complete ancillarity may be obtained by choosing $C = A$ and B trivial. Now choose M_1 in Theorem 6.2.3 so that it is trivial; the next result follows from the equivalence between 4 and 5 under the substitution $(M_2, F_n, G_n) \rightarrow (C, S_n \vee B, T_n)$ which involves, in particular, that Condition 3 becomes:

$$(6.5.2) \qquad\qquad T_{n+1} \perp\!\!\!\perp (C \vee S_n) \mid B \vee S_0 \vee T_n.$$

6.5.3 Proposition. C and (T_n) are $\mathcal{E}_{AVS_n}^{BVS_0}$-mutually ancillary and (T_n) is $(B \vee C \vee S_0)$-transitive for (S_n) if and only if C and (T_n) are $\mathcal{E}_{AVS_n}^{BVS_{n-1}}$-mutually ancillary and (T_n) is $(B \vee S_0)$-transitive for (S_n). ∎

Remark that we recover Proposition 6.5.1. by noting that the two transitivity conditions of Proposition 6.5.3 are trivial when $T_n = S_n$. The next examples provide insight into the necessity of the condition of transitivity in Proposition 6.5.3.

Example 1. Let us consider the two-dimensional process $x_n = (y_n, z_n)'$ defined by

$$(x_{n+1} \mid a, x_0^n) \sim N_2 \left[\begin{pmatrix} 0 \\ a\, y_n \end{pmatrix}; \begin{pmatrix} 1 & 0 \\ 0 & 1 - a^2 \end{pmatrix} \right]$$

where $a \in A = (-1, 1)$. Suppose that $(a \mid x_0)$ is uniformly distributed on A. Clearly, $y_{n+1} \perp\!\!\!\perp (A \vee X_0^n)$, $y_{n+1} \perp\!\!\!\perp z_{n+1} \mid A \vee X_0^n$, and $z_{n+1} \perp\!\!\!\perp X_0^n \mid A \vee y_n$. By Theorem 2.2.10 and Corollary 2.2.11, the two first conditional independences imply that $y_{n+1} \perp\!\!\!\perp (z_{n+1} \vee X_0^n) \mid A$. Therefore, for $n \geq 1$,

$$E(z_{n+1} \mid a, z_0^n, y_0) = a\, E(y_n \mid a, z_0^n, y_0) = a\, E(y_n \mid a) = 0,$$

and

$$\begin{aligned} V(z_{n+1} \mid a, z_0^n, y_0) &= (1 - a^2) + V(a\, y_n \mid a, z_0^n, y_0) \\ &= (1 - a^2) + a^2\, V(y_n \mid a) = 1. \end{aligned}$$

This means that, $\forall\, n \geq 1$,

$$(z_{n+1} \mid a, z_0^n, y_0) \sim N(0, 1) \text{ or, equivalently,}$$

$(z_n \mid a, x_0) \sim i.N(0, 1), \ \ \forall \, n \geq 2.$

We conclude that $\mathcal{A}$ and $\mathcal{Z}_2^n$ are mutually ancillary in $\mathcal{E}_{\mathcal{A} \vee \mathcal{X}_0^n}^{\mathcal{X}_0}$. However, $\mathcal{A}$ and $\mathcal{Z}_0^n$ are not mutually ancillary in $\mathcal{E}_{\mathcal{A} \vee \mathcal{X}_0^n}^{\mathcal{X}_0^{n-1}}$. Indeed, Proposition 6.5.3 is not applicable because $(\mathcal{Z}_0^n)$ is not $(\mathcal{A} \vee \mathcal{X}_0)$-transitive for $\mathcal{X}_0^n$. These two facts are evident since

$(z_{n+1} \mid a, x_0^n) \sim N(ay_n, 1 - a^2).$ ∎

Example 2. Consider the two-dimensional process $x_n = (y_n, z_n)'$ defined by

$$(x_{n+1} \mid a, x_0^n) \sim N_2 \left[\begin{pmatrix} y_n + c \\ b\, y_n \end{pmatrix}; I \right]$$

where $a = (b, c)' \in A = \mathbb{R}^2$. Clearly, because $(z_{n+1} \mid a, x_0^n) \sim N(b\, y_n, 1)$, $\mathcal{A} \perp\!\!\!\perp \mathcal{Z}_{n+1} \mid \mathcal{B} \vee \mathcal{X}_0^n$. In other words, $\mathcal{C}$ and $\mathcal{Z}_0^n$ are mutually ancillary in $\mathcal{E}_{\mathcal{A} \vee \mathcal{X}_0^n}^{\mathcal{B} \vee \mathcal{X}_0^{n-1}}$ but not in $\mathcal{E}_{\mathcal{A} \vee \mathcal{X}_0^n}^{\mathcal{B} \vee \mathcal{X}_0}$ because, e.g.,

$(\mathcal{A} \vee \mathcal{X}_0)z_n = b(\mathcal{A} \vee \mathcal{X}_0)y_{n-1} = b\{y_0 + (n-1)c\}, \ \forall \, n \geq 1.$

Thus Proposition 6.5.3 is not applicable because $(\mathcal{Z}_0^n)$ is not $(\mathcal{B} \vee \mathcal{X}_0)$-transitive for $(\mathcal{X}_0^n)$. Indeed $(\mathcal{A} \vee \mathcal{X}_0^n)z_{n+1} = b\, y_n = (\mathcal{B} \vee \mathcal{X}_0^n)z_{n+1}.$ ∎

6.5.3 Admissible Reductions in Joint Reductions

In this section the theorems provide transitivity conditions which guarantee the equivalence of initial and sequential admissibility in the case of mutual sufficiency, mutual exogeneity and Bayesian cuts.

A. Mutual Sufficiency

6.5.4 Theorem. $\mathcal{B}$ and $(\mathcal{T}_n)$ are $\mathcal{E}_{\mathcal{A} \vee \mathcal{S}_n}^{\mathcal{S}_0}$-mutually sufficient and $(\mathcal{T}_n)$ is $(\mathcal{A} \vee \mathcal{S}_0)$-transitive for $(\mathcal{S}_n)$ if and only if $\mathcal{B}$ and $(\mathcal{T}_n)$ are $\mathcal{E}_{\mathcal{A} \vee \mathcal{S}_n}^{\mathcal{S}_{n-1}}$-mutually sufficient and $(\mathcal{T}_n)$ is $(\mathcal{B} \vee \mathcal{S}_0)$-transitive for $(\mathcal{S}_n)$.

Proof. This theorem may be rephrased in terms of conditional independence as follows. Let:

(i) $\mathcal{A} \perp\!\!\!\perp \mathcal{T}_n \mid \mathcal{B} \vee \mathcal{S}_0$;

(ii) $\mathcal{B} \perp\!\!\!\perp \mathcal{S}_n \mid \mathcal{S}_0 \vee \mathcal{T}_n$;

(iii) $\mathcal{T}_{n+1} \perp\!\!\!\perp \mathcal{S}_n \mid \mathcal{A} \vee \mathcal{S}_0 \vee \mathcal{T}_n$;

(iv) $\mathcal{A} \perp\!\!\!\perp \mathcal{T}_n \mid \mathcal{B} \vee \mathcal{S}_{n-1}$;

(v) $\mathcal{B} \perp\!\!\!\perp \mathcal{S}_n \mid \mathcal{S}_{n-1} \vee \mathcal{T}_n$;

(vi) $\mathcal{T}_{n+1} \perp\!\!\!\perp \mathcal{S}_n \mid \mathcal{B} \vee \mathcal{S}_0 \vee \mathcal{T}_n$.

We need to prove that (i), (ii) and (iii) are equivalent to (iv), (v) and (vi). Indeed, when $\mathcal{C} = \mathcal{A}$, Proposition 6.5.3 shows that (i) and (iii) are equivalent to (iv) and (vi), and when $\mathcal{C}$ is trivial, Proposition 6.5.2 shows that under (vi), (ii) and (v) are equivalent.

It may be remarked, in view of Theorem 6.2.3, that the three conditions of Theorem 6.5.4, i.e., $\mathcal{B}$ and $(\mathcal{T}_n)$ are $\mathcal{E}^{\mathcal{S}_0}_{\mathcal{A} \vee \mathcal{S}_n}$-mutually sufficient and $(\mathcal{T}_n)$ is $(\mathcal{A} \vee \mathcal{S}_0)$-transitive for $(\mathcal{S}_n)$, are actually equivalent to the two following nonrelated conditional independence relations:

$$(6.5.3) \qquad\qquad \mathcal{T}_{n+1} \perp\!\!\!\perp (\mathcal{A} \vee \mathcal{S}_n) \mid \mathcal{B} \vee \mathcal{S}_0 \vee \mathcal{T}_n$$

$$(6.5.4) \qquad\qquad \mathcal{B} \perp\!\!\!\perp \mathcal{S}_n \mid \mathcal{S}_0 \vee \mathcal{T}_n.$$

Note also that the conditions of Theorem 6.5.4 imply (6.5.1) when $\mathcal{C}$ is trivial, i.e., $(\mathcal{B} \vee \mathcal{T}_{n+1}) \perp\!\!\!\perp \mathcal{S}_n \mid \mathcal{S}_0 \vee \mathcal{T}_n$, and this in turn implies that $(\mathcal{T}_n)$ is $\mathcal{S}_0$-transitive for $(\mathcal{S}_n)$, i.e., $\mathcal{T}_{n+1} \perp\!\!\!\perp \mathcal{S}_n \mid \mathcal{S}_0 \vee \mathcal{T}_n$.

B. Mutual Exogeneity

Concerning mutual exogeneity, we have the following theorem:

6.5.5 Theorem. If $(\mathcal{T}_n)$ is $(\mathcal{A} \vee \mathcal{S}_0)$-transitive for $(\mathcal{S}_n)$, then $\mathcal{C}$ and $(\mathcal{T}_n)$ are $\mathcal{E}^{\mathcal{S}_0}_{\mathcal{A} \vee \mathcal{S}_n}$-mutually exogenous if and only if $\mathcal{C}$ and $(\mathcal{T}_n)$ are $\mathcal{E}^{\mathcal{S}_{n-1}}_{\mathcal{A} \vee \mathcal{S}_n}$-mutually exogenous and $(\mathcal{T}_n)$ is $\mathcal{S}_0$-transitive for $(\mathcal{S}_n)$.

Proof. This theorem may be rephrased in terms of conditional independence as follows. Let:

(i) $C \perp\!\!\!\perp T_n \mid S_0$;

(ii) $A \perp\!\!\!\perp S_n \mid C \vee S_0 \vee T_n$;

(iii) $T_{n+1} \perp\!\!\!\perp S_n \mid A \vee S_0 \vee T_n$;

(iv) $C \perp\!\!\!\perp T_n \mid S_{n-1}$;

(v) $A \perp\!\!\!\perp S_n \mid C \vee S_{n-1} \vee T_n$;

(vi) $T_{n+1} \perp\!\!\!\perp S_n \mid S_0 \vee T_n$.

We need to prove that (i), (ii) and (iii) are equivalent to (iii), (iv), (v) and (vi). Indeed, when $B = A$, Proposition 6.5.2 shows that, under (iii), (ii) is equivalent to (v) and is equivalent to $(A \vee T_{n+1}) \perp\!\!\!\perp S_n \mid C \vee S_0 \vee T_n$. This implies that

(vii) $T_{n+1} \perp\!\!\!\perp S_n \mid C \vee S_0 \vee T_n$.

And when B is trivial, Proposition 6.5.3 shows that (i) and (vii) are equivalent to (iv) and (vi). ∎

With regards to mutual sufficiency, it may be shown, using Theorem 6.2.3, that the three conditions of Theorem 6.5.5, i.e., C and (T_n) are $\mathcal{E}_{A \vee S_n}^{S_0}$-mutually exogenous and (T_n) is $(A \vee S_0)$-transitive for (S_n), are equivalent to the two following nonrelated conditional independent relations:

$$(6.5.5) \qquad (A \vee T_{n+1}) \perp\!\!\!\perp S_n \mid C \vee S_0 \vee T_n$$

$$(6.5.6) \qquad C \perp\!\!\!\perp T_n \mid S_0.$$

Note that it follows from (6.5.5) that, under the conditions of Theorem 6.5.5, we also have that (T_n) is $(C \vee S_0)$-transitive for (S_n).

C. Bayesian Cuts

Combining these two theorems, we have the following theorem relating the initial and sequential cuts:

6.5.6 Theorem. $(\mathcal{B}, \mathcal{C}, (\mathcal{T}_n))$ operates a cut in $\mathcal{E}^{\mathcal{S}_0}_{\mathcal{A} \vee \mathcal{S}_n}$ and $(\mathcal{T}_n)$ is $(\mathcal{A} \vee \mathcal{S}_0)$-transitive for $(\mathcal{S}_n)$ if and only if $(\mathcal{B}, \mathcal{C}, (\mathcal{T}_n))$ operates a cut in $\mathcal{E}^{\mathcal{S}_{n-1}}_{\mathcal{A} \vee \mathcal{S}_n}$ and $(\mathcal{T}_n)$ is $(\mathcal{B} \vee \mathcal{S}_0)$-transitive for $(\mathcal{S}_n)$.

Proof. This theorem may be rephrased in terms of conditional independence, as follows. Let:

(i) $\mathcal{A} \perp\!\!\!\perp \mathcal{T}_n \mid \mathcal{B} \vee \mathcal{S}_0$;

(ii) $\mathcal{A} \perp\!\!\!\perp \mathcal{S}_n \mid \mathcal{C} \vee \mathcal{S}_0 \vee \mathcal{T}_n$;

(iii) $\mathcal{T}_{n+1} \perp\!\!\!\perp \mathcal{S}_n \mid \mathcal{A} \vee \mathcal{S}_0 \vee \mathcal{T}_n$;

(iv) $\mathcal{A} \perp\!\!\!\perp \mathcal{T}_n \mid \mathcal{B} \vee \mathcal{S}_{n-1}$;

(v) $\mathcal{A} \perp\!\!\!\perp \mathcal{S}_n \mid \mathcal{C} \vee \mathcal{S}_{n-1} \vee \mathcal{T}_n$;

(iv) $\mathcal{T}_{n+1} \perp\!\!\!\perp \mathcal{S}_n \mid \mathcal{B} \vee \mathcal{S}_0 \vee \mathcal{T}_n$.

Remark that (i), (ii) and (iii) are equivalent to (iv), (v) and (vi). Indeed, when $\mathcal{C} = \mathcal{A}$, Proposition 6.5.3 shows that (i) and (iii) are equivalent to (iv) and (vi); when $\mathcal{B} = \mathcal{A}$, Proposition 6.5.2 shows that under (iii), (ii) and (v) are equivalent.   ∎

Again, by Theorem 6.2.3, it may be shown that Conditions (i), (ii) and (iii) of the proof of Theorem 6.5.6 are equivalent to the two following non-related conditional independence conditions:

(6.5.7) $\mathcal{T}_{n+1} \perp\!\!\!\perp (\mathcal{A} \vee \mathcal{S}_n) \mid \mathcal{B} \vee \mathcal{S}_0 \vee \mathcal{T}_n,$

(6.5.8) $\mathcal{A} \perp\!\!\!\perp \mathcal{S}_n \mid \mathcal{C} \vee \mathcal{S}_0 \vee \mathcal{T}_n.$

Note also that the conditions of Theorem 6.5.6 imply (6.5.1) with $\mathcal{B} = \mathcal{A}$, i.e., $(\mathcal{A} \vee \mathcal{T}_{n+1}) \perp\!\!\!\perp \mathcal{S}_n \mid \mathcal{C} \vee \mathcal{S}_0 \vee \mathcal{T}_n$, and this clearly implies that we also have $(\mathcal{T}_n)$ is $(\mathcal{C} \vee \mathcal{S}_0)$-transitive for $(\mathcal{S}_n)$.

D. Initial and Sequential Cuts in Dominated Experiments

It may be illuminating to look at what is really the matter of Theorem 6.5.6 in the dominated case. We first need a definition.

6.5.7 Definition. $\mathcal{E} = (A \times S, \mathcal{A} \vee S, \Pi, (\mathcal{S}_n)_{n \geq 0})$ is a *dominated* sequential Bayesian experiment if for each $n \geq 0$, $\mathcal{E}_{\mathcal{A} \vee \mathcal{S}_n}$ is a dominated Bayesian experiment. ∎

According to Proposition 3.4.17, if $\mathcal{B} \perp \!\!\! \perp \mathcal{C} \mid \mathcal{S}_0, (\mathcal{B}, \mathcal{C}, (\mathcal{T}_n))$ operates a cut in $\mathcal{E}_{\mathcal{A} \vee \mathcal{S}_n}^{\mathcal{S}_0}$ if and only if

$$(6.5.9) \qquad g_{\mathcal{S}_n}^{\mathcal{A} \vee \mathcal{S}_0} = g_{\mathcal{T}_n}^{\mathcal{B} \vee \mathcal{S}_0} \cdot g_{\mathcal{S}_n}^{\mathcal{C} \vee \mathcal{S}_0 \vee \mathcal{T}_n}, \qquad \forall\, n \geq 1,$$

and $(\mathcal{B}, \mathcal{C}, (\mathcal{T}_n))$ operates a cut in $\mathcal{E}_{\mathcal{A} \vee \mathcal{S}_n}^{\mathcal{S}_{n-1}}$ if and only if

$$(6.5.10) \qquad g_{\mathcal{S}_n}^{\mathcal{A} \vee \mathcal{S}_{n-1}} = g_{\mathcal{T}_n}^{\mathcal{B} \vee \mathcal{S}_{n-1}} \cdot g_{\mathcal{S}_n}^{\mathcal{C} \vee \mathcal{S}_{n-1} \vee \mathcal{T}_n}, \qquad \forall\, n \geq 1.$$

But, by extending (1.4.29), we have

$$(6.5.11) \qquad g_{\mathcal{S}_n}^{\mathcal{A} \vee \mathcal{S}_0} = \prod_{1 \leq i \leq n} g_{\mathcal{S}_i}^{\mathcal{A} \vee \mathcal{S}_{i-1}}.$$

And so

$$(6.5.12) \qquad g_{\mathcal{S}_n}^{\mathcal{A} \vee \mathcal{S}_0} = \prod_{1 \leq i \leq n} g_{\mathcal{T}_i}^{\mathcal{B} \vee \mathcal{S}_{i-1}} \cdot \prod_{1 \leq i \leq n} g_{\mathcal{S}_i}^{\mathcal{C} \vee \mathcal{S}_{i-1} \vee \mathcal{T}_i}.$$

Consequently $g_{\mathcal{S}_n}^{\mathcal{A} \vee \mathcal{S}_0}$ may be written as the product of a $\mathcal{B}$-measurable function and a $\mathcal{C}$-measurable function. However, it should be realized that, unless certain additional conditions are satisfied, these two functions are not sampling densities. That condition $(\mathcal{T}_n)$ is $(\mathcal{B} \vee \mathcal{S}_0)$-transitive for $(\mathcal{S}_n)$ implies that

$$(6.5.13) \qquad \prod_{1 \leq i \leq n} g_{\mathcal{T}_i}^{\mathcal{B} \vee \mathcal{S}_{i-1}} = g_{\mathcal{T}_n}^{\mathcal{B} \vee \mathcal{S}_0}$$

and

$$(6.5.14) \qquad \prod_{1 \leq i \leq n} g_{\mathcal{S}_i}^{\mathcal{C} \vee \mathcal{S}_{i-1} \vee \mathcal{T}_i} = g_{\mathcal{S}_n}^{\mathcal{C} \vee \mathcal{S}_0 \vee \mathcal{T}_n}.$$

E. Some Examples

We conclude this section with some examples of cuts in initial and sequential experiments.

Example 1. Let x_n be a linear normal k-autoregressive process, i.e.,

$$(x_n \mid a, x_0^{n-1}) \sim N[P(L)x_n; \Sigma]$$

where $P(L)$ is a matrix whose elements are polynomials of degree k with no constant terms in the lag operator. By convention, $a = (P(L), \Sigma)$. If we partition x_n into $x_n = (y_n', z_n')'$ and $P(L)$ and Σ into

$$P(L) = \left(P_y'(L), P_z'(L) \right)' = \begin{pmatrix} P_{yy}(L) & P_{yz}(L) \\ P_{zy}(L) & P_{zz}(L) \end{pmatrix}$$

$$\Sigma = \begin{pmatrix} \Sigma_{yy} & \Sigma_{yz} \\ \Sigma_{zy} & \Sigma_{zz} \end{pmatrix}$$

then we obtain $(z_n \mid a, x_0^{n-1}) \sim N[P_z(L)x_n, \Sigma_{zz}]$; further $(y_n \mid a, x_0^{n-1}, z_n)$ is also normally distributed with

$$E(y_n \mid a, x_0^{n-1}, z_n) = \pi z_n + Q(L)x_n$$
$$V(y_n \mid a, x_0^{n-1}, z_n) = \Sigma_{yy.z}$$

where

$$\pi = \Sigma_{yz}\Sigma_{zz}^{-1}$$
$$Q(L) = P_y(L) - \Sigma_{yz}\Sigma_{zz}^{-1}P_z(L)$$
$$\Sigma_{yy.z} = \Sigma_{yy} - \Sigma_{yz}\Sigma_{zz}^{-1}\Sigma_{zy}.$$

Therefore, if one defines

$$b = (P_z(L), \Sigma_{zz})$$
$$c = (\pi, Q(L), \Sigma_{yy.z}),$$

and if $b \perp\!\!\!\perp c \mid x_0$, then $(b, c, (z_0^n))$ certainly operates a cut in the sequential experiment but, in general, does not operate a cut in the initial experiment. Note that the condition of prior independence is satisfied when $(P(L), \Sigma)$ are distributed a priori according to a "natural conjugate distribution", i.e., matrix normal inverted Wishart, (see, e.g., Drèze and Richard (1983)).

However, if we assume $P_{zy}(L) = 0$, $(z_n \mid a, x_0^{n-1}) \sim N[P_{zz}(L)z_n, \Sigma_{zz}]$. In terms of conditional independence, this implies that

$$z_{n+1} \perp\!\!\!\perp (a, x_0^n) \mid b, x_0, z_0^n.$$

Therefore (z_0^n) is both (b, x_0) and (a, x_0)-transitive for (x_0^n), and $(b, c, (z_0^n))$ operates a cut both in the initial and in the sequential experiment. ∎

Example 2. Now consider a dynamic version of Example 4 of Section 3.4.6, i.e., $(x_n \mid a, x_0^{n-1}) \sim N[\xi_n + P(L)(x_n - \xi_n), \Sigma]$ where $P(L)$ is defined as in Example 1, and $a = (\theta, \Sigma, \xi_i, 1 \leq i \leq \infty)$. Suppose that $A_\theta \xi_i = 0$, $\forall\ 1 \leq i \leq \infty$, where A_θ is a known matrix function of θ. With the same notation as in Example 3 and Example 4 of Section 3.4.6, we have:

$$(z_n \mid a, x_0^{n-1}) \sim N[\xi_{z,n} + P_z(L)(x_n - \xi_n), \Sigma_{zz}]$$
$$(y_n \mid a, x_0^{n-1}, z_n) \sim N[\xi_{y,n} + P_y(L)(x_n - \xi_n) +$$
$$\pi\{z_n - \xi_{z,n} - P_z(L)(x_n - \xi_n)\}, \Sigma_{yy \cdot z}]$$
$$B_\theta \xi_{y,n} + C_\theta \xi_{z,n} = 0.$$

Without placing restrictions on the parameter a, there is no useful cut nor transitivity property. The restriction $P_{zy}(L) = 0$ implies that (z_0^n) is (a, x_0)-transitive for (x_0^n). Assuming that B_θ is a (square) invertible matrix, an obviously useful choice of restrictions for a cut in the sequential experiment is $P_{zy}(L) = 0, \pi + B_\theta^{-1} C_\theta = 0$ and $P_{yz}(L) = \pi P_{zz}(L) - P_{yy}(L)\pi$. Under these restrictions $(b, c, (z_0^n))$ will operate a cut both in the initial and in the sequential experiments with

$$b = (P_{zz}(L), \Sigma_{zz}, \xi_{z,i}, 1 \leq i \leq \infty)$$

and

$$c = (P_{yy}(L), \pi, \Sigma_{yy \cdot z})$$

if the prior distribution is such that $b \perp\!\!\!\perp c \mid x_0$. Indeed, with these restrictions,

$$(z_n \mid a, x_0^{n-1}) \sim N[P_{zz}(L)z_n + (I - P_{zz}(L))\xi_{z,n}, \Sigma_{zz}]$$

and

$$(y_n \mid a, x_0^{n-1}, z_n) \sim N[P_{yy}(L)y_n + (I - P_{yy}(L))\pi z_n, \Sigma_{yy \cdot z}].$$

Note that, in this case, $\mathcal{E}_{a,z_0^n}^{x_0}$ captures all the incidental parameters; further, if we define $u_n = y_n - \pi z_n$, then $(u_n \mid a, x_0^{n-1}, z_n) \sim N[P_{yy}(L)u_n, \Sigma_{yy \cdot z}]$, i.e., u_n is an autoregressive process. ∎

Example 3. A second dynamic version of Example 4 of Section 3.4.6 is motivated by experiments where conditional expectations are unobservable functions of the process past. Using the same notation as in the preceding examples, consider the experiment $(x_n \mid a, x_0^{n-1}) \sim N(\xi_n, \Sigma)$ with $A_\theta \xi_n = P(L)x_n$. Clearly,

$$(z_n \mid a, x_0^{n-1}) \sim N(\xi_{z,n}, \Sigma_{zz}),$$
$$(y_n \mid a, x_0^{n-1}, z_n) \sim N[\xi_{y,n} + \pi(z_n - \xi_{z,n}), \Sigma_{yy \cdot z}]$$

and $B_\theta \xi_{y,n} + C_\theta \xi_{z,n} = P(L)x_n$. Using the same method as in Florens, Mouchart and Richard (1979, 1986, 1987) and in Engle, Hendry and Richard (1983) (see also Wu (1973)), a sequential cut may be obtained through the following restriction: $B_\theta \pi + C_\theta = 0$. Indeed, under these conditions, the conditional experiment may be written as $(y_n \mid a, x_0^{n-1}, z_n) \sim N(\eta_n, \Sigma_{yy.z})$ and $B_\theta \eta_n + C_\theta z_n = P(L)x_n$. Remark that $(b, c, (z_0^n))$ will operate a cut in the sequential experiment, with

$$b = (\Sigma_{zz}, \xi_{z,i}, \ 1 \le i < \infty)$$

and

$$c = (B_\theta, C_\theta, P(L), \ \eta_n : 1 \le n \le \infty)$$

if the prior distribution is such that $b \perp\!\!\!\perp c \mid x_0$. Remark also that if B_θ is (square) invertible the conditional experiment no longer has any incidental parameters, since $(y_n \mid a, x_0^{n-1}, z_n) \sim N[\pi z_n + B_\theta^{-1} P(L)x_n, \Sigma_{yy.z}]$. In this case we have a cut only in the sequential experiment. A cut in the initial experiment or, equivalently, a (b, x_0)-transitivity for (x_0^n) generally depends on specific assumptions regarding the process generating $(\xi_{z,n})$.   ∎

6.6 The Role of Transitivity: Further Results

Very often, particularly in discrete time series models, the specification of the model is sequential, i.e., defined by the sequence of experiments $\mathcal{E}_{\mathcal{A} \vee \mathcal{S}_n}^{\mathcal{S}_{n-1}}$. Sequentially admissible reductions are therefore easily checked, but what remains of interest is to establish whether such reductions are also initially admissible. As seen in Section 6.5, this requires verification of transitivity conditions. In this section, we look successively at weakening these transitivity conditions, and at their necessity.

6.6.1 Weakening of Transitivity Conditions

Even though transitivity conditions are theoretically relatively simple, verification may sometimes be difficult. In most cases, the $(\mathcal{A} \vee \mathcal{S}_0)$-transitivity is easy to verify since this depends on the sampling probabilities only. Indeed, this condition is equivalent to $P_{T_{n+1}}^{\mathcal{A} \vee \mathcal{S}_n} = P_{T_{n+1}}^{\mathcal{A} \vee \mathcal{S}_0 \vee T_n}$. However, the $(\mathcal{B} \vee \mathcal{S}_0)$-transitivity may be difficult to verify since it requires integration on

$\mathcal{A}$ conditionally on $\mathcal{B}$, which may pose considerable difficulty. And further, modelling specifications frequently presuppose $(\mathcal{A} \vee \mathcal{S}_0)$-transitivity without making this explicit completely. It is therefore helpful to be aware of certain weak additional assumptions which imply that other transitivities are verified.

In view of Proposition 6.5.2, this question is irrelevant in the comparison of sufficiency in initial and sequential analysis. But, in view of Proposition 6.5.3, this problem arises when trying to deduce ancillarity in initial analysis from ancillarity in sequential analysis. The following theorem shows that under a measurable separability condition, the same transitivity condition guarantees the equivalence between ancillarity in initial and sequential analysis.

6.6.1 Theorem. If $\mathcal{C} \parallel \mathcal{S}_n \mid \mathcal{B} \vee \mathcal{S}_0 \vee \mathcal{T}_n$ and if $(\mathcal{T}_n)$ is $(\mathcal{B} \vee \mathcal{C} \vee \mathcal{S}_0)$-transitive for $(\mathcal{S}_n)$, then $\mathcal{C}$ and $(\mathcal{T}_n)$ are $\mathcal{E}_{\mathcal{A} \vee \mathcal{S}_n}^{\mathcal{B} \vee \mathcal{S}_{n-1}}$-mutually ancillary if and only if they are $\mathcal{E}_{\mathcal{A} \vee \mathcal{S}_n}^{\mathcal{B} \vee \mathcal{S}_0}$-mutually ancillary. Moreover, in this case, $(\mathcal{T}_n)$ is also $(\mathcal{B} \vee \mathcal{S}_0)$-transitive for $(\mathcal{S}_n)$.

Proof. Let

(i) $\mathcal{C} \parallel \mathcal{S}_n \mid \mathcal{B} \vee \mathcal{S}_0 \vee \mathcal{T}_n$;

(ii) $\mathcal{T}_{n+1} \perp\!\!\!\perp \mathcal{S}_n \mid \mathcal{B} \vee \mathcal{C} \vee \mathcal{S}_0 \vee \mathcal{T}_n$;

(iii) $\mathcal{C} \perp\!\!\!\perp \mathcal{T}_{n+1} \mid \mathcal{B} \vee \mathcal{S}_n$;

(iv) $\mathcal{C} \perp\!\!\!\perp \mathcal{T}_n \mid \mathcal{B} \vee \mathcal{S}_0$;

(v) $\mathcal{T}_{n+1} \perp\!\!\!\perp \mathcal{S}_n \mid \mathcal{B} \vee \mathcal{S}_0 \vee \mathcal{T}_n$.

We must prove that under (i) and (ii), (iii) is both equivalent to (iv) and implies (v). From Proposition 6.5.3 we know that (ii) and (iv) imply (iii) and (v). By an application of Theorem 5.2.10, (i), (ii) and (iii) imply that $\mathcal{T}_{n+1} \perp\!\!\!\perp (\mathcal{C} \vee \mathcal{S}_n) \mid \mathcal{B} \vee \mathcal{S}_0 \vee \mathcal{T}_n$. Hence, $\mathcal{C} \perp\!\!\!\perp \mathcal{T}_{n+1} \mid \mathcal{B} \vee \mathcal{S}_0 \vee \mathcal{T}_n$, and so (iv) is obtained by making use of the equivalence between 1 and 2 in Theorem 6.2.3 with $\mathcal{M}_2$ trivial, and under the substitution

$$(\mathcal{M}_1, \mathcal{F}_n, \mathcal{G}_n) \rightarrow (\mathcal{C}, \mathcal{B} \vee \mathcal{S}_0 \vee \mathcal{T}_n, \mathcal{B} \vee \mathcal{S}_0). \qquad \blacksquare$$

We now apply this result to joint reductions. From the proof of Theorem 6.5.4 we obtain, as a corollary of Theorem 6.6.1, the following proposition:

6.6.2 Proposition. If $A \parallel S_n \mid B \vee S_0 \vee T_n$ and if (T_n) is $(A \vee S_0)$-transitive for (S_n), then B and (T_n) are $\mathcal{E}_{A \vee S_n}^{S_{n-1}}$-mutually sufficient if and only if they are $\mathcal{E}_{A \vee S_n}^{S_0}$-mutually sufficient. Moreover, in this case, (T_n) is both $(B \vee S_0)$- and (S_0)-transitive for (S_n).

Following the same steps, we obtain, for mutual exogeneity, the following proposition:

6.6.3 Proposition. If $C \parallel S_n \mid S_0 \vee T_n$ and if (T_n) is $(A \vee S_0)$-transitive for (S_n), then C and (T_n) are $\mathcal{E}_{A \vee S_n}^{S_{n-1}}$-mutually exogenous if and only if they are $\mathcal{E}_{A \vee S_n}^{S_0}$-mutually exogenous. Morevover, in this case, (T_n) is both $(C \vee S_0)$- and (S_0)-transitive for (S_n).   ∎

Finally, concerning cuts, we have the following theorem:

6.6.4 Theorem. If $A \parallel S_n \mid B \vee S_0 \vee T_n$ and if (T_n) is $(A \vee S_0)$-transitive for (S_n), then $(B, C, (T_n))$ operates a cut in $\mathcal{E}_{A \vee S_n}^{S_{n-1}}$ if and only if $(B, C, (T_n))$ operates a cut in $\mathcal{E}_{A \vee S_n}^{S_0}$. Moreover, in this case, (T_n) is both $(B \vee S_0)$- and $(C \vee S_0)$-transitive for (S_n).

Proof. Let

(i) $A \parallel S_n \mid B \vee S_0 \vee T_n$;

(ii) $T_{n+1} \perp\!\!\!\perp S_n \mid A \vee S_0 \vee T_n$;

(iii) $A \perp\!\!\!\perp T_n \mid B \vee S_{n-1}$;

(iv) $A \perp\!\!\!\perp S_n \mid C \vee S_{n-1} \vee T_n$;

(v) $A \perp\!\!\!\perp T_n \mid B \vee S_0$;

(vi) $A \perp\!\!\!\perp S_n \mid C \vee S_0 \vee T_n$.

We must prove that under (i) and (ii), (iii) and (iv) are equivalent to (v) and (vi). Proposition 6.5.2, with $\mathcal{B} = \mathcal{A}$, shows that under (ii), (iv) and (vi) are equivalent and imply that $(\mathcal{T}_n)$ is $(\mathcal{C} \vee \mathcal{S}_0)$-transitive for $(\mathcal{S}_n)$. Theorem 6.6.1, with $\mathcal{C} = \mathcal{A}$, shows that under (i) and (ii), (iii) and (v) are equivalent and imply that $(\mathcal{T}_n)$ is $(\mathcal{B} \vee \mathcal{S}_0)$-transitive for $(\mathcal{S}_n)$.   ∎

Note that all the measurable separability conditions appearing in Theorem 6.6.1, Propositions 6.6.2 and 6.6.3 and in Theorem 6.6.4, are readily satisfied when for each n the probability Π_n, that is the restriction of Π on $\mathcal{A} \vee \mathcal{S}_n$, is equivalent to a probability measure Π'_n such that $\mathcal{A} \perp\!\!\!\perp \mathcal{S}_n \mid \mathcal{S}_0; \Pi'_n$. This is a consequence of Corollary 5.2.11.

Theorem 6.6.4 is useful in ensuring that a cut in a sequential experiment is also a cut in the initial experiment, especially when the sequential experiment is specified in such a way that the conditional distributions of the process are not explicitly known. However, in some applications these distributions will be known (see Examples 1, 2 and 3 in Section 6.5). In such a context it is interesting to characterize a cut in $\mathcal{E}^{\mathcal{S}_{n-1}}_{\mathcal{A} \vee \mathcal{S}_n}$ and $\mathcal{E}^{\mathcal{S}_0}_{\mathcal{A} \vee \mathcal{S}_n}$ by properties of the prior distribution and of the conditional sampling distributions; such a characterization is provided by the following theorem.

6.6.5 Theorem. $(\mathcal{B}, \mathcal{C}, (\mathcal{T}_n))$ operates a cut both in $\mathcal{E}^{\mathcal{S}_0}_{\mathcal{A} \vee \mathcal{S}_n}$ and in $\mathcal{E}^{\mathcal{S}_{n-1}}_{\mathcal{A} \vee \mathcal{S}_n}$ if the following conditions hold:

1. $\mathcal{B} \perp\!\!\!\perp \mathcal{C} \mid \mathcal{S}_0$;

2. $\mathcal{T}_n \perp\!\!\!\perp (\mathcal{A} \vee \mathcal{S}_{n-1}) \mid \mathcal{B} \vee \mathcal{S}_0 \vee \mathcal{T}_{n-1}$; and

3. $\mathcal{A} \perp\!\!\!\perp \mathcal{S}_n \mid \mathcal{C} \vee \mathcal{S}_{n-1} \vee \mathcal{T}_n$.

Proof. Let

(i) $\mathcal{A} \perp\!\!\!\perp \mathcal{T}_n \mid \mathcal{B} \vee \mathcal{S}_0$;

(ii) $\mathcal{A} \perp\!\!\!\perp \mathcal{S}_n \mid \mathcal{C} \vee \mathcal{S}_0 \vee \mathcal{T}_n$;

(iii) $A \perp\!\!\!\perp T_n \mid B \vee S_{n-1}$;

(iv) $A \perp\!\!\!\perp S_n \mid C \vee S_{n-1} \vee T_n$;

(v) $T_{n+1} \perp\!\!\!\perp S_n \mid A \vee S_0 \vee T_n$;

(vi) $T_{n+1} \perp\!\!\!\perp S_n \mid B \vee S_0 \vee T_n$.

We must prove that 2 and 3 imply (i), (ii), (iii) and (iv). By (6.5.2) and Proposition 6.5.3 with $C = A$, 2 is equivalent to (i) and (v) and to (iii) and (vi). Now, 3 is equal to (iv) and , under (v), by Proposition 6.5.2 with $B = A$, (iv) is equivalent to (ii). ∎

Conditions 2 and 3 are easily interpreted in the sampling distributions of $\mathcal{E}_{A \vee S_n}^{S_{n-1}}$. Indeed, 2 means that the distribution of T_n, given $A \vee S_{n-1}$, depends on $B \vee S_0 \vee T_{n-1}$ only; 3 means that the distribution of S_n, given $A \vee S_{n-1} \vee T_n$, depends on $C \vee S_{n-1} \vee T_n$ only.

For applications it is interesting to rewrite Theorem 6.6.5 in terms of densities in a dominated sequential Bayesian experiment.

6.6.6 Corollary. Let $\mathcal{E} = (A \times S,\ A \vee S,\ \Pi, (S_n)_{n \geq 0})$ be a dominated sequential Bayesian experiment. Let $B \subset A, C \subset A$ and let (T_n) be a filtration adapted to (S_n). If

(i) $g_{BVC}^{S_0} = g_B^{S_0} \cdot g_C^{S_0}$,

(ii) $g_{T_n}^{A \vee S_{n-1}} = g_{T_n}^{B \vee S_0 \vee T_{n-1}} \quad \forall\, n \geq 0$,

(iii) $g_{S_n}^{A \vee S_{n-1} \vee T_n} = g_{S_n}^{C \vee S_{n-1} \vee T_n} \quad \forall\, n \geq 0$,

then $(B, C, (T_n))$ operates a cut both in $\mathcal{E}_{A \vee S_n}^{S_0}$ and in $\mathcal{E}_{A \vee S_n}^{S_{n-1}}$. ∎

6.6.2 Necessity of Transitivity Conditions

As we have seen transitivity conditions allow the comparison of the initial and sequential admissibility of reductions in sequential Bayesian experiments. It is natural to ask whether or not these conditions are relatively minimal in the sense that, under some other conditions on the experiment, the conditions of transitivity are really necessary. Along these lines, the main result is provided by the following theorem, which is to be compared to Proposition 6.5.3 and Theorem 6.6.1:

6.6.7 Theorem. If $S_n \ll C \mid B \vee S_0 \vee T_n$ and if C and (T_n) are both $\mathcal{E}_{A \vee S_n}^{S_0}$- and $\mathcal{E}_{A \vee S_n}^{S_{n-1}}$-mutually ancillary, then (T_n) is both $(B \vee S_0)$- and $(B \vee C \vee S_0)$-transitive for (S_n).

Proof. Let

(i) $S_n \ll C \mid B \vee S_0 \vee T_n$;

(ii) $C \perp\!\!\!\perp T_n \mid B \vee S_0$;

(iii) $C \perp\!\!\!\perp T_{n+1} \mid B \vee S_n$;

(iv) $T_{n+1} \perp\!\!\!\perp S_n \mid B \vee S_0 \vee T_n$;

(v) $T_{n+1} \perp\!\!\!\perp S_n \mid B \vee C \vee S_0 \vee T_n$.

We must prove that (i), (ii) and (iii) imply (iv) and (v). It follows from the proof of Theorem 6.2.3 (1 implies 2) that (ii) is equivalent to

(vi) $C \perp\!\!\!\perp T_{n+1} \mid B \vee S_0 \vee T_n$.

By an application of Theorem 5.4.13, (vi), (iii) and (i) imply that $T_{n+1} \perp\!\!\!\perp (C \vee S_n) \mid B \vee S_0 \vee T_n$, which implies (iv) and (v). ∎

This result is quite satisfactory but it must be recalled, in view of Section 5.5.2, that the assumption that $S_n \ll C \mid B \vee S_0 \vee T_n$ is relatively strong in this context. Indeed, along with $C \perp\!\!\!\perp T_n \mid B \vee S_0$, i.e., T_n is $\mathcal{E}_{C \vee S_n}^{B \vee S_0}$-ancillary, by Theorem 5.5.10, $S_n \ll C \mid B \vee S_0 \vee T_n$ implies that T_n is maximal ancillary in $\mathcal{E}_{C \vee S_n}^{B \vee S_0}$.

Applying Theorem 6.6.7 to joint reductions entails the following propositions:

6.6.8 Proposition. If $S_n \ll A \mid B \vee S_0 \vee T_n$ and if B and (T_n) are both $\mathcal{E}^{S_0}_{A \vee S_n}$- and $\mathcal{E}^{S_{n-1}}_{A \vee S_n}$-mutually sufficient, then (T_n) is (S_0)-, $(B \vee S_0)$- and $(A \vee S_0)$-transitive for (S_n).   ∎

6.6.9 Proposition. If $S_n \ll C \mid S_0 \vee T_n$ and if C and (T_n) are both $\mathcal{E}^{S_0}_{A \vee S_n}$- and $\mathcal{E}^{S_{n-1}}_{A \vee S_n}$-mutually exogenous, then (T_n) is (S_0)- and $(C \vee S_0)$-transitive for (S_n).   ∎

Remark that here we do not recover the necessity of (T_n) being $(A \vee S_0)$-transitive for (S_n).

6.6.10 Proposition. If $S_n \ll A \mid B \vee S_0 \vee T_n$ and if $(B, C, (T_n))$ operates a cut both in $\mathcal{E}^{S_0}_{A \vee S_n}$ and in $\mathcal{E}^{S_{n-1}}_{A \vee S_n}$, then (T_n) is $(C \vee S_0)$-, $(B \vee S_0)$- and $(A \vee S_0)$-transitive for (S_n).   ∎

We end this section with a proposition showing that under strong identifiability conditions, two transitivity conditions imply mutual ancillarity in the initial experiment and in the sequential experiment. Using Theorem 5.4.13 again, the proof of next proposition parallels the proof of Theorem 6.6.7.

6.6.11 Proposition. If $C \ll S_n \mid B \vee S_0 \vee T_n$ and if (T_n) is both $(B \vee S_0)$- and $(B \vee C \vee S_0)$-transitive for (S_n), then C and (T_n) are both $\mathcal{E}^{S_0}_{A \vee S_n}$- and $\mathcal{E}^{S_{n-1}}_{A \vee S_n}$-mutually ancillary.

Let us remark that the assumption $C \ll S_n \mid B \vee S_0 \vee T_n$ implies, by Theorem 5.5.4, that $C \vee B \vee S_0 \vee T_n$ is the minimal strong $\mathcal{E}^{B \vee S_0 \vee T_n}_{B \vee C \vee S_n}$-sufficient parameter. Note that this last property is frequently satisfied in applications.

7

Asymptotic Experiments

7.1 Introduction

In this chapter, we continue the analysis of sequential Bayesian experiments begun in Chapter 6. Our present concern is the relationship between the experiment stopped at time n and the asymptotic experiment generating infinite sequences of observations. Of course, asymptotic theory is a vast field in statistical theory, and cannot be treated exhaustively in a single chapter. Thus, we concentrate the attention on a particular series of problems related to almost sure convergence. In particular, asymptotic normality of posterior distributions is not considered here (interested readers should refer to LeCam (1986), Chap. 12; see also Dickey (1976), Hartigan (1983), and Walker (1969).

As in previous chapters, probabilistic tools are presented separately in Sections 7.2 (on limits of sequences of conditional independences) and 7.6 (on mutual conditional independences).

The first statistical topic compares admissibility conditions for global and asymptotic reduction. This problem was addressed in Chapter 6 with respect to filtrations (on the sample space). In contrast, in Section 7.3 the hypothesis of filtration is avoided as far as possible; this permits the treatment of a sequence of statistics rather than an increasing sequence relative to the accumulation of observations of some variables.

The second statistical topic, treated in Section 7.4, studies the existence of a consistent estimator for real-valued functions of parameters. In the present framework this is precisely the asymptotic aspect of exact estimability as presented in Section 4.7. We also examine, in some detail, the relationship between the Bayesian concept and various sampling-theory concepts of estimability. This second topic is illustrated, in Section 7.5, by the presentation of equivalent conditions for the exact estimability of discrete σ-fields (i.e., σ-fields generated by countable partitions).

Finally, Section 7.7 studies asymptotic properties in particular classes of experiments with special emphasis on conditional experiments; motivation for this analysis is found, in particular, in the desirability of obtaining results relevant for the asymptotic properties of regression-type models. Heuristically, an almost sure convergence of posterior expectations is often based on the existence of a sub-σ-field (on the sample space) that is both asymptotically sufficient and almost surely included in the σ-field of the parameters. Asymptotic sufficiency is established using techniques described in Section 7.3; the property of being almost surely included relies on slight extensions of 0-1 laws, i.e., laws for independent variables which are extended to conditionally independent variables. The relevant 0-1 laws are presented in a pure probabilistic set-up in Section 7.6; their statistical applications are in Section 7.7.

In the literature on time series analysis — particularly in econometrics — parameter estimability is generally obtained for stationary ergodic models. The results to be presented differ from the usual results in two respects. Firstly, the property of stationarity is weakened into a property of tail-sufficiency, which seems to be the weakest property which leads to suitable asymptotic behaviour. As this property may be difficult to verify directly we use, in Chapters 8 and 9, stationarity (in the sense of invariance for the shift) as a natural way of getting tail-sufficiency. Secondly, the property of ergodicity for the unreduced observations is weakened into a property of conditional ergodicity. In this chapter, ergodicity is in fact obtained through a condition of independence, to be weakened into conditional independence; because mixing properties require more structure, they are introduced in Chapter 8. In brief, a sequence of observations (y_n, z_n) is shown to exactly estimate the parameters characterizing the $(y \mid z)$-process under a condition of joint tail-sufficiency and of conditional independence.

This chapter seeks to provide the basic tools for the study of the asymptotic properties of a Bayesian experiment; in consequence, only a minimum of structure is introduced and as the necessary conditions are presented in what seems to be the weakest possible form, they are not verified directly; indeed the object of Chapters 8 and 9 is to use invariance arguments — such as stationarity or exchangeability — as efficient tools for checking the necessary conditions. This explains why this chapter has relatively few examples, and instead emphasizes methods; Chapter 8, and particularly Chapter 9, may be viewed as presenting examples and applications of the methods discussed in this chapter. Nonetheless the final section of this chapter presents a simple example which illustrates most of the concepts presented in the preceding sections.

Florens and Rolin (1984) reported the first steps of the work underlying this chapter; subsequent development of the asymptotic properties of conditional models and of finite parameter spaces is found in Florens, Mouchart and Rolin (1986) and Florens and Mouchart (1987c). However, this chapter considerably expands that material, and in particular incorporates a systematic study of asymptotically admissible reductions and of the correspondence between sampling theory and Bayesian results.

7.2 Limit of Sequences of Conditional Independences

The concern of this section is to describe those probabilistic tools which are most useful for the analysis of the properties of the limit of a sequence of conditional independence properties. We first recall the basic concepts of martingales and of tail σ-fields and then consider sequences of conditional independences, both with regards to filtrations and in general.

In this section, $(M, \mathcal{M}, P)$ is again an abstract probability space, $\mathcal{M}_i$ are sub-σ-fields of $\mathcal{M}$ (in particular $\mathcal{M}_0 = \{\phi, M\}$), $(\mathcal{F}_n)$ is a filtration in $(M, \mathcal{M})$ and $(\mathcal{G}_n)$ is a sequence of σ-fields adapted to $(\mathcal{F}_n)$ (see Definition 6.2.1). Also remember that $[\mathcal{M}_i]_p$ represents the class of p-integrable $\mathcal{M}_i$-measurable (real-valued) random variables defined on $\mathcal{M}$.

7.2.1 Definition. Consider a sequence (f_n) of random variables. (f_n) is a *martingale* adapted to a filtration $\mathcal{F}_n$ (or $(\mathcal{F}_n)$-martingale) if

(i) $f_n \in [\mathcal{F}_n]_1$

(ii) $\mathcal{F}_n f_{n+1} = f_n$.

7.2.2 Theorem. Let $\mathcal{F}_n \uparrow \mathcal{F}_\infty$ and (f_n) be an $(\mathcal{F}_n)$-martingale.

1.   If (f_n) is bounded in $\mathcal{L}_1$, i.e., $\sup_n \mathcal{M}_0(|f_n|) < \infty$, then there exists $g_1 \in [\mathcal{M}]_1$ such that
        (o) $f_n \to g_1$ a.s.P.

2.   The following conditions are equivalent:

        (i) $\exists g_2 \in [\mathcal{M}]_1$ such that $f_n \to g_2$ in L_1;
      (ii) $\exists g_3 \in [\mathcal{M}]_1$ such that $f_n = \mathcal{F}_n g_3$ a.s.P, $\forall\, n \in I\!N$;
    (iii) (f_n) is uniformly integrable,
        i.e., $\lim_{\alpha \to \infty} \sup_n \mathcal{M}_0[|f_n| 1_{\{|f_n| > \alpha\}}] = 0$.

Furthermore (iii) implies that (f_n) is bounded in L_1 and so if any one of these conditions is satisfied this implies (o) and $g_1 = g_2 = \mathcal{F}_\infty g_3$ a.s.P.

3.   If (f_n) is bounded in L_p with $p > 1$, i.e. $\sup_n \mathcal{M}_0(|f_n|^p) < \infty$,
    then 2.(iii) holds and therefore $f_n \to g_1$ a.s.P and furthermore in L_p.

For proofs and more details on martingales see, e.g., Neveu (1964) IV-5 (in particular Proposition IV-5-6), Neveu (1972), Chapter 4, or Dellacherie-Meyer (1980), Chapter V.

We now recall some facts on tail (or asymptotic) σ-fields.

7.2.3 Definition. Let $(\mathcal{G}_n)$ be a sequence of sub-σ-fields of $\mathcal{M}$. The *tail-σ-field* $\mathcal{G}_T$ is defined as

$$\mathcal{G}_T = \bigcap_{n \geq 0} \bigvee_{m \geq n} \mathcal{G}_m.$$

We also write $(\mathcal{G}_n)_T$ for $\mathcal{G}_T$ when such additional precision is required. When $\mathcal{G}_n$ is non-increasing (i.e., $\mathcal{G}_{n+1} \subset \mathcal{G}_n$ $\forall\, n \in I\!N$ or $\mathcal{G}_n \downarrow$),

$\mathcal{G}_T = \bigcap_{n\geq 0} \mathcal{G}_n$ whereas when $\mathcal{G}_n$ is non-decreasing — or equivalently a filtration — (i.e., $\mathcal{G}_n \subset \mathcal{G}_{n+1} \ \forall \ n \in \mathbb{N}$ or $\mathcal{G}_n \uparrow$), $\mathcal{G}_T = \mathcal{G}_\infty = \bigvee_{n\geq 0} \mathcal{G}_n$. When the sequence $(\mathcal{G}_n)$ is generated by a sequence of random variables g_n (i.e., $\mathcal{G}_n = \sigma(g_n)$), the tail σ-field $\mathcal{G}_T$ is the smallest σ-field that makes measurable the functions depending on the "last" coordinates g_n but not on the first m ones for any finite m. Note also that $\mathcal{G}_T$ makes $\liminf g_n$ and $\limsup g_n$ measurable for any sequence (g_n) such that $g_n \in [\mathcal{G}_n]$.

Note that Lemma 2.2.5(i) may be generalized into:

$$(7.2.1) \qquad\qquad \bigvee_{n\geq 0} \overline{\mathcal{G}_n} = \overline{\bigvee_{n\geq 0} \mathcal{G}_n}$$

for arbitrary sequences $(\mathcal{G}_n)$ whereas $\bigcap \overline{\mathcal{G}_n}$ is, in general, different from $\overline{\bigcap \mathcal{G}_n}$; the following properties are nonetheless observed:

7.2.4 Proposition. *(Elementary Properties of the Tail Operation)*

(i) $\overline{\bigcap_{n\geq 0} \mathcal{G}_n} \subset \bigcap_{n\geq 0} \overline{\mathcal{G}_n} \subset \overline{\mathcal{G}_T}$;

(ii) $(\overline{\mathcal{G}_n})_T = \overline{\mathcal{G}_T}$;

(iii)   If $\mathcal{F}_n \uparrow \mathcal{F}_\infty$ and $\mathcal{G}_n \uparrow \mathcal{G}_\infty$, then $(\mathcal{F}_n \vee \mathcal{G}_n)_\infty = \mathcal{F}_\infty \vee \mathcal{G}_\infty$;

(iv)   If $\mathcal{F}_n \downarrow \mathcal{F}_\infty$, then $\overline{\mathcal{F}_\infty} = (\overline{\mathcal{F}_n})_\infty$.

Proof. The first inclusion in (i) is trivial. For the second one, let us take $B \in \bigcap \overline{\mathcal{G}_n}$, i.e., $\forall \ n$, $\exists \ (B_n)$ such that $B_n \in \mathcal{G}_n$ and $B = B_n$ a.s. Because $\limsup B_n \in \mathcal{G}_T$ (by the definition of $\mathcal{G}_T$), and $B = \limsup B_n$ a.s. we have $B \in \overline{\mathcal{G}_T}$. Note that if $\mathcal{G}_n$ is decreasing, the three σ-fields in (i) are equal; this proves (ii) and (iv). Finally, (iii) derives directly from the associativity of $\vee$-operation once it is remarked that $\mathcal{F}_n \vee \mathcal{G}_n$ is also increasing. ∎

In the sequel we often rely on the following argument. Consider an integrable random variable defined on $M : m \in [\mathcal{M}]_1$ and a filtration in $\mathcal{M} : \mathcal{F}_n \uparrow \mathcal{F}_\infty$. The sequence $\mathcal{F}_n m$ is a $(\mathcal{F}_n)$-martingale which converges a.s. and in L_1 to $\mathcal{F}_\infty m$; this is an immediate application of Theorem 7.2.2. In particular, $\mathcal{F}_n m \to m$ once $m \in [\mathcal{F}_\infty]_1$. The next theorem illustrates another use of Theorem 7.2.2 when applied on a sequence of projections of σ-fields.

7.2.5 Theorem. For any sub σ-field $\mathcal{M}_1 \subset \mathcal{M}$ and any filtration $\mathcal{F}_n \uparrow \mathcal{F}_\infty$ in $\mathcal{M}$:

$$\overline{\mathcal{F}_\infty \mathcal{M}_1} \subset \overline{(\mathcal{F}_n \mathcal{M}_1)_T}.$$

Proof. It is sufficient to show that, for any $m \in [\mathcal{M}_1]_1$, $\mathcal{F}_\infty m$ is a.s. equal to a function measurable for the tail σ-field of the sequence $(\mathcal{F}_n \mathcal{M}_1)$. Indeed:

$$\forall\, m \in [\mathcal{M}_1]_1 : \mathcal{F}_n m \to \mathcal{F}_\infty m \quad \text{a.s.} P$$

by Theorem 7.2.2. As $\mathcal{F}_n m \in \mathcal{F}_n \mathcal{M}_1$, $\limsup \mathcal{F}_n m \in (\mathcal{F}_n \mathcal{M}_1)_T$. Therefore $\mathcal{F}_\infty m$ is a.s. equal to a function in $(\mathcal{F}_n \mathcal{M}_1)_T$. ∎

Reversing the orientation of the time index yields the following *reverse martingale theorem* (for proof, see, e.g., Dellacherie and Meyer (1980), Theorem V-33).

7.2.6 Theorem. Consider a decreasing sequence $\mathcal{F}_n \downarrow \mathcal{F}_T$ in $\mathcal{M}$. For any $m \in [\mathcal{M}]_1 : \mathcal{F}_n m \to \mathcal{F}_T m$ both a.s. and in L_1. In particular, $\mathcal{F}_n m \to m$ for $m \in [\mathcal{F}_T]_1$. ∎

We now present some results which combine conditional independence and martingale. The first result is similar to Theorems 6.2.2 and 6.2.3 where monotone class arguments are replaced by martingale arguments; this allows treatment of arbitrary sequences of conditioning σ-fields rather than filtrations.

7.2.7 Theorem. Consider three filtrations in $\mathcal{M} : \mathcal{F}_n \uparrow \mathcal{F}_\infty$, $\mathcal{G}_n \uparrow \mathcal{G}_\infty$ and $\mathcal{M}_n \uparrow \mathcal{M}_\infty$ along with a sequence $\mathcal{H}_n$ adapted to $(\mathcal{F}_n \vee \mathcal{M}_n)$, i.e.,

(o) $\mathcal{H}_n \subset \mathcal{F}_n \vee \mathcal{M}_n \quad \forall\, n \in I\!N$.

If

(i) $\mathcal{F}_n \perp\!\!\!\perp \mathcal{G}_n \mid \mathcal{M}_n \vee \mathcal{H}_n \quad \forall\, n \in I\!N$,

then

(ii) $\mathcal{F}_\infty \perp\!\!\!\perp \mathcal{G}_\infty \mid (\mathcal{M}_\infty \vee \mathcal{H}_n)_T$.

Proof. From (o) and (i) we have, for any $g \in [\mathcal{G}_k]_1, 0 \leq k \leq n$,

(iii) $(\mathcal{F}_n \vee \mathcal{M}_n)g = (\mathcal{H}_n \vee \mathcal{M}_n)g$.

By Theorem 7.2.2:

(iv) $(\mathcal{F}_n \vee \mathcal{M}_n)g \to (\mathcal{F}_\infty \vee \mathcal{M}_\infty)g$ a.s.

Taking the lim sup on both sides of (iii) we have, from (iv):

(v) $(\mathcal{F}_\infty \vee \mathcal{M}_\infty)g = \limsup_n (\mathcal{H}_n \vee \mathcal{M}_n)g$ a.s.

which may be written as

(vi) $(\mathcal{F}_\infty \vee \mathcal{M}_\infty)g \in \left[\overline{(\mathcal{H}_n \vee \mathcal{M}_n)_T}\right]$.

As $\mathcal{M}_n \uparrow \mathcal{M}_\infty$ implies $(\mathcal{H}_n \vee \mathcal{M}_n)_T = (\mathcal{H}_n \vee \mathcal{M}_\infty)_T$, (vi) implies:

(vii) $(\mathcal{F}_\infty \vee \mathcal{M}_\infty)g = (\mathcal{H}_n \vee \mathcal{M}_\infty)_T g$,

thus (ii) follows from the fact that (o) implies $(\mathcal{H}_n \vee \mathcal{M}_\infty)_T \subset \mathcal{F}_\infty \vee \mathcal{M}_\infty$ because of elementary Property 7.2.4. (iii) and from Monotone Class Theorem 0.2.21. ∎

Later on it will be useful to have a condition that allows one to replace $(\mathcal{M}_\infty \vee \mathcal{H}_n)_T$ in (ii) of Theorem 7.2.7 by $\mathcal{M}_\infty \vee \mathcal{H}_T$; describing such a condition is the object of the next theorem; its corollary provides an alternative description of a tail-σ-field.

7.2.8 Theorem. Let $\mathcal{M}_1 \subset \mathcal{M}$ and $(\mathcal{G}_n)$ be an arbitrary sequence in $\mathcal{M}$. Denote, as in Chapter 6, $\mathcal{G}_r^s = \bigvee_{r \leq n \leq s} \mathcal{G}_n$ (for $0 \leq r \leq s \leq \infty$). If

(i) $\mathcal{M}_1 \perp \!\!\! \perp \mathcal{G}_0^\infty \mid \mathcal{G}_T$,

then

(ii) $\overline{(\mathcal{M}_1 \vee \mathcal{G}_n)_T} = \overline{\mathcal{M}_1 \vee \mathcal{G}_T}$.

Proof. We have, by definition, that:

(iii) $(\mathcal{M}_1 \vee \mathcal{G}_n)_T = \bigcap_n \bigvee_{m \geq n} (\mathcal{M}_1 \vee \mathcal{G}_m) = \bigcap_n (\mathcal{M}_1 \vee \mathcal{G}_n^\infty);$

therefore, as $\mathcal{M}_1 \vee \mathcal{G}_T \subset \mathcal{M}_1 \vee \mathcal{G}_n^\infty \; \forall \, n$, we have $\mathcal{M}_1 \vee \mathcal{G}_T \subset (\mathcal{M}_1 \vee \mathcal{G}_n)_T$ where the inclusion may be strict; the hypothesis (i) will be used to prove the reverse inclusion.

Indeed, (i) implies that

(iv) $\mathcal{M}_1 \perp\!\!\!\perp \mathcal{G}_0^\infty \mid \mathcal{G}_n^\infty \; \forall \, n \in I\!N$

or, equivalently,

(v) $(\mathcal{M}_1 \vee \mathcal{G}_n^\infty) \perp\!\!\!\perp \mathcal{G}_0^\infty \mid \mathcal{G}_n^\infty \; \forall \, n \in I\!N.$

By (iii), this implies that

(vi) $(\mathcal{M}_1 \vee \mathcal{G}_n)_T \perp\!\!\!\perp \mathcal{G}_0^\infty \mid \mathcal{G}_n^\infty \; \forall \, n \in I\!N.$

By Theorem 7.2.7, (vi) implies that:

(vii) $(\mathcal{M}_1 \vee \mathcal{G}_n)_T \perp\!\!\!\perp \mathcal{G}_0^\infty \mid \mathcal{G}_T$

and, therefore, by Corollary 2.2.11 as $\mathcal{M}_1 \subset (\mathcal{M}_1 \vee \mathcal{G}_n)_T$:

(viii) $(\mathcal{M}_1 \vee \mathcal{G}_n)_T \perp\!\!\!\perp (\mathcal{M}_1 \vee \mathcal{G}_0^\infty) \mid \mathcal{M}_1 \vee \mathcal{G}_T.$

But (iii) also implies that $(\mathcal{M}_1 \vee \mathcal{G}_n)_T \subset \mathcal{M}_1 \vee \mathcal{G}_0^\infty$; therefore, by Corollary 2.2.8, (viii) implies that $(\mathcal{M}_1 \vee \mathcal{G}_n)_T \subset \overline{\mathcal{M}_1 \vee \mathcal{G}_T}.$ ∎

Combining twice Theorems 7.2.7 and 7.2.8 we obtain the following corollary, which is useful for the subsequent analysis:

7.2.9 Corollary. Under the conditions of Theorem 7.2.7 (namely, $\mathcal{F}_n \uparrow \mathcal{F}_\infty, \mathcal{G}_n \uparrow \mathcal{G}_\infty, \mathcal{M}_n \uparrow \mathcal{M}_\infty$ and $\mathcal{H}_n \subset \mathcal{F}_n \vee \mathcal{M}_n \; \forall \, n \in I\!N$), if:

(i) $\mathcal{F}_n \perp\!\!\!\perp \mathcal{G}_n \mid \mathcal{M}_n \vee \mathcal{H}_n \ \forall \, n \in \mathbb{N}$

(ii) $\mathcal{M}_n \perp\!\!\!\perp \mathcal{H}_0^n \mid \mathcal{H}_n,$

then

(iii) $\mathcal{F}_\infty \perp\!\!\!\perp \mathcal{G}_\infty \mid \mathcal{M}_\infty \vee \mathcal{H}_T.$ ∎

From Theorem 7.2.7 one may obtain the following dual result to Theorem 7.2.5, obtained by permuting projection and filtration.

7.2.10 Theorem. Let $\mathcal{M}_1 \subset \mathcal{M}$ and $\mathcal{F}_n \uparrow \mathcal{F}_\infty$ be a filtration in $\mathcal{M}$. Then $\mathcal{M}_1 \mathcal{F}_n \uparrow \mathcal{M}_1 \mathcal{F}_\infty.$

Proof. As $\mathcal{F}_n$ is a filtration, by elementary Property 4.3.2(iv), $\mathcal{M}_1 \mathcal{F}_n$ is also a filtration; therefore, $(\mathcal{M}_1 \mathcal{F}_n)_\infty \subset \mathcal{M}_1 \mathcal{F}_\infty$. By Theorem 4.3.3(i), $\mathcal{M}_1 \perp\!\!\!\perp \mathcal{F}_n \mid \mathcal{M}_1 \mathcal{F}_n \ \forall \, n$; thus by Theorem 7.2.7, $\mathcal{M}_1 \perp\!\!\!\perp \mathcal{F}_\infty \mid (\mathcal{M}_1 \mathcal{F}_n)_\infty$. By Theorem 4.3.3(ii), $\mathcal{M}_1 \mathcal{F}_\infty$ is the smallest sub-σ-field of $\mathcal{M}_1$ conditionally on which $\mathcal{M}_1$ and $\mathcal{F}_\infty$ are independent; this implies that $\mathcal{M}_1 \mathcal{F}_\infty \subset \overline{(\mathcal{M}_1 \mathcal{F}_n)_\infty}$. The equality is obtained once it is noted that both sides contain all the null sets of $\mathcal{M}_1$.   ∎

The next theorem is useful for the asymptotic study of dominated experiments.

7.2.11 Theorem. Let P and Q be two probabilities on $(M, \mathcal{M})$ and let $\mathcal{F}_n \uparrow \mathcal{F}_\infty$ be a filtration in $\mathcal{M}$. If

(i) $P_{\mathcal{F}_n} \ll Q_{\mathcal{F}_n} \ \forall \, n \geq 0$ and if $f_n = d \, P_{\mathcal{F}_n} / d \, Q_{\mathcal{F}_n},$
then
(ii) f_n is a martingale and $f_n \to f_\infty$ a.s.Q;

(iii) $f_n \to f_\infty$ in L_1 w.r.t. Q if and only if $P \ll Q$;

(iv) f_∞ is a generalized Radom-Nikodym derivative, i.e., f_∞ is characterized by: $\exists N \in \mathcal{F}_\infty$ such that $Q(N) = 0$ and

$$P(F) = \int_F f_\infty \, d \, Q + P(F \cap N) \quad \forall \, F \in \mathcal{F}_\infty.$$

Proof. (ii) is a simplified version of Proposition (III-2-7) in Neveu (1967); (iii) follows straightfowardly from Theorem 7.2.2 (see also Neveu (1964, ex: IV-5-3)).   ∎

7.3   Asymptotically Admissible Reductions

In this section we analyze the asymptotic admissibility of reductions. We first present some asymptotic properties of sequential experiments (Section 7.3.1) before considering the asymptotic admissibility of marginalization (Section 7.3.2) or of joint reductions (Section 7.3.3) in an unreduced experiment. We conclude this section by extending this analysis to conditional models.

7.3.1   Asymptotic Properties of Sequential Experiments

Let us consider a *sequential Bayesian experiment* as in Definition 6.3.1:

$$\mathcal{E} = (A \times S, \; \mathcal{A} \vee \mathcal{S}, \; \Pi, \; \mathcal{S}_n \uparrow \mathcal{S}_\infty).$$

In this chapter we implicitly assume, as in Chapter 6, that $n \in I\!N$ but we accept that $\mathcal{S}_\infty \subset \mathcal{S}$ is not necessarily equal to $\mathcal{S}$ (for instance, $\mathcal{S}$ may contain events due to randomization).

We first consider a limit property for posterior expectations. A direct application of Theorem 7.2.2 implies the following simple result:

7.3.1 Proposition. For any $a \in [\mathcal{A}]_1$, $\mathcal{S}_n a$ is a $(\mathcal{S}_n)$-martingale whose limit (a.s. and in L_1) is equal to $\mathcal{S}_\infty a$.   ∎

It should be stressed that Proposition 7.3.1. gives a genuinely Bayesian property in the sense that the martingale property of the Bayesian estimator $\mathcal{S}_n a$ — for an *arbitrary* (but integrable) function a of the parameters — is a property of the joint probability Π (more specifically, of the predictive probability P, i.e., the marginal of Π on $\mathcal{S}_\infty$); it differs from the usual asymptotic analysis of Bayesian estimators which is undertaken with respect to sampling probabilities. Connections between these approaches are considered in Section 7.4.

Theorem 7.2.11 permits the characterization of the limit of the sequences of densities associated to dominated experiments. As the limiting sampling probabilities tend to be undominated in most interesting cases, we assume that the probability Π is absolutely continuous with respect to the product of its margins in the finite horizon but not asymptotically.

7.3.2 Proposition. If

(i) $\Pi_{A \vee S_n} \ll \mu \otimes P_{S_n}, \quad \forall\, n \in \mathbb{N}$

then

(ii) $g_{A \vee S_n}$ is a positive $(A \vee S_n)$-martingale with respect to $\mu \otimes P$

which converges to g_∞ a.s.$\mu \otimes P$; the Lebesgue decomposition of Π with respect to $\mu \otimes P$ on $A \vee S$, is then given by

$$\Pi(M) = \int_M g_\infty\, d(\mu \otimes P) + \Pi(M \cap N) \quad \forall\, M \in A \vee S,$$

where $N \in A \vee S$ and $(\mu \otimes P)(N) = 0$.

(iii) $g_{A \vee S_n} \to g_\infty$ in L_1 if and only if $\Pi \ll \mu \otimes P$ on $A \vee S_\infty$;

in this case $g_\infty = g_{A \vee S_\infty}$ and $g_{A \vee S_n} = (\widetilde{A \vee S_n}) g_\infty$.

(iv) $g_\infty = 0$ a.s.$\mu \otimes P$ if and only if $\Pi \perp (\mu \otimes P)$. ∎

Note that (ii) is a simple consequence of (1.4.17) for $\mathcal{N} = A \vee S_n$, if we recall that $g_{A \vee S_{n'}} = (\widetilde{A \vee S_{n'}}) g_{A \vee S_n} \;\forall\, n' \leq n$ (see also in Section 6.3.1 the description of global analysis).

7.3.2 Asymptotic Sufficiency

We now investigate some relationships between sufficiency in finite samples and asymptotic sufficiency.

A. Sufficiency on the Sample Space

7.3.3 Definitions.

(a)   A sequence of statistics (T_n) is a *sufficient sequence* if:

  (i) (T_n) is (S_n)-adapted, i.e., $T_n \subset S_n \ \forall \, n \in \mathbb{N}$;
  (ii) T_n is sufficient in $\mathcal{E}_{A \lor S_n}$, the experiment stopped at time n, i.e.,
       $\mathcal{A} \perp\!\!\!\perp S_n \mid T_n \ \forall \, n \in \mathbb{N}$.

(b)   A statistic T is *asymptotically sufficient* if:

  (i) $T \subset S_\infty$;
  (ii) $\mathcal{A} \perp\!\!\!\perp S_\infty \mid T$.   ∎

Note that, in general (T_n) is not a filtration.

7.3.4 Proposition.

(i)   If (T_n) is a sufficient sequence of statistics its tail-σ-field, T_T, is asymptotically sufficient (i.e. $T_n \subset S_n$ and $\mathcal{A} \perp\!\!\!\perp S_n \mid T_n$ imply $T_T \subset S_\infty$ and $\mathcal{A} \perp\!\!\!\perp S_\infty \mid T_T$);

(ii) If T is an asymptotically sufficient σ-field, then $(S_n T)$ is a sufficient sequence of σ-fields.   ∎

Note that (i) is a straightforward application of Theorem 7.2.7 while (ii) is a simple consequence of Corollaries 2.2.4 (ii) and 4.3.5 (iii). Also, (i) can be viewed as a Bayesian version of similar results in a sampling theory framework (see Dynkin (1978), Diaconis and Freedman (1978), Lauritzen (1980), Section IV.4); result (ii) provokes the following question: Is the tail σ-field of the sequence $(S_n T)$ equal to T? The answer is no, in general, but the following inclusion obtains:

7.3.5 Proposition. Any asymptotically sufficient σ-field is included in the tail σ-field of the sufficient sequence formed by its projections on S_n, completed by the null sets of S_∞, i.e., $\mathcal{A} \perp\!\!\!\perp S_\infty \mid T \Rightarrow T \subset \overline{(S_n T)}_T \cap S_\infty$.   ∎

Indeed, the inclusion $T \subset \overline{(S_n T)}_T \cap S_\infty$ is a direct consequence of Theorem 7.2.5, after noting that $T \subset S_\infty$ implies $S_\infty T = \overline{T} \cap S_\infty$; Example 9.3.6.D shows why the inclusion is not, in general, an equality.

If we now start from a sufficient sequence (T_n) we know, by Proposition 7.3.4 (i), that its tail-σ-field T_T is asymptotically sufficient and by Proposition 7.3.4 (ii) that the sequence $(S_n T_T)$ is a sufficient sequence. Example 9.3.6.D also shows that there is, in general, no adaptation between the two sequences (T_n) and $(S_n T_T)$. This is, however, the case if (T_n) is minimal sufficient.

B. Minimal Sufficiency on the Sample Space

Recall, from Section 4.3, that $(S_n A)$ is a sequence of minimal sufficient statistics and that $S_\infty A$ is a minimal asymptotically sufficient statistic. We now investigate the links between $S_\infty A$ and the sequence $(S_n A)$. From Theorems 7.2.5 and 7.3.4 we have successively,

$$(7.3.1) \qquad S_\infty A \;\subset\; \overline{(S_n A)_T};$$

$$(7.3.2) \qquad S_n A \;\subset\; S_n(S_\infty A);$$

and, therefore,

$$(7.3.3) \qquad S_n A \;\subset\; S_n(S_\infty A) \subset S_n\{(S_n A)_T\}.$$

In (7.3.1) the inclusion is generally strict; in other words, the minimal asymptotically sufficient statistic is generally smaller that the tail-σ-field of the sequence of minimal sufficient statistics (see Example 9.3.6.D). Similarly, in (7.3.2), the inclusion is generally strict, i.e., if one projects the minimal asymptotically sufficient statistic $S_\infty A$ on the filtration (S_n) we get a sufficient sequence $S_n(S_\infty A)$ which is not necessarily minimal. However, this will be the case under a strong identification condition; the next theorem proves a slightly more general result:

7.3.6 Theorem. If $T \subset S_\infty$ is asymptotically sufficient and two-complete, then the sufficient sequence $(S_n T)$ is minimal, i.e., $S_n T$ is $\mathcal{E}_{A \vee S_n}$-minimal sufficient for all n.

Proof. In order to prove that $S_n T = S_n A$ we first note that, as $T \ll_2 A$ implies $T \ll A$, we have by Theorem 5.5.4 (i) that $\overline{T} = \overline{S_\infty A}$. Therefore, $S_\infty A \ll_2 A$ by Proposition 5.4.2. By the elementary Property 4.3.2. (vi), we then obtain that $S_n T = S_n \overline{T} = S_n(S_\infty A)$. Finally, Theorem 5.4.14 implies that $S_n(S_\infty A) = S_n A$. ∎

Note that, in view of Theorem 5.4.14, the condition of 2-completeness may be replaced by "p-complete for some p such that $1 < p < \infty$". Example 9.3.6.D also shows that 2-completeness of an asymptotically sufficient σ-field does not imply, in general, that its projections on the filtration S_n are 2-complete. However, the issue of whether the tail-σ-field of a sufficient sequence of 2-complete σ-fields is the minimal asymptotically sufficient σ-field seems unresolved.

C. Sufficiency on the Parameter Space

Sufficiency on the parameter space in sequential experiments is a particularly relevant topic in problems with incidental parameters such as $x_n \sim i.N(\mu_n, \sigma^2)$ (see Florens, Mouchart and Richard (1974, 1976, 1979, 1986 and 1987); see also Neyman (1951), Neyman and Scott (1948, 1951), Reiersøl (1950), Solari (1969), Sprent (1966, 1970), Anderson (1976) and Deistler (1976, 1986)). In this context identification (i.e., minimal sufficiency) and estimability, along with their connections between the sequence of finite sample experiments and the asymptotic experiment, are not only of interest in their own right, but also represent the first step in the analysis of sequences of conditional models (examined in Section 7.3.4).

Symmetrically to sufficiency on the sample space, we consider a sequence $(\mathcal{B}_n)$ of sub-σ-fields of $\mathcal{A}$:

7.3.7 Definitions.

(a)   a sequence of parameters $(\mathcal{B}_n)$ is a *sufficient sequence* if

   (i) $\mathcal{B}_n \subset \mathcal{A}, \quad \forall\, n \in I\!N$;
   (ii) $\mathcal{B}_n$ is sufficient in $\mathcal{E}_{\mathcal{A} \vee S_n}$, i.e., $\mathcal{A} \perp\!\!\!\perp S_n \mid \mathcal{B}_n \ \forall\, n \in I\!N$.

(b)   a parameter is *asymptotically sufficient* if

   (i) $\mathcal{B} \subset \mathcal{A}$;
   (ii) $\mathcal{B}$ is sufficient in the asymptotic experiment, i.e., $\mathcal{A} \perp\!\!\!\perp S_\infty \mid \mathcal{B}$. ∎

7.3.8 Proposition.

(i) If $(\mathcal{B}_n)$ is a sufficient sequence of parameters, then its tail-σ-field, $\mathcal{B}_T$, is asymptotically sufficient (i.e., $\mathcal{B}_n \subset \mathcal{A}$ and $\mathcal{A} \perp\!\!\!\perp S_n \mid \mathcal{B}_n$ imply $\mathcal{B}_T \subset \mathcal{A}$ and $\mathcal{A} \perp\!\!\!\perp S_\infty \mid \mathcal{B}_T$)

(ii) If $\mathcal{B}$ is an asymptotically sufficient parameter, then $\mathcal{B}$ is $\mathcal{E}_{\mathcal{A} \vee \mathcal{S}_n}$-sufficient for all n (i.e., $\mathcal{B} \subset \mathcal{A}$ and $\mathcal{A} \perp\!\!\!\perp \mathcal{S}_\infty \mid \mathcal{B}$ imply that $\mathcal{A} \perp\!\!\!\perp \mathcal{S}_n \mid \mathcal{B} \ \forall \ n$). ∎

Note that (i) is a direct application of Theorem 7.2.7 whereas (ii) is a trivial application of Corollary 2.2.4 (ii). In Proposition 7.3.8, the sequence $(\mathcal{B}_n)$ is, in general, not a filtration. The sequence of minimal sufficient parameters is nevertheless a filtration. More specifically, we have from Theorem 7.2.10:

7.3.9 Proposition. The sequence of minimal sufficient parameters is a filtration converging to the asymptotic minimal sufficient parameter (i.e., $\mathcal{AS}_n \uparrow \mathcal{AS}_\infty$). ∎

As far as identification is concerned, it may be noted that if an experiment $\mathcal{E}_{\mathcal{A} \vee \mathcal{S}_n}$ is identified (i.e., $\mathcal{AS}_n = \mathcal{A}$) for some finite n, then it is also identified for any $n' \geq n$ and therefore asymptotically identified (i.e., $\mathcal{AS}_\infty = \mathcal{A}$); of course, the converse does not hold, but we have the following result:

7.3.10 Proposition. Any $b \in [\mathcal{AS}_\infty]_1$ is a limit, a.s. and in L_1, of a sequence $b_n \in [\mathcal{AS}_n]_1$. In particular, if $\mathcal{E}_{\mathcal{A} \vee \mathcal{S}_\infty}$ is identified, this property holds for any $b \in [\mathcal{A}]_1$. ∎

By Proposition 7.3.9, this follows straightforwardly from Theorem 7.2.2 applied to the sequence $b_n = (\mathcal{AS}_n)b$. Note that if $\mathcal{AS}_\infty \neq \mathcal{AS}$, what is identified in $\mathcal{E}_{\mathcal{A} \vee \mathcal{S}}$ but not in $\mathcal{E}_{\mathcal{A} \vee \mathcal{S}_\infty}$ cannot be approximated by a sequence $b_n \in [\mathcal{AS}_n]_1$.

7.3.3 Asymptotic Admissibility of Joint Reductions

A. Mutual Ancillarity

Choosing $(\mathcal{M}_n)$ and $(\mathcal{H}_n)$ to be trivial in Theorem 7.2.7 gives a simple limit result for sequences of mutually ancillary σ-fields when their sequences are filtrations:

7.3.11 Proposition. If (C_n) and (T_n) are two filtrations (in A and in S) of mutually ancillary σ-fields, their limit is also mutually ancillary, i.e., $C_n \uparrow C_\infty, T_n \uparrow T_\infty$ and $C_n \perp\!\!\!\perp T_n$, $\forall n \in I\!\!N$, imply $C_\infty \perp\!\!\!\perp T_\infty$. ∎

Note that if either (C_n) or (T_n) were not a filtration, the conclusion might fail to hold for the tail σ-fields; this is shown in the following example:

Example. Let $s_n = (x_n, y_n)'$ and

$$(s_n \mid a) \sim i.N_2\left[0, \begin{pmatrix} 1 & a \\ a & 1 \end{pmatrix}\right];$$

let $S_n = \sigma(s_1, \ldots, s_n)$, $A = \sigma(a)$, $T_{2n-1} = \sigma(x_n)$, $T_{2n} = \sigma(y_n)$ and $C_n = A$. Clearly, the sequences (C_n) and (T_n) are mutually ancillary, however, (T_n) is not a filtration, (C_n) and $(\bigvee_{1 \leq i \leq n} T_i)$ are not mutually ancillary and, as will be shown in Chapter 9, $C_T = A$ and T_T are not mutually ancillary since $A \subset \overline{T_T}$. ∎

B. Mutual Exogeneity

With regards to mutual ancillarity, asymptotic properties are easily obtained for filtrations:

7.3.12 Proposition. If two filtrations $C_n \uparrow C_\infty$ and $T_n \uparrow T_\infty$ form a pair of mutually exogenous sequences (i.e., $C_n \perp\!\!\!\perp T_n$ and $A \perp\!\!\!\perp S_n \mid C_n \vee T_n$), then their increasing limits are also mutually exogenous (i.e., $C_\infty \perp\!\!\!\perp T_\infty$ and $A \perp\!\!\!\perp S_\infty \mid C_\infty \vee T_\infty$). ∎

This proposition is a simple consequence of Proposition 7.3.11 and Theorem 7.2.7, with $F_n = A$, $G_n = S_n$, $M_n = C_n \vee T_n$ and H_n trivial. If the sequence T_n is not a filtration, then a similar result holds, although somewhat restricted as it requires (C_n) to be a constant sequence and strenghtens the condition of mutual ancillarity.

7.3.13 Theorem. Let $C \subset A$ and (T_n) be an (S_n)-adapted sequence. If, in $\mathcal{E}_{A \vee S_n}$, C and T_n are mutually exogenous and C and T_0^n are mutually ancillary for any n, than C and T_T are mutually exogenous and C and T_0^∞ are mutually ancillary in $\mathcal{E}_{A \vee S_\infty}$.

Proof. After recalling that $T_T \subset T_0^\infty$, one may rewrite the theorem as: If

(i) $C \perp\!\!\!\perp T_0^n \ \forall\, n \in \mathbb{N}$

(ii) $A \perp\!\!\!\perp S_n \mid C \vee T_n$
     $\forall\, n \in \mathbb{N}$,

then

(iii) $C \perp\!\!\!\perp T_0^\infty$

(iv) $A \perp\!\!\!\perp S_\infty \mid C \vee T_T$.

That (i) implies (iii) is a simple consequence of Theorem 7.2.7; by this same theorem, (ii) implies that: $A \perp\!\!\!\perp S_\infty \mid (C \vee T_n)_T$ and by Theorem 7.2.8, (iii) implies that: $\overline{C \vee T_T} = \overline{(C \vee T_n)_T}$. ∎

C. Mutual Sufficiency

For mutual sufficiency, the result analogous to Theorem 7.3.13 is the following:

7.3.14 Theorem. Let (B_n) be a filtration of parameters, and let (T_n) be an arbitrary sequence of statistics. If B_n and T_n are mutually sufficient and B_n is $\mathcal{E}_{A \vee T_0^n}$-sufficient, then B_∞ and T_T are $\mathcal{E}_{A \vee S_\infty}$-mutually sufficient and B_∞ is $\mathcal{E}_{A \vee T_0^\infty}$-sufficient.

Proof. This theorem may be rewritten as: If

(i) $B_n \perp\!\!\!\perp S_n \mid T_n \ \forall\, n \in \mathbb{N}$

(ii) $A \perp\!\!\!\perp T_0^n \mid B_n \ \forall\, n \in \mathbb{N}$

then

(iii) $B_\infty \perp\!\!\!\perp S_\infty \mid T_T$

(iv) $A \perp\!\!\!\perp T_0^\infty \mid B_\infty,$

since (iv) trivially implies that $A \perp\!\!\!\perp T_T \mid B_\infty$. Now, by Theorem 7.2.7, (iii) is implied by (i) and (iv) is implied by (ii). ∎

D. Cut

With regards to mutual exogeneity and mutual sufficiency, Theorem 7.2.7 trivially implies that, for filtrations on both the parameter and the sample spaces, cuts for finite n produce asymptotic cuts. More interesting are the following results in the same spirit as Theorems 7.3.13 and 7.3.14:

7.3.15 Theorem. Let (B_n) be a filtration in A, $C \subset A$ and (T_n) an (S_n)-adapted sequence. If (B_n, C, T_n) operates a Bayesian cut in $\mathcal{E}_{A \vee S_n}$, and if B_n is $\mathcal{E}_{A \vee T_0^n}$-sufficient, then (B_∞, C, T_T) operates an asymptotic cut, i.e., a cut in $\mathcal{E}_{A \vee S_\infty}$, and B_∞ is $\mathcal{E}_{A \vee T_0^\infty}$-sufficient.

Proof. We first restate the theorem as: If

(i) $B_n \perp\!\!\!\perp C,$

(ii) $A \perp\!\!\!\perp T_0^n \mid B_n$

(iii) $A \perp\!\!\!\perp S_n \mid C \vee T_n,$

then

(iv) $B_\infty \perp\!\!\!\perp C,$

(v) $A \perp\!\!\!\perp T_0^\infty \mid B_\infty$

(vi) $A \perp\!\!\!\perp S_\infty \mid C \vee T_T.$

As before, (i) implies (iv) and (ii) implies (v). But, $A \perp\!\!\!\perp S_\infty \mid (C \vee T_n)_T$ is implied by (iii); and (i) and (ii) imply, by Theorem 2.2.10, $C \perp\!\!\!\perp (T_0^n \vee B_n)$, which yields $C \perp\!\!\!\perp T_0^\infty$. Under Theorem 7.2.8 this gives $\overline{(C \vee T_n)_T} = \overline{C \vee T_T}$. ∎

Note that when the sequence (T_n) is not increasing, condition (ii bis) — $\mathcal{A} \perp\!\!\!\perp T_n \mid \mathcal{B}_n$ — is weaker than condition (ii) which is required in order to ensure a sequence of cuts to have a suitable asymptotic property, i.e., an asymptotic cut with the tail σ-field of (T_n). From Theorem 6.2.3, condition (ii) may be rewritten in several equivalent forms. For instance, since (ii) is equivalent to (ii bis) and $\mathcal{A} \perp\!\!\!\perp T_0^n \mid \mathcal{B}_n \vee T_n$, then if $(\mathcal{B}_n, \mathcal{C}, T_n)$ operates a cut the condition $\mathcal{B}_n$ is $\mathcal{E}_{\mathcal{A} \vee T_0^n}$-sufficient is equivalent to T_n is $\mathcal{E}_{\mathcal{A} \vee T_0^n}^{\mathcal{B}_n}$-sufficient.

7.3.4 Asymptotically Admissible Reductions in Conditional Experiments

The asymptotic properties of regression-type models are based on the analysis of a sequence of conditional experiments such as $\mathcal{E}_{\mathcal{B} \vee T_n}^{\mathcal{M}_n}$, where both (T_n) and $(\mathcal{M}_n)$ are filtrations. More precisely, $T_n \uparrow T_\infty$ is $(\mathcal{S}_n)$-adapted and $\mathcal{M}_n \uparrow \mathcal{M}_\infty$ is $(\mathcal{A} \vee \mathcal{S}_n)$-adapted (typically, $(\mathcal{M}_n)$ is generated by a sequence of exogenous variables). In the spirit of Definition 6.3.1, such a structure is called a *sequential conditional Bayesian experiment*.

In this section, we analyse merely the problems of sufficiency and ancillarity. Joint reductions in conditional experiments do not raise problems specifically different from those encountered in the complete experiment.

A. Ancillarity

A simple consequence of Theorem 7.2.7 is the following proposition:

7.3.16 Proposition. A filtration $\mathcal{R}_n \uparrow \mathcal{R}_\infty$ of $\mathcal{E}_{\mathcal{B} \vee T_n}^{\mathcal{M}_n}$-ancillary statistics, i.e., $\mathcal{R}_n \subset \mathcal{M}_n \vee T_n$ and $\mathcal{B} \perp\!\!\!\perp \mathcal{R}_n \mid \mathcal{M}_n$, is asymptotically ancillary, i.e., $\mathcal{B} \perp\!\!\!\perp \mathcal{R}_\infty \mid \mathcal{M}_\infty$. ∎

B. Sufficiency on the Sample Space

A straightforward reinterpretation of Proposition 7.3.4 entails the following proposition:

7.3.17 Proposition.

(i) If $(\mathcal{N}_n)$ is a sequence of $\mathcal{E}_{\mathcal{B} \vee T_n}^{\mathcal{M}_n}$-sufficient statistics, i.e., $\mathcal{N}_n \subset \mathcal{M}_n \vee T_n$ and $\mathcal{B} \perp\!\!\!\perp T_n \mid \mathcal{M}_n \vee \mathcal{N}_n$, then $(\mathcal{M}_\infty \vee \mathcal{N}_n)_T$ is a strong $\mathcal{E}_{\mathcal{B} \vee T_\infty}^{\mathcal{M}_\infty}$-sufficient statistics, i.e., $\mathcal{M}_\infty \subset (\mathcal{M}_\infty \vee \mathcal{N}_n)_T$ and $\mathcal{B} \perp\!\!\!\perp T_\infty \mid (\mathcal{M}_\infty \vee \mathcal{N}_n)_T$;

(ii) If $\mathcal{N}$ is a $\mathcal{E}^{\mathcal{M}_\infty}_{\mathcal{B} \vee T_\infty}$-sufficient statistic, i.e., $\mathcal{B} \perp\!\!\!\perp T_\infty \mid \mathcal{M}_\infty \vee \mathcal{N}$, then $(\mathcal{M}_n \vee T_n)(\mathcal{M}_\infty \vee \mathcal{N})$ is a sequence of strong $\mathcal{E}^{\mathcal{M}_n}_{\mathcal{B} \vee T_n}$-sufficient statistics, i.e., $\mathcal{M}_n \subset (\mathcal{M}_n \vee T_n)(\mathcal{M}_\infty \vee \mathcal{N})$ and $\mathcal{B} \perp\!\!\!\perp T_n \mid (\mathcal{M}_n \vee T_n)(\mathcal{M}_\infty \vee \mathcal{N})$. ∎

With regards to minimal sufficiency on the sample space in conditional experiments, recall that

$$(7.3.4) \qquad \mathcal{N}^*_n = (\mathcal{M}_n \vee T_n)(\mathcal{B} \vee \mathcal{M}_n)$$

and

$$(7.3.5) \qquad \mathcal{N}^{**} = (\mathcal{M}_\infty \vee T_\infty)(\mathcal{B} \vee \mathcal{M}_\infty)$$

are, respectively, the minimal complete strong sufficient $\mathcal{E}^{\mathcal{M}_n}_{\mathcal{B} \vee T_n}$- and $\mathcal{E}^{\mathcal{M}_\infty}_{\mathcal{B} \vee T_\infty}$-statistics.

As in (7.3.1), (7.3.2) and (7.3.3) of Section 7.3.2.B, we have:

$$(7.3.6) \qquad \mathcal{N}^{**} \subset \mathcal{N}^*_T$$

$$(7.3.7) \qquad \mathcal{N}^*_n \subset (\mathcal{M}_n \vee T_n)\mathcal{N}^{**},$$

and this implies, in turn, that

$$(7.3.8) \qquad \mathcal{N}^*_n \subset (\mathcal{M}_n \vee T_n)\mathcal{N}^*_T.$$

In contrast with unreduced experiments, equality in (7.3.7), i.e., the minimal sufficiency of the projection of $\mathcal{N}^{**}$ on $(\mathcal{M}_n \vee T_n)$, is not assured by a strong identification condition only; it requires a supplementary transitivity hypothesis , as shown in the next theorem:

7.3.18 Theorem. If $\mathcal{N} \subset \mathcal{M}_\infty \vee T_\infty$ is a 2-complete $\mathcal{E}^{\mathcal{M}_\infty}_{\mathcal{B} \vee T_\infty}$-sufficient statistic, and if $(\mathcal{M}_n)$ is $\mathcal{B}$-transitive for $(\mathcal{M}_n \vee T_n)$, then the sequence $(\mathcal{M}_n \vee T_n)\mathcal{N}$ is a sequence of strong minimal $\mathcal{E}^{\mathcal{M}_n}_{\mathcal{B} \vee T_n}$-sufficient statistics.

Proof. By Definitions 5.5.1 and 6.4.1, and by Theorem 6.2.2, this theorem may be rephrased as: If

(i) $\mathcal{B} \perp\!\!\!\perp T_\infty \mid \mathcal{M}_\infty \vee \mathcal{N}$,

(ii) $\mathcal{N} \ll_2 \mathcal{B} \mid \mathcal{M}_\infty$, and

(iii) $\mathcal{M}_\infty \perp\!\!\!\perp T_n \mid \mathcal{B} \vee \mathcal{M}_n$,

then

(iv) $(\mathcal{M}_n \vee T_n)\mathcal{N} = \mathcal{N}_n^*$.

Since $\mathcal{N} \ll_2 \mathcal{B} \mid \mathcal{M}_\infty$ implies $\mathcal{N} \ll \mathcal{B} \mid \mathcal{M}_\infty$, and since $\mathcal{N} \vee \mathcal{M}_\infty$ is a strong $\mathcal{E}_{\mathcal{B} \vee T_\infty}^{\mathcal{M}_\infty}$-statistic, by Theorem 5.5.4 (i), that $\overline{\mathcal{N} \vee \mathcal{M}_\infty} = \overline{\mathcal{N}^{**}}$. Therefore, $\mathcal{N}^{**} \ll_2 \mathcal{B} \mid \mathcal{M}_\infty$. This is equivalent to $\mathcal{N}^{**} \ll_2 (\mathcal{B} \vee \mathcal{M}_\infty)$. Therefore, by Corollary 5.4.15,

$$\begin{aligned}
(\mathcal{M}_n \vee T_n)\mathcal{N}^{**} &= (\mathcal{M}_n \vee T_n)\{(\mathcal{M}_\infty \vee T_\infty)(\mathcal{B} \vee \mathcal{M}_\infty)\} \\
&= (\mathcal{M}_n \vee T_n)(\mathcal{B} \vee \mathcal{M}_\infty).
\end{aligned}$$

Now, since (iii) implies $\mathcal{M}_\infty \perp\!\!\!\perp (\mathcal{M}_n \vee T_n) \mid \mathcal{B} \vee \mathcal{M}_n$, we conclude, from Corollary 4.3.6(iv), that $(\mathcal{M}_n \vee T_n)(\mathcal{B} \vee \mathcal{M}_\infty) = (\mathcal{M}_n \vee T_n)(\mathcal{B} \vee \mathcal{M}_n) = \mathcal{N}_n^*$. ∎

Recall that Proposition 6.4.3 gives alternative characterizations of the transitivity condition (iii).

C. Sufficiency on the Parameter Space

Under supplementary conditions on the conditioning filtration, we now extend Proposition 7.3.8 to conditional experiments.

7.3.19 Theorem. If $(\mathcal{L}_n)$ is a sequence of $\mathcal{E}_{\mathcal{B} \vee T_n}^{\mathcal{M}_n}$-sufficient parameters such that $\mathcal{L}_0^n \perp\!\!\!\perp \mathcal{M}_n \mid \mathcal{L}_n \; \forall \, n \in I\!N$, then $\mathcal{L}_T$ is asymptotically sufficient. Moreover $\mathcal{L}_0^\infty \perp\!\!\!\perp \mathcal{M}_\infty \mid \mathcal{L}_T$.

Proof. This may be rephrased as: If

(o) $\mathcal{L}_n \subset \mathcal{B} \vee \mathcal{M}_n$,

(i) $\mathcal{B} \perp\!\!\!\perp T_n \mid \mathcal{L}_n \vee \mathcal{M}_n$

(ii) $\mathcal{L}_0^n \perp\!\!\!\perp \mathcal{M}_n \mid \mathcal{L}_n,$

then

(iii) $\mathcal{B} \perp\!\!\!\perp \mathcal{T}_\infty \mid \mathcal{L}_T \vee \mathcal{M}_\infty,$

(iv) $\mathcal{L}_0^\infty \perp\!\!\!\perp \mathcal{M}_\infty \mid \mathcal{L}_T.$

By Theorem 7.2.7, (i) implies $\mathcal{B} \perp\!\!\!\perp \mathcal{T}_\infty \mid (\mathcal{L}_n \vee \mathcal{M}_\infty)_T$ and (ii) implies $\mathcal{L}_0^\infty \perp\!\!\!\perp \mathcal{M}_\infty \mid \mathcal{L}_T$. It also implies $\overline{(\mathcal{L}_n \vee \mathcal{M}_\infty)_T} = \overline{\mathcal{L}_T \vee \mathcal{M}_\infty}$, by Theorem 7.2.8. ∎

7.3.20 Theorem. $\mathcal{L} \subset \mathcal{B}$ is a uniform $\mathcal{E}_{\mathcal{B} \vee \mathcal{T}_\infty}^{\mathcal{M}_\infty}$-sufficient parameter and $(\mathcal{M}_n)$ is $\mathcal{L}$-transitive for $(\mathcal{T}_n)$ if and only if $\mathcal{L}$ is a uniform $\mathcal{E}_{\mathcal{B} \vee \mathcal{T}_n}^{\mathcal{M}_n}$-sufficient parameter for each n and $(\mathcal{M}_n)$ is $\mathcal{B}$-transitive for $(\mathcal{T}_n)$.

Proof. This may be rephrased as: Let

(o) $\mathcal{L} \subset \mathcal{B}.$

Then

(i) $\mathcal{B} \perp\!\!\!\perp \mathcal{T}_\infty \mid \mathcal{L} \vee \mathcal{M}_\infty$
(ii) $\mathcal{M}_\infty \perp\!\!\!\perp \mathcal{T}_n \mid \mathcal{L} \vee \mathcal{M}_n$

are equivalent to

(iii) $\mathcal{B} \perp\!\!\!\perp \mathcal{T}_n \mid \mathcal{L} \vee \mathcal{M}_n$
(iv) $\mathcal{M}_\infty \perp\!\!\!\perp \mathcal{T}_n \mid \mathcal{B} \vee \mathcal{M}_n.$

Clearly (i) is equivalent, by Theorem 6.2.3, to:

(v) $\mathcal{B} \perp\!\!\!\perp \mathcal{T}_n \mid \mathcal{L} \vee \mathcal{M}_\infty.$

By Theorem 2.2.10, (v) and (ii) are equivalent to $(\mathcal{B} \vee \mathcal{M}_\infty) \perp\!\!\!\perp \mathcal{T}_n \mid \mathcal{L} \vee \mathcal{M}_n$ which is equivalent to (iii) and (iv). ∎

For minimal sufficiency on the parameter space in conditional experiments, Proposition 7.3.9 may be extended; however some care is required in treating null sets, and a transitivity condition is also required. Recall that

$$(7.3.9) \qquad \mathcal{L}_n^* = (\mathcal{B} \vee \mathcal{M}_n)(\mathcal{M}_n \vee \mathcal{T}_n)$$

and

$$(7.3.10) \qquad \mathcal{L}^{**} = (\mathcal{B} \vee \mathcal{M}_\infty)(\mathcal{M}_\infty \vee \mathcal{T}_\infty)$$

are, respectively, the minimal complete strong sufficient parameter in $\mathcal{E}_{\mathcal{B} \vee \mathcal{T}_n}^{\mathcal{M}_n}$ and in $\mathcal{E}_{\mathcal{B} \vee \mathcal{T}_\infty}^{\mathcal{M}_\infty}$.

7.3.21 Theorem. If $(\mathcal{M}_n)$ is $\mathcal{B}$-transitive for $(\mathcal{T}_n)$, then $(\mathcal{L}_n^*)$ is a filtration such that $\mathcal{L}^{**} = \overline{\mathcal{L}_\infty^*} \cap (\mathcal{B} \vee \mathcal{M}_\infty)$.

Proof. Since $\mathcal{M}_\infty \perp\!\!\!\perp \mathcal{T}_n \mid \mathcal{B} \vee \mathcal{M}_n$ implies that $(\mathcal{B} \vee \mathcal{M}_\infty) \perp\!\!\!\perp (\mathcal{M}_n \vee \mathcal{T}_n) \mid \mathcal{B} \vee \mathcal{M}_n$, we obtain, by Theorem 4.3.7 (iv), $\mathcal{L}_n^* = \{(\mathcal{B} \vee \mathcal{M}_\infty)(\mathcal{M}_n \vee \mathcal{T}_n)\} \cap (\mathcal{B} \vee \mathcal{M}_n)$. Now, clearly, $\mathcal{L}_n^*$ is a filtration since $(\mathcal{B} \vee \mathcal{M}_n)$ and $(\mathcal{B} \vee \mathcal{M}_\infty)(\mathcal{M}_n \vee \mathcal{T}_n)$ are filtrations. However, by Theorem 7.2.10, $\mathcal{L}^{**} = [(\mathcal{B} \vee \mathcal{M}_\infty)(\mathcal{M}_n \vee \mathcal{T}_n)]_\infty$. By Theorem 4.3.7 (v), $(\mathcal{B} \vee \mathcal{M}_\infty)(\mathcal{M}_n \vee \mathcal{T}_n) = \overline{(\mathcal{B} \vee \mathcal{M}_n)(\mathcal{M}_n \vee \mathcal{T}_n)} \cap (\mathcal{B} \vee \mathcal{M}_\infty) = \overline{\mathcal{L}_n^*} \cap (\mathcal{B} \vee \mathcal{M}_\infty)$. And by Lemma 2.2.5 (iii), $\overline{\mathcal{L}_n^*} \cap (\mathcal{B} \vee \mathcal{M}_\infty) = \mathcal{L}_n^* \vee \{(\mathcal{B} \vee \mathcal{M}_\infty) \cap \overline{\mathcal{I}}\}$ where, as usual, $\overline{\mathcal{I}} = \{M \in \mathcal{A} \vee \mathcal{S} : \Pi(M)^2 = \Pi(M)\}$. Hence

$$\mathcal{L}^{**} = \mathcal{L}_\infty^* \vee \{(\mathcal{B} \vee \mathcal{M}_\infty) \cap \overline{\mathcal{I}}\} = \overline{\mathcal{L}_\infty^*} \cap (\mathcal{B} \vee \mathcal{M}_\infty),$$

by another application of Lemma 2.2.5 (iii). ∎

As in the unreduced experiment, it is evident that if at some stage, the conditional experiment is identified (i.e., $\mathcal{L}_n^* = \mathcal{B} \vee \mathcal{M}_n$), then the asymptotic conditional experiment is identified (i.e., $\mathcal{L}^{**} = \mathcal{B} \vee \mathcal{M}_\infty$). However, by the martingale convergence theorem, using (2.2.3), any integrable $\mathcal{L}^{**}$-measurable function is the almost sure limit of the sequence of its projections on $\mathcal{L}_n^*$ which, by definition, are $\mathcal{L}_n^*$-measurable.

7.4   Asymptotic Exact Estimability

In this section we first examine the exact estimability from both a
Bayesian and an asymptotic point of view. More specifically, in a Bayesian
framework, consistency is exact estimability in an asymptotic experiment.
We apply the results of Section 4.7 (on exact estimability) to the particular
structure of an asymptotic experiment. For the general case, the first known
results probably date back to Doob (1949); further results were obtained by
Berk (1966, 1970) and LeCam (1986), Chap. 17. The finite (or countable)
case is treated in the following section. Some discussion of the connection
between Bayesian and sampling theory consistency, in the general case, con-
cludes this section.

7.4.1   Exact Estimability and Bayesian Consistency

We now examine concepts of exact estimability as introduced in Section
4.7, in the case of sequential Bayesian experiments, and relate this concept
to the more familiar concept of consistency. For expository reasons, we provide definitions and results for the general sequence of conditional Bayesian
experiments $\mathcal{E}^{\mathcal{M}_n}_{\mathcal{B} \vee \mathcal{T}_n}$, but we provide comments on these definitions and results for the sequence of unreduced experiment $\mathcal{E}_{\mathcal{A} \vee \mathcal{S}_n}$ only. In this section
we assume that $\mathcal{M}_n \uparrow \mathcal{M}_\infty$ and $\mathcal{T}_n \uparrow \mathcal{T}_\infty$.

The basic definition is based on the following theorem:

7.4.1 Theorem. Let $\mathcal{L}$ be an $\mathcal{E}^{\mathcal{M}_\infty}_{\mathcal{B} \vee \mathcal{T}_\infty}$-parameter (i.e., $\mathcal{L} \subset \mathcal{B} \vee \mathcal{M}_\infty$); the
following two conditions are then equivalent:

(i) $\mathcal{L} \subset \overline{\mathcal{M}_\infty \vee \mathcal{T}_\infty}$;

(ii) $\forall \, \ell \in [\mathcal{L}]_1, (\mathcal{M}_n \vee \mathcal{T}_n)\ell \to \ell$ a.s.Π.

Proof. By the martingale convergence theorem, $(\mathcal{M}_n \vee \mathcal{T}_n)\ell \to (\mathcal{M}_\infty \vee \mathcal{T}_\infty)\ell$
a.s.Π, and it is clear that $(\mathcal{M}_\infty \vee \mathcal{T}_\infty)\ell = \ell$ a.s.Π, $\forall \, \ell \in [\mathcal{L}]_1$ if and only if
$\mathcal{L} \subset \overline{\mathcal{M}_\infty \vee \mathcal{T}_\infty}$.   ∎

This motivates the next definition:

7.4.2 Definition. An $\mathcal{E}_{\mathcal{B}\vee\mathcal{T}_\infty}^{\mathcal{M}_\infty}$-parameter $\mathcal{L}$ satisfying any one of the two equivalent conditions of Theorem 7.4.1, is said to be *asymptotically exactly estimable in* $\mathcal{E}_{\mathcal{B}\vee\mathcal{T}_\infty}^{\mathcal{M}_\infty}$. ∎

For unreduced experiments $\mathcal{E}_{\mathcal{A}\vee\mathcal{S}_n}$ this definition becomes: $\mathcal{B}\subset\mathcal{A}$ is asymptotically exactly estimable if $\mathcal{S}_n b\to b$ a.s.Π for all $b\in[\mathcal{B}]_1$. Recall that $\mathcal{S}_n b$, the posterior expectation of b, is the usual Bayesian estimator corresponding to a quadratic loss function. The property "$\mathcal{S}_n b\to b$ a.s.Π" means that this sequence of estimators converges almost surely to the parameter b. This property thus represents the Bayesian concept of consistency. Thus a parameter $\mathcal{B}$ is asymptotically exactly estimable if every $\mathcal{B}$-measurable integrable function admits the conditional expectations with respect to $\mathcal{S}_n$ as a strongly consistent sequence of estimators. The Bayesian aspect of this analysis turns on two particularities: firstly, the estimator is the posterior expectation; secondly, the almost sure convergence is with respect to the joint probability Π. The links between these two particularities and the usual sampling theory concepts of consistency are examined in Section 7.4.2.

Using the theorems of Section 4.7, the study of exact estimability is usually structured as follows: It is known that the maximal exactly estimated parameter, $(\mathcal{B}\vee\mathcal{M}_\infty)\cap\overline{(\mathcal{M}_\infty\vee\mathcal{T}_\infty)}$, is contained in the minimal sufficient parameter $(\mathcal{B}\vee\mathcal{M}_\infty)(\mathcal{M}_\infty\vee\mathcal{T}_\infty)$ (Proposition 4.7.4(i)). The minimal sufficient parameter will be exactly estimable if and only if there exists a statistic $\mathcal{N}\subset\mathcal{M}_\infty\vee\mathcal{T}_\infty$ sufficient and exactly estimating $(\mathcal{N}\subset\overline{\mathcal{B}\vee\mathcal{M}_\infty})$ (Theorem 4.7.8). In view of Theorems 7.3.4 and 7.3.17, candidates for $\mathcal{N}$ are tail σ-fields of sequences of statistics sufficient in finite experiments. The verification of the exactly estimating character of such tail-σ-fields will rely on 0-1 laws which are presented in the sequel. Furthermore, in Chapters 8 and 9, sufficiency is also obtained using invariance arguments.

7.4.2 Sampling Theory and Bayesian Methods

As in Section 2.3.7, let us consider a statistical experiment $\mathcal{E}=\{(S,\mathcal{S}),P^a\ a\in A\}$, where a represents a point of A and P^a is the corresponding sampling probability. $\mathcal{A}$ is a σ-field on A such that $P^a(X)$ is $\mathcal{A}$-measurable, $\forall\ X\in\mathcal{S}$, and $\{\mathcal{S}_n,n\in I\!\!N\}$ is a filtration in $\mathcal{S}$ with

$\mathcal{S}_\infty = \bigvee_{0 \leq n \leq \infty} \mathcal{S}_n$. This is the usual framework for the analysis of asymptotic properties in sampling theory. In this context, a real parameter b is an $\mathcal{A}$-measurable real function of a, and $b(a)$ represents the values of b at the point a. We first recall the sampling theory definitions of estimability (see, for example, LeCam and Schwartz (1960), Breiman, LeCam and Schwartz (1964), Schwartz (1965), Skibinski (1967,1969), Martin and Vaguelsy (1969), Deistler and Seifert (1978).

7.4.3 Definition. A real-valued parameter b is *strongly* (respectively, *weakly*; respectively, *p-*) *sampling estimable* if there exists an $(\mathcal{S}_n)$-adapted sequence of estimators $\{s_n, n \in \mathbb{N}\}$ such that $s_n \to b(a)$ a.s.P^a (respectively, in probability with respect to P^a, respectively, in $L_p(P^a)$), $\forall\, a \in A$. The sequence $\{s_n, n \in \mathbb{N}\}$ is then said to be *strongly* (respectively, *weakly*, respectively, *p-*) *consistent* for b.   ∎

In a Bayesian framework, a probability μ on $(A, \mathcal{A})$ is introduced; its presence provides two ways of blending Bayesian and sampling theory arguments to relax Definition 7.4.3.

7.4.4 Definition. A real-valued parameter b is *μ-strongly* (respectively, *μ-weakly*; respectively, *μ-p-*) *sampling estimable* if there exists $E \in \mathcal{A}$ with $\mu(E) = 1$ such that b is strongly (respectively, weakly, respectively, *p-*) sampling estimable on the restricted parameter set E.   ∎

Let $\mathcal{E} = \{A \times S, \mathcal{A} \vee \mathcal{S}, \Pi = \mu \otimes P^{\mathcal{A}}\}$, the corresponding Bayesian experiment as defined in Section 1.2.1. In this framework, genuinely Bayesian definitions of estimability of a real parameter b are provided by the following definitions:

7.4.5 Definition. A real-valued parameter b is *strongly* (respectively, *weakly*; respectively, *p-*) *Bayesian estimable* if there exists an $(\mathcal{S}_n)$-adapted sequence of estimators $\{s_n, n \in \mathbb{N}\}$ such that $s_n \to b$ a.s.Π (respectively, in probability with respect to Π; respectively, in $L_p(\Pi)$).   ∎

We now examine the connections between the Definitions 7.4.3, 7.4.4 and 7.4.5. Two series of implications are straightforward:

(i) Each kind of estimability in Definition 7.4.3 implies the corresponding estimability in Definition 7.4.4 for any prior probability μ on $(A, \mathcal{A})$;

(ii) In each of Definitions 7.4.3, 7.4.4 and 7.4.5, strong estimability implies weak estimability, and p-estimability implies weak estimability (with the same adapted sequence of estimators).

Another implication is easily obtained when the parameter space A is countable:

(iii) If A is countable and if μ gives positive probability to each point of A, each μ-sampling estimability in Definition 7.4.4 implies the corresponding sampling estimability in Definition 7.4.3.

If A is uncountable, such an implication will not be true in general. Indeed, the proof that a property which is true almost everywhere is in fact true everywhere requires some kind of continuity (see Definition 2.3.13); but in an asymptotic experiment, the sampling probabilities are generally mutually singular (i.e., $P^a \perp P^{a'}$, $\forall\, a \neq a'$).

Other series of implications are less obvious and so are described by the next two theorems. The first of these relates various kinds of Bayesian estimability in Definitions 7.4.5 between them and to the concept of asymptotic exact estimability (Definition 7.4.2):

7.4.6 Theorem.

(i) Any real-valued weakly Bayesian estimable parameter b is strongly Bayesian estimable and $\mathcal{A} \cap \overline{S}_\infty$-measurable (i.e., is measurable with respect to the maximal exactly estimable parameter);

(ii) Any real-valued parameter $b \in [\mathcal{A} \cap \overline{S}_\infty]_p$ ($1 \leq p \leq \infty$) is strongly and p-Bayesian estimable. An adapted sequence of consistent estimators is given by the posterior expectations $\{\mathcal{S}_n b : n \in \mathbb{N}\}$.

Proof. (i) It suffices to note that if b is weakly Bayesian estimable and $\{s_n : n \in \mathbb{N}\}$ is an $(\mathcal{S}_n)$-adapted sequence of estimators such that $s_n \to b$ in probability with respect to Π, then there exists a subsequence $\{s_{n_k} : k \in \mathbb{N}\}$ such that $s_{n_k} \to b$ a.s. Π (see Neveu (1964), Proposition II.4.3), and so b is strongly Bayesian estimable and since $b = \limsup s_{n_k}$ a.s.Π, $b \in \overline{S}_\infty$.

(ii) If $b \in [\mathcal{A} \cap \overline{S}_\infty]_p$, then $b \in [\mathcal{A} \cap \overline{S}_\infty]_1$ and, by Theorem 7.2.2,

$\mathcal{S}_n b \to \mathcal{S}_\infty b$ a.s.Π, and in $L_1(\Pi)$. Now $\mathcal{S}_\infty b = b$ a.s.Π since $b \in \overline{\mathcal{S}}_\infty$. But $b \in [\mathcal{A}]_p$ implies that $\mathcal{S}_n b \to b$ in $L_p(\Pi)$ by Theorem 7.2.2.   ∎

Note that the assumption of p-integrability of b in Theorem 7.4.6 (ii), i.e., $b \in [\mathcal{A}]_p$ depends on the prior probability μ but not on the sampling probabilities. Theorem 7.4.6 is genuinely Bayesian in the following sense: In a sampling theory framework the existing subsequence, s_{n_k}, would depend on an unknown parameter and would therefore fail to be a sequence of estimators, whereas in a Bayesian framework this subsequence is based on the unique joint probability Π and therefore does not depend on the parameter.

The second theorem examines some links between each kind of estimability in 7.4.4 with the corresponding kind of estimability in 7.4.5:

7.4.7 Theorem.

(i)   A real parameter b strongly Bayesian estimable is μ-strongly sampling estimable;

(ii)   A real parameter b μ-weakly sampling estimable is weakly Bayesian estimable;

(iii) A bounded p-Bayesian estimable parameter b (i.e., $b \in [\mathcal{A}]_\infty$) is μ-p-sampling estimable.

Proof. (i) Let $s_- = \liminf s_n$ and $s_+ = \limsup s_n$. The event $\{s_n \to b\}$ may be written as $\{s_- = s_+ = b\}$. This event clearly belongs to $\mathcal{A} \vee \mathcal{S}$ and, by hypothesis, we have that $\Pi[\{s_+ = s_- = b\}] = 1$. This implies that $\mathcal{A}1_{\{s_+ = s_- = b\}} = 1$ a.s.μ. By definition of $\Pi = \mu \otimes P^{\mathcal{A}}$, and by the fact that $P^a[s_+ = s_- = b(a)]$ is $\mathcal{A}$-measurable (see, for instance, Neveu (1964), Proposition III-2-7), we obtain: $\mathcal{A}1_{\{s_+ = s_- = b\}} = P^a[s_+ = s_- = b(a)]$ a.s.μ. Hence there exists $E \in \mathcal{A}$ with $\mu(E) = 1$ such that $s_n \to b(a)$ a.s.P^a $\forall a \in E$.

(ii) By hypothesis, $\forall a \in E$ where $E \in \mathcal{A}$ and $\mu(E) = 1$, $\forall \varepsilon > 0$, we have:

$$\lim_{n \to \infty} P^a[|s_n - b(a)| > \varepsilon] = 0.$$

Now, as in (i), $P^a[|\,s_n - b(a)\,| > \varepsilon]$ is $\mathcal{A}$-measurable and bounded by 1. Since

$$\Pi[|s_n - b| > \varepsilon] = \int_A P^a[|s_n - b(a)| > \varepsilon]\mu(da),$$

an application of the Lebesgue dominated convergence theorem (see, for instance, Neveu (1964), II-3) entails the result, i.e.,

$$\lim_{n \to \infty} \Pi[|s_n - b| > \varepsilon] = 0.$$

(iii) By hypothesis, there exists an adapted sequence of estimators $\{s_n : n \in I\!\!N\}$ such that $s_n \to b$ in $L_p(\Pi)$. By Theorem 7.4.6(i), $b \in \overline{\mathcal{S}}_\infty$ and $b \in [\mathcal{A}]_p$, and so, by Theorem 7.4.6(ii), $\mathcal{S}_n b \to b$ a.s.Π. But if $b \in [\mathcal{A}]_\infty$, then there exists $\alpha \in I\!\!R^+$ such that $|b| \leq \alpha$, and so $|\mathcal{S}_n b - b|^p \leq (2\alpha)^p$. By the dominated convergence theorem for conditional expectations (see, for instance, Neveu (1964), IV-3), $\lim_{n \to \infty} \mathcal{A}(|\mathcal{S}_n b - b|^p) = 0$ a.s.Π or, equivalently, a.s.μ. Since, as in (i) of this theorem, $\mathcal{I}^a(|\mathcal{S}_n b - b(a)|^p)$ is $\mathcal{A}$-measurable, then $\mathcal{A}(|\mathcal{S}_n b - b|^p) = \mathcal{I}^a(|\mathcal{S}_n b - b(a)|^p)$ a.s.μ. So, there exists $E \in \mathcal{A}$ with $\mu(E) = 1$ such that $\forall\, a \in E$, $\lim_{n \to \infty} \mathcal{I}^a(|\mathcal{S}_n b - b(a)|^p) = 0$, i.e., $\mathcal{S}_n b \to b(a)$ in $L_p(P^a)$ or, equivalently, b is μ-p-sampling estimable. ∎

We conclude this section with some discussion of these results.

Remark 1. From the elementary properties and from Theorems 7.4.6(i) and 7.4.7(i) and (ii), μ-strong sampling, μ-weak sampling estimability, strong Bayesian estimability and weak Bayesian estimability are equivalent concepts, and so we will simply say Bayesian estimability. Moreover, from Theorems 7.4.6(ii) and 7.4.7(iii), if b is a bounded parameter, then Bayesian estimability, μ-p-sampling estimability and p-Bayesian estimability are equivalent concepts (see Table 1).

Remark 2. Due to the relations of hereby defined estimability concepts with the concept of exact estimability of σ-fields (Theorem 7.4.6), the σ-field generated by a family of Bayesian estimable real-valued parameters is exactly estimable. Thus every integrable parameter c measurable for this σ-field will be Bayesian estimable. Note that, in sampling theory, if $\{b_i : 1 \leq i \leq k\}$ is a family of real-valued parameters that are strongly (respectively, weakly) sampling estimable, then every continuous function c of the b_i's will be strongly (respectively, weakly) sampling estimable. But

Table 1. Estimability : Sampling Theory and Bayesian Concepts.

				Exact
Def. 7.4.3.	*Def. 7.4.4.*		*Def. 7.4.5.*	estimability
strongly $\overset{(1)}{\underset{(4)}{\overset{\Rightarrow}{\Leftarrow}}}$ sampling	μ-strongly sampling	$\underset{\text{Th. 7.4.7(i)}}{\Longleftarrow}$	strongly Bayesian	$\underset{\text{Th. 7.4.1}}{\Longleftrightarrow}$ $b \in A \cap \bar{S}_\infty$
$\Big\Downarrow$ (2)	$\Big\Downarrow$ (2)		$\Big\Downarrow$ (2) $\Big\Uparrow$ Th. 7.4.6(i)	$\Big\Downarrow$
weakly $\overset{(1)}{\underset{(4)}{\overset{\Rightarrow}{\Leftarrow}}}$ sampling	μ-weakly sampling	$\underset{\text{Th. 7.4.7(ii)}}{\Longrightarrow}$	weakly Bayesian	if $b \in [A]_p$
$\Big\Uparrow$ (3)	$\Big\Uparrow$ (3)		$\Big\Uparrow$ (3)	
p-sampling $\overset{(1)}{\underset{(4)}{\overset{\Rightarrow}{\Leftarrow}}}$	μ-p-sampling	$\underset{\text{Th.7.4.7(iii)}}{\Longleftarrow}$ if $b \in [A]_\infty$	p-Bayesian	$\underset{\text{Th. 7.4.6(ii)}}{\Longleftarrow}$ $b \in [A \cap \bar{S}_\infty]_p$

(1) $\forall$ μ, by definition.

(2) because almost sure convergence implies convergence in probability.

(3) because convergence in L_p implies convergence in probability.

(4) if A is countable and μ gives positive probabilities to each $a \in A$.

in this framework it seems difficult to extend this result to every Borel measurable function of the b_i's. The introduction of a prior probability μ on the parameters permits a step to be taken in that direction. Indeed, if c is any real-valued parameter integrable with respect to μ that is a (Borel) measurable function of the b_i's, then c is μ-strongly sampling estimable, i.e., there exists $E \in \mathcal{A}$ with $\mu(E) = 1$ such that c is strongly sampling estimable on the restricted parameter set E. Moreover, if c is bounded then it is also p-sampling estimable on this restricted parameter set for any $p \in [1, \infty[$. The adapted sequence of estimators strongly and, for any $p \in [1, \infty), p$-consistent for c may be chosen to be $\{S_n c : n \in I\!N\}$.

Remark 3. The above presentation of the relationships between the three kinds of estimability concepts is very similar to the presentation, in Section 4.6.2, of the relationships between different concepts of identification. The connections established in both analysis are also very similar: the "a.s.μ" concepts and the Bayesian concepts are equivalent and are implied by the sampling concepts. Note that this last implication does not require a hypothesis on $(A, \mathcal{A})$ for estimability but requires some regularity for identification (e.g., $(A, \mathcal{A})$ should be a Souslin space).

Moreover, it is known that estimability implies identification in each of the three approaches: for example, the sampling case is treated in LeCam and Schwartz (1960), the a.s.μ sampling case in Deistler and Seifert (1978) and the Bayesian case is Corollary 4.7.3. Let us observe that, in view of the equivalences of "a.s.μ" and the Bayesian concepts, Corollary 4.7.3. establishes that μ-sampling estimability implies a.s.s.-identification.

7.5   Estimability of Discrete σ-Fields

As an illustration of the above concepts we now consider the exact estimability of a σ-field $\mathcal{D}$ on the parameter space when it is generated by a measurable countable partition, i.e., $\mathcal{D} = \sigma\{E_d, d \in D \subset I\!N\}$ (where $E_d \neq \phi$, $E_d \in \mathcal{A}$, $\cup_d E_d = A$ and $d \neq d' \Rightarrow E_d \cap E_{d'} = \phi$); thus, E_d are the atoms of $\mathcal{D}$. This application is motivated, in particular, by the problems of discriminant analysis or of model choice when $\mathcal{D}$ is generated by the label of the sampling model (in such cases, the partition is typically finite) and the sampling probabilities are characterized by both a model label and a parameter that may be different for each model. In this situation, a fundamental issue is to establish whether, asymptotically, we "know the true model for sure", i.e., whether $\mathcal{D}$ is exactly estimable.

As $\mathcal{D}$ is generated by a (at most) countable partition of A, any $\mathcal{D}$-measurable function may be represented as a function of the canonical variable: $d : A \to D \subset I\!N$ defined by $d(a) = d \Leftrightarrow a \in E_d$ and to be considered as a model label. Hence the following notation: $\forall E \in \mathcal{A}$, $\forall x \in \mathcal{S}$,

$$\alpha(d) = \mu(E_d)$$

$$\alpha^S(d) = \mu^S(E_d)$$

$$\mu^d(E) = \mu(E \cap E_d)/\mu(E_d) = \mu(E \mid E_d)$$

$$\Pi^d(E \times X) = \Pi[(E \cap E_d) \times X]/\Pi(E_d \times S) = \Pi[E \times X \mid E_d \times S]$$

$$P^d(X) = \Pi^d(A \times X) = \Pi^d(E_d \times X)$$

Under the identification $E_d \subset A \leftrightarrow d \in D$ we may write the marginal model $\mathcal{E}_{\mathcal{D}\vee\mathcal{S}}$ as follows:

$$\mathcal{E}_{\mathcal{D}\vee\mathcal{S}} = (D \times S, \mathcal{D} \vee \mathcal{S}, \Pi = \alpha \otimes P^d = P \otimes \alpha^S)$$

where P^d may also be viewed as a predictive probability within the dth-model. This marginal model can be characterized by a finite (or countable parameter space); such a structure has been investigated, e.g., by De Groot (1970), Freedman (1963, 1965) or Freedman and Diaconis (1983).

The main consequence of the countable character of $\mathcal{D}$ is that we may suppose that $\alpha(d) > 0$ $\forall d \in D$, and consequently the family $(P^d)_{d\in D}$ (respectively, $(\Pi^d)_{d\in D}$) is equivalent to the corresponding marginal P (respectively, Π), i.e., $P^d(X) = 0$ $\forall d \Leftrightarrow P(X) = 0$). It will therefore be natural to consider, in particular, the Radon-Nikodym derivatives dP^d/dP and $d\Pi^d/d\Pi$. The first result gives equivalent characterizations of the exact estimability of $\mathcal{D}$ without reference to a filtration on $\mathcal{S}$.

7.5.1 Theorem. The following properties are equivalent:

(i) $\mathcal{D} \subset \bar{\mathcal{S}}$ (i.e., $\mathcal{D}$ is exactly estimable);

(ii) $\exists$ a partition of S, (X_d), such that $P^d(X_{d'}) = \mathbf{1}_{\{d=d'\}}$ (i.e., (P^d) are mutually singular);

(iii) $\exists$ a partition of S, (X_d), such that $d\Pi^d/d\Pi = [1/\alpha(d)]\mathbf{1}_{A\times X_d}$ a.s.Π;

(iv) $\exists$ a partition of S, (X_d), such that $\alpha^S(d) = \mathbf{1}_{X_d}$ a.s.P;

(v) $\exists$ a family in S, (X_d^*), such that $P^d(X_d^*) = P(\cup X_d^*) = 1$ and $P(X_d^* \cap X_{d'}^*) = 0, d \neq d'$;

(vi) $(dP^d/dP) \cdot (dP^{d'}/dP) = 0$ a.s.P, $\quad \forall d \neq d'$;

(vii) $\alpha^S(d) = 0$ a.s.$P^{d'}$, $\forall\, d \neq d'$.

Proof. We shall frequently appeal to the following version of Lemma 2.2.5 (iii):

(7.5.1) $\mathcal{M}_1 \subset \overline{\mathcal{M}_2} \Leftrightarrow \forall\, M_1 \in \mathcal{M}_1 \;\; \exists\, M_2 \in \mathcal{M}_2 \;:\; P(M_1 \triangle M_2) = 0$.

(i) ⇔ (ii)

$$\mathcal{D} \subset \overline{\mathcal{S}} \;\Leftrightarrow\; \exists (X_d^*)_{d \in D} \quad \text{such that} \quad X_d^* \in \mathcal{S} \;\; \text{and}$$
$$\Pi[(\{d\} \times S) \triangle (D \times X_d^*)] = 0 \;\; \forall\, d \in D \qquad \text{(by 7.5.1)}$$
$$\Leftrightarrow\; \exists (X_d^*) \quad \text{such that} \quad \alpha(d) P^d(S - X_d^*) + \sum_{d' \neq d} \alpha(d') P^{d'}(X_d^*) = 0$$
$$\Leftrightarrow\; \exists (X_d^*) \quad \text{such that} \quad P^d(S - X_d^*) = P^{d'}(X_d^*) = 0, d' \neq d$$
$$\Leftrightarrow\; \exists (X_d^*) \quad \text{such that} \quad P^d(X_{d'}^*) = \mathbf{1}_{\{d=d'\}}.$$

If (X_d^*) is not a partition, define an associated partition as follows:

$$X_1 \;=\; X_1^* \cup \left(S - \bigcup_d X_d^*\right),$$
$$X_d \;=\; X_d^* - \bigcup_{d' < d} X_{d'}^*, \quad \forall\, d \neq 1$$

we also have $P^d(X_{d'}) = \mathbf{1}_{\{d=d'\}}$.

(ii) ⇔ (iii)

$$\exists \text{ a partition } (X_d) \quad \text{such that} \quad P^d(X_{d'}) = \mathbf{1}_{\{d=d'\}}$$
$$\Leftrightarrow\; \exists \text{ a partition } (X_d) \quad \text{such that} \quad \Pi^d(A \times X_{d'}) = \mathbf{1}_{\{d=d'\}}$$
$$\Leftrightarrow\; \exists \text{ a partition } (X_d) \quad \text{such that}$$
$$\text{(7.5.2)} \qquad \Pi^d[M \cap (A \times X_d)] = \Pi^d(M) \qquad \forall\, M \in \mathcal{A} \vee \mathcal{S}$$
$$\Pi^d[M \cap (A \times X_{d'})] = 0 \qquad \forall\, d \neq d'.$$

Now, because $\Pi = \sum_d \alpha(d) \Pi^d$, we have

$$\text{(7.5.3)} \qquad \Pi[M \cap (A \times X_d)] = \alpha(d) \Pi^d(M), \text{i.e.,}$$

$$\frac{d\Pi^d}{d\Pi} \;=\; \frac{1}{\alpha(d)} \mathbf{1}_{A \times X_d} \quad \forall\, d \in D \quad \text{a.s.} \Pi.$$

(iv) ⇒ (ii)

(iv)⇒ $\alpha(d) = P(X_d)$ because
$$\alpha(d) = \int_S \alpha^S(d) \, dP = \int_S \mathbf{1}_{X_d} \quad dP = P(X_d)$$

⇒ $P^d(X_{d'}) = \mathbf{1}_{\{d=d'\}}$ because

$$
P^d(X_{d'}) = \frac{\Pi[\{d\} \times X_{d'}]}{\alpha(d)} \quad = \quad \frac{1}{\alpha(d)} \int_S \alpha^S(d) \cdot \mathbf{1}_{X_{d'}} dP
$$
$$
= \quad \frac{1}{\alpha(d)} \int_S \mathbf{1}_{X_d} \mathbf{1}_{X_{d'}} dP.
$$

(ii) ⇒ (iv)

Recall that, in general, $\alpha^S(d) = \alpha(d)(dP^d/dP)$. Therefore:

(ii) ⇒ ∀ $X \in S$

$$
P^d(X) = P^d(X \cap X_d) \quad = \quad \frac{1}{\alpha(d)} \sum \alpha(d') P^{d'}(X \cap X_d)
$$
$$
= \quad \frac{1}{\alpha(d)} P(X \cap X_d) = \frac{1}{\alpha(d)} \int_X \mathbf{1}_{X_d} dP
$$

$$
\Leftrightarrow \quad \frac{dP^d}{dP} = \frac{1}{\alpha(d)} \mathbf{1}_{X_d} \quad \text{a.s.} P.
$$

(ii) ⇒ (v) is trivial

(v) ⇒ (ii)

(v) ⇒ $P^d(X_{d'}^*) = 0 \quad d \neq d'$.

Indeed, $P^d(X_{d'}^*) \; = P^d(X_d^* \cap X_{d'}^*)$ because $P^d(X_d^*) = 1$
$$\qquad\qquad\quad = 0 \qquad\qquad\qquad \text{because } P(X_d^* \cap X_{d'}^*) = 0,$$

and note that the associated partition $X_1 = X_1^* \cup \{S - \bigcup_d X_d^*\}$,

$$
X_d = X_d^* - \bigcup_{d' < d} X_{d'}^*, \quad \forall \, d \neq 1,
$$

has the same properties.

(ii) ⇒ (vi)

(ii) ⇒ $\dfrac{dP^d}{dP} = \dfrac{1}{\alpha(d)} \mathbf{1}_{X_d}$ a.s.P (see (ii) ⇒(iv)

$$\Rightarrow \frac{dP^d}{dP} \cdot \frac{dP^{d'}}{dP} = \frac{1}{\alpha(d)} \frac{1}{\alpha(d')} \mathbf{1}_{X_d \cap X_{d'}} = 0 \text{ because } (X_d) \text{ is a partition.}$$

(vi) ⇒ (ii)

Define $X_d^* = \left\{ s \in S \mid \dfrac{dP^d}{dP} > 0 \right\}$.

Thus we obtain successively:

$$\begin{aligned} P^d(X_d^*) &= 1 \quad \text{because} \quad \frac{dP^d}{dP} = 0 \quad \text{on} \quad S - X_d^* \\ P(X_d^*) &\geq \alpha(d). \end{aligned}$$

Therefore

(vi) ⇒ $\dfrac{dP^{d'}}{dP} = 0$ on X_d^* because $\dfrac{dP^d}{dP} > 0$ on X_d^*,

$$\Rightarrow P^d(X_{d'}^*) = 0.$$

Thus (X_d^*) is an almost sure partition, to which one can associate, as above, a partition with the desired property.

(vi) ⇔ (vii)

(vii) ⇔ $\alpha(d) \dfrac{dP^d}{dP} = 0$ a.s.$P^{d'}$ $d \neq d'$

$$\Leftrightarrow \forall\, X \in \mathcal{S} \quad \int_X \frac{dP^d}{dP} \cdot dP^{d'} = 0 \quad d \neq d'$$

$$\Leftrightarrow \forall\, X \in \mathcal{S} \quad \int_X \frac{dP^d}{dP} \cdot \frac{dP^{d'}}{dP} dP = 0 \quad d \neq d'$$

$$\Leftrightarrow \frac{dP^d}{dP} \cdot \frac{dP^{d'}}{dP} = 0 \ \text{ a.s.}P \quad d \neq d'. \qquad\blacksquare$$

It is evident that the most important result in this theorem is the equivalence between the exact estimability of a finite-σ-field on the parameter space and the mutual singularity of the associated sampling probabilities marginalized with respect to that finite σ-field. The other characterizations shed more light on equivalent forms of this mutual singularity. The next corollary states an implication of such a context (the converse implication is obviously false).

7.5.2 Corollary. If $\mathcal{D} \subset \overline{\mathcal{S}}$, the family (Π^d) is mutually singular.   ∎

Theorem 7.5.1 is particularly useful in the asymptotic experiment. This is the focus of next corollary.

7.5.3 Corollary. Let us consider a filtration on $\mathcal{S} : \mathcal{S}_n \uparrow \mathcal{S}_\infty$. The following conditions are equivalent:

(i) $\mathcal{D} \subset \overline{\mathcal{S}}_\infty$;

(ii) $\exists$ a partition of $S, (X_d)$ such that $\alpha^{\mathcal{S}_n}(d) \to \mathbf{1}_{X_d}$ a.s.Π;

(iii) $\exists$ a partition of $S, (X_d)$ such that $\alpha^{\mathcal{S}_n}(d) \to \mathbf{1}_{X_d}$ in $L_1(\Pi)$;

(iv) $\alpha^{\mathcal{S}_n}(d) \to \mathbf{1}_{\{d=d'\}}$ a.s.$P^{d'}$.   ∎

In other words, if we interpret d as a model label, then its exact estimability is equivalent to a $0-1$ law on its posterior probability. Since the first results established by Kakutani (1948), the probability theory literature has developed a wealth of results in the form of conditions sufficient to ensure the mutual singularity of probabilities on filtration. Extensions to non i.i.d. sequences of Kakutani's original result are presented in a unified framework by Kabanov, Lipster and Shiryayev (1979-1980); for a shorter presentation, see Shiryayev (1981) or Memin (1985). It is worth mentioning that Theorem 7.5.1, in conjunction with that literature, has two different directions: *either* take two particular values $\{a_1, a_2\}$ in the parameter space, consider the conditional model $\mathcal{E}_{AVS}^{\{a_1,a_2\}}$, and note that Theorem 7.5.1 holds for arbitrary (strictly prositive) prior probability attached to those values, *or* consider d as a model label and integrate out any other parameters (see Zellner and Siow (1980)).

7.6   Mutual Conditional Independence and Conditional 0-1 Laws

As announced in concluding Section 7.4.1, the verification of the exactly estimating character of tail-σ-fields relies on 0-1 laws. A celebrated 0-1 law is the Kolmogorov 0-1 law (see, e.g., Chow and Teicher (1978)), which is deduced from the mutual independence of a sequence of σ-fields. In the next section, we define mutual conditional independence of a sequence of σ-fields on a general probability space and study some properties of this concept. The following section considers the case where the conditioning is operated with respect to a sequence of σ-fields.

7.6.1 Mutual Conditional Independence

In this section $(M, \mathcal{M}, P)$ is again an abstract probability space, $\mathcal{M}_i$ are sub-σ-fields of $\mathcal{M}$, in particular $\mathcal{M}_0 = \{M, \phi\}$. We consider an arbitrary family of sub-σ-fields of $\mathcal{M}$, $\{\mathcal{F}_i : i \in I\}$ where I is an arbitrary index set.

7.6.1 Definition. The family of sub-σ-fields of $\mathcal{M}$, $\{\mathcal{F}_i : i \in I\}$ is a family of *mutually conditionally independent σ-fields*, where the conditioning is with respect to a sub-σ-field $\mathcal{M}_1$ of $\mathcal{M}$ and we write $\perp\!\!\!\perp_{i \in I} \mathcal{F}_i \mid \mathcal{M}_1$ if, for every finite subset J of I and for every $f_i \in [\mathcal{F}_i]^+$, $\forall i \in J$,

$$\mathcal{M}_1 \left(\Pi_{i \in J} f_i \right) = \prod_{i \in J} \mathcal{M}_1(f_i). \qquad \blacksquare$$

Before stating the next theorem, which relates mutual conditional independence to the wedge operation, we first introduce some notation. For any subset J of I we will write

$$(7.6.1) \qquad \qquad \mathcal{F}_J = \bigvee_{i \in J} \mathcal{F}_i.$$

We will also write, when there is no ambiguity about I,

$$(7.6.2) \qquad \qquad \mathcal{F}_{]i[} = \mathcal{F}_{I-\{i\}} = \bigvee_{\substack{j \in I \\ j \neq i}} \mathcal{F}_j.$$

7.6.2 Theorem. Let $\{\mathcal{F}_i : i \in I\}$ be an arbitrary family of sub-σ-fields of $\mathcal{M}$, and let $\mathcal{M}_1$ be a sub-σ-field of $\mathcal{M}$. If $\perp\!\!\!\perp_{i \in I} \mathcal{F}_i \mid \mathcal{M}_1$, then $\forall \{I_j : j \in J\}$ partition of I where J is an arbitrary index set, we have:

$$\perp\!\!\!\perp_{j \in J} \mathcal{F}_{I_j} \mid \mathcal{M}_1.$$

Proof. Let K be a finite subset of J and L_j a finite subset of I_j, $\forall j \in K$. Then $L = \bigcup_{j \in K} L_j$ is a finite subset of I. If $f_i \in [\mathcal{F}_i]^+$ $\forall i \in L$, we have successively:

$$
\mathcal{M}_1 \left(\prod_{i \in L} f_i \right) = \mathcal{M}_1 \left[\prod_{j \in K} \left(\prod_{i \in L_j} f_i \right) \right]
$$

$$
= \prod_{j \in K} \prod_{i \in L_j} \mathcal{M}_1 (f_i)
$$

$$
= \prod_{j \in K} \mathcal{M}_1 \left(\prod_{i \in L_j} f_i \right).
$$

Define $h_j = \prod_{i \in L_j} f_i$. Clearly, $h_j \in [\mathcal{F}_{I_j}]^+$ and

$$
\mathcal{M}_1 \left(\prod_{j \in K} h_j \right) = \prod_{j \in K} \mathcal{M}_1 (h_j).
$$

Finally, using a monotone class argument, it is clear that the above relation is true for any $h_j \in [\mathcal{F}_{I_j}]^+$, $j \in K$.   ∎

In what follows, we often consider a countable family of sub-σ-fields of $\mathcal{M}, \{\mathcal{F}_n, n \in I\!N\}$. In this situation a set of equivalent definitions of mutual conditional independence is particularly useful.

7.6.3 Theorem. Let $\{\mathcal{F}_n; n \in I\!N\}$ be a sequence of sub-σ-fields of $\mathcal{M}$ and let $\mathcal{M}_1$ be a sub-σ-field of $\mathcal{M}$. The following conditions are equivalent:

(i) $\displaystyle \mathop{\perp\!\!\!\perp}_{n \in I\!N} \mathcal{F}_n \mid \mathcal{M}_1$;

(ii) $\displaystyle \mathop{\perp\!\!\!\perp}_{0 \le k \le n} \mathcal{F}_k \mid \mathcal{M}_1 \quad \forall \, n \in I\!N$;

(iii) $\mathcal{F}_0^n \perp\!\!\!\perp \mathcal{F}_{n+1}^\infty \mid \mathcal{M}_1 \quad \forall \, n \in I\!N$;

(iv) $\mathcal{F}_n \perp\!\!\!\perp \mathcal{F}_{]n[} \mid \mathcal{M}_1 \quad \forall \, n \in I\!N$;

(v) $\mathcal{F}_{n+1} \perp\!\!\!\perp \mathcal{F}_0^n \mid \mathcal{M}_1 \quad \forall \, n \in I\!N$.

Proof. By definition, (i) and (ii) are equivalent because any finite subset of $I\!N$ is included in an interval $[0, n]$ for some finite n. Now, by Theorem 7.6.2, (i) implies (iii) and (iv). Clearly, (iii) implies (v), and (iv) also implies (v).

Now, (v) implies (ii) by a recurrence argument. Indeed, let $f_k \in [\mathcal{F}_k]^+$, $k = 1, 2, \ldots, n$ and let us suppose that the following relation is true for ℓ:

$$\mathcal{M}_1 \left(\prod_{1 \leq k \leq n} f_k \right) = \mathcal{M}_1 \left(\prod_{1 \leq k \leq l} f_k \right) \times \prod_{\ell+1 \leq k \leq n} \mathcal{M}_1(f_k).$$

Then this reduction is true for $\ell - 1$, since by hypothesis,

$$\mathcal{M}_1 \left(\prod_{1 \leq k \leq \ell} f_k \right) = \mathcal{M}_1 \left(\prod_{1 \leq k \leq \ell-1} f_k \right) \cdot \mathcal{M}_1(f_\ell).$$

Hence $\mathcal{M}_1 \left(\prod_{1 \leq k \leq n} f_k \right) = \prod_{1 \leq k \leq n} \mathcal{M}_1(f_k)$ and this is (ii), by definition. ∎

Note that the equivalence between (i) and (ii) means that full mutual independence (i) is equivalent to (ii), i.e., the finite mutual independence at time n for any n. The equivalence between (i) and (v) says that full mutual independence (i) is equivalent to sequential independence at time n for any n, i.e., the $(n+1)^{th}$ σ-field is independent of the first n ones at any stage n.

An important corollary of Theorem 7.6.3 is the following:

7.6.4 Corollary. Let $\mathcal{M}_i, i = 1, 2, 3, 4$, be sub-$\sigma$-fields of $\mathcal{M}$. The following properties are equivalent:

(i) $\mathcal{M}_1 \perp\!\!\!\perp \mathcal{M}_2 \perp\!\!\!\perp \mathcal{M}_3 \mid \mathcal{M}_4$;

(ii) $\mathcal{M}_1 \perp\!\!\!\perp (\mathcal{M}_2 \vee \mathcal{M}_3) \mid \mathcal{M}_4$ and $\mathcal{M}_2 \perp\!\!\!\perp \mathcal{M}_3 \mid \mathcal{M}_4$. ∎

We now extend the Kolmogorov 0-1 law to mutual conditional independence:

7.6.5 Theorem. Let $\mathcal{M}_1$ be a sub-σ-field of $\mathcal{M}$ and let $\{\mathcal{F}_n, n \in I\!N\}$ be a sequence of sub-σ-fields of $\mathcal{M}$. If

(i) $\underset{n \in I\!N}{\perp\!\!\!\perp} \mathcal{F}_n \mid \mathcal{M}_1$,

then

(ii) $(\mathcal{M}_2 \vee \mathcal{F}_n)_T \subset \overline{\mathcal{M}_1} \quad \forall \, \mathcal{M}_2 \subset \mathcal{M}_1.$

In particular,

(iii) $\mathcal{F}_T \subset \overline{\mathcal{M}_1}.$

Proof. The proof is a slight extension of the proof given in Neveu (1964), Proposition IV-4-3. By Theorem 7.6.2, (i) implies $\mathcal{F}_0^n \perp\!\!\!\perp \mathcal{F}_{n+1}^\infty \mid \mathcal{M}_1$. Since $\mathcal{M}_2 \subset \mathcal{M}_1$ and $(\mathcal{M}_2 \vee \mathcal{F}_n)_T \subset \mathcal{M}_2 \vee \mathcal{F}_{n+1}^\infty$, this implies by Corollary 2.2.11, that $(\mathcal{M}_2 \vee \mathcal{F}_0^n) \perp\!\!\!\perp (\mathcal{M}_2 \vee \mathcal{F}_n)_T \mid \mathcal{M}_1$ and so by a monotone class argument (Theorem 6.2.3), $(\mathcal{M}_2 \vee \mathcal{F}_0^\infty) \perp\!\!\!\perp (\mathcal{M}_2 \vee \mathcal{F}_n)_T \mid \mathcal{M}_1$, which again implies that $(\mathcal{M}_2 \vee \mathcal{F}_n)_T \perp\!\!\!\perp (\mathcal{M}_2 \vee \mathcal{F}_n)_T \mid \mathcal{M}_1$ and so $(\mathcal{M}_2 \vee \mathcal{F}_n)_T \subset \overline{\mathcal{M}_1}$, by Corollary 2.2.8.   ∎

We conclude this section by extending some results of Section 4.3, related to the properties of projections of conditionally independent σ-fields, to arbitrary families.

7.6.6 Theorem. Let $\mathcal{M}_1 \subset \mathcal{M}$ and $\{\mathcal{F}_i : i \in I\}$ be an arbitrary family of sub$-\sigma$-fields of $\mathcal{M}$. If:

(i) $\displaystyle\perp\!\!\!\perp_{i \in I} \mathcal{F}_i \mid \mathcal{M}_1$

then, for any $J \subset I$, we have:

(ii) $\displaystyle \mathcal{M}_1 \mathcal{F}_J = \bigvee_{i \in J} \mathcal{M}_1 \mathcal{F}_i$

(iii) $\displaystyle \big(\bigvee_{i \in J} \mathcal{F}_i\big)\mathcal{M}_1 \subset \overline{\bigvee_{i \in J} \mathcal{F}_i \mathcal{M}_1}.$

Proof.

(ii) Since for any K finite subset of J and any $f_i \in [\mathcal{F}_i]^+$ $\; i \in K$, $\mathcal{M}_1(\prod_{i \in K} f_i) = \prod_{i \in K}(\mathcal{M}_1 f_i)$ we have: $\mathcal{M}_1(\prod_{i \in K} f_i) \in [\bigvee_{i \in K} \mathcal{M}_1 \mathcal{F}_i]$. By a monotone class argument: $\forall \, f \in [\mathcal{F}_J]^+ \, : \, \mathcal{M}_1 f \in [\bigvee_{i \in J} \mathcal{M}_1 \mathcal{F}_i]$ and, therefore, $\mathcal{M}_1 \mathcal{F}_J \subset [\bigvee_{i \in J} \mathcal{M}_1 \mathcal{F}_i]$. Furthermore, it is clear that $\mathcal{M}_1 \mathcal{F}_i \subset \mathcal{M}_1 \mathcal{F}_J \; \forall \, i \in J$ and, therefore, $\bigvee_{i \in J} \mathcal{M}_1 \mathcal{F}_i \subset \mathcal{M}_1 \mathcal{F}_J$.

(iii)   By Corollary 4.3.6 (ii) and by induction, we have that for any finite $K \subset J$: $\mathcal{F}_K \mathcal{M}_1 \subset \overline{\bigvee_{i \in K} \mathcal{F}_i \mathcal{M}_1}$ or, equivalently, because

$$\bigvee_{i \in K} \mathcal{F}_i \mathcal{M}_1 \subset \mathcal{F}_K, \qquad \mathcal{F}_K \perp\!\!\!\perp \mathcal{M}_1 \mid \bigvee_{i \in K} \mathcal{F}_i \mathcal{M}_1.$$

By Theorem 7.6.2 and by hypothesis, we have $\mathcal{F}_K \perp\!\!\!\perp \mathcal{F}_{J-K} \mid \mathcal{M}_1$ which implies, by Corollary 2.2.11, that $\mathcal{F}_K \perp\!\!\!\perp \mathcal{F}_{J-K} \mid \mathcal{M}_1 \vee (\bigvee_{i \in K} \mathcal{F}_i \mathcal{M}_1)$. Therefore, by Theorem 2.2.10, $\mathcal{F}_K \perp\!\!\!\perp (\mathcal{F}_{J-K} \vee \mathcal{M}_1) \mid \bigvee_{i \in K} \mathcal{F}_i \mathcal{M}_1$. By another application of Corollary 2.2.11, $\mathcal{F}_K \perp\!\!\!\perp (\mathcal{F}_{J-K} \vee \mathcal{M}_1) \mid \bigvee_{i \in J} \mathcal{F}_i \mathcal{M}_1$, and this implies that $\mathcal{F}_K \perp\!\!\!\perp \mathcal{M}_1 \mid \bigvee_{i \in J} \mathcal{F}_i \mathcal{M}_1$; by a monotone class argument, this in turn implies that $\mathcal{F}_J \perp\!\!\!\perp \mathcal{M}_1 \mid \bigvee_{i \in J} \mathcal{F}_i \mathcal{M}_1$ and, therefore, $\mathcal{F}_J \mathcal{M}_1 \subset \overline{\bigvee_{i \in J} \mathcal{F}_i \mathcal{M}_1}$.   ∎

Two important applications of (ii) are to the cases where J is a bounded interval (i.e., $J = \{0, 1, \ldots, n\}$) or where $J = I\!N$; this second situation is the focus of the next two corollaries (the proof of the second is trivial).

7.6.7 Corollary. Let $\mathcal{M}_1 \subset \mathcal{M}$ and $\{\mathcal{F}_n : n \in I\!N\}$ be an independent sequence of sub-σ-fields of $\mathcal{M}$ conditionally on $\mathcal{M}_1$. Then:

(i) $\mathcal{M}_1 \mathcal{F}_0^\infty = \bigvee_{n \in I\!N} \mathcal{M}_1 \mathcal{F}_n;$

(ii) $\overline{\mathcal{F}_T} \cap \mathcal{M}_1 \subset (\mathcal{M}_1 \mathcal{F}_n)_T.$

Proof. As (i) is a trivial consequence of the above theorem, we only prove (ii). By Theorem 7.6.6, $\mathcal{M}_1 \mathcal{F}_n^\infty = \bigvee_{m \geq n} \mathcal{M}_1 \mathcal{F}_m$. So $\mathcal{M}_1 \mathcal{F}_T \subset \bigvee_{m \geq n} \mathcal{M}_1 \mathcal{F}_m$, $\forall n \in I\!N$, and this implies $\mathcal{M}_1 \mathcal{F}_T \subset (\mathcal{M}_1 \mathcal{F}_n)_T$. By Theorem 7.6.5, $\mathcal{F}_T \subset \overline{\mathcal{M}_1}$ and so, by 4.3.2 (iii), $\mathcal{M}_1 \mathcal{F}_T = \overline{\mathcal{F}_T} \cap \mathcal{M}_1$.   ∎

7.6.8 Corollary. Under the conditions of Corollary 7.6.7, the sequence $(\overline{\mathcal{F}_0^n \mathcal{M}_1})$ is $(\mathcal{F}_n)$ recursive; more precisely:

$$\overline{\mathcal{F}_0^{n+1} \mathcal{M}_1} \subset \overline{\mathcal{F}_0^n \mathcal{M}_1} \vee \mathcal{F}_{n+1} \mathcal{M}_1 \subset \overline{\mathcal{F}_0^n \mathcal{M}_1} \vee \mathcal{F}_{n+1}.$$ ∎

7.6.2   Sifted Sequences of σ-Fields

Suppose that the $\mathcal{M}_1$-conditionally independent sequence of sub-σ-fields of $\mathcal{M}$, $\{\mathcal{H}_n : n \in I\!N\}$, is such that $\mathcal{H}_n = \mathcal{F}_n \vee \mathcal{G}_n$. In view of Theorem 7.6.3, this implies that $\perp\!\!\!\perp_{n \in I\!N} \mathcal{F}_n \mid \mathcal{M}_1$ and $\perp\!\!\!\perp_{n \in I\!N} \mathcal{G}_n \mid \mathcal{M}_1$. Consequently, we may apply the theorems of Section 7.6.1 to $(\mathcal{F}_n)$, $(\mathcal{G}_n)$ and $(\mathcal{H}_n)$. In this section, we seek to obtain similar results concerning the sequence $(\mathcal{H}_n)$ without making assumptions concerning the sequence $(\mathcal{G}_n)$ (such as $\perp\!\!\!\perp_{n \in I\!N} \mathcal{G}_n \mid \mathcal{M}_1$), but with assumptions on $(\mathcal{F}_n)$ conditionally on $\mathcal{G}_0^\infty$ (recall that our notation is such that $\mathcal{G}_0^\infty = \mathcal{G}_{I\!N}$ and, therefore, $\mathcal{G}_{]n[} = \mathcal{G}_{I\!N - \{n\}}$.

This requires some kind of transitivity condition, as is shown by the following theorem.

7.6.9 Theorem. Let $\mathcal{M}_1 \subset \mathcal{M}$ and let $\{\mathcal{F}_n : n \in I\!N\}$ and $\{\mathcal{G}_n : n \in I\!N\}$ be two sequences of sub-σ-fields of $\mathcal{M}$. Then:

(i) $\displaystyle \perp\!\!\!\!\perp_{0 \leq n < \infty} (\mathcal{G}_n \vee \mathcal{F}_n) \mid \mathcal{M}_1$

if and only if

(ii) $\displaystyle \perp\!\!\!\!\perp_{0 \leq n < \infty} \mathcal{G}_n \mid \mathcal{M}_1$,

(iii) $\displaystyle \perp\!\!\!\!\perp_{0 \leq n < \infty} \mathcal{F}_n \mid \mathcal{M}_1 \vee \mathcal{G}_0^\infty$,

(iv) $\mathcal{F}_n \perp\!\!\!\perp \mathcal{G}_0^\infty \mid \mathcal{M}_1 \vee \mathcal{G}_n, \quad \forall n \in I\!N$.

Moreover under (iii), (iv) is equivalent to

(v) $\mathcal{F}_I \perp\!\!\!\perp \mathcal{G}_0^\infty \mid \mathcal{M}_1 \vee \mathcal{G}_I, \quad \forall I \subset I\!N$.

Proof. By Theorem 7.6.3, (i) is equivalent to:

(vi) $(\mathcal{G}_n \vee \mathcal{F}_n) \perp\!\!\!\perp (\mathcal{G}_{]n[} \vee \mathcal{F}_{]n[}) \mid \mathcal{M}_1$.

By successive applications of Theorem 2.2.10, (vi) is equivalent to:

(vii) $\mathcal{G}_n \perp\!\!\!\perp \mathcal{G}_{]n[} \mid \mathcal{M}_1$;

(viii) $\mathcal{G}_n \perp\!\!\!\perp \mathcal{F}_{]n[} \mid \mathcal{M}_1 \vee \mathcal{G}_{]n[}$;

(ix) $\mathcal{F}_n \perp\!\!\!\perp \mathcal{F}_{]n[} \mid \mathcal{M}_1 \vee \mathcal{G}_0^\infty$;

(x) $\mathcal{F}_n \perp\!\!\!\perp \mathcal{G}_{]n[} \mid \mathcal{M}_1 \vee \mathcal{G}_n$.

Clearly, by Theorem 7.6.3, (vii) is equivalent to (ii) and (ix) is equivalent to (iii). Since (x) is of course equivalent to (iv), and, obviously (v) implies (viii), the proof will be completed if we can prove that, under (iii), (iv) implies (v). As it is obviously true for any I with one element, we prove by recurrence that if (v) is true for I, then it is true for $I \cup \{n\}$ $\forall\, n \notin I$, i.e.,

$$(\mathcal{F}_I \vee \mathcal{F}_n) \perp\!\!\!\perp \mathcal{G}_0^\infty \mid \mathcal{M}_1 \vee \mathcal{G}_I \vee \mathcal{G}_n,$$

which, by Theorem 2.2.10, is equivalent to:

(xi) $\mathcal{F}_I \perp\!\!\!\perp \mathcal{G}_0^\infty \mid \mathcal{M}_1 \vee \mathcal{G}_I \vee \mathcal{G}_n$,

(xii) $\mathcal{F}_n \perp\!\!\!\perp \mathcal{G}_0^\infty \mid \mathcal{M}_1 \vee \mathcal{G}_I \vee \mathcal{G}_n \vee \mathcal{F}_I$.

Now, clearly, (xi) is implied by (v). However, since $\mathcal{F}_n \perp\!\!\!\perp \mathcal{F}_I \mid \mathcal{M}_1 \vee \mathcal{G}_0^\infty$ is implied by (iii), (iii) and (iv) imply, by Theorem 2.2.10, that

$$\mathcal{F}_n \perp\!\!\!\perp (\mathcal{F}_I \vee \mathcal{G}_0^\infty) \mid \mathcal{M}_1 \vee \mathcal{G}_n,$$

and this implies (xii), by Corollary 2.2.11(ii). Hence (v) is true for any I finite subset of $\mathbb{N}$. If I is infinite, let (I_n) be a monotone sequence of finite subsets of $\mathbb{N}$ such that $I_n \uparrow I$. Since

$$\mathcal{F}_{I_n} \perp\!\!\!\perp \mathcal{G}_0^\infty \mid \mathcal{M}_1 \vee \mathcal{G}_{I_n}, \quad \forall\, n,$$

an application of Theorem 7.2.7 under the substitution $(\mathcal{F}_n, \mathcal{G}_n, \mathcal{M}_n, \mathcal{H}_n) \to (\mathcal{F}_{I_n}, \mathcal{G}_0^\infty, \mathcal{M}_1 \vee \mathcal{G}_{I_n}, \mathcal{M}_0)$ gives the result, i.e., $\mathcal{F}_I \perp\!\!\!\perp \mathcal{G}_0^\infty \mid \mathcal{M}_1 \vee \mathcal{G}_I$. ∎

It is obvious that, under the conditions of Theorem 7.6.9, Theorem 7.6.5 implies that

(7.6.3) $$(\mathcal{F}_n \vee \mathcal{G}_n)_T \subset \overline{\mathcal{M}_1}.$$

As stated at the beginning of this section, if one wishes to preserve most of the mutual conditional independences of the sequence $(\mathcal{F}_n)$ without making any assumption with respect to the distribution of the sequence $(\mathcal{G}_n)$, then more structure is required; making these structural requirements more explicit is the goal of the next theorem:

7.6.10 Theorem. Let $\mathcal{M}_1$ be a sub-σ-field of $\mathcal{M}$ and let $(\mathcal{F}_n)$ and $(\mathcal{G}_n)$ be two sequences of sub-σ-fields of $\mathcal{M}$. Then the following properties are equivalent:

I (i) $\displaystyle\mathop{\perp\!\!\!\perp}_{n \in I\!N} \mathcal{F}_n \mid \mathcal{M}_1 \vee \mathcal{G}_0^\infty$;

   (ii) $\mathcal{F}_n \perp\!\!\!\perp \mathcal{G}_0^\infty \mid \mathcal{M}_1 \vee \mathcal{G}_n$, $\quad \forall \, n \in I\!N$.

II (iii) $\displaystyle\mathop{\perp\!\!\!\perp}_{0 \leq k \leq n} \mathcal{F}_k \mid \mathcal{M}_1 \vee \mathcal{G}_0^n$, $\quad \forall \, n \in I\!N$;

   (iv) $\mathcal{F}_k \perp\!\!\!\perp \mathcal{G}_0^n \mid \mathcal{M}_1 \vee \mathcal{G}_k$, $\qquad \forall \, n, k \in I\!N, k \leq n$.

III (v) $\mathcal{F}_0^n \perp\!\!\!\perp \mathcal{G}_{n+1} \mid \mathcal{M}_1 \vee \mathcal{G}_0^n$, $\quad \forall \, n \in I\!N$;

   (vi) $\mathcal{F}_{n+1} \perp\!\!\!\perp \mathcal{G}_0^n \mid \mathcal{M}_1 \vee \mathcal{G}_{n+1}$, $\quad \forall \, n \in I\!N$;

   (vii) $\mathcal{F}_{n+1} \perp\!\!\!\perp \mathcal{F}_0^n \mid \mathcal{M}_1 \vee \mathcal{G}_0^{n+1}$, $\quad \forall \, n \in I\!N$.

7.6.11 Definition. Under any one of Property I, II or III of Theorem 7.6.10, we say that $(\mathcal{F}_n)$ *is $\mathcal{M}_1$-sifted by* $(\mathcal{G}_n)$. ∎

Before proving Theorem 7.6.10, let us make some remarks. Note that Property I consists of Conditions (iii) and (iv) of Theorem 7.6.9 and, therefore, may be understood as a weakening of the joint conditional independence as stated in (i) in Theorem 7.6.9. Property I is written in an asymptotic form (involving $\mathcal{G}_0^\infty$), whereas Property II is in a global form (i.e., involving $\mathcal{F}_0^n \vee \mathcal{G}_0^n$ for any n) and Property III is in a sequential one (i.e., assumptions on the probability on $\mathcal{F}_{n+1} \vee \mathcal{G}_{n+1}$ conditionally on $\mathcal{F}_0^n \vee \mathcal{G}_0^n$); in particular, (v) is a transitivity condition (see Chapter 6), namely, $(\mathcal{G}_0^n)$ is $\mathcal{M}_1$-transitive for $(\mathcal{F}_0^n \vee \mathcal{G}_0^n)$. In this regard, Theorems 7.6.3 and 7.6.10 fit into the same framework.

A weaker concept than $\mathcal{M}_1$-sifted sequences is the following:

7.6.12 Definition. The sequence $(\mathcal{G}_n)$ is $\mathcal{M}_1$-*allocated* to the sequence $(\mathcal{F}_n)$ if

(7.6.4) $$\mathcal{F}_I \perp\!\!\!\perp \mathcal{G}_0^\infty \mid \mathcal{M}_1 \vee \mathcal{G}_I \quad \forall\, I \subset I\!\!N \qquad \blacksquare$$

Here the basic idea is that, conditionally on $\mathcal{M}_1$, the dependence between the sequence $(\mathcal{F}_n)$ and the sequence $(\mathcal{G}_n)$ is limited to matching indices. In view of Theorem 7.6.9, (7.6.4) is clearly implied by Property I in Theorem 7.6.10, and (7.6.4) trivially implies the transitivity condition (v). Recalling the proof of Theorem 7.6.9, one should also note that if Condition (7.6.4) holds for finite subsets I only, then it also holds for any subset of $I\!\!N$.

Proof of Theorem 7.6.10

(I) $\Leftrightarrow$ (II)

By Theorem 7.6.3, (i) is equivalent to $\mathcal{F}_n \perp\!\!\!\perp \mathcal{F}_0^{n-1} \mid \mathcal{M}_1 \vee \mathcal{G}_0^\infty$, $\forall\, n \in I\!\!N$. So, by Theorem 2.2.10, (i) and (ii) are equivalent to

(viii) $\mathcal{F}_n \perp\!\!\!\perp (\mathcal{F}_0^{n-1} \vee \mathcal{G}_0^\infty) \mid \mathcal{M}_1 \vee \mathcal{G}_n \quad \forall\, n \in I\!\!N$.

Now, by Corollary 2.2.11 (ii), (viii) implies that

$$\mathcal{F}_n \perp\!\!\!\perp \mathcal{F}_0^{n-1} \mid \mathcal{M}_1 \vee \mathcal{G}_0^\ell, \quad \forall\, n \leq \ell,$$

and this is equivalent to (iii), by Theorem 7.6.3. Furthermore, by a monotone class argument, (ii) is equivalent to (iv). So (I) $\Rightarrow$ (II). To show that (II) $\Rightarrow$ (I), it remains to show that (iii) implies (i). But, by Theorem 7.6.3, (iii) is equivalent to $\mathcal{F}_k \perp\!\!\!\perp \mathcal{F}_0^{k-1} \mid \mathcal{M}_1 \vee \mathcal{G}_0^n$, $\forall\, k \leq n$. By Theorem 7.2.7, with the transposition $(\mathcal{F}_n, \mathcal{G}_n, \mathcal{M}_n, \mathcal{H}_n) \to (\mathcal{F}_k, \mathcal{F}_0^{k-1}, \mathcal{M}_1 \vee \mathcal{G}_0^n, \mathcal{M}_0)$, we obtain $\mathcal{F}_k \perp\!\!\!\perp \mathcal{F}_0^{k-1} \mid \mathcal{M}_1 \vee \mathcal{G}_0^\infty$, $\forall\, k$, which is equivalent to (i).

(I) $\Leftrightarrow$ (III)

Note that, by Theorem 2.2.10, (vi) and (vii) are equivalent to

(ix) $\mathcal{F}_{n+1} \perp\!\!\!\perp (\mathcal{F}_0^n \vee \mathcal{G}_0^n) \mid \mathcal{M}_1 \vee \mathcal{G}_{n+1} \quad \forall\, n \in I\!\!N$.

Clearly (viii) (equivalent to I) implies (ix), and (7.6.4) (implied by I) implies (v). Furthermore, (v) and (ix) imply, by induction, that

$$(\mathbf{x}) \quad \mathcal{F}_{n+1} \perp\!\!\!\perp (\mathcal{F}_0^n \vee \mathcal{G}_0^{n+k}) \mid \mathcal{M}_1 \vee \mathcal{G}_{n+1} \quad \forall\, n, k \in I\!\!N.$$

Indeed, (x) is clearly true for $k = 0$. Suppose now that (x) is true for a given k and remark that (x) for $k + 1$ is equivalent to (x) along with

$$\mathcal{F}_{n+1} \perp\!\!\!\perp \mathcal{G}_{n+k+1} \mid \mathcal{M}_1 \vee \mathcal{F}_0^n \vee \mathcal{G}_0^{n+k}.$$

By Corollary 2.2.11 (ii), this is implied by (v). The proof is concluded by noting that, by a monotone class argument, (x) is equivalent to (viii). ∎

We now extend the Kolmogorov 0-1 law so as to characterize the tail σ-field of the pair $(\mathcal{F}_n \vee \mathcal{G}_n)$ when $(\mathcal{F}_n)$ is $\mathcal{M}_1$-sifted by $(\mathcal{G}_n)$.

7.6.13 Theorem. If $(\mathcal{F}_n)$ is $\mathcal{M}_1$-sifted by $(\mathcal{G}_n)$, then

$$(7.6.5) \qquad\qquad (\mathcal{F}_n \vee \mathcal{G}_n)_T \subset \overline{(\mathcal{M}_1 \vee \mathcal{G}_n)_T}.$$

If, moreover, $\mathcal{M}_1 \perp\!\!\!\perp \mathcal{G}_0^\infty \mid \mathcal{G}_T$, then

$$(7.6.6) \qquad\qquad (\mathcal{F}_n \vee \mathcal{G}_n)_T \subset \overline{\mathcal{M}_1 \vee \mathcal{G}_T}.$$

And if, moreover, $\mathcal{G}_T \subset \overline{\mathcal{M}_1}$, then

$$(7.6.7) \qquad\qquad (\mathcal{F}_n \vee \mathcal{G}_n)_T \subset \overline{\mathcal{M}_1}.$$

Proof. By (7.6.4), we have

$$\mathcal{F}_n^\infty \perp\!\!\!\perp \mathcal{G}_0^\infty \mid \mathcal{M}_1 \vee \mathcal{G}_n^\infty, \quad \forall\, n \in I\!\!N.$$

But, by Theorem 7.6.3, (i) of Theorem 7.6.10 is equivalent to

$$\mathcal{F}_n^\infty \perp\!\!\!\perp \mathcal{F}_0^k \mid \mathcal{M}_1 \vee \mathcal{G}_n^\infty, \quad \forall\, k < n.$$

So, by Theorem 2.2.10 and Corollary 2.2.11, we have

$$(\mathcal{F}_n^\infty \vee \mathcal{G}_n^\infty) \perp\!\!\!\perp (\mathcal{F}_0^k \vee \mathcal{G}_0^\infty) \mid \mathcal{M}_1 \vee \mathcal{G}_n^\infty, \quad \forall\, k < n.$$

So

$$(\mathcal{F}_n \vee \mathcal{G}_n)_T \perp\!\!\!\perp (\mathcal{F}_0^k \vee \mathcal{G}_0^\infty) \mid \mathcal{M}_1 \vee \mathcal{G}_n^\infty, \quad \forall\, k < n.$$

Since $(\mathcal{M}_1 \vee \mathcal{G}_n^\infty)_T = (\mathcal{M}_1 \vee \mathcal{G}_n)_T$, we have, by Theorem 7.2.7 (with $\mathcal{M}_n = \mathcal{M}_1, \mathcal{H}_n = \mathcal{G}_n^\infty$),

$$(\mathcal{F}_n \vee \mathcal{G}_n)_T \perp\!\!\!\perp (\mathcal{F}_0^k \vee \mathcal{G}_0^\infty) \mid (\mathcal{M}_1 \vee \mathcal{G}_n)_T, \quad \forall\, k.$$

By a monotone class argument, this is equivalent to

$$(\mathcal{F}_n \vee \mathcal{G}_n)_T \perp\!\!\!\perp (\mathcal{F}_0^\infty \vee \mathcal{G}_0^\infty) \mid (\mathcal{M}_1 \vee \mathcal{G}_n)_T.$$

But $(\mathcal{F}_n \vee \mathcal{G}_n)_T \subset (\mathcal{F}_0^\infty \vee \mathcal{G}_0^\infty)$, and this is equivalent, by Corollary 2.2.8, to $(\mathcal{F}_n \vee \mathcal{G}_n)_T \subset \overline{(\mathcal{M}_1 \vee \mathcal{G}_n)_T}$. Now if $\mathcal{M}_1 \perp\!\!\!\perp \mathcal{G}_0^\infty \mid \mathcal{G}_T$ then, by Theorem 7.2.8,

$$\overline{(\mathcal{M}_1 \vee \mathcal{G}_n)_T} = \overline{\mathcal{M}_1 \vee \mathcal{G}_T}. \qquad \blacksquare$$

Note that in this theorem the result of Theorem 7.6.5, i.e., (7.6.3) or (7.6.7), is obtained under "milder" assumptions, i.e., $\mathcal{M}_1 \perp\!\!\!\perp \mathcal{G}_0^\infty \mid \mathcal{G}_T$ and $\mathcal{G}_T \subset \overline{\mathcal{M}_1}$ rather than $\perp\!\!\!\perp_{0 \leq n < \infty} \mathcal{G}_n \mid \mathcal{M}_1$, even if this last assumption does not imply the other two. Note also that under (i) in Theorem 7.6.10, which, by Corollary 2.2.11, implies $\perp\!\!\!\perp_{n \in I\!N}(\mathcal{F}_n \vee \mathcal{G}_n) \mid \mathcal{M}_1 \vee \mathcal{G}_0^\infty$, we obtain by Theorem 7.6.5,

$$(7.6.8) \qquad (\mathcal{F}_n \vee \mathcal{G}_n)_T \subset \overline{\mathcal{M}_1 \vee \mathcal{G}_0^\infty}.$$

7.7 Tail-Sufficient and Independent Bayesian Experiments

Two types of assumptions will be studied in this section. We first define Bayesian tail-sufficiency for non-conditional and for conditional experiments. In such experiments, asymptotically sufficient statistics are easily found. We next consider independent and conditional independent experiments, and determine asymptotically exactly estimating statistics. Finally, putting together these two sets of assumptions, we will deduce exact estimability properties.

7.7.1 Bayesian Tail-Sufficiency

Underlying the concept of Bayesian tail-sufficiency is the property that, the posterior expectation of any parameter when all the observations are known, does not depend on the first n observations regardless of n; in effect this means that the time origin is arbitrary.

It is useful to introduce the notion of successive observations:

7.7.1 Definition. *A dynamic Bayesian experiment* is a sequential Bayesian experiment

$$\mathcal{E} = \{A \times S, \mathcal{A} \vee \mathcal{S}, \Pi, \mathcal{S}_n \uparrow \mathcal{S}_\infty\}$$

for which there exists a sequence of statistics $\{\mathcal{X}_k : k \in I\!N\}$ (sub-σ-fields of $\mathcal{S}$) such that

(7.7.1) $$\mathcal{S}_n = \mathcal{X}_0^n = \bigvee_{0 \le k \le n} \mathcal{X}_k,$$

and so

(7.7.2) $$\mathcal{S}_\infty = \mathcal{X}_0^\infty = \bigvee_{0 \le k < \infty} \mathcal{X}_k.$$

In general, $\mathcal{X}_k$ will be interpreted as the information brought by the observation at time k, as is natural.

In this framework, Bayesian tail-sufficiency may be formalized as:

7.7.2 Theorem. In a dynamic Bayesian experiment, the following conditions are equivalent:

(7.7.3) $$A \perp\!\!\!\perp \mathcal{X}_0^\infty \mid \mathcal{X}_n^\infty, \quad \forall\, n \in I\!N;$$

(7.7.4) $$\lim_{k \to \infty} (\mathcal{X}_0^k a - \mathcal{X}_n^k a) = 0 \quad \text{a.s.} \quad \forall\, a \in [\mathcal{A}]_\infty^+, \quad \forall\, n \in I\!N;$$

(7.7.5) $$A \perp\!\!\!\perp \mathcal{X}_0^\infty \mid \mathcal{X}_T. \qquad \blacksquare$$

7.7.3 Definition. *A tail-sufficient Bayesian experiment* is a dynamic Bayesian experiment satisfying any one of the conditions of Theorem 7.7.2.   ∎

Before proving Theorem 7.7.2, let us provide some comment with regards to Definition 7.7.3. Bayesian tail-sufficiency appears as a genuinely asymptotic concept and may be seen either as the sufficiency of the tail statistic (7.7.5), or as the property that the posterior expectations, given an infinite trajectory, do not depend on the beginning of the trajectory (7.7.3), and (7.7.4) translates the very same idea in terms of approximating the asymptotic posterior expectations in terms of truncated ones (up to the first k observations). Note that Bayesian tail-sufficiency, which is both an easy-to-interpret and results-providing condition, may nevertheless be difficult to verify unless further structure is introduced; thus in Chapters 8 and 9 we show that a property such as stationarity implies the tail-sufficiency property.

Proof of Theorem 7.7.2: The equivalence of (7.7.3) and (7.7.4) is a simple application of Theorem 7.2.2 (i.e., the martingale convergence theorem). By Theorem 7.2.7, (7.7.3) clearly implies (7.7.5). Since $\mathcal{X}_T \subset \mathcal{X}_n^\infty \subset \mathcal{X}_0^\infty$, (7.7.5) also implies (7.7.3), by Corollary 2.2.11 (ii). ∎

We now extend this concept to conditional Bayesian experiments.

7.7.4 Definition. *A dynamic conditional Bayesian experiment* is defined by a conditional experiment, $\mathcal{E}_{\mathcal{B} \vee \mathcal{T}_\infty}^{\mathcal{M}_\infty}$, along with a sequence of conditional experiments, $\mathcal{E}_{\mathcal{B} \vee \mathcal{T}_n}^{\mathcal{M}_n}$, for which there exist two sequences of statistics (i.e., of sub-σ-fields of $\mathcal{M}_\infty \vee \mathcal{T}_\infty$) $(\mathcal{Y}_n)$ and $(\mathcal{Z}_n)$ such that:

$$(7.7.6) \qquad \mathcal{T}_n = \mathcal{Y}_0^n = \bigvee_{0 \leq k \leq n} \mathcal{Y}_k$$

and

$$(7.7.7) \qquad \mathcal{M}_n = \mathcal{Z}_0^n = \bigvee_{0 \leq k \leq n} \mathcal{Z}_k$$

and, therefore, $\mathcal{T}_n \uparrow \mathcal{T}_\infty = \mathcal{Y}_0^\infty$ and $\mathcal{M}_n \uparrow \mathcal{M}_\infty = \mathcal{Z}_0^\infty$; such an experiment will usually be written as $\mathcal{E}_{\mathcal{B} \vee \mathcal{Y}_0^\infty}^{\mathcal{Z}_0^\infty}$. ∎

We also use the following notation:

$$(7.7.8) \qquad \mathcal{X}_k = \mathcal{Y}_k \vee \mathcal{Z}_k \quad \text{and} \quad \mathcal{X}_0^n = \mathcal{M}_n \vee \mathcal{T}_n = \mathcal{Y}_0^n \vee \mathcal{Z}_0^n.$$

In this framework, conditional tail-sufficiency is introduced as follows:

7.7.5 Theorem. In a dynamic conditional Bayesian experiment $\mathcal{E}_{\mathcal{B} \vee \mathcal{Y}_0^\infty}^{\mathcal{Z}_0^\infty}$, the following conditions are equivalent:

$$(7.7.9) \qquad \mathcal{B} \perp\!\!\!\perp \mathcal{Y}_0^\infty \mid \mathcal{Z}_0^\infty \vee \mathcal{Y}_n^\infty, \quad \forall\, n \in I\!N;$$

$$(7.7.10) \qquad \lim_{k \to \infty} \left[(\mathcal{Z}_0^\infty \vee \mathcal{Y}_0^k) b - (\mathcal{Z}_0^\infty \vee \mathcal{Y}_n^k) b \right] = 0 \quad \text{a.s.}$$
$$\forall\, b \in [\mathcal{B}]_\infty^+, \quad \forall\, n \in I\!N;$$

(7.7.11) $\mathcal{B} \perp\!\!\!\perp \mathcal{Y}_0^\infty \mid (\mathcal{Z}_0^\infty \vee \mathcal{Y}_n)_T .$

Proof. The proof is essentially the same as for Theorem 7.7.2. Note, however, that the implication of (7.7.9) by (7.7.11) relies on the fact that $(\mathcal{Z}_0^\infty \vee \mathcal{Y}_n)_T \vee \mathcal{Y}_n^\infty = \mathcal{Z}_0^\infty \vee \mathcal{Y}_n^\infty .$ ∎

7.7.6 Definition. A dynamic conditional Bayesian experiment $\mathcal{E}_{\mathcal{B}\vee\mathcal{Y}_0^\infty}^{\mathcal{Z}_0^\infty}$ is *tail-sufficient* if any one of the conditions in Theorem 7.7.5 is satisfied.   ∎

Note that $\mathcal{Z}_0^\infty \subset (\mathcal{Z}_0^\infty \vee \mathcal{Y}_n)_T$, i.e., the asymptotic conditioning σ-field is contained in the asymptotically sufficient statistic. From the viewpoint of exact estimability, there is no hope, in such a case, of obtaining 0-1 type properties of the asymptotically sufficient statistic. This could be achieved if, in (7.7.11) one could replace $(\mathcal{Z}_0^\infty \vee \mathcal{Y}_n)_T$ by $(\mathcal{Z}_n \vee \mathcal{Y}_n)_T$. That would mean that $\mathcal{E}_{\mathcal{B}\vee\mathcal{X}_0^\infty}$ is a tail-sufficient Bayesian experiment, which suggests that tail-sufficiency in both the conditional experiment $\mathcal{E}_{\mathcal{B}\vee\mathcal{Y}_0^\infty}^{\mathcal{Z}_0^\infty}$ and in the marginal experiment $\mathcal{E}_{\mathcal{B}\vee\mathcal{Z}_0^\infty}$ would be equivalent to tail sufficiency in $\mathcal{E}_{\mathcal{B}\vee\mathcal{X}_0^\infty}$. But in fact this equivalence requires that a condition dealing with the elimination of irrelevant conditioning observations be satisfied (more precisely, this condition requires that the sequence $(\mathcal{Z}_n)$ be allocated to $(\mathcal{Y}_n)$ (see Definition 7.6.12), i.e., $\forall I \subset \mathbb{N} : \mathcal{Y}_I \perp\!\!\!\perp \mathcal{Z}_0^\infty \mid \mathcal{Z}_I$). Recall, from Section 7.6, that $(\mathcal{Y}_n)$ sifted by $(\mathcal{Z}_n)$ implies that $(\mathcal{Z}_n)$ is allocated to $(\mathcal{Y}_n)$, which implies that $\mathcal{Z}_0^n$ is transitive for $\mathcal{X}_0^n$, i.e., $\mathcal{Z}_0^{n+1} \perp\!\!\!\perp \mathcal{Y}_0^n \mid \mathcal{Z}_0^n$, $\forall\, n \in \mathbb{N}$. We may now state the following theorem.

7.7.7 Theorem.

(i)   If $(\mathcal{Z}_n)$ is allocated to $(\mathcal{Y}_n)$, and $\mathcal{E}_{\mathcal{B}\vee\mathcal{Y}_0^\infty \vee \mathcal{Z}_0^\infty}$ is tail-sufficient, then $\mathcal{E}_{\mathcal{B}\vee\mathcal{Z}_0^\infty}$ and $\mathcal{E}_{\mathcal{B}\vee\mathcal{Y}_0^\infty}^{\mathcal{Z}_0^\infty}$ are tail-sufficient.

(ii)   If $(\mathcal{Z}_n)$ is $\mathcal{B}$-allocated to $(\mathcal{Y}_n)$, and both $\mathcal{E}_{\mathcal{B}\vee\mathcal{Z}_0^\infty}$ and $\mathcal{E}_{\mathcal{B}\vee\mathcal{Y}_0^\infty}^{\mathcal{Z}_0^\infty}$ are tail-sufficient, then $\mathcal{E}_{\mathcal{B}\vee\mathcal{Y}_0^\infty \vee \mathcal{Z}_0^\infty}$ is tail-sufficient.

Proof. It suffices to prove that

(1) $\mathcal{Y}_n^\infty \perp\!\!\!\perp \mathcal{Z}_0^\infty \mid \mathcal{Z}_n^\infty$
(2) $\mathcal{B} \perp\!\!\!\perp (\mathcal{Z}_0^\infty \vee \mathcal{Y}_0^\infty) \mid \mathcal{Z}_n^\infty \vee \mathcal{Y}_n^\infty$

are equivalent to

(3) $\mathcal{Y}_n^\infty \perp\!\!\!\perp \mathcal{Z}_0^\infty \mid \mathcal{B} \vee \mathcal{Z}_n^\infty$,

(4) $\mathcal{B} \perp\!\!\!\perp \mathcal{Z}_0^\infty \mid \mathcal{Z}_n^\infty$,

(5) $\mathcal{B} \perp\!\!\!\perp \mathcal{Y}_0^\infty \mid \mathcal{Z}_0^\infty \vee \mathcal{Y}_n^\infty$.

By Theorem 2.2.10, (2) is equivalent to (5) and $\mathcal{B} \perp\!\!\!\perp \mathcal{Z}_0^\infty \mid \mathcal{Z}_n^\infty \vee \mathcal{Y}_n^\infty$. This, along with (1), are equivalent to $(\mathcal{B} \vee \mathcal{Y}_n^\infty) \perp\!\!\!\perp \mathcal{Z}_0^\infty \mid \mathcal{Z}_n^\infty$. By Theorem 2.2.10, this last conditional independence relation is equivalent to (3) and (4). ∎

Note the following robustness property of tail-sufficiency:

7.7.8 Proposition. If $\mathcal{E}_{\mathcal{B} \vee \mathcal{Y}_0^\infty}^{\mathcal{Z}_0^\infty}$ is tail-sufficient then, $\forall\, \mathcal{C} \subset \mathcal{B}$, $\mathcal{E}_{\mathcal{C} \vee \mathcal{Y}_0^\infty}^{\mathcal{Z}_0^\infty}$ is tail-sufficient.

Let us remark that the tail-sufficiency of the experiment $\mathcal{E}_{\mathcal{B} \vee \mathcal{Y}_0^\infty}^{\mathcal{Z}_0^\infty}$ may be implied by assumptions that have no relationship with the usual concept of stationarity, i.e., the invariance with respect to the shift operator. One rather trivial example is the following proposition:

7.7.9 Proposition. If $\mathcal{L}$ is an $\mathcal{E}_{\mathcal{B} \vee \mathcal{Y}_0^\infty}^{\mathcal{Z}_0^\infty}$-ancillary parameter (i.e., $\mathcal{L} \subset \mathcal{B} \vee \mathcal{Z}_0^\infty$ and $\mathcal{L} \perp\!\!\!\perp \mathcal{Y}_0^\infty \mid \mathcal{Z}_0^\infty$), then $\mathcal{E}_{\mathcal{L} \vee \mathcal{Y}_0^\infty}^{\mathcal{Z}_0^\infty}$ is tail-sufficient. ∎

7.7.2   Bayesian Independence

We use the definitions and results of Section 7.6 to present a Bayesian analysis of independent experiments. We show that this independence hypothesis provides exactly estimating statistics in asymptotic experiments.

7.7.10 Definition. An *independent Bayesian experiment* $\mathcal{E}_{\mathcal{A} \vee \mathcal{X}_0^\infty}$ is a dynamic Bayesian experiment such that

$$(7.7.12) \qquad\qquad \underset{0 \le k \le \infty}{\perp\!\!\!\perp} \mathcal{X}_k \mid \mathcal{A}. \qquad\qquad ∎$$

In an independent Bayesian experiment, the sequence $\overline{\mathcal{X}_0^n \mathcal{A}}$ of minimal sufficient statistics is $(\mathcal{X}_n)$-recursive, i.e., $\overline{\mathcal{X}_0^n \mathcal{A}} \subset \mathcal{X}_n \vee \overline{\mathcal{X}_0^{n-1} \mathcal{A}}$; this important property is a simple consequence of Corollary 7.6.8. From Theorem 7.6.5, we obtain the following proposition:

7.7.11 Proposition. In an independent Bayesian experiment, $\mathcal{X}_T$ is exactly estimating, i.e., $\mathcal{X}_T \subset \overline{\mathcal{A}}$. More generally, for any $\mathcal{B} \subset \mathcal{A}$,

$$(\mathcal{B} \vee \mathcal{X}_n)_T \subset \overline{\mathcal{A}}. \qquad \blacksquare$$

7.7.12 Definition. A *conditional independent Bayesian experiment* is a dynamic conditional experiment $\mathcal{E}_{\mathcal{B} \vee y_0^\infty}^{Z_0^\infty}$ such that $(\mathcal{Y}_n)$ is $\mathcal{B}$-sifted by (Z_n), i.e.,

(7.7.13) $$\underset{0 \leq n < \infty}{\perp\!\!\!\perp} \mathcal{Y}_n \quad | \quad \mathcal{B} \vee Z_0^\infty$$

(7.7.14) $$\mathcal{Y}_n \perp\!\!\!\perp Z_0^\infty \quad | \quad \mathcal{B} \vee Z_n. \qquad \blacksquare$$

Note the crucial role of condition (7.7.14); indeed, were it dropped there would be no formal reason to consider the fixed σ-field Z_0^∞ as a "parameter" rather than an "observation", and (7.7.13) alone would therefore be formally equivalent to (7.7.12). Thus (7.7.14) may be viewed as a characterization of the z-process. Note also that, under (7.7.13), the allocation assumption, i.e., (Z_n) is $\mathcal{B}$-allocated to $(\mathcal{Y}_n)$, is equivalent to (7.7.14), by Theorem 7.6.9. Theorem 7.6.13 entails the following proposition.

7.7.13 Proposition. In a conditional independent Bayesian experiment $\mathcal{X}_T$ is exactly estimating. More precisely:

$$\mathcal{X}_T \subset \overline{(\mathcal{B} \vee Z_n)_T} \subset \overline{\mathcal{B} \vee Z_0^\infty}. \qquad \blacksquare$$

From Theorem 7.6.13 we see that, under additional assumptions, Proposition 7.7.13 may be refined. Indeed, if $\mathcal{B} \perp\!\!\!\perp Z_0^\infty \mid Z_T$, i.e., $\mathcal{E}_{\mathcal{B} \vee Z_0^\infty}$ is tail-sufficient, then

(7.7.15) $$\mathcal{X}_T \subset \overline{\mathcal{B} \vee Z_T}.$$

Moreover, if Z_T is exactly estimating, i.e., $Z_T \subset \overline{\mathcal{B}}$, then so is $\mathcal{X}_T$, i.e.,

(7.7.16) $$\mathcal{X}_T \subset \overline{\mathcal{B}}.$$

Finally, note that without the assumption of the $\mathcal{B}$-allocation of (Z_n) to $(\mathcal{Y}_n)$, we obtain that $\mathcal{X}_T$ is exactly estimating only in $\mathcal{E}_{\mathcal{B} \vee y_0^\infty}^{Z_0^\infty}$, i.e.,

(7.7.17) $$\mathcal{X}_T \subset \overline{\mathcal{B} \vee Z_0^\infty}.$$

Finally, from Theorem 7.6.9, we obtain the following proposition:

7.7.14 Proposition. The experiment $\mathcal{E}_{\mathcal{B} \vee \mathcal{Y}_0^\infty \vee \mathcal{Z}_0^\infty}$ is independent if and only if $\mathcal{E}_{\mathcal{B} \vee \mathcal{Z}_0^\infty}$ is independent and $\mathcal{E}_{\mathcal{B} \vee \mathcal{Y}_0^\infty}^{\mathcal{Z}_0^\infty}$ is conditionally independent. ∎

7.7.3 Independent Tail-Sufficient Bayesian Experiments

Exact estimability results are now obtained by assuming both independence and tail-sufficiency.

7.7.15 Definition. An *independent tail-sufficient Bayesian* experiment is a dynamic Bayesian experiment that is both independent and tail-sufficient. ∎

We immediately obtain the following theorem:

7.7.16 Theorem. In an independent tail-sufficient Bayesian experiment $\mathcal{E}_{\mathcal{A} \vee \mathcal{X}_0^\infty}$ the minimal sufficient parameter is exactly estimable (i.e., $\mathcal{A} \mathcal{X}_0^\infty \subset \overline{\mathcal{X}_0^\infty}$). More precisely, $\overline{\mathcal{X}_T} = \overline{\mathcal{A}} \cap \overline{\mathcal{X}_0^\infty} = \overline{\mathcal{X}_0^\infty \mathcal{A}} = \overline{\mathcal{A} \mathcal{X}_0^\infty}$.

Proof. By Theorem 7.7.2, $\mathcal{X}_T$ is asymptotically sufficient. By Proposition 7.7.11, $\mathcal{X}_T$ is also exactly estimating, and, finally, by Theorem 4.7.8, the minimal sufficient parameter $\mathcal{A} \mathcal{X}_0^\infty$ is the maximal exactly estimable parameter $\mathcal{A} \cap \overline{\mathcal{X}_0^\infty}$. ∎

We now turn to conditional experiments. In order to extend Theorem 7.7.16 to the conditional case we must introduce a new definition:

7.7.17 Definition. A *conditional independent tail-sufficient Bayesian experiment* is a dynamic conditional Bayesian experiment $\mathcal{E}_{\mathcal{B} \vee \mathcal{Y}_0^\infty}^{\mathcal{Z}_0^\infty}$ that is both conditional independent and tail-sufficient. ∎

7.7.18 Theorem. Let us consider a dynamic Bayesian experiment $\mathcal{E}_{\mathcal{B} \vee \mathcal{X}_0^\infty}$, where $\mathcal{X}_k = \mathcal{Y}_k \vee \mathcal{Z}_k$ for any k. If:

(i) $\mathcal{E}_{\mathcal{B} \vee \mathcal{Z}_0^\infty}$ is a tail-sufficient Bayesian experiment,

(ii) $\mathcal{E}_{\mathcal{B} \vee \mathcal{Y}_0^\infty}^{\mathcal{Z}_0^\infty}$ is a conditional independent tail-sufficient Bayesian experiment,

(iii) $\mathcal{E}_{\mathcal{B} \vee \mathcal{Y}_0^\infty}^{\mathcal{Z}_0^\infty}$ is identified,

then $\mathcal{B}$ is exactly estimable. More precisely,

(7.7.18)
$$\overline{\mathcal{X}_T} = \overline{\mathcal{B} \vee \mathcal{Z}_T}.$$

Proof. 1) By Theorem 7.6.9 (v), (ii) implies that $(\mathcal{Z}_n)$ is $\mathcal{B}$-allocated to $(\mathcal{Y}_n)$. By Theorem 7.7.7 (ii), (i) and (ii) implies that $\mathcal{E}_{\mathcal{B} \vee \mathcal{X}_0^\infty}$ is tail-sufficient, i.e., $\mathcal{B} \perp\!\!\!\perp \mathcal{X}_0^\infty \mid \mathcal{X}_T$; and so, since $\mathcal{Z}_T \subset \mathcal{X}_T$, $\mathcal{X}_T$ is an $\mathcal{E}_{\mathcal{B} \vee \mathcal{X}_0^\infty}^{\mathcal{Z}_T}$-sufficient statistic.

2) By Proposition 7.7.13, (ii) implies that $\mathcal{X}_T \subset (\mathcal{B} \vee \mathcal{Z}_n)_T$. But, under (i), by (7.7.15), $\mathcal{X}_T \subset \overline{\mathcal{B} \vee \mathcal{Z}_T}$, i.e., $\mathcal{X}_T$ is $\mathcal{E}_{\mathcal{B} \vee \mathcal{X}_0^\infty}^{\mathcal{Z}_T}$-exactly estimating.

3) Hence, by Theorem 4.7.8,

$$\overline{\mathcal{X}_T} = \overline{(\mathcal{B} \vee \mathcal{Z}_T)\mathcal{X}_0^\infty} = \overline{\mathcal{X}_0^\infty (\mathcal{B} \vee \mathcal{Z}_T)} = \overline{(\mathcal{B} \vee \mathcal{Z}_T)} \cap \overline{\mathcal{X}_0^\infty}.$$

4) By Proposition 4.6.6 (ii), $(\mathcal{B} \vee \mathcal{Z}_0^\infty)\mathcal{X}_0^\infty = \mathcal{B} \vee \mathcal{Z}_0^\infty$, i.e., the identification of $\mathcal{E}_{\mathcal{B} \vee \mathcal{X}_0^\infty}^{\mathcal{Z}_0^\infty}$, implies $(\mathcal{B} \vee \mathcal{Z}_T)\mathcal{X}_0^\infty = \mathcal{B} \vee \mathcal{Z}_T$, the identification of $\mathcal{E}_{\mathcal{B} \vee \mathcal{X}_0^\infty}^{\mathcal{Z}_T}$. Therefore by 3), $\overline{\mathcal{X}_T} = \overline{\mathcal{B} \vee \mathcal{Z}_T}$ and, consequently, $\mathcal{B} \subset \overline{\mathcal{X}_0^\infty}$. ∎

Let us point out that the three assumptions of Theorem 7.7.18 may be interpreted as three steps of model building. The first assumption concerns relationships between the parameter $\mathcal{B}$ and the sequence of statistics $(\mathcal{Z}_n)$ only. Note that this assumption is, in particular, implied by the ancillarity of $(\mathcal{Z}_n)$ in $\mathcal{E}_{\mathcal{B} \vee \mathcal{X}_0^\infty}$. Note also that this assumption of ancillarity justifies the fact that, without losing information, we can consider the conditional experiment $\mathcal{E}_{\mathcal{B} \vee \mathcal{Y}_0^\infty}^{\mathcal{Z}_0^\infty}$. However, the assumption of ancillarity is not necessary for the exact estimability of $\mathcal{B}$.

The second assumption concerns the specification of the distribution of $\mathcal{Y}_0^\infty$ conditionally to the parameter $\mathcal{B}$ and the statistics $\mathcal{Z}_0^\infty$. We intended to express in terms of independence and tail-sufficiency the essential assumptions of the usual regression model. Our mode of presentation, however, is couched in terms of σ-fields and does not require any particular functional form.

Finally, the third assumption is simply a coherence assumption between the specification of the conditional model and the parameter of interest.

Note also that, in Theorem 7.7.18, the assumptions are asymptotic. But, in light of Theorem 7.6.10, Bayesian conditional independence assumptions are equivalent to global or sequential assumptions. If the assumption that $\mathcal{E}_{\mathcal{B} \vee \mathcal{Z}_0^\infty}$ is tail-sufficient, is provided by an ancillarity assumption ($\mathcal{B} \perp\!\!\!\perp \mathcal{Z}_0^\infty$), it is also equivalent to a global assumption of ancillarity ($\mathcal{B} \perp\!\!\!\perp \mathcal{Z}_0^n$) and to sequential assumptions of ancillarity ($\mathcal{Z}_n \perp\!\!\!\perp \mathcal{B} \mid \mathcal{Z}_0^{n-1}$, $\forall\, n \in I\!N$, with $\mathcal{Z}_0^{-1} = \mathcal{I}$) (see Chapter 6, Proposition 6.5.1). However, the conditional tail-sufficiency remains an asymptotic assumption. The introduction of the shift operator in Chapter 9 permits this property to be deduced from finite sample ones. From Theorem 7.3.21, and the comments which follow, the asymptotic property of identification may also be provided by an identification property of a finite conditional experiment.

To conclude this section, we consider the following problem. It is often the case in applications that one must establish whether the sufficient parameter $\mathcal{C}$ of a conditional experiment, obtained through a mutual exogeneity condition or through a Bayesian cut condition, is exactly estimable. In this context, assumptions of conditional independence and of conditional tail-sufficiency are made with respect to sampling probabilities, i.e., in the experiment $\mathcal{E}_{\mathcal{A} \vee \mathcal{X}_0^\infty}$ and not in the experiment $\mathcal{E}_{\mathcal{C} \vee \mathcal{X}_0^\infty}$. The following theorem addresses this problem in the case of mutual exogeneity; in light of Theorem 3.4.6, the Bayesian cut case receives the same answer.

7.7.19 Theorem. Let $\mathcal{E}_{\mathcal{A} \vee \mathcal{X}_0^\infty}$ be a dynamic Bayesian experiment, where $\mathcal{X}_k = \mathcal{Y}_k \vee \mathcal{Z}_k$ for any k and a parameter $\mathcal{C} \subset \mathcal{A}$. If:

(i) $\mathcal{C}$ and $\mathcal{Z}_0^n$ are $\mathcal{E}_{\mathcal{A} \vee \mathcal{X}_0^n}$-mutually exogeneous, $\quad \forall\, n \in I\!N$,

(ii) $\mathcal{E}_{\mathcal{A} \vee \mathcal{Y}_0^\infty}^{\mathcal{Z}_0^\infty}$ is a conditional independent tail-sufficient Bayesian experiment,

(iii) $\mathcal{C}$ is included in the minimal strong $\mathcal{E}_{\mathcal{A} \vee \mathcal{Y}_0^\infty}^{\mathcal{Z}_0^\infty}$-sufficient parameter,

(iv) $\mathcal{A}$ and $\mathcal{Z}_0^n$ are measurably separated conditionally on $\mathcal{C} \vee \mathcal{Z}_k$, $\forall\, k \leq n$,

then $\mathcal{C}$ is exactly estimable.

Proof. It suffices to show that $\mathcal{Z}_0^\infty$ is $\mathcal{E}_{\mathcal{C}\vee\mathcal{X}_0^\infty}$-ancillary and that $\mathcal{E}_{\mathcal{C}\vee\mathcal{Y}_0^\infty}^{\mathcal{Z}_0^\infty}$ is an identified conditional independent tail-sufficient Bayesian experiment and to then apply Theorem 7.7.18. In other words, using Theorem 7.6.10, we must show that:

(1) $\mathcal{C} \perp\!\!\!\perp \mathcal{Z}_0^n$, $\forall\, n \in \mathbb{N}$,

(2) $\mathcal{A} \perp\!\!\!\perp \mathcal{Y}_0^n \mid \mathcal{C} \vee \mathcal{Z}_0^n$, $\forall\, n \in \mathbb{N}$,

(3) $\mathcal{Y}_{n+1} \perp\!\!\!\perp \mathcal{Y}_0^n \mid \mathcal{A} \vee \mathcal{Z}_0^\infty$, $\forall\, n \in \mathbb{N}$,

(4) $\mathcal{Y}_k \perp\!\!\!\perp \mathcal{Z}_0^n \mid \mathcal{A} \vee \mathcal{Z}_k$, $\forall\, k, n \in \mathbb{N}$, $k \leq n$,

(5) $\mathcal{A} \perp\!\!\!\perp \mathcal{Y}_0^\infty \mid \mathcal{Z}_0^\infty \vee \mathcal{Y}_n^\infty$, $\forall\, n \in \mathbb{N}$,

(6) $\mathcal{C} \subset (\mathcal{A} \vee \mathcal{Z}_0^\infty)\mathcal{X}_0^\infty$,

(7) $\mathcal{A} \parallel \mathcal{Z}_0^n \mid \mathcal{C} \vee \mathcal{Z}_k$, $\forall\, k, n \in \mathbb{N}$, $k \leq n$
imply
(8) $\mathcal{C} \perp\!\!\!\perp \mathcal{Z}_0^\infty$,

(9) $\mathcal{Y}_{n+1} \perp\!\!\!\perp \mathcal{Y}_0^n \mid \mathcal{C} \vee \mathcal{Z}_0^\infty$, $\forall\, n \in \mathbb{N}$,

(10) $\mathcal{Y}_k \perp\!\!\!\perp \mathcal{Z}_0^n \mid \mathcal{C} \vee \mathcal{Z}_k$, $\forall\, k, n \in \mathbb{N}$, $k \leq n$,

(11) $\mathcal{C} \perp\!\!\!\perp \mathcal{Y}_0^\infty \mid \mathcal{Z}_0^\infty \vee \mathcal{Y}_n^\infty$, $\forall\, n \in \mathbb{N}$,

(12) $\mathcal{C} \vee \mathcal{Z}_0^\infty = (\mathcal{C} \vee \mathcal{Z}_0^\infty)\mathcal{X}_0^\infty$.

Clearly, (1) is equivalent to (8) and (5) implies (11), since $\mathcal{C} \subset \mathcal{A}$. Now (2), by Theorem 7.2.7, implies

(13) $\mathcal{A} \perp\!\!\!\perp \mathcal{Y}_0^\infty \mid \mathcal{C} \vee \mathcal{Z}_0^\infty$, which clearly implies

(14) $\mathcal{A} \perp\!\!\!\perp \mathcal{Y}_{n+1} \mid \mathcal{C} \vee \mathcal{Z}_0^\infty$.

By Theorem 2.2.10, (3) and (14) imply (9). Now, (2) implies

(15) $\mathcal{A} \perp\!\!\!\perp \mathcal{Y}_k \mid \mathcal{C} \vee \mathcal{Z}_0^n$

By Theorem 5.2.10, (15), (4) and (7) imply

(16) $\mathcal{Y}_k \perp\!\!\!\perp (\mathcal{A} \vee \mathcal{Z}_0^n) \mid \mathcal{C} \vee \mathcal{Z}_k$,

which clearly implies (10). Finally, $(\mathcal{A} \vee \mathcal{Z}_0^\infty) \perp\!\!\!\perp \mathcal{X}_0^\infty \mid \mathcal{C} \vee \mathcal{Z}_0^\infty$ follows from (13). So $\overline{(\mathcal{A} \vee \mathcal{Z}_0^\infty)\mathcal{X}_0^\infty} = \overline{(\mathcal{C} \vee \mathcal{Z}_0^\infty)\mathcal{X}_0^\infty}$. By (6), $\mathcal{C} \subset (\mathcal{A} \vee \mathcal{Z}_0^\infty)\mathcal{X}_0^\infty$ and this implies that $\mathcal{C} \vee \mathcal{Z}_0^\infty \subset \overline{(\mathcal{C} \vee \mathcal{Z}_0^\infty)\mathcal{X}_0^\infty} \cap (\mathcal{C} \vee \mathcal{Z}_0^\infty) = (\mathcal{C} \vee \mathcal{Z}_0^\infty)\mathcal{X}_0^\infty \subset \mathcal{C} \vee \mathcal{Z}_0^\infty$. This gives (12). ∎

Let us remark that the allocation assumption (4) and the measurable separability assumption (7) are used merely to obtain (16). So, with Assumption (16), the result is obtained without Assumption (iv). Note that Assumption (16) is a sampling property, i.e., the distribution of $\mathcal{Y}_k$ conditionally to $\mathcal{A} \vee \mathcal{Z}_0^n$ depends on $\mathcal{C} \vee \mathcal{Z}_k$ only; this is easily verified.

Note also that, even if the result is an asymptotic property, the measurable separability property is required only in finite sample (i.e., in $\mathcal{E}_{\mathcal{A} \vee \mathcal{Z}_0^n}$) in which it is generally a mild assumption. This is not at all the case in the asymptotic experiment because a typical situation is that in which the minimal sufficient parameter of the sequence $(\mathcal{Z}_n)$ is exactly estimable, i.e., is included in $\overline{\mathcal{Z}_0^\infty}$.

7.8 An Example

In this section we illustrate most of the concepts elaborated in this chapter by means of a simple example.

Let us consider $s = \{x_n : n \in I\!N\} \in S = I\!R^{I\!N}$ and $a \in A = I\!R$. Suppose that $(x_n \mid a) \sim i.N(\alpha_n a, \sigma^2)$ where $\sigma^2 \in I\!R_0^+$ is a known real positive number and $\{\alpha_n : n \in I\!N\}$ is a known sequence of non zero real numbers. Note that $\{\alpha_n : n \in I\!N\}$ may be, for instance, thought of as representing the shape of the trend of the dynamic process $\{x_n : n \in I\!N\}$. Suppose, furthermore, that $a \sim N(a_0, (\sigma^2/n_0))$, where $a_0 \in I\!R$ and $n_0 \in I\!R_0^+$ are known numbers.

7.8.1 Global and Sequential Analysis

Let $\mathcal{X}_0^n = \bigvee_{0 \le k \le n} \mathcal{X}_k$ where $\mathcal{X}_k = \sigma(x_k)$ and $\mathcal{A} = \sigma(a)$. Then, the usual computations on the normal distribution give:

$$(7.8.1) \qquad (a \mid x_0^n) \quad \sim \quad N(\mathcal{X}_0^n a, \frac{\sigma^2}{n_0 + \beta_n})$$

where

$$(7.8.2) \qquad \beta_n \quad = \quad \sum_{0 \le k \le n} \alpha_k^2$$

and

$$(7.8.3) \qquad \mathcal{X}_0^n a \quad = \quad \frac{n_0 a_0 + \beta_n t_n}{n_0 + \beta_n}$$

where

$$(7.8.4) \qquad t_n \quad = \quad \frac{1}{\beta_n} \sum_{0 \le k \le n} \alpha_k x_k.$$

Note also that

$$(7.8.5) \qquad (t_n \mid a) \quad \sim \quad N\left(a, \frac{\sigma^2}{\beta_n}\right)$$

and, therefore,

$$(7.8.6) \qquad t_n \quad \sim \quad N\left(a_0, \frac{\sigma^2}{n_0} + \frac{\sigma^2}{\beta_n}\right).$$

Hence,

$$(7.8.7) \qquad (\mathcal{X}_0^n a \mid a) \quad \sim \quad N\left(\frac{n_0 a_0 + \beta_n a}{n_0 + \beta_n}, \frac{\beta_n \sigma^2}{(n_0 + \beta_n)^2}\right)$$

and, therefore,

$$(7.8.8) \qquad \mathcal{X}_0^n a \quad \sim \quad N\left(a_0, \frac{\beta_n}{n_0 + \beta_n} \frac{\sigma^2}{n_0}\right).$$

Let

$$(7.8.9) \qquad \mathcal{T}_n \quad = \quad \sigma(t_n).$$

Note that

$$(7.8.10) \qquad \overline{\mathcal{T}_n} \quad = \quad \overline{\sigma(\mathcal{X}_0^n a)}$$

since t_n and $\mathcal{X}_0^n a$ are a.s. in bijection. Clearly, by (7.8.1) and (7.8.3), $\mathcal{T}_n$ is $\mathcal{E}_{A \vee \mathcal{X}_0^n}$-sufficient, i.e.,

$$(7.8.11) \qquad \qquad \mathcal{A} \perp\!\!\!\perp \mathcal{X}_0^n \mid \mathcal{T}_n$$

and, therefore,

$$(7.8.12) \qquad (a \mid t_n) \sim N\left(\frac{n_0 a_0 + \beta_n t_n}{n_0 + \beta_n}, \frac{\sigma^2}{n_0 + \beta_n}\right).$$

Using known results for the normal distribution, and noticing that the distributions of a and t_n are regular probabilities, an application of Theorem 5.5.17, shows that $\mathcal{T}_n$ is a 1-complete statistic and that $\mathcal{A}$ is a 1-complete parameter in $\mathcal{E}_{A \vee \mathcal{X}_0^n}$, i.e.,

$$(7.8.13) \qquad \qquad \mathcal{T}_n \ll_1 \mathcal{A}$$

and

$$(7.8.14) \qquad \qquad \mathcal{A} \ll_1 \mathcal{T}_n.$$

By Proposition 5.5.3 and Theorem 5.5.4, $\mathcal{A}$ and $\mathcal{T}_n$ are therefore identified and minimal sufficient, i.e.,

$$(7.8.15) \qquad \mathcal{A}\mathcal{X}_0^n = \mathcal{A}\mathcal{T}_n = \mathcal{A}$$

$$(7.8.16) \qquad \mathcal{X}_0^n \mathcal{A} = \overline{\mathcal{T}_n \cap \mathcal{X}_0^n}$$

$$(7.8.17) \qquad \mathcal{T}_n \mathcal{A} = \mathcal{T}_n.$$

Note also that, from Theorem 7.6.6,

$$(7.8.18) \qquad \mathcal{A}\mathcal{X}_0^n = \bigvee_{0 \le k \le n} \mathcal{A}\mathcal{X}_k$$

and $\mathcal{A}\mathcal{X}_k = \mathcal{A}$ as soon as $\alpha_k \ne 0$, as supposed.

Furthermore, let us remark that

$$(7.8.19) \qquad \mathcal{T}_{n+1} \subset \mathcal{T}_n \vee \mathcal{X}_{n+1}$$

i.e., $(\mathcal{T}_n)$ is $(\mathcal{X}_n)$-recursive. This is an example of Corollary 7.6.8, which states that under mutual conditional independence

$$(7.8.20) \qquad \mathcal{X}_0^{n+1} \mathcal{A} \subset \overline{\mathcal{X}_0^n \mathcal{A} \vee \mathcal{X}_{n+1}}.$$

Note also that, from (7.8.19), we easily obtain that

$$(7.8.21) \qquad \mathcal{T}_0^n = \mathcal{X}_0^n \quad \forall\, n \in I\!N$$

$$(7.8.22) \qquad \mathcal{T}_n^m = \mathcal{T}_n \vee \mathcal{X}_{n+1}^m \quad \forall\, n < m.$$

7.8.2 Asymptotic Analysis

According to Section 7.3.2, the above sequential analysis entails the following results. By Proposition 7.3.4 (i), we have:

$$(7.8.23) \qquad\qquad \mathcal{A} \perp\!\!\!\perp \mathcal{X}_0^\infty \mid \mathcal{T}_T$$

or, equivalently, using (7.8.16) and 7.2.4 (ii):

$$(7.8.24) \qquad\qquad \mathcal{X}_0^\infty \mathcal{A} \subset \overline{\mathcal{T}_T} = \overline{(\mathcal{X}_0^n \mathcal{A})_T}.$$

By Proposition 7.3.4 (ii) and (7.3.2),

$$(7.8.25) \qquad\qquad \mathcal{X}_0^n \mathcal{A} \subset \mathcal{X}_0^n (\mathcal{X}_0^\infty \mathcal{A}) \subset \mathcal{X}_0^n (\mathcal{T}_T)$$

and, according to Theorem 7.3.6., we have equalities rather than inclusions in (7.8.24) and (7.8.25) if we can show that $\mathcal{T}_T$ is 2-complete. Recall that this cannot be deduced from the fact that $(\mathcal{T}_n)$ is a sequence of 2-complete sufficient statistics.

In order to characterize $\mathcal{T}_T$ note first that, from (7.8.22),

$$(7.8.26) \qquad\qquad \mathcal{T}_n^\infty \;=\; \mathcal{T}_n \vee \mathcal{X}_{n+1}^\infty \quad \forall\, n \in I\!N$$

and, therefore,

$$(7.8.27) \qquad\qquad \mathcal{X}_T \subset \mathcal{T}_T \;=\; \bigcap_{n \in I\!N} (\mathcal{T}_n \vee \mathcal{X}_{n+1}^\infty).$$

The second step in the analysis is to use the results concerning the mutual conditional independence of $\{\mathcal{X}_n : n \in I\!N\}$, obtained in Section 7.6.1.

Note first that, using 7.3.22, repeated application of Theorem 2.2.10 and Corollary 2.2.11 establish that the following three sets of conditional independences are equivalent:

$$(7.8.28) \qquad \mathcal{X}_0^n \perp\!\!\!\perp \mathcal{X}_{n+1}^\infty \mid \mathcal{A} \qquad \text{and} \qquad \mathcal{X}_0^n \perp\!\!\!\perp \mathcal{A} \mid \mathcal{T}_n$$

$$(7.8.29) \qquad \mathcal{T}_n \perp\!\!\!\perp \mathcal{X}_{n+1}^\infty \mid \mathcal{A} \qquad \text{and} \qquad \mathcal{X}_0^n \perp\!\!\!\perp (\mathcal{A} \vee \mathcal{X}_{n+1}^\infty) \mid \mathcal{T}_n$$

$$(7.8.30) \qquad \mathcal{T}_n \perp\!\!\!\perp \mathcal{X}_{n+1}^\infty \mid \mathcal{A} \qquad \text{and} \qquad \mathcal{X}_0^n \perp\!\!\!\perp (\mathcal{A} \vee \mathcal{T}_n^\infty) \mid \mathcal{T}_n$$

According to Theorem 7.6.5 we have

$$(7.8.31) \qquad\qquad \mathcal{X}_T \subset \overline{\mathcal{A}}$$

and

(7.8.32) $$\overline{(\mathcal{A} \vee \mathcal{X}_n)_T} = \overline{\mathcal{A}}.$$

Since (7.8.23) implies that

(7.8.33) $$\mathcal{A} \perp \mathcal{T}_0^\infty \mid \mathcal{T}_T,$$

we also have, by Theorem 7.2.8, that

(7.8.34) $$\overline{(\mathcal{A} \vee \mathcal{T}_n)_T} = \overline{\mathcal{A} \vee \mathcal{T}_T}.$$

In order to obtain deeper results, we have to go further into the structure of the example. Since $\beta_n \to \beta$, where

(7.8.35) $$\beta = \sum_{0 \le k < \infty} \alpha_k^2,$$

and from the martingale convergence Theorem 7.2.2 and equation (7.8.3), we obtain the following result:

(7.8.36) $$t_n \to t \quad \text{a.s. and in} \quad L_p \quad \forall \, 1 \le p < \infty.$$

Moreover,

(7.8.37) $$\begin{aligned} t \;&=\; \mathcal{X}_0^\infty a & \text{if} \quad \beta = \infty \\ &=\; \left(1 + \tfrac{n_0}{\beta}\right) \mathcal{X}_0^\infty a - \tfrac{n_0}{\beta} a_0 & \text{if} \quad \beta < \infty. \end{aligned}$$

Let

(7.8.38) $$\mathcal{T} \;=\; \sigma(t).$$

Then, clearly, from (7.8.37)

(7.8.39) $$\overline{\mathcal{T}} \;=\; \overline{\sigma\{\mathcal{X}_0^\infty a\}}$$

and

(7.8.40) $$\mathcal{T} \subset \overline{\mathcal{T}_T}.$$

Let us show that $\mathcal{T}$ is also sufficient. More generally, (7.8.1) may be extended to

(7.8.41) $$(a \mid x_{n+1}^m) \;\sim\; N\left(\mathcal{X}_{n+1}^m a, \frac{\sigma^2}{n_0 + \beta_m - \beta_n} \right)$$

where

(7.8.42) $$\mathcal{X}^m_{n+1}a = \frac{n_0 a_0 + \beta_m t_m - \beta_n t_n}{n_0 + \beta_m - \beta_n}.$$

Hence, if we define

(7.8.43) $$v^m_n = a - \mathcal{X}^m_{n+1}a$$

(7.8.44) $$v_n = a - \mathcal{X}^\infty_{n+1}a,$$

then, clearly, if $\mathcal{V}^m_n = \sigma(v^m_n)$,

(7.8.45) $$\mathcal{V}^m_n \perp\!\!\!\perp \mathcal{X}^m_{n+1}$$

(7.8.46) $$v^m_n \sim N\left(0, \frac{\sigma^2}{n_0+\beta_m-\beta_n}\right)$$

and

(7.8.47) $$v^m_n \to v_n \quad \text{a.s. and in} \quad L_1 \quad \text{as } m \to \infty$$

by the martingale convergence Theorem 7.2.2. Now, for any $w \in [\mathcal{X}^k_{n+1}]_\infty$ with $k \le m$, and for any f bounded continuous function on $I\!R$,

(7.8.48) $$\mathcal{I}[f(v^m_n)w] = \mathcal{I}[f(v^m_n)]\mathcal{I}(w).$$

Passing through the limit by the dominated convergence theorem, this implies that

(7.8.49) $$\mathcal{I}[f(v_n)w] = \mathcal{I}[f(v_n)]\mathcal{I}(w).$$

Therefore, by a monotone class argument, if $\mathcal{V}_n = \sigma(v_n)$, then

(7.8.50) $$\mathcal{V}_n \perp\!\!\!\perp \mathcal{X}^\infty_{n+1} \quad \forall\, n \ge -1.$$

By the same argument, but using the reverse martingale Theorem 7.2.6,

(7.8.51) $$v_n \to a - \mathcal{X}_T a \quad \text{a.s. and in} \quad L_1,$$

and we also have that

(7.8.52) $$\sigma(a - \mathcal{X}_T a) \perp\!\!\!\perp \mathcal{X}_T.$$

Now, if we define

(7.8.53) $$\mathcal{V} = \sigma\{a - \mathcal{X}^\infty_0 a\},$$

by (7.8.39), we have that

(7.8.54) $$\mathcal{A} \subset \overline{\mathcal{V} \vee \mathcal{T}}$$

and

(7.8.55) $$\mathcal{V} \perp\!\!\!\perp \mathcal{X}_0^\infty.$$

Applying Corollary 2.2.11 shows that

(7.8.56) $$(\mathcal{V} \vee \mathcal{T}) \perp\!\!\!\perp \mathcal{X}_0^\infty \mid \mathcal{T},$$

since $\mathcal{T} \subset \mathcal{X}_0^\infty$, and therefore

(7.8.57) $$\mathcal{A} \perp\!\!\!\perp \mathcal{X}_0^\infty \mid \mathcal{T}.$$

7.8.3 The Case $\beta = \infty$

Since a.s. convergence implies convergence in distribution, it follows from (7.8.46) that

(7.8.58) $$\mathcal{X}_{n+1}^\infty a = a \quad \text{a.s.} \quad \forall\, n \geq -1.$$

Therefore,

(7.8.59) $$a = t \quad \text{a.s.}$$

and

(7.8.60) $$a \in \left[\overline{\mathcal{X}_T}\right].$$

Hence, clearly, $\mathcal{E}_{\mathcal{A} \vee \mathcal{X}_0^\infty}$ is tail-sufficient and a is exactly estimable. More precisely, from (7.8.59), (7.8.60) and (7.8.31) we obtain

(7.8.61) $$\overline{\mathcal{A}} = \overline{\mathcal{X}_T} = \overline{\mathcal{T}} \subset \overline{\mathcal{T}_T}$$

and, therefore,

(7.8.62) $$\overline{\mathcal{X}_0^\infty \mathcal{A}} = \overline{\mathcal{A} \mathcal{X}_0^\infty} = \overline{\mathcal{A}} \cap \overline{\mathcal{X}_0^\infty} = \overline{\mathcal{A}}.$$

7.8.4 The case $\beta < \infty$

As before, from (7.8.5) we obtain that

(7.8.63) $$(t \mid a) \sim \mathcal{N}\left(a, \frac{\sigma^2}{\beta}\right)$$

Therefore

(7.8.64) $$\begin{pmatrix} a \\ t \end{pmatrix} \sim \mathcal{N}_2\left[\begin{pmatrix} a_0 \\ a_0 \end{pmatrix}; \begin{pmatrix} \frac{\sigma^2}{n_0} & \frac{\sigma^2}{n_0} \\ \frac{\sigma^2}{n_0} & \frac{\sigma^2}{n_0} + \frac{\sigma^2}{\beta} \end{pmatrix}\right]$$

and this implies

$$(7.8.65) \qquad (a \mid t) \sim N\left(\frac{n_0 a_0 + \beta t}{n_0 + \beta}, \frac{\sigma^2}{n_0 + \beta}\right).$$

Note also that

$$(7.8.66) \qquad (\mathcal{X}_0^\infty a \mid a) \;\sim\; N\left(\frac{n_0 a_0 + \beta a}{n_0 + \beta}, \frac{\beta \sigma^2}{(n_0 + \beta)^2}\right)$$

and that

$$(7.8.67) \qquad \mathcal{X}_0^\infty a \;\sim\; N\left(a_0, \frac{\beta}{n_0 + \beta} \frac{\sigma^2}{n_0}\right).$$

This shows that $\Pi_{\mathcal{A} \vee \mathcal{T}} \sim (\mu \otimes P)_{\mathcal{A} \vee \mathcal{T}}$. Since $\mathcal{A} \perp\!\!\!\perp \mathcal{T}; \mu \otimes P$, from Proposition 5.2.3 and Theorem 5.2.7, we see that $\mathcal{T}$ and $\mathcal{A}$ are measurably separated, i.e.,

$$(7.8.68) \qquad \mathcal{A} \parallel \mathcal{T}.$$

Therefore, by Corollary 5.2.9 (iii),

$$(7.8.69) \qquad \mathcal{A} \parallel \mathcal{X}_0^\infty, \quad \text{i.e.,}$$

$$(7.8.70) \qquad \overline{\mathcal{A}} \cap \overline{\mathcal{X}_0^\infty} = \overline{\mathcal{I}},$$

which means that the maximal exactly estimable parameter is the trivial parameter.

On the other hand, an application of Theorem 5.5.17 shows that $\mathcal{T}$ is 1-complete, i.e.,

$$(7.8.71) \qquad \mathcal{T} \ll_1 \mathcal{A}$$

and therefore, by Theorem 7.3.6, we have

$$(7.8.72) \qquad \mathcal{X}_0^\infty \mathcal{A} \;=\; \overline{\mathcal{T}} \cap \mathcal{X}_0^\infty$$

$$(7.8.73) \qquad \mathcal{T} \mathcal{A} \;=\; \mathcal{T}$$

$$(7.8.74) \qquad \mathcal{A} \mathcal{X}_0^\infty \;=\; \mathcal{A} \mathcal{T} = \mathcal{A}$$

and

$$(7.8.75) \qquad \mathcal{X}_0^n \mathcal{A} \;=\; \overline{\mathcal{T}}_n \cap \mathcal{X}_0^n = \mathcal{X}_0^n \mathcal{T}.$$

Now from (7.8.70) and the fact that $\mathcal{X}_T \subset \overline{\mathcal{A}}$, we obtain that the tail σ-field is a trivial statistic, i.e.,

$$(7.8.76) \qquad \mathcal{X}_T \subset \overline{\mathcal{I}} \cap \mathcal{X}_0^\infty.$$

Note that this can be shown directly. Indeed, from (7.8.42), we obtain

$$(7.8.77) \qquad \mathcal{X}_{n+1}^{\infty} a \;=\; \frac{n_0 a_0 + \beta t - \beta_n t_n}{n_0 + \beta - \beta_n}$$

and, therefore,

$$(7.8.78) \qquad \mathcal{X}_T a \;=\; a_0.$$

Hence from (7.8.52),

$$(7.8.79) \qquad \mathcal{A} \perp\!\!\!\perp \mathcal{X}_T.$$

Since $\mathcal{X}_T \subset \overline{\mathcal{A}}$, we obtain, by Corollary 2.2.8,

$$(7.8.80) \qquad \mathcal{X}_T \subset \overline{\mathcal{I}} \cap \mathcal{X}_0^{\infty}.$$

Clearly, $\mathcal{E}_{\mathcal{A} \vee \mathcal{X}_0^{\infty}}$ is not tail-sufficient.

In fact, in this situation, the observation of the entire sequence $\{x_n : n \in I\!\!N\}$ is equivalent to the observation of the single statistic t, and this statistic is measurably separated from the parameter a.

8

Invariant Experiments

8.1 Introduction

In this chapter, the basic structure $\mathcal{E} = (A \times S, \, \mathcal{A} \vee \mathcal{S}, \Pi)$ is endowed with a set of transformations, and we investigate the consequences of invariance properties with respect to these transformations. Several related issues motivate an interest in invariance in statistics; some of these concerns are common to both the sampling theory and the Bayesian approaches, whereas others stem essentially from Bayesian thinking.

One motivation is the search for "impartial" statistical procedures, as suggested, e.g., in Lehmann (1959), or Arnold (1981). Thus, in a given situation it may seem desirable to look for procedures which behave identically over those parameter values for which the statistical model is invariant. Furthermore, in a Bayesian framework it has been argued that an invariant prior specification is a natural representation of a lack of prior information.

Interest in invariance very much reaches beyond such considerations; indeed, invariance often provides the basic tool for the reduction of statistical experiments because (conditional) invariance leads naturally to (conditional) independence properties. For instance, in an i.i.d. sampling from a normal process $s^2 = \Sigma(s_i - \bar{s})^2$ is an invariant statistic for the translation

349

group on the sample space and σ^2 is a maximal invariant parameter for those transformations on the parameter space and is a sufficient parameter for s^2; this is an example of the more general result that a maximal invariant parameter is sufficient for a maximal invariant statistic. For inference on σ^2 the converse question becomes: is it true that marginalizing the observation on s^2 would result in no loss of information for σ^2? In other words, is it true that a maximal invariant statistic is sufficient for a maximal invariant parameter? In a sampling theory framework the answer is, in general, "yes" under the heuristic proviso "in the absence of information about the sampling expectation". In a Bayesian framework, the answer is formally "yes" under a condition of invariance in the prior specification. Parenthetically, this is an interesting example of when mutual sufficiency may hold without having necessarily a cut. Invariance arguments will also provide criteria for ancillarity and exogeneity, because trivial invariant σ-fields transform mutual sufficiency into an ancillarity property.

From an asymptotic point of view, invariance arguments are also used for characterizing exactly estimable parameters; thus for instance, an important result is that in ergodic stationary processes identified parameters are exactly estimable where stationarity means invariance for the shift, which leaves the parameters unaffected; this is again an example of a more general result which asserts that, for those Bayesian experiments in which transformations on the product space $A \times S$ leave the joint probability invariant, the invariant functions of the parameters are exactly estimable under an ergodicity condition. Note that a theory which can treat transformations acting on both the parameters and the observations constitutes a particularly attractive tool for the treatment of incidental parameters.

A common structure clearly emerges from these different motivations, the study of which is the concern of this chapter. The next section, similarly to other chapters, tackles the abstract probability theory underlying these statistical problems. This section contains a review of the basic concepts of invariant sets and functions, of invariant measures viewed as a relationship between a set of transformations and a (sub-)σ-field, of ergodicity, mixing, and of the existence of invariant measures.

Let us briefly draw attention to some of the innovative aspects of this section. As throughout this monograph, emphasis is placed on the σ-

algebraic approach which provides a natural extension of the usual concepts in terms of marginal and conditional invariance; this section also presents in a systematic manner, the relations between joint, marginal and conditional invariance and their connection with (conditional) independence. The point properties of invariance are carefully displayed in order to obtain a (rigorous) basis for an operational version of those abstract concepts. It is also shown that almost sure invariance does require particular attention insofar as the almost sure invariant σ-field is, in general, different from the completed invariant σ-field. Finally, let us mention that most of the literature on invariance assumes that the set of transformations is endowed with a group structure, a natural example being the exchangeable processes (invariant under finite permutation) where, parenthetically, partial exchangeability is a natural example for conditional invariance. However, the statistical analysis of stationary processes (invariant under the shift operator) raises questions about the use of $I\!N$ or of $Z\!\!\!Z$ as the time index set: $Z\!\!\!Z$ generates a group of transformations but often requires the introduction of non-testable hypothesis (on the initial conditions at the left tail) while $I\!N$ generates only a monoïd (i.e., a semigroup with identity). For this reason, the theory first displays the results available in terms of a monoïd before endowing the set of transformations with a group structure. It should be mentioned that in Section 8.2.6 the set of transformations is also endowed with a probability measure; an interesting consequence of this addition is that it enables one to exhibit the equivalence of (conditional) invariance and (conditional) independence, up to some regularity conditions.

The last section introduces the idea of an invariant Bayesian experiment, i.e., invariance on the product space $A \times S$. The specification of invariant prior measure is not analyzed in detail; emphasis is given to sufficiency and exact estimability. This chapter deals with general models and examples; the next chapter applies this analysis to stochastic processes (time series) and exchangeable processes. This chapter draws upon and considerably extends Florens (1978, 1982) and (1986). In particular, results on the admissibility of reductions through invariance, characterization of a.s. invariant σ-fields, and randomization of transformations have been developed after the publication of these papers. The first two papers, however, develop the use of invariance arguments for the specification of prior measures in more detail than supplied here (see also Mouchart (1977)).

8.2   Invariance, Ergodicity, and Mixing

In this section $(M, \mathcal{M}, P)$ is, as usual, an abstract probability space, $\mathcal{M}_0 = \{\phi, M\}$ its trivial σ-field, and $\mathcal{M}_i$, $i \in I\!\!N$, are sub-σ-fields of $\mathcal{M}$.

8.2.1   Invariant Sets and Functions

Let us consider a measurable transformation φ of M, i.e., $\varphi : M \rightarrow M$ such that $\varphi^{-1}(\mathcal{M}) \subset \mathcal{M}$. We shall say that a measurable set $A \in \mathcal{M}$ is φ-invariant if $\varphi^{-1}(A) = A$ and that a (real-valued) measurable function $m \in [\mathcal{M}]$ is φ-invariant if $m = m \circ \varphi$. Note that these two definitions are coherent in the sense that $\varphi^{-1}(A) = A$ is equivalent to $\mathbf{1}_A = \mathbf{1}_A \circ \varphi$ since $\mathbf{1}_{\varphi^{-1}(A)} = \mathbf{1}_A \circ \varphi$.

8.2.1 Theorem. The collection of φ-invariant measurable sets is a σ-field called the φ-invariant σ-field and is denoted by $\mathcal{M}_\varphi$, i.e.,

$$(8.2.1) \qquad \mathcal{M}_\varphi = \{A \in \mathcal{M} : \varphi^{-1}(A) = A\},$$

and the set of real-valued $\mathcal{M}_\varphi$-measurable functions is equal to the set of φ-invariant measurable functions, i.e.,

$$(8.2.2) \qquad m \in [\mathcal{M}_\varphi] \Leftrightarrow m \in [\mathcal{M}] \quad \text{and} \quad m = m \circ \varphi.$$

Proof. $\mathcal{M}_\varphi$ is a σ-field because all set operations commute with inverse image operations. Recall that by definition, $\mathbf{1}_A = \mathbf{1}_A \circ \varphi$, $\forall A \in \mathcal{M}_\varphi$. Thus, by Theorem 0.2.21, $m \in [\mathcal{M}_\varphi] \Rightarrow m \in [\mathcal{M}]$ and $m = m \circ \varphi$. However, if $m \in [\mathcal{M}]$ and $m = m \circ \varphi$, then for any Borel set B, $m^{-1}(B) \in \mathcal{M}$ and $\varphi^{-1}[m^{-1}(B)] = (m \circ \varphi)^{-1}(B) = m^{-1}(B)$, i.e., $m^{-1}(B) \in \mathcal{M}_\varphi$. Thus $m \in [\mathcal{M}_\varphi]$.   ∎

More generally, we will consider a class Φ of measurable transformations of M. A measurable set $A \in \mathcal{M}$ will be said Φ-invariant if it is φ-invariant $\forall\, \varphi \in \Phi$ and a (real-valued) measurable function $m \in [\mathcal{M}]$ is Φ-invariant if it is φ-invariant $\forall\, \varphi \in \Phi$. In this situation we have the following straightforward corollary:

8.2.2 Corollary. The collection of Φ-invariant measurable sets is a σ-field called the Φ-*invariant σ-field* and is denoted by $\mathcal{M}_\Phi$, i.e.,

$$(8.2.3) \qquad\qquad \mathcal{M}_\Phi = \bigcap_{\varphi \in \Phi} \mathcal{M}_\varphi,$$

and the set of real-valued $\mathcal{M}_\Phi$-measurable functions is equal to the set of Φ-invariant measurable functions, i.e.,

$$(8.2.4) \qquad m \in [\mathcal{M}_\Phi] \Leftrightarrow m \in [\mathcal{M}] \quad \text{and} \quad m = m \circ \varphi, \quad \forall\, \varphi \in \Phi. \qquad \blacksquare$$

We extend the definition of Φ-invariance to not necessarily real-valued functions as:

8.2.3 Definition. Let $(N, \mathcal{N})$ be a measurable space and $h : M \to N$ with $h^{-1}(\mathcal{N}) \subset \mathcal{M}$.

(i) h is Φ-*invariant*, if $h^{-1}(\mathcal{N}) \subset \mathcal{M}_\Phi$;

(ii) h is *maximal Φ-invariant*, if $h^{-1}(\mathcal{N}) = \mathcal{M}_\Phi$. $\qquad \blacksquare$

Let us remark that, in view of (8.2.4), this definition of invariance is equivalent to the definition of invariance for real-valued functions. In fact, we have the same characterization as (8.2.4) under a minor assumption on $(N, \mathcal{N})$.

8.2.4 Theorem. Let $(N, \mathcal{N})$ be a measurable space and $h : M \to N$ with $h^{-1}(\mathcal{N}) \subset \mathcal{M}$. Then:

(i) If $h \circ \varphi = h, \quad \forall\, \varphi \in \Phi$, then h is Φ-invariant;

(ii) If h is Φ-invariant and $\mathcal{N}$ is separating, then $h \circ \varphi = h, \quad \forall\, \varphi \in \Phi$.

Proof.

(i) Let $B \in \mathcal{N}$. Then $h^{-1}(B) = (h \circ \varphi)^{-1}(B) = \varphi^{-1}[h^{-1}(B)] \ \forall\, \varphi \in \Phi$. Hence, $h^{-1}(B) \in \mathcal{M}_\Phi \ \forall\, B \in \mathcal{N}$, i.e., $h^{-1}(\mathcal{N}) \subset \mathcal{M}_\Phi$.

(ii) Since $h^{-1}(\mathcal{N}) \subset \mathcal{M}_\Phi$, we know that, $\forall B \in \mathcal{N}$, $\varphi^{-1}[h^{-1}(B)] = h^{-1}(B)$, $\forall \varphi \in \Phi$. Therefore, $\forall x \in M$, $\forall \varphi \in \Phi$, $\forall B \in \mathcal{N}$,

$$1_{\varphi^{-1}[h^{-1}(B)]}(x) = 1_B[h[\varphi(x)]] = 1_{h^{-1}(B)}(x) = 1_B[h(x)].$$

Hence, $\forall x \in M$, $\forall \varphi \in \Phi$, $h[\varphi(x)] \in A^{\mathcal{N}}_{h(x)}$. But, since $\mathcal{N}$ is separating, $A^{\mathcal{N}}_{h(x)} = \{h(x)\}$ (see Section 0.2.1). Therefore, $(h \circ \varphi)(x) = h(x)$, $\forall x \in M$, $\forall \varphi \in \Phi$.   ∎

As a corollary, we obtain the following characterization of a maximal Φ-invariant function:

8.2.5 Corollary. Let $(N, \mathcal{N})$ be a measurable space with $\mathcal{N}$ separating, and let $h : M \to N$ with $h^{-1}(\mathcal{N}) \subset \mathcal{M}$.

(i) h is Φ-invariant, if and only if $h \circ \varphi = h$, $\forall \varphi \in \Phi$;

(ii) h is maximal-Φ-invariant if and only if:
    1. $h \circ \varphi = h$ $\forall \varphi \in \Phi$, and
    2. $\forall m \in [\mathcal{M}_\Phi]$, $\exists f_m \in [\mathcal{N}]$ such that $m = f_m \circ h$.

Proof. In view of Theorem 8.2.4, it suffices to show that $\mathcal{M}_\Phi \subset h^{-1}(\mathcal{N})$ if and only if (ii.2) holds. This is obviously true, since, by Theorem 0.2.11, $m \in [h^{-1}(\mathcal{N})]$ if and only if $\exists f_m \in [\mathcal{N}]$ such that $m = f_m \circ h$.   ∎

Let us remark that the first part of Corollary 8.2.5 is the extension of (8.2.4), in Corollary 8.2.2, to non-necessarily real-valued functions; it merely requires the minor assumption that $\mathcal{N}$ is separating.

It is natural to endow Φ with a certain structure. Indeed, remark that if A is a φ-invariant measurable set, then A is also φ^n-invariant $\forall n \in \mathbb{N}$ (where φ^n is the nth composition of φ with the convention that $\varphi^0 = i$ is the identity on M). Therefore, $\mathcal{M}_\varphi \subset \mathcal{M}_{\varphi^n}$ $\forall n \in \mathbb{N}$, and thus

$$(8.2.5) \qquad\qquad \Phi = \{\varphi^n : n \in \mathbb{N}\} \Rightarrow \mathcal{M}_\Phi = \mathcal{M}_\varphi.$$

Further, if A is both φ_1-invariant and φ_2-invariant, then A is also $\varphi_1 \circ \varphi_2$-invariant, i.e., $\mathcal{M}_{\varphi_1} \cap \mathcal{M}_{\varphi_2} \subset \mathcal{M}_{\varphi_1 \circ \varphi_2}$. Therefore, concerning the

Φ-invariant σ-field, there is no loss of generality in supposing that Φ is closed under composition of measurable transformations. Consequently, in what follows we assume that Φ has at least the structure of a monoïd (i.e., a semigroup with a unit; for this concept, see, e.g., Cohn (1974), Chapter 3).

Finally, it often happens that Φ also has a group structure. Indeed, if Φ consists of bijective and bimeasurable transformations (i.e., such that φ^{-1} is also a measurable transformation on M) and if A is a φ-invariant measurable set, then it is also a φ^{-1}-invariant measurable set. In this context, there is no loss of generality in assuming that Φ is closed under inversion; Φ is then a group of measurable transformations on M. Conversely, if Φ is a group of measurable transformations on M, Φ consists of bijective and bimeasurable transformations on M.

Let us remark that the characterization of Φ-invariance and maximal Φ-invariance given in Theorem 8.2.4 and Corollary 8.2.5 did not require Φ to exhibit any particular structure. When Φ is a group it is possible to obtain another characterization of these invariance properties as point properties. These point properties, presented below, are in fact the usual definitions of Φ-invariance and maximal Φ-invariance in elementary textbooks.

8.2.2 Invariance as Point Properties

We first define a Φ-maximal character of a function h in terms of point properties.

8.2.6 Definition. Let $(N, \mathcal{N})$ be a measurable space and $h\colon M \to N$ with $h^{-1}(\mathcal{N}) \subset \mathcal{M}$. We say that h is Φ-*adapted* if

$$(8.2.6) \qquad h(x') = h(x) \Rightarrow \exists\, \varphi \in \Phi \text{ such that } x' = \varphi(x)$$

or, equivalently, if

$$(8.2.7) \qquad h^{-1}[\{h(x)\}] \subset \Phi_x, \quad \forall\, x \in M,$$

where

$$(8.2.8) \qquad \Phi_x = \{\varphi(x) : \varphi \in \Phi\}. \qquad \blacksquare$$

The next two theorems link this definition with the inclusion of $\mathcal{M}_\Phi$ in $h^{-1}(\mathcal{N})$. Let us first denote by $A_x^{\mathcal{M}_\Phi}$ and $A_x^{h^{-1}(\mathcal{N})}$ the atoms of x in $\mathcal{M}_\Phi$ and in $h^{-1}(\mathcal{N})$, i.e.,

$$(8.2.9) \qquad A_x^{\mathcal{M}_\Phi} \;=\; \{x' : \mathbf{1}_A(x') = \mathbf{1}_A(x) \quad \forall\, A \in \mathcal{M}_\Phi\}$$

and

$$(8.2.10) \quad A_x^{h^{-1}(\mathcal{N})} \;=\; \{x' : \mathbf{1}_A(x') = \mathbf{1}_A(x) \quad \forall\, A \in h^{-1}(\mathcal{N})\}.$$

By the definition of $\mathcal{M}_\Phi$, and noting that $\mathbf{1}_A[\varphi(x)] = \mathbf{1}_{\varphi^{-1}(A)}(x)$, we have:

$$(8.2.11) \qquad\qquad\qquad \Phi_x \;\subset\; A_x^{\mathcal{M}_\Phi} \qquad \forall\, x \in M.$$

Clearly,

$$(8.2.12) \qquad\qquad h^{-1}[\{h(x)\}] \;\subset\; A_x^{h^{-1}(\mathcal{N})} \qquad \forall\, x \in M$$

and if $\mathcal{N}$ is separating since by (0.2.12), $A_x^{h^{-1}(\mathcal{N})} = h^{-1}[A_{h(x)}^{\mathcal{N}}]$, we also have:

$$(8.2.13) \qquad\qquad h^{-1}[\{h(x)\}] = A_x^{h^{-1}(\mathcal{N})}, \quad \forall\, x \in M.$$

8.2.7 Theorem. If $\mathcal{M}$ is a Blackwell σ-field, and if $\mathcal{N}$ is both separating and separable, then h Φ-adapted implies that $\mathcal{M}_\Phi \subset h^{-1}(\mathcal{N})$.

Proof. By (8.2.7), (8.2.8), and (8.2.13) we see that

$$A_x^{h^{-1}(\mathcal{N})} \subset A_x^{\mathcal{M}_\Phi}, \quad \forall\, x \in M.$$

Note that $h^{-1}(\mathcal{N})$ is separable since $\mathcal{N}$ is separable. Therefore, by Blackwell Theorem 0.2.16, $\mathcal{M}_\Phi \subset h^{-1}(\mathcal{N})$. ∎

Note that $m = m \circ \varphi$, $\forall\, \varphi \in \Phi$, if and only if m is constant on Φ_x, $\forall\, x \in M$. So, by (8.2.4), $m \in [\mathcal{M}_\Phi]$ if and only if $m \in [\mathcal{M}]$ and m is constant on Φ_x, $\forall\, x \in M$, i.e., $\Phi_x \subset m^{-1}[\{m(x)\}]$, $\forall\, x \in M$. Recall that $m \in [\mathcal{M}_\Phi]$ is constant on the atoms of $\mathcal{M}_\Phi$, i.e., $A_x^{\mathcal{M}_\Phi} \subset m^{-1}[\{m(x)\}]$, $\forall\, x \in M$. Therefore, in view of (8.2.11), without appealing to Blackwell Theorem, $m \in [\mathcal{M}_\Phi]$ if and only if $m \in [\mathcal{M}]$ and m is constant on the atoms of $\mathcal{M}_\Phi$.

The converse of Theorem 8.2.7 is obtained when Φ is a group.

First, note that, when Φ is group, Φ_x is the equivalence class of x for the equivalence relation on M defined as

$$(8.2.14) \qquad\qquad x' \underset{\Phi}{\sim} x \quad \text{if } \exists\, \varphi \in \Phi \text{ such that } x' = \varphi(x),$$

and Φ_x is then called *the orbit under Φ of x*. (When Φ is merely a monoïd, the relation given by (8.2.14) is reflexive and transitive only.)

Second, note that when Φ is a group,

$$\varphi^{-1}(\Phi_x) = \Phi_x, \quad \forall \, \varphi \in \Phi, \forall \, x \in M$$

(when Φ is a monoïd, we only have that $\Phi_x \subset \varphi^{-1}(\Phi_x)$, $\forall \, \varphi \in \Phi, \forall \, x \in M$). Therefore, when Φ is a group we have:

(8.2.15) $\Phi_x \in \mathcal{M} \Rightarrow \Phi_x \in \mathcal{M}_\Phi$

and, in view of (8.2.9), we have that

(8.2.16) $\Phi_x \in \mathcal{M} \Rightarrow \Phi_x = A_x^{\mathcal{M}_\Phi}.$

We are now ready to state the converse of Theorem 8.2.7.

8.2.8 Theorem. If Φ is a group and if the orbits under Φ are measurable, i.e., $\Phi_x \in \mathcal{M}$, $\forall \, x \in M$, then $\mathcal{M}_\Phi \subset h^{-1}(\mathcal{N})$ implies that h is Φ-adapted.

Proof. In view of Formulae (8.2.9) and (8.2.10), $\mathcal{M}_\Phi \subset h^{-1}(\mathcal{N})$ implies that $A_x^{h^{-1}(\mathcal{N})} \subset A_x^{\mathcal{M}_\Phi}$. Therefore, by Equations (8.2.12) and (8.2.16), $h^{-1}[\{h(x)\}] \subset \Phi_x$, i.e., (8.2.7) holds.   ∎

As a corollary of Theorems 8.2.7 and 8.2.8, we have:

8.2.9 Corollary. Let $\mathcal{M}$ be a Blackwell σ-field, Φ a group of measurable transformations of M such that the orbits under Φ are measurable, $\mathcal{N}$ a separable and separating σ-field on N, and $h : M \to N$ with $h^{-1}(\mathcal{N}) \subset \mathcal{M}$. Then the following conditions are equivalent:

(i) h is a maximal Φ-invariant (i.e. $\mathcal{M}_\Phi = h^{-1}(\mathcal{N})$);

(ii)   a) h is Φ-invariant (i.e., $h = h \circ \varphi \; \forall \, \varphi \in \Phi$);
       b) h is Φ-adapted (i.e. $h(x') = h(x) \Rightarrow \exists \, \varphi \in \Phi : x' = \varphi(x)$);

(iii) $\Phi_x = h^{-1}[\{h(x)\}] \; \forall \, x \in M$.   ∎

Let us remark that (ii) is the usual textbook definition of a maximal Φ-invariant function.

The existence of a maximal Φ-invariant function remains an open question. However, when considering the characterization given by (ii) there is a natural candidate for this function, namely, the canonical map from M to the quotient set of M under the equivalence relation induced by the group Φ. More precisely, let us define

(8.2.17)
$$\tilde{M} = M / \underset{\Phi}{\sim}$$

and
(8.2.18)
$$\tilde{\Phi} : M \to \tilde{M} \quad : \quad \tilde{\Phi}(x) = \Phi_x.$$

As σ-field on $\tilde{M}$, it is natural to choose the largest σ-field with respect to which $\tilde{\Phi}$ is measurable, i.e.,

(8.2.19)
$$\tilde{\mathcal{M}} = \{C \subset \tilde{M} : \tilde{\Phi}^{-1}(C) \in \mathcal{M}\}.$$

This clearly implies that
$$\tilde{\Phi}^{-1}(\tilde{\mathcal{M}}) \subset \mathcal{M}.$$

According to Corollary 8.2.9, the maximal Φ-invariant character of $\tilde{\Phi}$ would normally require assumptions on $(M, \mathcal{M})$, Φ and $(\tilde{M}, \tilde{\mathcal{M}})$. However, as shown in the next theorem, the canonical character of $\tilde{\Phi}$ allows us to drop such hypotheses. This does not mean, however, that these assumptions are irrelevant; indeed, the constructions of $\tilde{M}, \tilde{\mathcal{M}}$ and $\tilde{\Phi}$ are rather artificial, and are often difficult to characterize in applications.

8.2.10 Theorem. Let Φ be a group. Then:

(i) $\tilde{\Phi}$ is Φ-invariant and Φ-adapted, i.e., $\tilde{\Phi}^{-1}[\{\tilde{\Phi}(x)\}] = \Phi_x, \quad \forall\, x \in M,$

(ii) $\tilde{\Phi}$ is maximal Φ-invariant, i.e., $\mathcal{M}_\Phi = \tilde{\Phi}^{-1}(\tilde{\mathcal{M}}).$

Moreover

(iii) $\tilde{\mathcal{M}} = \tilde{\Phi}(\mathcal{M}_\Phi).$

Proof.

(i) $x' \in \tilde{\Phi}^{-1}[\{\tilde{\Phi}(x)\}]$ if and only if $\tilde{\Phi}(x') = \tilde{\Phi}(x)$ i.e. $\Phi_{x'} = \Phi_x$ and, therefore, if and only if $x' \in \Phi_x$.

(ii) From (i) $\tilde{\Phi} \circ \varphi = \tilde{\Phi}$, $\forall \varphi \in \Phi$, and by Theorem 8.2.4(i), we obtain $\tilde{\Phi}^{-1}(\tilde{\mathcal{M}}) \subset \mathcal{M}_\Phi$. We next show that, $\forall X \in \mathcal{M}_\Phi$, $\tilde{\Phi}^{-1}[\tilde{\Phi}(X)] = X$. If this is true then, by definition of $\tilde{\mathcal{M}}$, $\tilde{\Phi}(X) \in \tilde{\mathcal{M}}$ and, therefore, $\mathcal{M}_\Phi \subset \tilde{\Phi}^{-1}(\tilde{\mathcal{M}})$. Now, clearly, $X \subset \tilde{\Phi}^{-1}[\tilde{\Phi}(X)]$. However, $x \in \tilde{\Phi}^{-1}[\tilde{\Phi}(X)]$ if and only if $\exists \; x' \in X$ such that $\tilde{\Phi}(x) = \tilde{\Phi}(x')$, i.e., $\Phi_x = \Phi_{x'}$. Now, if $x' \in X$ and $X \in \mathcal{M}_\Phi$, then, by (8.2.9), $A_{x'}^{\mathcal{M}_\Phi} \subset X$ and, by (8.2.11), $\Phi_{x'} \subset X$. Therefore $x \in X$, and $\tilde{\Phi}^{-1}[\tilde{\Phi}(X)] \subset X$.

(iii) We saw in (ii) that $\tilde{\Phi}(\mathcal{M}_\Phi) \subset \tilde{\mathcal{M}}$. Clearly $\tilde{\Phi} \circ \tilde{\Phi}^{-1}$ is the identity on $\tilde{\mathcal{M}}$. Therefore $\forall \, C \in \tilde{\mathcal{M}}$, $C = \tilde{\Phi}[\tilde{\Phi}^{-1}(C)]$; but $\tilde{\Phi}^{-1}(C) \in \mathcal{M}_\Phi$ and therefore, $\tilde{\mathcal{M}} \subset \tilde{\Phi}(\mathcal{M}_\Phi)$ ∎

8.2.3 Invariance and Conditional Invariance of σ-Fields

It is usual to consider situations where $P[\varphi^{-1}(A)] = P(A)$, $\forall \, A \in \mathcal{M}$, i.e., $P \circ \varphi^{-1} = P$ on $\mathcal{M}$, and to say in this context that φ is a *measure-preserving transformation* of $(M, \mathcal{M}, P)$ or that P is φ-*invariant* on $\mathcal{M}$. For later use it will be more convenient to define φ-invariance of a sub-σ-field of $\mathcal{M}$ in the sense of the φ-invariance of the probability restricted to a sub-σ-field.

8.2.11 Definition. $\mathcal{M}_1 \subset \mathcal{M}$ is Φ-*invariant* or Φ *invariates* $\mathcal{M}_1$, denoted as $\Phi \, \mathrm{I} \, \mathcal{M}_1$, if

$$(8.2.20) \qquad \mathcal{M}_0(m_1 \circ \varphi) = \mathcal{M}_0 m_1 \quad \forall \, \varphi \in \Phi, \quad \forall m_1 \in [\mathcal{M}_1]^+.$$

If we want to stress the role of the probability P in this concept, we write $\Phi \, \mathrm{I} \, \mathcal{M}_1; P$. ∎

We first remark that, by Theorem 0.2.21, this definition is equivalent to the usual definition restricted to $\mathcal{M}_1$, i.e., $P \circ \varphi^{-1} = P$ on $\mathcal{M}_1$. It is then sufficient to check the definition for m_1 equal to the indicator functions $\mathbf{1}_A$ with $A \in \mathcal{M}_1$.

The next two propositions state elementary but important properties of the Φ-invariance of a σ-field, viz., its stability for the inclusion and the fact that, for any probability P on $(M, \mathcal{M})$, Φ invariates the invariant σ-field $\mathcal{M}_\Phi$.

8.2.12 Proposition. If $\Phi \text{ I } \mathcal{M}_1$, then, $\forall \, \mathcal{M}_2 \subset \mathcal{M}_1, \Phi \text{ I } \mathcal{M}_2$. In particular, $\Phi \text{ I } (\mathcal{M}_1 \cap \overline{\mathcal{M}}_0)$. ∎

While $\Phi \text{ I } \mathcal{M}_1$ means $P \circ \varphi^{-1} = P$ on $\mathcal{M}_1$, $\forall \, \varphi \in \Phi$, it should be noted that $\Phi \text{ I } (\mathcal{M}_1 \cap \overline{\mathcal{M}}_0)$ is in fact equivalent to $P \circ \varphi^{-1} \ll P$ on $\mathcal{M}_1$, $\forall \, \varphi$ on Φ, i.e., the restriction to $\mathcal{M}_1$ of $P \circ \varphi^{-1}$ is absolutely continuous with respect to the restriction to $\mathcal{M}_1$ of P; in particular, $\Phi \text{ I } \overline{\mathcal{M}}_0$, being equivalent to $P \circ \varphi^{-1} \ll P$, $\forall \, \varphi \in \Phi$, means that transforming P by φ does not create new null sets.

8.2.13 Proposition. For any probability P on $(M, \mathcal{M})$: $\Phi \text{ I } \mathcal{M}_\Phi$. As a consequence $\Phi \text{ I } (\mathcal{M}_\Phi \cap \overline{\mathcal{M}}_0)$. ∎

For later use, it is interesting to consider sub-σ-fields for which the measurability of the transformation φ is preserved.

8.2.14 Definition.

(i) $\mathcal{M}_1 \subset \mathcal{M}$ is φ-*stable* if

$$(8.2.21) \qquad\qquad \varphi^{-1}(\mathcal{M}_1) \subset \mathcal{M}_1,$$

i.e., φ is measurable when considered as $\varphi : (M, \mathcal{M}_1) \to (M, \mathcal{M}_1)$ or, alternatively.

$$(8.2.22) \qquad\qquad m_1 \in [\mathcal{M}_1] \Rightarrow m_1 \circ \varphi \in [\mathcal{M}_1].$$

(ii) $\mathcal{M}_1 \subset \mathcal{M}$ is Φ-*stable* if $\mathcal{M}_1$ is φ-stable $\forall \, \varphi \in \Phi$. ∎

The main consequence of invariance is provided by the ergodic theorem, which will now be stated in a slightly more general context. For a proof see, e.g., Breiman (1968), Proposition 6.21 and Corollary 6.25.

8.2.15 Proposition. If $\mathcal{M}_1$ is Φ-stable and if Φ I $\mathcal{M}_1$, then $\forall\, m_1 \in [\mathcal{M}_1]_1$, $\forall \varphi \in \Phi$,

$$(8.2.23) \qquad \frac{1}{n} \sum_{0 \le k \le n-1} m_1 \circ \varphi^k \to (\mathcal{M}_\varphi \cap \mathcal{M}_1) m_1$$

a.s. and in L_1 as $n \to \infty$. ∎

As usual, in the presence of a probability P, it is interesting to enlarge the notion of invariant sets to almost sure invariant sets; this will be useful when linking invariance with conditional independence.

We say that a measurable set $A \in \mathcal{M}$ is *almost surely φ-invariant* if $\varphi^{-1}(A) = A$ a.s. or, equivalently, if $P[A \triangle \varphi^{-1}(A)] = 0$ and a measurable function $m \in [\mathcal{M}]$ is *almost surely φ-invariant* if $m = m \circ \varphi$ a.s. Analogously to Theorem 8.2.1, we have:

8.2.16 Proposition. The collection of almost surely φ-invariant measurable sets is a σ-field called the *almost surely-φ-invariant σ-field* and is denoted by $\mathcal{M}_\varphi^*$, i.e.,

$$(8.2.24) \qquad \mathcal{M}_\varphi^* = \{A \in \mathcal{M} : \varphi^{-1}(A) = A \quad \text{a.s.}\},$$

and a measurable function $m \in [\mathcal{M}]$ is $\mathcal{M}_\varphi^*$-measurable if and only if it is almost surely φ-invariant, i.e.,

$$(8.2.25) \qquad m \in [\mathcal{M}_\varphi^*] \Leftrightarrow m \in [\mathcal{M}] \text{ and } m = m \circ \varphi \quad \text{a.s.} \qquad \blacksquare$$

The following theorem relates the almost sure invariant measurable sets to the completion by null sets of the invariant σ-field:

8.2.17 Theorem. If $\mathcal{M}_1$ is φ-stable and $\{\varphi\}$ I $(\mathcal{M}_1 \cap \overline{\mathcal{M}_0})$ then

$$(8.2.26) \qquad \mathcal{M}_1 \cap \mathcal{M}_\varphi^* = \overline{\mathcal{M}_1 \cap \mathcal{M}_\varphi} \cap \mathcal{M}_1$$

or, equivalently,

$$(8.2.27) \qquad \text{for } m_1 \in [\mathcal{M}_1], \ m_1 = m_1 \circ \varphi \quad \text{a.s.}$$
$$\Leftrightarrow \exists\, m_1' \in [\mathcal{M}_1], \text{ with } m_1' = m_1' \circ \varphi, \text{ such that } m_1 = m_1' \text{ a.s..}$$

In particular, if $\{\varphi\}$ I $\overline{\mathcal{M}_0}$, then

(8.2.28) $$\mathcal{M}_\varphi^* = \overline{\mathcal{M}}_\varphi.$$

Proof. Clearly, $\mathcal{M}_1 \cap \mathcal{M}_\varphi \subset \mathcal{M}_1 \cap \mathcal{M}_\varphi^*$. However, $\mathcal{M}_1 \cap \overline{\mathcal{M}_0} \subset \mathcal{M}_1 \cap \mathcal{M}_\varphi^*$, since $A \in \mathcal{M}_1$, $P(A) = 0$ and $\{\varphi\}$ I $(\mathcal{M}_1 \cap \overline{\mathcal{M}_0})$ imply $P[\varphi^{-1}(A)] = 0$. Hence, $A = \varphi^{-1}(A)$ a.s.P. Therefore, $\overline{\mathcal{M}_1 \cap \mathcal{M}_\varphi} \cap \mathcal{M}_1 \subset \mathcal{M}_1 \cap \mathcal{M}_\varphi^*$. If $A \in \mathcal{M}_1$ and $A = \varphi^{-1}(A)$ a.s., then $A = \varphi^{-n}(A)$ a.s. Indeed, if $P[A \Delta \varphi^{-n}(A)] = 0$, then $P[\varphi^{-1}[A \Delta \varphi^{-n}(A)]] = 0$, i.e., $\varphi^{-1}(A) = \varphi^{-(n+1)}(A)$ a.s. Hence, if $A = \varphi^{-1}(A)$ a.s. and $A = \varphi^{-n}(A)$ a.s., then $A = \varphi^{-(n+1)}(A)$ a.s. Now, let $B = \cap_n \cup_{m \geq n} \varphi^{-m}(A)$. By the φ-stability of $\mathcal{M}_1$, $B \in \mathcal{M}_1$ and, clearly, $\varphi^{-1}(B) = B$ and $A = B$ a.s. Hence, $B \in \mathcal{M}_1 \cap \mathcal{M}_\varphi$ and $A \in \overline{\mathcal{M}_1 \cap \mathcal{M}_\varphi} \cap \mathcal{M}_1$.   ∎

The notion of almost sure invariance may be extended to a family Φ of measurable transformations. We say that a measurable set $A \in \mathcal{M}$ is *almost surely Φ-invariant* if it is almost surely φ-invariant $\forall \varphi \in \Phi$, and a measurable function $m \in [\mathcal{M}]$ is *almost surely Φ-invariant* if it is almost surely φ-invariant $\forall \varphi \in \Phi$. Analogously to Corollary 8.2.2, we have:

8.2.18 Corollary. The collection of almost surely-Φ- invariant measurable sets is a σ-field denoted $\mathcal{M}_\Phi^*$ and is called the *almost surely Φ-invariant σ-field*,

(8.2.29) $$\mathcal{M}_\Phi^* = \bigcap_{\varphi \in \Phi} \mathcal{M}_\varphi^*$$

and a measurable function $m \in [\mathcal{M}]$ is $\mathcal{M}_\Phi^*$-measurable if and only if it is almost surely Φ-invariant, i.e.,

(8.2.30) $m \in [\mathcal{M}_\Phi^*] \Leftrightarrow m \in [\mathcal{M}]$ and $m = m \circ \varphi$ a.s. $\forall \varphi \in \Phi$.   ∎

Remark that Φ I $(\mathcal{M}_1 \cap \overline{\mathcal{M}_0})$ is not only equivalent to $P \circ \varphi^{-1} \ll P$ on $\mathcal{M}_1$, $\forall \varphi \in \Phi$, but is also equivalent to $\mathcal{M}_1 \cap \overline{\mathcal{M}_0} \subset \mathcal{M}_1 \cap \mathcal{M}_\Phi^*$; indeed, consider any $A \in \mathcal{M}_1$ such that $P(A) = P(A)^2$; then $A = \varphi^{-1}(A)$ a.s. is equivalent to saying that $P(A) = P(\varphi^{-1}(A))$. We can now extend Theorem 8.2.17 as follows:

8.2.19 Corollary. If $\mathcal{M}_1$ is Φ-stable and Φ I $\mathcal{M}_1 \cap \overline{\mathcal{M}_0}$, then

(8.2.31) $$\mathcal{M}_1 \cap \mathcal{M}_\Phi^* = \bigcap_{\varphi \in \Phi} \overline{\mathcal{M}_1 \cap \mathcal{M}_\varphi} \cap \mathcal{M}_1$$

or, equivalently,

(8.2.32) $\quad$ for $m \in [\mathcal{M}_1], m = m \circ \varphi$ a.s. $\quad \forall \, \varphi \in \Phi$

$$\Leftrightarrow \forall \, \varphi \in \Phi, \ \exists \ m_\varphi \in [\mathcal{M}_1], \text{ with } m_\varphi = m_\varphi \circ \varphi,$$

$$\text{such that } m = m_\varphi \text{ a.s.}$$

In particular,

(8.2.33) $\qquad \overline{\mathcal{M}_1 \cap \mathcal{M}_\Phi} \cap \mathcal{M}_1 \subset \mathcal{M}_1 \cap \mathcal{M}_\Phi^*.$

Consequently, if Φ I $\overline{\mathcal{M}_0}$, then

(8.2.34) $\qquad \overline{\mathcal{M}_\Phi} \subset \mathcal{M}_\Phi^* = \bigcap_{\varphi \in \Phi} \overline{\mathcal{M}_\varphi}.$ $\qquad$ ∎

With some additional conditions on Φ, Corollary 8.2.19 may be strengthened:

8.2.20 Theorem. If

(i) $\quad \Phi$ is a countable group, $\mathcal{M}_1$ is Φ-stable and Φ I $(\mathcal{M}_1 \cap \overline{\mathcal{M}_0})$,

or if

(ii) $\quad \Phi = \{\varphi^n : n \in I\!N\}$, $\mathcal{M}_1$ is φ-stable and $\{\varphi\}$ I $(\mathcal{M}_1 \cap \overline{\mathcal{M}_0})$,

then

(8.2.35) $\qquad \mathcal{M}_1 \cap \mathcal{M}_\Phi^* = \overline{\mathcal{M}_1 \cap \mathcal{M}_\Phi} \cap \mathcal{M}_1$

or, equivalently,

(8.2.36) $\quad$ for $m_1 \in [\mathcal{M}_1]$, $m_1 = m_1 \circ \varphi$ a.s. $\quad \forall \, \varphi \in \Phi$

$$\Leftrightarrow \exists \ m_1' \in [\mathcal{M}_1] \text{ with } m_1' = m_1' \circ \varphi \quad \forall \varphi \in \Phi,$$

$$\text{such that } m_1 = m_1' \text{ a.s.}$$

In particular, if Φ is a countable group and Φ I $\overline{\mathcal{M}_0}$ or if $\Phi = \{\varphi^n : n \in I\!N\}$ and $\{\varphi\}$ I $\overline{\mathcal{M}_0}$, then

(8.2.37) $\qquad \mathcal{M}_\Phi^* = \overline{\mathcal{M}_\Phi}.$

Proof.

(i) Clearly $\overline{\mathcal{M}_1 \cap \mathcal{M}_\Phi} \cap \mathcal{M}_1 \subset \mathcal{M}_1 \cap \mathcal{M}_\Phi^*$. However, if $A \in \mathcal{M}_1 \cap \mathcal{M}_\Phi^*$, let $B = \cup_{\varphi \in \Phi} \varphi^{-1}(A)$. Then, $\varphi^{-1}(B) = B$, $\forall\, \varphi \in \Phi$, since Φ is a group and, since Φ is countable, $B \in \mathcal{M}_1$, i.e., $B \in \mathcal{M}_1 \cap \mathcal{M}_\Phi$. But $A = B$ a.s. Hence, $\mathcal{M}_1 \cap \mathcal{M}_\Phi^* \subset \overline{\mathcal{M}_1 \cap \mathcal{M}_\Phi} \cap \mathcal{M}_1$.

(ii) By (8.2.5), $\mathcal{M}_\Phi = \mathcal{M}_\varphi$. Clearly, by (8.2.29), $\mathcal{M}_\Phi^* \subset \mathcal{M}_\varphi^*$. Now, if $m \in [\mathcal{M}_1]$ and $m \circ \varphi^n = m$ a.s., then by the φ-stability of $\mathcal{M}_1$, $m \circ \varphi^n \in \mathcal{M}_1$. Since $\{\varphi\}$ I $\mathcal{M}_1 \cap \overline{\mathcal{M}_0}$, this implies $m \circ \varphi^{n+1} = m \circ \varphi$ a.s. Therefore, $m = m \circ \varphi^n$ a.s. $\forall\, n \in I\!N$ if $m \in \mathcal{M}_1 \cap \mathcal{M}_\varphi^*$ and so $\mathcal{M}_1 \cap \mathcal{M}_\varphi^* = \mathcal{M}_1 \cap \mathcal{M}_\Phi^*$. ∎

The extension to the case of Φ being uncountable requires that more structure be placed on Φ; this case is treated in Section 8.2.6.

Proposition 8.2.13 may be extended to almost sure invariance:

8.2.21 Proposition. For any probability P on $(M, \mathcal{M})$, Φ I $\mathcal{M}_\Phi^*$. Consequently, if Φ I $\overline{\mathcal{M}_0}$, then Φ I $\overline{\mathcal{M}_\Phi}$. ∎

We now extend Definition 8.2.11 to the conditional case:

8.2.22 Definition. Let $\mathcal{M}_1$ and $\mathcal{M}_2$ be sub-σ-fields of $\mathcal{M}$. $\mathcal{M}_1$ *is Φ-invariant conditionally on $\mathcal{M}_2$ or Φ invariates $\mathcal{M}_1$ conditionally on $\mathcal{M}_2$* — denoted as Φ I $\mathcal{M}_1 \mid \mathcal{M}_2$ — if:

(i) Φ I $(\mathcal{M}_2 \cap \overline{\mathcal{M}_0})$,

(ii) $\varphi^{-1}(\mathcal{M}_2)(m_1 \circ \varphi) = (\mathcal{M}_2 m_1) \circ \varphi \quad \forall\, \varphi \in \Phi, \forall\, m_1 \in [\mathcal{M}_1]^+$. ∎

In the above definition, assumption (ii) is essential. Note that condition (i) ensures that, if m_2 and m_2' are two versions of $\mathcal{M}_2 m_1$, $m_2 \circ \varphi$ and $m_2' \circ \varphi$ are two versions of $\varphi^{-1}(\mathcal{M}_2)(m_1 \circ \varphi)$. Condition (i) is therefore introduced in the definition for sake of coherence with the definition of conditional expectation since, according to our convention, (ii) is an almost sure equality.

It is worthwhile rewriting condition (ii) of Definition 8.2.22 for two particular cases. Firstly, if Φ is a group and $\mathcal{M}_2$ is Φ-stable, then by bijectivity we in fact have that, $\varphi^{-1}(\mathcal{M}_2) = \mathcal{M}_2$, and so condition (ii), i.e.,

$$(8.2.38) \qquad \varphi^{-1}(\mathcal{M}_2)(m_1 \circ \varphi) = (\mathcal{M}_2 m_1) \circ \varphi$$

is equivalent to the simpler condition:

$$(8.2.39) \quad \mathcal{M}_2(m_1 \circ \varphi) = (\mathcal{M}_2 m_1) \circ \varphi \quad \forall\, \varphi \in \Phi,\ \forall\, m_1 \in [\mathcal{M}_1]^+.$$

Secondly, if $\mathcal{M}_2 \subset \mathcal{M}_\Phi^*$, $\varphi^{-1}(A) = A$ a.s. $\forall\, A \in \mathcal{M}_2$, so that (i) is trivially satisfied, $\overline{\varphi^{-1}(\mathcal{M}_2)} = \overline{\mathcal{M}}_2$. Since $(\mathcal{M}_2 m_1) \circ \varphi = \mathcal{M}_2 m_1$ a.s., using (2.2.3), condition (ii) is therefore equivalent to:

$$(8.2.40) \quad \mathcal{M}_2(m_1 \circ \varphi) = \mathcal{M}_2 m_1 \quad \forall\, \varphi \in \Phi,\ \forall\, m_1 \in [\mathcal{M}_1]^+.$$

By taking $\mathcal{M}_2 = \mathcal{M}_0$ this shows, in particular, that Definition 8.2.22 is coherent with Definition 8.2.11, since $\Phi\ \mathrm{I}\ \mathcal{M}_1 \mid \mathcal{M}_0$ is equivalent to $\Phi\ \mathrm{I}\ \mathcal{M}_1$.

It is important to remark that, if $\mathcal{M}_2$ is Φ-stable, then (8.2.39) is actually equivalent to (8.2.38) along with

$$(8.2.41) \quad \mathcal{M}_2(m_1 \circ \varphi) = \varphi^{-1}(\mathcal{M}_2)(m_1 \circ \varphi) \quad \forall\, \varphi \in \Phi,\ \forall\, m_1 \in [\mathcal{M}_1]^+,$$

and this is equivalent to

$$\varphi^{-1}(\mathcal{M}_1) \perp\!\!\!\perp \mathcal{M}_2 \mid \varphi^{-1}(\mathcal{M}_2) \quad \forall\, \varphi \in \Phi.$$

It is worth noting that it will often be necessary to introduce this further assumption in order to obtain interesting consequences of conditional invariance in more general situations than those where either Φ is a group, or $\mathcal{M}_2$ is included in $\mathcal{M}_\Phi^*$.

Example. Let $x = \binom{y}{z} \sim N_2(\mu, \Sigma)$, $\varphi_a(x) = \binom{y+a}{z+a}$, $\Phi = \{\varphi_a : a \in \mathbb{R}\}$, $\mathcal{M}_1 = \sigma(y)$ and $\mathcal{M}_2 = \sigma(z)$. Let us write $f_N(y \mid \cdot)$ for the density function (with respect to the Lebesgue measure) of a normal distribution on the y-space. We may show that $\Phi\ \mathrm{I}\ \mathcal{M}_1 \mid \mathcal{M}_2$, i.e., $\mathcal{M}_2(m_1 \circ \varphi) = (\mathcal{M}_2 m_1) \circ \varphi$, for any function m_1 defined on the y-space, provided $\sigma_{yz} = \sigma_{zz}$. Indeed,

$$\mathcal{M}_2(m_1 \circ \varphi_a) = \int m_1(y + a) f_N(y \mid \mu_y - \mu_z + z, \sigma_{yy} - \sigma_{zz})\, dy$$

$$\mathcal{M}_2 m_1 = \int m_1(y) f_N(y \mid \mu_y - \mu_z + z, \sigma_{yy} - \sigma_{zz})\, dy$$

and, therefore,

$$(\mathcal{M}_2 m_1) \circ \varphi_a = \int m_1(y) f_N(y \mid \mu_y - \mu_z + z + a, \sigma_{yy} - \sigma_{zz})\, dy$$

$$= \mathcal{M}_2(m_1 \circ \varphi_a)$$

where the last equality is obtained by a change of variable $y \rightarrow y - a$ and follows from the translation invariance of the Lebesgue measure. ∎

Note that, as shown in the above example, $\Phi \, I \, \mathcal{M}_1 \mid \mathcal{M}_2$ does not, in general, imply either $\Phi \, I \, (\mathcal{M}_1 \vee \mathcal{M}_2)$ nor $\Phi \, I \, \mathcal{M}_1$.

8.2.23 Theorem.

(i)   If $\Phi \, I \, \mathcal{M}_1 \mid \mathcal{M}_2$, then $\forall \, \mathcal{M}_3 \subset \mathcal{M}_1$, $\Phi \, I \, \mathcal{M}_3 \mid \mathcal{M}_2$;

(ii) $\Phi \, I \, \mathcal{M}_1 \mid \mathcal{M}_2 \Leftrightarrow \Phi \, I \, (\mathcal{M}_1 \vee \mathcal{M}_2) \mid \mathcal{M}_2$;

(iii) If $\Phi \, I \, \overline{\mathcal{M}_0}$, then $\Phi \, I \, \mathcal{M}_1 \mid \mathcal{M}_2 \Leftrightarrow \Phi \, I \, \overline{\mathcal{M}_1} \mid \mathcal{M}_2 \Leftrightarrow \Phi \, I \, \mathcal{M}_1 \mid \overline{\mathcal{M}_2}$;

(iv) If $\Phi \, I \, \mathcal{M}_1 \mid \mathcal{M}_2$, then $\Phi \, I \, (\mathcal{M}_1 \vee \mathcal{M}_2) \cap \overline{\mathcal{M}_0}$.

Proof.

(i) Follows from the definition.

(ii) Take $m_i \in [\mathcal{M}_i]^+, i = 1, 2$.

$$
\begin{aligned}
\varphi^{-1}(\mathcal{M}_2)[(m_1 m_2) \circ \varphi] &= \varphi^{-1}(\mathcal{M}_2)[m_1 \circ \varphi \cdot m_2 \circ \varphi] \\
&= m_2 \circ \varphi \cdot \varphi^{-1}(\mathcal{M}_2)(m_1 \circ \varphi) \\
&\quad (\text{since } m_2 \circ \varphi \in [\varphi^{-1}(\mathcal{M}_2)]) \\
&= m_2 \circ \varphi \cdot (\mathcal{M}_2 m_1) \circ \varphi \\
&= (m_2 \cdot \mathcal{M}_2 m_1) \circ \varphi \quad (\text{by Definition 8.2.22 (ii)}) \\
&= \mathcal{M}_2(m_1 m_2) \circ \varphi,
\end{aligned}
$$

for $\mathcal{M}_2(m_1 m_2) = m_2 \mathcal{M}_2 m_1$ implies $\mathcal{M}_2(m_1 m_2) \circ \varphi = (m_2 \mathcal{M}_2 m_1) \circ \varphi$ by Definition 8.2.22(i), and the result follows by Theorem 0.2.21.

(iii) If $m \in [\overline{\mathcal{M}_1}]$, then there exists $m_1 \in [\mathcal{M}_1]$ such that $m = m_1$ a.s. Therefore, $\mathcal{M}_2 m = \mathcal{M}_2 m_1$ a.s. and, since $\Phi \, I \, \overline{\mathcal{M}_0}$, $m \circ \varphi = m_1 \circ \varphi$ a.s. and $(\mathcal{M}_2 m) \circ \varphi = (\mathcal{M}_2 m_1) \circ \varphi$. This shows that $\Phi \, I \, \mathcal{M}_1 \mid \mathcal{M}_2$ implies that $\Phi \, I \, \overline{\mathcal{M}_1} \mid \mathcal{M}_2$ when $\Phi \, I \, \overline{\mathcal{M}_0}$. However, $\Phi \, I \, \overline{\mathcal{M}_0}$ implies that $\varphi^{-1}(\overline{\mathcal{M}_0}) \subset \overline{\mathcal{M}_0}$. Therefore, $\varphi^{-1}(\mathcal{M}_2) \subset \varphi^{-1}(\overline{\mathcal{M}_2}) \subset \overline{\varphi^{-1}(\mathcal{M}_2)}$. By (2.2.3), $\varphi^{-1}(\overline{\mathcal{M}_2})(m_1 \circ \varphi) = \varphi^{-1}(\mathcal{M}_2)(m_1 \circ \varphi)$ and $\mathcal{M}_2 m_1 = \overline{\mathcal{M}_2} m_1$ a.s. Under $\Phi \, I \, \overline{\mathcal{M}_0}$ this shows that $\Phi \, I \, \mathcal{M}_1 \mid \mathcal{M}_2$ if and only if $\Phi \, I \, \mathcal{M}_1 \mid \overline{\mathcal{M}_2}$.

(iv) If $A \in \mathcal{M}_1 \vee \mathcal{M}_2$ and $P(A) = 0$, then

$$P[\varphi^{-1}(A)] = \mathcal{M}_0[\mathbf{1}_A \circ \varphi] = \mathcal{M}_0[\varphi^{-1}(\mathcal{M}_2)(\mathbf{1}_A \circ \varphi)] = \mathcal{M}_0[(\mathcal{M}_2\mathbf{1}_A) \circ \varphi] = 0$$

by (ii) and Definition 8.2.22(i), since $\mathcal{M}_2\mathbf{1}_A = 0$ a.s. ■

The next theorem shows that conditional invariance generates some conditional independence properties. More precisely, if we recall that $\mathcal{M}_2\mathcal{M}_1$ is the smallest sub-σ field of $\mathcal{M}_2$ conditionally on which $\mathcal{M}_1$ and $\mathcal{M}_2$ are independent then, when $\Phi \; \mathrm{I} \; \mathcal{M}_1 \mid \mathcal{M}_2$, this property still holds for σ-fields obtained under the inverse image of φ for any $\varphi \in \Phi$.

8.2.24 Theorem. If $\Phi \; \mathrm{I} \; \mathcal{M}_1 \mid \mathcal{M}_2$, then, $\forall \; \varphi \in \Phi$,

(i) $\quad \varphi^{-1}(\mathcal{M}_2)[\varphi^{-1}(\mathcal{M}_1)] = \overline{\varphi^{-1}(\mathcal{M}_2\mathcal{M}_1)} \cap \varphi^{-1}(\mathcal{M}_2),$

and this implies that

(ii) $\quad \varphi^{-1}(\mathcal{M}_1) \perp\!\!\!\perp \varphi^{-1}(\mathcal{M}_2) \mid \varphi^{-1}(\mathcal{M}_2\mathcal{M}_1).$

Proof. Since $\forall \; \varphi \in \Phi$, $\forall \; m_1 \in [\mathcal{M}_1]^+$,
$\varphi^{-1}(\mathcal{M}_2)(m_1 \circ \varphi) = (\mathcal{M}_2m_1) \circ \varphi$, $\overline{\varphi^{-1}(\mathcal{M}_2)[\varphi^{-1}(\mathcal{M}_1)]} = \overline{\varphi^{-1}(\mathcal{M}_2\mathcal{M}_1)}$.
It now suffices to recall that, by definition of the projection of σ-fields,
$\overline{\varphi^{-1}(\mathcal{M}_2)[\varphi^{-1}(\mathcal{M}_1)]} \cap \varphi^{-1}(\mathcal{M}_2) = \varphi^{-1}(\mathcal{M}_2)[\varphi^{-1}(\mathcal{M}_1)].$ ■

The main property of conditional invariance is provided by the following theorem which analyses the relationships between marginal, conditional and joint Φ-invariance.

8.2.25 Theorem. Let $\mathcal{M}_i$, $i = 1, 2, 3$ be sub-σ-fields of $\mathcal{M}$. Then the following properties are equivalent:

(i) $\quad \Phi \; \mathrm{I} \; (\mathcal{M}_1 \vee \mathcal{M}_2) \mid \mathcal{M}_3$

(ii) $\quad \Phi \; \mathrm{I} \; \mathcal{M}_1 \mid \mathcal{M}_3$ and $\Phi \; \mathrm{I} \; \mathcal{M}_2 \mid \mathcal{M}_3 \vee \mathcal{M}_1.$

Proof. (i) and (ii) entails that Φ I $(\mathcal{M}_1 \vee \mathcal{M}_2 \vee \mathcal{M}_3) \cap \overline{\mathcal{M}_0}$ by Theorem 8.2.23(iv). If $m_i \in [\mathcal{M}_i]^+, i = 1, 2$, then:

$$\varphi^{-1}(\mathcal{M}_3)[(m_1 m_2) \circ \varphi] = \varphi^{-1}(\mathcal{M}_3)[m_1 \circ \varphi \cdot m_2 \circ \varphi]$$
$$= \varphi^{-1}(\mathcal{M}_3)[m_1 \circ \varphi \cdot \varphi^{-1}(\mathcal{M}_1 \vee \mathcal{M}_3)(m_2 \circ \varphi)]$$
$$= \varphi^{-1}(\mathcal{M}_3)[m_1 \circ \varphi \cdot (\mathcal{M}_1 \vee \mathcal{M}_3)m_2 \circ \varphi]$$

if Φ I $\mathcal{M}_2 \mid \mathcal{M}_3 \vee \mathcal{M}_1$

$$= \varphi^{-1}(\mathcal{M}_3)[\{m_1 \cdot (\mathcal{M}_1 \vee \mathcal{M}_3)m_2\} \circ \varphi]$$
$$= \mathcal{M}_3[m_1 \cdot (\mathcal{M}_1 \vee \mathcal{M}_3)m_2] \circ \varphi$$

if Φ I $(\mathcal{M}_1 \vee \mathcal{M}_3) \mid \mathcal{M}_3$, which is implied by Φ I $\mathcal{M}_1 \mid \mathcal{M}_3$ (Theorem 8.2.23(ii)),

$$= \mathcal{M}_3(m_1 m_2) \circ \varphi.$$

Therefore, (ii) implies (i) by Theorem 0.2.21. Now, by Theorem 8.2.23(i) and (ii), Φ I $(\mathcal{M}_1 \vee \mathcal{M}_2) \mid \mathcal{M}_3$ implies that Φ I $(\mathcal{M}_1 \vee \mathcal{M}_3) \mid \mathcal{M}_3$. Therefore, under (i),

$$\varphi^{-1}(\mathcal{M}_3)[m_1 \circ \varphi \cdot \varphi^{-1}(\mathcal{M}_1 \vee \mathcal{M}_3)(m_2 \circ \varphi)]$$
$$= \varphi^{-1}(\mathcal{M}_3)[m_1 \circ \varphi \cdot (\mathcal{M}_1 \vee \mathcal{M}_3)m_2 \circ \varphi] \quad \forall \, m_i \in [\mathcal{M}_i]^+, \quad i = 1, 2$$

or, equivalently, $\forall \, m_i \subset [\mathcal{M}_i]^+, i = 1, 2, 3$,

$$\mathcal{M}_0[m_1 \circ \varphi \cdot m_3 \circ \varphi \cdot \varphi^{-1}(\mathcal{M}_1 \vee \mathcal{M}_3)(m_2 \circ \varphi)]$$
$$= \mathcal{M}_0[m_1 \circ \varphi \cdot m_3 \circ \varphi \cdot (\mathcal{M}_1 \vee \mathcal{M}_3)m_2 \circ \varphi].$$

This shows that

$$\varphi^{-1}(\mathcal{M}_1 \vee \mathcal{M}_3)(m_2 \circ \varphi) = (\mathcal{M}_1 \vee \mathcal{M}_3)m_2 \circ \varphi,$$

i.e, Φ I $\mathcal{M}_2 \mid \mathcal{M}_1 \vee \mathcal{M}_3$, and so (i) implies (ii).   ∎

The next theorem shows that conditional invariance provides conditional independence between a.s. invariant σ-fields. This is in fact the most useful consequence of the concept of invariance:

8.2.26 Theorem. Let $\mathcal{M}_i$, $i = 1, 2$, be sub-σ-fields of $\mathcal{M}$. If

(i) Φ I $\mathcal{M}_1 \mid \mathcal{M}_2$,

(ii) $\mathcal{M}_2$ is Φ-stable,

(iii) $\mathcal{M}_1 \cap \mathcal{M}_{\Phi}^* \perp\!\!\!\perp \mathcal{M}_2 \mid \varphi^{-1}(\mathcal{M}_2)$ $\forall \varphi \in \Phi$,

then

(iv) $\mathcal{M}_1 \cap \mathcal{M}_{\Phi}^* \perp\!\!\!\perp \mathcal{M}_2 \mid \mathcal{M}_2 \cap \mathcal{M}_{\Phi}^*$.

Proof. Let $m_1 \in [\mathcal{M}_1 \cap \mathcal{M}_{\Phi}^*]^+$. Then $m_1 \circ \varphi = m_1$ a.s. and therefore, under (i), $\varphi^{-1}(\mathcal{M}_2)m_1 = \mathcal{M}_2 m_1 \circ \varphi$. Under (ii) and (iii), by Corollary 2.2.3(ii), $\mathcal{M}_2 m_1 = \varphi^{-1}(\mathcal{M}_2)m_1$. Therefore, $\mathcal{M}_2 m_1 = \mathcal{M}_2 m_1 \circ \varphi$, i.e., $\mathcal{M}_2 m_1 \in [\mathcal{M}_2 \cap \mathcal{M}_{\Phi}^*]$ and this is equivalent to (iv) by Theorem 2.2.6. ∎

Let us remark that, under (i) and (ii), it follows from the proof that (iii) and (iv) are actually equivalent.

However, as remarked after introducing the definition of conditional invariance, when Φ is a group, then $\mathcal{M}_2$ Φ-stable is in fact equivalent to $\varphi^{-1}(\mathcal{M}_2) = \mathcal{M}_2$, $\forall \varphi \in \Phi$. Therefore, in this situation, (iii) is readily satisfied. This gives the following corollary:

8.2.27 Corollary. If Φ is a group, and if:

(i) Φ I $\mathcal{M}_1 \mid \mathcal{M}_2$

(ii) $\mathcal{M}_2$ is Φ-stable,

then

(iii) $\mathcal{M}_1 \cap \mathcal{M}_{\Phi}^* \perp\!\!\!\perp \mathcal{M}_2 \mid \mathcal{M}_2 \cap \mathcal{M}_{\Phi}^*$. ∎

There is another important situation where condition (iii) of Theorem 8.2.26 is readily satisfied. This is when $\mathcal{M}_1 \subset \mathcal{M}_{\Phi}^*$ and Φ I $\mathcal{M}_2$. More precisely, we have the following theorem:

8.2.28 Theorem. If $\mathcal{M}_1 \subset \mathcal{M}_{\Phi}^*$ and $\mathcal{M}_2$ is Φ-stable, then the following statements are equivalent:

(i) $\Phi \; I \; \mathcal{M}_2 \mid \mathcal{M}_1$

(ii) $\Phi \; I \; (\mathcal{M}_1 \vee \mathcal{M}_2)$

(iii) $\Phi \; I \; \mathcal{M}_2$ and $\Phi \; I \; \mathcal{M}_1 \mid \mathcal{M}_2$

(iv) $\Phi \; I \; \mathcal{M}_2 \mid \mathcal{M}_2 \cap \mathcal{M}_\Phi^*$ and $\mathcal{M}_1 \perp\!\!\!\perp \mathcal{M}_2 \mid \mathcal{M}_2 \cap \mathcal{M}_\Phi^*$.

Proof. By Proposition 8.2.21 and Theorem 8.2.23(i), $\Phi \; I \; \mathcal{M}_1$. Therefore, by Theorem 8.2.25, (i) is equivalent to (ii) and (ii) is equivalent to (iii). Clearly, (ii) implies $\Phi \; I \; \mathcal{M}_2 \mid \mathcal{M}_2 \cap \mathcal{M}_\Phi^*$ by Theorems 8.2.23(i) and 8.2.25. Now, if $m_1 \in [\mathcal{M}_1]_\infty$, then $m_1 \circ \varphi = m_1$ a.s. and therefore, under (iii), $\varphi^{-1}(\mathcal{M}_2)m_1 = \mathcal{M}_2 m_1 \circ \varphi$. But:

$$
\begin{aligned}
\mathcal{M}_0[\{\mathcal{M}_2 m_1 - (\mathcal{M}_2 m_1)\circ\varphi\}^2] &= \mathcal{M}_0[\{\mathcal{M}_2 m_1 - \varphi^{-1}(\mathcal{M}_2)m_1\}^2] \\
&= \mathcal{M}_0[m_1 \cdot \mathcal{M}_2 m_1] - \mathcal{M}_0[m_1 \cdot \varphi^{-1}(\mathcal{M}_2)m_1] \\
&= \mathcal{M}_0[m_1 \cdot \mathcal{M}_2 m_1] - \mathcal{M}_0[m_1 \cdot (\mathcal{M}_2 m_1)\circ\varphi]
\end{aligned}
$$

since $\mathcal{M}_2$ is Φ-stable, i.e., $\varphi^{-1}(\mathcal{M}_2) \subset \mathcal{M}_2$. But under (ii), we have

$$
\begin{aligned}
\mathcal{M}_0[m_1 \cdot (\mathcal{M}_2 m_1) \circ \varphi] &= \mathcal{M}_0[m_1 \circ \varphi \cdot (\mathcal{M}_2 m_1) \circ \varphi] \\
&= \mathcal{M}_0[m_1 \cdot \mathcal{M}_2 m_1].
\end{aligned}
$$

Therefore,

$$
\mathcal{M}_2 m_1 = (\mathcal{M}_2 m_1) \circ \varphi, \quad \text{i.e.,} \quad \mathcal{M}_2 m_1 \in \mathcal{M}_2 \cap \mathcal{M}_\Phi^*
$$

and, by Theorem 2.2.6, this is equivalent to $\mathcal{M}_1 \perp\!\!\!\perp \mathcal{M}_2 \mid \mathcal{M}_2 \cap \mathcal{M}_\Phi^*$. Now, by Corollary 2.2.3(iii) and the Φ-stability of $\mathcal{M}_2$, $\mathcal{M}_1 \perp\!\!\!\perp \mathcal{M}_2 \mid \mathcal{M}_2 \cap \mathcal{M}_\Phi^*$ implies that

$$
\begin{aligned}
\mathcal{M}_1[(\mathcal{M}_2 \cap \mathcal{M}_\Phi^*)(m_2 \circ \varphi)] &= \mathcal{M}_1(m_2 \circ \varphi) \\
\mathcal{M}_1[(\mathcal{M}_2 \cap \mathcal{M}_\Phi^*)m_2] &= \mathcal{M}_1 m_2 \quad \forall \, \varphi \in \Phi \quad \forall \, m_2 \in [\mathcal{M}_2]^+.
\end{aligned}
$$

But, by (8.2.40), $\Phi \; I \; \mathcal{M}_2 \mid \mathcal{M}_2 \cap \mathcal{M}_\Phi^*$ implies that
$$
(\mathcal{M}_2 \cap \mathcal{M}_\Phi^*)(m_2 \circ \varphi) = (\mathcal{M}_2 \cap \mathcal{M}_\Phi^*)m_2.
$$

Therefore, we obtain $\mathcal{M}_1(m_2 \circ \varphi) = \mathcal{M}_1 m_2 \; \forall \, \varphi \in \Phi$ and $\forall \, m_2 \in [\mathcal{M}_2]^+$ and, by (8.2.40), this is equivalent to (i), since $\mathcal{M}_1 \subset \mathcal{M}_\Phi^*$.   ■

This theorem leads to the following result analogous to Theorem 8.2.26:

8.2.29 Theorem. Let $\mathcal{M}_1$ be a sub-σ-field of $\mathcal{M}$. If

(i) $\Phi \, I \, \mathcal{M}_1$,

then $\forall \, \mathcal{M}_2 \subset \mathcal{M}_1$ Φ-stable

(ii) $\mathcal{M}_1 \cap \mathcal{M}_\Phi^* \perp\!\!\!\perp \mathcal{M}_2 \mid \mathcal{M}_2 \cap \mathcal{M}_\Phi^*$.

Proof. By Theorem 8.2.23 (i), $\Phi \, I \, \mathcal{M}_1$ implies that

$$\Phi \, I \, \{\mathcal{M}_2 \vee (\mathcal{M}_1 \cap \mathcal{M}_\Phi^*)\}.$$

By Theorem 8.2.28, this implies (ii).   ∎

The next theorem gives a condition under which the conditional σ-field in the conditioning Φ-invariance may be reduced.

8.2.30 Theorem. Let $\mathcal{M}_i$, $i = 1, 2, 3$ be sub-σ-fields of $\mathcal{M}$. Then the following statements are equivalent:

(i)   1. $\mathcal{M}_1 \perp\!\!\!\perp \mathcal{M}_2 \mid \mathcal{M}_3$
       2. $\Phi \, I \, \mathcal{M}_1 \mid \mathcal{M}_2 \vee \mathcal{M}_3$

(ii)   1. $\varphi^{-1}(\mathcal{M}_1) \perp\!\!\!\perp \varphi^{-1}(\mathcal{M}_2) \mid \varphi^{-1}(\mathcal{M}_3) \ \ \forall \, \varphi \in \Phi$
       2. $\Phi \, I \, \mathcal{M}_1 \mid \mathcal{M}_3$
       3. $\Phi \, I \, (\mathcal{M}_2 \vee \mathcal{M}_3) \cap \overline{\mathcal{M}_0}$.

Proof. Clearly, (i) 2 implies (ii) 3 by definition. Under (i) 2, we have that

$$\varphi^{-1}(\mathcal{M}_2 \vee \mathcal{M}_3)(m_1 \circ \varphi) = (\mathcal{M}_2 \vee \mathcal{M}_3) m_1 \circ \varphi, \quad \forall \, m_1 \in [\mathcal{M}_1]^+.$$

But, under (i) 1, by Theorem 2.2.1(ii), $(\mathcal{M}_2 \vee \mathcal{M}_3)m_1 = \mathcal{M}_3 m_1$. Therefore, under (i),

$$\{\varphi^{-1}(\mathcal{M}_2) \vee \varphi^{-1}(\mathcal{M}_3)\}(m_1 \circ \varphi) = (\mathcal{M}_3 m_1) \circ \varphi.$$

This is clearly equivalent to the two following identities:

$$\{\varphi^{-1}(\mathcal{M}_2) \vee \varphi^{-1}(\mathcal{M}_3)\}(m_1 \circ \varphi) \ = \ \varphi^{-1}(\mathcal{M}_3)(m_1 \circ \varphi)$$
$$\varphi^{-1}(\mathcal{M}_3)(m_1 \circ \varphi) \ = \ (\mathcal{M}_3 m_1) \circ \varphi,$$

i.e., (ii) 1 and (ii) 2. However, (ii) 1 with $\varphi = i$ is equivalent to (i) 1 and

$$\varphi^{-1}(\mathcal{M}_2 \vee \mathcal{M}_3)(m_1 \circ \varphi) = (\mathcal{M}_3 m_1) \circ \varphi,$$

under (ii) 1 and (ii) 2. Under (i) 1 and (ii) 3,

$$(\mathcal{M}_2 \vee \mathcal{M}_3) m_1 \circ \varphi = \mathcal{M}_3 m_1 \circ \varphi.$$

Therefore,

$$\varphi^{-1}(\mathcal{M}_2 \vee \mathcal{M}_3)(m_1 \circ \varphi) = (\mathcal{M}_2 \vee \mathcal{M}_3) m_1 \circ \varphi$$

and this, along with (ii) 3, is equivalent to (i) 2.   ∎

8.2.4   Ergodicity and Mixing

We now extend the concept of ergodic transformations slightly.

8.2.31 Definition. Φ is *ergodic on* $\mathcal{M}_1$ *conditionally on* $\mathcal{M}_2$ if

(i) $\Phi \; I \; (\mathcal{M}_1 \vee \mathcal{M}_2) \cap \overline{\mathcal{M}_0}$

(ii) $\mathcal{M}_1 \cap \mathcal{M}_\Phi^* \subset \overline{\mathcal{M}_2}$

If $\mathcal{M}_2$ is trivial, we simply say that Φ is *ergodic on* $\mathcal{M}_1$.   ∎

Note that when $\mathcal{M}_2$ is trivial the ergodicity of Φ means that the Φ-invariant measurable sets have probability 0 or 1. The following lemma shows that when Φ is ergodic on $\mathcal{M}_1$ conditionally on $\mathcal{M}_2$ the a.s.Φ-invariant $\mathcal{M}_1$-measurable functions are almost surely equal to a.s.Φ-invariant $\mathcal{M}_2$-measurable functions.

8.2.32 Lemma.   If Φ is ergodic on $\mathcal{M}_1$ conditionally on $\mathcal{M}_2$, then

$$\mathcal{M}_1 \cap \mathcal{M}_\Phi^* \subset \overline{\mathcal{M}_2 \cap \mathcal{M}_\Phi^*}.$$

Proof. If $m_1 \in [\mathcal{M}_1 \cap \mathcal{M}_\Phi^*]$, $m_1 = m_1 \circ \varphi$ a.s. $\forall \, \varphi \in \Phi$. By definition 8.2.31(ii), there exists $m_2 \in [\mathcal{M}_2]$ such that $m_1 = m_2$ a.s. However, since

Φ I $(\mathcal{M}_1 \vee \mathcal{M}_2) \cap \overline{\mathcal{M}}_0$, we have $m_1 \circ \varphi = m_2 \circ \varphi$ a.s. Hence, $m_2 = m_2 \circ \varphi$ a.s. and $m_1 = m_2$ a.s., i.e., $m_1 \in [\overline{\mathcal{M}_2 \cap \mathcal{M}_\Phi^*}]$. ∎

A stronger but often more easily checked condition for ergodicity is that of mixing, which we now define precisely:

8.2.33 Definition. Φ is *mixing on $\mathcal{M}_1$ conditionally on $\mathcal{M}_2$* if

(i) Φ I $(\mathcal{M}_1 \vee \mathcal{M}_2) \cap \overline{\mathcal{M}}_0$

(ii) $\forall\, m_1, m_1' \in [\mathcal{M}_1]_\infty$, $\forall\, \varphi \in \Phi$,
  $\mathcal{M}_2[m_1' \cdot m_1 \circ \varphi^n] \to \mathcal{M}_2 m_1' \cdot \mathcal{M}_2 m_1$ a.s. $n \to \infty$.

If $\mathcal{M}_2$ is trivial, we simply say that Φ *is mixing on $\mathcal{M}_1$*. ∎

When $\mathcal{M}_2$ is trivial, the mixing property of Φ may be equivalently stated on the indicator functions of $\mathcal{M}_1$, as follows:

  $\forall\, A, B \in \mathcal{M}_1, \forall\, \varphi \in \Phi, \quad P[A \cap \varphi^{-n}(B)] \to P(A)P(B)$ a.s. $n \to \infty$.

8.2.34 Theorem. If Φ is mixing on $\mathcal{M}_1$ conditionally on $\mathcal{M}_2$, then Φ is ergodic on $\mathcal{M}_1$ conditionally on $\mathcal{M}_2$.

Proof. If $m_1 \in [\mathcal{M}_1 \cap \mathcal{M}_\Phi^*]$, then $m_1 = m_1 \circ \varphi^n$ a.s. $\forall\, n \in I\!N$. Therefore condition (ii) of mixing implies that

  $\mathcal{M}_2(m_1' m_1) = \mathcal{M}_2(m_1') \mathcal{M}_2(m_1) \quad \forall\, m_1' \in [\mathcal{M}_1]^+ \quad \forall\, m_1 \in [\mathcal{M}_1 \cap \mathcal{M}_\Phi^*]^+.$

This is equivalent to $\mathcal{M}_1 \perp\!\!\!\perp \mathcal{M}_1 \cap \mathcal{M}_\Phi^* \mid \mathcal{M}_2$, which by Corollary 2.2.8, implies that $\mathcal{M}_1 \cap \mathcal{M}_\Phi^* \subset \overline{\mathcal{M}}_2$. ∎

8.2.5 Existence of Invariant Measure

In Section 8.2.3, we defined the Φ-invariance of $\mathcal{M}_1$ (i.e. Φ I $\mathcal{M}_1$) for a given probability space $(M, \mathcal{M}, P)$. One may also look for the existence of a probability measure P on a given measurable space $(M, \mathcal{M})$, which would give Φ I $\mathcal{M}_1$. In general, if Φ has only one element or, equivalently, has the form $\{\varphi^n : n \in I\!N\}$, there may be a very large number of invariant probability measures. But once we add new elements to Φ, the set of Φ-invariant measures decreases and eventually will possibly be empty. Heuristically, the theory of Haar measures shows that even if Φ becomes a "large enough" group, one may show both that there still exists such a Φ-invariant measure and that it may be characterized.

Unfortunately, the theory of Haar measures requires several mathematical preliminaries which have not been necessary up to now. In this section we briefly sketch the main results, to be used in the sequel, and refer to Halmos (1950) Chapter X to XII, or Nachbin (1965) for a more systematic exposition.

When Φ is a group, it is natural to associate to Φ two groups of transformations:

$$\Phi_L = \{\varphi_L(\varphi') = \varphi \circ \varphi' \mid \varphi \in \Phi\}$$
$$\Phi_R = \{\varphi_R(\varphi') = \varphi' \circ \varphi \mid \varphi \in \Phi\}.$$

These two groups clearly coincide if Φ is a commutative group. In order to define Haar measures in such a case, we need to assume that Φ is a locally compact topological group endowed with its Borel σ-field $\mathcal{F}$. Remember that a Radon measure on a topological space is a measure on the Borel sets which gives finite measure on the compact sets. In this case, we define:

8.2.35 Definition. A *left* (respectively, *right*) *Haar measure* Q on $(\Phi, \mathcal{F})$ is a nonidentically zero Radon measure such that Φ_L I $\mathcal{F}$ (respectively, Φ_R I $\mathcal{F}$), i.e., such that $\forall \varphi \in \Phi$, $\forall A \in \mathcal{F}$, $Q(\varphi \circ A) = Q(A)$ (respectively, $Q(A \circ \varphi) = Q(A)$). ■

The Haar Theorem is proved in Halmos (1950) (Section 58, Theorem B and Section 60, Theorem C) and in Nachbin (1965) (Theorems 1 and 4 in Chapter 2).

8.2.36 Proposition. For every locally compact topological group Φ,
(i)   there exists a left (respectively, right) Haar measure which is unique up to a strictly positive factor of proportionality;

(ii)   this invariant measure is finite if and only if Φ is compact;

(iii)   if Φ is compact there is a unique probability which is both a right and
left Haar measure.   ∎

The particular case of Φ being compact is especially interesting. Indeed,
if $\mathcal{M}_1 \subset \mathcal{M}$ is such that the map $\varphi(x) : (\Phi \times M, \mathcal{F} \otimes \mathcal{M}_1) \to (M, \mathcal{M}_1)$
is measurable, i.e., $\forall\, A \in \mathcal{M}_1$, $\{(\varphi, x) : \varphi(x) \in A\} \in \mathcal{F} \otimes \mathcal{M}_1$, then
any probability measure P such that $\Phi\,\mathrm{I}\,\mathcal{M}_1;\ P$ is uniquely and entirely
determined by its trace on $\mathcal{M}_1 \cap \mathcal{M}_\Phi^*$, i.e., $P_{\mathcal{M}_1 \cap \mathcal{M}_\Phi^*}$, and by its conditional
probabilities to $\mathcal{M}_1$, i.e. $P_{\mathcal{M}}^{\mathcal{M}_1}$. This may be seen as follows: if $\Phi\,\mathrm{I}\,\mathcal{M}_1;\ P$,
then for any $m_1 \in [\mathcal{M}_1]^+$ and $m_1' \in [\mathcal{M}_1 \cap \mathcal{M}_\Phi^*]^+$,

$$\int_M m_1[\varphi(x)]m_1'(x)P(dx) = \int_M m_1(x)m_1'(x)P(dx).$$

Therefore, by Fubini Theorem, if Q is the Haar probability on $(\Phi, \mathcal{F})$

$$\int_M m_1(x)m_1'(x)P(dx) = \int_M \overline{m_1}(x)m_1'(x)P(dx)$$

where

$$\overline{m_1}(x) = \int_\Phi m_1[\varphi(x)]Q(d\varphi).$$

Clearly, $\overline{m_1} \in [\mathcal{M}_1 \cap \mathcal{M}_\Phi]$ and, therefore, $(\mathcal{M}_1 \cap \mathcal{M}_\Phi^*)m_1 = \overline{m_1}$,
$\forall\, m_1 \in [\mathcal{M}_1]^+$. Note that the computation of $\overline{m_1}(x)$ does not involve P.

Therefore, in the decomposition of P on $\mathcal{M}$ as

$$P = P_{\mathcal{M}_1 \cap \mathcal{M}_\Phi^*} \otimes P_{\mathcal{M}_1}^{\mathcal{M}_1 \cap \mathcal{M}_\Phi^*} \otimes P_{\mathcal{M}}^{\mathcal{M}_1},$$

$P_{\mathcal{M}_1}^{\mathcal{M}_1 \cap \mathcal{M}_\Phi^*}$ is unique and $P_{\mathcal{M}_1 \cap \mathcal{M}_\Phi^*}$ and $P_{\mathcal{M}}^{\mathcal{M}_1}$ are arbitrary when $\Phi\,\mathrm{I}\,\mathcal{M}_1;\ P$.

In particular, if Φ is a finite group, then for any $\mathcal{M}_1 \subset \mathcal{M}$ Φ-stable
and for any probability measure P on $(M, \mathcal{M})$ such that $\Phi\,\mathrm{I}\,\mathcal{M}_1;\ P$,

$$(\mathcal{M}_1 \cap \mathcal{M}_\Phi^*)m_1 = \overline{m_1}, \ \forall\, m_1 \in [\mathcal{M}_1]^+$$

where

$$\overline{m_1} = \frac{1}{|\Phi|} \sum_{\varphi \in \Phi} m_1 \circ \varphi$$

and $|\Phi|$ is the cardinality of Φ.

8.2.6   Randomization of the Set of Transformations

Let $(M, \mathcal{M}, P)$ be an abstract probability space and Φ a monoïd of measurable transformations on M. Let us take $\mathcal{F}$ a σ-field on Φ compatible with the product of composition on Φ and the action of Φ on $\mathcal{M}$, i.e.,

$\circ$: $(\Phi \times \Phi, \mathcal{F} \otimes \mathcal{F}) \to (\Phi, \mathcal{F})$ defined as $\circ(\varphi, \varphi') = \varphi \circ \varphi'$ is measurable, i.e., $\{(\varphi, \varphi') : \varphi \circ \varphi' \in B\} \in \mathcal{F} \otimes \mathcal{F} \quad \forall B \in \mathcal{F}$

$\star$: $(\Phi \times M, \mathcal{F} \otimes \mathcal{M}) \to (M, \mathcal{M})$ defined as $\star(\varphi, x) = \varphi(x)$ is measurable, i.e., $\{(\varphi, x) : \varphi(x) \in A\} \in \mathcal{F} \otimes \mathcal{M} \quad \forall A \in \mathcal{M}$.

Let us note that, with these properties, if we define for each $\varphi \in \Phi$,

$$\varphi_R : \Phi \to \Phi \quad : \quad \varphi_R(\varphi') = \varphi' \circ \varphi$$
$$\varphi_L : \Phi \to \Phi \quad : \quad \varphi_L(\varphi') = \varphi \circ \varphi'$$

then φ_R and φ_L are measurable, i.e.,

$$\varphi_R^{-1}(\mathcal{F}) \subset \mathcal{F} \quad \text{and} \quad \varphi_L^{-1}(\mathcal{F}) \subset \mathcal{F}.$$

Similarly, if for each $x \in M$ we define

$$\hat{x} : \Phi \to M : \hat{x}(\varphi) = \varphi(x)$$

then $\hat{x}$ is measurable, i.e., $\hat{x}^{-1}(\mathcal{M}) \subset \mathcal{F}$.

It is often necessary to make another assumption concerning the action of $(\Phi, \mathcal{F})$ on $(M, \mathcal{M})$:

8.2.37 Definition. The action of $(\Phi, \mathcal{F})$ on $(M, \mathcal{M})$ is *progressively measurable* if $\forall \, \mathcal{M}_1 \subset \mathcal{M}$ Φ-stable, the map:

$\star$: $(\Phi \times M, \mathcal{F} \otimes \mathcal{M}_1) \to (M, \mathcal{M}_1)$ is measurable, i.e., $\{(\varphi, x) : \varphi(x) \in A\} \in \mathcal{F} \otimes \mathcal{M}_1 \quad \forall \, A \in \mathcal{M}_1$.   ■

Note that this assumption is not too restrictive since we already have that, $\forall \, m_1 \in [\mathcal{M}_1]$, the function $m_1[\varphi(x)] : \Phi \times M \to \mathbb{R}$ considered as a function of φ, i.e., $m_1 \circ \hat{x}$, is $\mathcal{F}$-measurable, and considered as a function of x, i.e., $m_1 \circ \varphi$, is $\mathcal{M}_1$-measurable by the Φ-stability of $\mathcal{M}_1$. Thus this definition amounts to saying that a function measurable in each of its components is jointly measurable.

Now let Q be a probability measure on $(\Phi, \mathcal{F})$. By the presence of this structure we may hope, with some hypotheses on Q, to characterize as in Theorem 8.2.20, the almost surely Φ-invariant σ-field by relaxing the hypothesis of countability of Φ. This will be the next theorem which slightly generalizes a similar theorem in Lehmann (1959) (Theorem 4, Chapter 6).

8.2.38 Theorem. Let $(M, \mathcal{M}, P)$ a probability space and Φ a set of measurable transformations on $\mathcal{M}$. If:

(i) Φ is a group,

(ii) $\mathcal{F}$ is a σ-field on Φ compatible with the structure such that the action of Φ on M is progressively measurable,

(iii) Q is a probability on $(\Phi, \mathcal{F})$ such that Φ_R I $\overline{\mathcal{F}_0}; Q$, i.e.,
$Q \circ \varphi_R^{-1} \ll Q, \quad \forall \varphi \in \Phi$,

(iv) $\mathcal{M}_1$ is Φ-stable and Φ I $\mathcal{M}_1 \cap \overline{\mathcal{M}_0}; P$, i.e.,
$P \circ \varphi^{-1} \ll P$ on $\mathcal{M}_1 \; \forall \varphi \in \Phi$,

then

(v) $\mathcal{M}_1 \cap \mathcal{M}_\Phi^* = \overline{\mathcal{M}_1 \cap \mathcal{M}_\Phi} \cap \mathcal{M}_1$, i.e.,
for $m_1 \in [\mathcal{M}_1]$, $m_1 = m_1 \circ \varphi$ a.s.$P \; \forall \varphi \in \Phi$ if and only if
$\exists \; m_1' \in [\mathcal{M}_1 \cap \mathcal{M}_\Phi]$, i.e, $m_1' \circ \varphi = m_1' \; \forall \varphi \in \Phi$, such that $m_1 = m_1'$
a.s.P.

Proof.

(i) By (8.2.33), we already know that $\overline{\mathcal{M}_1 \cap \mathcal{M}_\Phi} \cap \mathcal{M}_1 \subset \mathcal{M}_1 \cap \mathcal{M}_\Phi^*$.

(ii) Let $m_1 \in [\mathcal{M}_1 \cap \mathcal{M}_\Phi^*]_\infty$. Let us define

$$\overline{m}_1(x) = \int_\Phi m_1[\varphi(x)] Q(d\varphi).$$

By (ii), (iv) and Fubini Theorem, $\overline{m}_1 \in [\mathcal{M}_1]$. Let us successively define

$$
\begin{aligned}
N &= \{(\varphi, x) : m_1[\varphi(x)] \neq m_1(x)\}, \\
N_\varphi &= \{x : m_1[\varphi(x)] \neq m_1(x)\}, \\
M_x &= \{\varphi : m_1[\varphi(x)] \neq m_1(x)\}.
\end{aligned}
$$

Then by (ii) and (iv) $N \in \mathcal{F} \otimes \mathcal{M}_1$, $N_\varphi \in \mathcal{M}_1$, and $M_x \in \mathcal{F}$. Since, by Fubini Theorem, $P(N_\varphi) = 0 \quad \forall \, \varphi \in \Phi$,

$$(Q \otimes P)(N) = \int_M Q(M_x) P(dx) = \int_\Phi P(N\varphi) Q(d\varphi) = 0$$

and $Q(M_x)$ is $\mathcal{M}_1$-measurable. Therefore, if $A = \{x : Q(M_x) > 0\}$, $A \in \mathcal{M}_1$ and $P(A) = 0$. Moreover, $\forall \, x \in A^c$, $m_1[\varphi(x)] = m_1(x)$ a.s.Q and, therefore, $\overline{m}_1(x) = m_1(x)$, i.e., $m_1 = \overline{m}_1$ a.s.P. Let us similarly define

$$
\begin{aligned}
\overline{N} &= \{(\varphi, x) : m_1[\varphi(x)] \neq \overline{m}_1(x)\}, \\
\overline{N}_\varphi &= \{x : m_1[\varphi(x)] \neq \overline{m}_1(x)\}, \\
\overline{M}_x &= \{\varphi : m_1[\varphi(x)] \neq \overline{m}_1(x)\}, \\
\overline{A} &= \{x : Q(\overline{M}_x) > 0\}.
\end{aligned}
$$

Now, if $m_1' = \overline{m}_1 \mathbf{1}_{\overline{A}^c}$, i.e. $m_1'(x) = \overline{m}_1(x)$ for $x \in \overline{A}^c$ and 0 for $x \in \overline{A}$, then clearly $m_1' \in [\mathcal{M}_1]$ since, as before, $\overline{A} \in \mathcal{M}_1$. We now show that $m_1 = m_1'$ a.s.P and that $m_1' \circ \varphi = m_1' \; \forall \, \varphi \in \Phi$. Since $m_1 \circ \varphi = m_1$ a.s.P and $m_1 = \overline{m}_1$ a.s.P, $m_1 \circ \varphi = \overline{m}_1$ a.s.P, i.e., $P(\overline{N}_\varphi) = 0 \; \forall \, \varphi \in \Phi$. As before, by Fubini Theorem, $P(\overline{A}) = 0$ and, therefore, $m_1' = \overline{m}_1$ a.s.P and $m_1' = m_1$ a.s.P. Now, if $x \in \overline{A}^c$, then $Q(\overline{M}_x) = 0$ and by (iii), $Q[\varphi_R^{-1}(\overline{M}_x)] = 0 \; \forall \, \varphi \in \Phi$, but

$$\varphi_R^{-1}(\overline{M}_x) = \{\varphi' : m_1[\varphi'[\varphi(x)]] \neq \overline{m}_1(x)\}.$$

Therefore,
$$m_1[\varphi'[\varphi(x)]] = \overline{m}_1(x) \text{ a.s.}Q \text{ in } \varphi'.$$

This implies that
$$\overline{m}_1[\varphi(x)] = \overline{m}_1(x) \quad \text{and}$$
$$m_1[\varphi'[\varphi(x)]] = \overline{m}_1[\varphi(x)] \quad \text{a.s.}Q \text{ in } \varphi', \text{ i.e.,}$$
$$Q(\overline{M}_{\varphi(x)}) = 0.$$

Therefore, $\forall \, x \in \overline{A}^c$, $\forall \, \varphi \in \Phi$,
$$\overline{m}_1[\varphi(x)] = \overline{m}_1(x) \quad \text{and} \quad \varphi(x) \in \overline{A}^c.$$

However, by (i) if $\varphi(x) \in \overline{A}^c$, then $x = \varphi^{-1}[\varphi(x)] \in \overline{A}^c$. Therefore if $x \in \overline{A}$, then $\varphi(x) \in \overline{A} \; \forall \, \varphi \in \Phi$. This shows that $m_1'[\varphi(x)] = m_1'(x) \; \forall \, x \in M$, $\forall \, \varphi \in \Phi$. ∎

Taking $\mathcal{M}_1 = \mathcal{M}$ entails the following corollary:

8.2.39 Corollary. If:

(i) Φ is a group,

(ii) $\mathcal{F}$ is a σ-field on Φ compatible with the structure,

(iii) Q a probability on $(\Phi, \mathcal{F})$ such that $\Phi_R \text{ I } \overline{\mathcal{F}_0}; Q$

(iv) $\Phi \text{ I } \overline{\mathcal{M}_0}; P,$

then

(v) $\mathcal{M}_{\Phi}^* = \overline{\mathcal{M}_{\Phi}}.$ ∎

This probability structure on Φ, may be used to characterize Φ-invariance as almost surely equivalent to independence on the product probability space. More precisely, under the assumption of compatibility of $\mathcal{F}$ with the action of Φ on $\mathcal{M}$, the map $\varphi \to \mathcal{M}_0(m \circ \varphi) = \int_M m[\varphi(x)] P(dx)$ is $\mathcal{F}$-measurable, for any $m \in [\mathcal{M}]$. Under this weaker assumption alone, Q and P induces a unique probability measure R on $(\Phi \times M, \mathcal{F} \otimes \mathcal{M})$ defined by extending

$$(8.2.42) \quad R(B \times A) = \int_B P[\varphi^{-1}(A)] Q(d\varphi) \quad \forall\, B \in \mathcal{F},\ \forall\, A \in \mathcal{M}.$$

More generally, $\forall\, f \in [\mathcal{F}]^+, m \in [\mathcal{M}]^+,$

$$(8.2.43) \qquad \mathcal{I}_R(fm) = \int_{\Phi} f(\varphi) \mathcal{M}_0(m \circ \varphi) Q(d\varphi),$$

where $\mathcal{I}_R$ denotes the expectation with respect to R.

However, the stronger assumption of the compatibility of $\mathcal{F}$ with the action of Φ on M permits the use of Fubini Theorem, since $m[\varphi(x)]$ is $\mathcal{F} \otimes \mathcal{M}$-measurable. Therefore, under this stronger assumption, $\forall\, r \in [\mathcal{F} \otimes \mathcal{M}]^+$

$$
\begin{aligned}
(8.2.44) \qquad \mathcal{I}_R(r) &= \int_{\Phi} Q(d\varphi) \int_M r(\varphi, \varphi(x)) P(dx) \\
&= \int_M P(dx) \int_{\Phi} r(\varphi, \varphi(x)) Q(d\varphi).
\end{aligned}
$$

Let us note that (8.2.44) may also be written as

(8.2.45) $$\mathcal{I}_R(r) = \mathcal{I}_{R_0}(r \circ h),$$

where $R_0 = Q \otimes P$ and $h : (\Phi \times M, \mathcal{F} \otimes \mathcal{M}) \to (\Phi \times M, \mathcal{F} \otimes \mathcal{M})$ is defined as $h(\varphi, x) = (\varphi, \varphi(x))$ and is measurable since $\forall\, B \in \mathcal{F}$ and $A \in \mathcal{M}$, $h^{-1}(B \times A) = (B \times M) \cap \{(\varphi, x) : \varphi(x) \in A\} \in \mathcal{F} \otimes \mathcal{M}$.

If we identify, as we usually do, sub-σ-fields of $\mathcal{F}$ (respectively, $\mathcal{M}$) with the corresponding sub-σ-fields on $\Phi \times M$, and measurable functions on Φ (respectively, M) with measurable functions on $\Phi \times M$, then $\mathcal{F} \otimes \mathcal{M} = \mathcal{F} \vee \mathcal{M}$ and we may notice that $f \circ h = f \; \forall\, f \in [\mathcal{F}]$ and that $m \circ h = m \circ \varphi$ $\forall\, m \in [\mathcal{M}]$. With this identification, Definition (8.2.42) is in fact equivalent to saying that the conditional probability on $\mathcal{F}$ with respect to R admits $P \circ \varphi^{-1}$ as a regular version. So, $\forall\, m \in [\mathcal{M}]^+$,

(8.2.46) $$\mathcal{F}m = \mathcal{M}_0(m \circ \varphi) \quad \text{a.s.}R.$$

This reinterpretation entails the following proposition:

8.2.40 Proposition. Let $\mathcal{M}_1$ be a sub-σ-field of $\mathcal{M}$. Then
(i) $\mathcal{F} \perp\!\!\!\perp \mathcal{M}_1; R$
if and only if
(ii) $\mathcal{M}_0(m_1 \circ \varphi) = \mathcal{M}_0(m_1)$ a.s.Q $\quad \forall\, m_1 \in [\mathcal{M}_1]^+$.   ∎

If Φ is countable and Q gives positive probability on any $\varphi \in \Phi$, or if Φ is a topological space such that $\forall\, m \in [\mathcal{M}]_\infty$, $\mathcal{M}_0(m \circ \varphi)$ is a continuous function of φ and Q gives positive probability to any open subset of φ, we have already seen that, in such situations, Q is regular for Φ in the following sense:

8.2.41 Definition. Q is *regular* for Φ if $m \in [\mathcal{M}]_\infty$ and $\mathcal{M}_0(m \circ \varphi) = 0$ a.s.Q implies $\mathcal{M}_0(m \circ \varphi) = 0 \quad \forall\, \varphi \in \Phi$.   ∎

Such a regularity property of Q entails the interpretation of conditional Φ-invariance as a conditional independence property, as is shown in the next theorem:

8.2.42 Theorem. If $\Phi\, \mathrm{I}\, (\mathcal{M}_2 \cap \overline{\mathcal{M}}_0); P$ and if Q is regular for Φ, then

(i) $\Phi \text{ I } \mathcal{M}_1 \mid \mathcal{M}_2; P$

if and only if

(ii) $\mathcal{F} \perp\!\!\!\perp \mathcal{M}_1 \mid \mathcal{M}_2; R.$

Proof. Using condition (iii) of Theorem 2.2.1 on the equivalent characterizations of conditional independence we have that:

$$\mathcal{F} \perp\!\!\!\perp \mathcal{M}_1 \mid \mathcal{M}_2; R$$

if and only if

(8.2.47) $\mathcal{F}(m_1 m_2) = \mathcal{F}(m_2 \mathcal{M}_2 m_1) \quad \forall \, m_i \in [\mathcal{M}_i]^+, i = 1, 2.$

By (8.2.46), this is equivalent to

(8.2.48) $\mathcal{M}_0[m_1 \circ \varphi \cdot m_2 \circ \varphi] = \mathcal{M}_0[m_2 \circ \varphi \cdot (\mathcal{M}_2 m_1) \circ \varphi]$ a.s.$Q.$

By the regularity of Q for Φ, (8.2.48) is then true $\forall \, \varphi \in \Phi$, and is therefore equivalent to

(8.2.49) $\varphi^{-1}(\mathcal{M}_2)(m_1 \circ \varphi) = (\mathcal{M}_2 m_1) \circ \varphi$ a.s.P, $\forall \, \varphi \in \Phi$, $\forall \, m_1 \in [\mathcal{M}_1]^+,$

which is condition (ii) of Definition 8.2.22 of conditional invariance.   ∎

Thus (conditional) invariance may be interpreted as a (conditional) independence property between the transformations φ's and the sets in $\mathcal{M}_1$, once Φ has been endowed with a regular probability Q on a suitable σ-field $\mathcal{F}$. As this underlying probability is arbitrary (but regular), this equivalence shows that both properties are structurally similar and that in fact conditional invariance and conditional independence will share similar properties as, e.g., Theorem 2.2.10 and Theorem 8.2.25.

8.3   Invariant Experiments

8.3.1   Construction and Definition of an Invariant Bayesian Experiment

The structure of an invariant statistical experiment is usually presented as follows. In the statistical experiment $\mathcal{E}$ defined as

(8.3.1) $\mathcal{E} = \{(S, \mathcal{S}); P^a : a \in A\}$

there exists a set of measurable transformations G; this means that each $g \in G$ is a function $g : S \to S$ such that $g^{-1}(S) \subset S$. The usual invariance condition is the following: $\forall g \in G, \forall a \in A, \exists$ a unique $a' \in A$ such that

$$(8.3.2) \qquad P^a[g^{-1}(X)] = P^{a'}(X) \quad \forall X \in S.$$

In other words, the image under g of any sampling probability P^a, i.e., $P^a \circ g^{-1}$, is another sampling probability $P^{a'}$. Thus, to each $g \in G$ is associated a transformation $\bar{g} : A \to A$ defined by $\bar{g}(a) = a'$. Therefore, if $\bar{G} = \{\bar{g} : g \in G\}$, it is shown that if G is a monoid (resp. a group), $\overline{G}$ is a monoïd (respectively, a group) and (8.3.2) may be rewritten as

$$(8.3.3) \qquad \mathcal{I}^a(t \circ g) = \mathcal{I}^{\bar{g}(a)}(t), \ \forall \, a \in A, \ \forall \, g \in G, \ \forall \, t \in [S]^+.$$

Let us define $\varphi : A \times S \to A \times S$ by

$$(8.3.4) \qquad \varphi(a, s) = (\bar{g}(a), g(s)).$$

Under the hypothesis that the σ-field $\mathcal{A}$ on A is such that $\forall g \in G, \bar{g} : A \to A$ is a measurable transformation, i.e., $\bar{g}^{-1}(\mathcal{A}) \subset \mathcal{A}$, then $\Phi = \{\varphi : g \in G\}$ is a set of measurable transformations on $A \times S$, i.e., $\varphi^{-1}(\mathcal{A} \otimes S) \subset \mathcal{A} \otimes S$ and Φ is a monoïd (respectively, a group) if G is a monoïd (respectively, a group). Now, if μ is a probability measure on $(A, \mathcal{A})$, then recall that P^a is by construction a regular version of the restriction to S of the conditional probability Π on $\mathcal{A}$.

So, if we identify, as before, sub-σ-fields of $\mathcal{A}$ (respectively, S) with sub-σ-fields of the corresponding cylinders on $A \times S$, and measurable functions on A (respectively, S) with measurable functions on $A \times S$, then with this identification (8.3.3) may be rewritten in the Bayesian Experiment

$$(8.3.5) \qquad \mathcal{E} \ = \ (A \times S, \mathcal{A} \vee S, \Pi)$$

as

$$(8.3.6) \qquad \mathcal{A}(t \circ \varphi) \ = \ (\mathcal{A}t) \circ \varphi \qquad \forall \, t \in [S]^+.$$

Note that, under (8.3.4),

$$(8.3.7) \qquad \varphi^{-1}(\mathcal{A}) \ \subset \ \mathcal{A}$$

and

(8.3.8) $\qquad \varphi^{-1}(\mathcal{S}) \subset \mathcal{S}.$

Now, if the prior probability μ on $(A, \mathcal{A})$ is such that

(8.3.9) $\qquad \mu \circ \overline{g}^{-1} \ll \mu \quad \forall g \in G$

or, equivalently, $A \in \mathcal{A}$, and $\Pi(A) = 0$ implies that $\Pi[\varphi^{-1}(A)] = 0$, so Φ invariates the null sets of $\mathcal{A}$. Therefore, under (8.3.2), (8.3.9) and mesurability conditions on the transformations in G and $\overline{G}$, we see that, with (8.3.4), the invariance in sampling theory framework is translated in the Bayesian experiment by: Φ is a set of measurable transformations on $A \times S$ such that $\mathcal{A}$ and $\mathcal{S}$ are Φ-stable, $\Phi \text{ I } \mathcal{S} \mid \mathcal{A}$ and, moreover, $\mathcal{A} \perp\!\!\!\perp \varphi^{-1}(\mathcal{S}) \mid \varphi^{-1}(\mathcal{A})$ $\forall \; \varphi \in \Phi$. More generally, we will define:

8.3.1 Definition. Let $\mathcal{E} = (A \times S, \mathcal{A} \vee \mathcal{S}, \Pi)$ be an unreduced Bayesian experiment and Φ a monoïd of measurable transformations on $A \times S$. Then $\mathcal{E}$ is Φ-*invariant* (respectively, *sampling* Φ-*invariant*) (respectively, *posterior* Φ-*invariant*) if $\mathcal{A}$ and $\mathcal{S}$ are Φ-stable and $\Phi \text{ I } (\mathcal{A} \vee \mathcal{S})$ (respectively, $\Phi \text{ I } \mathcal{S} \mid \mathcal{A}$) (respectively, $\Phi \text{ I } \mathcal{A} \mid \mathcal{S}$). ∎

These definitions may be extended quite naturally to general reduced Bayesian experiment $\mathcal{E}^{\mathcal{M}}_{\mathcal{B}\vee\mathcal{T}}$ where $\mathcal{B} \subset \mathcal{A}, \mathcal{T} \subset \mathcal{S}$ and $\mathcal{M} \subset \mathcal{A} \vee \mathcal{S}$:

8.3.2 Definition. Let Φ be a monoïd of measurable transformations on $A \times S$. Then $\mathcal{E}^{\mathcal{M}}_{\mathcal{B}\vee\mathcal{T}}$ is Φ-*invariant* (respectively, *sampling* Φ-*invariant*) (respectively, *posterior* Φ-*invariant*) if $\mathcal{M}$, $\mathcal{B} \vee \mathcal{M}$ and $\mathcal{M} \vee \mathcal{T}$ are Φ-stable and $\Phi \text{ I } (\mathcal{B} \vee \mathcal{T}) \mid \mathcal{M}$ (respectively, $\Phi \text{ I } \mathcal{T} \mid \mathcal{B} \vee \mathcal{M}$) (respectively, $\Phi \text{ I } \mathcal{B} \mid \mathcal{M} \vee \mathcal{T}$). ∎

Theorem 8.2.25 allows us to link Φ-invariance and conditional Φ-invariance.

8.3.3 Proposition. The following statements are equivalent:

(i) $\mathcal{E}^{\mathcal{M}}_{\mathcal{B}\vee\mathcal{T}}$ is Φ-*invariant*;

(ii) $\mathcal{E}^{\mathcal{M}}_{\mathcal{B}\vee\mathcal{T}}$ is *sampling* Φ-*invariant* and Φ I $\mathcal{B}$ | $\mathcal{M}$ (i.e., *prior* Φ-*invariant*);

(iii) $\mathcal{E}^{\mathcal{M}}_{\mathcal{B}\vee\mathcal{T}}$ is *posterior* Φ-*invariant* and Φ I $\mathcal{T}$ | $\mathcal{M}$ (i.e., *predictively* Φ-*invariant*).   ∎

The next proposition shows that, when $\mathcal{E}$ is Φ-invariant, any reduction on stable sub-σ-fields remains Φ-invariant; the proof is a simple consequence of Theorems 8.2.23(i) and 8.2.25:

8.3.4 Proposition. If $\mathcal{E}$ is Φ-invariant, if $\mathcal{B} \subset \mathcal{A}$, $\mathcal{T} \subset \mathcal{S}$ and $\mathcal{M} \subset \mathcal{A} \vee \mathcal{S}$ are such that $\mathcal{M}$, $\mathcal{B}\vee\mathcal{M}$ and $\mathcal{M}\vee\mathcal{T}$ are Φ-stable, then $\mathcal{E}^{\mathcal{M}}_{\mathcal{B}\vee\mathcal{T}}$ is Φ-invariant. ∎

In the unreduced Bayesian experiment $\mathcal{E}$, the most usual use of the Proposition 8.3.3 runs as follows: the sampling model suggests a set Φ of measurable transformations such that $\mathcal{E}$ is sampling Φ-invariant. Let us suppose that there exists a probability on $(A, \mathcal{A})$ such that Φ I $\mathcal{A}$ and that this probability is chosen as the prior probability. Then $\mathcal{E}$ will be Φ-invariant. From this it can be deduced that Φ I $\mathcal{S}$, i.e. the predictive probability is invariant under Φ and Φ I $\mathcal{A}$ | $\mathcal{S}$, i.e. the posterior probabilities are invariant under Φ.

This procedure raises a major problem related to the choice of an invariant prior probability. If $\mathcal{A}$ is almost surely invariant, i.e., if a is unchanged under φ, $\forall \varphi \in \Phi$, then this condition may sometimes be trivially satisfied for any prior probability. In terms of the introduction to this section, φ is obtained by g, this means that the sampling probabilities are invariant under g, i.e., $\bar{g}$ is the identity on A, $\forall g \in G$. Some examples, involving stationary and exchangeable processes, are presented in the next chapter. The opposite situation occurs when Φ is sufficiently rich. In this case, the condition Φ I $\mathcal{A}$ may characterize a unique invariant measure (up to a multiplicative function of φ). To preserve the invariance structure and, in particular, to obtain an invariant estimate (i.e., a posterior invariance) it is natural to choose this measure as a prior one. Unfortunately in general this measure is not a probability, and some care is needed in using it; we omit the study of this situation in this book. As indicated in Chapter 1, recall that a prior measure may be used as soon as it induces a σ-finite predictive measure. In this case, the posterior probabilities are well defined. Additional problems ermerge when decomposing a measure into marginals and conditionals, for the latter exercise makes sense only when the marginal is σ-finite. Consequently, extensions of the results obtained in this book to the case of an unbounded prior measure are rather delicate. The existence of a bounded invariant prior measure is generally asserted in compact parameter spaces, but this situation is not particularly standard. One can imagine intermediate situations in which the choice of the prior probability is free with respect to some of the parameters, typically on the invariant sub-σ-field, and is determined on the remaining parameters conditionally on this σ-field. The same difficulties arise due to the fact that this conditional prior is often unbounded.

We now turn to some examples.

Example 1.

Let $s = (x_1, x_2, \ldots, x_n)$, $x_i = (y_i, z_i)' \in \mathbb{R}^2$, $a \in [0, 2\pi[= A$ with $(x_i \mid a) \sim i.N_2\left[\left(\begin{smallmatrix} \cos a \\ \sin a \end{smallmatrix}\right); I\right]$. Let $R_\alpha = \left(\begin{smallmatrix} \cos \alpha & -\sin \alpha \\ \sin \alpha & \cos \alpha \end{smallmatrix}\right)$ be the rotation of angle α. Clearly, if $g_\alpha(x_i) = R_\alpha x_i$, then $\bar{g}_\alpha(a) = a + \alpha$. Therefore, $\varphi_\alpha(a, s) = (a + \alpha, \{R_\alpha x_i; 1 \le i \le n\})$ where $a + \alpha$ is understood as modulo 2π, i.e., in $\mathbb{R}/2\pi\mathbb{Z}$. Let μ be the uniform distribution on $[0, 2\pi[$, i.e., with a density with respect to Lebesgue measure equal to $\frac{1}{2\pi} 1_{[0, 2\pi[}(a)$. Obviously, if $\Phi = \{\varphi_\alpha : \alpha \in [0, 2\pi[\}$, then Φ is a group and Φ I A and Φ I $S \mid A$. Hence, Φ I S, Φ I $A \mid S$, and Φ I $(A \vee S)$. Note that, by direct computation, the joint density is given by

$$\pi(a, s) = (2\pi)^{-\frac{n+1}{2}} e^{-\frac{1}{2}\Sigma_{1 \le i \le n}\{(y_i - \cos a)^2 + (z_i - \sin a)^2\}}$$
$$= (2\pi)^{-\frac{n+1}{2}} e^{-\frac{1}{2}(y'y + z'z + n)} e^{nt \cos(u - a)}$$

where

$$\left(\begin{matrix} \bar{y} \\ \bar{z} \end{matrix}\right) = t\left(\begin{matrix} \cos u \\ \sin u \end{matrix}\right).$$

Hence the predictive density is given by

$$p(s) = (2\pi)^{-\frac{n}{2}} I_0(nt) e^{-\frac{1}{2}(y'y + z'z + n)}.$$

Hence Φ I $\mathcal{S}$ since $R'_\alpha R_\alpha$ is the identity matrix. The posterior density is

$$\mu(a \mid s) = (2\pi)^{-\frac{1}{2}} I_0(nt)^{-1} e^{nt\cos(a-u)},$$

where $I_0(x)$ is the incomplete Bessel function (see Watson (1983), appendix A, and Abramowitz and Stegun (1984)). Therefore Φ I $\mathcal{A} \mid \mathcal{S}$, since $(a + \alpha) - (u + \alpha) = a - u$.   ∎

Example 2.

Let $s = (s_1, \ldots, s_n)$, $s_i \in \mathbb{R}$, $a \in \mathbb{R}$ with $(s_i \mid a) \sim i.N(a, 1)$. Let $g_\alpha(s_i) = s_i + \alpha$, hence $\bar{g}_\alpha(a) = a + \alpha$ and therefore $\varphi_\alpha(a, s) = (a + \alpha, s_1 + \alpha, \ldots, s_n + \alpha)$. Hence Φ I $\mathcal{S} \mid \mathcal{A}$ with $\Phi = \{\varphi_\alpha : \alpha \in \mathbb{R}\}$, provided that the prior measure is equivalent to the Lebesgue measure on $\mathbb{R}$; here, because of the obtained conditional invariance, the translation group appears as a "natural" group of transformations operating on both the sample and the parameter space. But in this situation, the Haar measure is the Lebesgue measure on $\mathbb{R}$. If this is chosen as a prior measure, then the joint density is given by

$$\pi(a, s) = (2\pi)^{-\frac{n}{2}} e^{-\frac{1}{2}\Sigma_{1 \leq i \leq n}(s_i - a)^2}.$$

It is evident that this is invariant for Φ, i.e., Φ I $(\mathcal{A} \vee \mathcal{S})$. The predictive measure is σ-finite and its density is given by

$$p(s) = (2\pi)^{-\frac{n-1}{2}} n^{-\frac{1}{2}} e^{-\frac{1}{2}\Sigma_{1 \leq i \leq n}(s_i - \bar{s})^2},$$

where $\bar{s} = \frac{1}{n}\Sigma_{1 \leq i \leq n} s_i$. Clearly, $p(s)$ is invariant under Φ. The posterior probabilities are given by

$$\mu(a \mid s) = (2\pi)^{-\frac{1}{2}} n^{\frac{1}{2}} e^{-\frac{n}{2}(a - \bar{s})^2},$$

which are obviously conditionally invariant under Φ. This situation illustrates that Proposition 8.3.3 may hold even with improper prior measure.

Example 3. Multivariate Regression

Let $s = (x_1, x_2, \ldots, x_n)$, $x_i = (y_i, z_i)'$, $y_i \in \mathbb{R}^p$, $z_i \in \mathbb{R}^k$, $a = (B', \Sigma)$ where B is a $k \times p$-matrix and Σ a $p \times p$-symmetric positive definite matrix with

$$(y_i \mid a, z_1^n) \sim i.N_p(B'z_i, \Sigma).$$

Let $g_{M,N}(x_i) = (My_i + Nz_i, z_i)'$ where M is a $p \times p$-nonsingular matrix, and N a $p \times k$-matrix, and Φ the set of such transformations. It follows that $\bar{g}_{M,N}(a) = (MB' + N, M \Sigma M')$. Let the prior density with respect to the Lebesgue measure on $\mathbb{R}^{p+k} \times C_p$, where $C_p \subset \mathbb{R}^{\frac{p(p+1)}{2}}$ corresponds to the cone of symmetric positive definite $p \times p$-matrices, be given by

$$\mu(B', \Sigma) = |\Sigma|^{-\frac{\nu+p+1}{2}}.$$

Then,

$$\Phi \text{ I } y_1^n \mid a \vee z_1^n.$$

The joint density is given by

$$\pi(y_1^n, a \mid z_1^n) = |\Sigma|^{-\frac{\nu+p+n+1}{2}} (2\pi)^{-\frac{np}{2}} \times \exp\{-\frac{1}{2}\text{tr}[\Sigma^{-1}(Y - ZB)'(Y - ZB)]\},$$

where $Y' = (y_1, \ldots, y_n)$ and $Z' = (z_1, \ldots, z_n)$. The predictive measure is σ-finite and its density with respect to the Lebesgue measure is given by

$$p(y_1^n \mid z_1^n) = K \mid Y'M_Z Y|^{-\frac{n+\nu-k}{2}} |Z'Z|^{-\frac{p}{2}}$$

where $M_Z = I - Z(Z'Z)^{-1}Z'$ and K is an appropriate constant. Now, since $g_{M,N}(Y, Z) = (YM' + ZN', Z)$ and the Jacobian of this transformation is $|M|^n$ the predictive measure will be invariant if $\nu = k$. The posterior probability is characterized by

$$(B \mid \Sigma, Y, Z) \quad \sim \quad MN(\hat{B}, (Z'Z)^{-1} \otimes \Sigma),$$

where $\hat{B} = (Z'Z)^{-1}Z'Y$ and

$$(\Sigma \mid Y, Z) \quad \sim \quad IW(Y'M_Z Y, n).$$

The density of the Inverted Wishart with respect to the trace of the Lebesgue measure on C_p is given by

$$\mu(\Sigma \mid Y, Z) = \left\{ 2^{\frac{np}{2}} \pi^{\frac{p(p-1)}{4}} \prod_{i=1}^{p} \Gamma(\frac{1}{2}(n + 1 - i)) \right\}^{-1}$$
$$|Y' M_Z Y|^{\frac{n}{2}} |\Sigma|^{-\frac{n+p+1}{2}} \exp -\frac{1}{2}\text{tr}\Sigma^{-1}(Y' M_Z Y).$$

The density of the matrix normal distribution of B with respect to the Lebesgue measure is given by

$$\mu(B \mid \Sigma, Y, Z) = \left[(2\pi)^{pk} |\Sigma|^k |Z'Z|^{-p} \right]^{-\frac{1}{2}} \exp -\frac{1}{2} \text{tr}\Sigma^{-1}(B - \hat{B})'Z'Z(B - \hat{B}).$$

It can be verified that Φ I A | S. (For more details on the computation of the posterior and the predictive probabilities, see Tiao and Zellner (1964) and Ando and Kaufmann (1965).)   ∎

Example 4. Pitman Parametrization.

In the spirit of Pitman (1939), we can consider a statistical experiment defined as follows. Let $(S, \mathcal{S}, Q)$ be a probability space and A a group acting on S by the operation $\star : A \times S \to S$, i.e., if e is the identity of the group A, $e \star s = s$ $\forall s \in S$ and $a_1 \star (a_2 \star s) = (a_1 a_2) \star s$ $\forall a_1, a_2 \in A$, $\forall s \in S$. Let $\mathcal{A}$ be a σ-field on A compatible with the structure of the group and with the action of the group on S, i.e., $a_1 a_2^{-1} : A \times A \to A$ is measurable and $a \star s : A \times S \to S$ is measurable. For $\alpha \in A$, let us define $g_\alpha : S \to S$ by $g_\alpha(s) = \alpha \star s$. Then g_α is a measurable transformation on S, $g_e = i$, the identity on S, and $g_{\alpha_1} \circ g_{\alpha_2} = g_{\alpha_1 \alpha_2}$ so that $g_\alpha^{-1} = g_{\alpha^{-1}}$. For $a \in A$ we define the sampling probabilities by $P^a = Q \circ g_a^{-1}$. By Fubini Theorem, $\forall X \in \mathcal{S}$, $P^a(X) : A \to R$ is measurable and, for any $t \in [\mathcal{S}]^+$, $\mathcal{I}^a(t) = \int_S t(a \star s) Q(ds)$. For $\alpha \in A$ let us define $\overline{g}_\alpha : A \to A$ by $\overline{g}_\alpha(a) = \alpha a$ and $\varphi_\alpha : A \times S \to A \times S$ by $\varphi_\alpha(a, s) = (\overline{g}_\alpha(a), g_\alpha(s)) = (\alpha a, \alpha \star s)$. Then, clearly, $\overline{g}_e$ is the identity on A, $\overline{g}_{\alpha_1} \circ \overline{g}_{\alpha_2} = \overline{g}_{\alpha_1 \alpha_2}$ $\forall \alpha_1, \alpha_2 \in A$ and so, $\forall \alpha \in A$, $\overline{g}_\alpha^{-1} = \overline{g}_{\alpha^{-1}}$. Similarly, φ_e is the identity on $A \times S$, $\varphi_{\alpha_1} \circ \varphi_{\alpha_2} = \varphi_{\alpha_1 \alpha_2}$ $\forall \alpha_1, \alpha_2 \in A$ and $\varphi_\alpha^{-1} = \varphi_{\alpha^{-1}}$. If $G = \{g_\alpha : \alpha \in A\}$, $\overline{G} = \{\overline{g}_\alpha : \alpha \in A\}$ and $\Phi = \{\varphi_\alpha : \alpha \in A\}$, then $\overline{G}$ and Φ are groups of transformations isomorphic to A, whereas the map $\alpha \to g_\alpha$ from A to G is only a homomorphism, since $g_{\alpha_1} = g_{\alpha_2} \not\Rightarrow \alpha_1 = \alpha_2$. Let μ be a probability measure on $(A, \mathcal{A})$ and let Π be the joint probability on $(A \times S, \mathcal{A} \vee \mathcal{S})$. Then, for any $m \in [\mathcal{A} \vee \mathcal{S}]^+$,

$$\mathcal{I}m = \int_{A \times S} m(\alpha, \alpha \star \sigma) Q(d\sigma) \mu(d\alpha)$$

and

$$\mathcal{A}m = \int_S m(a, a \star \sigma) Q(d\sigma).$$

Therefore, for any $t \in [\mathcal{S}]^+$,

$$\mathcal{A}(t \circ \varphi_\alpha) = \int_S t(\alpha a \star \sigma) Q(d\sigma) \quad \text{and}$$

$$(\mathcal{A}t) \circ \varphi_\alpha = \{\int_S t(a \star \sigma) Q(d\sigma)\} \circ \varphi_\alpha$$

$$= \int_S t(\alpha a \star \sigma) Q(d\sigma).$$

Since $\forall b \in [\mathcal{A}]$, $b \circ \varphi_\alpha = b(\alpha a)$ and $\forall t \in [\mathcal{S}]$, $t \circ \varphi_\alpha = t(\alpha \star s)$. Obviously, $\mathcal{A}$ and $\mathcal{S}$ are Φ-stable. Now, Φ I $(\mathcal{A} \cap \overline{\mathcal{I}})$ if and only if μ is such that $\mu \circ \overline{g}_\alpha^{-1} \ll \mu \, \forall \alpha \in A$ or, equivalently, if $\mu \circ \overline{g}_\alpha^{-1} \sim \mu \, \forall \alpha \in A$ since $\overline{G}$ is a group. Therefore, for such a prior, Φ I $\mathcal{S} \mid \mathcal{A}$. If there exists a prior probability μ on $(A, \mathcal{A})$ such that $\mu \circ \overline{g}_\alpha^{-1} = \mu \, \forall \alpha \in A$, then Φ I $\mathcal{A}$. Therefore Φ I $(\mathcal{A} \vee \mathcal{S})$, Φ I $\mathcal{S}$ and Φ I $\mathcal{A} \mid \mathcal{S}$. This may be checked directly since, $\forall m \in [\mathcal{A} \vee \mathcal{S}]^+$,

$$
\begin{aligned}
\mathcal{I}(m \circ \varphi_\alpha) &= \int_{A \times S} m[\alpha \alpha', (\alpha \alpha') \star \sigma] Q(d\sigma) \mu(d\alpha) \\
&= \int_{A \times S} m[\alpha'', \alpha'' \star \sigma] Q(d\sigma) \mu \circ \overline{g}_\alpha^{-1}(d\alpha'') \\
&= \mathcal{I}(m).
\end{aligned}
$$

Therefore, if μ^s is a regular version of the posterior probability then, $\forall b \in [\mathcal{A}]^+$, $\mathcal{S}(b \circ \varphi_\alpha) = (\mathcal{S}b) \circ \varphi_\alpha$, i.e.,

$$
\int_A b(\alpha \alpha') \mu^s(d\alpha') = \int_A b(\alpha') \mu^{\alpha \star s}(d\alpha').
$$

This example covers the cases where the parameter is either a location parameter or a scale parameter. Note that these results may be thought of as conditional on Q. If Q is also considered as a parameter then for any prior on Q, such as a Dirichlet process (see the next example), we will have that Φ I (Q, a, s) when defining $\varphi_\alpha(Q, a, s) = (Q, \alpha a, \alpha \star s)$. Therefore we will also have that Φ I $a \mid s$, i.e., after the integration of the nuisance parameter Q. ∎

Example 5. Nonparametric Bayesian Experiment.

Let $s = (x_1, \ldots, x_n) \in [0, 1]^n = S$, $a \in A$, where A is the set of probability measures on $[0, 1]$ with its Borel σ-field. We suppose that $\underset{1 \le i \le n}{\parallel} x_i \mid a$ and $P(x_i \in B \mid a) = a(B)$. Let $g : [0, 1] \to [0, 1]$ strictly increasing continuous with $g(0) = 0$ and $g(1) = 1$, and let G be the set of such functions. If $g(s) = (g(x_1), \ldots, g(x_n))$, then $\overline{g}(a) = a \circ g^{-1}$. Therefore, let us define $\varphi(a, s) = (\overline{g}(a), g(s))$; if $\Phi = \{\varphi : g \in G\}$, then $\mathcal{A}$ and $\mathcal{S}$ are Φ-stable and $\mathcal{A}(t \circ \varphi) = (\mathcal{A}t) \circ \varphi$ a.s. for any prior μ on A. If we take as a prior $a \sim Di(n_0 a_0)$, i.e., the Dirichlet Process (see, for instance, Ferguson (1973, 1974) or Rolin (1983)) where $n_0 \in]0, \infty[$ and a_0 is a probability measure on $[0, 1]$ — recall that, by definition, this means that for every measurable partition $(B_1, B_2, \ldots, B_k)$ of $[0, 1]$, the vector $(a(B_1), a(B_2), \ldots, a(B_k))$

is distributed as a k-dimensional Dirichlet distribution with parameters $(n_0 a_0(B_1), n_0 a_0(B_2), \ldots, n_0 a_0(B_k))$ —. It has been shown that $(a \mid s) \sim Di(n_* a_*)$ where $n_* = n_0 + n$ and $a_* = (n_0/n_*)a_0 + (n/n_*)\hat{F}(s)$, where $\hat{F}(s)$ is the empirical distribution of s, i.e., $\hat{F}(s)(B) = \frac{1}{n}\sum_{1 \leq i \leq n} 1_B(x_i)$. In the class of Dirichlet priors, Φ-invariant prior probabilities are characterized by $a_0 = p\varepsilon_0 + (1 - p)\varepsilon_1$ where $p \in [0,1]$, and ε_x is the Dirac measure at point x since, clearly, $a_0 \circ g^{-1} = a_0$ and if $a \sim Di(n_0 a_0)$, then $a \circ g^{-1} \sim Di(n_0 a_0 \circ g^{-1})$. Hence Φ I a. Moreover, if this is true, then Φ I (a,s) and Φ I $s \mid a$. This last conditional invariance can be verified directly, since $a_* \circ g^{-1} = (n_0/n_*)a_0 + (n/n_*)\hat{F}[g(s)]$. The coherence of this example is troubled by the following problem: if $a \sim Di(n_0 a_0)$ with $a_0 = p\varepsilon_0 + (1-p)\varepsilon_1$, then $a = b\varepsilon_0 + (1-b)\varepsilon_1$ a.s. where $b \sim B(n_0 p, n_0(1-p))$, i.e. b has a beta distribution. Therefore, in the joint probability, this implies that almost surely x_i is equal to 0 or 1, which may contradict the observations. More precisely, conditionally on a, $x_i = 0$ or 1 with probability b and $1 - b$. This implies that $x_i = 0$ or 1 with probability p, $1 - p$. Another version of the posterior distribution, i.e., conditionally on s, is therefore given by $a = b_*\varepsilon_0 + (1 - b_*)\varepsilon_1$ a.s. where $b_* \sim B[n_*t, n_*(1 - t)]$ and

$$t = \frac{n_0}{n_*}p + \frac{n}{n_*}u(s) \quad \text{where } u(s) = \frac{1}{n}\sum_{1 \leq i \leq n} 1_{\{x_i = 0\}}.$$

8.3.2   Invariance and Reduction

As seen in Section 8.2.3, invariance properties imply conditional independence relations. These can be used in invariant Bayesian experiments to obtain admissible reductions. We first introduce the following notation. For $\mathcal{M} \subset \mathcal{A} \vee \mathcal{S}$

$$(8.3.10) \qquad\qquad \mathcal{M}_\Phi = (\mathcal{A} \vee \mathcal{S})_\Phi \cap \mathcal{M}$$

is the Φ-invariant sub-σ-field of $\mathcal{M}$ and

$$(8.3.11) \qquad\qquad \mathcal{M}_\Phi^* = (\mathcal{A} \vee \mathcal{S})_\Phi^* \cap \mathcal{M}$$

is the almost surely Φ-invariant sub-σ-field of $\mathcal{M}$.

In particular, we define:

8.3.5 Definition. Let $\mathcal{E} = \{A \times S, \mathcal{A} \vee \mathcal{S}, \Pi\}$ be an unreduced Bayesian experiment and Φ a monoid of measurable transformations on $A \times S$ such that $\mathcal{A}$ is Φ-stable and $\mathcal{S}$ is Φ-stable. Then

(8.3.12) $\mathcal{A}_\phi$ *is the maximal invariant parameter,*

(8.3.13) $\mathcal{A}_\phi^*$ *is the almost surely maximal invariant parameter,*

(8.3.14) $\mathcal{S}_\phi$ *is the maximal invariant statistic, and*

(8.3.15) $\mathcal{S}_\phi^*$ *is the almost surely maximal invariant statistic.*   ∎

It is often useful to make the following assumptions on Φ:

8.3.6 Definition. Φ is regular for $\mathcal{E}$ if

(8.3.16) $\Phi \perp\!\!\!\perp \mathcal{A} \cap \overline{\mathcal{I}}$ and $\mathcal{A}_\Phi^* = \overline{\mathcal{A}_\Phi} \cap \mathcal{A}$

(8.3.17) $\Phi \perp\!\!\!\perp \overline{\mathcal{I}} \cap \mathcal{S}$ and $\mathcal{S}_\Phi^* = \overline{\mathcal{S}_\Phi} \cap \mathcal{S}.$ ∎

This means that the almost surely maximal invariant parameter (respectively, statistic) is the maximal invariant parameter (respectively, statistic) completed by the null sets of $\mathcal{A}$ (respectively, $\mathcal{S}$) i.e., the trivial parameter (respectively, statistic); we have already seen, in Sections 8.2.3 and 8.2.6, several situations in which Φ is regular for $\mathcal{E}$.

We now turn to the main result of this section, the proof of which follows directly from Theorem 8.2.29:

8.3.7 Proposition. If $\mathcal{E}$ is Φ-invariant, the almost surely maximal invariant parameter and statistic are mutually sufficient, i.e.,

(8.3.18) $\mathcal{A}_\Phi^* \quad \perp\!\!\!\perp \quad \mathcal{S} \mid \mathcal{S}_\Phi^*$

(8.3.19) $\mathcal{A} \quad \perp\!\!\!\perp \quad \mathcal{S}_\Phi^* \mid \mathcal{A}_\Phi^*.$

Moreover, if Φ is regular for $\mathcal{E}$, the maximal invariant parameter and statistic are mutually sifficient, i.e.,

(8.3.20) $\mathcal{A}_\Phi \quad \perp\!\!\!\perp \quad \mathcal{S} \mid \mathcal{S}_\Phi$

(8.3.21) $\mathcal{A} \quad \perp\!\!\!\perp \quad \mathcal{S}_\Phi \mid \mathcal{A}_\Phi.$ ∎

As a corollary of this proposition, let us consider the case where $\mathcal{E}$ is only sampling Φ-invariant.

8.3.8 Corollary. If Φ is a group, and if $\mathcal{E}$ is sampling Φ-invariant, then the almost surely maximal Φ-invariant parameter is sufficient in the experiment marginalized on both the parameter and the almost surely maximal Φ-invariant statistic, i.e.,

$$(8.3.22) \qquad\qquad \mathcal{A} \perp\!\!\!\perp S_\Phi^* \mid \mathcal{A}_\Phi^*. \qquad\qquad\qquad \blacksquare$$

Since $\Phi \mathrm{I} \mathcal{S} \mid \mathcal{A}$, this corollary is a direct consequence of Corollary 8.2.27.

Barnard (1963) defined the principle of Φ-sufficiency based on Property (8.3.22). This principle says that, since the distribution of S_Φ^* depends only on $\mathcal{A}_\Phi^*$, then in the absence of prior information on $\mathcal{A}_\Phi^*$, S_Φ^* contains all the available information on $\mathcal{A}_\Phi^*$. In the Bayesian framework this means that S_Φ^* is sufficient for $\mathcal{A}_\Phi^*$, i.e., $\mathcal{A}_\Phi^* \perp\!\!\!\perp \mathcal{S} \mid S_\Phi^*$. Therefore, in view of Theorem 8.3.7, the most natural assumption implying these properties is that the prior probability is such that $\Phi \mathrm{I} \mathcal{A}$; consequently, this invariance assumption seems to be the most appropriate way to represent absence of prior information.

This approach can be used, in particular, if $\mathcal{B} \subset \mathcal{A}$ represents a parameter of interest. Then one may look for a family $\overline{G}$ of transformations on $\mathcal{A}$ such that $\mathcal{B} = \mathcal{A}_\Phi$. One then looks for a family G of transformations on S such that $\Phi \mathrm{I} \mathcal{S} \mid \mathcal{A}$. With an invariant prior probability, i.e., $\Phi \mathrm{I} \mathcal{A}$, this ensures that $\Phi \mathrm{I} (\mathcal{A} \vee \mathcal{S})$ and, therefore, $\mathcal{B}$ and S_Φ^* are mutually sufficient; in this context, as shown in Section 3.4.1, inference on $\mathcal{B}$ enjoys both good robustness properties and is easily computed (see Section 3.4.1), even when no cut is available.

It is interesting to consider special cases of this theorem for particular structures of Φ. Recall that $\varphi(a, s) = (\overline{g}(a), g(s))$. When $\overline{g}(a) = a \quad \forall g \in G$ (as, e.g., in stationary process with invariance with respect to the shift-operator), then $\mathcal{A} \subset \mathcal{A}_\Phi$ and, therefore, $\mathcal{A}_\Phi^* = \mathcal{A}_\Phi = \mathcal{A}$. In this case, (8.3.19) becomes trivial and (8.3.18) says that S_Φ^* is a sufficient statistic in $\mathcal{E}$. The opposite situation is when $\mathcal{A}_\Phi^* = \mathcal{A} \cap \overline{\mathcal{I}}$, i.e., the trivial parameter. Then 8.3.18 becomes trivial and (8.3.19) says that S_Φ^* is an ancillary statistic in $\mathcal{E}$. Therefore, invariance may be seen as a tool for obtaining ancillary and sufficient statistics in a Bayesian experiment.

8.3.9 Corollary.

(i) If $\mathcal{E}$ is sampling Φ-invariant and $\mathcal{A}_\Phi^* = \mathcal{A}$, then $\mathcal{A} \perp\!\!\!\perp \mathcal{S} \mid \mathcal{S}_\Phi^*$.

(ii) If Φ is a group, $\mathcal{E}$ is sampling Φ-invariant and $\mathcal{A}_\Phi^* = \mathcal{A} \cap \overline{\mathcal{I}}$, then $\mathcal{A} \perp\!\!\!\perp \mathcal{S}_\Phi^*$.

(iii) If $\mathcal{E}$ is Φ-invariant and $\mathcal{A}_\Phi^* = \mathcal{A} \cap \overline{\mathcal{I}}$, then $\mathcal{A} \perp\!\!\!\perp \mathcal{S}_\Phi^*$.   ∎

The proof of (i) follows directly from Theorem 8.2.28; (ii) is a consequence of Corollary 8.3.8, and (iii) of Proposition 8.3.7.

Theorem 8.3.7 and Corollary 8.3.8 are also useful in searching for pivotal functions, i.e., random variables functions of both the parameter and the observation whose distribution does not depend on the parameter.

8.3.10 Definition. A sub-σ-field $\mathcal{M} \subset \mathcal{A} \vee \mathcal{S}$ is called a *pivotal σ-field* if $\mathcal{M} \perp\!\!\!\perp \mathcal{A}$.   ∎

The usual procedure for finding a pivotal σ-field is to look for a group of transformation $\overline{G}$ on A such that $\mathcal{A}_\Phi$ is trivial. One then determines G, a group of transformations on S such that $\Phi \, \mathrm{I} \, \mathcal{S} \mid \mathcal{A}$. Finally, one need merely exhibit a random variable (i.e., function of both the parameter and the observation) that is Φ-invariant. More precisely:

8.3.11 Corollary.

(i) If Φ is a group and if $\mathcal{E}$ is sampling Φ-invariant, then
$(\mathcal{A} \vee \mathcal{S})_\Phi^* \perp\!\!\!\perp \mathcal{A} \mid \mathcal{A}_\Phi^*$.
Therefore, if $\mathcal{A}_\Phi^* = \mathcal{A} \cap \overline{\mathcal{I}}$, then $(\mathcal{A} \vee \mathcal{S})_\Phi^*$ is a pivotal σ-field.

(ii) If $\mathcal{E}$ is Φ-invariant, then
$(\mathcal{A} \vee \mathcal{S})_\Phi^* \perp\!\!\!\perp \mathcal{A} \mid \mathcal{A}_\Phi^*, \quad (\mathcal{A} \vee \mathcal{S})_\Phi^* \perp\!\!\!\perp \mathcal{S} \mid \mathcal{S}_\Phi^*$.

Proof.

(i) $\Phi \, \mathrm{I} \, \mathcal{S} \mid \mathcal{A} \Rightarrow \Phi \, \mathrm{I} \, (\mathcal{A} \vee \mathcal{S}) \mid \mathcal{A}$, by Theorem 8.2.23(ii) and the result follows from Corollary 8.2.27.

(ii) follows directly from Theorem 8.2.29.   ∎

In applications it is often the case that $\mathcal{A}_\Phi$ is trivial; it is rarely the case that $\mathcal{S}_\Phi$ is trivial. However, if the experiment is reduced to a sufficient statistic, this may well be the case. This situation is of some interest in that, when a pivotal function is independent of the observation, the posterior distribution is the same as the joint and the sampling distributions.

We now turn to the problem of reducing an invariant experiment on a sufficient statistic. Similarly, a reduction may be made on a sufficient parameter. Recall that, by Theorem 8.2.25, $\Phi \text{ I } \mathcal{S} \mid \mathcal{A}$ is equivalent to $\Phi \text{ I } \mathcal{T} \mid \mathcal{A}$ and that $\Phi \text{ I } \mathcal{S} \mid \mathcal{A} \vee \mathcal{T}$, i.e., $\mathcal{E}$ is sampling Φ-invariant if and only if $\mathcal{E}_{\mathcal{A}\vee\mathcal{T}}$ and $\mathcal{E}^{\mathcal{T}}_{\mathcal{A}\vee\mathcal{S}}$ are sampling Φ-invariant.

The next results are Bayesian versions of the main results of Hall, Wijman and Ghosh (1965) (Theorems 3.1 and 3.2) (see also Petit (1970)):

8.3.12 Theorem. Let Φ be a group and let $\mathcal{E}$ be sampling Φ-invariant. If $\mathcal{T} \subset \mathcal{S}$ is a Φ-stable $\mathcal{E}$-sufficient statistic, then $\mathcal{T}^*_\Phi$ is $\mathcal{E}_{\mathcal{A}\vee\mathcal{S}^*_\Phi}$-sufficient and $\mathcal{S}^*_\Phi$ and $\mathcal{T}$ are sampling independent conditionally on $\mathcal{T}^*_\Phi$, i.e.,

$$\mathcal{A} \perp\!\!\!\perp \mathcal{S}^*_\Phi \mid \mathcal{T}^*_\Phi \quad \text{and} \quad \mathcal{T} \perp\!\!\!\perp \mathcal{S}^*_\Phi \mid \mathcal{A} \vee \mathcal{T}^*_\Phi.$$

Proof. Note that by Theorem 2.2.10, $\mathcal{A} \perp\!\!\!\perp \mathcal{S}^*_\Phi \mid \mathcal{T}^*_\Phi$ and $\mathcal{T} \perp\!\!\!\perp \mathcal{S}^*_\Phi \mid \mathcal{A} \vee \mathcal{T}^*_\Phi$ is equivalent not only to $(\mathcal{A} \vee \mathcal{T}) \perp\!\!\!\perp \mathcal{S}^*_\Phi \mid \mathcal{T}^*_\Phi$ but also to $\mathcal{A} \perp\!\!\!\perp \mathcal{S}^*_\Phi \mid \mathcal{T}$ and to $\mathcal{T} \perp\!\!\!\perp \mathcal{S}^*_\Phi \mid \mathcal{T}^*_\Phi$. Now, $\mathcal{T}$ $\mathcal{E}$-sufficient, i.e., $\mathcal{A} \perp\!\!\!\perp \mathcal{S} \mid \mathcal{T}$ clearly implies that $\mathcal{A} \perp\!\!\!\perp \mathcal{S}^*_\Phi \mid \mathcal{T}$. Since $\Phi \text{ I } \mathcal{S} \mid \mathcal{A}$ implies that $\Phi \text{ I } \mathcal{S} \mid \mathcal{A} \vee \mathcal{T}$, and since $\mathcal{A} \perp\!\!\!\perp \mathcal{S} \mid \mathcal{T}$, we also obtain, by Theorem 8.2.30, that $\Phi \text{ I } \mathcal{S} \mid \mathcal{T}$. Therefore, by Corollary 8.2.27, $\mathcal{S}^*_\Phi \perp\!\!\!\perp \mathcal{T} \mid \mathcal{T}^*_\Phi$. ∎

The same result holds when the assumption that Φ is a group is replaced by the assumption that $\Phi \text{ I } \mathcal{A}$.

8.3.13 Theorem. Let $\mathcal{E}$ be a Φ-invariant experiment. If $\mathcal{T} \subset \mathcal{S}$ is a Φ-stable $\mathcal{E}$-sufficient statistic then $(\mathcal{A} \vee \mathcal{T}) \perp\!\!\!\perp \mathcal{S}^*_\Phi \mid \mathcal{T}^*_\Phi$.

Proof. The proof is the same as above, since $\Phi \text{ I } (\mathcal{A}\vee\mathcal{S})$ implies that $\Phi \text{ I } \mathcal{S}$. Therefore $\mathcal{S}^*_\Phi \perp\!\!\!\perp \mathcal{T} \mid \mathcal{T}^*_\Phi$ by Theorem 8.2.29. ∎

The next result shows that, when $\mathcal{E}$ is sampling Φ-invariant and

$\mathcal{A}_\Phi^* = \mathcal{A}$, a Φ-stable $\mathcal{E}$-sufficient statistic $\mathcal{T}$ may be reduced to its almost surely Φ-invariant sub-σ-field.

8.3.14 Theorem. Let $\mathcal{E}$ be sampling Φ-invariant and $\mathcal{A}_\Phi^* = \mathcal{A}$. If $\mathcal{T}$ is a Φ-stable $\mathcal{E}$-sufficient statistic, then $\mathcal{T}_\Phi^*$ is also $\mathcal{E}$-sufficient. In particular, the minimal $\mathcal{E}$-sufficient statistic $\mathcal{SA}$ is included in $\mathcal{S}_\Phi^*$. Therefore $(\mathcal{SA})_\Phi^* = \mathcal{SA}$.

Proof. Since $\Phi \mathbin{I} \mathcal{S} \mid \mathcal{A}$ implies that $\Phi \mathbin{I} \mathcal{T} \mid \mathcal{A}$, $\mathcal{A} \perp\!\!\!\perp \mathcal{T} \mid \mathcal{T}_\Phi^*$ by Corollary 8.3.9(i). This, along with $\mathcal{T}$ $\mathcal{E}$-sufficient, i.e., $\mathcal{A} \perp\!\!\!\perp \mathcal{S} \mid \mathcal{T}$, is equivalent to $\mathcal{A} \perp\!\!\!\perp \mathcal{S} \mid \mathcal{T}_\Phi^*$ by Theorem 2.2.10. By Corollary 8.3.9(i), $\mathcal{A} \perp\!\!\!\perp \mathcal{S} \mid \mathcal{S}_\Phi^*$. Therefore, by Theorem 4.3.3, $\mathcal{SA} \subset \overline{\mathcal{S}_\Phi^*} \cap \mathcal{S} = \mathcal{S}_\Phi^*$ since $\overline{\mathcal{I}} \cap \mathcal{S} \subset \mathcal{S}_\Phi^*$. ∎

In this kind of reduction, there is also a connection between pivotal σ-fields.

8.3.15 Corollary. Let Φ be a group and let $\mathcal{E}$ be a sampling Φ-invariant experiment. Then $(\mathcal{A} \vee \mathcal{S})_\Phi^* \perp\!\!\!\perp (\mathcal{A} \vee \mathcal{T}) \mid (\mathcal{A} \vee \mathcal{T})_\Phi^*$ for any Φ-stable statistic $\mathcal{T} \subset \mathcal{S}$.

Proof. $\Phi \mathbin{I} \mathcal{S} \mid \mathcal{A}$ is, as before, equivalent to $\Phi \mathbin{I} \mathcal{T} \mid \mathcal{A}$ and $\Phi \mathbin{I} \mathcal{S} \mid \mathcal{A} \vee \mathcal{T}$. By Theorem 8.2.23(ii) and (i), $\Phi \mathbin{I} (\mathcal{A} \vee \mathcal{S}) \mid \mathcal{A} \vee \mathcal{T}$ and the result follows by Corollary 8.2.27.

We now extend the result of Corollary 8.3.11.

8.3.16 Corollary. Let $\mathcal{E}$ be a Φ-invariant experiment. If $\mathcal{T} \subset \mathcal{S}$ is a Φ-stable $\mathcal{E}$-sufficient statistic,
then

(i) $(\mathcal{A} \vee \mathcal{T})_\Phi^* \perp\!\!\!\perp \mathcal{A} \mid \mathcal{A}_\Phi^*$

(ii) $(\mathcal{A} \vee \mathcal{T})_\Phi^* \perp\!\!\!\perp \mathcal{S} \mid \mathcal{T}_\Phi^*$.

Proof. Since $\Phi \mathbin{I} (\mathcal{A} \vee \mathcal{S})$ implies $\Phi \mathbin{I} (\mathcal{A} \vee \mathcal{T})$, by Corollary (8.3.11)(ii), we have (i) and $(\mathcal{A} \vee \mathcal{T})_\Phi^* \perp\!\!\!\perp \mathcal{T} \mid \mathcal{T}_\Phi^*$. Now, $\mathcal{A} \perp\!\!\!\perp \mathcal{S} \mid \mathcal{T}$ implies $(\mathcal{A} \vee \mathcal{T}) \perp\!\!\!\perp \mathcal{S} \mid \mathcal{T}$ by Corollary 2.2.11. Therefore $(\mathcal{A} \vee \mathcal{T})_\Phi^* \perp\!\!\!\perp \mathcal{S} \mid \mathcal{T}$. But this, along with $(\mathcal{A} \vee \mathcal{T})_\Phi^* \perp\!\!\!\perp \mathcal{T} \mid \mathcal{T}_\Phi^*$, is equivalent to $(\mathcal{A} \vee \mathcal{T})_\Phi^* \perp\!\!\!\perp \mathcal{S} \mid \mathcal{T}_\Phi^*$ by Theorem 2.2.10. ∎

Theorem 8.3.12 calls for two remarks. Firstly, if Φ is regular for $\mathcal{E}$, then $\mathcal{S}_\Phi^*$ and $\mathcal{T}_\Phi^*$ may be replaced by $\mathcal{S}_\Phi$; $\mathcal{T}_\Phi$ and this is precisely the results of Hall, Wijsman and Ghosh up to the null sets of $\mathcal{A}$ for the prior probability. Secondly, the only place where the group structure of Φ is used is in the application of Corollary 8.2.27. If Φ is a monoïd, the result will be obtained by Theorem 8.2.26 under the further assumption that $\mathcal{S}_\Phi^* \perp\!\!\!\perp \mathcal{T} \mid \varphi^{-1}(\mathcal{T})$, and this is implied by $\varphi^{-1}(\mathcal{S}) \perp\!\!\!\perp \mathcal{T} \mid \varphi^{-1}(\mathcal{T})$, since $\mathcal{S}_\Phi^* \subset \overline{\varphi^{-1}(\mathcal{S})}$.

We conclude this section by looking for invariant σ-fields in the examples presented in Section 8.3.1.

Example 6. In this example, Φ is regular for $\mathcal{E}$ because the uniform distribution on $[0, 2\pi[$ is invariant under Φ. $\mathcal{A}_\Phi = \mathcal{I}$ and $\mathcal{S}_\Phi$ is somewhat difficult to describe. Note that $\overline{x} = (\overline{y}, \overline{z})'$ is sufficient for a. If $\mathcal{T} = \sigma(\overline{x})$, then $\mathcal{A} \perp\!\!\!\perp \mathcal{S} \mid \mathcal{T}$. According to Corollary 8.3.8, for the invariant prior probability $\mathcal{A} \perp\!\!\!\perp \mathcal{S}_\Phi$, i.e., $\mathcal{S}_\Phi$ is ancillary. Now, $\mathcal{T}_\Phi = \sigma\{\overline{x}'\overline{x}\} = \sigma\{t^2\}$, and therefore, by Theorem 8.3.12, $\mathcal{A} \perp\!\!\!\perp \mathcal{S}_\Phi \mid \mathcal{T}_\Phi$ and $\mathcal{S}_\Phi \perp\!\!\!\perp \mathcal{T} \mid \mathcal{A} \vee \mathcal{T}_\Phi$. These two relationships do not depend on the prior probability.   ∎

Example 7. In this example, Φ is regular for $\mathcal{E}$. $\mathcal{A}_\Phi = \mathcal{I}$, by Corollary 8.3.8. In fact, $\mathcal{S}_\Phi = \sigma\{s_i - \overline{s} : 1 \leq i \leq n\}$ where $\overline{s} = \frac{1}{n}\sum_{1 \leq i \leq n} s_i$, since $u = (s_1 - \overline{s}, \ldots, s_n - \overline{s})'$ is maximal invariant and $(u \mid a) \sim N_n(0, I - \frac{1}{n}ee')$ where $e = (1, 1, \ldots, 1)'$. Clearly, $\mathcal{T} = \sigma\{\overline{s}\}$ is sufficient for a, i.e., $\mathcal{A} \perp\!\!\!\perp \mathcal{S} \mid \mathcal{T}$, and $\mathcal{T}_\Phi = \mathcal{I}$. Therefore, by Theorem 8.3.12, $\mathcal{A} \perp\!\!\!\perp \mathcal{S}_\Phi$ and $\mathcal{S}_\Phi \perp\!\!\!\perp \mathcal{T} \mid \mathcal{A}$ for any prior probability, i.e., $u \perp\!\!\!\perp a$ and $u \perp\!\!\!\perp \overline{s} \mid a$.   ∎

Example 8. In this example, let us choose $z_i = e = (1, 1, \ldots, 1)'$ and set $\eta = B'e$. Then, if $a = (\eta, \Sigma)$, $(y_i \mid a) \sim i.N_p(\eta, \Sigma)$, $g_{M,\nu}(y_i) = My_i + \nu$ where $\nu = N'e$, M is a nonsingular $p \times p$-matrix, and $\overline{g}_{M,\nu}(\eta, \Sigma) = (M\eta + \nu, M\Sigma M')$. Let us further simplify this example by taking $M = I$. Then Φ is the translation group on $\mathbb{R}^p$. Let us define $\overline{y} = \frac{1}{n}\sum_{1 \leq i \leq n} y_i$, $u = (y_1 - \overline{y}, \ldots, y_n - \overline{y})'$ and $v = \frac{1}{n}\sum_{1 \leq i \leq n}(y_i - \overline{y})(y_i - \overline{y})' = \frac{1}{n}u'u$. Now $\mathcal{S}_\Phi = \sigma\{u\}$ and $\mathcal{A}_\Phi = \sigma\{\Sigma\}$. For any prior probability which is the product of the Lebesgue measure on

$I\!\!R^p$ with any probability on C_p, by Proposition 8.3.7 and since Φ is regular for $\mathcal{E}$, $\mathcal{A}_\Phi \perp\!\!\!\perp \mathcal{S} \mid \mathcal{S}_\Phi$ and $\mathcal{A} \perp\!\!\!\perp \mathcal{S}_\Phi \mid \mathcal{A}_\Phi$, i.e., $\Sigma \perp\!\!\!\perp \overline{y} \mid u$ and $\mu \perp\!\!\!\perp u \mid \Sigma$. Now $\mathcal{T} = \sigma\{\overline{y}, v\}$ is the minimal sufficient statistic for (η, Σ). $\mathcal{T}_\Phi = \sigma(v)$ and, by Theorem 8.3.12, for any prior probability $\mathcal{A} \perp\!\!\!\perp \mathcal{S}_\Phi \mid \mathcal{T}_\Phi$ and $\mathcal{S}_\Phi \perp\!\!\!\perp \mathcal{T} \mid \mathcal{A} \vee \mathcal{T}_\Phi$, i.e., $(\eta, \Sigma) \perp\!\!\!\perp u \mid v$ and $u \perp\!\!\!\perp \overline{y} \mid \eta, \Sigma, v$. Let us examine another special case of this example. Let us take $\Sigma = \sigma^2 I$, $M = Q$, $\nu = 0$, where $Q'Q = I$. Now $\overline{g}_Q(\eta, \sigma^2) = (Q\eta, \sigma^2)$. Therefore $\mathcal{A}_\Phi = (\eta'\eta, \sigma^2)$ for $\eta'\eta = \xi'\xi$ if and only if there exists Q orthogonal such that $\eta = Q\xi$. $\mathcal{S}_\Phi$ is difficult to describe. If $w = t(v) = \frac{1}{n}\sum_{1 \le i \le n}(y_i - \overline{y})'(y_i - \overline{y})$, then clearly $\mathcal{T} = \sigma\{\overline{y}, w\}$ is minimal sufficient for a proper prior probability and $\mathcal{T}_\Phi = \sigma\{\overline{y}'\overline{y}, w\}$. By Theorem 8.3.12, $\mathcal{A} \perp\!\!\!\perp \mathcal{S}_\Phi \mid \mathcal{T}_\Phi$ and $\mathcal{S}_\Phi \perp\!\!\!\perp \mathcal{T} \mid \mathcal{A} \vee \mathcal{T}_\Phi$, i.e., $\eta \perp\!\!\!\perp \mathcal{S}_\Phi \mid \overline{y}'\overline{y}, w$ and $\mathcal{S}_\Phi \perp\!\!\!\perp \overline{y} \mid \eta, \sigma^2, \overline{y}'\overline{y}, w$. With an invariant prior, $\mathcal{A}_\Phi \perp\!\!\!\perp \mathcal{T} \mid \mathcal{T}_\Phi$ and $\mathcal{A} \perp\!\!\!\perp \mathcal{T}_\Phi \mid \mathcal{A}_\Phi$, by Proposition 8.3.7, i.e., $(\eta'\eta, \sigma^2) \perp\!\!\!\perp \overline{y} \mid \overline{y}'\overline{y}, w$ and $\eta \perp\!\!\!\perp (\overline{y}'\overline{y}, w) \mid \eta'\eta, \sigma^2$. ∎

Example 9. In this example, Φ is a group isomorphic to A and μ is a prior probability such that $\mu \circ \overline{g}_\alpha^{-1} \ll \mu$ $\forall \alpha \in A$. Let us introduce the "inverse" of a measure on a group A, by $\mu^{-1}(B) = \mu(B^{-1})$. Then for any $f \in [\mathcal{A}]^+$, $\int_A f(\alpha)\mu^{-1}(d\alpha) = \int_A f(\alpha^{-1})\mu(d\alpha)$. Moreover, if $h_\alpha : A \to A$ is defined as $h_\alpha(\alpha') = \alpha'\alpha$, then

$$\int_A f(\alpha)(\mu^{-1} \circ h_a^{-1})(d\alpha) = \int_A f(\alpha a)\mu^{-1}(d\alpha)$$

$$= \int_A f[(a^{-1}\alpha)^{-1}]\mu(d\alpha)$$

$$= \int_A f(\alpha^{-1})\{\mu \circ (\overline{g}_{a^{-1}})^{-1}\}(d\alpha)$$

Therefore, if $\mu \circ \overline{g}_\alpha^{-1} \sim \mu$ $\forall \alpha \in A$, then $\mu^{-1} \circ h_\alpha^{-1} \sim \mu^{-1}$ $\forall \alpha \in A$. Therefore, using Theorem 8.2.38, for any $\mathcal{M} \subset \mathcal{A} \vee \mathcal{S}$ such that, $\forall M \in \mathcal{M}$ the inverse image of M under the map

$$(\alpha, \alpha', \sigma) \in A \times (A \times S) \to (\alpha\alpha', \alpha \star \sigma) \in A \times S$$

belongs to $\mathcal{A} \otimes \mathcal{M}$, $\mathcal{M}_\phi^* = \overline{\mathcal{M}_\phi} \cap \mathcal{M}$. One is thus interested only in Φ-invariant σ-fields. Observe that $\mathcal{A}_\Phi = \mathcal{I}$ because $\overline{g}_\alpha(a) = \alpha a$ and A is group. There is no interesting characterization of $\mathcal{S}_\Phi$ but, by Corollary 8.3.8, $\mathcal{S}_\Phi$ is ancillary, i.e., $\mathcal{A} \perp\!\!\!\perp \mathcal{S}_\Phi$. Now if $m \in [(\mathcal{A} \vee \mathcal{S})_\Phi]$, then

$m(\alpha a, \alpha \star s) = m(a, s)$ $\forall\,\alpha, a \in A$ and $\forall\,s \in S$. Taking $\alpha = a^{-1}$ shows that $m(a, s) = m(e, a^{-1} \star s)$. Now, $u = a^{-1} \star s : A \times S \to S$ is maximal invariant. Indeed, if $a^{-1} \star s = \alpha^{-1} \star \sigma$, then $\sigma = (\alpha a^{-1}) \star s$,; i.e., $\sigma = \alpha' \star s$ and $\alpha = \alpha' a$, and therefore $(\alpha, \sigma) = \varphi_{\alpha'}(a, s)$. By Corollary 8.3.11, $u \perp\!\!\!\perp a$, i.e., u is a pivotal function. By direct computation, $\forall\,f$ positive Borel functions on S, $\mathcal{I}[f(u)] = \int_S f(\sigma)Q(d\sigma)$. Hence the joint distribution and the sampling distribution of u is given by Q. If μ is such that Φ I $\mathcal{A}$ then, by Corollary 8.3.11, we also know that $u \perp\!\!\!\perp s \mid \mathcal{S}_\Phi$. A most interesting situation is when there exists a Φ-stable sufficient statistic with values in the group A, i.e., in the parameter space. More precisely, let $t : A \times S \to A$, $\mathcal{T} = t^{-1}(\mathcal{A}) \subset \mathcal{S}$, such that t is sufficient, i.e., $a \perp\!\!\!\perp s \mid t$. We suppose that there exists $\beta : A \to A$, with $\mathcal{B} = \beta^{-1}(\mathcal{A}) \subset \mathcal{A}$ such that $t(\alpha \star s) = \beta(\alpha)t(s)$ $\forall\,\alpha \in A$, $\forall\,s \in S$. Since $t \circ \varphi_\alpha = t(\alpha \star s)$, this ensures that $\mathcal{T}$ is Φ-stable. However, it is evident that $\beta(\alpha_1\alpha_2) = \beta(\alpha_1)\beta(\alpha_2)$ and β is an endomorphism on A. We first remark that $\mathcal{B}$ is sufficient for $\mathcal{T}$ and then for $\mathcal{S}$, since $\mathcal{T}$ is sufficient. Indeed,

$$
\begin{aligned}
\mathcal{A}[f(t)] &= \int_S f[t(a \star \sigma)]Q(d\sigma) \\
&= \int_A f[\beta(a)\tau]\tilde{Q}(d\tau)
\end{aligned}
$$

where $\tilde{Q} = Q \circ t^{-1}$. Therefore $\mathcal{A}[f(t)]$ is $\mathcal{B}$-measurable. If we suppose that $\mathcal{E}$ is identified: $\mathcal{A} = \mathcal{B}$, and if the points of A are measurable, then β is bijective. By replacing t by $\beta^{-1} \circ t$, it is readily seen that we may suppose, without loss of generality, that $t(\alpha \star s) = \alpha t(s)$. Therefore $\mathcal{T}_\Phi = \mathcal{I}$, and if $f[\alpha a, t(\alpha \star s)] = f[a, t(s)]$ $\forall\,\alpha \in A$, then $f(a, t(s)) = f(e, a^{-1}t(s))$. This shows, by Corollary 8.2.5, that $v = a^{-1}t$ is maximal invariant, i.e., $(\mathcal{A}\vee\mathcal{T})_\Phi = v^{-1}(\mathcal{A})$. By Corollary 8.3.11(i), $v \perp\!\!\!\perp a$; this may be seen directly as follows:

$$
\begin{aligned}
\mathcal{I}[g(a)h(v)] &= \int_{A \times S} g(\alpha)h[\alpha^{-1}t(\alpha \star \sigma)]\mu(d\alpha)Q(d\sigma) \\
&= \int_A g(\alpha)\mu(d\alpha)\int_A h(\tau)\tilde{Q}(d\tau).
\end{aligned}
$$

Hence $a \perp\!\!\!\perp v$ and $\tilde{Q}$ is the distribution of v. If the prior probability μ is invariant then, by Corollary 8.3.15, $v \perp\!\!\!\perp s$ and, therefore, $v \perp\!\!\!\perp t$. This is not readily seen since

$$
\mathcal{I}[f(t)h(v)] = \mathcal{I}[f(av)h(v)]
$$

$$= \int_{A \times A} f(\alpha\tau) h(\tau) \mu(d\alpha) \tilde{Q}(d\tau)$$

$$= \int_A h(\tau) \left\{ \int_A f(\alpha\tau) \mu(d\alpha) \right\} \tilde{Q}(d\tau).$$

It will nonetheless be verified if we can show that

$$\int_A f(\alpha\tau) \mu(d\alpha) = \int_A f(\chi) \tilde{P}(d\chi),$$

where $\tilde{P}$ is the predictive distribution of t. In fact, if μ is a left-invariant probability on A this is also right-invariant and the above relation is then true with $\tilde{P} = \mu$. Indeed,

$$\int_A f[\alpha^{-1}\tau\beta] \mu(d\alpha) = \int_A f(\alpha^{-1}) \mu(d\alpha)$$

since μ is left-invariant. Therefore, $\forall \beta \in A$,

$$\int_A f(\alpha^{-1}) \mu(d\alpha) = \int_A \mu(d\tau) \int_A f[\alpha^{-1}\tau\beta] \mu(d\alpha)$$

$$= \int_A \mu(d\alpha) \int_A f[\alpha^{-1}\tau\beta] \mu(d\tau)$$

$$= \int_A \mu(d\alpha) \int_A f[\tau\beta] \mu(d\tau)$$

$$= \int_A f(\tau\beta) \mu(d\tau).$$

Hence,

$$\int_A f(\tau\beta) \mu(d\tau) = \int_A f(\tau) \mu(d\tau) = \int_A f(\alpha^{-1}) \mu(d\alpha),$$

i.e., μ is right invariant and $\mu^{-1} = \mu$.

It is now easy to compute the posterior distribution of the parameter. Indeed,

$$S[g(a)] = T[g(a)] = T[g[tv^{-1}]]$$

$$= \int_A g(t\tau^{-1}) \tilde{Q}(d\tau)$$

$$= \int_A g(t\tau) \tilde{Q}^{-1}(d\tau)$$

$$= \int_A g(\alpha)(\tilde{Q}^{-1} \circ \overline{g}_t^{-1})(d\alpha).$$

This shows that $\tilde{Q}^{-1} \circ \overline{g}_t^{-1}$ is the posterior distribution of a. ∎

Remark. The above example illustrates an important point of view in statistics which is midway between sampling and Bayesian analysis. Formally, the model is Bayesian, since there is a prior probability on the parameters. However, this prior probability is in fact straightforwardly derived from the sampling process, due to an invariance property with respect to a group naturally acting on this process. This point of view bears several names in the literature. It is sometimes called "fiducial" or "'structural" (Fraser (1968, 1972)) or "inner inference" (Villegas (1977(a), 1981)). In this approach, the most important technical problem is normally the unboundedness of the invariant measure, and is the source of paradoxes (Stone (1976) and Dawid, Stone and Zidek (1973)). Unboundedness was avoided above either by considering only a sampling invariance property (where only null sets of the prior probability are taken into account) or by assuming the existence of an invariant prior probability. In this latter case the probability is both left- and right-invariant, which greatly simplified the proofs. The extension to the unbounded case requires further conditions , and these are not examined in this book. Let us remark however, that the main results presented in this chapter have been established in a more general setting than that of Pitman's experiment, as analyzed in the above example, by relaxing the group assumption, in particular. This approach satisfactorily relates invariance and the reduction of an experiment.

8.3.3   Invariance and Exact Estimability

We now turn to the analysis of the role of Φ-invariance in the study of exact estimability. Recall that the maximal exactly estimable parameter is $\mathcal{A} \cap \overline{\mathcal{S}}$, which may be difficult to characterize. It is therefore interesting to identify sub-σ-fields of $\mathcal{A} \cap \overline{\mathcal{S}}$. Ergodicity and Φ-invariance give rise to the following kind of results:

8.3.17 Theorem. If $\mathcal{E}$ is Φ-invariant and Φ is ergodic on $\mathcal{S}$ conditionally on $\mathcal{A}$, then $\mathcal{A}_\Phi^* \mathcal{S}$ is exactly estimable.

Proof. By Theorem 8.3.7, $\mathcal{A}_\Phi^* \perp\!\!\!\perp \mathcal{S} \mid \mathcal{S}_\Phi^*$, i.e., $\mathcal{S}_\Phi^*$ is an $\mathcal{E}_{\mathcal{A}_\Phi^* \vee \mathcal{S}}$- sufficient statistic. By Lemma 8.2.26, $\mathcal{S}_\phi^* \subset \overline{\mathcal{A}_\Phi^*}$ and hence the result follows by Theorem 4.7.8, with $\mathcal{M} = \mathcal{I}$, $\mathcal{T} = \mathcal{S}$, $\mathcal{B} = \mathcal{A}_\Phi^*$ and $\mathcal{N} = \mathcal{S}_\Phi^*$.   ∎

Remark that it follows from the proof of Theorem 4.7.8 that, under the hypotheses of this theorem, we have

$$(8.3.23) \qquad\qquad A_\Phi^* S = A_\Phi^* \cap \overline{S}$$

$$(8.3.24) \qquad\qquad S_\Phi^* = S A_\Phi^* = \overline{A_\Phi^*} \cap S.$$

Note that $A_\Phi^* S$ is not necessarily the maximal exactly estimable parameter in $\mathcal{E}_{A \vee S}$, i.e., $A \cap \overline{S}$. However, any exactly estimable subparameter B for which S_Φ^* is sufficient is necessarily included in $A_\Phi^* S$. Indeed, $B \perp\!\!\!\perp S \mid S_\Phi^*$ and $B \subset \overline{S}$ imply $B \subset \overline{S_\Phi^*}$. Hence $B \subset \overline{S_\Phi^*} \cap A$. But $\overline{S_\Phi^*} \cap A = \overline{A_\Phi^*} \cap \overline{S} \cap A = A_\Phi^* \cap \overline{S} = A_\Phi^* S$, by (8.2.23) and (8.2.24), since $\overline{A_\Phi^*} \cap A = A_\Phi^*$.

The extension of Theorem 8.3.17 from the unreduced experiment $\mathcal{E}$ to the general reduced experiment $\mathcal{E}_{B \vee T}^{\mathcal{M}}$ requires a further assumption:

8.3.18 Theorem. Let $B \subset A$, $T \subset S$, $\mathcal{M} \subset A \vee S$. If $\mathcal{E}_{B \vee T}^{\mathcal{M}}$ is Φ-invariant, $(B \vee \mathcal{M})_\Phi^* \perp\!\!\!\perp \mathcal{M} \vee T \mid \varphi^{-1}(\mathcal{M} \vee T) \; \forall \; \varphi \in \Phi$, and Φ is ergodic on $\mathcal{M} \vee T$ conditionally on $B \vee \mathcal{M}$, then $(B \vee \mathcal{M})_\Phi^*(\mathcal{M} \vee T)$ is exactly estimable.

Proof. Since $\Phi \mathrm{I} (B \vee T) \mid \mathcal{M}$ then, $\Phi \mathrm{I} (B \vee \mathcal{M}) \mid \mathcal{M} \vee T$ by Theorem 8.2.25. This invariance and $(B \vee \mathcal{M})_\Phi^* \perp\!\!\!\perp (\mathcal{M} \vee T) \mid \varphi^{-1}(\mathcal{M} \vee T) \; \forall \; \varphi \in \Phi$, entail, by Theorem 8.2.26, that $(B \vee \mathcal{M})_\Phi^* \perp\!\!\!\perp \mathcal{M} \vee T \mid (\mathcal{M} \vee T)_\Phi^*$. The rest of the proof is identical to the proof of Theorem 8.3.15, setting $A = B \vee \mathcal{M}$ and $S = \mathcal{M} \vee T$. ∎

The further assumption, $(B \vee \mathcal{M})_\Phi^* \perp\!\!\!\perp (\mathcal{M} \vee T) \mid \varphi^{-1}(\mathcal{M} \vee T) \; \forall \varphi \in \Phi$, is readily satisfied when Φ is a group. By Theorems 8.2.25 and 8.2.29, this assumption is also redundant if $\Phi \mathrm{I} \mathcal{M}$.

In the last chapter the general theory of invariant experiments, as expanded in this chapter, will be applied to the analysis of stochastic processes; particular emphasis is placed on stationary and exchangeable processes.

9

Invariance in Stochastic Processes

9.1 Introduction

In Chapter 8 we presented a general theory for invariant experiments, characterizing mutually sufficient σ-fields, exactly estimating and estimable σ-fields. In this chapter those results are used to analyze two classes of models which are particularly important for empirical work, namely, stationary and exchangeable processes. In contrast to Chapter 8, in this chapter the invariant transformations $\varphi(a, s) = (\bar{g}(a), g(s))$ are such that $\bar{g}$ is always the identity on A.

We first consider experiments constructed from stationary or exchangeable sampling processes, viewed as processes invariant with respect to shifts or to finite permutations. These sampling invariance properties lead to the invariance of the joint probability characterizing the Bayesian experiment, which is, in turn, equivalent to invariance properties of the predictive process and of the posterior expectations. Thus stationarity or exchangeability involve the asymptotic sufficiency of the invariant σ-field; these important results follow directly from the general results obtained in Chapter 8.

The analysis is pursued by considering the properties of exact estimability for these experiments; clearly, such an investigation requires additional assumptions. The simplest case is the i.i.d. case, where identified parameters are exactly estimable. Furthermore, the i.i.d. case makes possible a complete characterization of exactly estimable exchangeable models.

For stationary processes, the case of Moving Average of order q processes (MA (q)) is straightforward because their q-dependent character implies that their tail-σ-field is exactly estimating. Markovian stationary processes are then studied and we present, in terms of measurable separability, a version of Doeblin's condition which is sufficient for their exact estimability. We also give a general presentation of ARMA (p, q) models; their exact estimability is derived from the preceding analysis of Markov processes. Note that this presentation is in σ-algebraic terms, and does not rely on linear representations. Finally, the last section extends some of the results to conditional models.

This chapter follows a slightly different presentation schema. As usual, we start from a Bayesian Experiment $\mathcal{E} = (A \times S, \mathcal{A} \vee \mathcal{S}, \Pi)$ on which is defined a stochastic process (x_n) with values in a measurable space $(U, \mathcal{U})$. It is natural to introduce the canonical representation of this process, viz., a new Bayesian Experiment $\widehat{\mathcal{E}}$ with the same sample parameter space A but with $U^{I\!N}$ for the sample space; on this new sample space is defined the coordinate representation process on which the shift or the permutation operators operate. But we do not wish to lose from view the original experiment $\mathcal{E}$; thus we shall systematically apply to $\mathcal{E}$ the results derived more naturally on the representation experiment $\widehat{\mathcal{E}}$. This double analysis in terms of both $\mathcal{E}$ and $\widehat{\mathcal{E}}$ may seem somewhat cumbersome but permits a change in the process, and therefore on the representation experiment, without affecting the original experiment. And in fact the analysis in terms of representation is particularly suited to making use of the results, obtained in Chapter 8, for stochastic processes, although it may be bypassed for the study of exact estimability.

The material developed in this chapter is based on (but considerably extends) Rolin (1986) for Section 9.3, and Florens, Mouchart and Rolin (1986) for Section 9.4.

9.2 Bayesian Stochastic Processes and Representations

9.2.1 Introduction

Consider the following simple experiment: $S = I\!R^n$, $A = I\!R$, $s = (s_1, \ldots, s_n)$ with $(s_i \mid a) \sim i.N(a, 1)$ along with a group of translations on $A \times S$ defined as $\varphi_\alpha(a, s) = (a + \alpha, s_i + \alpha, \ldots, s_n + \alpha)$, $\alpha \in \mathbb{R}$. Clearly, the translated distribution is $i.N(a + \alpha, 1)$, so that we obtain $\Phi \; I \; s \mid a$ for $\Phi = \{\varphi_\alpha : \alpha \in \mathbb{R}\}$. Here, because of the obtained conditional invariance, the group of translations appears as a "natural" group of transformations operating on both the sample and the parameter spaces. One should nevertheless remark that, until now, and with the exception of Chapter 8, we have been reluctant to assume any algebraic or topological structure for $(A \times S, \mathcal{A} \vee S)$: we wanted these spaces to be as "abstract" as possible and merely required a measurable structure, i.e., a possibility of assigning probabilities to well-defined families of sets. It is therefore not natural to introduce groups of operators acting on such spaces; and in the above example one may observe that the structure $(s_i \mid a) \sim i.N(a, 1)$ implicitly refers to some representation of the observation — the translation group would appear much less natural for $\exp s_i$, for instance. Furthermore s_i is likely to be one of many possible random variables that may be defined from an actual experimental design. These considerations suggest that in actual statistical modelling one will introduce representations $\widehat{\mathcal{E}} = (\widehat{A} \times \widehat{S}, \widehat{A} \vee \widehat{S}, \widehat{\Pi})$ of an abstract experiment $\mathcal{E} = (A \times S, \mathcal{A} \vee S, \Pi)$ by means of transformations $f : A \times S \to \widehat{A} \times \widehat{S}$ which have the property that $f(a, s) = (f_1(a), f_2(s))$, where each f_k is measurable, and such that in $\widehat{\mathcal{E}}$ one may exhibit, for instance, interesting invariance properties. We shall see subsequently that, for example, the shift operator is not naturally defined on an abstract sample space S although it can be naturally defined on processes defined on S, e.g., $\widehat{S} = \mathbb{R}^{I\!N}$, $f_2(s) = (x_n)_{n \in I\!N}$ and $x_n \in \mathbb{R}$.

Insofar as this section is meant to be rather general, the introduction of representations may appear as arbitrary. In fact, there may be interest in considering several representations of the same experiment, translating to the abstract experiment $\mathcal{E}$ the results obtained for the different representations $\widehat{\mathcal{E}}$ (such as minimal sufficient σ-fields etc.).

9.2.2 Representation of Experiments

In this section we make the concept of representation precise and then present a basic result that allows one to translate results obtained with respect to a representation to the original experiment. We first state two theorems in the context of a general abstract probability space and then apply them to a representation of a Bayesian experiment.

9.2.1 Theorem. Let $(M, \mathcal{M}, P)$ be a probability space, $(\widehat{M}, \widehat{\mathcal{M}})$ be a measurable space, and let $f : M \to \widehat{M}$ be such that $f^{-1}(\widehat{\mathcal{M}}) \subset \mathcal{M}$. If $\widehat{P} = P \circ f^{-1}$, $\widehat{\mathcal{N}} \subset \widehat{\mathcal{M}}$ and $\mathcal{N} = f^{-1}(\widehat{\mathcal{N}})$, then $\forall\, m \in [f^{-1}(\widehat{\mathcal{M}})]^+$, $\exists\, \widehat{m} \in [\widehat{\mathcal{M}}]^+$ such that

(i) $m = \widehat{m} \circ f$

(ii) $\mathcal{N}m = (\widehat{\mathcal{N}}\widehat{m}) \circ f$.

Proof. As before, $\mathcal{M}_0 = \{\phi, M\}$ and $\widehat{\mathcal{M}}_0 = \{\phi, \widehat{M}\}$; projections on $\mathcal{M}_0$ and on $\widehat{\mathcal{M}}_0$ are mathematical expectations with respect to P and $\widehat{P}$ respectively. The existence of $\widehat{m} \in [\widehat{\mathcal{M}}]^+$ satisfying (i) follows from Chapter 0, Theorem 0.2.11. The proof of (ii) results from simple manipulations of conditional expectations and of image probabilities (see Proposition 0.3.8). Thus consider arbitrary $\widehat{n} \in [\widehat{\mathcal{N}}]^+$ and $\widehat{m} \in [\widehat{\mathcal{M}}]^+$; we then have that:

$$
\begin{aligned}
\mathcal{M}_0[\widehat{n} \circ f \cdot \mathcal{N}(\widehat{m} \circ f)] &= \mathcal{M}_0[\widehat{n} \circ f \cdot \widehat{m} \circ f] \\
&= \widehat{\mathcal{M}}_0(\widehat{n}\widehat{m}) = \widehat{\mathcal{M}}_0[\widehat{n}(\widehat{\mathcal{N}}\widehat{m})] \\
&= \mathcal{M}_0[\widehat{n} \circ f \cdot (\widehat{\mathcal{N}}\widehat{m}) \circ f]
\end{aligned}
$$

This shows that

$$
\mathcal{N}(\widehat{m} \circ f) = (\widehat{\mathcal{N}}\widehat{m}) \circ f. \qquad \blacksquare
$$

Theorem 9.2.1. permits properties of the representation experiment $\widehat{\mathcal{E}}$ to be viewed as properties of the original experiment $\mathcal{E}$. The next theorem illustrates as an example, the translation of a conditional independence property.

9.2.2 Theorem. Let $(M, \mathcal{M}, P)$ a probability space and $(\widehat{M}, \widehat{\mathcal{M}}, \widehat{P})$ a representation of this probability space through $f : M \to \widehat{M}$ as defined in Theorem 9.2.1. Let $\widehat{\mathcal{M}}_i$ $(i = 1, 2, 3)$ be sub-σ-fields of $\widehat{\mathcal{M}}$ and let $\mathcal{M}_i = f^{-1}(\widehat{\mathcal{M}}_i)$ $(i = 1, 2, 3)$. Then:

(i) $\widehat{\mathcal{M}}_1 \perp\!\!\!\perp \widehat{\mathcal{M}}_2 \mid \widehat{\mathcal{M}}_3;\ \widehat{P}$

if and only if

(ii) $\mathcal{M}_1 \perp\!\!\!\perp \mathcal{M}_2 | \mathcal{M}_3; \; P.$

Proof. Let $\widehat{m}_i \in [\widehat{\mathcal{M}_i}]^+ \quad (i = 1, 2)$. Then, by Theorem 9.2.1,

$$\mathcal{M}_3(\widehat{m}_1 \circ f \cdot \widehat{m}_2 \circ f) = \widehat{\mathcal{M}}_3(\widehat{m}_1 \widehat{m}_2) \circ f \qquad \text{a.s.} P.$$
$$\mathcal{M}_3(\widehat{m}_1 \circ f) = (\widehat{\mathcal{M}}_3 \widehat{m}_1) \circ f \qquad \text{a.s.} P.$$
$$\mathcal{M}_3(\widehat{m}_2 \circ f) = (\widehat{\mathcal{M}}_3 \widehat{m}_2) \circ f \qquad \text{a.s.} P.$$

Therefore

$$\mathcal{M}_3(\widehat{m}_1 \circ f \cdot \widehat{m}_2 \circ f) = \mathcal{M}_3(\widehat{m}_1 \circ f) \cdot \mathcal{M}_3(\widehat{m}_2 \circ f) \quad \text{a.s.} P.$$

if and only if

$$\widehat{\mathcal{M}}_3(\widehat{m}_1 \widehat{m}_2) = \widehat{\mathcal{M}}_3(\widehat{m}_1) \cdot \widehat{\mathcal{M}}_3(\widehat{m}_2) \qquad \text{a.s.} \widehat{P}. \qquad \blacksquare$$

We now provide a precise definition of a representation experiment. This is a preliminary step before using the general results of Theorems 9.2.1. and 9.2.2. on Bayesian experiments.

9.2.3 Definition. Let $\mathcal{E} = (A \times S, \mathcal{A} \vee \mathcal{S}, \Pi)$ be a Bayesian experiment. Then $\widehat{\mathcal{E}} = (\widehat{A} \times \widehat{S}, \widehat{\mathcal{A}} \vee \widehat{\mathcal{S}}, \widehat{\Pi})$ is said to be a *representation of $\mathcal{E}$ (through f)* if there exists a measurable function $f : A \times S \to \widehat{A} \times \widehat{S}$ such that

(i) $f = (f_1, f_2)$ with $f_1 : A \to \widehat{A}$ and $f_2 : S \to \widehat{S}$ both being measurable and, as usual, also considered as defined on $A \times S$.

(ii) $\widehat{\Pi} = \Pi \circ f^{-1}$ $\qquad\qquad\qquad\qquad\qquad\qquad\qquad\qquad$ $\blacksquare$

Note that, given this definition, any parameter (respectively, statistic) in $\widehat{\mathcal{E}}$ is sent to a parameter (respectively, a statistic) in $\mathcal{E}$ by the inverse image of f_1 (respectively f_2).

Theorem 9.2.1. may be further refined in the context of a Bayesian experiments as is illustrated by the following proposition:

9.2.4 Proposition. Let $\widehat{\mathcal{E}} = (\widehat{A} \times \widehat{S}, \; \widehat{\mathcal{A}} \vee \widehat{\mathcal{S}}, \; \widehat{\Pi})$ be a representation of $\mathcal{E} = (A \times S, \; \mathcal{A} \vee \mathcal{S}, \; \Pi)$. For any $\widehat{\mathcal{B}} \subset \widehat{\mathcal{A}}$ and $\widehat{\mathcal{T}} \subset \widehat{\mathcal{S}}$, let us define

$\mathcal{B} = f_1^{-1}(\widehat{\mathcal{B}})$ and $\mathcal{T} = f_2^{-1}(\widehat{\mathcal{T}})$. Then:

(i) $\forall\, b \in [\mathcal{B}]^+$ $\exists\, \widehat{b} \in [\widehat{\mathcal{B}}]^+$ such that $b = \widehat{b} \circ f_1$ and $Tb = (\widehat{T}\widehat{b}) \circ f_2$.

(ii) $\forall\, t \in [\mathcal{T}]^+$ $\exists\, \widehat{t} \in [\widehat{\mathcal{T}}]^+$ such that $t = \widehat{t} \circ f_2$ and $Bt = (\widehat{B}\widehat{t}\,) \circ f_1$.   ∎

As an example of how to use both this result and Theorem 9.2.2., we draw attention to the following simple corollary.

9.2.5 Corollary. Let $\mathcal{A}' = f_1^{-1}(\widehat{\mathcal{A}})$ and $\mathcal{S}' = f_2^{-1}(\widehat{\mathcal{S}})$. Let $\widehat{\mathcal{T}} \subset \widehat{\mathcal{S}}$ and $\mathcal{T} = f_2^{-1}(\widehat{\mathcal{T}})$. Then

$$\widehat{\mathcal{A}} \perp \widehat{\mathcal{S}} \mid \widehat{\mathcal{T}}; \widehat{\Pi} \;\Leftrightarrow\; \mathcal{A}' \perp \mathcal{S}' \mid \mathcal{T}; \Pi,$$

i.e., any sufficient statistic in $\widehat{\mathcal{E}}$ is translated, through f_2, into a sufficient statistic in $\mathcal{E}_{\mathcal{A}' \vee \mathcal{S}'}$.   ∎

9.2.3 Bayesian Stochastic Processes

In the Bayesian experiment $\mathcal{E} = (A \times S, \mathcal{A} \vee \mathcal{S}, \Pi)$ a *Bayesian stochastic process* is a sequence of random variables $\{x_n : n \in I\!\!N\}$ defined on $(S, \mathcal{S})$ with values in some measurable space $(U, \mathcal{U})$; this measurable space is called the *state space* of the process. We have already studied sequences of random variables in Chapter 6. However, at that stage, we did not require that each random variable x_n has values in the same measurable space, which is essential when considering shift and permutation operators.

For the *representation of a Bayesian stochastic process*, we define:

(9.2.1) $\widehat{A} = A, \quad \widehat{\mathcal{A}} = \mathcal{A}, \quad f_1 = i,$

i.e., f_1 is the identity map on A;

(9.2.2) $\widehat{S} = U^{I\!\!N}, \qquad \widehat{\mathcal{S}} = \mathcal{U}^{I\!\!N},$

i.e., $\widehat{\mathcal{S}}$ is the product σ-field of infinitely many copies of $\mathcal{U}$; and

(9.2.3) $f_2 = x = (x_n, n \in I\!\!N) = (x_0, x_1, x_2, \ldots).$

$\widehat{\Pi}$ is defined as in Definition 9.2.3, namely, $\widehat{\Pi} = \Pi \circ f^{-1}$ and, in particular, $\widehat{\mu} = \mu$ and $\widehat{P} = P \circ x^{-1}$.

We denote, as usual, the sub-σ-fields of $\mathcal{S}$ generated by the stochastic process as follows:

(9.2.4) $$\mathcal{X}_n = \sigma\{x_n\} = x_n^{-1}(\mathcal{U})$$

Similarly, for any J subset of $I\!N$:

(9.2.5) $$\mathcal{X}_J = \bigvee_{n \in J} \mathcal{X}_n = \sigma\{x_n : n \in J\}$$

In particular we denote

(9.2.6) $$\mathcal{X}_n^m = \bigvee_{n \leq p \leq m} \mathcal{X}_p$$

(9.2.7) $$\mathcal{X}_n^\infty = \bigvee_{m \geq n} \mathcal{X}_m;$$

$\mathcal{X}_n^\infty$ is called the *future after n of the process*. $\mathcal{X}_0^n$ is the *past up to time n of the process* and is the canonical filtration associated to the stochastic process x. Note that

(9.2.8) $$\mathcal{X}_0^\infty = x^{-1}(\mathcal{U}^{I\!N}) = \mathcal{S}'.$$

As in Chapter 7, the tail-σ-field of the process x is given by

(9.2.9) $$\mathcal{X}_T = \bigcap_n \mathcal{X}_n^\infty = \bigcap_n \bigvee_{m \geq n} \mathcal{X}_m$$

It will on occasion be interesting to introduce the *coordinate process* (respectively, the *parameter random variable*) defined on the representation space $(\widehat{S}, \widehat{\mathcal{S}})$ (respectively, the parameter space $(\widehat{A}, \widehat{\mathcal{A}})$) or, as usual, also considered as defined on $(\widehat{A} \times \widehat{S}, \widehat{\mathcal{A}} \vee \widehat{\mathcal{S}})$, as follows:

(9.2.10) $$\widehat{x}_n(\alpha, u) \quad = \widehat{x}_n(u) \quad = u_n$$
(9.2.11) $$\widehat{a}(\alpha, u) \quad = \widehat{a}(\alpha) \quad = \alpha$$
$$\forall\, \alpha \in \widehat{A}, \quad \forall\, u \in \widehat{S}, \quad \forall\, n \in I\!N$$

To this coordinate process we may associate, using the same definitions, the sub-σ-fields of $\widehat{S}$. For instance:

(9.2.12) $$\widehat{\mathcal{X}}_n = \widehat{x}_n^{-1}(\mathcal{U}),$$

and similarly for $\widehat{\mathcal{X}}_J, \widehat{\mathcal{X}}_n^n, \widehat{\mathcal{X}}_n^\infty, \widehat{\mathcal{X}}_0^\infty$ and so on. Note that for the coordinate process

(9.2.13) $$\widehat{\mathcal{X}}_0^\infty = \widehat{\mathcal{S}} = \widehat{x}^{-1}(\mathcal{U}^{I\!N}).$$

The notation for the coordinate process is motivated by the following identity:

(9.2.14) $$\widehat{x}_n \circ x = x_n \quad \forall\, n \in I\!N.$$

Note also that the coordinate process allows us to write the following identity:

(9.2.15) $$\mathcal{I}[b\, f \circ x] = \widehat{\mathcal{I}}[b\, f \circ \widehat{x}], \quad \forall\, b \in [A]^+, \quad \forall\, f \in [\mathcal{U}^{I\!N}]^+.$$

9.2.4 Shift and Permutations

We now provide a standard definition of the shift operator and the finite permutation operators (see, for instance, Chung (1968), 8.1). These operators lead to the most typical invariances used in the statistical analysis of stochastic processes. For example, the shift operator is used in time-series analysis whereas the permutation operator is used in the analysis of survey data.

9.2.6 Definition. The *shift operator*, denoted by τ is the mapping from $U^{I\!N}$ to $U^{I\!N}$ defined by

(9.2.16) $$\tau(u)_n = u_{n+1} \quad \forall\, n \in I\!N. \qquad\qquad \blacksquare$$

Note that τ is always a measurable transformation on $(U^{I\!N}, \mathcal{U}^{I\!N})$. Clearly τ is a surjective map but is not injective. Successive compositions of τ give the iterates of τ, i.e., $\tau^k : U^{I\!N} \to U^{I\!N}$

(9.2.17) $$\tau^k(u)_n = u_{n+k} \quad \forall\, k, n \in I\!N.$$

If τ^0 is defined as the identity map on $U^{I\!N}$, then the set of measurable transformations $\Gamma = \{\tau^k : k \in I\!N\}$ is a monoïd but not a group. This fact is sometimes troublesome. One way to overcome this difficulty is to consider τ as a map from $U^{Z\!\!Z}$ to $U^{Z\!\!Z}$. Clearly, τ is now a bijective measurable transformation on $U^{Z\!\!Z}$, and would be quite natural if the Bayesian stochastic process were indexed by $Z\!\!Z$ instead of $I\!N$. Let us note that this is often done in stationary processes and in time series analysis. For the same reason as in Chapter 6, this case will not be studied in detail in this chapter, which is why we do not introduce this representation formally.

The shift operator also acts by composition on the Bayesian stochastic process x. Namely,

$$(9.2.18) \qquad (\tau^k \circ x)_n = x_{n+k} \quad \forall\, k, n \in I\!N.$$

Thus $\tau^k \circ x$ is the kth times shifted stochastic process, i.e., starting at x_k.

In general, it is difficult to define τ as a measurable transformation on the abstract measurable space $(A \times S,\ \mathcal{A} \vee \mathcal{S})$ while under our usual conventions τ may also be considered as a measurable transformation on the representation space $(\widehat{A} \times \widehat{S},\ \widehat{A} \vee \widehat{S},\ \widehat{\Pi})$ and may afterwards be translated in the framework of the original experiment $\mathcal{E} = (A \times S,\ \mathcal{A} \vee \mathcal{S},\ \Pi)$ through (i, x).

Let us note that, with these conventions on τ and the definition of the coordinate process, we have the following identities:

$$(9.2.19) \qquad \widehat{x}_{n+k} = \widehat{x}_n \circ \tau^k = (\tau^k \circ \widehat{x})_n \quad \forall\, k, n \in I\!N$$

$$(9.2.20) \qquad \widehat{a} \circ \tau = \widehat{a}.$$

This implies, in particular, that

$$(9.2.21) \qquad \widehat{\mathcal{X}}_n = \tau^{-n}(\widehat{\mathcal{X}}_0).$$

As in Chapter 8, an invariant sub-σ-field is associated to the shift operator. More precisely, we define the *σ-field of shift-invariant sets* in $\mathcal{U}^{I\!N}$:

$$(9.2.22) \qquad \mathcal{U}_\Gamma = \{B \in \mathcal{U}^{I\!N} : \tau^{-1}(B) = B\},$$

and, according to Theorem 8.2.1,

$$(9.2.23) \qquad f \in [\mathcal{U}_\Gamma] \Leftrightarrow f \in [\mathcal{U}^{I\!N}] \quad \text{and} \quad f \circ \tau = f.$$

The *shift-invariant events of the stochastic process* x are then defined as

$$(9.2.24) \qquad \mathcal{X}_\Gamma = x^{-1}(\mathcal{U}_\Gamma).$$

Similarly, we define the *shift-invariant events of the coordinate process* $\widehat{x}$ as

$$(9.2.25) \qquad \widehat{\mathcal{X}}_\Gamma = \widehat{x}^{-1}(\mathcal{U}_\Gamma).$$

As in Chapter 8, we consider the Γ-invariant σ-field in $\widehat{\mathcal{E}}$, $(\widehat{\mathcal{A}} \vee \widehat{\mathcal{S}})_\Gamma$, and its traces either on sub-σ-fields of $\widehat{\mathcal{A}}$ or on sub-σ-fields of $\widehat{\mathcal{S}}$. The structure of these invariant σ-fields are given in the following lemma.

9.2.7 Lemma. In the representation experiment $\widehat{\mathcal{E}} = (\widehat{A} \times \widehat{S}, \; \widehat{\mathcal{A}} \vee \widehat{\mathcal{S}}, \; \widehat{\Pi})$ of the experiment $\mathcal{E} = (A \times S, \; \mathcal{A} \vee \mathcal{S}, \Pi)$, we have

(i) $\forall \, \widehat{\mathcal{B}} \subset \widehat{\mathcal{A}} \qquad \widehat{\mathcal{B}}_\Gamma = \widehat{\mathcal{B}}$

(ii) $\forall \, \widehat{\mathcal{T}} \subset \widehat{\mathcal{S}} \qquad \widehat{\mathcal{T}}_\Gamma = \widehat{\mathcal{X}}_\Gamma \cap \widehat{\mathcal{T}}$

and, in particular,

(iii) $\widehat{\mathcal{S}}_\Gamma = \widehat{\mathcal{X}}_\Gamma$.

Moreover, in the original experiment,

(iv) $\mathcal{X}_\Gamma = x^{-1}(\widehat{\mathcal{S}}_\Gamma)$.

Proof. To prove (i), it suffices to use (9.2.20), which shows that we have $\widehat{\mathcal{A}} \subset (\widehat{\mathcal{A}} \vee \widehat{\mathcal{S}})_\Gamma$. Now, by the definition of $\widehat{\mathcal{T}}_\Gamma$, (ii) is a direct consequence of (iii). If $\widehat{m} \in \widehat{\mathcal{X}}_\Gamma$, $\widehat{m} = f \circ \widehat{x}$ with $f \in [\mathcal{U}_\Gamma]$. Therefore, by (9.2.19), $\widehat{m} \circ \tau = f \circ \widehat{x} \circ \tau = f \circ \tau \circ \widehat{x} = f \circ \widehat{x} = \widehat{m}$, and so $\widehat{m} \in [(\widehat{\mathcal{A}} \vee \widehat{\mathcal{S}})_\Gamma \cap \widehat{\mathcal{S}}] = [\widehat{\mathcal{S}}_\Gamma]$. However, if $\widehat{m} \in [\widehat{\mathcal{S}}_\Gamma]$, then there exists $f \in [\mathcal{U}^{I\!N}]$ such that $\widehat{m} = f \circ \widehat{x}$ and $\widehat{m} \circ \tau = \widehat{m}$. This implies that $f \circ \widehat{x} = f \circ \tau \circ \widehat{x}$. Therefore, by the definition of the coordinate process, $f \circ \tau = f$, i.e., $f \in [\mathcal{U}_\Gamma]$, and so $\widehat{m} \in [\widehat{\mathcal{X}}_\Gamma]$. Finally, (iv) is a direct consequence of (iii) and (9.2.14). ∎

The shift operator may be used to represent the tail σ-field of the Bayesian process x, and of the coordinate process $\widehat{x}$. Indeed let us define the σ-fields of sets in $\mathcal{U}^{I\!N}$ whose definition does not depend on the first n coordinates for any finite n, i.e.,

$$(9.2.26) \qquad\qquad \mathcal{U}_T = \bigcap_n \tau^{-n}(\mathcal{U}^{I\!N})$$

or equivalently, by Theorem 0.2.11,

(9.2.27)
$$f \in [\mathcal{U}_T] \Leftrightarrow f \in [\mathcal{U}^{I\!N}] \quad \text{and} \quad \forall\, n \in I\!N$$
$$\exists f_n \in [\mathcal{U}^{I\!N}] \quad \text{such that} \quad f = f_n \circ \tau^n.$$

Although $\mathcal{U}_T$ is not really a tail σ-field, we nevertheless have:

(9.2.28)
$$\widehat{\mathcal{X}}_T = \widehat{x}^{-1}(\mathcal{U}_T)$$

and

(9.2.29)
$$\mathcal{X}_T = x^{-1}(\widehat{\mathcal{X}}_T) = x^{-1}(\mathcal{U}_T).$$

We now turn to the same construction for the group of finite permutations.

9.2.8 Definition. To a permutation of the first k integers we associate the *finite permutation operator σ* from $U^{I\!N}$ to $U^{I\!N}$ defined by:

(9.2.30)
$$\sigma(u)_n = u_{\sigma(n)} \quad \forall\, n < k$$
$$= u_n \quad \forall\, n \geq k$$

We denote by Σ_k the set of all finite permutation operators satisfying (9.2.30); $\Sigma = \cup_{k \in I\!N} \Sigma_k$ is the set of finite permutation operators.

Note that any $\sigma \in \Sigma$ is a bijective bimeasurable transformation on $(U^{I\!N}, \mathcal{U}^{I\!N})$. Hence Σ, as well as Σ_k for any k, is a group of measurable transformations on $(U^{I\!N}, \mathcal{U}^{I\!N})$.

As for the shift operator, Σ acts by composition on the process x. Indeed, if $\sigma \in \Sigma_k$,

(9.2.31)
$$(\sigma \circ x)_n = x_{\sigma(n)} \quad \forall\, n < k$$
$$= x_n \quad \forall\, n \geq k$$

Thus $\sigma \circ x$ is the transformed stochastic process obtained by permuting, according to σ, the first k coordinates of the stochastic process x.

As for the shift operator, Σ may be considered as a group of measurable transformations on the representation space $(\widehat{A} \times \widehat{S}, \widehat{A} \vee \widehat{S})$. Note that, with this convention,

(9.2.32)
$$\sigma \circ \widehat{x} = \widehat{x} \circ \sigma$$

(9.2.33)
$$\widehat{a} \circ \sigma = \widehat{a}$$

As before, we define the *σ-field of symmetric sets* in $\mathcal{U}^{I\!N}$ as

$$(9.2.34) \qquad \mathcal{U}_\Sigma = \{B \in \mathcal{U}^{I\!N} : \sigma^{-1}(B) = B \quad \forall \, \sigma \in \Sigma\},$$

and note that, according to Theorem 8.2.1,

$$(9.2.35) \quad f \in [\mathcal{U}_\Sigma] \Leftrightarrow f \in [\mathcal{U}^{I\!N}] \quad \text{and} \quad f \circ \sigma = f \quad \forall \, \sigma \in \Sigma.$$

The *symmetric events of the stochastic process* x is then defined as

$$(9.2.36) \qquad\qquad \mathcal{X}_\Sigma = x^{-1}(\mathcal{U}_\Sigma)$$

and, similarly, we defined the *symmetric events of the coordinate process* $\widehat{x}$ as

$$(9.2.37) \qquad\qquad \widehat{\mathcal{X}}_\Sigma = \widehat{x}^{-1}(\mathcal{U}_\Sigma).$$

Considering, as in Chapter 8, the Σ-invariant σ-field in $\widehat{\mathcal{E}}, (\widehat{\mathcal{A}} \vee \widehat{\mathcal{S}})_\Sigma$ and its traces on sub-σ-fields of $\widehat{\mathcal{A}}$ or on sub-σ-fields of $\widehat{\mathcal{S}}$, we may reproduce the proof of Lemma 9.2.7 to obtain the following lemma:

9.2.9 Lemma. In the representation experiment $\widehat{\mathcal{E}} = (\widehat{A} \times \widehat{S}, \widehat{\mathcal{A}} \vee \widehat{\mathcal{S}}, \widehat{\Pi})$ of the experiment $\mathcal{E} = (A \times S, \mathcal{A} \vee \mathcal{S}, \Pi)$, then

(i) $\forall \, \widehat{\mathcal{B}} \subset \widehat{\mathcal{A}} \qquad \widehat{\mathcal{B}}_\Sigma = \widehat{\mathcal{B}}$

(ii) $\forall \, \widehat{\mathcal{T}} \subset \widehat{\mathcal{S}} \qquad \widehat{\mathcal{T}}_\Sigma = \widehat{\mathcal{X}}_\Sigma \cap \widehat{\mathcal{T}},$

and, in particular,

(iii) $\widehat{\mathcal{S}}_\Sigma = \widehat{\mathcal{X}}_\Sigma.$

Moreover, in the original experiment,

(iv) $\mathcal{X}_\Sigma = x^{-1}(\widehat{\mathcal{S}}_\Sigma).$ ■

Now, it follows from (9.2.22), (9.2.26) and (9.2.34), that

$$(9.2.38) \qquad\qquad \mathcal{U}_\Gamma \subset \mathcal{U}_T \subset \mathcal{U}_\Sigma \subset \mathcal{U}^{I\!N}.$$

Therefore, for the coordinate process $\hat{x}$,

(9.2.39) $$\hat{\mathcal{X}}_\Gamma \subset \hat{\mathcal{X}}_T \subset \hat{\mathcal{X}}_\Sigma \subset \hat{\mathcal{X}}_0^\infty,$$

and the same inclusions hold for the Bayesian stochastic process x, i.e.,

(9.2.40) $$\mathcal{X}_\Gamma \subset \mathcal{X}_T \subset \mathcal{X}_\Sigma \subset \mathcal{X}_0^\infty.$$

Until now, all these asymptotic and invariant σ-fields have been analyzed without any use of the probability measure. Invariance properties of the probability measure will allow us to describe a.s. equivalent σ-fields and to refine the inclusion properties obtained above.

9.3 Standard Bayesian Stochastic Processes

A natural application of invariant parameter experiments is the analysis of discrete time invariant stochastic processes such as stationary or exchangeable processes (see, e.g., Anderson (1971)). In what follows these processes are analyzed both from the point of view of invariance, in light of the general theory of Chapter 8, and from the point of view of exact estimability, as developed in Chapter 7. This exact estimability will be obtained for the particular cases of i.i.d. and of Markovian stationary processes.

Both stationary and exchangeable processes are generally defined by properties of their sampling distributions. In the framework of a Bayesian experiment, this will be extended, in a natural way, to properties of the joint distribution; these experiments will consequently be characterized by properties of the predictive probability and of the posterior expectations. This is useful for finding asymptotically sufficient σ-fields and, therefore, conditions for exact estimability of the minimal sufficient parameter.

9.3.1 Stationary Processes

The structure of a stationary process is usually presented as follows. In the statistical experiment $\mathcal{E}$ defined as

(9.3.1) $$\mathcal{E} = \{(S, \mathcal{S}); \ P^a \quad a \in A\}$$

the stochastic process $x = \{x_n : n \in I\!\!N\}$ defined on $(S, \mathcal{S})$ with values in $(U, \mathcal{U})$ is stationary if

$$(9.3.2) \qquad \mathcal{I}^a [f(x_1, x_2, \ldots, x_{n+1})] = \mathcal{I}^a [f(x_0, x_1, \ldots, x_n)]$$
$$\forall\, n \in I\!\!N, \quad \forall\, a \in A, \quad \forall\, f \in [\mathcal{U}^{n+1}]^+.$$

Using a monotone class argument and the definition of the shift operator in Section 9.2.4, (9.3.2) is actually equivalent to

$$(9.3.3) \qquad \mathcal{I}^a (f \circ \tau^n \circ x) = \mathcal{I}^a (f \circ x)$$
$$\forall\, n \in I\!\!N, \quad \forall\, a \in A, \quad \forall\, f \in [\mathcal{U}^{I\!\!N}]^+.$$

In the Bayesian experiment

$$(9.3.4) \qquad\qquad \mathcal{E} = (A \times S, \mathcal{A} \vee \mathcal{S}, \Pi)$$

it implies that, for any prior probability,

$$(9.3.5) \qquad A[f(x_1, x_2, \ldots, x_{n+1})] = A[f(x_0, x_1, \ldots, x_n)]$$
$$\forall\, n \in I\!\!N, \quad \forall\, f \in [\mathcal{U}^{n+1}]^+,$$

and this is also equivalent to

$$(9.3.6) \qquad A(f \circ \tau^n \circ x) = A(f \circ x) \qquad \forall\, n \in I\!\!N, \quad \forall\, f \in [\mathcal{U}^{I\!\!N}]^+.$$

Now, in the representation $\widehat{\mathcal{E}}$ of $\mathcal{E}$ as defined in Section 9.2.3, (9.3.5) becomes

$$(9.3.7) \qquad \widehat{A}(\hat{t} \circ \tau) = \widehat{A}(\hat{t}) \qquad \forall\, n \in I\!\!N \quad \forall\, \hat{t} \in [\widehat{\mathcal{X}}_0^n]^+,$$

and (9.3.6) becomes

$$(9.3.8) \qquad \widehat{A}(\hat{t} \circ \tau^n) = \widehat{A}(\hat{t}) \qquad \forall\, n \in I\!\!N \quad \forall\, \hat{t} \in [\widehat{\mathcal{X}}_0^\infty]^+.$$

Now, recalling, by Lemma 9.2.7, that $\widehat{A}_\Gamma = \widehat{A}$, one sees that (9.3.7) is actually equivalent, by Definition 8.2.22, to

$$(9.3.9) \qquad\qquad \{\tau\} \, \mathrm{I} \, \widehat{\mathcal{X}}_0^n \mid \widehat{A} \quad \forall\, n \in I\!\!N,$$

and that (9.3.8), equivalent to (9.3.9), amounts to

$$(9.3.10) \qquad\qquad \Gamma \, \mathrm{I} \, \widehat{\mathcal{X}}_0^\infty \mid \widehat{A}.$$

This leads to the following definition of Bayesian stationarity:

9.3.1 Definition. In the Bayesian experiment $\mathcal{E} = (A \times S, \mathcal{A} \vee S, \Pi)$, the *Bayesian stochastic process* $x = \{x_n : n \in I\!\!N\}$ *is* $\mathcal{M}$*-stationary*, where $\mathcal{M} \subset \mathcal{A} \vee S$, if

$$\mathcal{M}(f \circ \tau \circ x) = \mathcal{M}(f \circ x) \quad \forall f \in [\mathcal{U}^{I\!\!N}]^+. \qquad \blacksquare$$

This definition, specialized on a sub-σ-field $\mathcal{B}$ of $\mathcal{A}$, and reinterpreted in the representation $\widehat{\mathcal{E}}$ of $\mathcal{E}$ gives the following theorem, which is expressed using the notation of Section 9.2:

9.3.2 Theorem. In the Bayesian experiment $\mathcal{E} = (A \times S, \mathcal{A} \vee S, \Pi)$, let $\mathcal{B} \subset \mathcal{A}$ and $x = \{x_n : n \in I\!\!N\}$ be a Bayesian stochastic process. Then the following properties are equivalent:

(i) x is $\mathcal{B}$-stationary

(ii) $\Gamma \perp\!\!\!\perp \widehat{\mathcal{S}} \mid \widehat{\mathcal{B}}$

(iii) $\Gamma \perp\!\!\!\perp (\widehat{\mathcal{B}} \vee \widehat{\mathcal{S}})$

(iv) $\Gamma \perp\!\!\!\perp \widehat{\mathcal{S}}$ and $\Gamma \perp\!\!\!\perp \widehat{\mathcal{B}} \mid \widehat{\mathcal{S}}$

(v) $\Gamma \perp\!\!\!\perp \widehat{\mathcal{S}} \mid \widehat{\mathcal{S}_\Gamma}$ and $\widehat{\mathcal{B}} \perp\!\!\!\perp \widehat{\mathcal{S}} \mid \widehat{\mathcal{S}_\Gamma}$.

Moreover, in this situation,

(vi) $\widehat{\mathcal{S}_\Gamma^\bullet} = \overline{\widehat{\mathcal{S}_\Gamma} \cap \widehat{\mathcal{S}}}$.

Proof. Clearly, in the representation $\widehat{\mathcal{E}}$ of $\mathcal{E}$, $\mathcal{B}(f \circ \tau \circ x) = \mathcal{B}(f \circ x)$ is equivalent, by (9.2.15), to $\widehat{\mathcal{B}}(f \circ \tau \circ \widehat{x}) = \widehat{\mathcal{B}}(f \circ \widehat{x})$; since $\tau \circ \widehat{x} = \widehat{x} \circ \tau$, this is equivalent to $\widehat{\mathcal{B}}(f \circ \widehat{x} \circ \tau) = \widehat{\mathcal{B}}(f \circ \widehat{x})$. By Definition 8.2.22, this is equivalent to $\Gamma \perp\!\!\!\perp \widehat{\mathcal{S}} \mid \widehat{\mathcal{B}}$, because $\widehat{\mathcal{B}_\Gamma} = \widehat{\mathcal{B}}$ by Lemma 9.2.7. Since $\widehat{\mathcal{B}} \subset (\widehat{\mathcal{A}} \vee \widehat{\mathcal{S}})_\Gamma$ by Theorem 8.2.28, (ii), (iii), (iv) are equivalent and are equivalent to $\Gamma \perp\!\!\!\perp \widehat{\mathcal{S}} \mid \widehat{\mathcal{S}_\Gamma^\bullet}$ and $\mathcal{B} \perp\!\!\!\perp \widehat{\mathcal{S}} \mid \widehat{\mathcal{S}_\Gamma^\bullet}$. By Theorem 8.2.20, $\widehat{\mathcal{S}_\Gamma^\bullet} = \overline{\widehat{\mathcal{S}_\Gamma} \cap \widehat{\mathcal{S}}}$ and therefore, by Theorem 8.2.23 (iii) and Corollary 2.2.7, this is equivalent to (v). $\qquad \blacksquare$

Note that, by a monotone class argument, $\Gamma \perp\!\!\!\perp \widehat{\mathcal{S}} \mid \widehat{\mathcal{B}}$ is actually equivalent to $\Gamma \perp\!\!\!\perp \widehat{\mathcal{X}_0^n} \mid \widehat{\mathcal{B}} \ \forall n \in I\!\!N$. If the equivalences between (ii), (iii), and

(iv) remain true when $\widehat{S}$ is replaced by $\widehat{\mathcal{X}}_0^n$, then (v) is no longer equivalent because $\widehat{\mathcal{X}}_0^n$ is not Γ-stable.

It is whorthwhile rewriting (iv) and (v) in terms of posterior expectations. Let us first note that Γ I $\widehat{S}$ amounts to saying that, in $\widehat{S}$, the predictive probability is shift-invariant, whereas Γ I $\widehat{B}$ | $\widehat{S}$ is equivalent to saying that

$$(9.3.11) \qquad \tau^{-k}(\widehat{S})b = (\widehat{S}b) \circ \tau^k \qquad \forall\, b \in [\widehat{B}]^+, \quad \forall\, k \in I\!N$$

or, equivalently,

$$(9.3.12) \qquad \widehat{\mathcal{X}}_k^\infty b = (\widehat{\mathcal{X}}_0^\infty b) \circ \tau^k \qquad \forall\, b \in [\widehat{B}]^+, \quad \forall\, k \in I\!N.$$

Now, $\widehat{B} \perp\!\!\!\perp \widehat{S} \mid \widehat{S}_\Gamma$ is equivalent to saying that $\widehat{S}_\Gamma = \widehat{\mathcal{X}}_\Gamma$ is an $\widehat{\mathcal{E}}_{\widehat{B}\vee\widehat{S}}$-sufficient statistic or, equivalently,

$$(9.3.13) \qquad \widehat{\mathcal{X}}_0^\infty b = \widehat{\mathcal{X}}_\Gamma b \quad \forall\, b \in [\widehat{B}]^+.$$

Therefore, since $\widehat{\mathcal{X}}_\Gamma$ is shift-invariant,

$$(9.3.14) \qquad \widehat{\mathcal{X}}_0^\infty b = \widehat{\mathcal{X}}_k^\infty b = \widehat{\mathcal{X}}_\Gamma b.$$

Then, according to Theorem 7.7.2, the dynamic Bayesian experiment $\widehat{\mathcal{E}} = (\widehat{A} \times \widehat{S}, \widehat{A} \vee \widehat{S}, \widehat{\Pi}, \widehat{\mathcal{X}}_0^n)$ is tail-sufficient, i.e., $\widehat{\mathcal{X}}_T$ is also an $\widehat{\mathcal{E}}_{\widehat{B}\vee\widehat{S}}$-sufficient statistic, but this is of course evident since by (9.2.39), $\widehat{\mathcal{X}}_\Gamma \subset \widehat{\mathcal{X}}_T$.

Theorem 9.3.2 induces the following corollary on the original Bayesian experiment:

9.3.3 Corollary. In the Bayesian experiment $\mathcal{E} = (A \times S, \mathcal{A} \vee \mathcal{S}, \Pi)$, let $x = \{x_n : n \in I\!N\}$ be a Bayesian stochastic process with state space $(U, \mathcal{U})$ and let $B \subset \mathcal{A}$. Then

(i) x is B-stationary

if and only if

(ii) x is $\mathcal{X}_\Gamma$-stationary

and

(iii) $\mathcal{X}_\Gamma$ is an $\mathcal{E}_{\mathcal{B}\vee\mathcal{X}_0^\infty}$-sufficient statistic.

Moreover,

(iv) $\mathcal{X}_\Gamma^* = \overline{\mathcal{X}_\Gamma} \cap \mathcal{X}_0^\infty$.

Proof. By Lemma 9.2.7, (iv), $\mathcal{X}_\Gamma = x^{-1}(\widehat{\mathcal{S}_\Gamma})$. So, by Theorem 9.2.2, $\widehat{\mathcal{B}} \perp \widehat{\mathcal{S}} \mid \widehat{\mathcal{S}_\Gamma}; \widehat{\Pi}$ is equivalent to $\mathcal{B} \perp \mathcal{X}_0^\infty \mid \mathcal{X}_\Gamma; \Pi$. However, by (8.2.40), $\Gamma \mathrm{I} \widehat{\mathcal{S}} \mid \widehat{\mathcal{S}_\Gamma}$ is equivalent to $\widehat{\mathcal{S}_\Gamma}(f \circ \widehat{x} \circ \tau^k) = \widehat{\mathcal{S}_\Gamma}(f \circ \widehat{x}) \ \forall k \in I\!N, \forall f \in [\mathcal{U}^{I\!N}]^+$. Since, by (9.2.19), $\widehat{x} \circ \tau^k = \tau^k \circ \widehat{x}$, this is equivalent to $\mathcal{X}_\Gamma(f \circ \tau^k \circ x) = \mathcal{X}_\Gamma(f \circ x) \ \forall f \in [\mathcal{U}^{I\!N}]^+, \forall k \in I\!N$, i.e., x is $\mathcal{X}_\Gamma$-stationary. ∎

Note that (ii) is a property depending on the predictive probability only, whereas (iii) is a property of the posterior expectations.

By Theorem 4.7.8, the minimal sufficient parameter of a $\mathcal{B}$-stationary process will be exactly estimable if $\mathcal{X}_\Gamma$ is exactly estimating, i.e., $\mathcal{X}_\Gamma \subset \overline{\mathcal{B}}$.

Note that in terms of Chapter 8, this may also be interpreted as Γ is ergodic on $\mathcal{X}_0^\infty$ conditionally on $\mathcal{B}$. As shown in Theorem 8.2.34, a sufficient condition for this ergodic property is the mixing condition given in Definition 8.2.33. For the special case of a $\mathcal{B}$-stationary process x, this mixing condition may be written as

$$(9.3.15) \quad \forall f, g \in [\mathcal{U}^{I\!N}]^+ \quad \mathcal{B}(f \circ x \cdot g \circ \tau^n \circ x) \longrightarrow \mathcal{B}(f \circ x)\mathcal{B}(g \circ x)$$
$$\mathrm{a.s.}\Pi \quad \text{as } n \to \infty$$

Note that under such a mixing condition we obtain from Theorem 4.7.8, or from (8.3.23) and (8.3.24),

$$(9.3.16) \qquad \overline{\mathcal{X}_\Gamma} = \overline{\mathcal{B}\mathcal{X}_0^\infty} = \overline{\mathcal{X}_0^\infty \mathcal{B}} = \overline{\mathcal{B}} \cap \overline{\mathcal{X}_0^\infty},$$

i.e., in $\mathcal{E}_{\mathcal{B}\vee\mathcal{X}_0^\infty}$ there exist almost sure equalities between the shift-invariant σ-field of the process x, the minimal sufficient parameter, the minimal sufficient statistic, and the maximal exactly estimating statistic and estimable parameter.

The following proposition follows directly from Theorems 8.2.23(i) and 8.2.25.

9.3.4 Proposition. If x is a $\mathcal{B}$-stationary Bayesian process then, for any $\mathcal{C} \subset \mathcal{B}$, x is a $\mathcal{C}$-stationary Bayesian process. In particular, x is an $\mathcal{I}$-stationary Bayesian process, i.e., a predictively stationary Bayesian process. ∎

Note, however, that although Γ is ergodic on $\mathcal{X}_0^\infty$ conditionally on $\mathcal{B}$, Γ is not necessarily ergodic on $\mathcal{X}_0^\infty$ conditionally on $\mathcal{C}$. We will see in the next sections that the condition of ergodicity becomes redundant once more structure, such as conditional independence properties, is imposed on the stochastic process since the mixing Condition (9.3.15) may be interpreted as an asymptotic conditional independence.

9.3.2 Exchangeable and i.i.d. Processes

The structure of an exchangeable process is usually presented as follows. In the statistical experiment $\mathcal{E}$, defined as

(9.3.17) $$\mathcal{E} = \{(S, \mathcal{S}); \ P^a \quad a \in A\},$$

the stochastic process $x = \{x_n : n \in I\!N\}$ defined on $(S, \mathcal{S})$ with values in $(U, \mathcal{U})$ is exchangeable if

(9.3.18) $$\mathcal{I}^a\big[f[x_{\sigma(0)}, x_{\sigma(1)}, \dots, x_{\sigma(k)}]\big] = \mathcal{I}^a[f(x_0, x_1, \dots, x_k)]$$
$$\forall \, k \in I\!N, \quad \forall \, \sigma \in \Sigma_{k+1}, \quad \forall \, a \in A, \quad \forall \, f \in [\mathcal{U}^{k+1}]^+.$$

Using the definitions of the finite permutation operators presented in Section 9.2, in the Bayesian experiment

(9.3.19) $$\mathcal{E} = \{A \times S, \mathcal{A} \vee \mathcal{S}, \Pi\}$$

(9.3.18) implies that, for any prior probability,

(9.3.20) $$\mathcal{A}(f \circ \sigma \circ x) = \mathcal{A}(f \circ x) \quad \forall \, f \in [\mathcal{U}^{I\!N}]^+, \quad \forall \, \sigma \in \Sigma.$$

In the representation $\widehat{\mathcal{E}}$ of $\mathcal{E}$ defined in Section 9.2.3, and using (9.2.15) and (9.2.32), (9.3.20) becomes:

(9.3.21) $$\widehat{\mathcal{A}}(\widehat{t} \circ \sigma) = \widehat{\mathcal{A}}(\widehat{t}) \quad \forall \, \widehat{t} \in [\widehat{\mathcal{S}}]^+, \quad \forall \, \sigma \in \Sigma.$$

According to Definition (8.2.22), and since $\widehat{\mathcal{A}}_\Sigma = \widehat{\mathcal{A}}$ by Lemma 9.2.9, (9.3.21) amounts to saying that

$$(9.3.22) \qquad\qquad \Sigma \; \mathbb{I} \; \widehat{\mathcal{S}} \mid \widehat{\mathcal{A}}.$$

Note also that, since Σ is a countable group, by Theorem 8.2.20,

$$(9.3.23) \qquad\qquad (\widehat{\mathcal{A}} \vee \widehat{\mathcal{S}})^*_\Sigma = \overline{(\widehat{\mathcal{A}} \vee \widehat{\mathcal{S}})_\Sigma}$$

and the same relation holds for any Σ-stable sub-σ-field of $\widehat{\mathcal{A}} \vee \widehat{\mathcal{S}}$.

This leads to the following definition of Bayesian exchangeability:

9.3.5 Definition. In the Bayesian experiment $\mathcal{E} = (A \times S, \mathcal{A} \vee \mathcal{S}, \Pi)$ *the Bayesian stochastic process* $x = \{x_n : n \in I\!N\}$ *is* $\mathcal{M}$-*exchangeable*, where $\mathcal{M} \subset \mathcal{A} \vee \mathcal{S}$, if

$$\mathcal{M}(f \circ \sigma \circ x) = \mathcal{M}(f \circ x) \quad \forall \, f \in [\mathcal{U}^{I\!N}]^+, \quad \forall \, \sigma \in \Sigma. \qquad \blacksquare$$

Taking conditional expectations entails the following proposition:

9.3.6 Proposition. If $x = \{x_n : n \in I\!N\}$ is an $\mathcal{M}$-exchangeable Bayesian stochastic process, then $\forall \, \mathcal{N} \subset \mathcal{M}$, x is $\mathcal{N}$-exchangeable. In particular, x is $\mathcal{I}$-exchangeable. $\qquad \blacksquare$

Since, conditionally on $\mathcal{M}$, $(x_0, \ldots x_{n-1}, x_n)$ and $(x_1, \ldots, x_n, x_0)$ have the same distribution for any $n \in I\!N$, this is true for $(x_0, \ldots, x_{n-1})$ and $(x_1, \ldots, x_n)$. This entails the following proposition:

9.3.7 Proposition. If $x = \{x_n : n \in I\!N\}$ is a $\mathcal{M}$-exchangeable Bayesian stochastic process, then x is $\mathcal{M}$-stationary. $\qquad \blacksquare$

By duplicating the proofs of Theorem 9.3.2 and Corollary 9.3.3, we can reproduce Theorem 9.3.2 replacing Γ by Σ; we then obtain the following proposition:

9.3.8 Proposition. In the Bayesian experiment $\mathcal{E} = (A \times S, \mathcal{A} \vee \mathcal{S}, \Pi)$, let $\mathcal{B} \subset \mathcal{A}$, and $x = \{x_n : n \in I\!N\}$ be a Bayesian stochastic process then:

(i) x is $\mathcal{B}$-exchangeable

if and only if

(ii) x is $\mathcal{X}_\Sigma$-exchangeable

and

(iii) $\mathcal{X}_\Sigma$ is an $\mathcal{E}_{\mathcal{B} \vee \mathcal{X}_0^\infty}$-sufficient statistic.

Moreover,

(iv) $\mathcal{X}_\Sigma^* = \overline{\mathcal{X}}_\Sigma \cap \mathcal{X}_0^\infty.$ ∎

Note that, once again, (ii) is a property of the predictive probability only, whereas (iii) is a property of the posterior expectations. Indeed, (ii) is equivalent to

$$(9.3.24) \qquad \mathcal{X}_\Sigma(f \circ \sigma \circ x) = \mathcal{X}_\Sigma(f \circ x) \quad \forall \, \sigma \in \Sigma, \quad \forall \, f \in [\mathcal{U}^{I\!N}]^+,$$

whereas (iii) is equivalent to

$$(9.3.25) \qquad\qquad\qquad \mathcal{X}_0^\infty b = \mathcal{X}_\Sigma b \qquad \forall \, b \in [\mathcal{B}]^+.$$

This property reveals, in particular, that the posterior expectations are invariant under permutations of the components of the process x.

Characterizing exchangeable processes as the marginalization of i.i.d. processes is a well-established tradition in probability theory. The pioneering paper of de Finetti (1937) for $\{0, 1\}$ valued processes was first generalized by Hewitt and Savage (1955) before subsequently being extended in several ways. (For a good exposition see Chow and Teicher (1978), 7.3. Th. 2 and Dellacherie and Meyer (1980), Ch. 5; see also Freedman (1962, 1963), Pitman (1978), Aldous and Pitman (1979), Diaconis and Freedman (1984), and Aldous (1985)). The goal of Theorem 9.3.11 is to present a general Bayesian version of these results; this theorem also provides another complete dual characterization of a $\mathcal{B}$-exchangeable Bayesian process. Before stating the theorem, some additional definitions are required.

9.3.9 Definition. In the Bayesian experiment $\mathcal{E} = (A \times S, \mathcal{A} \vee \mathcal{S}, \Pi)$, let $\mathcal{M} \subset \mathcal{A} \vee \mathcal{S}$ and $x = \{x_n : n \in I\!\!N\}$ be a Bayesian stochastic process. Then

(i) x is $\mathcal{M}$-*identically distributed* if
$$\mathcal{M} \, g(x_k) = \mathcal{M} \, g(x_0) \quad \forall \, k \in I\!\!N, \quad \forall \, g \in [\mathcal{U}]^+;$$

(ii) x is $\mathcal{M}$-*independent* if
$$\underset{0 \leq n < \infty}{\perp\!\!\!\perp} \mathcal{X}_n \mid \mathcal{M};$$

(iii) x is $\mathcal{M}$-*i.i.d.* if it is $\mathcal{M}$-independent and $\mathcal{M}$-identically distributed. $\blacksquare$

Let us remark that, in the representation $\widehat{\mathcal{E}}$ of $\mathcal{E}$, that x is $\mathcal{B}$-identically distributed is actually equivalent to

(9.3.26) $$\Gamma \, I \, \widehat{\mathcal{X}_0} \mid \widehat{\mathcal{B}}.$$

Indeed, since $\widehat{\mathcal{B}}_\Gamma = \widehat{\mathcal{B}}$, $\Gamma \, I \, (\widehat{\mathcal{B}} \cap \overline{\widehat{\mathcal{I}}})$ by Proposition 8.2.13. Now,

$$\mathcal{B} \, g(x_k) = \mathcal{B} \, g(x_0) \quad \forall \, k \in I\!\!N, \, \forall \, g \in [\mathcal{U}]^+$$

is equivalent, by (9.2.15), to

$$\widehat{\mathcal{B}}(g \circ \widehat{x}_k) = \widehat{\mathcal{B}}(\widehat{g} \circ \widehat{x}_0) \quad \forall \, k \in I\!\!N, \, \forall \, g \in [\mathcal{U}]^+,$$

and since $\widehat{x}_k = \widehat{x}_0 \circ \tau^k$ this is equivalent to

$$\widehat{\mathcal{B}}(\widehat{t} \circ \tau^k) = \widehat{\mathcal{B}}(\widehat{t}) \quad \forall \, k \in I\!\!N, \, \forall \, \widehat{t} \in [\widehat{\mathcal{X}_0}]^+.$$

Note the following straightforward proposition.

9.3.10 Proposition. In the Bayesian experiment $\mathcal{E} = (A \times S, \mathcal{A} \vee \mathcal{S}, \Pi)$ let $\mathcal{M} \subset \mathcal{A} \vee \mathcal{S}$ and let $x = \{x_n : n \in I\!\!N\}$ be a Bayesian stochastic process. Then:

(i) if x is $\mathcal{M}$-stationary, then x is $\mathcal{M}$-identically distributed;

(ii) if x is $\mathcal{M}$-i.i.d., then x is $\mathcal{M}$-exchangeable. $\blacksquare$

Let us remark that, as a corollary, a $\mathcal{M}$-stationary and $\mathcal{M}$-independent process is a $\mathcal{M}$-i.i.d. process, and conversely. However, any marginalization on $\mathcal{N} \subset \mathcal{M}$ of an $\mathcal{M}$-i.i.d. process is $\mathcal{N}$-exchangeable, by Proposition 9.3.6, since it is $\mathcal{M}$-exchangeable.

The next theorem proves that any exchangeable process is a marginalization of an i.i.d. process.

9.3.11 Theorem. In the Bayesian experiment $\mathcal{E} = (A \times S, \mathcal{A} \vee \mathcal{S}, \Pi)$ let $\mathcal{B} \subset \mathcal{A}$ and let $x = \{x_n : n \in I\!N\}$ be a Bayesian stochastic process. Then

(i) x is $\mathcal{B}$-exchangeable

if and only if

(ii) x is $\mathcal{X}_\Gamma$-i.i.d.

and

(iii) $\mathcal{X}_\Gamma$ is an $\mathcal{E}_{\mathcal{B} \vee \mathcal{X}_0^\infty}$-sufficient statistic.

Moreover if x is $\mathcal{B}$-exchangeable, then

(iv) $\overline{\mathcal{X}}_\Gamma = \overline{\mathcal{X}}_T = \overline{\mathcal{X}}_\Sigma.$

Proof. (i) $\Rightarrow$ (ii) and (iii): Since $\mathcal{B}$-exchangeability implies $\mathcal{B}$-stationarity, we already know, by Corollary 9.3.3, that (i) implies (iii) and that x is $\mathcal{X}_\Gamma$-stationary. We also know, by Proposition 9.3.8(iii), that x is also $\mathcal{X}_\Sigma$-exchangeable.

Now, by the ergodic theorem (Proposition 8.2.15), $\forall\, k \in I\!N\ \forall\, g \in [\mathcal{U}]_\infty$, as $n \to \infty$,

$$\frac{1}{n} \sum_{0 \le \ell \le n-1} g(x_{k+\ell+1}) \to \mathcal{X}_\Gamma g(x_{k+1}) \quad \text{a.s.}\Pi \text{ and in } L_1.$$

Therefore, by the conditional dominated convergence theorem (see, for instance, Neveu (1970), Section IV-3), $\forall\, t \in [\mathcal{X}_0^k]_\infty$, as $n \to \infty$

$$
\frac{1}{n} \sum_{0 \leq \ell \leq n-1} \mathcal{X}_\Sigma[t\, g(x_{k+\ell+1})] \quad = \mathcal{X}_\Sigma \left[t\, \frac{1}{n} \sum_{0 \leq \ell \leq n-1} g(x_{k+\ell+1}) \right]
$$

$$
\to \quad \mathcal{X}_\Sigma[t\, \mathcal{X}_\Gamma[g(x_{k+1})]] \quad = \mathcal{X}_\Sigma(t) \cdot \mathcal{X}_\Gamma[g(x_{k+1})].
$$

But, by $\mathcal{X}_\Sigma$-exchangeability, $\forall\, \ell \geq 0$,

$$
\mathcal{X}_\Sigma[t\, g(x_{k+\ell+1})] = \mathcal{X}_\Sigma[t\, g(x_{k+1})].
$$

Therefore, $\forall\, k \in \mathbb{N}$, $\forall\, g \in [\mathcal{U}]_\infty$, $\forall\, t \in [\mathcal{X}_0^k]_\infty$,

$$
\mathcal{X}_\Sigma[t\, g(x_{k+1})] = \mathcal{X}_\Sigma t \cdot \mathcal{X}_\Gamma[g(x_{k+1})].
$$

Taking $t \equiv 1$ and conditional expectation with respect to $\mathcal{X}_\Gamma$ shows that $\mathcal{X}_\Sigma[g(x_{k+1})] = \mathcal{X}_\Gamma[g(x_{k+1})]$ and that

$$
\mathcal{X}_{k+1} \perp\!\!\!\perp \mathcal{X}_0^k \mid \mathcal{X}_\Gamma \quad \text{and} \quad \mathcal{X}_{k+1} \perp\!\!\!\perp \mathcal{X}_0^k \mid \mathcal{X}_\Sigma.
$$

By Theorem 7.6.3 and Proposition 9.3.10, x is $\mathcal{X}_\Gamma$-i.i.d. and $\mathcal{X}_\Sigma$-i.i.d. Hence (i) implies (ii). However, if $f_k \in [\mathcal{U}]_\infty$, $0 \leq k \leq n$,

$$
\mathcal{X}_\Sigma \left[\prod_{0 \leq k \leq n} f_k(x_k) \right] \quad = \prod_{0 \leq k \leq n} \mathcal{X}_\Sigma[f_k(x_k)]
$$

$$
= \prod_{0 \leq k \leq n} \mathcal{X}_\Gamma[f_k(x_k)]
$$

$$
= \mathcal{X}_\Gamma \left[\prod_{0 \leq k \leq n} f_k(x_k) \right].
$$

Therefore by Theorem 0.2.21, $\forall\, t \in [\mathcal{X}_0^\infty]_\infty$ $\mathcal{X}_\Sigma t = \mathcal{X}_\Gamma t$. Since clearly, $\mathcal{X}_\Gamma \subset \mathcal{X}_T \subset \mathcal{X}_\Sigma$, this shows that $\mathcal{X}_\Sigma \subset \overline{\mathcal{X}}_\Gamma$; this implies (iv).

(ii) and (iii) $\Rightarrow$ (i): By (iii), $\mathcal{B} \perp\!\!\!\perp \mathcal{X}_0^\infty \mid \mathcal{X}_\Gamma$, and (ii) is equivalent to $\mathcal{X}_0^n \perp\!\!\!\perp \mathcal{X}_{n+1}^\infty \mid \mathcal{X}_\Gamma$ $\forall\, n \in \mathbb{N}$ by Theorem 7.6.3. Therefore, by Theorem 7.6.4, $\mathcal{B} \perp\!\!\!\perp \mathcal{X}_0^n \perp\!\!\!\perp \mathcal{X}_{n+1}^\infty \mid \mathcal{X}_\Gamma$ and, by the same theorem, $\mathcal{X}_0^n \perp\!\!\!\perp (\mathcal{B} \vee \mathcal{X}_{n+1}^\infty) \mid \mathcal{X}_\Gamma$. By Corollary 2.2.11, $\mathcal{X}_0^n \perp\!\!\!\perp \mathcal{X}_{n+1}^\infty \mid \mathcal{B} \vee \mathcal{X}_\Gamma$ and this is equivalent, by Theorem 7.6.3, to saying that x is $(\mathcal{B} \vee \mathcal{X}_\Gamma)$-independent.

Now, by (ii) and (iii),

$$(\mathcal{B} \vee \mathcal{X}_\Gamma)[g(x_k)] = \mathcal{X}_\Gamma[g(x_k)] = \mathcal{X}_\Gamma[g(x_0)] = (\mathcal{B} \vee \mathcal{X}_\Gamma)[g(x_0)].$$

Therefore (ii) and (iii) imply that x is $(\mathcal{B} \vee \mathcal{X}_\Gamma)$-i.i.d. It is therefore $(\mathcal{B} \vee \mathcal{X}_\Gamma)$-exchangeable (Proposition 9.3.10(ii)) and hence $\mathcal{B}$-exchangeable (Proposition 9.3.6).   ∎

The mixing condition given in (9.3.15) is in fact an asymptotic conditional independence. It thus seems quite natural to expect that, for a $\mathcal{B}$-i.i.d. process, $\mathcal{X}_\Gamma$ would be exactly estimating and we indeed have the following theorem:

9.3.12 Theorem. In the Bayesian experiment $\mathcal{E} = (A \times S, \mathcal{A} \vee \mathcal{S}, \Pi)$, let $\mathcal{B} \subset \mathcal{A}$ and $x = \{x_n : n \in I\!N\}$ be a Bayesian stochastic process. Then

(i) x is $\mathcal{B}$-i.i.d.

if and only if

(ii) x is $\mathcal{X}_\Gamma$-i.i.d.

and

(iii) $\mathcal{X}_\Gamma$ is an $\mathcal{E}_{\mathcal{B} \vee \mathcal{X}_0^\infty}$-sufficient and exactly estimating statistic.

Moreover , if x is $\mathcal{B}$-i.i.d.,

(iv) $\mathcal{B}\mathcal{X}_0^\infty = \mathcal{B}\mathcal{X}_0$

(v) $\overline{\mathcal{X}_\Gamma} = \overline{\mathcal{X}_T} = \overline{\mathcal{X}_\Sigma} = \overline{\mathcal{B}\mathcal{X}_0} = \overline{\mathcal{X}_0^\infty \mathcal{B}} = \overline{\mathcal{B}} \cap \overline{\mathcal{X}_0^\infty}.$

Proof. Since (i) implies $\mathcal{B}$-exchangeability by Proposition 9.3.10(ii), (i) implies (ii) and $\mathcal{B} \perp\!\!\!\perp \mathcal{X}_0^\infty \mid \mathcal{X}_\Gamma$ by Theorem 9.3.11. Now, by Theorem 7.6.5, x $\mathcal{B}$-independent implies $\mathcal{X}_T \subset \overline{\mathcal{B}}$ which implies $\mathcal{X}_\Gamma \subset \overline{\mathcal{B}}$, i.e., $\mathcal{X}_\Gamma$ is exactly estimating. Now x $\mathcal{B}$-independent implies, by Corollary 7.6.7(i), that

$\mathcal{B}\mathcal{X}_0^\infty = \bigvee_{0 \le n \le \infty} \mathcal{B}\mathcal{X}_n$; x $\mathcal{B}$-identically distributed implies $\mathcal{B}\mathcal{X}_n = \mathcal{B}\mathcal{X}_0$. This shows (iv). Since $\mathcal{X}_\Gamma \subset \overline{\mathcal{B}}$, Theorems 9.3.11(iv) and (9.3.16) imply (v). However, the proof of Theorem 9.3.11 shows that (ii) and $\mathcal{B} \perp\!\!\!\perp \mathcal{X}_0^\infty \mid \mathcal{X}_\Gamma$ imply that x is $(\mathcal{B} \vee \mathcal{X}_\Gamma)$-i.i.d. It is therefore $\mathcal{B}$-i.i.d. under $\mathcal{X}_\Gamma \subset \overline{\mathcal{B}}$. ∎

This theorem shows that if x is $\mathcal{B}$-i.i.d. the minimal sufficient parameter $\mathcal{B}\mathcal{X}_0^\infty$ is exactly estimable, i.e., $\mathcal{B}\mathcal{X}_0^\infty \subset \overline{\mathcal{X}_0^\infty}$ and, in particular, if $\mathcal{B}$ is identified, i.e., $\mathcal{B} = \mathcal{B}\mathcal{X}_0^\infty$, then $\mathcal{B}$ is exactly estimable. Note that, since $\mathcal{B}\mathcal{X}_0^\infty = \mathcal{B}\mathcal{X}_0$, $\mathcal{B}$ will be identified by the process if and only if it is identified by the first observation.

It is interesting to remark that, by a slight conditional extension of Hewitt-Savage Theorem, x $\mathcal{B}$-i.i.d. implies that $\mathcal{X}_\Sigma \subset \overline{\mathcal{B}}$. This result has been obtained here without using the techniques of the proof of the Hewitt-Savage Theorem.

Let us remark that, comparing Corollary 9.3.3 with Theorem 9.3.11, the stronger condition of $\mathcal{B}$-exchangeability rather than $\mathcal{B}$-stationarity merely implies a stronger property of the predictive probability, namely, $\mathcal{X}_\Gamma$-i.i.d. instead of $\mathcal{X}_\Gamma$-stationarity. Consequently this cannot greatly simplify the conditions under which $\mathcal{X}_\Gamma$ is exactly estimating. On the contrary, the next corollary show that, if x is not $\mathcal{B}$-i.i.d, $\mathcal{X}_\Gamma$ is not exactly estimating.

9.3.13 Corollary. Let $x = \{x_n : n \in I\!N\}$ be a $\mathcal{B}$-exchangeable Bayesian stochastic process. Then $\mathcal{X}_\Sigma$ is exactly estimating if and only if x is $\mathcal{B}$-i.i.d.

Proof. As shown in the proof of Theorem 9.3.11, x $\mathcal{B}$-exchangeable is equivalent to x $(\mathcal{B} \vee \mathcal{X}_\Gamma)$-i.i.d. Therefore $\mathcal{X}_\Gamma \subset \overline{\mathcal{B}}$ implies x $\mathcal{B}$-i.i.d. and, by Theorem 9.3.12, x $\mathcal{B}$-i.i.d. implies $\mathcal{X}_\Gamma \subset \overline{\mathcal{B}}$. ∎

This corollary shows that a stochastic process which is $\mathcal{B}$-exchangeable but not $\mathcal{B}$-i.i.d., and for which the minimal sufficient parameter $\mathcal{B}\mathcal{X}_0^\infty$ is exactly estimable ($\subset \overline{\mathcal{X}_0^\infty}$), has a minimal sufficient statistic, $\mathcal{X}_0^\infty \mathcal{B} = \mathcal{X}_\Sigma \mathcal{B}$, strictly smaller than $\mathcal{X}_\Sigma$, i.e., $\mathcal{X}_\Sigma \mathcal{B} \ne \mathcal{X}_\Sigma$. This means that in such a process one must seek a sufficient and exactly estimating statistic inside $\mathcal{X}_\Sigma$, since $\mathcal{X}_\Sigma$ is too large. Therefore, unlike $\mathcal{B}$-stationary processes, the exact estimability of the minimal sufficient parameter of a $\mathcal{B}$-exchangeable process cannot rely on the mixing Condition (9.3.15) unless it is $\mathcal{B}$-i.i.d.

Nonetheless, since any B-exchangeable process is a marginalization of an i.i.d. process, it is worthwhile to studying these processes in the hope of obtaining more tractable conditions for the exact estimability of the minimal sufficient parameter.

9.3.14 Definition. In the Bayesian experiment $\mathcal{E} = (A \times S, \mathcal{A} \vee \mathcal{S}, \Pi)$, a Bayesian stochastic process $x = \{x_n : n \in I\!N\}$ with state space $(U, \mathcal{U})$ is a *B-generalized i.i.d. process* with $B \subset \mathcal{A}$ *with respect to* $\mathcal{M} \subset \mathcal{A} \vee \mathcal{S}$ if

(i) x is $\mathcal{M}$-i.i.d.

(ii) $B \perp\!\!\!\perp \mathcal{X}_0^\infty \mid \mathcal{M}.$ ∎

Note that, from Theorem 9.3.11, if x is B-exchangeable, then it is a B-generalized i.i.d. process with respect to $\mathcal{X}_\Gamma$. However, reproducing the end of the proof of Theorem 9.3.11 shows that if x is a B-generalized i.i.d. process with respect to $\mathcal{M}$, then it is a $(B \vee \mathcal{M})$-i.i.d. process and, consequently, a B-exchangeable process by Propositions 9.3.6 and 9.3.10(ii).

The advantage of this definition derives from the fact that the experiment $\mathcal{E}_{B \vee \mathcal{X}_0^\infty}$ is entirely characterized by the experiment $\mathcal{E}_{B \vee \mathcal{M} \vee \mathcal{X}_0}$. Therefore the condition for exact estimability of the minimal sufficient parameter relies merely on $\mathcal{E}_{B \vee \mathcal{M} \vee \mathcal{X}_0}$, where $\mathcal{M}$ (in contrast with $\mathcal{X}_\Gamma$) will generally be nonasymptotic. Indeed, we have in fact the following result:

9.3.15 Theorem. Let $x = \{x_n : n \in I\!N\}$ be a B-generalized i.i.d. process with respect to $\mathcal{M}$. The minimal sufficient parameter is exactly estimable if and only if

$$\overline{B(\mathcal{M}\mathcal{X}_0)} = \overline{(\mathcal{M}\mathcal{X}_0)B} = \overline{B} \cap \overline{\mathcal{M}\mathcal{X}_0}.$$

Proof. By Proposition 4.7.7, the minimal sufficient parameter will be exactly estimable in $\mathcal{E}_{B \vee \mathcal{X}_0^\infty}$ if and only if $\overline{\mathcal{X}_0^\infty B} = \overline{B \mathcal{X}_0^\infty}$. But $\mathcal{X}_\Gamma$ is sufficient in $\mathcal{E}_{B \vee \mathcal{X}_0^\infty}$, by Theorem 9.3.11, since x is B-exchangeable. Therefore, by Theorem 4.3.7, $\overline{\mathcal{X}_0^\infty B} = \overline{\mathcal{X}_\Gamma B}$ and $B \mathcal{X}_0^\infty = B \mathcal{X}_\Gamma$. Now by Theorem 9.3.12, x $\mathcal{M}$-i.i.d. implies that $\overline{\mathcal{X}_\Gamma} = \overline{\mathcal{M}\mathcal{X}_0}$. Therefore the minimal sufficient parameter will be exactly estimable, i.e., $B \mathcal{X}_0^\infty \subset \overline{\mathcal{X}_0^\infty}$, if and only if $\overline{(\mathcal{M}\mathcal{X}_0)B} = \overline{B(\mathcal{M}\mathcal{X}_0)}$. It follows from Section 4.7 that this implies that

$$\overline{(\mathcal{M}\mathcal{X}_0)B} = \overline{B(\mathcal{M}\mathcal{X}_0)} = \overline{B} \cap \overline{\mathcal{M}\mathcal{X}_0}$$

and, by Proposition 4.7.7, this relation is in fact equivalent to

$$\mathcal{B}(\mathcal{M}\mathcal{X}_0) \subset \overline{\mathcal{M}\mathcal{X}_0},$$

and is also equivalent to

$$(\mathcal{M}\mathcal{X}_0)\mathcal{B} \subset \overline{\mathcal{B}}. \qquad \blacksquare$$

Let us remark that the condition of exact estimability of $\mathcal{B}$-generalized i.i.d. processes with respect to $\mathcal{M}$ does not imply that $\mathcal{X}_\Gamma$ is exactly estimating but in fact requires that $\mathcal{X}_\Gamma\mathcal{B}$ is exactly estimating.

Theorem 9.3.15 leads to the following important corollary:

9.3.16 Corollary. In any marginalization on the parameter space of an identified i.i.d. process the minimal sufficient parameter is exactly estimable. In other words, if $x = \{x_n : n \in I\!N\}$ is a $\mathcal{B}$-i.i.d. process with $\mathcal{B}\mathcal{X}_0 = \mathcal{B}$, then for any $\mathcal{C} \subset \mathcal{B}$, $\mathcal{C}\mathcal{X}_0^\infty \subset \overline{\mathcal{X}_0^\infty}$.

Proof. Since $\mathcal{C} \subset \mathcal{B}$, Condition (ii) of Definition 9.3.14, i.e., $\mathcal{C} \perp\!\!\!\perp \mathcal{X}_0^\infty \mid \mathcal{B}$ is trivially satisfied, and x is therefore a $\mathcal{C}$-generalized i.i.d. process with respect to $\mathcal{B}$. Since $\mathcal{B}\mathcal{X}_0 = \mathcal{B}$, $\mathcal{C}[\mathcal{B}\mathcal{X}_0] = \mathcal{C}\mathcal{B} = \mathcal{C} \subset \mathcal{B} = \mathcal{B}\mathcal{X}_0$. Therefore $\mathcal{C}[\mathcal{B}\mathcal{X}_0] \subset \overline{\mathcal{B}\mathcal{X}_0}$ and the condition of Theorem 9.3.15 is satisfied. $\qquad \blacksquare$

9.3.3 Moving Average Processes

In time series analysis moving average processes are defined by representing a stationary process in terms of an i.i.d. process. More precisely, let $x = \{x_n : n \in I\!N\}$ be a $\mathcal{B}$-i.i.d. process with state space $(I\!R^d, \mathcal{B}^d)$. The stationary process $y = \{y_n : q \leq n < \infty\}$ with state space $(I\!R^d, \mathcal{B}^d)$ is defined in terms of x and matrix parameters $(C_k : 1 \leq k \leq q)$, where $C_k \in [\mathcal{B}]$, by

$$(9.3.27) \qquad y_n = x_n + \sum_{1 \leq k \leq q} C_k x_{n-k}, \quad \forall\, n \geq q.$$

The study of the exact estimability of such a process does not rely on the linear structure (9.3.27). This structure may be extended, for instance, to

$$(9.3.28) \qquad y_n = f(c, x_n, x_{n-1}, \dots x_{n-q}).$$

Note that (9.3.28) implies that $\mathcal{Y}_n \subset \mathcal{C} \vee \mathcal{X}_{n-q}^n$ and, therefore,

$$\mathcal{Y}_{n+q+1} \perp\!\!\!\perp \mathcal{Y}_q^n \mid \mathcal{B}, \qquad \forall\, n \geq q,$$

i.e., y is a q-dependent process. Thus, in a Bayesian framework, a very general notion of moving average process may be considered:

9.3.17 Definition. In the Bayesian experiment $\mathcal{E} = (A \times S, \mathcal{A} \vee \mathcal{S}, \Pi)$ a $\mathcal{B}$-stationary process $y = \{y_n : n \in I\!\!N\}$ with state space $(U, \mathcal{U})$ is a *moving-average process of order q* if there exist $\mathcal{C} \subset \mathcal{B}$ and a $\mathcal{B}$-i.i.d. process $x = \{x_n : n \in I\!\!N\}$ with state space $(V, \mathcal{V})$ such that

$$\mathcal{Y}_n \subset \mathcal{C} \vee \mathcal{X}_{n-q}^n, \qquad \forall\, n \geq q.$$

This leads to the following theorem:

9.3.18 Theorem. In a moving average of order q $\mathcal{B}$-stationary process y, the minimal sufficient parameter is exactly estimable. Moreover,

$$\overline{\mathcal{Y}_\Gamma} = \overline{\mathcal{Y}_T} = \overline{\mathcal{B}\mathcal{Y}_0^\infty} = \overline{\mathcal{Y}_0^\infty \mathcal{B}} = \overline{\mathcal{B}} \cap \overline{\mathcal{Y}_0^\infty}.$$

Proof. By $\mathcal{B}$-stationarity (Corollary 9.3.3), we know that $\mathcal{Y}_\Gamma$ and $\mathcal{Y}_T$ are sufficient statistics and $\mathcal{Y}_\Gamma \subset \mathcal{Y}_T$. By Definition 9.3.17, it is easily seen that $\mathcal{Y}_T \subset \bigcap_n (\mathcal{C} \vee \mathcal{X}_n^\infty)$. Now, since $\mathcal{C} \subset \mathcal{B}$ and $\mathcal{X}_{n+1}^\infty \perp\!\!\!\perp \mathcal{X}_0^n \mid \mathcal{B}$, by Theorem 7.6.5, $(\mathcal{C} \vee \mathcal{X}_n)_T \subset \overline{\mathcal{B}}$. Hence $\mathcal{Y}_\Gamma$ and $\mathcal{Y}_T$ are exactly estimating and the result follows from Section 4.7.   ∎

9.3.4   Markovian Stationary Processes

The structure of a Markovian process is usually presented as follows. In the statistical experiment $\mathcal{E}$, defined as

$$(9.3.29) \qquad\qquad \mathcal{E} = \{(S, \mathcal{S}); P^a\ a \in A\},$$

the stochastic process $x = \{x_n : n \in I\!\!N\}$ defined on $(S, \mathcal{S})$ with values in $(U, \mathcal{U})$ is Markovian if:

$$(9.3.30) \qquad\qquad (\mathcal{X}_0^n)^a\ f(x_{n+1}) = (\mathcal{X}_n)^a\ f(x_{n+1})$$
$$\forall\, n \in I\!\!N, \quad \forall\, a \in A, \quad \forall\, f \in [\mathcal{U}]^+.$$

Recall that, for $s \in [\mathcal{S}]^+$ and $\mathcal{T} \subset \mathcal{S}$, $\mathcal{T}^a s$ denotes the expectation of s conditional on $\mathcal{T}$ with respect to P^a. In the Bayesian experiment

$$(9.3.31) \qquad \mathcal{E} = \{A \times S, \mathcal{A} \vee \mathcal{S}, \Pi\}$$

this implies that, for any prior probability,

$$(9.3.32) \qquad \mathcal{X}_{n+1} \perp \mathcal{X}_0^n \mid \mathcal{A} \vee \mathcal{X}_n.$$

Recall that such a relation was analyzed in Section 6.4.2. We are now led to the following definition:

9.3.19 Definition. In the Bayesian experiment $\mathcal{E} = (A \times S, \mathcal{A} \vee \mathcal{S}, \Pi)$, the Bayesian stochastic process $x = \{x_n : n \in I\!\!N\}$ is $\mathcal{M}$-*Markovian*, where $\mathcal{M} \subset \mathcal{A} \vee \mathcal{S}$, if:

$$(9.3.33) \qquad \mathcal{X}_{n+1} \perp \mathcal{X}_0^n \mid \mathcal{M} \vee \mathcal{X}_n$$

or, equivalently,

$$(9.3.34) \qquad \mathcal{X}_{n+1}^\infty \perp \mathcal{X}_0^n \mid \mathcal{M} \vee \mathcal{X}_n.$$

For the equivalence of (9.3.33) and (9.3.34) see Definition 6.4.4 or Theorem 6.2.2.

It is well known that the stationarity of a Markovian process is equivalent to the stationarity of the transitions combined with an initial condition; this fact motivates the definition of the following concept:

9.3.20 Definition. In the Bayesian experiment $\mathcal{E} = (A \times S, \mathcal{A} \vee \mathcal{S}, \Pi)$, the Bayesian stochastic process $x = \{x_n : n \in I\!\!N\}$ is $\mathcal{B}$-*homogeneous*, where $\mathcal{B} \subset \mathcal{A}$, if $\forall f \in [\mathcal{U}]^+$ $\exists h \in [\mathcal{B} \otimes \mathcal{U}]^+$ such that,

$$(9.3.35) \quad (\mathcal{B} \vee \mathcal{X}_n)(f \circ x_{n+1}) = h(a, x_n) \quad \text{a.s.}\Pi \quad \forall\, n \in I\!\!N. \qquad \blacksquare$$

In the representation $\widehat{\mathcal{E}}$ of $\mathcal{E}$, defined in Section 9.2.3, and using the properties of the Shift operator as seen in Section 9.2.4, (9.3.35) becomes

$$(9.3.36) \qquad \tau^{-n}(\widehat{\mathcal{B}} \vee \widehat{\mathcal{X}}_0)(\widehat{t} \circ \tau^n) = \{(\widehat{\mathcal{B}} \vee \widehat{\mathcal{X}}_0)\widehat{t}\} \circ \tau^n$$
$$\forall\, \widehat{t} \in [\widehat{\mathcal{X}}_1]^+, \forall\, n \in I\!\!N.$$

By Definition 8.2.22(ii), this means that

$$(9.3.37) \qquad \Gamma \text{ I } \widehat{\mathcal{X}}_1 \mid \widehat{\mathcal{B}} \vee \widehat{\mathcal{X}}_0$$

as soon as

$$(9.3.38) \qquad \Gamma \text{ I } (\widehat{\mathcal{B}} \vee \widehat{\mathcal{X}}_0) \cap \overline{\overline{\mathcal{I}}}.$$

For a $\mathcal{B}$-Markovian process, (9.3.36) may actually be extended to any $\widehat{t} \in [\widehat{\mathcal{X}}_1^\infty]^+$, as shown in the following lemma:

9.3.21 Lemma. Let $x = \{x_n : n \in I\!N\}$ be a $\mathcal{B}$-Markovian process. Then

(i) $\Gamma \text{ I } \widehat{\mathcal{X}}_1 \mid \widehat{\mathcal{B}} \vee \widehat{\mathcal{X}}_0$

if and only if

(ii) $\Gamma \text{ I } \widehat{\mathcal{X}}_0^\infty \mid \widehat{\mathcal{B}} \vee \widehat{\mathcal{X}}_0.$

Proof. By Theorem 0.2.21 it suffices to show that, $\forall n \in I\!N$, $\Gamma \text{ I } \widehat{\mathcal{X}}_0^n \mid \widehat{\mathcal{B}} \vee \widehat{\mathcal{X}}_0$. Now, by Theorem 8.2.23(ii), this is true for $n = 1$ and, by an induction argument, it suffices to show that if it is true for n, then it is true for $n+1$. By Theorem 8.2.23(iv), $\Gamma \text{ I } (\widehat{\mathcal{X}}_0^n \vee \widehat{\mathcal{B}}) \cap \overline{\overline{\mathcal{I}}}$. Therefore, since $\tau^{n+k} = \tau^n \circ \tau^k$, $\Gamma \text{ I } \widehat{\mathcal{X}}_1 \mid \widehat{\mathcal{B}} \vee \widehat{\mathcal{X}}_0$ implies that $\Gamma \text{ I } \widehat{\mathcal{X}}_{n+1} \mid \widehat{\mathcal{B}} \vee \widehat{\mathcal{X}}_n$. Indeed, $\forall f \in [\mathcal{U}]^+$,

$$\tau^{-k}(\widehat{\mathcal{B}} \vee \widehat{\mathcal{X}}_n)(f \circ \widehat{x}_{n+1} \circ \tau^k) = (\widehat{\mathcal{B}} \vee \widehat{\mathcal{X}}_0)(f \circ \widehat{x}_1) \circ \tau^n \circ \tau^k \quad \forall k \in I\!N$$

and

$$(\mathcal{B} \vee \widehat{\mathcal{X}}_n)(f \circ \widehat{x}_{n+1}) = (\mathcal{B} \vee \widehat{\mathcal{X}}_0)(f \circ \widehat{x}_1) \circ \tau^n.$$

By Theorem (8.2.30),

$$\widehat{\mathcal{X}}_{n+k+1} \perp\!\!\!\perp \widehat{\mathcal{X}}_k^{n+k} \mid \widehat{\mathcal{B}} \vee \widehat{\mathcal{X}}_{n+k}$$

(implied by the $\mathcal{B}$-Markovian property),

$$\Gamma \text{ I } \widehat{\mathcal{X}}_{n+1} \mid \widehat{\mathcal{B}} \vee \widehat{\mathcal{X}}_n \quad \text{and} \quad \Gamma \text{ I } (\widehat{\mathcal{B}} \vee \widehat{\mathcal{X}}_0^n) \cap \overline{\overline{\mathcal{I}}}$$

is equivalent to

$$\widehat{\mathcal{X}}_{n+1} \perp\!\!\!\perp \widehat{\mathcal{X}}_0^n \mid \widehat{\mathcal{B}} \vee \widehat{\mathcal{X}}_n \quad \text{and} \quad \Gamma \text{ I } \widehat{\mathcal{X}}_{n+1} \mid \widehat{\mathcal{B}} \vee \widehat{\mathcal{X}}_0^n.$$

But, by Theorem 8.2.25, this along with $\Gamma \text{ I } \widehat{\mathcal{X}}_0^n \mid \widehat{\mathcal{B}} \vee \widehat{\mathcal{X}}_0$ is equivalent to $\Gamma \text{ I } \widehat{\mathcal{X}}_0^{n+1} \mid \widehat{\mathcal{B}} \vee \widehat{\mathcal{X}}_0$. ∎

By Theorem 8.2.25, we know that

$$(9.3.39) \qquad\qquad \Gamma \text{ I } \widehat{\mathcal{X}}_0^\infty \mid \widehat{\mathcal{B}},$$

is equivalent to $\Gamma \text{ I } \widehat{\mathcal{X}}_0^\infty \mid \widehat{\mathcal{B}} \vee \widehat{\mathcal{X}}_0$ along with

$$(9.3.40) \qquad\qquad \Gamma \text{ I } \widehat{\mathcal{X}}_0 \mid \widehat{\mathcal{B}},$$

which means that $\widehat{x}$ is $\widehat{\mathcal{B}}$-identically distributed (see Definition 9.3.9(i)). Therefore a $\mathcal{B}$-Markovian process is $\mathcal{B}$-stationary if and only if it is $\mathcal{B}$-homogeneous and $\mathcal{B}$-identically distributed. More precisely, we have the following lemma:

9.3.22 Lemma. A $\mathcal{B}$-Markovian process $x = \{x_n : n \in I\!\!N\}$ is $\mathcal{B}$-stationary if and only if it is $\mathcal{B}$-homogeneous and

$$(9.3.41) \qquad\qquad \mathcal{B}(f \circ x_1) = \mathcal{B}(f \circ x_0) \quad \forall\, f \in [\mathcal{U}]^+.$$

Proof. Clearly, $\mathcal{B}$-stationarity implies (9.3.41) and $\mathcal{B}$-homogeneity. Now we show that (9.3.35) and (9.3.41) imply that x is $\mathcal{B}$-identically distributed. Indeed, by an induction argument,

$$\begin{aligned}
\mathcal{B}(f \circ x_{n+1}) &= \mathcal{B}[(\mathcal{B} \vee \mathcal{X}_n)(f \circ x_{n+1})] = \mathcal{B}[h(a, x_n)] \\
&= \mathcal{B}[h(a, x_0)] = \mathcal{B}[(\mathcal{B} \vee \mathcal{X}_0)(f \circ x_1)] = \mathcal{B}(f \circ x_1),
\end{aligned}$$

i.e., if x_n is $\mathcal{B}$-distributed as x_0, x_{n+1} is $\mathcal{B}$-distributed as x_1. Now in the representation $\widehat{\mathcal{E}}$ of $\mathcal{E}$, since $\widehat{\mathcal{B}}_\Gamma = \widehat{\mathcal{B}}$, by Proposition 8.2.13, $\Gamma \text{ I } (\widehat{\mathcal{B}} \cap \widehat{\overline{\mathcal{I}}})$ and therefore, by (8.2.40), $\Gamma \text{ I } \widehat{\mathcal{X}}_0 \mid \widehat{\mathcal{B}}$. By Theorem 8.2.23(iv), this implies that $\Gamma \text{ I } (\widehat{\mathcal{B}} \vee \widehat{\mathcal{X}}_0) \cap \widehat{\overline{\mathcal{I}}}$ and, therefore, $\mathcal{B}$-homogeneity is equivalent to $\Gamma \text{ I } \widehat{\mathcal{X}}_1 \mid \mathcal{B} \vee \widehat{\mathcal{X}}_0$. By Lemma 9.3.21, this is equivalent to $\Gamma \text{ I } \widehat{\mathcal{X}}_0^\infty \mid \widehat{\mathcal{B}} \vee \widehat{\mathcal{X}}_0$ and, therefore, $\Gamma \text{ I } \widehat{\mathcal{X}}_0^\infty \mid \widehat{\mathcal{B}}$. ∎

The exactly estimating property of $\mathcal{X}_\Gamma$ for a $\mathcal{B}$-stationary process has been easily demonstrated under the additional structure of $\mathcal{B}$-independence. One way to relax this independence hypothesis is to suppose a $\mathcal{B}$-Markovian property where independence is replaced by conditional independence. The following theorem represents a crucial step into this direction:

9.3.23 Theorem. In the Bayesian experiment $\mathcal{E} = (A \times S, \mathcal{A} \vee \mathcal{S}, \Pi)$, let $\mathcal{B} \subset \mathcal{A}$ and $x = \{x_n : n \in I\!N\}$ be a Bayesian stochastic process. Then

(o) x is $\mathcal{B}$-Markovian and $\mathcal{B}$-stationary

if and only if

(i) x is $\mathcal{X}_\Gamma$-Markovian and $\mathcal{X}_\Gamma$-stationary,

(ii) $\mathcal{X}_\Gamma$ is an $\mathcal{E}_{\mathcal{B} \vee \mathcal{X}_0^\infty}$-sufficient statistic,

(iii) $\mathcal{X}_\Gamma \subset \overline{\mathcal{B} \vee \mathcal{X}_0}$.

Moreover, if x is $\mathcal{B}$-Markovian and $\mathcal{B}$-stationary,

(iv) $\forall\, f \in [\mathcal{U}_\Gamma]^+\ \exists\, h \in [\mathcal{B} \otimes \mathcal{U}]^+$ such that
        $f \circ x = h(a, x_n)$ a.s. $\Pi\quad \forall\, n \in I\!N$

which implies, in particular, that

(v) $\mathcal{X}_\Gamma \subset \underset{0 \le n < \infty}{\cap} \overline{\mathcal{B} \vee \mathcal{X}_n}$.

Proof. (o) $\Rightarrow$ (i) (ii) (iii). By Corollary 9.3.3., we know that (o) implies (ii) and that x is $\mathcal{X}_\Gamma$-stationary. To prove that (o) implies (iv) and thus (iii), note first that, by the $\mathcal{B}$-Markov property for any $f \in [\mathcal{U}_\Gamma]_\infty$

$$(\mathcal{B} \vee \mathcal{X}_n)(f \circ x) = (\mathcal{B} \vee \mathcal{X}_n)(f \circ \tau^n \circ x)$$
$$= (\mathcal{B} \vee \mathcal{X}_0^n)(f \circ \tau^n \circ x) = (\mathcal{B} \vee \mathcal{X}_0^n)(f \circ x).$$

Therefore, by Theorem 7.2.2,

$$(\mathcal{B} \vee \mathcal{X}_n)(f \circ x) \to f \circ x \quad \text{a.s.}\Pi \text{ and in } L_1.$$

By Lemma 9.3.21, let $h \in [\mathcal{B} \otimes \mathcal{U}]_\infty$ such that

$$(\mathcal{B} \vee \mathcal{X}_n)(f \circ x) = h(a, x_n) \quad \text{a.s.}\Pi \ \forall\, n \in I\!N.$$

Hence, by $\mathcal{B}$-stationarity, $\forall\ k \in I\!N$ and $\forall\ n \geq k$,

$$
\begin{aligned}
\mathcal{B}[|(\mathcal{B} \vee \mathcal{X}_n)(f \circ x) - f \circ x|] &= \mathcal{B}[|h(a, x_n) - f \circ x|] \\
&= \mathcal{B}[|h(a, (\tau^{n-k} \circ x)_k) - f \circ \tau^{n-k} \circ x|] \\
&= \mathcal{B}[|h(a, x_k) - f \circ x|].
\end{aligned}
$$

Therefore, by the conditional dominated convergence theorem,

$$
\mathcal{B}[|h(a, x_k) - f \circ x|] = 0 \quad \forall\ k \in I\!N,
$$

i.e., $f \circ x = h(a, x_k)$ a.s. Π, $\forall\ k \in I\!N$ and this implies (v).

Now, since $\mathcal{X}_\Gamma \subset \overline{\mathcal{B} \vee \mathcal{X}_n}$, by the $\mathcal{B}$-Markov property,

$$
\mathcal{X}_{n+1} \perp\!\!\!\perp \mathcal{X}_0^n \mid \mathcal{B} \vee \mathcal{X}_\Gamma \vee \mathcal{X}_n.
$$

However, (ii) implies, by Corollary 2.2.11, that

$$
\mathcal{B} \perp\!\!\!\perp \mathcal{X}_0^n \mid \mathcal{X}_\Gamma \vee \mathcal{X}_n.
$$

By Theorem 2.2.10, these two conditional independence relations are equivalent to

$$
(\mathcal{B} \vee \mathcal{X}_{n+1}) \perp\!\!\!\perp \mathcal{X}_0^n \mid \mathcal{X}_\Gamma \vee \mathcal{X}_n.
$$

This clearly implies that

$$
\mathcal{X}_{n+1} \perp\!\!\!\perp \mathcal{X}_0^n \mid \mathcal{X}_\Gamma \vee \mathcal{X}_n,
$$

i.e., x is $\mathcal{X}_\Gamma$-Markovian.

(i) (ii) (iii) $\Rightarrow$ (o). By Corollary 9.3.3., we already know that (i) and (ii) imply that x is $\mathcal{B}$-stationary. Now, by (iii), for any $f \in [\mathcal{U}_\Gamma]_\infty$ there exists $h \in [\mathcal{B} \otimes \mathcal{U}]_\infty$ such that $f \circ x = h(a, x_0)$ a.s. Π. But by $\mathcal{B}$-stationarity

$$
\begin{aligned}
\mathcal{B}[|h(a, x_n) - f \circ x|] &= \mathcal{B}[|h[a, (\tau^n \circ x)_0] - f \circ \tau^n \circ x|] \\
&= \mathcal{B}[|h(a, x_0) - f \circ x|] = 0.
\end{aligned}
$$

Therefore, $\forall\ n \in I\!N$, $f \circ x = h(a, x_n)$ a.s. Π and (iv) is satisfied.

By Proposition 7.6.4, since x $\mathcal{X}_\Gamma$-Markovian is equivalent to

$$
\mathcal{X}_0^n \perp\!\!\!\perp \mathcal{X}_{n+1}^\infty \mid \mathcal{X}_\Gamma \vee \mathcal{X}_n, \quad \forall\ n \in I\!N,
$$

and since (ii) implies $\mathcal{B} \perp\!\!\!\perp \mathcal{X}_0^\infty \mid \mathcal{X}_\Gamma \vee \mathcal{X}_n$, we have

$$\mathcal{X}_0^n \perp\!\!\!\perp \mathcal{X}_{n+1}^\infty \perp\!\!\!\perp \mathcal{B} \mid \mathcal{X}_\Gamma \vee \mathcal{X}_n.$$

This implies in turn that

$$(\mathcal{B} \vee \mathcal{X}_0^n) \perp\!\!\!\perp \mathcal{X}_{n+1}^\infty \mid \mathcal{X}_\Gamma \vee \mathcal{X}_n$$

and that

$$\mathcal{X}_0^n \perp\!\!\!\perp \mathcal{X}_{n+1}^\infty \mid \mathcal{B} \vee \mathcal{X}_\Gamma \vee \mathcal{X}_n.$$

Since $\mathcal{X}_\Gamma \subset \overline{\mathcal{B} \vee \mathcal{X}_n}$, this is equivalent to $\mathcal{X}_0^n \perp\!\!\!\perp \mathcal{X}_{n+1}^\infty \mid \mathcal{B} \vee \mathcal{X}_n$, i.e., x is $\mathcal{B}$-Markovian.   ∎

In view of Theorem 9.3.23, it is very easy to find a sufficient condition for $\mathcal{X}_\Gamma$ to be exactly estimating in a $\mathcal{B}$-Markovian and $\mathcal{B}$-stationary process:

9.3.24 Proposition. Let x be a $\mathcal{B}$-Markovian and $\mathcal{B}$-stationary process. If $\mathcal{X}_0 \parallel \mathcal{X}_1 \mid \mathcal{B}$, then the minimal sufficient parameter is exactly estimable. Moreover,

$$\overline{\mathcal{X}_\Gamma} = \overline{\mathcal{B}\mathcal{X}_0^\infty} = \overline{\mathcal{X}_0^\infty \mathcal{B}} = \overline{\mathcal{B}} \cap \overline{\mathcal{X}_0^\infty}.   ∎$$

The condition $\mathcal{X}_0 \parallel \mathcal{X}_1 \mid \mathcal{B}$ is in fact slightly too strong. Indeed, it means that if there exist $B, C \in \mathcal{B} \otimes \mathcal{U}$ such that $\mathbf{1}_B(a, x_0) = \mathbf{1}_C(a, x_1)$ a.s. Π, then $\exists\, E \in \mathcal{B}$ such that $\mathbf{1}_B(a, x_0) = \mathbf{1}_E(a)$ a.s. Π. But, by Theorem 9.3.23(iv), if $A \in \mathcal{X}_\Gamma$ and if $B \in \mathcal{B} \otimes \mathcal{U}$ is such that $\mathbf{1}_A = \mathbf{1}_B(a, x_0)$ a.s. Π, then $\mathbf{1}_A = \mathbf{1}_B(a, x_1)$ a.s. Π. Therefore $\mathcal{X}_\Gamma$ will be included in $\overline{B}$ if the following condition holds: if $B \in \mathcal{B} \otimes \mathcal{U}$ is such that $\mathbf{1}_B(a, x_0) = \mathbf{1}_B(a, x_1)$ a.s. Π, then there exists $E \in \mathcal{B}$ such that $\mathbf{1}_B(a, x_0) = \mathbf{1}_E(a)$ a.s. Π. Equivalently $\mathcal{X}_\Gamma \subset \overline{B}$ under the following condition

$$(9.3.42) \qquad B \in \mathcal{B} \otimes \mathcal{U}, \quad (\mathcal{B} \vee \mathcal{X}_0)\mathbf{1}_B(a, x_1) = \mathbf{1}_B(a, x_0)$$
$$\Rightarrow \{\mathcal{B}\mathbf{1}_B(a, x_0)\}^2 = \mathcal{B}\mathbf{1}_B(a, x_0).$$

Under regularity assumptions such as the existence of a regular transition function $p(a, u, B)$ satisfying $\forall\, a \in A$, $\forall\, B \in \mathcal{U}$,

$$(9.3.43) \qquad\qquad (\mathcal{B} \vee \mathcal{X}_0)\mathbf{1}_B(x_1) = p(a, x_0, B).$$

Condition (9.3.42) is the Bayesian equivalent of the fact that $\forall\ a \in A$, the Markov process x is irreducible in the sense that there do not exist two nonempty closed disjoint sets where C is a closed set if $p(a, u, C) = 1$, $\forall\ u \in C$ (this is Doeblin's condition; see for instance Stout (1974), Section 3.6, or Breiman (1968), 7.3,). More details on Markov chains may be found, e.g., in Cohn (1965, 1974), and Orey (1971).

9.3.5 Autoregressive Moving Average Processes

Autoregressive moving average processes are defined through a representation of a stationary process in terms of an i.i.d. process. More precisely, let $x = \{x_n : n \in I\!N\}$ be a $\mathcal{B}$-i.i.d. process with state space $(I\!R^d, \mathcal{B}^d)$. The autoregressive moving average $\mathcal{B}$-stationary process $y = \{y_n : n \in I\!N\}$ of order (p, q) with state space $(I\!R^d, \mathcal{B}^d)$ is defined through x and matrix parameters $\{A_k : 1 \leq k \leq p)$ and $\{B_\ell : 1 \leq \ell \leq q\}$, where $A_k \in [\mathcal{B}]$ and $B_\ell \in [\mathcal{B}]$, by

$$(9.3.44) \qquad y_n = - \sum_{1 \leq k \leq p} A_k y_{n-k} + \sum_{1 \leq \ell \leq q} B_\ell x_{n-\ell} + x_n.$$

Once again, this linear structure is irrelevant for the study of exact estimability and we may consider the extended structure

$$(9.3.45) \qquad y_n = f(c, y_{n-1}, y_{n-2}, \ldots, y_{n-p}; x_n, x_{n-1}, \ldots, x_{n-q}).$$

Note that (9.3.45) implies that $\mathcal{Y}_n \subset C \vee \mathcal{Y}_{n-p}^{n-1} \vee \mathcal{X}_{n-q}^n$.

This leads to a very general notion of autoregressive moving average process in a Bayesian experiment.

9.3.25 Definition. In the Bayesian experiment $\mathcal{E} = (A \times S, \mathcal{A} \vee \mathcal{S}, \Pi)$ a $\mathcal{B}$-stationary process $y = \{y_n : n \in I\!N\}$ with state space $(U, \mathcal{U})$ is an *autoregressive moving average process* of order (p, q) if there exist $C \subset \mathcal{B}$ and a $\mathcal{B}$-i.i.d. process $x = \{x_n : n \in I\!N\}$ with state space $(V, \mathcal{V})$ such that the process $\{(y_n, x_n) : n \in I\!N\}$ with state space $(U \times V, \mathcal{U} \otimes \mathcal{V})$ is $\mathcal{B}$-stationary and such that

$$\mathcal{Y}_n \subset C \vee \mathcal{Y}_{n-p}^{n-1} \vee \mathcal{X}_{n-q}^n, \quad \forall\ n \geq r = \max(p, q).$$

This leads to the following theorem.

9.3.26 Theorem. In an autoregressive moving average $\mathcal{B}$-stationary process y of order (p, q), the minimal sufficient parameter is exactly estimable if:

$$(9.3.46) \qquad \mathcal{Y}_{r-p}^{r-1} \quad \perp\!\!\!\perp \quad \mathcal{X}_r^\infty \mid \mathcal{B} \vee \mathcal{X}_{r-q}^{r-1},$$

$$(9.3.47) \qquad (\mathcal{Y}_{r-p}^{r-1} \vee \mathcal{X}_{r-q}^{r-1}) \quad \| \quad (\mathcal{Y}_{2r-p}^{2r-1} \vee \mathcal{X}_{2r-q}^{2r-1}) \mid \mathcal{B},$$

and, moreover,

$$(9.3.48) \qquad \overline{\mathcal{Y}_\Gamma} = \overline{\mathcal{B}\mathcal{Y}_0^\infty} = \overline{\mathcal{Y}_0^\infty \mathcal{B}} = \overline{\mathcal{B} \cap \overline{\mathcal{Y}_0^\infty}}.$$

Proof. Let us first show that under (9.3.46), $\forall\, n \geq r$ x_n is independent of $\mathcal{Y}_{r-p}^{n-1} \vee \mathcal{X}_{r-q}^{n-1}$ conditionally on $\mathcal{B}$. Indeed, since x is a $\mathcal{B}$-i.i.d. process,

$$(9.3.49) \qquad \mathcal{X}_{r-q}^{r-1} \perp\!\!\!\perp \mathcal{X}_r^\infty \mid \mathcal{B},$$

and, by Theorem 2.2.10, (9.3.46) and (9.3.49) are equivalent to

$$(9.3.50) \qquad \mathcal{X}_r^\infty \perp\!\!\!\perp (\mathcal{X}_{r-q}^{r-1} \vee \mathcal{Y}_{r-p}^{r-1}) \mid \mathcal{B}.$$

Since $\mathcal{X}_n^\infty \perp\!\!\!\perp \mathcal{X}_r^{n-1} \mid \mathcal{B}$, we also have, by Theorem 7.6.4,

$$(9.3.51) \qquad \mathcal{X}_n^\infty \perp\!\!\!\perp \mathcal{X}_r^{n-1} \perp\!\!\!\perp (\mathcal{X}_{r-q}^{r-1} \vee \mathcal{Y}_{r-p}^{r-1}) \mid \mathcal{B},$$

and so

$$(9.3.52) \qquad \mathcal{X}_n^\infty \perp\!\!\!\perp (\mathcal{Y}_{r-p}^{r-1} \vee \mathcal{X}_{r-q}^{n-1} \vee \mathcal{C}) \mid \mathcal{B}$$

by Corollary 2.2.11, since $\mathcal{C} \subset \mathcal{B}$. But, by a recurrence argument, $\mathcal{Y}_n \subset \mathcal{C} \vee \mathcal{Y}_{n-p}^{n-1} \vee \mathcal{X}_{n-q}^n$ implies that

$$(9.3.53) \qquad \mathcal{Y}_r^{n-1} \subset \mathcal{C} \vee \mathcal{Y}_{r-p}^{r-1} \vee \mathcal{X}_{r-q}^{n-1}.$$

Therefore,

$$(9.3.54) \qquad \mathcal{X}_n^\infty \perp\!\!\!\perp (\mathcal{Y}_{r-p}^{n-1} \vee \mathcal{X}_{r-q}^{n-1}) \mid \mathcal{B}.$$

Now let us consider the process $z = \{z_n : n \geq r - 1\}$ defined by

$$(9.3.55) \qquad z_n = (y_n, y_{n-1}, \ldots, y_{n-p+1}, x_n, x_{n-1}, \ldots, x_{n-q+1}).$$

This process is $\mathcal{B}$-stationary since $\{(y_n, x_n) : n \in I\!N\}$ is $\mathcal{B}$-stationary. Clearly, in our usual notation,

$$(9.3.56) \qquad Z_n = \mathcal{Y}_{n-p+1}^n \vee \mathcal{X}_{n-q+1}^n,$$

and so

$$(9.3.57) \qquad Z_{r-1}^{n-1} = \mathcal{Y}_{r-p}^{n-1} \vee \mathcal{X}_{r-q}^{n-1}.$$

However, since $\mathcal{Y}_n \subset \mathcal{C} \vee \mathcal{Y}_{n-p}^{n-1} \vee \mathcal{X}_{n-q}^n$,

$$(9.3.58) \quad Z_n = \mathcal{Y}_{n-p+1}^{n-1} \vee \mathcal{X}_{n-q+1}^n \vee \mathcal{Y}_n \subset \mathcal{C} \vee \mathcal{Y}_{n-p}^{n-1} \vee \mathcal{X}_{n-q}^{n-1} \vee \mathcal{X}_n.$$

Therefore,

$$(9.3.59) \qquad Z_n \subset \mathcal{C} \vee Z_{n-1} \vee \mathcal{X}_n.$$

But, by (9.3.54),

$$(9.3.60) \qquad \mathcal{X}_n \perp\!\!\!\perp Z_{r-1}^{n-1} \mid \mathcal{B},$$

and so, since $\mathcal{C} \subset \mathcal{B}$, by Corollary 2.2.11

$$(9.3.61) \qquad (\mathcal{C} \vee Z_{n-1} \vee \mathcal{X}_n) \perp\!\!\!\perp Z_{r-1}^{n-1} \mid \mathcal{B} \vee Z_{n-1}.$$

Using (9.3.59), we obtain

$$(9.3.62) \qquad Z_n \perp\!\!\!\perp Z_{r-1}^{n-1} \mid \mathcal{B} \vee Z_{n-1}.$$

The process z is therefore $\mathcal{B}$-Markovian and $\mathcal{B}$-stationary. By Theorem 9.3.23 (v), Z_Γ will be exactly estimating (i.e., $Z_\Gamma \subset \overline{\mathcal{B}}$) if

$$(9.3.63) \qquad \overline{\mathcal{B} \vee Z_{r-1}} \cap \overline{\mathcal{B} \vee Z_{2r-1}} = \overline{\mathcal{B}},$$

and, in view of (9.3.57), this is precisely (9.3.47) . To complete the proof, it suffices to note that
$$(9.3.64) \qquad \mathcal{Y}_\Gamma \vee \mathcal{X}_\Gamma \subset Z_\Gamma$$

since, for instance, the process y may be seen as a function of the process z which commutes with the properly defined shift operators. Therefore, $\mathcal{Y}_\Gamma$ is exactly estimating. ∎

Before concluding this section, it is worthwhile commenting on conditions (9.3.46) and (9.3.47). Condition (9.3.47) is a very mild measurability assumption since, heuristically, it says that if $f(a, z_{r-1}) = g(a, z_{2r-1})$ a.s. Π, then $f(a, z_{r-1}) = h(a)$ a.s. Π. Condition (9.3.46) is in fact a very natural initial condition. Indeed, when the processes are indexed by $\mathbb{Z}$, as is usually done in time series analysis, (9.3.46) is trivially satisfied under the commonly invoked assumption that

$$(9.3.65) \qquad\qquad \mathcal{Y}_n \subset \overline{\mathcal{C} \vee \mathcal{X}_{-\infty}^n}.$$

9.3.6   An Example

We now provide an example which illustrates many of the concepts expanded in Chapters 7 and 9.

Let $A = \mathbb{R}_0^+$ and $\mathcal{A} = \mathcal{B}_0^+ = \mathcal{B} \cap \mathbb{R}_0^+$. Take a prior probability μ on $(A, \mathcal{A})$ which admits a strictly positive density on A with respect to Lebesgue measure. Let m be a real random variable with $\mathcal{M} = \sigma(m)$ such that

$$(9.3.66) \qquad\qquad \mathcal{M} \quad \perp\!\!\!\perp \quad \mathcal{A}$$

and

$$(9.3.67) \qquad\qquad m \quad \sim \quad N(0, 1).$$

Consider $x = \{x_n : n \in \mathbb{N}\}$ a real valued $(\mathcal{A} \vee \mathcal{M})$-i.i.d. stochastic process normally distributed with expectation m and variance a, i.e.,

$$(9.3.68) \qquad\qquad \underset{0 \leq n < \infty}{\perp\!\!\!\perp} \mathcal{X}_n \mid \mathcal{A} \vee \mathcal{M}$$

$$(9.3.69) \qquad\qquad (x_n \mid a, m) \sim N(m, a).$$

In this context, according to Sections 2.3.7 and 5.5.4, μ is a regular probability, and so Bayesian and sampling theory concepts of sufficiency and completion are equivalent. Remark also that, for the experiment $\mathcal{E}_{\mathcal{A} \vee \mathcal{M} \vee \mathcal{X}_0^n}$, there exists $\Pi_0 \sim \Pi$ such that

$$(9.3.70) \qquad\qquad \mathcal{A} \perp\!\!\!\perp \mathcal{M} \perp\!\!\!\perp \mathcal{X}_0^n; \Pi_0 \quad \forall n \in \mathbb{N}.$$

This fact will prove useful in establishing measurable separability using Corollary 5.2.11.

We now analyze the Bayesian experiments $\mathcal{E}_{(\mathcal{A}\vee\mathcal{M})\vee\mathcal{X}_0^\infty}$, $\mathcal{E}^{\mathcal{A}}_{\mathcal{M}\vee\mathcal{X}_0^\infty}$, $\mathcal{E}^{\mathcal{M}}_{\mathcal{A}\vee\mathcal{X}_0^\infty}$, and $\mathcal{E}_{\mathcal{A}\vee\mathcal{X}_0^\infty}$.

A. The Experiment $\mathcal{E}_{(\mathcal{A}\vee\mathcal{M})\vee\mathcal{X}_0^\infty}$.

If we define

$$(9.3.71) \qquad u_n \;=\; \frac{1}{n+1} \sum_{0 \le k \le n} x_k, \quad \mathcal{U}_n = \sigma\left(u_n\right)$$

$$(9.3.72) \qquad v_n \;=\; \frac{1}{n+1} \sum_{0 \le k \le n} \left(x_k - u_n\right)^2, \quad \mathcal{V}_n = \sigma\left(v_n\right)$$

then, by standard results, $\mathcal{U}_n \vee \mathcal{V}_n$ is sufficient and 1-complete in $\mathcal{E}_{(\mathcal{A}\vee\mathcal{M})\vee\mathcal{X}_0^n}$, i.e.,

$$(9.3.73) \qquad (\mathcal{A} \vee \mathcal{M}) \perp\!\!\!\perp \mathcal{X}_0^n \mid \mathcal{U}_n \vee \mathcal{V}_n$$

$$(9.3.74) \qquad (\mathcal{U}_n \vee \mathcal{V}_n) \ll_1 (\mathcal{A} \vee \mathcal{M}),$$

and is therefore minimal sufficient, i.e.,

$$(9.3.75) \qquad \mathcal{X}_0^n \left(\mathcal{A} \vee \mathcal{M}\right) = \overline{\left(\mathcal{U}_n \vee \mathcal{V}_n\right)} \cap \mathcal{X}_0^n.$$

Note also that $\mathcal{E}_{(\mathcal{A}\vee\mathcal{M})\vee\mathcal{X}_0^n}$ is measurably separated and identified, i.e.,

$$(9.3.76) \qquad (\mathcal{A} \vee \mathcal{M}) \parallel \mathcal{X}_0^n$$

$$(9.3.77) \qquad (\mathcal{A} \vee \mathcal{M}) \mathcal{X}_0^n = (\mathcal{A} \vee \mathcal{M}) \mathcal{X}_0 = \mathcal{A} \vee \mathcal{M}.$$

Indeed, $(\mathcal{A} \vee \mathcal{M}) x_0 = m$ and $(\mathcal{A} \vee \mathcal{M}) x_0^2 = m^2 + a$. Therefore $(\mathcal{A} \vee \mathcal{M}) \subset (\mathcal{A} \vee \mathcal{M}) \mathcal{X}_0$.

Let us also recall that

$$(9.3.78) \qquad \mathcal{U}_n \perp\!\!\!\perp \mathcal{V}_n \mid \mathcal{A} \vee \mathcal{M}$$

$$(9.3.79) \qquad (u_n \mid a, m) \;\sim\; N\left(m, \frac{a}{n+1}\right)$$

$$(9.3.80) \qquad (v_n \mid a, m) \;\sim\; \frac{a}{n+1}\chi_n^2,$$

i.e., the chi-square distribution with n degrees of freedom. Therefore

$$(9.3.81) \qquad \mathcal{V}_n \perp\!\!\!\perp \mathcal{M} \mid \mathcal{A},$$

and by Theorem 2.2.10

(9.3.82) $V_n \perp\!\!\!\perp (\mathcal{M} \vee \mathcal{U}_n) \mid \mathcal{A}.$

In $\mathcal{E}_{(\mathcal{A} \vee \mathcal{M}) \vee \mathcal{X}_0^\infty}$, since x is $(\mathcal{A} \vee \mathcal{M})$-i.i.d. and $\mathcal{E}_{(\mathcal{A} \vee \mathcal{M}) \vee \mathcal{X}_0^\infty}$ is identified, by Theorem 9.3.12, $\mathcal{X}_\Gamma$ is sufficient and exactly estimating, i.e.,

(9.3.83) $\overline{\mathcal{X}_\Gamma} = \overline{\mathcal{X}_T} = \overline{\mathcal{X}_\Sigma} = \overline{\mathcal{A} \vee \mathcal{M}},$

and therefore

(9.3.84) $\mathcal{X}_0^\infty (\mathcal{A} \vee \mathcal{M}) = \overline{\mathcal{X}_\Gamma} \cap \mathcal{X}_0^\infty.$

Clearly, $\mathcal{X}_\Gamma$ is 1-complete, i.e.,

(9.3.85) $\mathcal{X}_\Gamma \ll_1 (\mathcal{A} \vee \mathcal{M}).$

According to Theorem 7.3.6, $\mathcal{X}_0^n \mathcal{X}_\Gamma$ is $\mathcal{E}_{(\mathcal{A} \vee \mathcal{M}) \vee \mathcal{X}_0^n}$-minimal sufficient. This is clear since

(9.3.86) $\mathcal{X}_0^n \mathcal{X}_\Gamma = \mathcal{X}_0^n (\mathcal{A} \vee \mathcal{M}) = \overline{(\mathcal{U}_n \vee \mathcal{V}_n)} \cap \mathcal{X}_0^n.$

Note that, in this case, $\mathcal{X}_0^n \mathcal{X}_\Gamma$ is also 1-complete in $\mathcal{E}_{(\mathcal{A} \vee \mathcal{M}) \vee \mathcal{X}_0^n}$.

Now, by Proposition 7.3.4, $(\mathcal{U}_n \vee \mathcal{V}_n)_T$ is also $\mathcal{E}_{(\mathcal{A} \vee \mathcal{M}) \vee \mathcal{X}_0^\infty}$-sufficient and, by (7.3.1),

$$\mathcal{X}_0^\infty (\mathcal{A} \vee \mathcal{M}) \subset \overline{(\mathcal{U}_n \vee \mathcal{V}_n)}_T,$$

and therefore

(9.3.87) $\overline{\mathcal{A} \vee \mathcal{M}} \subset \overline{(\mathcal{U}_n \vee \mathcal{V}_n)}_T.$

Note that this inclusion can also be deduced from the strong law of large numbers, since $u_n \longrightarrow m$ a.s.Π and $v_n \longrightarrow a$ a.s. Π. However, u_n and v_n are invariant under any permutation of the $(n+1)$-first coordinates; consequently,

(9.3.88) $(\mathcal{U}_n \vee \mathcal{V}_n)_T \subset \mathcal{X}_\Sigma,$

and so

(9.3.89) $\overline{(\mathcal{U}_n \vee \mathcal{V}_n)}_T = \overline{\mathcal{A} \vee \mathcal{M}}.$

In this situation the tail σ-field of a sufficient sequence of 2-complete statistics is minimal asymptotically sufficient, i.e.,

(9.3.90) $\overline{\mathcal{X}_0^\infty (\mathcal{A} \vee \mathcal{M})} = \bigcap_n \overline{\bigcup_{m \geq n} \mathcal{X}_0^m (\mathcal{A} \vee \mathcal{M})}.$

Note that, if

(9.3.91) $$y_n = x_n - m,$$

then

(9.3.92) $$\mathcal{M} \perp \mathcal{Y}_0^\infty \mid \mathcal{A},$$

and y is an $\mathcal{A}$-i.i.d. process. Note also that $\mathcal{E}_{\mathcal{A} \vee \mathcal{Y}_0^\infty}$ is identified, i.e.,

(9.3.93) $$\mathcal{A} \mathcal{Y}_0^\infty = \mathcal{A} \mathcal{Y}_0 = \mathcal{A}.$$

Indeed, $\mathcal{A} y_0^2 = a$. Therefore, by Theorem 9.3.12,

(9.3.94) $$\overline{\mathcal{Y}_\Gamma} = \overline{\mathcal{Y}_T} = \overline{\mathcal{Y}_\Sigma} = \overline{\mathcal{A}} = \overline{\mathcal{Y}_0^\infty \mathcal{A}}.$$

Now, considering the definition of y_n and v_n, $v_n \in [\mathcal{Y}_0^n]$ and v_n is clearly invariant under permutations of the first $(n+1)$-coordinates of the process y. Therefore $\mathcal{V}_T \subset \mathcal{Y}_\Sigma$, and so $\mathcal{V}_T \subset \overline{\mathcal{A}}$. Now, $\mathcal{A} \subset \overline{\mathcal{V}_T}$ since $v_n \longrightarrow a$ a.s. II, and thus

(9.3.95) $$\overline{\mathcal{V}_T} = \overline{\mathcal{A}}.$$

B. The Experiment $\mathcal{E}_{\mathcal{M} \vee \mathcal{X}_0^\infty}^{\mathcal{A}}$.

In this experiment (with known variance) $\mathcal{U}_n$ is sufficient and 1-complete in $\mathcal{E}_{\mathcal{M} \vee \mathcal{X}_0^n}^{\mathcal{A}}$, i.e.,

(9.3.96) $$\mathcal{M} \perp \mathcal{X}_0^n \mid \mathcal{A} \vee \mathcal{U}_n$$

(9.3.97) $$\mathcal{U}_n \ll_1 \mathcal{M} \mid \mathcal{A}.$$

$\mathcal{U}_n$ is therefore minimal sufficient, i.e.,

(9.3.98) $$(\mathcal{A} \vee \mathcal{X}_0^n)(\mathcal{A} \vee \mathcal{M}) = \overline{(\mathcal{A} \vee \mathcal{U}_n)} \cap (\mathcal{A} \vee \mathcal{X}_0^n).$$

By Corollary 5.2.11, $\mathcal{E}_{\mathcal{M} \vee \mathcal{X}_0^n}^{\mathcal{A}}$ is measurably separated, i.e.,

(9.3.99) $$\mathcal{M} \parallel \mathcal{X}_0^n \mid \mathcal{A}.$$

By Proposition 4.6.6, $\mathcal{E}_{\mathcal{M} \vee \mathcal{X}_0^n}^{\mathcal{A}}$ is also identified, i.e.,

(9.3.100) $$(\mathcal{A} \vee \mathcal{M})(\mathcal{A} \vee \mathcal{X}_0^n) = \mathcal{A} \vee \mathcal{M}.$$

This is straightforward to verify since

$$(\mathcal{A} \vee \mathcal{M})(\mathcal{A} \vee \mathcal{X}_0^n) = \mathcal{A} \vee \{(\mathcal{A} \vee \mathcal{M}) \mathcal{X}_0^n\} = \mathcal{A} \vee \mathcal{M}.$$

Now, recalling (9.3.81), $V_n \perp\!\!\!\perp \mathcal{M} \mid \mathcal{A}$, V_n is $\mathcal{E}^{\mathcal{A}}_{\mathcal{M} \vee \mathcal{X}_0^\infty}$-ancillary and, by Basu First Theorem (5.5.9), we obtain

$$(9.3.101) \qquad\qquad V_n \perp\!\!\!\perp (\mathcal{M} \vee \mathcal{U}_n) \mid \mathcal{A},$$

which is (9.3.82), and a fundamental property of normal populations; consequently, this result may be viewed as a corollary of Basu First Theorem.

In $\mathcal{E}^{\mathcal{A}}_{\mathcal{M} \vee \mathcal{X}_0^\infty}$, the process $\{(a, x_n) : n \in I\!\!N\}$ is $(\mathcal{A} \vee \mathcal{M})$-i.i.d. Since $\mathcal{E}^{\mathcal{A}}_{\mathcal{M} \vee \mathcal{X}_0^\infty}$ is identified, by Theorem (9.3.12) we know that

$$(9.3.102) \qquad \overline{(\mathcal{A} \vee \mathcal{X}_n)_\Gamma} = \overline{(\mathcal{A} \vee \mathcal{X}_n)_T} = \overline{(\mathcal{A} \vee \mathcal{X}_n)_\Sigma}$$
$$= \overline{\mathcal{A} \vee \mathcal{M}} = \overline{(\mathcal{A} \vee \mathcal{X}_0^\infty)(\mathcal{A} \vee \mathcal{M})}.$$

By (9.3.83), $\overline{\mathcal{X}_T} = \overline{\mathcal{A} \vee \mathcal{M}}$. Therefore

$$(9.3.103) \qquad\qquad \overline{(\mathcal{A} \vee \mathcal{X}_n)_T} = \overline{\mathcal{A} \vee \mathcal{X}_T} = \overline{\mathcal{A} \vee \mathcal{M}}.$$

Hence $\mathcal{X}_T$ is minimal sufficient and 1-complete in $\mathcal{E}^{\mathcal{A}}_{\mathcal{M} \vee \mathcal{X}_0^\infty}$. By Theorem 7.3.6, $(\mathcal{A} \vee \mathcal{X}_0^n)(\mathcal{A} \vee \mathcal{X}_T)$ is $\mathcal{E}^{\mathcal{A}}_{\mathcal{M} \vee \mathcal{X}_0^n}$-minimal sufficient and is also 1-complete since, by (9.3.98),

$$(9.3.104) \qquad\qquad \overline{(\mathcal{A} \vee \mathcal{X}_0^n)(\mathcal{A} \vee \mathcal{X}_T)} = \overline{\mathcal{A} \vee \mathcal{U}_n}.$$

By Proposition 7.3.4 and expression (7.3.1),

$$(9.3.105) \qquad\qquad \overline{(\mathcal{A} \vee \mathcal{X}_0^\infty)(\mathcal{A} \vee \mathcal{M})} \subset \overline{(\mathcal{A} \vee \mathcal{U}_n)_T}.$$

However, u_n is invariant under any permutation of the $(n+1)$ first coordinates, so $(\mathcal{A} \vee \mathcal{U}_n)_T \subset (\mathcal{A} \vee \mathcal{X}_n)_\Sigma$. Therefore, by (9.3.102),

$$(9.3.106) \qquad\qquad \overline{(\mathcal{A} \vee \mathcal{U}_n)_T} = \overline{\mathcal{A} \vee \mathcal{M}}.$$

In this experiment it is also the case that the tail σ-field of a 1-complete minimal sufficient sequence is the 1-complete minimal asymptotically sufficient statistic, i.e.,

$$(9.3.107) \qquad \overline{(\mathcal{A} \vee \mathcal{X}_0^\infty)(\mathcal{A} \vee \mathcal{M})} = \bigcap_n \overline{\bigcup_{k \geq n} (\mathcal{A} \vee \mathcal{X}_0^k)(\mathcal{A} \vee \mathcal{M})}.$$

Note that, from (9.3.106), we obtain that $\mathcal{U}_T \subset \overline{\mathcal{A} \vee \mathcal{M}}$. By the strong law of large numbers, $u_n \to m$ a.s. II, and $(1/n+i+1) \sum_{0 \leq k \leq n} x_{i+k}^2 \to m^2 + a$

a.s.Π $\forall$ $i \in I\!N$. But $x_{i+k} = (i+k+1)u_{i+k} - (i+k)u_{i+k-1}$. Hence $\mathcal{M} \subset \overline{\mathcal{U}_T}$ and $\mathcal{A} \subset \overline{\mathcal{U}_{i-1}^{\infty}}$ $\forall i \in I\!N$. Therefore

$$(9.3.108) \qquad \overline{\mathcal{U}_T} \;=\; \overline{\mathcal{A} \vee \mathcal{M}},$$

and

$$(9.3.109) \qquad \overline{(\mathcal{A} \vee \mathcal{U}_n)_T} \;=\; \overline{\mathcal{A} \vee \mathcal{U}_T} = \overline{\mathcal{A} \vee \mathcal{M}}.$$

C. The Experiment $\mathcal{E}_{\mathcal{A} \vee \mathcal{X}_0^{\infty}}^{\mathcal{M}}$.

This experiment is very similar to the experiment $\mathcal{E}_{\mathcal{M} \vee \mathcal{X}_0^{\infty}}^{\mathcal{A}}$. But, whereas in $\mathcal{E}_{\mathcal{M} \vee \mathcal{X}_0^{\infty}}^{\mathcal{A}}$ the minimal sufficient statistic $\mathcal{U}_n$ was uniform, i.e., $\mathcal{U}_n \subset \mathcal{X}_0^n$, this is not the case in $\mathcal{E}_{\mathcal{A} \vee \mathcal{X}_0^n}^{\mathcal{M}}$. Indeed, if we define

$$(9.3.110) \qquad t_n = \frac{1}{n+1} \sum_{0 \le k \le n} (x_k - m)^2 \,, \mathcal{T}_n = \sigma\left(t_n\right),$$

then $\mathcal{T}_n$ is sufficient and 1-complete in $\mathcal{E}_{\mathcal{A} \vee \mathcal{X}_0^n}^{\mathcal{M}}$ i.e.,

$$(9.3.111) \qquad \mathcal{A} \perp\!\!\!\perp \mathcal{X}_0^n \mid \mathcal{M} \vee \mathcal{T}_n,$$

$$(9.3.112) \qquad \mathcal{T}_n \ll_1 \mathcal{A} \mid \mathcal{M},$$

but $\mathcal{T}_n$ is not uniform.

Let us recall that

$$(9.3.113) \qquad t_n = v_n + (u_n - m)^2 \,,$$

and

$$(9.3.114) \qquad (t_n \mid a, m) \sim \frac{a}{n+1} \chi_{n+1}^2.$$

This implies that

$$(9.3.115) \qquad \mathcal{M} \perp\!\!\!\perp \mathcal{T}_n \mid \mathcal{A},$$

and shows that $\mathcal{T}_n$ is 1-complete. Just as in the above section, $\mathcal{E}_{\mathcal{A} \vee \mathcal{X}_0^n}^{\mathcal{M}}$ is measurably separated and identified, and $\mathcal{T}_n$ is minimal sufficient, i.e.,

$$(9.3.116) \qquad \mathcal{A} \parallel \mathcal{X}_0^n \mid \mathcal{M},$$

$$(9.3.117) \qquad (\mathcal{A} \vee \mathcal{M})(\mathcal{M} \vee \mathcal{X}_0^n) = \mathcal{A} \vee \mathcal{M},$$

$$(9.3.118) \qquad (\mathcal{M} \vee \mathcal{X}_0^n)(\mathcal{A} \vee \mathcal{M}) = \overline{(\mathcal{M} \vee \mathcal{T}_n)} \cap (\mathcal{M} \vee \mathcal{X}_0^n),$$

and in $\mathcal{E}_{\mathcal{A} \vee \mathcal{X}_0^\infty}^{\mathcal{M}}$, $\mathcal{X}_T$ is minimal sufficient and exactly estimating (and therefore 1-complete), i.e.,

(9.3.119) $\overline{(\mathcal{M} \vee \mathcal{X}_n)_T} = \overline{\mathcal{M} \vee \mathcal{X}_T} = \overline{\mathcal{A} \vee \mathcal{M}} = \overline{(\mathcal{M} \vee \mathcal{X}_0^\infty)(\mathcal{A} \vee \mathcal{M})}.$

Now,

(9.3.120) $\overline{(\mathcal{M} \vee \mathcal{X}_0^n)(\mathcal{M} \vee \mathcal{X}_T)} = \overline{\mathcal{M} \vee \mathcal{T}_n}$

and

(9.3.121) $\overline{(\mathcal{M} \vee \mathcal{T}_n)_T} = \overline{\mathcal{A} \vee \mathcal{M}}.$

But by (9.3.91) and (9.3.94), $\mathcal{T}_n$ is minimal sufficient and 1-complete in $\mathcal{E}_{\mathcal{A} \vee \mathcal{Y}_0^n}$. So

(9.3.122) $\overline{\mathcal{T}_T} = \overline{\mathcal{A}},$

and, therefore,

(9.3.123) $\overline{(\mathcal{M} \vee \mathcal{T}_n)_T} = \overline{\mathcal{M} \vee \mathcal{T}_T}.$

D. The Experiment $\mathcal{E}_{\mathcal{A} \vee \mathcal{X}_0^\infty}$.

This experiment presents far more pathological features than the three preceding ones, and illustrates results obtained in Chapter 9. It also provides counterexamples to false conjonctures which might have been advanced in Chapter 7.

In $\mathcal{E}_{\mathcal{A} \vee \mathcal{X}_0^n}$, by standard computations,

(9.3.124) $((x_0, x_1, \ldots, x_n)' \mid a) \sim N_{n+1}(0, aI + ee')$

where I is the identity matrix, and $e = (1, 1, \ldots, 1)'$. The density function only depends on

$$(x_0, x_1, \ldots, x_n)(aI + ee')^{-1}(x_0, x_1, \ldots, x_n)'$$
$$= \tfrac{1}{a}(x_0, x_1, \ldots, x_n)(I - \tfrac{1}{a+n+1}ee')(x_0, x_1, \ldots, x_n)'$$
$$= \tfrac{n+1}{a}(v_n + u_n^2) - \tfrac{(n+1)^2}{a(a+n+1)}u_n^2$$
$$= \tfrac{n+1}{a}v_n + \tfrac{n+1}{a+n+1}u_n^2,$$

where v_n and u_n are given in (9.3.71) and (9.3.72).

If $\mathcal{W}_n = \sigma(u_n^2) \subset \mathcal{U}_n$ then, according to Section 4.4.4, $\mathcal{V}_n \vee \mathcal{W}_n$ is $\mathcal{E}_{\mathcal{A} \vee \mathcal{X}_0^n}$-minimal sufficient, i.e.,

(9.3.125) $\mathcal{A} \perp\!\!\!\perp \mathcal{X}_0^n \mid \mathcal{V}_n \vee \mathcal{W}_n.$

$\mathcal{E}_{\mathcal{A} \vee \mathcal{X}_0^n}$ is measurably separated and identified, i.e.,

(9.3.126) $$\mathcal{A} \parallel \mathcal{X}_0^n$$

and

(9.3.127) $$\mathcal{A}\mathcal{X}_0^n = \mathcal{A}\mathcal{X}_0 = \mathcal{A}.$$

Indeed, $\mathcal{A}x_0^2 = 1 + a$.

From (9.3.82), (9.3.79) and (9.3.80) we deduce that:

(9.3.128) $$\mathcal{V}_n \quad \perp\!\!\!\perp \quad \mathcal{W}_n \mid \mathcal{A}$$

(9.3.129) $$(v_n \mid a) \quad \sim \quad \frac{a}{n+1}\chi_n^2$$

(9.3.130) $$(u_n \mid a) \quad \sim \quad N\left(0; \frac{a+n+1}{n+1}\right).$$

Therefore,

(9.3.131) $$(u_n^2 \mid a) \sim \frac{a+n+1}{n+1}\chi_1^2.$$

Note also that

(9.3.132) $$\begin{aligned}(u_n^2 \mid a, m) \quad &\sim \quad \frac{a}{n+1}\left\{N\left[\sqrt{\frac{n+1}{a}}m, 1\right]\right\}^2 \\ &= \quad \frac{a}{n+1}\chi_1^2\left[\frac{(n+1)\,m^2}{a}\right],\end{aligned}$$

which is a non central chi-square distribution.

Consequently, if

(9.3.133) $$\mathcal{N} = \sigma\left\{m^2\right\} \subset \mathcal{M},$$

then

(9.3.134) $$\mathcal{W}_n \perp\!\!\!\perp \mathcal{M} \mid \mathcal{A} \vee \mathcal{N}.$$

From (9.3.129) and (9.3.131) we deduce that $\mathcal{V}_n \vee \mathcal{W}_n$ is not complete, since

(9.3.135) $$\mathcal{A}\left[v_n - n\left(u_n^2 - 1\right)\right] = \frac{a}{n+1}n - n\left(\frac{a+n+1}{n+1} - 1\right) = 0.$$

Now, in $\mathcal{E}_{\mathcal{A} \vee \mathcal{X}_0^\infty}$ the process x is $\mathcal{A}$-exchangeable, according to Definition 9.3.5. By Theorem 9.3.11, $\overline{\mathcal{X}_\Gamma} = \overline{\mathcal{X}_T} = \overline{\mathcal{X}_\Sigma}$, x is $\mathcal{X}_\Gamma$-i.i.d., and $\mathcal{X}_\Gamma$ is $\mathcal{E}_{\mathcal{A} \vee \mathcal{X}_0^\infty}$-sufficient. Now, by Corollary 9.3.13, since x is not $\mathcal{A}$-i.i.d., $\mathcal{X}_\Gamma$ is not exactly estimating and we must search for a smaller sufficient and exactly estimating statistic.

Now, by Definition 9.3.14, x is an $\mathcal{A}$-generalized i.i.d. process with respect to $\mathcal{A} \vee \mathcal{M}$ and is also a marginalization on the parameter space since x is $\mathcal{A} \vee \mathcal{M}$-i.i.d. Recall that $\mathcal{E}_{(\mathcal{A} \vee \mathcal{M}) \vee \mathcal{X}_0^\infty}$ and $\mathcal{E}_{\mathcal{A} \vee \mathcal{X}_0^\infty}$ are identified. Therefore, by Corollary 9.3.16, $\mathcal{A}$ is exactly estimable, i.e., $\mathcal{A} \subset \overline{\mathcal{X}_0^\infty}$. More precisely, since $\overline{\mathcal{X}_\Gamma} = \overline{\mathcal{A} \vee \mathcal{M}}$, by Theorem 9.3.15, we have $\overline{\mathcal{X}_0^\infty \mathcal{A}} = \overline{\mathcal{A} \mathcal{X}_0^\infty} = \overline{\mathcal{A}}$. Now, by (9.3.95), $\overline{\mathcal{V}_T} = \overline{\mathcal{A}}$, and so $\mathcal{V}_T$ is sufficient and exactly estimable and therefore 1-complete. By Theorem 7.3.6,

$$\mathcal{X}_0^n \mathcal{V}_T = \overline{(\mathcal{V}_n \vee \mathcal{W}_n)} \cap \mathcal{X}_0^n,$$

and this shows that the projection on $\mathcal{X}_0^n$ of a complete asymptotically sufficient statistic is not complete, although it is minimal sufficient.

Finally, by the strong law of large numbers in $\mathcal{E}_{(\mathcal{A} \vee \mathcal{M}) \vee \mathcal{X}_0^\infty}$, $u_n^2 \to m^2$ a.s. Π as $n \to \infty$. It is then also true in $\mathcal{E}_{\mathcal{A} \vee \mathcal{X}_0^\infty}$. Therefore, using (9.3.89), we have the following sequence of results:

(9.3.136) $\overline{\mathcal{V}_T} = \overline{\mathcal{A}} \subset \overline{\mathcal{A} \vee \mathcal{N}} \subset \overline{(\mathcal{V}_n \vee \mathcal{W}_n)_T} \subset \overline{(\mathcal{V}_n \vee \mathcal{U}_n)_T} = \overline{\mathcal{A} \vee \mathcal{M}}.$

This shows that, in $\mathcal{E}_{\mathcal{A} \vee \mathcal{X}_0^\infty}$,

(9.3.137) $\overline{\mathcal{X}_0^\infty \mathcal{A}} \neq \overline{\bigcap_n \bigcup_{m \geq n} \mathcal{X}_0^m \mathcal{A}},$

i.e., the minimal asymptotically sufficient statistic is strictly included in the tail of the sequence of the minimal sufficient statistics.

9.4 Conditional Stochastic Processes

9.4.1 Introduction

A useful motivation for this section is the analysis of the regression model $y_n = f(z_n, a) + \varepsilon_n$ where, conditionally on z, the ε's are i.i.d. (with 0 expectation). We shall say that the sampling process generating y_n is conditionally independent and stationary, in a sense to be made precise.

It is well known that the existence of consistent estimators for a in such a (nonlinear) regression model requires more than assumptions on the conditional distributions generating $(y \mid z)$; it also imposes restrictions on the behaviour of the z-process. In the linear case:

$$y_n = \sum_{1 \leq i \leq k} a_i z_{ni} + \varepsilon_n,$$

convergence of the least squares (or maximum likelihood) estimator of a is generally obtained by assuming the convergence of $n^{-1} Z_n' Z_n$ (where $Z_n = [z_{ti}], 1 \leq t \leq n, 1 \leq i \leq k$).

A preliminary question concerns the nature of z_n. Sometimes, z_n is considered as "nonrandom", as in the analysis of variance; in other cases, it may be more natural to view the z_n's as having been randomly generated by some unspecified process. Thus, in the linear model, the convergence of $n^{-1} Z_n' Z_n$ may be as a sequence of either nonrandom or random matrices, and it is easily shown that the convergence of the least squares estimator will be either in probability or almost sure according to the type of convergence of $n^{-1} Z_n' Z_n$. When f is not linear and z_n is nonrandom, Malinvaud [1970] proposes an assumption on the asymptotic behaviour of the empirical distribution function. More recently, Burguete, Gallant, and Souza [1982] proposed for a random z_n, an assumption on the convergence of the sequence $n^{-1} \sum_{1 \leq i \leq n} \varphi(\varepsilon_i, z_i)$ for every φ dominated by a given integrable function. Note that a trivial extension for the random regressor case would be to assume that z_n and ε_n are independent, both i.i.d., and that the distribution of z_n is characterized by parameters *a priori* independent of both a and the parameter characterizing the distribution of ε_n; it is then easy to show that, in this context, (y_n, z_n) are jointly i.i.d. and the estimation of a (i.e., the maximum likelihood estimator or posterior expectation) is the same in the joint and the conditional models. Since the model is embedded in an i.i.d. framework, identification is a sufficient condition for exact estimability, and the assumption of convergence of $n^{-1} Z_n' Z_n$ or of $n^{-1} \Sigma_{1 \leq i \leq n} \varphi(\varepsilon_i, z_i)$ is a consequence of the law of large numbers.

It should be noted that the assumptions that z_n is i.i.d. and that there is a cut (prior independence of the parameters), are very constraining; this section seeks to obtain consistency of posterior expectations from weaker assumptions. The main results of this section may heuristically be viewed as follows: in an identified conditional i.i.d. process, a condition of stationarity on the conditioning process is a sufficient condition for the convergence of posterior expectations.

In Chapter 7, Section 7.7.3, this problem was treated using an "arbitrary time origin", i.e., a tail-sufficiency condition; heuristically, this means that the posterior expectation of an arbitrary integrable function of the parameter given all the process $\{x_n : n \in I\!N\}$ is not affected by deleting the first k observations, i.e., is equal to the posterior expectation given $\{x_n : k \leq n < \infty\}$. As shown in Theorem 7.7.2, this property is equivalent to the sufficiency of the tail σ-field. In this section we analyze this problem using the concept of the conditional invariance for the shift operator as studied in Chapter 8, since it was shown in Section 9.3 that shift invariance implies the tail-sufficiency property.

In Chapter 7, a first kind of result, namely, Theorems 7.7.18 and 7.7.19, was obtained, based on the weaker assumption of the tail-sufficiency property; however, these results relied on asymptotic assumptions. The results of this section will be less general since based on the stronger assumption of shift invariance, but they make use of assumptions verifiable on finite sample size experiments. Note that for the marginal process generating z_n, that is, the conditioning variable, we do not assume ergodicity but merely stationarity. This feature leads to results for the conditional models different from those for the joint models. It may indeed be shown that if the joint process generating $x_n = (y_n, z_n)$ is stationary and ergodic, then the identified parameters are exactly estimable. In this section we show that if the joint process is stationary (which is implied by the stationarity of $(y_n \mid z_n)$ and of (z_n)), and if the conditional model generating $(y_n \mid z_n)$ is ergodic, then its identified parameters are exactly estimable. In this section, the ergodicity of the conditional model is obtained through an assumption of conditional (mutual) independence. It is natural to obtain conditional ergodicity through a weaker conditional mixing assumption. Such an approach is the natural framework for analyzing models such as $y_n = f(y_{n-1}, z_n, a) + \varepsilon_n$.

Another feature of our results concerns the treatment of nuisance parameters. In a Bayesian framework it is natural to integrate out nuisance parameters, and to then analyze the experiments marginalized on the space (parameters of interest × observations). In the main theorem, a general situation is examined, whereas in its corollary it is assumed that the parameter of interest is a sufficient parameter for the conditional sampling process.

Our results, however, do not depend on whether or not the marginal process provides information about the parameter of interest. This is so because we are concerned with consistency but not with asymptotic efficiency — our results implicitly assume a complete specification of the joint model. Thus a supplementary assumption of a cut or of mutual ancillarity is not necessary for obtaining exact estimability, although such an assumption would allow one to simplify the specification and, in general, obtain more robust results.

9.4.2 Shift in Conditional Stochastic Processes

In the Bayesian experiment $\mathcal{E} = (A \times S, \mathcal{A} \vee \mathcal{S}, \Pi)$ the Bayesian stochastic process $x = \{x_n : n \in \mathbb{N}\}$ defined on $(S, \mathcal{S})$ with values in the measurable space $(U, \mathcal{U})$ is decomposed into

$$(9.4.1) \qquad x_n = (y_n, z_n),$$

where y_n and z_n are valued in $(V, \mathcal{V})$ and $(W, \mathcal{W})$ respectively. Thus

$$(9.4.2) \qquad U = V \times W \quad \text{and } \mathcal{U} = \mathcal{V} \otimes \mathcal{W}.$$

The sub-σ-fields generated by this process are

$$(9.4.3) \qquad \mathcal{Y}_n = y_n^{-1}(\mathcal{V})$$

and

$$(9.4.4) \qquad \mathcal{Z}_n = z_n^{-1}(\mathcal{W})$$

so that

$$(9.4.5) \qquad \mathcal{X}_n = \mathcal{Y}_n \vee \mathcal{Z}_n = x_n^{-1}(\mathcal{U}).$$

Other related σ-fields are defined as in Section 9.2.3. As in that section, it is useful to introduce the coordinate process defined on the representation space $\widehat{\mathcal{E}} = (\widehat{A} \times \widehat{S}, \widehat{\mathcal{A}} \vee \widehat{\mathcal{S}}, \widehat{\Pi})$ by

$$(9.4.6) \qquad \widehat{y}_n(u) \;=\; v_n$$

$$(9.4.7) \qquad \widehat{z}_n(u) \;=\; w_n$$

since $u_n = (v_n, w_n)$.

To this coordinate process we associate the sub-σ-fields of $\widehat{A} \vee \widehat{S}$ defined by

$$(9.4.8) \qquad\qquad \widehat{\mathcal{Y}}_n \;=\; \widehat{y}_n^{-1}(\mathcal{V})$$

and
$$(9.4.9) \qquad\qquad \widehat{\mathcal{Z}}_n \;=\; \widehat{z}_n^{-1}(\mathcal{W})$$

so that
$$(9.4.10) \qquad\qquad \widehat{\mathcal{X}}_n \;=\; \widehat{\mathcal{Y}}_n \vee \widehat{\mathcal{Z}}_n = \widehat{x}_n^{-1}(\mathcal{U}).$$

As in Section 9.2.4, if the shift operator acts by composition on the stochastic processes y and z as

$$(9.4.11) \qquad\qquad \begin{aligned} (\tau^k \circ y)_n &= y_{n+k} \\ (\tau^k \circ z)_n &= z_{n+k} \end{aligned}$$

it may be defined as acting on the representation space $(\widehat{A} \times \widehat{S}, \widehat{A} \vee \widehat{S})$ in such a way that

$$(9.4.12) \qquad\qquad \begin{aligned} \widehat{y}_n \circ \tau^k &= \widehat{y}_{n+k} \\ \widehat{z}_n \circ \tau^k &= \widehat{z}_{n+k} \\ \widehat{a} \circ \tau^k &= \widehat{a}. \end{aligned}$$

This permits the basic sub-σ-fields to be represented as:

$$(9.4.13) \qquad\qquad \begin{aligned} \widehat{\mathcal{Y}}_n &= \tau^{-n}(\widehat{\mathcal{Y}}_0), \\ \widehat{\mathcal{Z}}_n &= \tau^{-n}(\widehat{\mathcal{Z}}_0). \end{aligned}$$

Clearly,
$$(9.4.14) \qquad\qquad \widehat{\mathcal{B}}_\Gamma \;=\; \widehat{\mathcal{B}} \qquad \forall\, \widehat{\mathcal{B}} \subset \widehat{\mathcal{A}}.$$

Finally, let us recall that
$$(9.4.15) \qquad\qquad \widehat{\mathcal{S}} = \widehat{\mathcal{X}}_0^\infty.$$

9.4.3 Conditional Shift-Invariance

In the Bayesian experiment $\mathcal{E} = (A \times S, \mathcal{A} \vee \mathcal{S}, \Pi)$, let $\mathcal{B} \subset \mathcal{A}$ be a subparameter and let x be a Bayesian stochastic process as defined in Section 9.4.2. Consider the representation of the experiment $\mathcal{E}$ marginalized on $\widehat{\mathcal{B}} \vee \widehat{\mathcal{X}}_0^\infty$, i.e.,

$$(9.4.16) \qquad\qquad \widehat{\mathcal{E}}_{\widehat{\mathcal{B}} \vee \widehat{\mathcal{X}}_0^\infty} = \{A \times U^{I\!N}, \widehat{\mathcal{B}} \vee \widehat{\mathcal{X}}_0^\infty, \widehat{\Pi}\},$$

and let us recall that $\widehat{\mathcal{E}}_{\widehat{\mathcal{B}} \vee \widehat{\mathcal{X}}_0^\infty}$ is sampling shift-invariant if x is $\mathcal{B}$-stationary or, equivalently,

(9.4.17) $\qquad\qquad\qquad \Gamma \text{ I } \widehat{\mathcal{X}}_0^\infty \mid \widehat{\mathcal{B}}.$

Since $\widehat{\mathcal{B}}_\Gamma = \widehat{\mathcal{B}}$, this is in fact equivalent to saying, by Theorem 8.2.28, that $\mathcal{E}_{\widehat{\mathcal{B}} \vee \widehat{\mathcal{X}}_0^\infty}$ is shift-invariant, i.e.,

(9.4.18) $\qquad\qquad\qquad \Gamma \text{ I } (\widehat{\mathcal{B}} \vee \widehat{\mathcal{X}}_0^\infty).$

As shown in Section 9.3.1, (9.4.17) is actually equivalent to

(9.4.19) $\qquad\qquad\qquad \Gamma \text{ I } \widehat{\mathcal{X}}_0^n \mid \widehat{\mathcal{B}} \qquad \forall\, n \in I\!N,$

i.e., $\widehat{\mathcal{E}}_{\widehat{\mathcal{B}} \vee \widehat{\mathcal{X}}_0^\infty}$ is sampling shift-invariant if and only if $\widehat{\mathcal{E}}_{\widehat{\mathcal{B}} \vee \widehat{\mathcal{X}}_0^n}$ is sampling shift-invariant $\forall\, n \in I\!N$; this means that an asymptotic property of shift-invariance is in fact equivalent to a finite horizon property of shift invariance. We now consider a less general version of Proposition 8.3.3, more suited to the present analysis.

9.4.1 Proposition.

(i) $\widehat{\mathcal{E}}_{\widehat{\mathcal{B}} \vee \widehat{\mathcal{X}}_0^n}$ is (sampling) shift-invariant if and only if $\widehat{\mathcal{E}}_{\widehat{\mathcal{B}} \vee \widehat{\mathcal{Z}}_0^n}$ and $\widehat{\mathcal{E}}^{\widehat{\mathcal{Z}}_0^n}_{\widehat{\mathcal{B}} \vee \widehat{\mathcal{Y}}_0^n}$ are (sampling) shift-invariant.

(ii) $\widehat{\mathcal{E}}_{\widehat{\mathcal{B}} \vee \widehat{\mathcal{X}}_0^\infty}$ is (sampling) shift-invariant if and only if $\widehat{\mathcal{E}}_{\widehat{\mathcal{B}} \vee \widehat{\mathcal{Z}}_0^\infty}$ and $\widehat{\mathcal{E}}^{\widehat{\mathcal{Z}}_0^\infty}_{\widehat{\mathcal{B}} \vee \widehat{\mathcal{Y}}_0^\infty}$ are (sampling) shift-invariant. $\qquad\blacksquare$

Let us recall the definitions of sampling shift-invariance conditionally on the z-process. $\widehat{\mathcal{E}}^{\widehat{\mathcal{Z}}_0^n}_{\widehat{\mathcal{B}} \vee \widehat{\mathcal{Y}}_0^n}$ is sampling shift-invariant if

(9.4.20) $\qquad\qquad\qquad \Gamma \text{ I } \widehat{\mathcal{Y}}_0^n \mid \widehat{\mathcal{B}} \vee \widehat{\mathcal{Z}}_0^n,$

i.e., $\Gamma \text{ I } (\widehat{\mathcal{B}} \vee \widehat{\mathcal{Z}}_0^n) \cap \overline{\widehat{\mathcal{J}}}$, and $\forall\, k, n \in I\!N$,

(9.4.21) $\quad \tau^{-k}(\widehat{\mathcal{B}} \vee \widehat{\mathcal{Z}}_0^n)(\widehat{t} \circ \tau^k) = \{(\widehat{\mathcal{B}} \vee \widehat{\mathcal{Z}}_0^n)\widehat{t}\} \circ \tau^k \quad \forall\, \widehat{t} \in \left[\widehat{\mathcal{Y}}_0^n\right]^+,$

or, equivalently, if

(9.4.22) $\left(\widehat{\mathcal{B}} \vee \widehat{\mathcal{Z}}_k^{n+k}\right)\left(\widehat{t} \circ \tau^k\right) = \left\{\left(\widehat{\mathcal{B}} \vee \widehat{\mathcal{Z}}_0^n\right)\widehat{t}\right\} \circ \tau^k \quad \forall\, \widehat{t} \in \left[\widehat{\mathcal{Y}}_0^n\right]^+.$

Similarly, $\widehat{\mathcal{E}}_{\widehat{\mathcal{B}} \vee \widehat{\mathcal{Y}}_0^\infty}^{\widehat{\mathcal{Z}}_0^\infty}$ is sampling shift invariant if

(9.4.23) $\Gamma \mathrm{I} \, \widehat{\mathcal{Y}}_0^\infty \mid \widehat{\mathcal{B}} \vee \widehat{\mathcal{Z}}_0^\infty,$

i.e., $\Gamma \mathrm{I} \left(\widehat{\mathcal{B}} \vee \widehat{\mathcal{Z}}_0^\infty \right) \cap \overline{\widehat{\mathcal{I}}}$ and $\forall \, k \in I\!N,$

(9.4.24) $\left(\widehat{\mathcal{B}} \vee \widehat{\mathcal{Z}}_k^\infty \right) \left(\widehat{t} \circ \tau^k \right) = \left\{ \left(\widehat{\mathcal{B}} \vee \widehat{\mathcal{Z}}_0^\infty \right) \widehat{t} \right\} \circ \tau^k \quad \forall \, \widehat{t} \in \left[\widehat{\mathcal{Y}}_0^\infty \right]^+.$

Let us note that, contrary to the marginal experiment $\widehat{\mathcal{E}}_{\widehat{\mathcal{B}} \vee \widehat{x}_0^n}$, if $\widehat{\mathcal{E}}_{\widehat{\mathcal{B}} \vee \widehat{\mathcal{Y}}_0^n}^{\widehat{\mathcal{Z}}_0^n}$ is shift invariant, i.e., $\Gamma \mathrm{I} \left(\widehat{\mathcal{B}} \vee \widehat{\mathcal{Y}}_0^n \right) \mid \widehat{\mathcal{Z}}_0^n$, this is not equivalent to $\widehat{\mathcal{E}}_{\widehat{\mathcal{B}} \vee \widehat{\mathcal{Y}}_0^n}^{\widehat{\mathcal{Z}}_0^n}$ sampling shift invariant, i.e., $\Gamma \mathrm{I} \, \widehat{\mathcal{Y}}_0^n \mid \widehat{\mathcal{B}} \vee \widehat{\mathcal{Z}}_0^n$, since it also requires $\widehat{\mathcal{E}}_{\widehat{\mathcal{B}} \vee \widehat{\mathcal{Y}}_0^n}^{\widehat{\mathcal{Z}}_0^n}$ to be prior-shift invariant, i.e., $\Gamma \mathrm{I} \, \widehat{\mathcal{B}} \mid \widehat{\mathcal{Z}}_0^n$, by Proposition 8.3.3 (ii). This is, however, a consequence of $\widehat{\mathcal{E}}_{\widehat{\mathcal{B}} \vee \widehat{\mathcal{Z}}_0^n}$ sampling shift invariant, i.e., $\Gamma \mathrm{I} \, \widehat{\mathcal{Z}}_0^n \mid \widehat{\mathcal{B}}$. But it remains true that asymptotic sampling shift invariance is implied by a finite horizon sampling shift invariance in the experiment conditioned on the z-process. More precisely, we have the following theorem:

9.4.2 Theorem.

(i)   If $\Gamma \mathrm{I} \left(\widehat{\mathcal{B}} \vee \widehat{\mathcal{Z}}_0^\infty \right) \cap \overline{\widehat{\mathcal{I}}}$, and if $\widehat{\mathcal{E}}_{\widehat{\mathcal{B}} \vee \widehat{\mathcal{Y}}_0^n}^{\widehat{\mathcal{Z}}_0^n}$ is sampling shift invariant $\forall n \in I\!N$, then $\widehat{\mathcal{E}}_{\widehat{\mathcal{B}} \vee \widehat{\mathcal{Y}}_0^\infty}^{\widehat{\mathcal{Z}}_0^\infty}$ is sampling shift invariant.

(ii) If $\left(\widehat{\mathcal{Z}}_n \right)$ is $\widehat{\mathcal{B}}$-allocated to $\left(\widehat{\mathcal{Y}}_n \right)$ and if $\widehat{\mathcal{E}}_{\widehat{\mathcal{B}} \vee \widehat{\mathcal{Y}}_0^\infty}^{\widehat{\mathcal{Z}}_0^\infty}$ is sampling shift invariant, then $\widehat{\mathcal{E}}_{\widehat{\mathcal{B}} \vee \widehat{\mathcal{Y}}_0^n}^{\widehat{\mathcal{Z}}_0^n}$ is sampling shift invariant $\forall n \in I\!N$.

Proof.

(i) We have to show that (9.4.22) implies (9.4.24).

For $\widehat{t} \in \left[\widehat{\mathcal{Y}}_0^\ell \right]_\infty$ as $n \to \infty$, $\left(\widehat{\mathcal{B}} \vee \widehat{\mathcal{Z}}_k^{n+k} \right) \left(\widehat{t} \circ \tau^k \right) \to \left(\widehat{\mathcal{B}} \vee \widehat{\mathcal{Z}}_k^\infty \right) \left(\widehat{t} \circ \tau^k \right)$ and $\left(\widehat{\mathcal{B}} \vee \widehat{\mathcal{Z}}_0^n \right) \widehat{t} \to \left(\widehat{\mathcal{B}} \vee \widehat{\mathcal{Z}}_0^\infty \right) \widehat{t}$ a.s. (and in L_1) by the martingale convergence theorem, Theorem 7.7.2. Since $\Gamma \mathrm{I} \left(\widehat{\mathcal{B}} \vee \widehat{\mathcal{Z}}_0^\infty \right) \cap \overline{\widehat{\mathcal{I}}},$

$\left\{\left(\hat{\mathcal{B}} \vee \hat{\mathcal{Z}}_0^n\right) \hat{\imath}\right\} \circ \tau^k \;\rightarrow\; \left\{\left(\hat{\mathcal{B}} \vee \hat{\mathcal{Z}}_0^\infty\right) \hat{\imath}\right\} \circ \tau^k$ a.s. Therefore (9.2.24) holds $\forall \hat{\imath} \in \left[\hat{\mathcal{Y}}_0^\ell\right]_\infty$ $\forall \ell \in I\!N$ and, by Theorem 0.2.21, (9.2.24) holds $\forall \hat{\imath} \in \left[\hat{\mathcal{Y}}_0^\infty\right]_\infty$.

(ii) If $\left(\hat{\mathcal{Z}}_n\right)$ is $\hat{\mathcal{B}}$-allocated to $\left(\hat{\mathcal{Y}}_n\right)$ and if $\hat{\imath} \in \left[\hat{\mathcal{Y}}_0^n\right]_\infty$,

$$\left(\hat{\mathcal{B}} \vee \hat{\mathcal{Z}}_k^\infty\right)\left(\hat{\imath} \circ \tau^k\right) \;=\; \left(\hat{\mathcal{B}} \vee \hat{\mathcal{Z}}_k^{n+k}\right)\left(\hat{\imath} \circ \tau^k\right)$$
$$\left(\hat{\mathcal{B}} \vee \hat{\mathcal{Z}}_0^\infty\right)\hat{\imath} \;=\; \left(\hat{\mathcal{B}} \vee \hat{\mathcal{Z}}_0^n\right)\hat{\imath}.$$

Therefore, since $\Gamma\,I\left(\hat{\mathcal{B}} \vee \hat{\mathcal{Z}}_0^\infty\right) \cap \overline{\overline{\mathcal{I}}}$, (9.4.24) implies (9.4.22). Hence (9.4.23) implies (9.4.20). ∎

As recalled in the introduction, the ergodicity of $\hat{\mathcal{E}}_{\hat{\mathcal{B}} \vee \hat{\mathcal{Y}}_0^\infty}^{\hat{\mathcal{Z}}_0^\infty}$ is obtained through the usual assumption that $\hat{\mathcal{E}}_{\hat{\mathcal{B}} \vee \hat{\mathcal{Y}}_0^\infty}^{\hat{\mathcal{Z}}_0^\infty}$ is conditionally independent or equivalently that the process $(\hat{\mathcal{Y}}_n)$ is $\hat{\mathcal{B}}$-sifted by $(\hat{\mathcal{Z}}_n)$, i.e.,

$$(9.4.25) \qquad \underset{0 \le n < \infty}{\perp\!\!\!\perp} \hat{\mathcal{Y}}_n \mid \hat{\mathcal{B}} \vee \hat{\mathcal{Z}}_0^\infty$$

along with

$$(9.4.26) \qquad \hat{\mathcal{Y}}_n \perp\!\!\!\perp \hat{\mathcal{Z}}_0^\infty \mid \hat{\mathcal{B}} \vee \hat{\mathcal{Z}}_n \quad \forall n \in I\!N.$$

Recall that, by Theorem 7.6.10, this asymptotic assumption is equivalent to a sequence of finite sample assumptions and that it implies that $\left(\hat{\mathcal{Z}}_n\right)$ is $\hat{\mathcal{B}}$-allocated to $\left(\hat{\mathcal{Y}}_n\right)$, i.e.,

$$(9.4.27) \qquad \hat{\mathcal{Y}}_I \perp\!\!\!\perp \hat{\mathcal{Z}}_0^\infty \mid \hat{\mathcal{B}} \vee \hat{\mathcal{Z}}_I \quad \forall I \subset I\!N.$$

The assumption that $\hat{\mathcal{E}}_{\hat{\mathcal{B}} \vee \hat{\mathcal{Y}}_0^\infty}^{\hat{\mathcal{Z}}_0^\infty}$ is conditionally independent permits the characterization of the sampling shift invariance of $\hat{\mathcal{E}}_{\hat{\mathcal{B}} \vee \hat{\mathcal{Y}}_0^\infty}^{\hat{\mathcal{Z}}_0^\infty}$ to be simplified as is shown in the following theorem:

9.4.3 Theorem. If $\hat{\mathcal{E}}_{\hat{\mathcal{B}} \vee \hat{\mathcal{Y}}_0^\infty}^{\hat{\mathcal{Z}}_0^\infty}$ is conditionally independent, then $\hat{\mathcal{E}}_{\hat{\mathcal{B}} \vee \hat{\mathcal{Y}}_0^\infty}^{\hat{\mathcal{Z}}_0^\infty}$ is sampling shift invariant if and only if $\hat{\mathcal{E}}_{\hat{\mathcal{B}} \vee \hat{\mathcal{Y}}_0}^{\hat{\mathcal{Z}}_0^\infty}$ is sampling shift invariant and $\Gamma\,I\left(\hat{\mathcal{B}} \vee \hat{\mathcal{Z}}_0^\infty\right) \cap \overline{\overline{\mathcal{I}}}$.

Proof. By Theorem 9.4.2 (ii) and (9.4.27) the condition is necessary. By Theorem 9.4.2 (i), the condition will also be sufficient if it implies that $\widehat{\mathcal{E}}^{\widehat{\mathcal{Z}}^n_0}_{\widehat{\mathcal{B}} \vee \widehat{\mathcal{Y}}^n_0}$ is sampling shift invariant $\forall n \in I\!N$. Thus we have to prove that $\Gamma \text{ I } \widehat{\mathcal{Y}}_0 \mid \widehat{\mathcal{B}} \vee \widehat{\mathcal{Z}}_0$ and $\Gamma \text{ I } \left(\widehat{\mathcal{B}} \vee \widehat{\mathcal{Z}}^\infty_0 \right) \cap \overline{\overline{\mathcal{I}}}$ implies $\Gamma \text{ I } \widehat{\mathcal{Y}}^n_0 \mid \widehat{\mathcal{B}} \vee \widehat{\mathcal{Z}}^n_0 \; \forall n \in I\!N$. Since $\Gamma \text{ I } \left(\widehat{\mathcal{B}} \vee \widehat{\mathcal{Z}}^\infty_0 \right) \cap \overline{\overline{\mathcal{I}}}$ implies that $\Gamma \text{ I } \left(\widehat{\mathcal{B}} \vee \widehat{\mathcal{Z}}^n_0 \right) \cap \overline{\overline{\mathcal{I}}}$ by Proposition 8.2.12, it suffices to prove (9.4.22). Let $f_m \in [\mathcal{V}]^+, 0 \leq m \leq n$. Then for any $k, n \in I\!N$:

$$\left(\widehat{\mathcal{B}} \vee \widehat{\mathcal{Z}}^{n+k}_k \right) \left[\prod_{0 \leq m \leq n} f_m \circ \widehat{y}_{m+k} \right]$$

$$= \left(\widehat{\mathcal{B}} \vee \widehat{\mathcal{Z}}^\infty_0 \right) \left[\prod_{0 \leq m \leq n} f_m \circ \widehat{y}_{m+k} \right] \qquad \text{by (9.4.27)}$$

$$= \prod_{0 \leq m \leq n} \left(\widehat{\mathcal{B}} \vee \widehat{\mathcal{Z}}^\infty_0 \right) [f_m \circ \widehat{y}_{m+k}] \qquad \text{by (9.4.25)}$$

$$= \prod_{0 \leq m \leq n} \left(\widehat{\mathcal{B}} \vee \widehat{\mathcal{Z}}_{m+k} \right) [f_m \circ \widehat{y}_{m+k}] \qquad \text{by (9.4.26)}$$

Now, if $\Gamma \text{ I } \widehat{\mathcal{Y}}_0 \mid \widehat{\mathcal{B}} \vee \widehat{\mathcal{Z}}_0$, then

$$\left(\widehat{\mathcal{B}} \vee \widehat{\mathcal{Z}}_{m+k} \right) [f_m \circ \widehat{y}_{m+k}] \; = \; \left(\widehat{\mathcal{B}} \vee \widehat{\mathcal{Z}}_0 \right) [f_m \circ \widehat{y}_0] \circ \tau^{m+k}$$

$$= \; \left\{ \left(\widehat{\mathcal{B}} \vee \widehat{\mathcal{Z}}_0 \right) [f_m \circ \widehat{y}_0] \circ \tau^m \right\} \circ \tau^k$$

$$= \; \left(\widehat{\mathcal{B}} \vee \widehat{\mathcal{Z}}_m \right) [f_m \circ \widehat{y}_m] \circ \tau^k.$$

since $\Gamma \text{ I } \left(\widehat{\mathcal{B}} \vee \widehat{\mathcal{Z}}_m \right) \cap \overline{\overline{\mathcal{I}}}$. Therefore,

$$\left(\widehat{\mathcal{B}} \vee \widehat{\mathcal{Z}}^{n+k}_k \right) \left[\left(\prod_{0 \leq m \leq n} f_m \circ \widehat{y}_m \right) \circ \tau^k \right]$$

$$= \left(\widehat{\mathcal{B}} \vee \widehat{\mathcal{Z}}^{n+k}_k \right) \left[\prod_{0 \leq m \leq n} f_m \circ \widehat{y}_{m+k} \right]$$

$$= \prod_{0 \leq m \leq n} \left(\widehat{\mathcal{B}} \vee \widehat{\mathcal{Z}}_m \right) [f_m \circ \widehat{y}_m] \circ \tau^k$$

$$= \left\{ \prod_{0 \leq m \leq n} \left(\widehat{\mathcal{B}} \vee \widehat{\mathcal{Z}}_m \right) [f_m \circ \widehat{y}_m] \right\} \circ \tau^k$$

$$= \left\{ \left(\widehat{\mathcal{B}} \vee \widehat{\mathcal{Z}}^n_0 \right) \left[\prod_{0 \leq m \leq n} f_m \circ \widehat{y}_m \right] \right\} \circ \tau^k$$

since $\Gamma \, I \left(\widehat{\mathcal{B}} \vee \widehat{\mathcal{Z}}_0^n \right) \cap \overline{\overline{\mathcal{I}}}$. By Theorem 0.2.21, $\forall \, \widehat{t} \in \left[\widehat{\mathcal{Y}}_0^n \right]^+$,

$$\left(\widehat{\mathcal{B}} \vee \widehat{\mathcal{Z}}_k^{n+k} \right) \left(\widehat{t} \circ \tau^k \right) = \left\{ \left(\widehat{\mathcal{B}} \vee \widehat{\mathcal{Z}}_0^n \right) \widehat{t} \right\} \circ \tau^k,$$

i.e., $\Gamma \, I \, \widehat{\mathcal{Y}}_0^n \mid \widehat{\mathcal{B}} \vee \widehat{\mathcal{Z}}_0^n$. ∎

We can now state the main result of this section which, as indicated in the introduction, is analogous to Theorem 7.7.18. Although the assumptions used here are stronger than those underlying Theorem 7.7.18, they provide the same result: exact estimability of the parameter, but with the added feature that these assumptions are finite sample properties only. Since we are dealing with shift invariance, these assumptions provided more natural characterization of invariant σ-fields while Theorem 7.7.18 characterized tail σ-fields.

9.4.4 Theorem. Let $\widehat{\mathcal{E}}_{\widehat{\mathcal{B}} \vee \widehat{\mathcal{X}}_0^\infty}$ be a representation of a Bayesian experiment $\mathcal{E}_{\mathcal{A} \vee \mathcal{X}_0^\infty}$ marginalized on a subparameter $\mathcal{B} \subset \mathcal{A}$. Let us suppose that

(i) $\widehat{\mathcal{E}}_{\widehat{\mathcal{B}} \vee \widehat{\mathcal{Z}}_0^n}$ is sampling shift invariant $\forall n \in I\!N$;

(ii) $\widehat{\mathcal{E}}_{\widehat{\mathcal{B}} \vee \widehat{\mathcal{Y}}_0^\infty}^{\widehat{\mathcal{Z}}_0^\infty}$ is conditionally independent;

(iii) $\widehat{\mathcal{E}}_{\widehat{\mathcal{B}} \vee \widehat{\mathcal{Y}}_0}^{\widehat{\mathcal{Z}}_0}$ is sampling shift invariant;

(iv) $\widehat{\mathcal{E}}_{\widehat{\mathcal{B}} \vee \widehat{\mathcal{Y}}_0^\infty}^{\widehat{\mathcal{Z}}_0^\infty}$ is identified;

then $\widehat{\mathcal{B}}$ is exactly estimable in $\widehat{\mathcal{E}}_{\widehat{\mathcal{B}} \vee \widehat{\mathcal{X}}_0^\infty}$. More precisely,

(v) $\overline{\widehat{\mathcal{X}}_\Gamma} = \overline{\widehat{\mathcal{B}} \vee \widehat{\mathcal{Z}}_\Gamma}$.

Proof.

1. As shown before, (i) is equivalent to $\Gamma \, I \, \widehat{\mathcal{Z}}_0^\infty \mid \widehat{\mathcal{B}}$. By Theorem 8.2.23 (iv), (i) implies

(vi) $\Gamma \, I \, (\widehat{\mathcal{B}} \vee \widehat{\mathcal{Z}}_0^\infty) \cap \overline{\overline{\mathcal{I}}}$.

However, by Theorems 8.2.28 and 8.2.29, since $\widehat{\mathcal{B}}_\Gamma = \widehat{\mathcal{B}}$ and $\widehat{\mathcal{Z}}_0^\infty$ is Γ-stable, (i) implies $(\widehat{\mathcal{B}} \vee \widehat{\mathcal{Z}}_0^\infty)_\Gamma^* \perp\!\!\!\perp \widehat{\mathcal{Z}}_0^\infty \mid \widehat{\mathcal{Z}}_\Gamma^*$. But under (vi), by Theorem 8.2.17,

$$(\widehat{\mathcal{B}} \vee \widehat{\mathcal{Z}}_0^\infty)_\Gamma^* = \overline{(\widehat{\mathcal{B}} \vee \widehat{\mathcal{Z}}_0^\infty)_\Gamma \cap (\widehat{\mathcal{B}} \vee \widehat{\mathcal{Z}}_0^\infty)}$$

$$\widehat{\mathcal{Z}}_\Gamma^* = \overline{\widehat{\mathcal{Z}}_\Gamma \cap \widehat{\mathcal{Z}}_0^\infty}.$$

Therefore (i) implies $(\widehat{\mathcal{B}} \vee \widehat{\mathcal{Z}}_0^\infty)_\Gamma \perp\!\!\!\perp \widehat{\mathcal{Z}}_0^\infty \mid \widehat{\mathcal{Z}}_\Gamma$. By Corollary 2.2.11, since

$$\widehat{\mathcal{B}} \vee \widehat{\mathcal{Z}}_\Gamma \subset (\widehat{\mathcal{B}} \vee \widehat{\mathcal{Z}}_0^\infty)_\Gamma,$$

(i) implies

$$(\widehat{\mathcal{B}} \vee \widehat{\mathcal{Z}}_0^\infty)_\Gamma \perp\!\!\!\perp (\widehat{\mathcal{B}} \vee \widehat{\mathcal{Z}}_0^\infty) \mid \widehat{\mathcal{B}} \vee \widehat{\mathcal{Z}}_\Gamma.$$

By Corollary 2.2.8, this implies that $(\widehat{\mathcal{B}} \vee \widehat{\mathcal{Z}}_0^\infty)_\Gamma \subset \overline{\widehat{\mathcal{B}} \vee \widehat{\mathcal{Z}}_\Gamma}$. Therefore (i) implies

(vii) $\overline{(\widehat{\mathcal{B}} \vee \widehat{\mathcal{Z}}_0^\infty)_\Gamma} = \overline{\widehat{\mathcal{B}} \vee \widehat{\mathcal{Z}}_\Gamma}.$

2. Now, by Theorem 9.4.3, and under (ii), (iii) and (iv) are equivalent to $\widehat{\mathcal{E}}_{\widehat{\mathcal{B}} \vee \widehat{\mathcal{Y}}_0^\infty}^{\widehat{\mathcal{Z}}_0^\infty}$ being sampling shift invariant, i.e., $\Gamma \mathrel{\mathrm{I}} \widehat{\mathcal{Y}}_0^\infty \mid \widehat{\mathcal{B}} \vee \widehat{\mathcal{Z}}_0^\infty$. Along with (i), this is equivalent to

(viii) $\Gamma \mathrel{\mathrm{I}} \widehat{\mathcal{X}}_0^\infty \mid \widehat{\mathcal{B}}.$

As in 1, (viii) implies

(ix) $\Gamma \mathrel{\mathrm{I}} \left(\widehat{\mathcal{B}} \vee \widehat{\mathcal{X}}_0^\infty \right) \cap \overline{\overline{\mathcal{I}}},$

and

(x) $\overline{(\widehat{\mathcal{B}} \vee \widehat{\mathcal{X}}_0^\infty)_\Gamma} = \overline{\widehat{\mathcal{B}} \vee \widehat{\mathcal{X}}_\Gamma}.$

However, by Theorem 8.2.28, (viii) is equivalent to $\Gamma \mathrel{\mathrm{I}} \widehat{\mathcal{X}}_0^\infty \mid \widehat{\mathcal{X}}_\Gamma^*$ and $\widehat{\mathcal{B}} \perp\!\!\!\perp \widehat{\mathcal{X}}_0^\infty \mid \widehat{\mathcal{X}}_\Gamma^*$. But under (ix), by Theorem 8.2.17, $\widehat{\mathcal{X}}_\Gamma^* = \overline{\widehat{\mathcal{X}}_\Gamma \cap \widehat{\mathcal{X}}_0^\infty}$. Therefore (viii) implies

(xi) $\hat{B} \perp \hat{\mathcal{X}}_0^\infty \mid \hat{\mathcal{X}}_\Gamma$,

i.e., $\hat{\mathcal{X}}_\Gamma$ is an $\mathcal{E}_{\hat{B} \vee \hat{\mathcal{X}}_0^\infty}$-sufficient statistic.

3. Now (ii) implies (9.4.25), i.e., $\perp_{0 \leq n < \infty} \hat{\mathcal{Y}}_n \mid \hat{B} \vee \hat{\mathcal{Z}}_0^\infty$ which, by Corollary 2.2.11, is equivalent to $\perp_{0 \leq n < \infty} \hat{\mathcal{X}}_n \mid \hat{B} \vee \hat{\mathcal{Z}}_0^\infty$. Hence, by Theorem 7.6.5, $\left(\hat{B} \vee \hat{\mathcal{X}}_n \right)_T \subset \overline{\hat{B} \vee \hat{\mathcal{Z}}_0^\infty}$ and, therefore, $\left(\hat{B} \vee \hat{\mathcal{X}}_0^\infty \right)_\Gamma \subset \overline{\hat{B} \vee \hat{\mathcal{Z}}_0^\infty}$. But under (ix), by Definition 8.2.31, this means that Γ is ergodic on $\hat{B} \vee \hat{\mathcal{X}}_0^\infty$ conditionally on $\hat{B} \vee \hat{\mathcal{Z}}_0^\infty$. Hence, by Lemma 8.2.32, $\left(\hat{B} \vee \hat{\mathcal{X}}_0^\infty \right)_\Gamma \subset \overline{\left(\hat{B} \vee \hat{\mathcal{Z}}_0^\infty \right)_\Gamma}$. But under (viii) and (ix), and since $\hat{\mathcal{Z}}_\Gamma \subset \hat{\mathcal{X}}_\Gamma$, this is equivalent to

(xii) $\overline{\hat{B} \vee \hat{\mathcal{X}}_\Gamma} = \overline{\left(\hat{B} \vee \hat{\mathcal{X}}_0^\infty \right)_\Gamma} = \overline{\left(\hat{B} \vee \hat{\mathcal{Z}}_0^\infty \right)_\Gamma} = \overline{\hat{B} \vee \hat{\mathcal{Z}}_\Gamma}$.

Therefore $\hat{\mathcal{X}}_\Gamma \subset \overline{\hat{B} \vee \hat{\mathcal{Z}}_\Gamma}$, i.e., $\hat{\mathcal{X}}_\Gamma$ is exactly estimating in $\mathcal{E}_{\hat{B} \vee \hat{\mathcal{X}}_0^\infty}^{\hat{\mathcal{Z}}_\Gamma}$ and in $\mathcal{E}_{\hat{B} \vee \hat{\mathcal{X}}_0^\infty}^{\hat{\mathcal{Z}}_0^\infty}$.

4. Since $\hat{\mathcal{X}}_\Gamma$ is sufficient and exactly estimating in $\mathcal{E}_{\hat{B} \vee \hat{\mathcal{X}}_0^\infty}^{\hat{\mathcal{Z}}_\Gamma}$, this experiment is exact and, by Theorem 4.7.8 and Proposition 4.7.7,

(xiii) $\overline{\hat{\mathcal{X}}_\Gamma} = \overline{\left(\hat{B} \vee \hat{\mathcal{Z}}_\Gamma \right) \hat{\mathcal{X}}_0^\infty} = \overline{\hat{\mathcal{X}}_0^\infty \left(\hat{B} \vee \hat{\mathcal{Z}}_\Gamma \right)} = \overline{\left(\hat{B} \vee \hat{\mathcal{Z}}_\Gamma \right) \cap \hat{\mathcal{X}}_0^\infty}$.

5. Finally, by Proposition 4.6.6 (ii), (iv) implies that $\mathcal{E}_{\hat{B} \vee \hat{\mathcal{X}}_0^\infty}^{\hat{\mathcal{Z}}_\Gamma}$ is identified, i.e., $\left(\hat{B} \vee \hat{\mathcal{Z}}_\Gamma \right) \hat{\mathcal{X}}_0^\infty = \hat{B} \vee \hat{\mathcal{Z}}_\Gamma$. Therefore

(xiv) $\overline{\hat{\mathcal{X}}_\Gamma} = \overline{\hat{B} \vee \hat{\mathcal{Z}}_\Gamma}$,

and this shows that $\hat{B} \subset \overline{\hat{\mathcal{X}}_0^\infty}$, i.e., $\hat{B}$ is exactly estimable. ∎

Remark that in Theorem 9.4.4, assumption (iv) is the only asymptotic assumption. But recall that assumption (ii) implies the allocation Property (9.4.27) which implies, in particular, that $\hat{\mathcal{Y}}_0^n \perp \hat{\mathcal{Z}}_0^\infty \mid \hat{B} \vee \hat{\mathcal{Z}}_0^n$, i.e., $(\hat{\mathcal{Z}}_0^n)$ is $\hat{B}$-transitive for $(\hat{\mathcal{Y}}_0^n)$. Now, by Theorem 7.3.21, this implies that

(9.4.28) $\left(\hat{B} \vee \hat{\mathcal{Z}}_0^\infty \right) \hat{\mathcal{X}}_0^\infty = \overline{\bigvee_{0 \leq n < \infty} \left(\hat{B} \vee \hat{\mathcal{Z}}_0^n \right) \hat{\mathcal{X}}_0^n \cap \left(\hat{B} \vee \hat{\mathcal{Z}}_0^\infty \right)}$.

Therefore $\widehat{\mathcal{E}}_{\widehat{\mathcal{B}}\vee\widehat{\mathcal{Y}}_0^\infty}^{\widehat{\mathcal{Z}}_0^\infty}$ is identified once $\widehat{\mathcal{E}}_{\widehat{\mathcal{B}}\vee\widehat{\mathcal{Y}}_0^n}^{\widehat{\mathcal{Z}}_0^n}$ is identified for some $n \in I\!N$.

Let us compare Theorem 9.4.4 and Theorem 7.7.18. Firstly, assumption (i) of 9.4.4, i.e., $\Gamma \text{ I } \widehat{\mathcal{Z}}_0^\infty \mid \widehat{\mathcal{B}}$ implies, by Theorems 8.2.28 and 8.2.17, that $\widehat{\mathcal{B}} \perp\!\!\!\perp \widehat{\mathcal{Z}}_0^\infty \mid \widehat{\mathcal{Z}}_\Gamma$. Since $\widehat{\mathcal{Z}}_\Gamma \subset \widehat{\mathcal{Z}}_T$, by Corollary 2.2.11, this implies that $\widehat{\mathcal{E}}_{\widehat{\mathcal{B}}\vee\widehat{\mathcal{Z}}_\infty}$ is tail-sufficient; this is assumption (i) of Theorem 7.7.18. Secondly, (xi) in the proof of Theorem 9.4.4 implies that $\widehat{\mathcal{E}}_{\widehat{\mathcal{B}}\vee\widehat{\mathcal{X}}_\infty}$ is tail-sufficient or, equivalently, $\widehat{\mathcal{B}} \perp\!\!\!\perp \widehat{\mathcal{X}}_0^\infty \mid \widehat{\mathcal{X}}_n^\infty \; \forall\, n \in I\!N$ by Theorem 7.7.2. But this is equivalent to $\widehat{\mathcal{B}} \perp\!\!\!\perp \left(\widehat{\mathcal{Y}}_0^\infty \vee \widehat{\mathcal{Z}}_0^\infty \right) \mid \widehat{\mathcal{Y}}_n^\infty \vee \widehat{\mathcal{Z}}_0^\infty$ and, by Corollary 2.2.11, this implies that $\widehat{\mathcal{B}} \perp\!\!\!\perp \widehat{\mathcal{Y}}_0^\infty \mid \widehat{\mathcal{Z}}_0^\infty \vee \widehat{\mathcal{Y}}_n^\infty$, i.e., $\widehat{\mathcal{E}}_{\widehat{\mathcal{B}}\vee\widehat{\mathcal{Y}}_0^\infty}^{\widehat{\mathcal{Z}}_0^\infty}$ is tail-sufficient. As the other assumptions are the same, this shows that assumptions of Theorem 9.4.4 are stronger than the assumptions of Theorem 7.7.18 and therefore, as a by-product, Theorem 9.4.4 also entails that

(9.4.29) $\overline{\widehat{\mathcal{X}}_T} = \overline{\widehat{\mathcal{B}} \vee \widehat{\mathcal{Z}}_T}.$

Hence $\widehat{\mathcal{X}}_T$ is both sufficient and exactly estimating in $\widehat{\mathcal{E}}_{\widehat{\mathcal{B}}\vee\widehat{\mathcal{X}}_0^\infty}^{\widehat{\mathcal{Z}}_T}$. Note that, in this set-up, nothing shows that $\overline{\widehat{\mathcal{Z}}_\Gamma} = \overline{\widehat{\mathcal{Z}}_T}$ or that $\overline{\widehat{\mathcal{X}}_\Gamma} = \overline{\widehat{\mathcal{X}}_T}$; this suggests that Theorems 7.7.18 and 9.4.4 may in fact be different in nature.

As in Section 7.7, we now present a version of Theorem 9.4.4, analogous to Theorem 7.7.19, that is based on properties of the sampling process before integrating out nuisance parameters.

9.4.5 Theorem. Let $\widehat{\mathcal{E}}_{\widehat{\mathcal{A}}\vee\widehat{\mathcal{X}}_0^\infty}$ be a representation of $\mathcal{E}_{\mathcal{A}\vee\mathcal{X}_0^\infty}$ and let $\mathcal{B} \subset \mathcal{A}$ be a subparameter such that:

(i) $\widehat{\mathcal{E}}_{\widehat{\mathcal{B}}\vee\widehat{\mathcal{Z}}_\infty}$ is sampling shift invariant;

(ii) $\widehat{\mathcal{E}}_{\widehat{\mathcal{A}}\vee\widehat{\mathcal{Y}}_0^\infty}^{\widehat{\mathcal{Z}}_0^\infty}$ is conditionally independent;

(iii) $\widehat{\mathcal{E}}_{\widehat{\mathcal{A}}\vee\widehat{\mathcal{Y}}_0}^{\widehat{\mathcal{Z}}_0}$ is sampling shift invariant;

(iv) $\widehat{\mathcal{B}}$ is an $\widehat{\mathcal{E}}_{\widehat{\mathcal{A}}\vee\widehat{\mathcal{Y}}_0}^{\widehat{\mathcal{Z}}_0}$-sufficient parameter;

(v) $\widehat{\mathcal{B}}$ is included in the minimal strong $\widehat{\mathcal{E}}_{\widehat{\mathcal{A}}\vee\widehat{\mathcal{Y}}_0^\infty}^{\widehat{\mathcal{Z}}_0^\infty}$-sufficient parameter;

then $\widehat{B}$ is exactly estimable.

Proof. By Theorems 7.7.19 and 9.4.4, it suffices to prove that (iii) and (iv) imply

(vi) $\widehat{B}$ is $\mathcal{E}^{\widehat{\widehat{Z}}_n}_{\widehat{A} \vee \widehat{\mathcal{Y}}_n}$ -sufficient $\forall\, n \in I\!N$;

(vii) $\mathcal{E}^{\widehat{\widehat{Z}}_0}_{\widehat{B} \vee \widehat{\mathcal{Y}}_0}$ is sampling shift invariant.

Indeed, under (vi), by Theorem 7.7.19, (ii) implies that $\mathcal{E}^{\widehat{\widehat{Z}}_\infty^0}_{\widehat{B} \vee \widehat{\mathcal{Y}}_0^\infty}$ is conditionally independent and (v) implies that $\mathcal{E}^{\widehat{\widehat{Z}}_\infty^0}_{\widehat{B} \vee \widehat{\mathcal{Y}}_0^\infty}$ is identified; the result then follows from Theorem 9.4.4. Thus, we have to prove that

(1) $\Gamma \mathrel{I} \widehat{\mathcal{Y}}_0 \mid \widehat{A} \vee \widehat{Z}_0$,

and

(2) $\widehat{A} \perp\!\!\!\perp \widehat{\mathcal{Y}}_0 \mid \widehat{B} \vee \widehat{Z}_0$

imply

(3) $\Gamma \mathrel{I} \widehat{\mathcal{Y}}_0 \mid \widehat{B} \vee \widehat{Z}_0$,

and

(4) $\widehat{A} \perp\!\!\!\perp \widehat{\mathcal{Y}}_n \mid \widehat{B} \vee \widehat{Z}_n$.

But this is an immediate consequence of Theorem 8.2.30 under the substitution $(\mathcal{M}_1, \mathcal{M}_2, \mathcal{M}_3) \rightarrow \left(\widehat{\mathcal{Y}}_0, \widehat{A}, \widehat{B} \vee \widehat{Z}_0 \right)$. ∎

Both Theorems 9.4.4 and 9.4.5 assume shift invariance of $\widehat{\mathcal{E}}_{\widehat{B} \vee \widehat{Z}_\infty^0}$ and, therefore, its tail sufficiency. Thus $\widehat{\mathcal{E}}_{\widehat{A} \vee \widehat{Z}_\infty^0}$ is allowed to not enjoy such a property provided that the prior distribution conditional on the parameter of interest is such that, after integrating out the nuisance parameter, $\widehat{\mathcal{E}}_{\widehat{B} \vee \widehat{Z}_\infty^0}$ is shift invariant, i.e., the process z will be $\mathcal{B}$-stationary whithout necessarily being $\mathcal{A}$-stationary. A simple example of such a situation is a process characterized, amongst other, by *a priori* i.i.d. incidental parameters. If shift invariance of $\widehat{\mathcal{E}}_{\widehat{B} \vee \widehat{Z}_\infty^0}$ is implied by the shift invariance of $\widehat{\mathcal{E}}_{\widehat{A} \vee \widehat{Z}_\infty^0}$, neither shift invariance nor tail sufficiency of $\widehat{\mathcal{E}}_{\widehat{A} \vee \widehat{Z}_\infty^0}$ is a necessary condition for shift invariance of $\widehat{\mathcal{E}}_{\widehat{B} \vee \widehat{Z}_\infty^0}$.

If $\widehat{B}$ and $\widehat{Z}_0^\infty$ are mutually ancillary (i.e., $\widehat{B} \perp\!\!\!\perp \widehat{Z}_0^\infty$), then by Theorem 8.2.30 the shift invariance of $\mathcal{E}_{\widehat{B} \vee \widehat{Z}_0^\infty}$ is equivalent to the shift invariance of $\mathcal{E}_{\widehat{Z}_0^\infty}$, i.e., of the predictive distribution of the process z. This supplementary condition of mutual ancillarity makes the reduction of $\mathcal{E}_{\widehat{B} \vee \widehat{X}_0^\infty}$ into $\mathcal{E}_{\widehat{B} \vee \widehat{Y}_0^\infty}^{\widehat{Z}_0^\infty}$ admissible but leaves open the possibility that the transition $\widehat{P}_{\widehat{Y}_0^\infty}^{\widehat{B} \vee \widehat{Z}_0^\infty}$ underlying the conditional experiment may not be sampling probabilities but a conditional probability obtained after integrating out the nuisance parameters of the sampling probabilities. In Theorem 9.4.5 the role of assumption (iv), which under assumption (ii) implies the sufficiency of $\widehat{B}$ in $\mathcal{E}_{\widehat{A} \vee \widehat{Y}_0^\infty}^{\widehat{Z}_0^\infty}$, is precisely to preclude such a possibility and, therefore, to let the other assumptions bear on the sampling probabilities. These assumptions of mutual ancillarity and of sufficiency of $\widehat{B}$ are implied, by Theorem 3.3.6 (iii), by an assumption of a cut in $\mathcal{E}_{\widehat{A} \vee \widehat{X}_0^\infty}$, i.e., when the parameter characterizing the distribution of the process z is *a priori* independent of the parameters characterizing the distribution of the process y conditional to the process z.

Thus the main difference between Theorems 9.4.4 and 9.4.5 lies in the transition $\widehat{P}_{\widehat{Y}_0^\infty}^{\widehat{B} \vee \widehat{Z}_0^\infty}$ characterizing the conditional experiment $\mathcal{E}_{\widehat{B} \vee \widehat{Y}_0^\infty}^{\widehat{Z}_0^\infty}$: in Theorem 9.4.4 the transition is obtained after integrating out the nuisance parameters whereas in Theorem 9.4.5 the transition is the sampling probabilities with $\widehat{B}$ as a sufficient parameter.

As the exact estimability of $\widehat{B}$ implies the exact estimability of any $\widehat{C} \subset \widehat{B}$, and as the shift invariance of $\mathcal{E}_{\widehat{B} \vee \widehat{Z}_0^\infty}$ implies the shift invariance of $\mathcal{E}_{\widehat{C} \vee \widehat{Z}_0^\infty}$ for any $\widehat{C} \subset \widehat{B}$ (by Theorem 8.2.23.(i), and since $\widehat{B}_\Gamma = \widehat{B}$ implies $\Gamma \perp\!\!\!\perp \left(\widehat{B} \vee \widehat{Z}_0^\infty \right)$ by Theorem 8.2.28), we may conclude that the main difference, for practical purposes, finally resides in whether or not the parameter of interest to be exactly estimated is, or is not, included in a parameter characterizing a conditional sampling process.

Bibliography

Abramowitz, M. and I.A. Stegun (1964), *Handbook of Mathematical Functions*. National Bureau of Standards, New York: Dover Publications.

Akaike, H. (1974), A new look at the statistical model identification. *IEEE Transactions on Automatic Control*, **19**, 716–723.

Aldous, D.J. (1985), Exchangeability and related topics. (Lecture Notes in Mathematics, 1117). New York: Springer-Verlag.

Aldous, D.J. and J.W. Pitman (1979), On the zero-one law for exchangeable events. *Annals of Probability*, **7**, 704–723.

Anderson, T.W. (1971), *The Statistical Analysis of Time Series*. New York: John Wiley.

Anderson, T.W. (1976), Estimation of linear functional relationships: approximate distributions and connexions with simultaneous equations in econometrics (with discussion). *Journal of the Royal Statistical Society*, Series B , **38**, 1–36.

Anderson, T.W. (1984), Estimating linear statistical relationships. *The Annals of Statistics*, **12**, 1–45.

Ando, A. and G.M. Kaufman (1965), Bayesian analysis of the independent multinormal process, neither mean nor precision known. *Journal of the American Statistical Association*, **60**, 347–358.

Arnold, S. (1981), *The Theory of Linear Models and Multivariate Analysis*. New York: John Wiley.

Bahadur, R.R. (1954), Sufficiency and statistical decision functions. *The Annals of Mathematical Statistics*, **25**, 423–462.

Bahadur, R.R. (1955a), A characterization of sufficiency. *The Annals of Mathematical Statistics*, 26(2), 286–293.

Bahadur, R.R. (1955b), Statistics and subfields. *The Annals of Mathematics*, **26**(3), 490–497.

Barankin, E.W. (1961), Sufficient parameters: solution of the minimal dimensionality problem. *Annals of the Institute of Statistical Mathematics*, **12**, 91–118.

Barnard, G.A. (1963), Some aspects of the fiducial argument. *Journal of the Royal Statistical Society*, Series B, **25**, 111–114.

Barnard, G.A. and D.A. Sprott (1971), A note on Basu's example of anomalous ancillary statistics. In: *Foundations of Statistical Inference. A Symposium.* See Godambe and Sprott (1971), 163–176.

Barndorff-Nielsen, O. (1973), On M-ancillarity. *Biometrika*, **60**, 447–455.

Barndorff-Nielsen, O. (1978), *Information and Exponential Families in Statistical Theory.* New York: John Wiley.

Barra, J.R. (1971), *Notions fondamentales de statistique mathématique.* Paris: Dunod. (English translation: *Mathematical Basis of Statistics.* New York: Academic Press (1981)).

Basu, D. (1955), On statistics independent of a complete sufficient statistic. *Sankhyā*, **15**, 377–380.

Basu, D. (1958), On statistics independent of a sufficient statistics. *Sankhyā*, **20**, 223–226.

Basu, D. (1959), The family of ancillary statistics. *Sankhyā*, **21**, 247–256.

Basu, D. (1964), Recovery of ancillary information. *Sankhyā A*, Series A, **26**, 3–16.

Basu, D. (1975), Statistical information and likelihood. Part I. *Sankhyā A*, **37**, 1–71.

Basu, D. (1977), On the elimination of nuisance parameters. Part I. *Journal of the American Statistical Association*, **72**, 355–366.

Basu, D. and T.P. Speed (1974), Bibliography of sufficiency. Unpublished.

Becker, N. and I. Gordon (1983), On Cox's criterion for discriminating between alternative ancillary statistics. *International Statistical Review*, **51**, 89–92.

Berger, J. (1980), *Statistical Decision Theory: Foundations, Concepts and Methods*. New York: Springer-Verlag.

Berk, R.H. (1966), Limiting behavior of posterior distributions when the model is incorrect. *The Annals of Mathematical Statistics*, **37**, 51–58.

Berk, R.H. (1970), Consistency a posteriori. *The Annals of Mathematical Statistics*, **41**, 894–906.

Bernardo, J.M. (1979), Reference posterior distributions for Bayesian inferences (with discussion). *Journal of the Royal Statistical Society*, Series B,**41**, 113–147.

Bernardo, J.M., De Groot, M.H., Lindley, D.V. and A.F.M. Smith (editors), (1980), *Bayesian Statistics. Proceedings of the First International Meeting on Bayesian Statistics*, held in Valencia (Spain), May 28-June 2, 1979. Valencia: University Press.

Bernardo, J.M., De Groot, M.H., Lindley, D.V. and A.F.M. Smith (editors) (1985), *Bayesian Statistics 2. Proceedings of the Second Valencia International Meeting*, September 6-10, 1983. Amsterdam: North-Holland.

Bernardo, J.M., De Groot, M.H., Lindley, D.V. and A.F.M. Smith (editors) (1988), *Bayesian Statistics 3. Proceedings of the Third Valencia International Meeting*, 1987. Oxford: Clarendom Press.

Bhaskara Rao, K.P.S. and B.V.Rao (1981), *Borel Spaces*. (Dissertationes Mathematicae CXC). Warsaw: Polska Akademia Nauk, Instytut Matematyczny.

Billingsley, P.(1979), *Probability and Measures*. New York: John Wiley.

Blackwell, D. (1951), Comparison of experiments. *Proceedings of the Second Berkeley Symposium on Mathematical Statistics and Probabilities*. Berkeley: University of California Press, 93–102.

Blackwell, D. (1953), Equivalent comparison of experiments. *The Annals of Mathematical Statistics*, **24**, 265–272.

Blackwell, D. (1956), On a class of probability spaces. *Proceedings of the Third Berkeley Symposium on Mathematical Statistics and Probabilities*, **2**. Berkeley: University of California Press, 1–6.

Blackwell, D. and L.E. Dubins (1975), On existence and non-existence of proper, regular conditional distribution. *The Annals of Probability*, **3**, 741–752.

Blumenthal, R.M. and R.K. Getoor (1968), *Markov Processes and Potential Theory*. New York: Academic Press.

Box, G.E.P. and G.C. Tiao (1973), *Bayesian Inference in Statistical Analysis*. Reading, (Mass.): Addison-Welsey.

Breiman, L. (1968), *Probability*. London: Addison-Welsey.

Breiman, L., LeCam, L. and L. Schwartz (1964), Consistent estimates and zero-one sets. *The Annals of Mathematical Statistics*, **35**, 157–161.

Bunke, H. and O. Bunke (1974), Identifiability and estimability. *Mathematische Operationsforschung und Statistik*, **5**, 223–233.

Burguete, J.-F., A.R. Gallant and G. Souza (1982), On unification of the asymptotic theory of nonlinear econometric models. *Econometric Review*, **1**, 151–190.

Burkholder, D.L. (1961), Sufficiency in the undominated case. *The Annals of Mathematical Statistics*, **32**, 1191–1200.

Cano, J.A., Hernandez, A., and E. Moreno (1988), On Kolmogorov partial sufficiency. In: *Bayesian Statistics 3*. See J.M. Bernardo, M.H. De Groot, D.V. Lindley, and A.F.M. Smith, (1988), 553–556.

Cassel, C.M., Särndal, C.E. and J.H. Wretman (1977), *Foundations of Inference in Survey Sampling*. New York: John Wiley.

Chamberlain, G. (1982), On the general equivalence of Granger and Sims causality. *Econometrica*, **50**, 569–581.

Chandra, S. (1977), On the mixtures of probablity distributions. *Scandinavian Journal of Statistics*, **4**, 105–112.

Chow, Y.S. and H. Teicher (1978), *Probability Theory*. Berlin: Springer Verlag.

Chung, K.L. (1968), *A Course in Probability Theory*. New York: Harcourt-Brace.

Cifarelli, D.M., Muliere, P. and M. Scarsini (1981), Il Modello Lineare nell'Approccio Bayesiano non Parametrico. *Quaderni dell'Istituto Matematico G. Castlenuovo*, Roma.

Cocchi, D. and M. Mouchart (1986), Linear Bayes estimation in finite populations with a categorical auxiliary variable. CORE Discussion Paper 8615, Université Catholique de Louvain, Louvain-la-Neuve, Belgium.

Cocchi, D. and M. Mouchart (1989), Approximations of Bayesian solutions in finite population models. CORE Discussion Paper 8905, Université Catholique de Louvain, Louvain-la-Neuve, Belgium.

Cohn, H. (1965), On a class of dependent random variables. *Revue Roumaine de Mathématiques Pures et Appliquées*, **10**, 1593–1606.

Cohn, H. (1974), On the tail events of a Markov chain. *Zeitschrift für Wahrscheinlichkeitstheorie und Verwandte Gebiete*, **29**, 65–72.

Cornet, B. and H. Tulkens (editors) (1989), *Contributions to Operations Research and Economics. Proceedings of the CORE XXth Anniversary Symposium*. Cambridge (Mass.): The MIT Press, forthcoming.

Cox, D.R. (1961), Tests of separate families of hypotheses. *Proceedings of the Fourth Berkeley Symposium on Mathematical Statistics and Probability*, **1**. Berkeley: The University of California Press, 105–123.

Cox, D.R. (1962), Further results on tests of separate families of hypotheses. *Journal of the Royal Statistical Society*, Series B, **24**, 406–424.

Cox, D.R. (1971), The choice between alternative ancillary statistics. *Journal of the Royal Statistical Society*, Series B, **33**, 251–255.

Cox, D.R. and D.V. Hinkley (1974), *Theoretical Statistics*. London: Chapman and Hall.

Csiszar, I. (1967a), Information type measures of differences of probability distributions and indirect observations. *Studia Scientiarum Mathematicarum Hungria*, **2**, 299–318.

Csiszar, I. (1967b), On topological properties of f-divergences. *Studia Scientiarum Mathematicarum Hungria*, **2**, 319–329.

Davidson, J.E.H., Hendry, D.F., Srba, F., and S. Yeo (1978), Econometric modelling of the aggregate time-series relationship between consumers expenditure and income in United Kingdom. *Economic Journal*, **88**, 661–692.

Dawid, A.P. (1979a), Conditional independence in statistical theory (with discussion). *The Journal of the Royal Statistical Society*, Series B, **41**(1), 1–31.

Dawid, A.P. (1979b), Some misleading arguments involving conditional independence. *Journal of the Royal Statistical Society*, Series B, **41**, 249–252.

Dawid, A.P. (1980a), A Bayesian look at nuisance parameters. In: *Bayesian Statistics*. See J.M. Bernardo, M.H. de Groot, D.V. Lindley, and A.F.M. Smith, (1980), 167–184.

Dawid, A.P. (1980b), Conditional independence for statistical operations. *The Annals of Statistics*, **8**, 598–617.

Dawid, A.P., Stone, M., and J.V. Zidek (1973), Marginalization paradoxes in Bayesian and structural inference. *Journal of The Royal Statistical Society*, Series B, **35** 189–233.

de Finetti, B. (1937), La prévision: ses lois logiques, ses sources subjectives. *Annales de l'Institut Henri Poincaré*, **7**. (English translation: Foresight: its logical laws, its subjective sources. In Kyburg and Smokler (1964)).

de Finetti, B. (1974), *Probability, Induction and Statistics*. New York: John Wiley.

De Groot, M. (1970), *Optimal Statistical Decisions*. New York: McGraw Hill.

Deistler, M. (1976), The identifiability of linear econometric models with autocorrelated errors. *International Economic Review*, **17**, 26–46.

Deistler, M. (1986), Identifiability and causality in linear dynamic errors-in-variables systems. In: *Asymptotic Theory for Non i.i.d. Processes*. See J.-P. Florens, M. Mouchart, J.-P. Raoult, J.-M. Rolin, and L. Simar (1986), 145–154.

Deistler, M. and H.G. Seifert (1978), Identifiability and consistent estimability in dynamic econometric models. *Econometrica*, **46**, 969–980.

Dellacherie, C. and P.A. Meyer (1975), *Probabilité et potentiel*. Paris: Hermann.

Dellacherie, C. and P.A. Meyer (1980), *Probabilité et potentiel (B), Théorie des martingales*. Paris: Hermann.

Diaconis, P. and D. Freedman (1979), Sufficiency and exchangeability. Unpublished.

Diaconis, P. and D. Freedman (1984) Partial exchangeability and sufficiency. In: *Statistics: Applications and New Directions*, edited by J. K. Ghosh and J. Roy. Calcutta: Indian Statistical Institute, 205–236.

Diaconis, P. and D. Freedman (1986), On the consistency of Bayes estimates (with discussion). *The Annals of Statistics*, **14**, 1–67.

Diaconis, P. and D. Ylvisaker (1979), Conjugate priors for exponential families. *The Annals of Statistics*, **7**(2), 269–281.

Diaconis, P. and S. Zabell (1982), Updating subjective probability. *Journal of the American Statistical Association*, **77**, 822–830. (With a more extensive discussion in their Stanford technical report n° 136, 1979.)

Dickey, J.M. (1976), Approximate posterior distributions. *Journal of the American Statistical Association*, **71**, 680–689.

Doob, J.L. (1949), Applications of the theory of martingales. *Colloques Internationaux du C.N.R.S.*. Paris, C.N.R.S., 22–28.

Doob, J.L. (1953), *Stochastic Processes*. New York: John Wiley.

Drèze, J.H. (1974), Bayesian theory of identification in simultaneous equations models. In: *Studies of Bayesian Econometrics and Statistics*, edited by S. Fienberg and A. Zellner. Amsterdam: North-Holland, 159–174.

Drèze, J.H. and J.-F. Richard (1983), Bayesian analysis of simultaneous equation systems. In: *Handbook of Econometrics*, edited by Z. Griliches and M. Intriligator. Amsterdam: North-Holland, 517–598.

Dynkin, E.B. (1978), Sufficient statistics and extreme points. *Annals of Probability*, **5**, 705–730.

Elbers, C. and G. Ridder (1982), True and spurious duration dependence: the identifiability of the proportional hazard model. *Review of Economic Studies*, **49**, 403–410.

Engle, R.F., Hendry, D.F., and J.-F. Richard (1983), Exogeneity. *Econometrica*, **51**(2), 277–304.

Ferguson, T.S. (1973), A Bayesian analysis of some nonparametric problems. *Annals of Statistics*, **1**, 209–230.

Ferguson, T.S. (1974), Prior distribution on spaces of probability measures. *Annals of Statistics*, **2**, 615–629.

Fiori, G., Florens, J.-P., and H.W. Lai Tong (1982), Analyse des innovations dans un processus multivarié: application à des données françaises. *Annales de l'INSEE*, **46**, 3–24.

Fisher, F.M. (1966), *The Identification Problem in Econometrics*. New York: McGraw Hill.

Florens, J.-P. (1974), Contributions aux applications des statistiques bayésiennes à l'économétrie. Thèse de doctorat de troisième cycle, Université de Provence, Marseille, France.

Florens, J.-P. (1978), Mesures a priori et invariance dans une expérience bayésienne. *Publications de l'Institut de Statistique de l'Université de Paris*, **23**, 29–56.

Florens, J.-P. (1979), Exogénéité et non causalité dans un modèle bayésien. Communications aux Journées de l'Association des Statisticiens Universitaires, Paris.

Florens, J.-P. (1980), Spécification et réduction des expériences bayésiennes. Application au modèle économétrique linéaire. Thèse de doctorat d'Etat. Université de Rouen, France.

Florens, J.-P. (1982), Expériences bayésiennes invariantes. *Annales de l'Institut Henri Poincaré*, **18**(1-2), 305-317.

Florens, J.-P. (1983), Approximate reductions of Bayesian experiments. In: *Specifying Statistical Models*. See J.-P. Florens, M. Mouchart, J.-P. Raoult, L. Simar, and A. F. M. Smith (1983), 85-92.

Florens, J.-P. (1986), Consistency and invariance in Bayesian experiment. Document de Travail 8601. GREQE, Marseille.

Florens, J.P. (1988), Exhaustivité paramétrique et enveloppement de modèles. In : *Mélanges économiques, Essais en l'honneur d'Edmond Malinvaud*. Paris: Economica.

Florens, J.P., Hendry, D., and J.-F. Richard (1989), Encompassing and specificity. Cahier 8904, GREMAQ, Université des Sciences Sociales, Toulouse, France. (Also available as a discussion paper from Duke University).

Florens, J.-P. and M. Mouchart (1977), Reduction of Bayesian experiments. CORE Discussion Paper 7737, Université Catholique de Louvain, Louvain-la-Neuve, Belgium (revised July 1979).

Florens, J.-P. and M. Mouchart (1980), Initial and sequential reduction of Bayesian experiments. CORE Discussion Paper 8015, Université Catholique de Louvain, Louvain-la-Neuve, Belgium.

Florens, J.-P. and M. Mouchart (1982), A note on non causality. *Econometrica*, **50**(3), 583-591.

Florens, J.-P. and M. Mouchart (1985a), Conditioning in dynamic models. *Journal of Time Series Analysis*, **53**(1), 15-35.

Florens, J.-P. and M. Mouchart (1985b), A linear theory for non causality. *Econometrica*, **53**(1), 157-175.

Florens, J.-P. and M. Mouchart (1985c), Model selection: some remarks from a Bayesian viewpoint. In: *Model Choice*. See J.-P. Florens, M. Mouchart, J.-P. Raoult and L. Simar (1985), 27–44.

Florens, J.-P. and M. Mouchart (1986a), Exhaustivité, ancillarité et identification en statistique bayésienne. *Annales d'Economie et de Statistique*, 4, 63–93.

Florens, J.-P. and M. Mouchart (1986b), Some examples of Bayesian experiments. *Statistica*, 46(4), 439–448.

Florens, J.-P. and M. Mouchart (1989), Bayesian specification test. In: *Contributions to Operations Research and Economics*. See B. Cornet and H. Tulkens (editors) (1989).

Florens, J.-P., Mouchart, M., Raoult, J.-P., and L. Simar (1984), *Alternative Approaches to Time Series Analysis*. Proceedings of the 3rd Franco-Belgian Meeting of Statisticians, held in Louvain-la-Neuve, November 25-26, 1982. Bruxelles: Publications des Facultés Universitaires Saint-Louis.

Florens, J.-P., Mouchart, M., Raoult, J.-P., and L. Simar (1985), *Model Choice*. Proceedings of the 4th Franco-Belgian Meeting of Statisticians, held in Louvain-la-Neuve, November 24-25, 1983. Bruxelles: Publications des Facultés Universitaires Saint-Louis.

Florens, J.-P., Mouchart, M., Raoult, J.-P., Rolin, J.-M., and L. Simar (1986), *Asymptotic Theory for Non i.i.d. Processes*. Proceedings of the 5th Franco-Belgian Meeting of Statisticians, held at C.I.R.M., Marseille, November 23-24, 1984. Bruxelles: Publications des Facultés Universitaires Saint-Louis.

Florens, J.-P., Mouchart, M., Raoult, J.-P., Simar, L., and A.F.M. Smith (1983), *Specifying Statistical Models, from Parametric to Non-Parametric Using Bayesian Approaches*. Proceedings of the 2nd Franco-Belgian Meeting of Statisticians, held in Louvain-la-Neuve, October 15-16, 1981. (Lecture Notes in Statistics, 16). New York: Springer-Verlag.

Florens, J.-P., Mouchart, M., and J.-F. Richard (1974), Bayesian inference in error-in-variables models. *Journal of Multivariate Analysis*, **4**, 419–452.

Florens, J.-P., Mouchart, M., and J.-F. Richard (1976), Likelihood analysis of linear models. CORE Discussion Paper 7619, Université Catholique de Louvain, Louvain-la-Neuve, Belgium.

Florens, J.-P., Mouchart, M., and J.-F. Richard (1979), Specification and inference in linear models. CORE Discussion Paper 7943, Université Catholique de Louvain, Louvain-la-Neuve, Belgium.

Florens, J.-P., Mouchart, M., and J.-F. Richard (1986), Structural time series modelling: a Bayesian approach. *Applied Mathematics and Computation*, **20**(3–4), 365–400.

Florens, J.-P., Mouchart, M., and J.-F. Richard (1987), Dynamic error-in-variables models and limited information analysis. *Annales d'Economie et de Statistique*, **6/7**, 289–310.

Florens, J.-P., Mouchart, M., and J.-M. Rolin (1980), Réductions dans les expériences bayésiennes séquentielles. *Cahiers du Centre d'Etudes de Recherche Opérationnelle*, **22**(3–4), 353–362.

Florens, J.-P., Mouchart, M. and J.-M. Rolin (1985), On two definitions of identification. *Statistics*, **16**(2), 213–218.

Florens, J.-P., Mouchart, M., and J.-M. Rolin (1986), Exact estimability in conditional models. In: *Asymptotic Theory for Non i.i.d. Processes*. See J.-P. Florens, M. Mouchart, J.-P. Raoult, J.-M. Rolin, and L. Simar (1986), 121–144.

Florens, J.-P., Mouchart, M., and J.-M. Rolin (1987), Non causality and marginal Markov processes. CORE Discussion Paper 8706, Université Catholique de Louvain, Louvain-la-Neuve, Belgium.

Florens, J.-P., Mouchart, M., and S. Scotto (1983), Approximate sufficiency on the parameter space and model selection. In: *44th Session of the International Statistical Institute: Contributed Papers*, **2**, 763–766.

Florens, J.-P. and J.-F. Richard (1989), Encompassing in finite parameter space. Cahier 8905, GREMAQ, Université des Sciences Sociales, Toulouse, France. (Also available as a discussion paper from Duke University).

Florens, J.-P. and J.-M. Rolin (1984), Asymptotic sufficiency and exact estimability in Bayesian experiments. In: *Alternative Approaches to Time Series Analysis*. See J.-P. Florens, M. Mouchart, J.-P. Raoult, and L. Simar (1984), 121–142.

Florens, J.P. and S. Scotto (1984), Information value and econometric modelling. D.P. 17, Southern European Economics Discussion Series, ASSET, I.E.P., Universidad del Pais Vasco, Bilbao (Spain).

Fraser, D.A.S., 1968, *The Structure of Inference*. New York: John Wiley.

Fraser, D.A.S., (1972), Bayes, likelihood, or structural. *The Annals of Mathematical Statistics*, **43**, 777–790.

Fraser, D.A.S. (1973), The elusive ancillary. In: *Multivariate Statistical Inference*, edited by D.G. Kabe and R.P. Gupta. Amsterdam: North-Holland, 41–48.

Freedman, D.A. (1962), Invariants under mixing which generalize de Finetti's theorem. *The Annals of Mathematical Statistics*, **33**(3), 916–923.

Freedman, D.A. (1963), Invariants under mixing which generalize de Finetti's theorem. Continuous time parameter. *The Annals of Mathematical Statistics*, **34**(4), 1194–1216.

Freedman, D.A. (1963), On the asymptotic behavior of Bayes estimates in the discrete case. *The Annals of Mathematical Statistics*, **34**, 1386–1403.

Freedman, D.A. (1965), On the asymptotic behavior of Bayes estimates in the discrete case II. *The Annals of Mathematical Statistics*, **36**, 454–456.

Freedman, D.A. and P. Diaconis (1983), On inconsistent Bayes estimates in the discrete case. *The Annals of Statistics*, **11**, 1109–1118.

Godambe, V.P. (1976), Conditional likelihood and unconditional optimum estimating equations. *Biometrika*, **63**(2), 277–284.

Godambe, V.P. (1980), On sufficiency and ancillarity in the presence of a nuisance parameter. *Biometrika*, **67**(1), 155–162.

Godambe, V.P. and D.A. Sprott (1971), *Foundations of Statistical Inference. A Symposium.* Toronto: Holt, Reinehart, and Winston.

Goel, P.K. and M.H. De Groot (1979), Comparison of experiments and information measures. *The Annals of Statistics*, **7**, 1066–1077.

Goldberger, A.S. (1964), *Econometric Theory.* New York: John Wiley.

Good, I.J. (1950), *Probability and the Weighing of Evidence.* London: Griffin.

Good, I.J. (1965) *The Estimation of Probabilities. An Essay on Modern Bayesian Methods.* Cambridge (Mass.): M.I.T. Press.

Gouriéroux, C., Monfort, A., and A. Trognon (1983), Testing nested and non nested hypotheses. *Journal of Econometrics*, **21**, 83–115.

Gouriéroux, C., Monfort, A., and A. Trognon (1984), Pseudo-maximum likelihood methods: theory. *Econometrica*, **52**, 681–700.

Granger, C.W.J. (1969), Investigating causal relations by econometric models and cross-spectral methods. *Econometrica*, **37**, 424–438.

Griliches, Z. (1974), Errors in variables and other unobservables. *Econometrica*, **42**(6), 971–988.

Hajek, J. (1965), On basic concepts of statistics. *Proceedings of the Fifth Berkeley Symposium on Mathematical Statistics and Probabilities*, **1**, Berkeley: The University of California Press, 139–162.

Hall, W.J., Wijsman, R.A., and J.K. Ghosh (1965), The relationship between sufficiency and invariance with applications in sequential analysis. *The Annals of Mathematical Statistics*, **36**, 575–614.

Halmos, P.R. (1950), *Measure Theory.* New York: Van Nostrand Reinhold.

Halmos, P. and L.J. Savage (1949), Application of the R.N. theorem to the theory of sufficient statistics. *The Annals of Mathematical Statistics*, **20**, 225–241.

Hannan, E.J. (1971), The identification problem for multiple equation systems with moving average errors. *Econometrica* **39**, 751–767.

Hartigan, J.A. (1964), Invariant prior distributions. *The Annals of Mathematical Statistics*, **35**, 836–845.

Hartigan, J.A. (1983), *Bayes Theory*. New York: Springer-Verlag.

Haussman, J.A. (1978), Specification tests in econometrics. *Econometrica*, **46**, 1251–1271.

Haussman, J.A. and W. Taylor (1981), A generalized specification test. *Economic Letters*, **8**, 239–245.

Hendry, D.F. and G.J. Anderson (1977), Testing dynamic specification in small simultaneous systems: an application to a model of building society behavior in the United Kingdom. In: *Frontiers in Quantitative Economics*, Vol. 3A, edited by M. D. Intriligator. Amsterdam: North-Holland, 361–383.

Hendry, D. and J.-F. Richard (1982), On the formulation of empirical models in dynamic econometrics. *Journal of Econometrics*, **20**, 3–33.

Hendry, D. and J.F. Richard (1983), The econometric analysis of economic time series. *International Statistical Review*, **51**, 111–163.

Hendry, D. and J.-F. Richard (1987), Recent developments in the theory of encompassing. CORE DP 8722, Université Catholique de Louvain, Louvain-la-Neuve, Belgium. In: *Contributions to Operations Research and Economics*. See B. Cornet and H. Tulkens, (editors) (1989).

Hewitt, E. and L.J. Savage (1955), Symmetric measures on Cartesian products. *Transactions of the American Mathematical Society*, **80**, 470–501.

Heyer, H. (1982), *Theory of Statistical Experiments*. New York: Springer-Verlag.

Hoffmann-Jørgensen, J. (1971), Existence of Conditional Probabilities. *Mathematica Scandinavia*, **28**, 257–264.

Hosoya, Y. (1977), On the Granger condition for non-causality. *Econometrica*, **45**, 1735–1736.

Huber, P.J. (1967), The behavior of maximum likelihood estimates under non-standard conditions. *Proceedings of the Fifth Berkeley Symposium on Mathematical Statistics and Probability*, 1, Berkeley: The University of California Press, 221–233.

Hunt, G.A. (1957), Markov processes and potentials I and II. *Illinois Journal of Mathematics*, 1, 44–93 and 316–369.

Jeffreys, H. (1961), *Theory of Probability*. Third edition. London: Oxford University Press.

Kabanov, Y., Lipster, A. Ch., and A.N. Shiryayev (1979-1980), Absolute continuity and singularity of locally absolute continuous probability distributions, I, II. Translated from *Math. USSR Sbornisk*, 35, 631–680 (1979), and 36, 31–58 (1980).

Kadane, J. (1974), The role of identification in Bayesian theory. In: *Studies in Bayesian Econometrics and Statistics*, edited by S. Fienberg and A. Zellner. Amsterdam: North-Holland.

Kagan, A., Linnik, Y.V., and C.R. Rao (1973), *Characterization Problems in Mathematical Statistics*. New York: John Wiley.

Kakutani, S. (1948), On equivalence of infinite product measures. *Annals of Mathematics*, 49, 214–224.

Kalbfleisch, J. D. (1975), Sufficiency and conditionality (with discussion). *Biometrika*, 62, 251–259.

Kendall, M.G. and A. Stuart (1952), *The Advanced Theory of Statistics. Vol. 1 : Distribution Theory*. London: Griffin.

Kendall, M.G. and A. Stuart (1961), *The Advanced Theory of Statistics. Vol. 2 : Inference and Relationship*. London: Griffin.

Kendall, M.G. and A. Stuart (1966), *The Advanced Theory of Statistics. Vol. 3: Design and Analysis, and Time Series*. London: Griffin.

Koehn, U. and D.L. Thomas (1975), On statistics independent of a sufficient statistic: Basu's lemma. *The American Statistician*, 39, 40–42.

Kolmogorov, A.N. (1942), Determination of the center of dispersion and degree of accuracy for a limited number of observations (in Russian). *Izvestija Akademii Nauk SSSR*, 6, 3–32.

Kolmogorov, A.N. (1950), *Foundations of the Theory of Probability*. New York: Chelsea. (Translated from: *Grundbegriffe der Wahrscheinlichkeitsrechnung*, (1933). Berlin: Springer-Verlag).

Koopmans, T.C. (1950), When is an equations system complete for statistical purposes? In: *Statistical Inference in Dynamic Economic Models*, edited by T.C. Koopmans. New York: John Wiley, 393–490.

Koopmans, T.C. and O. Reiersøl (1950), The identification of structural characteristics. *Annals of Mathematical Statistics*, **21**, 165–181.

Kyburg, H.E. and H.E. Smokler (editors), (1964), *Studies in Subjective Probability*. New York: John Wiley.

Lauritzen, S.L. (1982), *Statistical Models as Extremal Families*. Aalborg: Aalborg University Press.

Leamer, E. (1978), *Specification Searches: Ad Hoc Inference with Non Experimental Data*. New York: John Wiley.

Leamer, E.,(1983), Model choice and specification analysis. In: *Handbook of Econometrics*, edited by Z. Grilickes and M.D. Intriligator, Vol. 1. Amsterdam: North-Holland.

LeCam, L., (1958) Les propriétés asymptotiques de solutions de Bayes. *Publications de l'Institut de Statistique de l'Université de Paris*, **7**, 17–35.

LeCam, L. (1964), Sufficiency and approximate sufficiency. *The Annals of Mathematical Statistics*, **35**, 1419–1455.

LeCam, L. (1966), Likelihood functions for large numbers of independent observations. In: *Festschrift for J. Neyman*, edited by F. N. David. New York: John Wiley. 167–187.

LeCam, L. (1969), Théorie asymptotique de la décision statistique. Montréal: Les Presses de l'Université de Montréal.

LeCam, L. (1970), On the assumptions used to prove asymptotic normality of maximum likelihood estimates. *The Annals of Mathematical Statistics*, **41**(3), 802–828.

LeCam, L. (1986) *Asymptotic Methods in Statistical Decision Theory*. New York: Springer-Verlag.

LeCam, L. and L. Schwartz (1960), A necessary and sufficient condition for the existence of consistent estimates. *The Annals of Mathematical Statistics*, **31**, 140–150.

Lehmann, E.L. (1959), *Testing Statistical Hypothesis*. New York: John Wiley.

Lehmann, E.L. and H. Scheffé (1950), Completeness, similar regions and unbiased tests, Part I. *Sankhyā*, **10**, 305–340.

Lehmann, E.L. and H. Scheffé (1955), Completeness, similar regions and unbiased tests. Part II. *Sankhyā*, **15**, 219–236.

Lindley, D.V. (1961), The use of prior probability distributions in statistical inference and decisions. *Proceedings of the Fourth Berkeley Symposium on Probability and Statistics*, **1**. Berkeley: The University of California Press, 453–468.

Lindley, D.V. and G.M. El-Sayyad (1968), The Bayesian estimation of a linear functional relationship. *Journal of the Royal Statistical Society*, Series B, **30**(1), 190–202.

Lindley, D.V. and A.F.M. Smith (1972), Bayes estimates for the linear model. *Journal of the Royal Statistical Society*, Series B, **34**(1), 1–41.

Lindquist, A. and G. Picci (1979), On stochastic realization problem. *SIAM Journal on Control and Optimization*, **17**, 365–389.

Lindquist, A. and G. Picci (1982), On a condition for minimality of Markovian splitting subspaces. *Systems and Control letters*, **1**, 264–269.

Lindquist, A., Picci,G., and G. Ruckebusch (1979), On minimal splitting subspaces and Markovian representations. *Mathematical Systems Theory*, **12**, 271–279.

Littaye-Petit, M., Piednoir, J., and B. Van Cutsem (1969), Exhaustivité. *Annales de l'Institut Henri Poincaré*, **5**, 289–322.

Mac Kean, H.P. Jr. (1963), Brownian motion with a several-dimensional time. *Theory of Probability and its Applications*, **8**, 335–354.

Malinvaud, E. (1978), *Méthodes statistiques de l'économétrie*. Third edition, Paris: Dunod.

Marsaglia, G. (1964), Conditional means and covariances of normal variables with singular covariance matrix. *Journal of the American Statistical Association*, **49**, 1203–1204.

Martin, F., Petit, J.-L., and H. Littaye (1971), Comparaison des expériences. *Annales de l'Institut Henri Poincaré*, **7**, 145–179.

Martin, F., Petit, J.-L., and M. Littaye (1973), Indépendance conditionnelle dans le modèle statistique bayésien. *Annales de l'Institut Henri Poincaré*, **9**, 19–40.

Martin, F. and D. Vaguelsy (1969), Propriétés asymptotiques du modèle statistique. *Annales de l'Institut Henri Poincaré*, **5**(4), 355–384.

Memin, J. (1985), Contiguity, absolute continuity and tightness. In: *Model choice*. See J.-P. Florens, M. Mouchart, J.-P. Raoult, and L. Simar (1985), 11–25.

Metivier, M. (1968), *Notions fondamentales de la théorie des probabilités*. Second edition. Paris: Dunod.

Mizon, G.E. and J.-F. Richard (1986), The encompassing principle and its application to testing non-nested hypotheses. *Econometrica*, **54**, 657–678.

Monfort, A. (1982), *Cours de Statistique mathématique*. Paris: Economica.

Mouchart, M. (1976), A note on Bayes theorem. *Statistica*, **36**(2), 349–357.

Mouchart, M. and R. Orsi (1976), Polynomial approximation of distributed lags and linear restrictions: a Bayesian approach. *Empirical Economics*, **1**, 129–152.

Mouchart, M. and R. Orsi (1986), A note on price adjustment models in disequilibrium econometrics. *Journal of Econometrics*, **31**, 209–217.

Mouchart, M. and H. Roche (1987), Bayesian analysis of load curves through spline functions. *The Statistician*, **36**(2), 289–296.

Mouchart, M. and J.-M. Rolin (1984a), On k-sufficiency. *Statistica*, **44**(3), 367–371.

Mouchart, M. and J.-M. Rolin (1984b), A note on conditional independence. *Statistica*, **44**(4), 557–584.

Mouchart, M. and J.-M. Rolin (1985), Letter to the Editor. *Statistica*, **45**(3), 427–430.

Mouchart, M. and J.-M. Rolin (1986), On the σ-algebraic realization problem. CORE Discussion Paper 8604, Université Catholique de Louvain, Louvain-la-Neuve, Belgium.

Mouchart, M. and J.-M. Rolin (1989), On maximal ancillarity. *Statistica*,49(1).

Mouchart, M. and L. Simar (1984), Bayesian predictions: non parametric methods and least squares approximations. In: *Alternative Approaches to Time Series Analysis*. See J.-P. Florens, M. Mouchart, J.-P. Raoult, and L. Simar (1984), 11–28.

Nachbin, L. (1965), *The Haar Integral*. New York: Van Nostrand Reinhold.

Neveu, J. (1964), *Bases mathématiques du calcul des probabilités*. Paris: Masson. (Second edition: 1970). English translation: *Mathematical Foundations of the Calculus of Probability* (1965). San Francisco: Holden-Day.

Neveu, J. (1972), *Martingales en Temps Discret*. Paris: Masson.

Neyman, J. (1951), Existence of a consistent estimates of the directional parameter in a linear structural relation between two variables. *Annals of Mathematical Statistics*, **22**, 497–512.

Neyman, J. and E. Scott (1948), Consistent estimates based on partially consistent observations. *Econometrica*, **16**, 1–12.

Neyman, J. and E. Scott (1951), On certain methods of estimating the linear structural relationship. *The Annals of Mathematical Statistics*, **22**, 352–361. (Corrections: **23** (1952), 135.)

Orey, S. (1971), *Limit Theorems for Markov Chain Probabilities*. London: Van Nostrand.

Parthasarathy, K.R. (1977), *Introduction to Probability and Measure*. Delhi: The MacMillan Company of India Ltd.

Pesaran, M. H., (1974), On the general problem of model selection. *Review of Economic Studies*, **41**, 153–171.

Petit, J.-L. (1970), Exhaustivité, ancillarité et invariance. *Annales de l'Institut Henri Poincaré*, **6**(4), 327–334.

Picci, G. (1976), Stochastic realization of Gaussian processes. *Proceedings of the IEEE*, **64**(1), 112–122.

Picci, G. (1977), Some connections between the theory of sufficient statistics and the identifiability problem. *SIAM Journal on Applied Mathematics*, **33**, 383–398.

Pierce, D.A. and L.D. Haugh (1977), Causality in temporal systems: characterizations and a survey. *Journal of Econometrics*, **5**, 265–293.

Pitcher, T.S. (1957), Sets of measures not admitting necessary and sufficient statistics or subfield. *The Annals of Mathematical Statistics*, **28**, 267–268.

Pitcher, T.S. (1965), A more general property than domination for sets of probability measures. *Pacific Journal of Statistics*, **15**, 597–611.

Pitman, E.J.C. (1939), Location and scale parameters. *Biometrika*, **30**, 391–421.

Pitman, E.J.C. (1979), *Some Basic Theory for Statistical Inference*. London: Chapman and Hall.

Pitman, J.W. (1978), An extension of de Finetti's theorem. *Advances in Applied Probability*, **10**, 268–269.

Pitman, J.W. and T.S. Speed (1973), A note on random times. *Stochastic Processes and Applications*, **1**, 369–374.

Press, S.J. (1972), *Applied Multivariate Analysis*. New York: Holt, Rinehart, and Winston.

Raiffa, H. and R. Schlaifer (1961), *Applied Statistical Decision Theory*. Boston: Division of Research, Harvard Business School.

Ramamoorthi, R.V. (1980a), Sufficiency, pairwise sufficiency and Bayes sufficiency in undominated experiments. Ph.D. Thesis submitted to the Indian Statistical Institute, Calcutta.

Ramamoorthi, R.V. (1980b), On pairwise sufficiency and sufficiency in standard Borel spaces. *Sankhyā*, Series A, **47**, 139–145.

Ramsey, F.P. (1926), Truth and probability. In: *The Foundations of Mathematics and Other Logical Essays*, edited by R.B. Braithwaite (1950). New York: The Humanities Press. Also reprinted in Kyburg and Smokler (1964), 61–92.

Raoult, J.-P. (1975), *Structures statistiques*. Paris: Presses Universitaires de France.

Reiersøl, O. (1950), Identifiability of a linear relation between variables which are subject to error. *Econometrica*, **18**, 375–389.

Rolin, J.-M. (1975), The inverse of a continuous additive functional. *Pacific Journal of Mathematics*, **2**, 585–604.

Rolin, J.-M. (1983), Non parametric Bayesian statistics: a stochastic process approach. In: *Specifying Statistical Models, from Parametric to Non-Parametric Using Bayesian Approaches.* See J.-P. Florens, M. Mouchart, J.-P. Raoult, and L. Simar (1983), 108–133.

Rolin, J.-M. (1985), Selection of variables in discriminant analysis. In: *Model Choice.* See J.-P. Florens, M. Mouchart, J.-P. Raoult, and L. Simar (1985), 103–120.

Rolin, J.-M. (1986), Asymptotic behaviour of posterior expectations. In: *Asymptotic Theory for Non i.i.d. Processes.* See J.-P. Florens, M. Mouchart, J.-P. Raoult and L. Simar (1986), 93–120.

Romier, G. (1969), Modèle d'expérimentation statistique. *Annales de l'Institut Henri Poincaré*, **5**, 275–288.

Rosenblatt, M.R. (1962), *Random Processes*. New York: Oxford University Press.

Rothenberg, T.J. (1971), Identification in parametric models. *Econometrica*, **39**, 577–591.

Roy, K.V. and R.V. Ramamoorthi (1979), Relationship between Bayes, classical and decision-theoretic sufficiency. *Sankhyā*, Series A, **41**, 48–58.

Sacksteder, R. (1967), A note on statistical equivalence. *The Annals of Mathematical Statistics*, **38**, 784–794.

Savage, L.J. (1954), *The Foundations of Statistics*. New York: John Wiley.

Sawa, T. (1978), Information criteria for discriminating among alternative regression models. *Econometrica*, **46**, 1273–1292.

Schönfeld, P. (1975), A survey of recent concepts of identification. CORE Discussion Paper 7515, Université Catholique de Louvain, Louvain, Belgium.

Schwartz, L. (1960), Consistency of Bayes procedures. Unpublished Ph. D. Thesis, University of California, Berkeley.

Schwartz, L. (1965), On Bayes procedures. *Zeitschrift für Wahrscheinlichkeitstheorie und verwandte Gebiete*, **4**, 10-26.

Scott, A. and T.M.F. Smith (1971), Bayes estimates for subclasses in stratified sampling. *Journal of the American Statistical Association*, **66**, 834–836.

Scott, A. and T.M.F. Smith (1973), Survey design, symmetry, and posterior distributions. *Journal of the Royal Statistical Society*, Series B, **35**, 57–60.

Shafer, G. (1986), Savage revisited. *Statistical Science*, 1(4), 463–501.

Shiryayev, A.N. (1981), Martingales: recent developments, results and applications. *International Statistical Review*, 49(3), 199–234.

Sims, C.A. (1972), Money, income, and causality. *American Economic Review*, **62**, 540–552.

Sims, C.A. (1980), Macroeconomics and reality. *Econometrica*, 48(1), 1–48.

Skibinski, M. (1967), Adequate subfields and sufficiency. *The Annals of Mathematical Statistics*, **38**, 155–161.

Skibinski, M. (1969), Some known results concerning zero-one sets. *Annals of the Institute of Statistical Mathematics*, **21**, 541–545.

Smith, T.M.F. (1983), On the validity of inferences from non-random samples. *Journal of the Royal Statistical Society*, Series A, **146**, 394–403.

Solari, M.E. (1969), The "maximum likelihood solution" of the problem of estimating a linear functional relationship. *Journal of the Royal Statistical Society*, Series B, **31**, 372–375.

Soler, J.-L. (1970), Notions de liberté en statistique mathématique. Thèse de doctorat de 3ème cycle, Université de Grenoble, France.

Speed, T.P. (1976), A note on pairwise sufficiency and completions. *Sankhyā*, Series A, **38**, 194–196.

Sprent, P. (1966), A generalized least squares approach to linear functional relationships. *Journal of the Royal Statistical Society*, Series B, **28**, 278–297.

Sprent, P. (1970), The saddle point of the likelihood surface for a linear functional relationship. *Journal of the Royal Statistical Society*, Series B, **32**, 432–434.

Sprott, D.A. (1975), Marginal and conditional sufficiency. *Biometrika*, **62**, 599–605.

Stone, M. (1976), Strong inconsistency from uniform priors. *Journal of the American Statistical Association*, **71**, 114–125.

Stout, W.F. (1974), *Almost Sure Convergence*. New York: Academic Press.

Strassen, V. (1965), The existence of probability measures with given marginals. *The Annals of Mathematical Statistics*, **36**, 423–439.

Sugden, R.A. (1979), Inference on symmetric functions of exchangeable populations. *Journal of the Royal Statistical Society*, Series B, **41**, 269–273.

Sugden, R.A. (1985), A Bayesian view of ignorable designs in survey sampling inference. In: *Bayesian Statistics 2*. See J.M. Bernardo, M.H. de Groot, D.V. Lindley, and A.F.M. Smith (1985), 751–754.

Sugden, R.A. and T.M.F. Smith (1984), Ignorable and informative designs in survey sampling inference. *Biometrika*, **71**, 495–506.

Teicher, H. (1960), On the mixture of distributions. *The Annals of Mathematical Statistics*, **31**, 55–73.

Teicher, H. (1961), Identifiability of mixtures. *The Annals of Mathematical Statistics*, **32**, 244–248.

Teicher, H. (1967), Identifiability of mixtures of product measures. *The Annals of Mathematical Statistics*, **38**, 1300–1302.

Tiao, G. and A. Zellner (1964), On the Bayesian estimation of multivariate regression. *Journal of the Royal Statistical Society*, Series B, **26**, 277–285.

Torgersen, E.N. (1972), Comparison of translation experiments. *The Annals of Mathematical Statistics*, **43**, 1383–1399.

Torgersen, E.N. (1976), Comparison of statistical experiments. *Scandinavian Journal of Statistics*, **3**, 186–208.

Torgersen, E.N. (1981), Measures of information based on comparison with total information and with total ignorance. *The Annals of Statistics*, **9**, 638–657.

Van Putten, C. and J.H. Van Schuppen (1985), Invariance properties of the conditional independence relation. *The Annals of Probablity*, **13**(3), 934–945.

Villegas, C. (1971), On Haar priors. In: *Foundations of Statistical Inference*, edited by V.P. Godambe and D.A. Sproot. Toronto: Holt, Rinehart, and Winston.

Villegas, C. (1972), Bayesian inference in linear relations. *Annals of Mathematical Statistics*, **43**, 1767–1791.

Villegas, C. (1977a), Inner statistical inference. *Journal of the American Statistical Association*, **72**, 453–458.

Villegas, C. (1977b), On the representation of ignorance. *Journal of the American Statistical Association*, **72**, 653–654.

Villegas, C. (1981), Inner Statistical Inference II. *Annals of Statistics*, **9**, 768–776.

Walker, A.M. (1969), On the asymptotic behaviour of posterior distributions. *Journal of the Royal Statistical Society*, Series B, **31**(1), 80–88.

Watson, G.S. (1983), *Statistics on Sphere*. New York: John Wiley.

White, H. (1982), Maximum likelihood estimation of misspecified models. *Econometrica*, **50**, 1–26.

Wu, D.L. (1973), Alternative tests of independence between stochastic regressors and disturbances. *Econometrica*, **41**, 733–750.

Zacks, S. (1971), *Theory of Statistical Inference*. New York: John Wiley.

Zellner, A. (1971), *An Introduction to Bayesian Inference in Econometrics*. New York: John Wiley.

Zellner, A. and A. Siow (1980), Posterior odds ratios for selected regression hypothesis. In: *Bayesian Statistics*. See J.M. Bernardo, M.H. de Groot, D.V. Lindley, and A.F.M. Smith, (1980), 585–603.

/

Author Index

Subject Index